FORAMINIFERA

FORAMINIFERA

THEIR CLASSIFICATION AND ECONOMIC USE

FOURTH EDITION, REVISED AND ENLARGED, WITH
AN ILLUSTRATED KEY TO THE GENERA

BY
JOSEPH A. CUSHMAN
LATE RESEARCH ASSOCIATE IN MICROPALAEONTOLOGY, HARVARD UNIVERSITY

HARVARD UNIVERSITY PRESS
CAMBRIDGE, MASSACHUSETTS
LONDON, ENGLAND

SIXTH PRINTING OF THE FOURTH EDITION, 1980

PRINTED IN THE UNITED STATES OF AMERICA

ISBN 0-674-30801-8

PREFACE

THE FIRST EDITION of this work was published in 1928 and the second in 1933. The second edition was accompanied by a second volume with forty plates and a Key to the families and genera as then used. In the third edition, published in 1940, the Key and plates were added to and included with the text in a single volume.

In this fourth edition, as much as possible of the work published through 1947 and some of that published during 1948 has been included. Many genera have been erected since the third edition and a considerable number of them are here described and figured; others, which must await more detailed study, have been omitted, as it is impossible to place them with any degree of certainty.

Interest in the Foraminifera has grown greatly in the last few years, partly because of their great value in connection with the petroleum industry and in general geologic correlation. Many problems remain to be solved, however, in spite of the great amount of research that has been done on the group.

The classification adopted here is that of the earlier editions, with such changes as can be made by new evidence that has come to light. As before, it is based upon the known geologic history of the genera, upon the phylogenetic characters through a study of much fossil material from all continents, and finally upon a study of the ontogeny in many microspheric specimens, which show generic relationships much more definitely than megalospheric specimens of the same species.

It is impossible to acknowledge here all the help that I have had from institutions and individuals. That such help and friendly criticism have made possible improvement over the first edition is gratefully acknowledged.

To my friends, Dr. Carl O. Dunbar of Yale University, Dr. T. Wayland Vaughan of the United States Geological Survey, and Dr. W. Storrs Cole of Cornell University, I am deeply indebted for their work on the difficult groups of the fusulinids and orbitoids, two groups of great importance, which need in each case a specialist to understand fully their unique structures and rapidly increasing literature.

To my daughters, Alice E. Cushman for her painstaking work on the proofs, and Ruth Cushman Hill for lettering the plates, I am much indebted. To Ruth Todd for help in revising the manuscript I owe my grateful thanks.

J. A. C.

SHARON, MASSACHUSETTS
August 1948

CONTENTS

FINDING LIST OF FIGURES AND PLATES

TEXT FIGURES

TEXT PLATES

KEY PLATES

FORAMINIFERA

CHAPTER I

THE LIVING ANIMAL

THE FORAMINIFERA are almost entirely marine animals, though a very few live in brackish or even fresh water. They are single-celled animals belonging to the Protozoa. Except in a few of the simplest types, there is developed a test, either of agglutinated foreign material, or of chitin, or of calcareous material secreted by the animal itself. These tests are preserved as fossils in many of the geologic formations since Cambrian time. In the existing oceans foraminifera occur in enormous numbers, and in water from the continental shelf out to about 2000 fathoms or more their tests form the thick *Globigerina*-ooze of the ocean floor. As fossils, they are often very abundant, and in the Paleozoic as well as in the younger formations they have formed thick limestones. The great pyramids of Egypt are constructed from limestones made largely of fossil foraminifera. In the tropics, the sands of the beaches are often largely composed of the tests of foraminifera, and in shallow water their great numbers actually form obstructing shoals.

Most of the abundant living species are of small size, except in the shallow waters of the tropics, where numerous species of considerable size still exist. In the Eocene and Oligocene, species with large tests several inches across were developed. There is but a single living species, now found in the Indo-Pacific, which has as large a diameter as those of the earlier fossils, and this species has a very thin discoid form.

Species have definite geologic and geographic ranges, and when these are known in detail they become of importance in determining the age of sediments and the conditions under which they were deposited.

In spite of the fact that so much has been written in regard to the foraminifera, a literature now including several thousand papers, comparatively little has been written about the animal itself. Although foraminifera may be obtained on almost any sea-coast, and it has been shown that they may live and develop for years in a balanced aquarium, not until very recent years has the complete life history been known, and even now there seems to be much controversy as to the development in different groups. The habits and physiologic characters of the animal and its relationships to the environment are very incompletely known. Studies along these lines will add materially to our knowledge of the group.

Very little is as yet known about the growth rate of the individual chambers in the foraminifera, or the length of time between the various portions of the life cycle. The work of Hofker has shown that under certain adverse conditions a stage similar to encysting in other groups may take place, and the individual be practically dormant for considerable periods.

The animal in the foraminifera should be considered as living both inside and outside of its test. In other words, the test in many groups at least is an internal one. Even in the imperforate groups, the ornamentation of the exterior is added to after the test is completed, and in such forms, even when disturbed, the protoplasm is not wholly withdrawn into the test.

The living animal consists of a mass of protoplasm with a nucleus, the latter inside the test, while the thinner protoplasm in finely reticulate pseudopodia streams out from the test in all directions. In these streams there are usually two currents, the central portion moving outward and the periphery moving inward toward the test. This is not a steady movement, but has definite rhythms. The form of the pseudopodia in the different families is not the same, and these should be further studied. The incoming currents bear food particles and débris of various sorts; thus in attached specimens the outer surface often becomes more or less covered with such débris. The pseudopodia in the moving or feeding individual often extend out very many times the diameter of the test. In the free forms, the whole animal, with its test, may move forward over the surface at a slow pace.

It may be rather definitely stated that individuals, of different species at least, are repellent to one another. Numerous examples of this were observed at the Tortugas laboratory. There is certainly no fusing of individuals of different species or genera to form tests of an intermediate character, as has been sometimes conjectured.

LIFE HISTORY

An alternation of generations in the foraminifera has been known for some time. Two forms of the species are usually found together, one small in size but with a large proloculum or initial chamber, the other of larger size but starting with a much smaller proloculum. From the size of the proloculum these two forms have been named, respectively, *megalospheric* and *microspheric*. The megalospheric form is usually much more abundant. The two forms are to be looked for in all species, and it is important in descriptive work to distinguish them. There may be a whole series of megalospheric forms with a varying size of proloculum, and to certain phases of this development Hofker has given the name *Trimorphism*. This will be discussed in a later chapter.

The microspheric form has a number of nuclei, often a larger number than there are chambers, scattered irregularly through the protoplasm of the body. There seems to be a rather definite relation between the size of the nuclei and the size of the chamber in which they occur, the larger

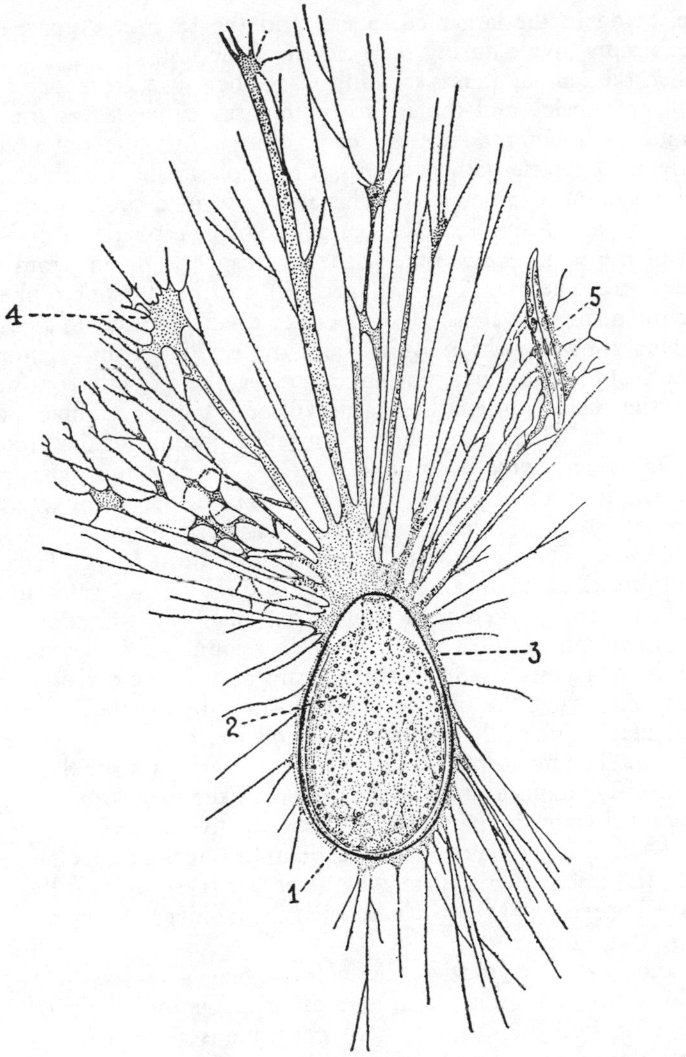

LIVING ANIMAL

DARK LINE SHOWS WALL OF TEST (1), FILLED WITH PROTOPLASM (2), AND ALSO COMPLETELY COVERED (3). ANASTOMOSING PSEUDOPODIA (4), WITH A DIATOM (5).
(AFTER SCHULTZE)

PLATE I

nuclei being in the larger chambers, and the reverse. Apparently these nuclei simply divide during the growth of the test.

When the animal attains its adult stage there is a great increase in the number of nuclei, and the entire protoplasm either leaves the test and accumulates about the exterior, or is drawn into the outer chambers. Finally, each nucleus gathers a mass of protoplasm about itself and secretes the proloculum of a new test. This newly formed proloculum is of the larger type, and is the first chamber of the megalospheric form, instead of being of the same size as that of the microspheric parent from which it was derived. The megalospheric form differs from the microspheric form in having a single nucleus. This does not divide, but moves along as new chambers are added, keeping in about the middle chamber numerically. Nucleoli appear in increasing numbers as the growth continues, and finally the whole nucleus breaks down, and a great number of minute nuclei appear. These draw about themselves portions of the protoplasmic mass, and then divide by mitotic division. Finally, the mass leaves the test in the form of flagellated zoospores. These then conjugate and give rise to the small proloculum of the microspheric form, thus completing the life cycle. The empty test thus left behind must form a large proportion of the dredged foraminifera, accounting for the great number of adult forms always present in both recent and fossil collections.

The microspheric form is the result of a conjugation or sexual process, while the megalospheric form is the result of a simple division or asexual process. As a rule, the megalospheric form is by far the more common, and in many species the microspheric form is very rare, or even as yet unknown. The microspheric form, while it starts as a smaller individual, in most cases attains a much larger size than the megalospheric, as might be suspected from the nature of the reproductive processes by which it is formed. In species where there are definite stages in development, it is usually the microspheric form which repeats these more fully, the early stages being reduced or entirely skipped in the megalospheric form of the species.

In some cases the megalospheric form may give rise to a group of megalospheric young instead of to zoospores. On the whole, the life cycle agrees well with the alternation of generations as seen in certain other groups of animals.

The two forms are of great importance in the understanding of the phylogeny of the group. In general the microspheric form may be considered as conservative, reviewing as it does the early stages which indicate the history of its development. On the other hand, the megalospheric form is progressive, quickly passing over or even leaving out these early ancestral stages, and pushing forward to new developments often not attained at all in the microspheric form. Thus from the early stages of the microspheric form it is possible to find the ancestral forms from which

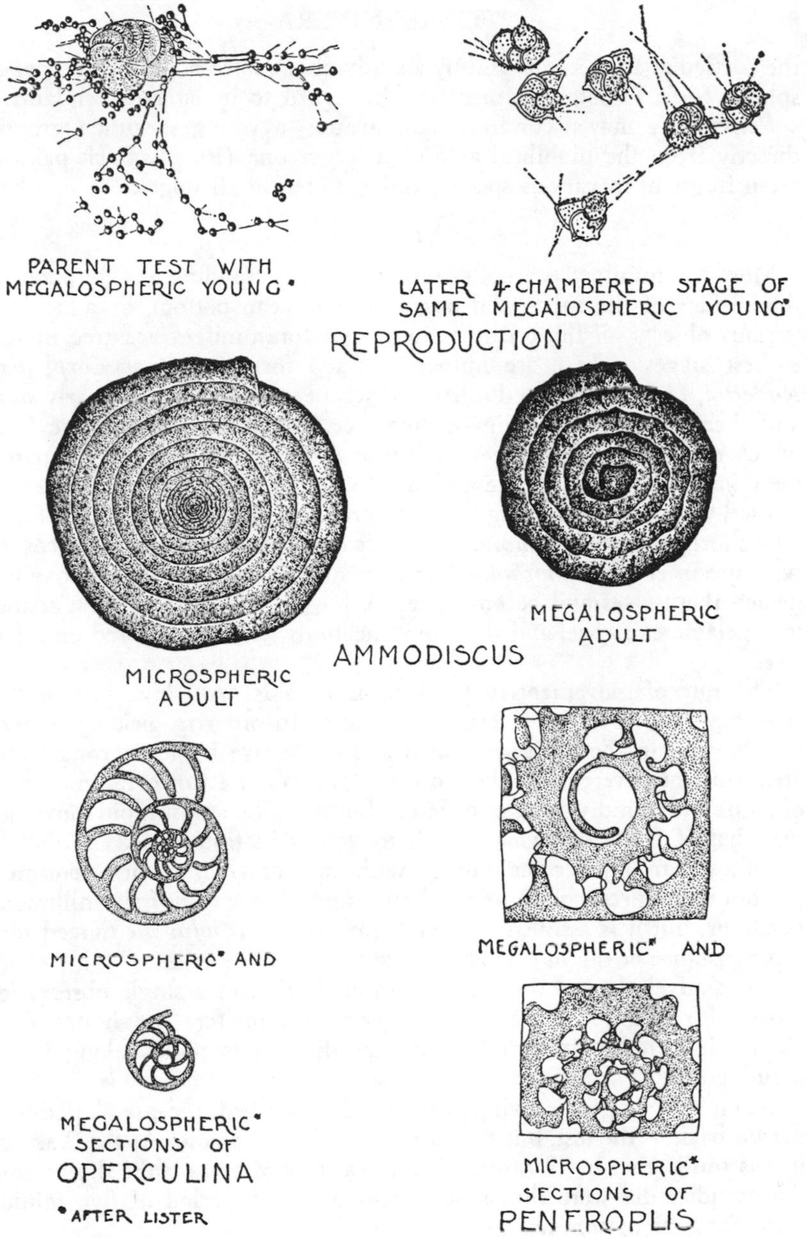

MEGALOSPHERIC AND MICROSPHERIC FORMS

Plate 2

the particular form came, and by a study of the later stages of the megalospheric form to obtain information in regard to its later descendants.

Plastogamy may occur in certain groups, a young specimen growing directly from the umbilical area of a larger one (Pl. 3). Such pairs are often frequent in various species, and are seen in all stages.

HABITS

Most foraminifera are bottom-living forms, crawling slowly about on the surface of the muds and oozes of the ocean bottom, or attached to various objects on the ocean bottom. All foraminifera are free in their earliest stages. There are numerous fixed forms, such as *Carpenteria*, *Rupertia*, *Homotrema*, and others, which become fixed in the early stages and thereafter do not again become free. There are many more forms which may attach themselves loosely to other objects for all or a part of their life history, but are never truly fixed. Hydroid stems are often encrusted with a mass of living foraminifera of the Rotaliidae and Miliolidae. The short eel grass, *Posidonia*, of tropical shallow waters is often covered with specimens of *Planorbulina*, *Sorites*, and other forms. These may later detach themselves and become free. A few species have become adapted to a pelagic existence, and their modifications will be discussed on a later page.

The rate of movement in the foraminifera is very slow, yet for their size they may cover considerable distances. In order to pick up material for the test, in those species which show selective habits, a considerable area must be covered. At the Tortugas laboratory I timed the movement of a number of individuals of different families. The most rapid movement was that of *Iridia diaphana*, which moved at its fastest speed at the rate of almost a millimeter per minute, with an average of about a centimeter per hour. A species of *Discorbis* had an average speed of six millimeters per hour, but it is a much smaller form. *Archaias aduncus* moved for a short distance at the rate of about five millimeters per hour. This species has a relatively heavy, thick test. *Sorites duplex*, in a single observation, moved for a time at the rate of twelve millimeters per hour. From observations made, it would seem that the test is pulled along by the pseudopodia.

When a feeding or moving specimen is disturbed, the pseudopodia are drawn back to the test, but they are soon sent out again. At the Tortugas it was found that in all cases, if no actual injury to the animal was done, the pseudopodia were thrust out again within a period of five minutes from the time of contraction.

Individuals, particularly of different species or genera, have a decided repellent relation when their pseudopodia meet. It was noted in several instances that such specimens, when their pseudopodia came in contact, changed their direction to avoid one another. On the other hand, when

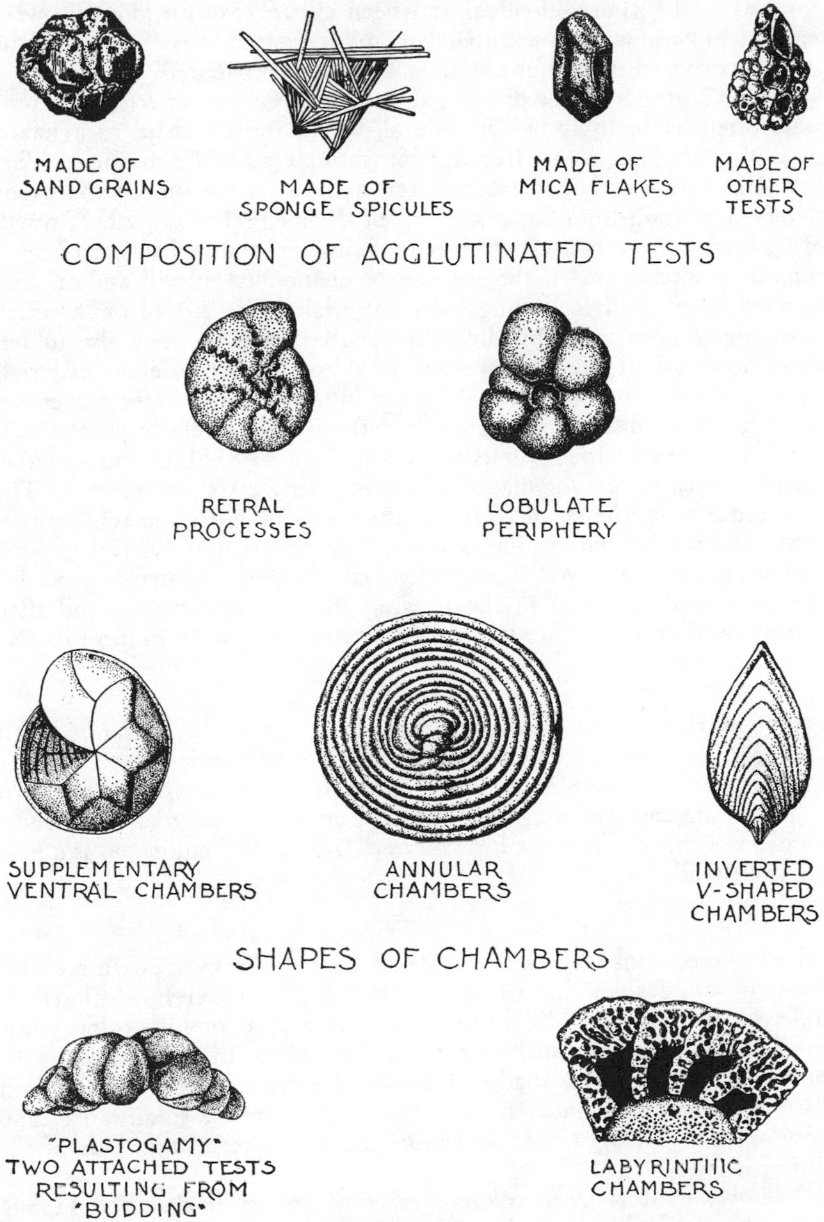
MADE OF SAND GRAINS
MADE OF SPONGE SPICULES
MADE OF MICA FLAKES
MADE OF OTHER TESTS
COMPOSITION OF AGGLUTINATED TESTS
RETRAL PROCESSES
LOBULATE PERIPHERY
SUPPLEMENTARY VENTRAL CHAMBERS
ANNULAR CHAMBERS
INVERTED V-SHAPED CHAMBERS
SHAPES OF CHAMBERS
"PLASTOGAMY" TWO ATTACHED TESTS RESULTING FROM "BUDDING"
LABYRINTHIC CHAMBERS

PLATE 3

portions of the same individual had been cut in two, pseudopodia were rapidly thrown out from each portion, and in contact these fused, and the two parts moved toward one another and finally coalesced.

At the Tortugas it was discovered that in specimens of *Iridia* the tests were often left entirely by the animal, which moved about as a naked mass of protoplasm with a free and comparatively rapid movement. That the animal may withdraw from its test and pass some time without one is very significant from the standpoint of the method of growth. Growth of the test, in those species which have a single chamber, has often been a subject of speculation. If the test can be abandoned at will and another secreted, or made by collecting more material in the case of those which have agglutinated tests, this difficulty is solved, and we may also understand how various sedentary species are able to collect different materials which are not common, and make them into tests. Given free movement and a power of discrimination, which certain species seem to possess, it is not difficult to explain how tests are made of one sort of building material, spicules, mica flakes, ambulacral plates of brittle stars, etc. (Pl. 3). The occurrence of specimens of *Iridia* on the leaves of *Posidonia* some inches above the sea bottom, while its test is composed of material from the bottom, is also explained. The material could easily be carried up to this elevation while ingested in the moving mass of protoplasm, and then concentrated on the surface when the animal settled down to form its test.

FOOD

In general, so far as known, the usual food of the foraminifera consists of vegetable material, diatoms and various other algae furnishing the greater part. In some of the pelagic forms it has been observed that copepods are captured and eaten, as well as other protozoa. The whole problem of the food of these forms is very largely conjecture, and is a field for investigation.

COLOR

Living specimens often are beautifully colored. From my own observations, it would seem that the protoplasm iself is differently colored in different groups. The Miliolidae often have a light pinkish color, while the Peneroplidae are usually of a salmon color. In other groups the protoplasm is of various shades of brown, and in some it is wholly or in part colorless. This matter of color may be of greater significance than now appears, if it is studied in connection with the relationships of the different groups.

The test itself is often colored, especially when fresh. The Homotremidae are brilliantly colored, pink, red or orange. Some of the Calcarinidae also show marked color when fresh. *Rotalia rubra*, a common species of the West Indian region, is a deep red or pink, as is also

Globigerinoides rubra. Many of the Rotaliidae have a brownish color, especially in the young, due to the chitinous inner layer of the test. Nearly all the arenaceous forms are reddish- or yellowish-brown.

COMMENSAL ALGAE

It has been found that the larger living foraminifera of the tropics, particularly those living at depths of less than thirty fathoms in coral-reef regions, have algae associated with them. That there is a commensal relationship similar to that known to exist between algae and corals seems very probable. Owing to the comparatively shallow depth of penetration of sunlight, even in clear tropical waters, such algae are limited to about thirty fathoms, as photosynthesis cannot apparently take place much below that level.

For the same reason, the statements sometimes given in regard to the habitat of certain foraminifera that they were "attached to algae" should be corrected, especially when the shallowest records for these forms may be in hundreds of fathoms, representing depths where algal life is unknown.

CHAPTER II

THE TEST

THE COVERING of the animal in the foraminifera is usually referred to as a test, rather than as a shell such as is secreted by special organs in the Mollusca, etc. In the very simplest forms of the Allogromiidae the animal is naked and without a distinct test. In other primitive forms there is a more definite cell wall, but this may be broken through at any point.

There is a definite sequence in the development of the test, now almost universally accepted among workers on the foraminifera. The earliest definite tests were chitinous; these were followed by arenaceous or agglutinated tests developed by the addition of foreign material cemented to the exterior of the chitinous test. The cementing material may be of various sorts, chitinous, "ferruginous," siliceous, or calcareous. Little is really known of the chemical composition of these cements, although they have been often written about with much assurance. It is a field for exploration. As the cement becomes the dominating feature, its secretion by the animal renders the foreign material unnecessary, and smooth tests result: of "ferruginous" cement (*Trochammina* and many genera); of siliceous cement (Silicinidae); imperforate, bluish-white, calcareous forms (Miliolidae and Ophthalmidiidae); or clear, finely perforate, calcareous forms (Rotaliidae, etc.). All stages in this development may be observed, and the change from a chitinous and arenaceous test to a purely calcareous one may be seen in the development of a single individual.

CHITINOUS TESTS

The most primitive sort of definite test that is developed is a chitinous one, often thin and transparent, but with a definite shape and aperture. Many of the families of the foraminifera, whatever the later character of the test, have the earliest portion with a thin, chitinous, inner layer, representing the primitive test of the group. Purely chitinous tests occur in the Allogromiidae, and in primitive genera of other groups, *Leptodermella*, *Nodellum*, etc.

In brackish water it has been found that the absence or scarcity of calcareous material results in the development of purely chitinous tests by some species which otherwise might have a certain amount of calcareous material in their normal tests.

ARENACEOUS TESTS

In most of the earlier and more primitive groups of the foraminifera there is an arenaceous or agglutinated test, made up of foreign material, sand grains, sponge spicules, mica flakes, etc., loosely or firmly cemented together over a thin, chitinous, inner layer, representing the primitive chitinous test of the still simpler groups. This outer material of the agglutinated tests is of various sorts. In some of the most primitive forms, *Astrorhiza* for example (Key, pl. 1, fig. 1), the mud or sand of the ocean bottom in which the animal lives is loosely cemented about channels leading to the central chamber. Foreign bodies of all sorts, sponge spicules, other foraminiferal tests, etc., are included indiscriminately, and only the inner portion is at all firmly cemented. The only purpose seems to be to form a somewhat rigid protecting wall about the softer protoplasmic body.

Other less primitive forms show power of selection in varying degrees. *Rhabdammina* in its various species usually uses sand grains, while *Marsipella*, in the same bottom sample, will have mostly sponge spicules. A certain general constituent of the bottom material is selected and others are discarded.

In some species with agglutinated tests there is a high degree of selectivity. *Psammosphaera fusca* uses only sand grains of various sizes, those of one color sometimes being used to the exclusion of others. *P. parva* uses sand grains of more or less uniform size, and usually adds a single, large, acerose, sponge spicule built into the wall and projecting far at either end. This is not accidental, for specimens without the spicules are few, and I have rarely seen one with a short or broken spicule, but almost always with a very long uninjured one. *P. bowmanni* uses only mica flakes cemented by the edges, forming a weak and irregular test, and the selective power must be great as the bottom samples in which it occurs often have few mica flakes present. *P. rustica* uses acerose sponge spicules for the framework of the test, fitting smaller pieces of broken spicules into the polygonal areas so that they completely fill the openings (Pl. 3). If the material is ingested in the protoplasmic body, and later carried to the surface and cemented, it is not difficult to account for the apparent mechanical ability of the organism. That this selection occurs in single-celled forms, which are but a speck of protoplasmic material, is the great wonder.

The cement of the test in the most primitive forms is apparently chitinous like the inner wall of the test, and the foreign material is simply included in the outer portion. In the majority of arenaceous forms, from the Paleozoic to the present oceans, the cement is "ferruginous" and has a yellowish- or reddish-brown color, commonly seen in living forms and in well-preserved specimens throughout the fossil series. This "ferruginous" cement may be in small amount, as in *Rhabdammina*, or make up almost the entire test, as in *Ammolagena* and *Glomospira*, where the arenaceous particles are very inconspicuous. All gradations are present in many of the families, and individuals of the same species, and even different parts of the same specimen, may show very considerable differences in the relative proportion of cement and foreign materials.

It is apparent from a study of the arenaceous group that the ferruginous cement may, in shallow, warm waters, be largely replaced by calcareous cement. From this stage it is a simple step to an entirely calcareous test, a development which may be seen actually taking place in a single individual, in living and fossil species. Distinctly arenaceous walls change to calcareous ones in the same individual, but the opposite process is unknown, although sometimes theoretically claimed.

In a few groups siliceous cement is developed, and many forms develop calcareous cement.

SILICEOUS TESTS

Truly siliceous tests are developed only in a few groups, particularly in the family Silicinidae, from the Jurassic to the present oceans. These are probably developed directly from the Ammodiscidae, through *Ammodiscus* and *Glomospira*. Siliceous tests are recorded among the Miliolidae, but this is probably due to the mistaken placing of such forms as *Silicosigmoilina* (Key, pl. 13, fig. 21) and *Miliammina* (Key, pl. 13, figs. 23–26) among the Miliolidae, which they superficially resemble. In brackish or other acid waters the Miliolidae often develop a much thicker chitinous layer which withstands the acid action, as do Mollusca in similar environments. It may be stated that siliceous tests in the foraminifera are unusual.

CALCAREOUS TESTS

It may be easily and completely demonstrated that calcareous tests develop directly from arenaceous ones. The change from the typical ferruginous to calcareous cement in agglutinated forms is well known, but the change from a calcareous cemented arenaceous form to one that is wholly calcareous can be easily demonstrated. This may take place gradually in the fossil series, as shown by a study of related species, or better still, in the stages of a single individual. In *Cymbalopora* (Key, pl. 32, figs. 13, 14) the early chambers are arenaceous — calcareous cement with agglutinated particles, with a chitinous lining — but the later chambers gradually become entirely calcareous. The same development may be seen even more clearly in living species of *Eggerella*, *Dorothia*, etc.

Of the two types of calcareous tests, imperforate and perforate, the former is the more primitive. *Cornuspira* (Key, pl. 16, figs. 1–3) and many of its relatives occur in the later Paleozoic, and the change from such forms as *Ammodiscus* and *Glomospira* is a simple one. The bluish-white appearance of these forms, which are also often referred to in the literature as "porcellanous" forms, is characteristic. The peculiar calcareous cement is distinctly different from that of those forms which develop into the perforate calcareous group, and the two need study from a chemical standpoint.

In the Miliolidae the arenaceous ancestral character, as well as the inner chitinous wall, is kept in the more primitive genera, but in the Ophthalmidiidae this character is lost. The two families are distinct in many respects, and the more they are studied the greater are the differences that appear.

In the calcareous perforate tests, which make up such a large proportion of the Post-Paleozoic foraminifera, the development from arenaceous forms with calcareous cement has already been noted. They have not been studied from a chemical standpoint, as they deserve. The relationships of the groups with calcareous perforate tests are much less simple than has been commonly thought, and it is quite probable that such forms have arisen from the arenaceous forms several times, in different places in the geologic record. This is one field which will repay careful research.

The perforations in the wall of the test are very variable. This difference in size has sometimes been used as a definite character for various groups, but the size varies in different species of the same genus, and particularly at different stages in the same individual. Some forms with coarse perforations in the young may have fine ones in the adult chambers, and the reverse is also true.

FORMS OF THE TEST

The most primitive form of the test is that of the Astrorhizidae, where there is a central body with numerous channels out to the surface formed by the material collected about the pseudopodia. In the Rhizamminidae there are two open ends. In the Saccamminidae, the chamber becomes more definite and a single aperture is developed. This latter is geometrically the simplest test, although not the most primitive.

After the single chamber, the next stage in development is the formation of an elongate, tubular chamber, usually coiled about the proloculum or first chamber. This may then be broken up into irregular chambers, and finally into the definitely chambered forms. It might be supposed, as some authors have done, that a repetition of simple, globular chambers in various geometric forms would give rise to the definitely chambered forms, but the primitive forms, except in a few groups, seem to have arisen from coiled, tubular forms, the evidence of which is seen in the early stages of various primitive forms. Uncoiling, as elsewhere in the animal kingdom, is often a sign of gerontic or old age characters. In the foraminifera, as in other groups, this may finally result in specialized, almost or wholly uncoiled forms. The primitive coiled forms may be planispiral, or may develop conical or even cylindrical spirals.

From the planispiral types (Fig. 2), by division into chambers, are derived many forms in the different groups. From the conical spiral types have developed great groups, such as the Rotaliidae and the families derived from it. Such forms are spoken of as trochoid tests, after the genus *Trochus* in the Mollusca. In such forms the chambers of all the whorls are visible from the dorsal side, but usually only those of the last-formed chamber may be seen from the ventral side (Fig. 1). There is usually an umbilical area in the central portion of the ventral side, which may be either open or variously filled.

CHAMBERS

The initial chamber in the foraminifera is known as the *proloculum*, and may be either small and the result of the union of zoospores, the microspheric proloculum, or larger and the result of asexual division, the *megalospheric* proloculum (Pl. 2). In primitive forms the second chamber is often elongate and tubular, either straight or variously coiled. Chambers may be closely coiled, loosely coiled, or uncoiled. In many coiled forms the chambers remain evolute, showing those of the earlier coils, or closely involute, covering the earlier coils. Many of the terms used in the description of chambers need no explanation, as they are common terms and easily understood. Some of the more commonly used terms are illustrated

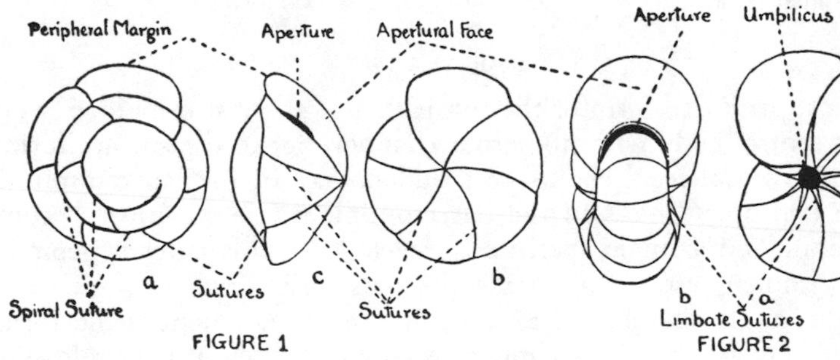

FIGURE 1

FIGURE 2

on Plates 3 and 4. The portions of the chamber are named similarly to those of the walls and are usually thought of in connection with the wall. Such terms as peripheral, proximal, dorsal, ventral, etc., will be readily understood. In trochoid forms (Fig. 1), the side which shows all the whorls of the spire is the dorsal side, and the one showing only the last-formed whorl is the ventral side. The dorsal side is usually more or less convex, but may be flat, or even slightly concave, in forms which are attached by the dorsal side. That portion of the chamber which is adjacent to or contains the aperture is spoken of as the apertural face.

Chambers are usually simple, that is, undivided. In many of the higher

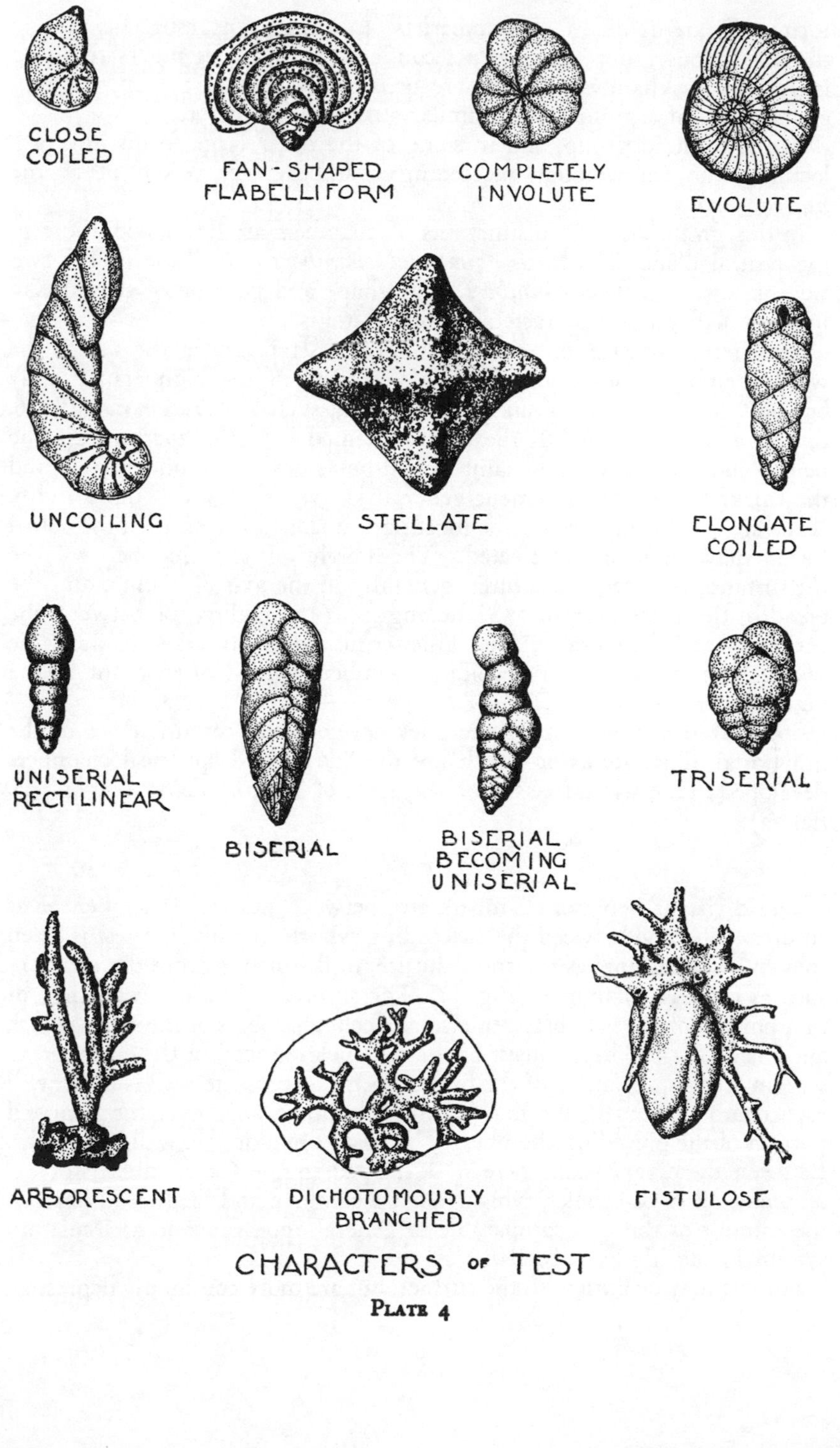

CHARACTERS OF TEST

PLATE 4

forms they are divided into chamberlets. In some forms, especially among the arenaceous group, there is a secondary filling of chambers resulting in *labyrinthic* chambers. These are usually the sign of the approaching culmination of a group, as are similar structures elsewhere.

In the orbitoid group, and in some of the other large forms, the proloculum and immediately succeeding chambers are spoken of as the *nucleoconch*.

In this group also two distinct sets of chambers are developed, those of the central plane, known as *equatorial chambers*, and those of the two sides, known as *lateral chambers*. The shape and position of these make important diagnostic characters in the orbitoids.

In different descriptions there is a great discrepancy in the use of the words height, breadth, and thickness as applied to chambers. It may be convenient to think of an uncoiled form, such as *Saracenaria* (Pl. 16, fig. 2), as a simple one. If the test is oriented with the aperture up, the height and breadth of the chamber will be as ordinarily understood, and the thickness, the measurement generally at right angles to the breadth. If the straight forms are, for the most part, uncoiled forms, then the coiled forms may be similarly treated. The height of the chamber will be the distance between the sutures, generally in the axis of coiling, and the breadth, the measurement at right angles to the coiling, or between the periphery and the spiral suture. This terminology will allow of the same relative use in coiled and uncoiling chambers, which often occur in the same species.

Specialized chambers are sometimes developed at certain stages of the individual, illustrations of which are the "float" and spherical chambers developed in the adult, free-swimming stage of *Tretomphalus* (Key, pl. 32, fig. 22).

SUTURES

The divisions between chambers and between whorls are spoken of as sutures. The line between the succeeding whorls or coils in a test is often spoken of by authors as the spiral suture, in distinction from the ordinary sutures between chambers (Fig. 1). The suture, ordinarily, is the line of that portion of the wall between two adjacent chambers of the same whorl, and inside the test may consist of the old enclosed wall of the chamber, as seen in section. In many of the higher forms a complete wall is built with each chamber, that is, the new chamber builds a floor over the included portion of the preceding chamber. This results in a double wall in sections. Between these two walls there is developed in the Camerinidae, for example, a series of tubules, which may be complex and lead to all parts of the interior of the test, comparable in general appearance to a circulatory system in the higher animals.

Sutures may be flush with the surface, but are more commonly depressed, and in some cases are raised above the general surface. They are often very much thickened, and are then spoken of as *limbate*, a character of importance in descriptive work.

WALL

The materials of which the wall of the test is made have already been discussed. In most foraminifera the wall is perforate. These openings, either large or small, allow the fine protoplasmic materials access to the exterior, as well as through the larger aperture. Occasionally the perforations may be very definitely placed, as in the higher forms such as the Globigerinidae (Pl. 5), where each is in a polygonal area of the surface and forms a cancellated surface.

The wall may be locally thickened, giving rise to a pattern, as is seen in some species of *Lagena*, in *Epistomina*, and elsewhere. The thickening gives whiter areas against the darker background of the thinner portions. In the Globigerinidae a highly specialized character is found in the development of fine spines clothing the test. These are outgrowths of the wall, and in these specialized pelagic forms may have a function in supporting the protoplasm, which often becomes attenuated and many times the diameter of the test.

In some of the higher groups, especially the orbitoids, there are solid masses called "pillars" developed in the wall. These may start very early, and be continued to the surface throughout further development. The presence or absence of these is of diagnostic importance in several groups. Definite thickenings of the umbilical region in *Rotalia* and allied genera may form definite "ventral or umbilical plugs" (Pl. 5). In *Elphidium* the wall of the chamber is extended across the preceding suture to form "retral processes" (Pl. 3).

ORNAMENTATION

The calcareous foraminifera, especially, are often highly ornamented. This seems to be due in many cases to an excessive amount of calcareous material. The ornamentation may involve various parts of the test in varying degrees, such as the general surface of the chamber, the sutures, the periphery, the spiral suture, or the aperture (Pl. 5).

Patterns in the chamber wall due to difference in thickness have already been noted. There may also be pits, either irregularly distributed or making up more or less geometrical patterns in the wall. These depressions in the higher forms, like the Globigerinidae, the Rupertiidae, etc., may produce a definite, cancellated appearance of the wall because of the greater thickening along the sides of polygonal areas with the perforations in the center.

The most common ornamentation of the general chamber wall is in the form of raised areas. These may take the form of raised costae, knobs,

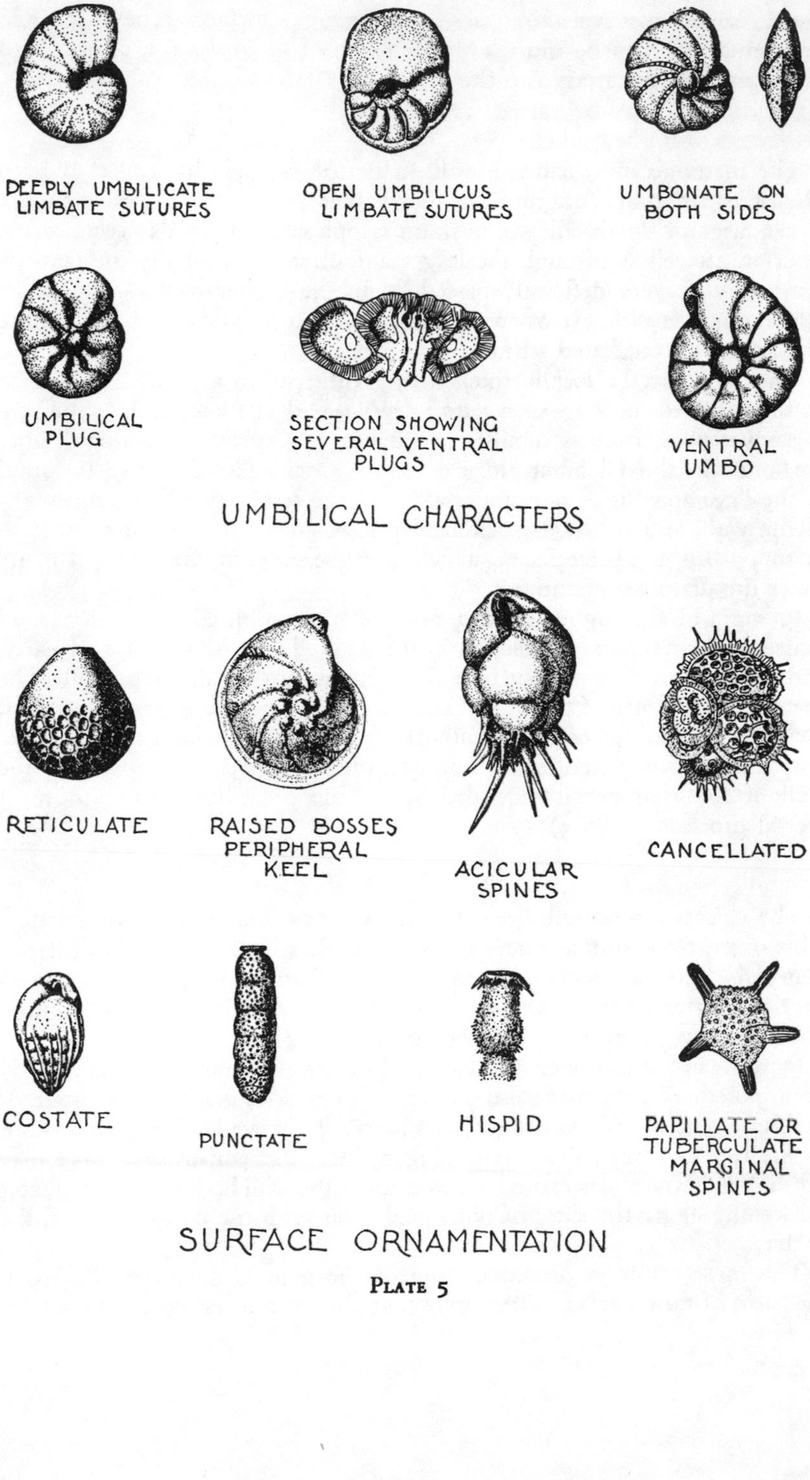
DEEPLY UMBILICATE
LIMBATE SUTURES
OPEN UMBILICUS
LIMBATE SUTURES
UMBONATE ON
BOTH SIDES
UMBILICAL
PLUG
SECTION SHOWING
SEVERAL VENTRAL
PLUGS
VENTRAL
UMBO
UMBILICAL CHARACTERS
RETICULATE
RAISED BOSSES
PERIPHERAL
KEEL
ACICULAR
SPINES
CANCELLATED
COSTATE
PUNCTATE
HISPID
PAPILLATE OR
TUBERCULATE
MARGINAL
SPINES
SURFACE ORNAMENTATION

Plate 5

spines, or various types of raised meshwork. These raised forms of ornamentation may be generally scattered, form characteristic patterns, or be confined to certain definite portions of the chamber surface.

The sutures may be raised into definite ridges or costae, and these costae may again be broken up into rows of bead-like knobs, or may even develop spines. The sutures are often thick and limbate, of clear shell material, appearing darker against the lighter areas between the sutures. Where the sutures meet the periphery, spines are often developed, or peripheral keels connecting with the sutures.

The periphery of the test in spiral forms is a place where ornamentation is often greatly developed. It may consist of a thickened keel, or one that is thin, and becomes a broad carina in several families. Spines, either in relation to the ends of the sutures, the keel, or the general surface, are often developed to large size. Costae of the general surface may be developed into spines at their outer edge. In some forms, large stout spines are developed, independent of the individual chambers, and are added to throughout the life of the animal.

The spiral suture is itself often thickened, standing above the general surface, or broken into knobs. Its ornamentation is usually more closely related to that of the sutures and periphery than to that of the general chamber surface.

About the aperture, spinose or granular surfaces are often developed in several different families. Where a neck and lip are developed, there are often spines about the outer surface. In forms like *Buliminoides*, or some species of *Discorbis*, etc., where there is a depressed umbilical area or apertural face, specialized costae running in toward the aperture are developed, often becoming highly ornate.

In the old age of individuals, or in the later development of species, the ornamentation is often lost, as in other animals, and the test that was highly ornamented in the early stages becomes gradually less so, and in the adult entirely smooth (Pl. 5). Such a disappearance of characters does not take place equally in all parts of the test. In coiled forms of some of the Lagenidae, for example, the ornamentation on the general chamber surface becomes smooth, first on the proximal side, and the ornamentation is held longest on the peripheral part. In the same specimen, the ornamentation of the suture lines disappears first in the peripheral part of the suture, and is retained longest on the proximal end, an exact reverse of the conditions on the chamber wall between sutures.

In many species, the ornamentation is closely identified with the individual chambers and sutures, or patterns are broken at the edges of the chambers. In others, the ornamentation belongs to the test as a whole, and is continued unbroken from chamber to chamber. The large spines may be added to as each chamber is built, so that they have a laminated appearance in section (Pl. 6).

APERTURE

The aperture of the test is one of its most important parts from the standpoint of relationships and descriptive work. It is the opening through which the main body of protoplasm has its chief egress to the exterior. The aperture in the adult of a given species is rather constant, when its development is known. The fact that the aperture changes as the test develops has been held to be a feature which would deprive the aperture of much consideration as a feature of systematic, descriptive worth. The fact that its changes are usually in a logical sequence, and that these very changes are capable of interpreting relationships, has often been overlooked. The young of a complex form will have in its early stages the simple aperture that is characteristic of the genus related to that stage in its development. Thus in *Pyrgo*, the early quinqueloculine stage will have the simple tooth characteristic of *Quinqueloculina*, in its next, or triloculine stage, the bifid tooth characteristic of *Triloculina*, and in the adult, the broad tooth characteristic of that species of *Pyrgo*.

In the more primitive forms the aperture is but a simple opening in the wall of the test. In most coiled forms it is a simple opening between the base of the chamber wall and the preceding whorl. In uncoiled forms the aperture tends to become terminal and to appear in the apertural face itself, finally becoming entirely terminal. The same is true in the biserial forms, such as the Textulariidae, and in the uniserial forms that have developed from them.

In forms that become much compressed so that the aperture tends to become a very long, narrow opening, there are partitions built across the opening, as in *Peneroplis*, even though the chamber itself is not divided into chamberlets. In complex types in which the chambers are subdivided into chamberlets, the apertures are usually multiple. Toward the end of development, in many different groups, there is a decided tendency to have multiple apertures, instead of a single one. This can be seen by referring to the various plates showing development (Pls. 10, 11). Many species, in their adult chamber, have numerous large pores or supplementary apertures, especially those that have become pelagic in habit.

The forms which have become uniserial, as a rule, have terminal apertures. These typically develop a cylindrical neck, and often a phialine lip, characters more closely associated with mechanical form than with systematic relationships. Parallelisms of this general type are abundant in the foraminifera.

In many foraminifera there is developed a tooth of some sort in the aperture. This is usually an outgrowth of the side and many become very complex. In the Miliolidae (Pl. 14) and the Valvulinidae (Pl. 12) some of these forms will be found. In the Buliminidae and the Ellipsoidinidae, there are frequently tooth-like structures in the aperture that are related

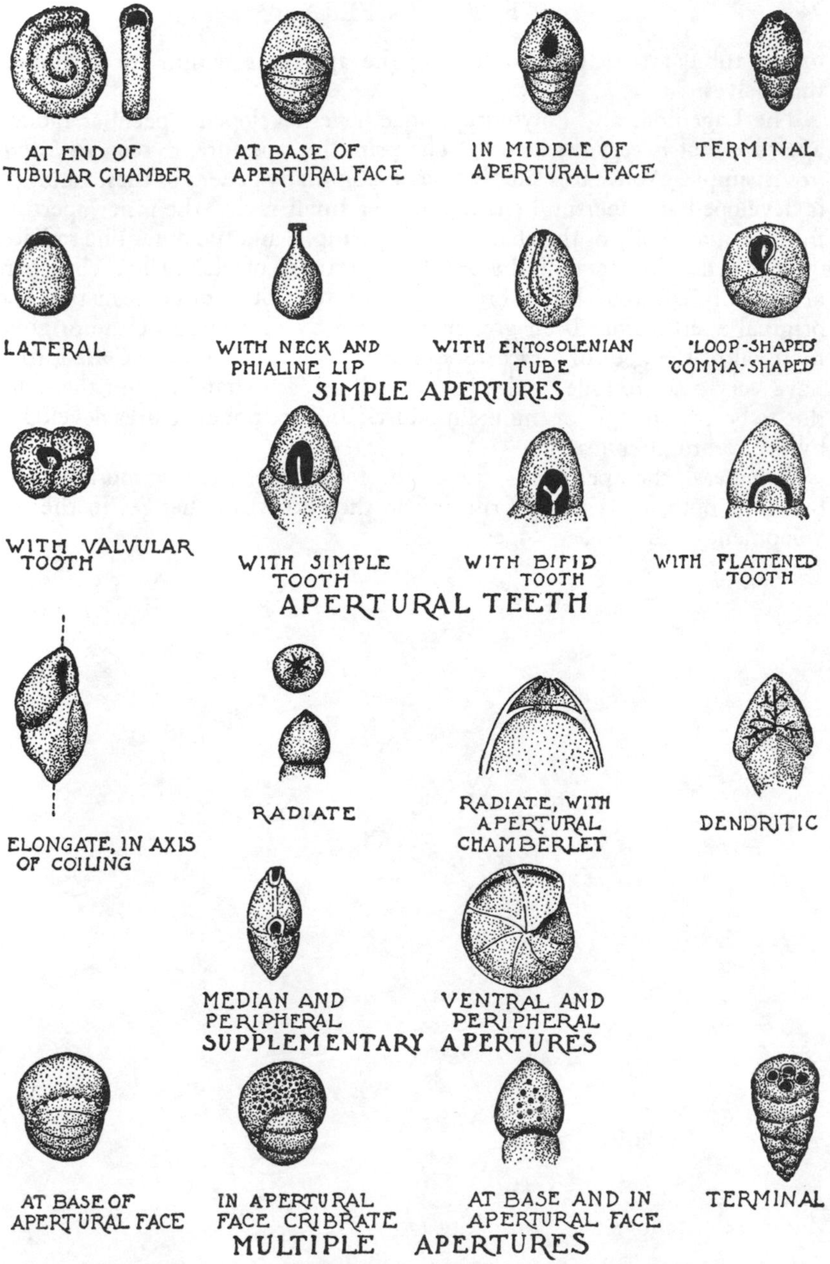
AT END OF TUBULAR CHAMBER
AT BASE OF APERTURAL FACE
IN MIDDLE OF APERTURAL FACE
TERMINAL
LATERAL
WITH NECK AND PHIALINE LIP
WITH ENTOSOLENIAN TUBE
'LOOP-SHAPED' 'COMMA-SHAPED'
SIMPLE APERTURES
WITH VALVULAR TOOTH
WITH SIMPLE TOOTH
WITH BIFID TOOTH
WITH FLATTENED TOOTH
APERTURAL TEETH
ELONGATE, IN AXIS OF COILING
RADIATE
RADIATE, WITH APERTURAL CHAMBERLET
DENDRITIC
MEDIAN AND PERIPHERAL
VENTRAL AND PERIPHERAL
SUPPLEMENTARY APERTURES
AT BASE OF APERTURAL FACE
IN APERTURAL FACE CRIBRATE
AT BASE AND IN APERTURAL FACE
TERMINAL
MULTIPLE APERTURES

VARIOUS TYPES OF APERTURES

PLATE 6

to the tubular structures connecting the apertures within the chambers themselves.

The Lagenidae and Polymorphinidae have developed a peculiar radiate aperture that is very distinctive. The primitive aperture, as seen in *Robulus*, is simple. Above this simple aperture, in many genera of these families, is developed an "apertural chamberlet" or small cavity, the inner aperture into the main body of the chamber being simple, and the outer one radiate.

In the development of the test the apertures of the earlier chambers are usually covered. Upon breaking these tests, it is often seen that the original aperture has been greatly enlarged by resorption of the original material of the apertural face. Some forms, notably *Ceratobulimina*, may have very neat, rounded holes in the wall of each chamber, but these are due to boring sponges or minute molluscs, and are not apertures developed by the foraminifer itself.

In general, the aperture is one of the features of the test most worthy of study, both in its adult form and in the successive changes in the development of the earlier stages.

CHAPTER III

COLLECTING AND PREPARING MATERIAL

Recent material may be collected in many localities along beaches. On coasts with shallow waters which have many algae or hydroids the tests of foraminifera often come in on the beach in enormous numbers. As these tests are very light, each wave carries them up and deposits them at its highest point. When the wave recedes, a whitish line, often largely composed of foraminifera, will be found. This may be scraped up carefully, and later prepared for study. Such deposits at Rimini, on the Adriatic, furnished the earlier authors with many of their species, and the deposits there today, when the waves are breaking, are very rich. Where there are deposits of fossil foraminiferal material along the shore, the fossil species may be mingled with the recent ones, making the study of such collections especially difficult.

In deeper water, of a few feet or fathoms, it is often possible to make good collections by means of a simple dredge. In Jamaica, with a bright, new, tin pail, which could be seen in several fathoms through a water glass, I found it possible to get samples of the bottom, even along coral reefs, in ten fathoms or more. A much better means is the "bull-dog" snapper, which is so devised that it shuts on contact with the bottom and brings up a teacupful of the bottom material. By adding a weight to this apparatus, samples can be obtained even in two or three hundred fathoms.

An apparatus designed by Dr. C. S. Piggott of the Carnegie Institution is capable of taking cores ten feet or more in length from the ocean bottom even at great depths. Studies of the foraminifera of these cores have proved very interesting in comparison with the present fauna of the same spot. Changes in sea level and in temperature during late geologic time have been definitely shown by such comparative studies. In a similar manner, sections of Eocene and Cretaceous sediments have been obtained from the ocean bottom, extending our knowledge of geologic structures into areas otherwise inaccessible. Work along these lines has but barely begun.

PRESERVING

For studying the animal itself nothing can take the place of living material. Material collected in alcohol is of some use, but formaldehyde should not be used, as it frequently destroys the lime of the tests. If only the tests are to be studied, it is by far the best to wash the material in fresh water and then dry it. It can then be packed in bottles or boxes and await future study without any deterioration.

WASHING

For the examination of the foraminifera clean tests are necessary. In order to get these, the dredged material which contains mud and fine sand should be washed. This is best done by means of nested sieves such as are obtainable at most laboratory supply houses. Brass sieves with meshes of 200, 120, 80, 40, etc., to the inch can be obtained. For most practical purposes sieves with 40-, 80-, and 200-mesh to the inch are sufficient. The Cushman Laboratory now uses exclusively the sieve frames with collars, which take silk bolting cloth for the screen. This may be more easily washed and cleaned, and is easily replaced. The mud is placed directly in the top sieve and a stream of water with a fine spray played upon the material. If the sieves are shaken, so that the material is kept in motion, the finer particles will be washed through readily. The resulting clean foraminifera can then be dried. It is sometimes more satisfactory to wash material through the coarsest sieve first, into some sort of retainer, and then again pass this through the finer sieves. By this means the finer meshed sieves are not clogged with the material. If sufficient water is used constantly, the tests will be practically in suspension and will not be damaged, even with a rather strong current of water. It is sometimes best to wash very delicate material in a muslin bag, simply agitating the bag in a container of water until all fine mud is washed away. Material may be dried in the open air, or if speed is necessary, by artificial means, either on a hot plate or on some sort of stove. With very delicate material it is often best not to dry the sample too quickly, or to have it get too hot in the process.

Some workers prefer to wash all their material by a decanting process, pouring off the muddy water until the whole becomes entirely clear. This takes much more time than the use of sieves, but may result in a larger number of the most delicate tests being perfectly preserved for study.

SORTING

After the material is washed, it often helps in the examination if preliminary sorting can be done. There are different methods of doing this. One is that called "spinning." By this method the material is put with clean water in a plate, or watch glass, or in any dish with water, so that a circular motion can be set up. This is the old method by which gold was "panned" by the miners. The gold dust was heavier than the sand and came to the middle of the pan. The foraminifera, however, being lighter material than the sand, accumulate on the outer edges and can be washed off in the process into a larger receptacle below.

Another method by which rough sorting can be done is "decanting." If the material is shaken up in a tall vessel of some sort, the lighter specimens will stay in suspension for a short period and can be poured off, leaving the heavier ones on the bottom. Successive stages will separate most of the calcareous tests from the sand and the heavier foraminifera.

One of the most useful methods is "floating." The washed material, after drying, is slightly heated and thrown upon cold water, when those smaller tests which are filled with air will float on the surface, and can be poured off. In this way beautiful material can be prepared which consists very largely of pure foraminiferal tests. This last method combined with decanting will give the best results.

By the use of heavy liquids, such as bromoform, etc., it is possible to greatly concentrate the tests of foraminifera and separate them from the remainder of the material. It is often possible in this way to save hours of work that would be involved in looking over a sample and picking out the individual specimens. All such work should be done with the use of a ventilating hood.

The most satisfactory method of mechanical sorting is the use of carbon tetrachloride worked out by the late Doctor Ozawa. This is an excellent, speedy, and very cheap method for Recent, Tertiary, and some Cretaceous and Jurassic material, wherever the tests are hollow. The material should be very carefully washed, and then any dry, fine material taken out with a 200-mesh sieve. The sample is then placed in a dish and ordinary commercial carbon tetrachloride poured over it. The foraminifera, fine mollusca, and some ostracods will instantly float to the surface and may be retained by pouring the surface liquid through ordinary fine cotton cloth. The separate on the cloth will be entirely dry and ready for examination in a minute or two, and the residue may be allowed to dry, or thrown away if there is no further use for it. This gives an excellent, clean, quantitative sample, and many such samples can be prepared in an hour's work, and the tetrachloride may be used over and over again.

STORAGE OF MATERIAL

It is possible to store washed material in a variety of ways. Bottles or boxes of convenient size may be used. The various sizes of jars with snap covers make very convenient containers, as they may be opened with a simple pressure on the center of the cap, and closed tightly by a pressure at the sides. For small samples, the folded papers used by chemists are convenient. The data can be written on them before they are filled, and they can be easily filed in a very small space. They are also very light and if carefully packed will stand shipment well. Simple cases with uniform compartments fitted with bottles are now on the market.

FOSSIL MATERIAL

Fossil material may be collected from many kinds of sediments. For the study of free forms, it is best to collect from the shaly or marly partings frequently found alternating with harder beds. Unconsolidated fossil material may be treated by the means already described for recent collections.

With harder material, various methods of preparation may be used. By boiling with alkaline solutions, soda or potash, the specimens may frequently be gotten out in clean condition for study. The use of the autoclave has been described by Driver (1928). The use of the oxy-acetylene torch has been described by G. D. Hanna and by F. and H. Hodson. In the case of hard material, it has been found that the use of a grinding machine will often break down the sample in such a way that the individual specimens will crack out along the surface, and a considerable percentage of specimens with a well-preserved surface for study may be obtained.

With very hard, compact material, such as limestone, the only resort is a study of sectioned specimens. This method is of use with the orbitoids, the Fusulinidae, and some others, but is not as valuable for many of the Rotaliidae or other groups where the exterior is necessary for specific determinations. Weathered specimens will often show good exteriors, and after their determination it is possible to recognize the same species in sections from the same horizon.

CHAPTER IV

METHODS OF STUDY

THE TREATMENT of material will vary much according to the use to be made of it. If it has been collected and prepared purely for the purpose of making a determination of age for stratigraphic correlation, it may be of no further value to the worker. If the sample has been collected for scientific study, it will take a very different course. In the next chapter the methods used in economic work will be discussed, and here the treatment for scientific study only.

SELECTING AND MOUNTING

If a fauna is to be worked up for a scientific collection or for publication it will be necessary to pick out the specimens that are desired. In this work a binocular microscope with a large field and plenty of light is almost a necessity. As most recent foraminifera and many fossil ones are light colored, the material should be loosely scattered over a darker surface. Black is the most commonly used, although at least one of my students much prefers green. Blackened trays may be made by blacking a shallow pasteboard tray with waterproof ink. When this is scratched another coat can be easily applied. An excellent tray may be made by placing a plate of glass over a piece of black velvet. This gives an intense black surface against which specimens stand out with great distinctness. It is best to have the specimens so scattered that individuals stand out clearly. If the material is put on thickly, it is difficult to distinguish forms clearly and the eye strain is much greater. Lines in white may be ruled so that the black surface is divided into squares, if one wishes to search the whole area, and a mechanical stage is of still greater help. One of the simplest backgrounds may be made by exposing a photographic plate and developing it so that a uniform black surface results. This may be used for the bottom of a tray, or if corners are to be eliminated, with their danger of retaining specimens, the plate alone may be used on the stage of the microscope. Trays of dark glass with white lines for this special purpose may now be purchased from dealers in microscopical supplies.

The selected specimens may be mounted on slides directly, or may be sorted into families and genera for further study. In picking out the specimens, the best method is to use a moistened brush. The finest brushes obtainable, oo size, made of red sable bristles, are by far the best. These make a very fine point indeed, and if moistened and touched to the specimen can be used to carry it to the slide. Needles with wax are sometimes used, but they are not elastic, and easily break delicate specimens. Camel's hair brushes are not sufficiently elastic to be of much service. The sable brushes, if left moistened and the tip drawn to a point, will last for a long time.

It is necessary to mount the selected material for safety, for further study, and for permanent preservation. Slides of various sorts are used (Pl. 7). One of the simplest, cheapest, and most generally satisfactory types of slide is made of two pieces of pasteboard, the upper of good grade and thick, the bottom one of cheaper grade and thinner. The upper one is punched with holes of the desired size and number, and between the two a piece of black paper is inserted. The two then are pasted with Map Mounter's Paste which is very strong. To avoid the necessity for black paper, the lower piece may be made of black surfaced heavy paper, or be covered with various black waterproof substances.

For slides of individual species, it is best to use a slide with a single opening, preferably centered and not over a half-inch in diameter. This leaves plenty of room for labeling and for cover. A very simple method is to place an ordinary cover glass above the opening with a tiny drop of glue, or better, Dupont's Household Cement, at three or four places near the edge. The point of a knife or needle will quickly snap off the cover if it is necessary to change the position of the specimens for further study. A neater method of cover is that provided by a sheath made of a glass slide and a base of a thin strip of cardboard, the two attached by strips of gummed paper at the sides.

A newer method is to use a metal sheath which may be stamped out and has the edges bent over to hold the glass slide. These are easily removed if it is necessary to remount or turn the specimens. Manufactured slides of many kinds for different purposes are on the market. There is a wide range in the quality of such slides, however.

Doctor Zinndorf of Germany has produced a slide which is very excellent in many ways. It is made of black or dark blue celluloid, with a central cavity and a square glass cover held in place by two clips cemented to the slide. This slide has the advantage of being very firmly closed yet easily opened, and specimens can be kept loose in the slide with a large margin of safety. The main objection is the danger of celluloid, and slides of other substances such as bakelite can be made as cheaply. Pasteboard slides with a slit left for the insertion of a cover glass are also very useful, and are much used in Europe.

Another type of pasteboard slide that is used to good advantage is one with a large rectangular opening, the background made of double-thick photographic paper in black with white lines, forming 10, 20, 50, or 100 squares, each with a white number. When these are covered with a glass sheath they make excellent slides for the rapid study of faunas.

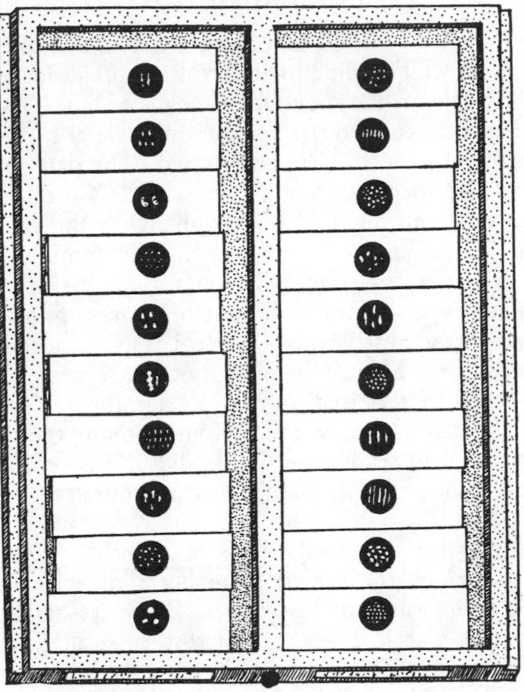

UNIT TRAY WITH SLIDES

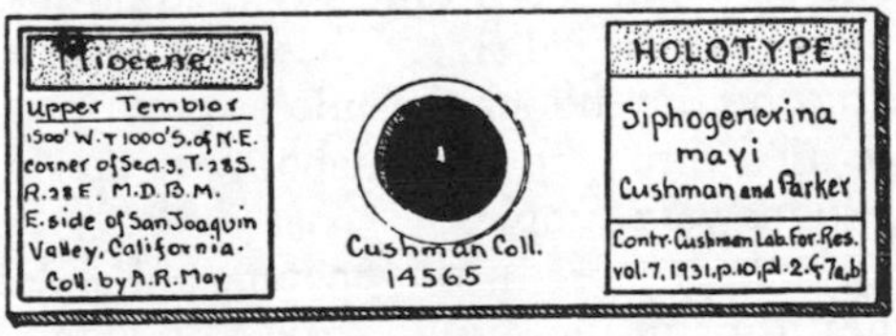

SPECIES SLIDE

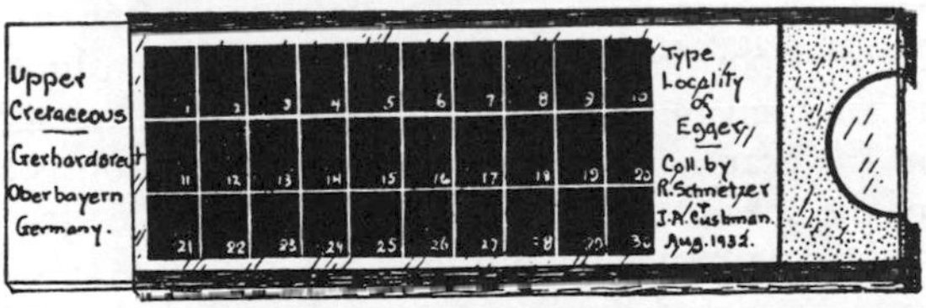

FAUNAL OR GROUP SLIDE

PLATE 7

Two of the standard types of slide used at the Cushman Laboratory are illustrated on Plate 7. The slide for individual species has the outlines for labels printed directly on the slide when they are made. The small oblong on the upper left has a colored strip of gummed paper for fossil specimens, yellow for Tertiary, green for Cretaceous, etc. The data for the sample is on the left and the identification on the right. Various type slides are marked in red or green on the upper right, with the reference to place of publication placed at the lower right. The faunal slide has thirty squares of good size so that they will contain many small specimens and several of ordinary size. Slides with 20–100 squares are often used, but as a practical matter the thirty-square slide has been adopted as the one serving our work best. Material is picked out in separate slides by groups and then mounted in systematic order by families. If one is studying Textulariidae, for example, they will be found among the earlier numbers, and there is no need to search the whole slide. Where hundreds of such faunal or locality slides are available, this preliminary work saves much time later.

All the slides so far described are 3 × 1 inches. Dr. A. Franke of Arnstadt has advocated the use of a smaller slide which is much more compact. This is described in his papers. They are arranged in narrow trays which may be placed under the microscope so that all the slides of a tray may be examined one by one without removing the slide from the tray. Doctor Franke showed me his collection mounted in these slides and they certainly have the advantage of taking up very little space. A figure is given here of the standard unit tray used in this laboratory, made of plywood for 20 slides of usual size. These can all be examined without removal by passing the tray under the microscope. Unit cases holding 24 of these trays are used throughout the laboratory and make the arrangement of the collections very simple.

Specimens may be kept loose in the openings of slides that are covered with the glass sheaths or in the Zinndorf type of slide. With most slides, and for greater safety, specimens should be attached. For this purpose gum of some sort is necessary. Ordinary glue will crack on drying and frequently break the specimen. It is difficult to soften glue when it is desired to alter the position of a specimen. In the Cushman Laboratory the gum that has been found most satisfactory is one made by dissolving ordinary gum tragacanth in warm water to form a consistency desirable for the size of the specimens to be mounted. A drop or two of formaldehyde can be added to prevent moulding. This gum is transparent, and a thin coating can be placed directly on the slide and allowed to dry. The moisture from the brush, as the specimen is transferred, will soften the gum sufficiently for the specimen to become attached, and before drying the specimen can be oriented to the desired position. The use of this gum will not cover important structures, and it is very easy to moisten and remove the specimen if desired.

Some workers prefer to use glass slides with concave cells, placing the specimens under an ordinary cover glass. An objection to these slides is that the specimens often get lodged between the cover and the slide and are broken. The slides themselves, also, are easily broken; furthermore, they must have some sort of label attached, whereas the cardboard slides can be written upon directly.

In studying and drawing specimens it is often necessary to place them in different positions. Ordinary plasticene can be used, preferably of a dark color. This is sticky, however, and it is often difficult to clean specimens. A better material is the black wax used in biological laboratories for making models from serial sections. It is easily softened with the warmth of the hand, and is not sticky. A hole may be made with the point of a needle and the specimen set in it, or often simply stood on end on the wax itself. Far better than either of the above in my own work I find the prism rotator. With this simple device it is possible to study the specimen without moving it, and to get views from above and below. The entire periphery may be viewed from every angle by rotating the prism. It is invaluable for full study of specimens, especially if one wishes to view the aperture from all angles. For drawing, the prism rotator may be used with the ordinary camera-lucida, and all views made without touching the specimen. For measuring specimens the prism rotator is also invaluable, for with an ocular micrometer the various dimensions may be obtained without disturbing the specimen.

In the study of dry specimens it will often be found that moistening the exterior will bring out the structure very distinctly. This may be done with water or, if long study is needed in drawing, with glycerin, which will dry less rapidly.

Photography is very useful in the study of specimens and in making permanent records. The methods used in the Cushman Laboratory have been described elsewhere (Contr. Cushman Lab. Foram. Res., vol. 2, pt. 1, 1926, pp. 1–3), but are repeated here. Various methods have been tried out for a number of years, but always difficulty was encountered in getting sufficient depth of focus with the desired magnification. Several years ago a method was worked out by the writer, and later put into definite form. It has continued to give excellent results. The first step is to obtain a negative of the greatest possible depth of focus regardless of magnification, and then an enlargement is made from this negative to the desired size. In this way the details are kept with the deep focus.

For the actual photographing a vertical camera is used. The particular camera used in the laboratory is a type "H," Bausch and Lomb, with the camera parts to take 5 × 7 plates. Kits are used in the plate holders for smaller sizes. Any good compound microscope may be used. For objectives the Micro-Tessars of Bausch and Lomb are used. The 30-millimeter equivalent focus has been found to be of greatest value, although those of greater focal length are excellent for large specimens. For focusing, the diaphragm should be wide open or nearly so, but for the actual exposure the stop should be cut down at least to 11 of the scale, or even to 22 to get the greatest possible sharpness and depth. An ocular may be used if desired, but much the best results are obtained without it. With the type "H" camera, with the bellows extended to full length, there is a magnification of the image on the plate to about 18 to 20 diameters. This will give an excellent depth. By all means, a focusing glass of some form should be used, and every change of specimens very carefully refocused. The entire result depends on this point of very exact focus. With the combination of a 32-millimeter objective, no ocular, and bellows full length, all the specimens in a circle of about five millimeters may be photographed at once.

For the lighting, one of the regular Bausch and Lomb illuminating outfits with a 6-volt, 108-watt lamp is used. This is placed in the most advantageous position, and then screwed to the table to form a permanent fixture, with a table switch. With the microscope and camera fixed, the only thing needed is to get the slide in position. The light from this unit will be found very intense and the shadows it casts very dark. In the laboratory here a counter-light is used. This is a 250-watt, 115-volt "flood-light" bulb in an ordinary, pliable gooseneck which can be quickly bent to any desired position.

In actual operation a table was built in across the end of the "exposure room" of the laboratory, the center of the top removable, with a second solid shelf below. In this lower part the camera and microscope are placed, also the counter-light. On the top, at the left, is the strong light as already noted. This arrangement brings the top of the camera, even when extended, low enough to permit focusing to be done from the floor. It is unnecessary to stoop to place the slides in position or change the diaphragm, or to stand on a higher level for focusing.

When these factors already mentioned have become fixed, a table of variant factors should be worked out. The intensity of the strong light may be varied by focusing, and this will greatly change the time of exposure. The greatest speed can be obtained with the filament just out of focus on the slide. Each new bulb will be found to vary — the variation is sometimes as much as fifty per cent — and the intensity is apt to decrease with use. This factor must be constantly checked. Specimens will require very different exposures. White foraminifera of tropical reefs, containing chalk-white Miliolidae, will need much less exposure than the gray dull material of much of the American Cretaceous, for example. Length of exposure should be increased in preference to opening the iris diaphragm.

The best results will be obtained from fairly slow plates, and any good plate will give excellent results. The normal development time for the plate should be taken as the base, and length of exposure, intensity of light, etc., varied until the desired sharpness of the developed plate results. Tank development should be used entirely.

When good negatives are obtained, the next step is to get the size best adapted to the purpose desired. In this work an Eastman auto-focus enlarging camera is used. With this camera working in a vertical position no focusing is necessary. The enlarging paper is placed directly on the table in the red light. With a smaller auto-focus enlarger a magnification of 3½ times may be obtained, making the final print 60 to 75 diameters, a size sufficient for all practical purposes. Where records are desired for filing in the laboratory, printing is done on double-thick 4 × 6 paper, which can then be used as 4 × 6 file card with any notes that may be necessary.

By the methods outlined here we have photographed 2500 specimens of foraminifera in an afternoon on less than a hundred slides, and had the negatives ready for printing in the evening.

The longest time is spent in mounting the slides for photographing, but if flat slides are used with a black background, gummed ready for use, one quickly becomes very expert in placing the specimen in position and arranging a number in the 5-millimeter circle. Specimens of fairly uniform size should of course be mounted together to insure uniform focus.

This method, proved by several years of constant usage in the laboratory, will be found a valuable one wherever numbers of foraminifera are handled, and where permanent records are desired.

The only change that has been made in the apparatus described is to increase the length of the bellows to gain a greater magnification, and therefore to obviate so much secondary enlargement. With longer bellows the camera is used in a horizontal position. The use of dark-field illumination will be found very useful in photographing, particularly in the study of sections.

Recently an adapter has been introduced with which a Leica camera may be used, and very good results are being obtained in this way. The field is much smaller, but for individual specimens it is usually sufficient. With a combination of 72 mm., 48 mm., and 32 mm. Microtessar lenses with different tube lengths to the camera all necessary magnifications can be obtained, and the resulting negatives on very fine emulsion film may be enlarged to a considerable size.

Sections in many specimens are very necessary for the study of the structure of the wall and of the early stages. These sections may be made by infiltrating the specimen with balsam, and when hard, grinding it down upon an ordinary hone until the desired plane is reached. It may be left in this condition, or softened, turned over, and ground down to a thin section, according to the particular need. Larger specimens in matrix may be treated the same as in the making of ordinary rock sections. Specimens may be infiltrated with balsam or other material and the test itself dissolved to give the shape of the cavity and the connections between chambers.

In describing foraminifera the following outline is offered as that which the writer has long used and which is now generally accepted: general appearance, chambers, sutures, wall, aperture, and color. These five or six distinctive groups of descriptive characters are set off from one another by semicolons for clearness.

1. General appearance will include: relative size, proportions, characters of the periphery, changes in plan of development, condition of attachment, and such other general points as are not included in the following more detailed characters.

2. Chambers, including number, relative size and shape, arrangement and characters of the interior.

3. Sutures, including amount of depression or elevation; clearness; amount of limbation, if any; changes in various parts or sides of the test; direction; straight, or, if curved, the relative amount of curvature; relations to the ornamentation.

4. Wall, including relative thickness; materials of which composed; kind and relative amount of cement; finish of the exterior; relative size of perforations, if a perforate form; ornamentation, especially changes in different stages of development or in different parts of the wall.

5. Aperture, including changes of position at different stages of development; relation to peculiar structures or modifications of the chamber, development of neck or lip; and ornamentation directly connected with the aperture itself.

6. Color — usually not evident in fossil forms, although occasionally of decided importance.

CHAPTER V

ECONOMIC USE

For the use of foraminifera in determining geologic correlation, especially in checking well drillings, methods different from those used for purely scientific study are necessary. Speed in handling material is often one of the most important factors. Anything that will allow of the greatest speed and at the same time the greatest accuracy is essential, and other things may be sacrificed to these.

Avoidance of scientific names other than that of the genus has been used by the writer in all his economic work. It has been found sufficient to place a specimen in its proper genus, then to give it a number with the formation character. Thus *Nodosaria* Tm–1 will stand in the work as a definite species of *Nodosaria* from the Tertiary and the Miocene Monterey shale; *Nodosaria* Cv–1 a particular species from the Cretaceous, Velasco shale of Mexico. A "type specimen" is selected and mounted as a permanent reference slide for each "species" or number. A short description on a catalog card is made and filed for reference under its number, and camera-lucida drawings of different views are made and filed. As *Nodosaria* Tm–2 is found, it finds its proper place, and is likewise permanently recorded. For ease in reference, plates are built up of the figures, so that all numbered Nodosarias, for instance, are quickly seen on one plate. No occasion arises for consulting published figures or the general literature, which, unless time and a large library are available, would probably result in a wrong specific name being given to the specimen. The abbreviated form of letter and number makes recording simple, makes columns in charts less difficult to handle, and is a great time saver. It also places the species in its general geologic position for future work. A worker can mount, draw, and briefly describe many numbered forms while he might be trying to run down the form to a satisfactory name, even if plenty of literature were at hand.

VERTICAL RANGES

As material is studied from a section, from a well core, or even from carefully collected samples from a standard tool well, the vertical distribution becomes known. With the quick reference to the figures as the samples are studied, changes are easily seen, and new numbered "species" are added to the growing list. Occurrences are noted with records as to relative abundance.

From this data a chart may be constructed showing the vertical distribution of the numbered "species" in the particular section or well. It will be found in any section that there are species which are so rare that they are of little use in determining the position of samples. Even if their ranges are short and accurate, the time consumed in finding them in a sample, and the possibility of missing them, make their value very slight. On the other hand, there will be found species whose ranges are long, and which may be present through too much of the section to be of use in detailed work. In every section, however, there are species whose ranges are relatively short, and which are abundant enough to be quickly found in a sample if they are present at all. Such species are ideal for correlation purposes. If, then, these key species are selected, a chart may be built up from them that will make the placing of unknown samples in the section a matter of comparison. The "tops" and "bottoms" of ranges may be used. If two species of different but overlapping ranges are both present, it is at once apparent that the sample that contains them came from the zone of their overlapping. By using many species the zone of overlap becomes narrowed, until, with rich faunas carefully worked out, the accuracy of placing samples becomes one of close discrimination of specific characters. Even if the zone of overlap may be of considerable amount, in a case where but a few species are available, the relative abundance of species may serve as an additional check and narrow down the limits. Even in similar sediments, the relative abundance of species at different horizons is often widely different.

Charts may be made and duplicated to give a series of workers the data that have been worked out by the persons specializing in different parts of a section. One of the best of these is a visual chart which is really a key to the particular section (see figure). For the first division, two very different forms, each of which is common, yet limited to its own part of the section, should be chosen. The left side of the chart may thus be divided into Divisions 1 and 2, those of *Nodosaria* Xy–1 and *Discorbis* Xy–1. The upper portion, Division 1, may be further divided by *Nonion* Xy–1 and *Textularia* Xy–1. The zone of *Nonion* Xy–1 may be subdivided into three horizons represented by *Bolivina* Xy–1, *Uvigerina* Xy–1, and *Cassidulina* Xy–1, and so on with the other main divisions. The figures of each of these key species for the different divisions may be placed in the proper position on the charts, and it becomes entirely a matter of discrimination of characters to find the particular horizon where an unknown sample should be placed. From the rough, first plotting (see figure) of the occurrence of these and other species will be seen the original data from the core of 290 feet of section from which the visual chart is made. Other species had too few records or had too long a range to be useful in this particular problem. Endless variations in making charts of this sort will suggest themselves to workers engaged in such problems.

With drilling wells where ranges are known, it is very simple to keep up with the horizons if charts of the sections from near-by wells or sec-

	Nodosaria Xy-1	Nodosaria Xy-2	Nonion Xy-1	Nonion Xy-2	Bolivina Xy-1	Verneuilina Xy-1	Textularia Xy-1	Spirillina Xy-1	Clavulina Xy-1	Robulus Xy-1	Globigerina Xy-1	Pyrgo Xy-1	Cassidulina-Xy-1	Eponides Xy-1	Orbulina Xy-1	Ammobaculites Xy-1	Dentalina Xy-1	Dentalina Xy-2	Glandulina Xy-1	Globulina Xy-1	Frondicularia Xy-1	Lagena Xy-1	Uvigerina Xy-1	Pyrulina Xy-1
210'	F		C		C					F	C									F				
220	C		C		C						C													F
230	C		C		C						C													
240	F		F							C	F			F				R					C	
250	R		F							C	F												C	
260	F	R	R			R				R	R						R			C			C	
270	R		F							R	F			R									F	
280	C		F							F	F		C							C				
290	F		F							C	C		C											
300	C	F					F			F	C										F			F
310							F			R	C						R				F			
320	R			R			R			F	F			R							C			
330	C	F					F			C	F										F			
340	C	F					C			F	C					F								
350	C			R			C			F	C			R		F		R						R
360	F						C			C	F					F								
370	F			R			F				C					F								
380								C	F		F									F				
390								C	F	F	F													
400		R				R			C	F	F								F					
410				R					F	R	R			R			R		F					R
420		R									R				F					F				
430				R							R	F			F									
440				R						R		F			C									
450											R	F			C									
460															C							F		
470										R	R				F							F		

C = Common. F = Frequent. R = Rare.

Rough tabulation of occurrence of species in 260-foot section of core giving data for accompanying idealized chart. Other species not used have too long range or too few records.

FIGURE 3

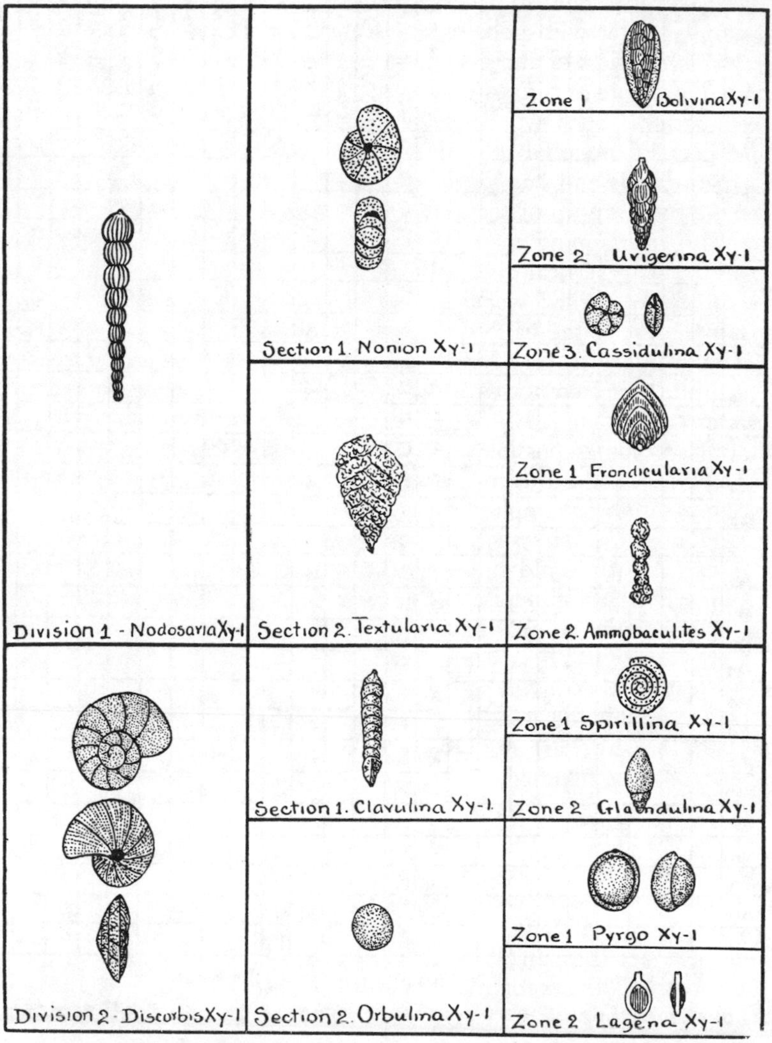

FIGURE 4. IDEALIZED CHART OF SECTION ZONED BY FORAMINIFERA

tions are worked out. Various horizons will be recognized, so that they may be picked up even in rotary drillings, when material from such horizons comes to the surface. The main use is in the case of core samples or bit samples where the exact position in the well is known, but the problem becomes one of placing the samples exactly in the section. It is a simple matter with good samples to tell when passage is made from one clear formation to another if the section has been worked out, but to place random core or bit samples accurately means a greater refinement of detailed ranges. Such work can be done accurately, however, with sufficiently detailed material to start with, by a worker who can recognize valid distinctions and has keen powers of discrimination.

For the building up of sections, continuous core samples are by far the best. With overlapping cores it becomes possible to build up continuous sections for long vertical distances. Next to cores, samples carefully taken from cable tool drilled wells are useful in working up sections. If the first sample out of the bailer is taken, and samples are taken at frequent and regular intervals, very good working sections may be built up. Rotary well samples are often not worth the time spent on them, as far as gaining information for detailed work is concerned. After the section has been built up it becomes possible to use rotary samples, but only to use the "tops" of ranges and thus to tell when a horizon has been reached.

In examination of samples, the "wet method" advocated by Driver may be found useful. Many species show their characters much better when wet than when dry, and this method is useful from this point of view, and also takes less time than that which requires the material to be dried before examination. The appearance of specimens is very different in liquid from what it is when dry, and a change of samples from one method to the other is not advocated.

In speeding up the handling of specimens the procedure adopted in this laboratory by Earl A. Trager will be found useful. Where many samples are to be handled, the containers are of the same size and shape, so as to pack to take up as little space as possible. Each container has its own number. This "pan number" becomes the key to the sample. The data from each sample are written on a sheet and each sample given its pan number. The specimen is then put to soak in its pan, samples are washed as they are ready, and are transferred to a filter paper on which the pan number is again written. When dry, the pan number is inserted in the container, and comparison with the original sheets gives the person making the examination the full data to substitute for the pan number, which is then ready for the next lot. It has been found possible by this method to have a great many samples under preparation at one time without fear of loss of data, or of trouble in keeping labels dry or with their proper sample. It saves much clerical labor and makes for speed in handling material.

CHAPTER VI

GEOGRAPHIC DISTRIBUTION

NEARLY ALL of the many papers on the Recent foraminifera are faunal papers taking up limited areas. In the case of the *Challenger* Report which covered all the seas, Brady had such a wide latitude in his conception of species that faunal limits were not distinguished except in a general way. It has been possible in the studies of all the living species of certain genera, as well as a study of faunas from various parts of the world, to arrive at a fairly comprehensive idea of the living foraminiferal fauna. That there are very definite faunas is apparent, and many of them may be subdivided. These faunas are of interest to the worker on fossil foraminifera as well as to one working on the living faunas, for the later Tertiary and Recent faunas have been established for a considerable period. The Pliocene faunas of Florida and California are very dissimilar, but each is very close to the Recent fauna now living off the coast of the respective regions.

The migrations of faunas in Tertiary times have been marked. For example, the Eocene (Lutetian) fauna of southern England and the Paris Basin migrated gradually through the Mediterranean region to Australia. Many of these species of the European Eocene are found but little changed in the Miocene of Australia, and some of them are found with but little modification in the recent material from the Australian coast. The Miocene warm-water faunas of the Austrian and Hungarian regions also migrated to the Indo-Pacific, and many of the species are still living in that area. So in the Lower Oligocene of the southeastern United States, the species became extinct at the end of the Lower Oligocene in that region, but had migrated to the Pacific, where some of them, or closely allied species, still persist. Many of the large forms now living only in the Indo-Pacific, *Operculina*, *Siderolites*, *Baculogypsina*, *Calcarina*, etc., were very widely distributed in the Tertiary, but have now become extinct except in these restricted areas.

These migrations of the Tertiary account in a large measure for some of the peculiar distributions that are found today. The living fauna of the Caribbean and West Indian region is much more like that of the Australian region than any other. The cold-water fauna of the west coast of South America is more like that of northern Europe than it is like that of the islands of the Oceanic groups, a comparatively short distance away.

In general, the bottom-living Recent foraminifera may be divided into cold-water and warm-water faunas. These are shown crudely on the two accompanying maps. The cold-water faunas are also found to some extent in the deeper, colder waters of the oceans. The Arctic fauna covers the cold waters of the Arctic region and comes southward along both sides of the Atlantic, much farther to the south on the western side. Likewise, it works southward into the Pacific, especially on the western side, but some species work southward along the eastern side.

The Antarctic fauna is very similar in many of its groups to the Arctic one, but has many characteristic species. It covers the general Antarctic

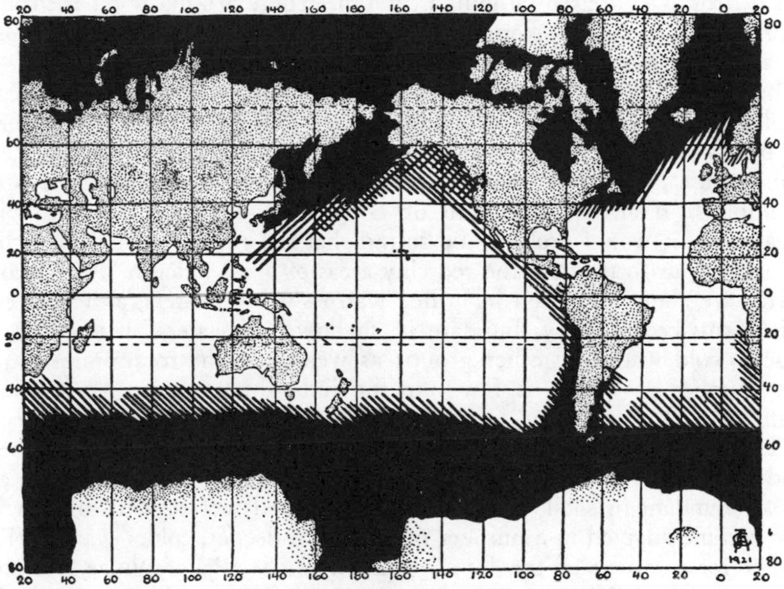

FIGURE 5. DISTRIBUTION OF COLD-WATER FAUNAS

regions, and works northward, especially on the eastern side of the Pacific and also to some extent along the coast of Africa.

The warm-water faunas are more easily subdivided, and may have many more subdivisions than those shown on the accompanying map. What may be termed the Mediterranean fauna works out into the eastern Atlantic, and also into the Red Sea and Indian Ocean. The West Indian fauna extends from the coast of Florida to southern Brazil, northward to Bermuda in lesser degree, and in a still less characteristic form across the ridges of the Atlantic to the coast of Ireland. In the Pacific there is a general fauna covering the region from Africa to the Polynesian Islands. Many things are lacking in the western portion, which may be separated as an East African fauna, while the other is kept as an East Indian fauna, including Polynesia, running northward to southern Japan, and including Hawaii, where many of the characteristic things are again lacking. It is possible to draw lines more closely, and find a number of smaller faunas in this general area.

With the foraminifera, it may be said that temperature is the great controlling force, and depth, except as controlled by temperature, a much smaller factor.

The most abundant single deposit of foraminiferal origin today is *Globigerina*-ooze. It is made up of the tests of the pelagic foraminifera of the Globigerinidae and Globorotaliidae, and is formed in the ocean basins from 500 to 2500 fathoms in depth. There are many other groups represented, especially from 1000 fathoms up, but these two families greatly predominate.

In abyssal depths, where it is difficult for carbonate of lime to accumulate, many arenaceous forms are found. This is true, not, as seems to be a current opinion, because they are found only in such habitats, but because they are the only ones that can persist under the particular conditions. If an equal amount of *Globigerina*-ooze were treated with weak acid, it would probably be found that the residue would be richer in arenaceous forms than the red clay areas of greater depth. Arenaceous forms are abundant, often in rather warm shallow waters, although certain forms become very abundant in shallow cold waters. It is generally true in cold waters, in other groups as well as in the foraminifera, that the number of species is few but the individuals occur in enormous numbers.

The Lagenidae, as a rule, are characteristic of the continental shelf, and between 50 and 500 fathoms are very abundant. The Miliolidae are most abundant in shallow, warm water of coral reef regions, but *Pyrgo* has become adapted in a number of species to deeper, colder waters. The larger foraminifera of the families Camerinidae, Peneroplidae, Alveolinellidae, Calcarinidae, etc., are almost exclusively tropical in waters under 30 fathoms. This, as already noted, is largely due to the restricted limits at which the commensal algae associated with them are able to live. The approximate depth of many Tertiary sediments may be definitely determined from what is known of the distribution of their contained genera in the present oceans.

As already noted, the study of core samples from the present ocean bottom has made possible much data for the observation of changes in faunas in very late geologic time, and also the change in ecologic conditions.

In studying fossil faunas it must be constantly borne in mind that changed ecologic conditions will mean a certain amount of change in the faunas. Studies now under way on both our coasts, on lines of carefully collected samples from shore out into deep water, are giving very definite data for the determination of conditions under which late Tertiary sediments of these same regions were deposited.

PELAGIC FORAMINIFERA

The deposits of *Globigerina*-ooze, as has already been mentioned, are composed very largely of pelagic foraminifera. These are distributed by warm ocean currents, and their tests fall to the bottom in great numbers, gradually building up a sticky, calcareous mud. Fossil oozes of this type are found especially in tropical islands that have apparently been raised

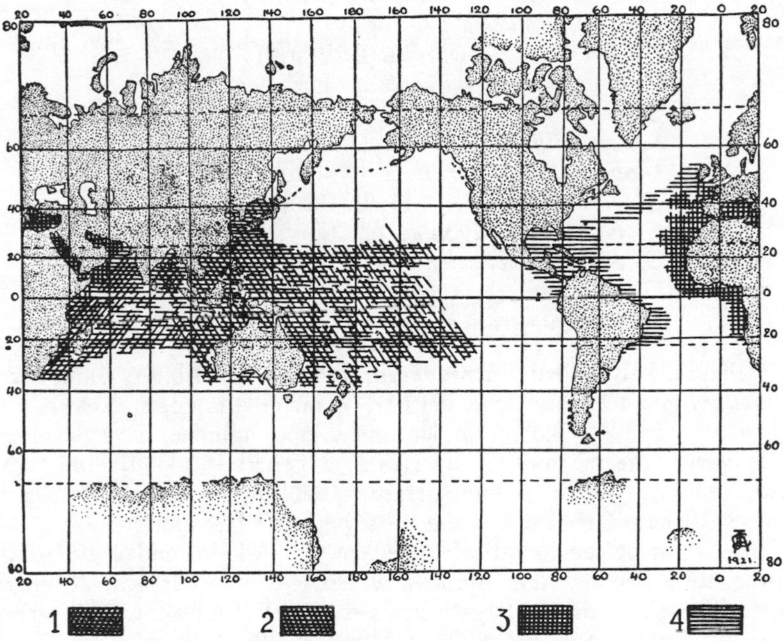

FIGURE 6. DISTRIBUTION OF WARM-WATER FAUNAS
1. East African. 2. Indo-Pacific. 3. Mediterannean. 4. West Indian.

during the Tertiary. The twenty-five species mentioned below have been recorded as pelagic. All but one of these, *Tretomphalus bulloides*, belong to the two families Globigerinidae and Globorotaliidae. *Tretomphalus* becomes pelagic only in the adult phase, and is mainly found in tow net collecting about coral islands. The early stages are apparently bottom living, and the species is not therefore to be grouped with the oceanic pelagic group. The following species are noted as pelagic:

Globigerina bulloides d'Orbigny.
Globigerina dubia Egger.
Globigerina inflata d'Orbigny.
Globigerina cretacea d'Orbigny (?).

Globigerina digitata H. B. Brady.
Globigerina dutertrei d'Orbigny.
Globigerina helicina d'Orbigny.
Globigerina pachyderma (Ehrenberg).
Globigerinoides rubra (d'Orbigny).
Globigerinoides sacculifera (H. B. Brady).
Globigerinoides conglobata (H. B. Brady).
Globigerinella aequilateralis (H. B. Brady).
Orbulina universa d'Orbigny.
Hastigerina pelagica (H. B. Brady).
Pulleniatina obliquiloculata (Parker and Jones).
Candeina nitida d'Orbigny.
Sphaeroidinella dehiscens (Parker and Jones).
Globorotalia menardii (d'Orbigny).
Globorotalia tumida (H. B. Brady).
Globorotalia patagonica (d'Orbigny).
Globorotalia canariensis (d'Orbigny).
Globorotalia crassa (d'Orbigny).
Globorotalia truncatulinoides (d'Orbigny).

To these may be added *Tretomphalus bulloides* (d'Orbigny), and *Hastigerinella digitata* (Rhumbler), neither, so far as known, of oceanic type.

The two families, Globigerinidae and Globorotaliidae, have developed either very large or multiple apertures, a generally globular or tumid form, and a rugose or spinose surface, often clothed with long, slender spines. Of these *Orbulina* has the most perfect adaptation.

While most of the foraminifera except the Allogromiidae are strictly marine, there are a certain number which live in water away from the ocean. These records rest largely upon the works of Daday, who records the genus *Entzia* from salt pools of Hungary; Brodsky, who records the genera *Spiroloculina*, "*Biloculina*," *Lagena*, *Nodosaria*, *Textularia*, and *Cornuspira*, from springs in the Desert of Kara-kum in central Asia; and Asano, who has recorded numerous species from brackish water in Japan. These waters are supposed to be the remnants of an old Miocene sea. It has been thought that the foraminifera found therein might have been transferred by birds from the ocean, but Brodsky is very emphatic that this was not the case. Other records not yet published seem to make it probable that foraminifera will be found in salt or brackish lakes in many parts of the world.

CHAPTER VII

GEOLOGIC DISTRIBUTION

IT IS NOT POSSIBLE in a short chapter to take up the detailed geologic distribution of any but the larger groups of the foraminifera. The nude types must be very old, and of course have left no traces in the rocks. Likewise those forms which have developed chitin alone are preserved only under the most favorable conditions. The arenaceous group, especially those which have very definite tests, are capable of good preservation if conditions are favorable. By the newer methods of dissolving limestones, a greatly increased knowledge of the earlier forms of the arenaceous group is being rapidly accumulated. The chitinous base and more or less elastic cementing materials make the arenaceous tests much less capable of holding their form under pressure than is the case with the calcareous forms. In Cretaceous and even Tertiary sediments, which, although unconsolidated, have been subjected to stresses and shearing, the calcareous foraminifera will be found to have held their shape well, while the arenaceous forms with them will often be greatly twisted out of shape. The calcareous forms are well preserved as a rule, unless there has been leaching, or the material turned to greensand before it was fully fossilized.

In the early rocks, microscopic objects have been found which have been assigned to the foraminifera, usually to the genus *Orbulina*, which is really a specialized end form, in no sense a primitive one, and whose geologic history is not long. In the pre-Cambrian (?) quartzites of France, Cayeux has recorded microscopic objects (*Cayeuxina* Galloway) which he has referred to the foraminifera, although noting that the structure is obliterated and that some of them may be equally well referred to the radiolaria. The size of the chambers, 0.01 mm. in diameter, is much smaller than the microspheric proloculum of known forms, and there is nothing about the specimens as figured to suggest foraminifera as much as yeast cells, or many other microscopic objects. Dawson figures very similar objects which occur in great numbers in association with *Eozoon*, often apparently occurring in the chambers of that form. These he designates as Archaeosphaerinae. His figures and description certainly indicate a very close relationship to *Eozoon*, a form which has long since lost its claim to be considered as belonging to the foraminifera. The very close similarity to Dawson's and Cayeux's figures would seem to indicate a similar source for the two sets of objects, and that neither can have any valid claim as foraminifera.

Matthew records from the Cambrian of New Brunswick both *Orbulina* and *Globigerina* (*Matthewina* Galloway), giving several specific names under both genera. I have examined specimens of both genera as named by Matthew in his collections, but have been unable to make out anything of true foraminiferal nature in either. They very strongly resemble minute concretions. This opinion has been concurred in by others who have examined the originals. The so-called Orbulinas are merely small globular bodies, and the so-called Globigerinas, aggregates of these in various groupings. That they are in any sense related to the highly specialized and much later appearing genera to which they are assigned is hardly tenable, and in my own mind, from the material I have studied, that they are foraminifera at all is equally questionable.

Chapman has recorded from the Cambrian of the Malverns in England well-distinguished foraminifera which he refers to the genera *Lagena*, *Nodosaria*, *Marginulina*, *Cristellaria*, and *Spirillina*. These are rather abundant, especially the *Spirillina*.

A recent study of this material has been made by Dr. Wood, who shows that this material of supposed Cambrian age is really lower Liassic (Wood, The supposed Cambrian foraminifera from the Malverns.—Quart. Journ. Geol. Soc. London, vol. 102, pt. 4, 1947, pp. 447–460, pls. 26–28, text fig. 1). This has a good result in helping to eliminate various inconsistencies in ranges of various genera and families and to validate the classification as recorded here. From the Cambrian of Russia, Ehrenberg records numerous casts which are referred to *Verneuilina*, *Bolivina*, *Nodosaria*, *Pulvinulina*, and *Rotalia*. These might equally well be casts of *Verneuilina*, *Textularia*, *Nodosinella* or *Reophax*, and *Trochammina*. The originals show very little, and the geologic age is open to question. Brady records *Lagena* from the Silurian, which, however, may be *Archaelagena*. Terquem describes four species referred to *Placopsilina*, on crinoids from the Upper Silurian, three of which, at first glance, would appear to be Nodosarias and in section might easily be mistaken for this genus. According to Terquem, however, they were attached forms. He also refers Devonian specimens questionably to *Orbulina*, *Lagenulina*, *Cristellaria*, *Fusulina*, and *Globigerina*. Terquem figures peculiar globular bodies (*Terquemina* Galloway) which are only casts and not the original tests, the wall characters of which are indeterminable as the wall is not present. Few of these various groups of spherical bodies resemble foraminifera, and it is very doubtful if they have any relation to the group. Keeping refers material from the Silurian of Central Wales to *Dentalina*, *Rotalia* (?), and *Textularia*. The figures of the last-mentioned genus, at least, seem to be correctly identified. Recently, by the method of dissolving limestones, excellent faunas of beautifully preserved arenaceous foraminifera have been found by Moreman, Dunn, Croneis, and others in the Ordovician and Silurian, and continuation of this work will undoubtedly add greatly to our knowledge of this pre-Carboniferous portion of the geologic column, which has hitherto been largely without foraminiferal records, probably because search was largely made for the larger calcareous forms which were not yet developed at this time. Eisenack, in the Baltic region, has found in the early Paleozoic members of the Astrorhizidae and Saccamminidae showing the early primitive character of these families.

Beginning with the Carboniferous, foraminifera are often abundant. In the Pennsylvanian of America there are rich faunas with abundant specimens. Arenaceous forms almost completely predominate. The Saccamminidae, Hyperamminidae, Reophacidae, Ammodiscidae, Lituolidae, Textulariidae, Verneuilinidae, Fusulinidae, Trochamminidae, Orbitolinidae, and Placopsilinidae, a majority of the arenaceous group, are present and some of them well developed in the Pennsylvanian. Some of the Lagenidae are possibly present, and the beginnings of the Miliolidae and Camerinidae. There are a few other groups recorded, but they should be carefully checked, since in some cases they rest upon single specimens. Altogether, the Paleozoic foraminifera are predominantly or almost exclusively arenaceous, and this condition continues on into the Permian, where some of the lines of development reach their climax and become extinct. In the Permian, particularly, the imperforate calcareous forms often become abundant. The change from the arenaceous Ammodiscidae to the imperforate calcareous Ophthalmidiidae is a slight one. This family lived on and became very abundant in the Jurassic as well as later.

Triassic foraminifera are extremely rare, although I have had the privilege of examining a rich fauna, beautifully preserved, which when fully studied will yield an excellent amount of data for this period. This is one of the portions of the geologic column that greatly needs further study.

The Jurassic is characterized by the dominance of the Lagenidae. These become very abundant, and develop many species. While many of the forms are like the modern ones, there are many primitive forms among them which need detailed studies. Arenaceous groups persist, and at some horizons are abundant. The Silicinidae appear in the Lias, more of the Miliolidae, and the Ophthalmidiidae are often abundant. The earliest, unmistakable, primitive genera of the Buliminidae and Rotaliidae also appear, but not the more specialized families derived from the Rotaliidae.

There are very few large species of foraminifera in the Jurassic, and these are mainly specialized arenaceous forms, but the faunas have many smaller species. For the first time, the perforate calcareous types, as represented particularly by the Lagenidae, predominate over the arenaceous forms, although at some horizons the latter are abundant.

With the Lower Cretaceous there is a great development of new forms. Arenaceous forms continue as they do to the Recent oceans, but they are largely overshadowed by the calcareous forms. More of the genera of the Buliminidae appear, such as *Virgulina* and *Bolivina*. In the Rotaliidae a few more of the primitive genera appear. The Globigerinidae make their first appearance with undoubted species, and the Heterohelicidae appear, although better developed in the Upper Cretaceous. The simpler forms of the Anomalinidae appear. Of the large forms, *Orbitolina* becomes very abundant.

The Upper Cretaceous greatly increases the number of genera that are developed. Many of them such as *Pseudotextularia*, *Planoglobulina*, *Eouvigerina*, etc., became extinct in America with the Cretaceous, although in Europe they apparently persisted into the Lower Eocene. With the Upper Cretaceous there are new families appearing, especially the more specialized ones derived from the Rotaliidae, such as the Calcarinidae, Chilostomellidae, Globorotaliidae, Peneroplidae, and the Orbitoididae. Some of these may be found to have had their real beginnings in the Lower Cretaceous. Pelagic foraminifera appear to have been first definitely developed in the Cretaceous. The Globigerinidae and Globorotaliidae are well developed and abundant, and the Gümbelinas seem to have been pelagic at this time, although none of their group is pelagic today. There is a great development of the Valvulinidae in the Upper Cretaceous, particularly of France, and also of specialized Lituolidae in the same region. Many of these large, highly specialized forms did not persist into the Tertiary. In general, the Upper Cretaceous takes on many modern aspects in the foraminifera, although in many respects it is primitive. The specialized groups of this period have mainly disappeared.

With the Eocene, modern forms are more abundant. Large species are developed in the warm seas of this time. The genera *Orbitolites*, *Opertorbitolites*, *Alveolinella*, *Flosculina*, *Borelis*, *Operculina*, *Heterostegina*, *Camerina*, *Asterigerina*, *Amphistegina*, *Lepidocyclina*, *Discocyclina*, and others, develop large species. The higher groups of the Rotaliidae and derived families appear. While many of these had specialized species, the general character of the fauna has persisted, and many of the genera are now found living in the shallow waters of the Indo-Pacific. As the Tertiary progresses, species are replaced by more modern ones, and gradually the fauna comes to that of the present oceans. Outside of the Allogromiidae, which naturally would not be found in the fossil state, none of the families, except the Neusinidae and Keramosphaeridae, is wholly represented by recent species. These two are very specialized families with very few genera and species, and very limited in their distribution at the present time.

A chart showing the geologic distribution of the Polymorphinidae will be found under that family. It would be instructive to have more of these to show that the grouping in the various families has followed the geologic sequence, and the student can easily construct similar charts for the various families from the data given with each genus.

As more is known of the distribution of various genera of the foraminifera, it would be of great value to study the early stages of development in successive geologic formations of those genera which have a long stratigraphic range. This should give basic data for showing the development and relationships of genera and families and make the classification more accurate.

CHAPTER VIII

TRIMORPHISM

IN THE LAST FEW YEARS, Hofker has applied the name Trimorphism to certain phenomena relating to the foraminifera. The application of the existence of trimorphism has met with varied results. Some authors have denied that it existed in their material, and only a few have given it very serious consideration. If it is entirely true, and as widely applicable as its proponents claim, then our present classification is entirely to be made over to fit the conditions. The whole basis of trimorphism is that there are three forms in every species, A_1, A_2, and B. The first two represent the megalospheric forms, and the last the microspheric form. As the megalospheric generations may, apparently, be many, with a constantly diminishing size to the proloculum, the three definite forms are perhaps not so accurate as to express it in some sort of series, and the whole might better be called Polymorphism, if that term were not already overworked in other ways.

The microspheric and megalospheric forms have already been discussed in some detail. There may be more than one development of the megalospheric form between two generations of microspheric ones. The size of the megalospheric proloculum in these intermediate generations is not constant, and the resulting test lacks some of the early stages. With the largest megalospheric proloculum, the adult characters are taken on almost at once. With the smaller megalospheric proloculum, some of the early stages skipped in the preceding are now present. With the microspheric proloculum, the greatest number of early stages are present.

The problem involved would not be so perplexing, nor so troublesome, if it did not involve our nomenclature. In the figures are shown three forms from the Miocene of the Vienna Basin. The first of these was named by d'Orbigny *Nodosaria aculeata*. It has a large proloculum, even larger than the succeeding chamber. There are but four chambers in the test, arranged in a straight line, and it might be called *Nodosaria* with little question. The second was named *Dentalina floscula* by d'Orbigny. It has a smaller megalospheric proloculum, six chambers in gradually increasing size, and the axis of the test is curved. The third was named *Marginulina hirsuta* by d'Orbigny. It has a microspheric proloculum, nine chambers or more, the early ones somewhat compressed, but the final ones exactly like those in the other two forms. The ornamentation in the adult of all three is the same. All three forms occur together, and it is evident from a study of a suite of specimens from the type locality that all three forms belong to a single species. The microspheric form is the only one that has the full characters. The three forms may be graphically represented, using P for the proloculum, M for the *Marginulina* stage, D for the *Dentalina* stage, and N for the *Nodosaria* stage, as follows:

$P + M + D + N$ for the microspheric form.
$P + D + N$ for that with a smaller megalospheric form.
$P + N$ for that with a larger megalospheric form.

It becomes difficult to name such forms. There are three generic names involved as well as three specific ones. The specific names can be disposed of on the basis of the usual application of the Rules of Nomenclature, the

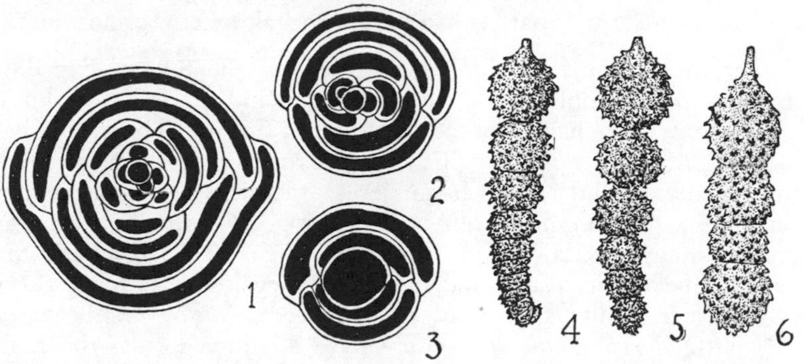

FIGURE 7

FIGS. 1–3. *Idalina antiqua* (d'Orbigny). (After Munier-Chalmas and Schlumberger.) Fig. 1. Section of microspheric specimen with quinqueloculine early stage followed by triloculine and biloculine stages. Fig. 2. Section of megalospheric specimen with triloculine early stage followed by a biloculine stage, the quinqueloculine stage skipped. Fig. 3. Section of a megalospheric specimen with the biloculine stage following the proloculum, the triloculine and quinqueloculine stages skipped. Figs. 4–6. (After d'Orbigny.) Fig. 4. Microspheric form "*Marginulina hirsuta* d'Orbigny." Fig. 5. Megalospheric form "*Dentalina floscula* d'Orbigny." Fig. 6. Megalospheric form "*Nodosaria aculeata* d'Orbigny.

first described one holding its specific name, and dropping the other two. With the generic names the difficulty is increased. As the microspheric form is the only one that shows the full characters, it would seem best to call all the specimens *Marginulina aculeata* d'Orbigny, and consider the two megalospheric forms, and others of the same sort, as specimens with incomplete stages. It is obvious that this could be done only with the knowledge of all the forms of the species.

Another example may be seen in *Idalina antiqua* (d'Orbigny). In the section of a specimen with a microspheric proloculum, there are fifteen chambers developed after the proloculum. The early chambers I–VIII are quinqueloculine, IX–XII triloculine, and the remainder biloculine, the normal sequence in microspheric forms of this group. Figure 2 is a section of a specimen with a small megalospheric proloculum, and having but nine chambers after the proloculum. There is no truly quinqueloculine stage, chambers I–VI showing the triloculine stage and the remainder biloculine. In figure 3 is a section of a specimen with a very large megalospheric proloculum and with but four chambers after it, all of them biloculine. If this were not an involute species covering the early stages, a case exactly parallel to that previously given would obtain. Graphically the stages may be represented as follows:

$P + Q + T + B$ for the microspheric form.
$P + T + B$ for that with a smaller megalospheric proloculum.
$P + B$ for that with a larger megalospheric proloculum.

Hofker in his work on the Foraminifera of the Siboga Expedition gives numerous examples along this same line, and they are well known to all students who have worked with many forms. There are many species of *Quinqueloculina* in which the microspheric and megalospheric forms are known, and in which no development goes beyond the quinqueloculine stage. *Quinqueloculina*, then, is a good genus for such forms. There are many species of *Triloculina* in which the microspheric form has a quinqueloculine young and the larger megalospheric forms do not go beyond a triloculine stage. For all such forms, *Triloculina* becomes a good genus. *Pyrgo* (*Biloculina*) may have all three forms: the microspheric quinqueloculine, then triloculine, and finally biloculine; the form with the smaller megalospheric proloculum, triloculine, then biloculine, and lastly, individuals with a very large megalospheric proloculum, that become at once biloculine. For all these, *Pyrgo* (*Biloculina*) becomes a perfectly good genus.

So in the case of other groups, if all three forms seem to show different generic characters there is difficulty, but if one generic character is involved the generic problem is not difficult. It may be shown that when all three forms occur without the addition of a new character, a primitive form is under observation. In the microspheric form which shows several stages it may be safely assumed that they are taken on in the order in which they once developed, and that they represent more primitive genera which are already known, or are to be looked for, in either the recent or fossil series. A classification built on this basis, as is the one here, must be close to the truth of the actual development of the different groups.

A closer understanding of the results of the trimorphism must lead to a simplification of our treatment of species. It is very evident that it is unsafe to describe a species entirely from the megalospheric form, even though that may be the more common one. Sections should be obtained, if necessary, to know whether the worker is dealing with a microspheric or a megalospheric form, so that he may search his material for the microspheric form if he does not have it. It will undoubtedly be possible to unite species under a single name which now may be placed as different species and under different genera. This task of simplification and grouping together of forms does not mean that there are not very many species and genera of the foraminifera, but that the known facts of development have not been taken into account in naming forms, or in grouping them.

Perhaps the most serious difficulty in the application of trimorphism is the too wide application of it. For example, three generic forms have sometimes been placed together as one which do not have the same geographical distribution. If the three forms are really part of one species they cannot have different distributions. The various facts in regard to the application of trimorphism must be worked out in greater detail, and, if possible, under experimental control with living material in the laboratory, to see what actually takes place. This should not be difficult for workers near marine laboratories, and it is hoped that more definite and detailed facts may soon be forthcoming on this important phase of work with the foraminifera.

CHAPTER IX

CLASSIFICATION

NUMEROUS ATTEMPTS at classification of the foraminifera have been made from the time of the appearance of d'Orbigny's classic Tableau Méthodique published in 1826. Nearly all the classifications since d'Orbigny's time have been based upon the resemblance in form of the adult test. The classification adopted in Brady's *Challenger* Report of 1884, with its ten families, has been the one in most general use since that date. It, however, places together forms which are now known to have very different beginnings, although their adult forms may have certain points in common. The classification used here is, with few changes, the same as that in the earlier editions. It is based upon the best thought developed by many authors on the relationships of the foraminifera since the publication of Brady's Monograph in 1884. Since Brady's time much has been learned in regard to the development of foraminifera, and many new genera have been erected. The classification given here has been wholly, or in large part, adopted by workers on the foraminifera throughout the world.

While our knowledge of the foraminifera has increased greatly in the last few years, very much has yet to be learned before any really final classification of the group can be made. The earliest Paleozoic has yet to be fully investigated, and the Permian and Triassic should yield a wealth of material for the necessary filling in of very important gaps that are now largely blank. It is very probable that similar groups have arisen, not only more than once, but several times, from the same or different sources, and only an intensive study of the development and relationships of fossil faunas will give the true picture. It seems very safe to predict that the final classification of the foraminifera will be found to be much more complex than most workers have heretofore thought. Parallelisms are very numerous. For example, the biserial forms that Brady included under the Textulariidae now are placed in at least seven different families.

An ideal classification should be based upon the known phylogeny of a group as shown by the fossil record, and coupled with the ontogeny of the individual as shown in its complete development, together with what may be learned of the morphology and physiology of the group.

In the foraminifera, as has been mentioned, there are at least two distinct forms, one the result of the fusion of gametes after mitotic division, the microspheric form, the other the result of simple division, the megalospheric form. In the first of these forms, the early stages are more nearly complete, while in the second, the early stages may be skipped, and the adult characters taken on almost at once. It is very evident that any classification must be based upon the relationships shown in this microspheric form of the species. Unfortunately, the microspheric form is often rare, and the megalospheric form the common one. Acceleration of development takes place in various groups, so that there are some species which hold to ancestral characters and show several developmental stages, whereas others are much more specialized, and skip, or greatly reduce, these stages. Also, parallelisms, or the development of similar structural forms in the adult, are very common in the foraminifera.

The simplest foraminifera are the Allogromiidae, where the test is wanting or consists of a thin, chitinous wall. This latter may be more or less globular or elongate, and open at both ends. With the advent of the arenaceous or agglutinated test on the outside of the thin, chitinous layer, a structure of more or less permanency is established, and in the strongly cemented tests is capable of preservation in the fossil series. One of the simplest of this form of test is that seen in the Astrorhizidae (Key, pl. 1), where there is a central chamber with irregular arms, the whole test rather loosely cemented. Another simple form is the agglutinated test, open at both ends, in the Rhizamminidae (Key, pl. 1), a form easily derived from such primitive forms as *Shepheardella* in the Allogromiidae (Pl. 8, fig. 9). The simple, single-chambered forms of the Saccamminidae (Key, pl. 2) may be directly derived from such primitive forms as *Allogromia* (Pl. 8, figs. 6, 7).

As the next simplest stage in development, there is the initial chamber, or proloculum, followed by a long, undivided, tubular chamber. This may be straight, as in *Hyperammina* (Key, pl. 3, fig. 1), or become coiled in various types of spirals, as in *Ammodiscus* and its allies (Pl. 9). The planispiral form is the simplest of these, and occurs in different types, the arenaceous in *Ammodiscus* (Pl. 9, fig. 1), in *Cornuspira* in the imperforate calcareous group (Pl. 15, fig. 1), and in *Spirillina* in the perforate calcareous group (Key, pl. 29, figs. 1–4). All the evidence points to the development of *Cornuspira* from *Ammodiscus* by the development of the calcareous cement, and in a similar way *Spirillina* has developed. The long chamber next becomes divided into divisions or chambers, and the various families with chambered tests gradually make their appearance in the fossil series.

The Lituolidae (Pl. 10) represent, in their primitive forms, the simple stages from *Ammodiscus* to *Trochamminoides* and thence to *Haplophragmoides* and *Endothyra*. These forms developed at least in the Devonian. From the simple forms, very complex ones develop gradually in the Cretaceous and Tertiary. In the Fusulinidae, derived from

Endothyra, are some of the largest of the Paleozoic foraminifera, which become complex in their internal structure. *Loftusia* is one of the largest of all the foraminifera, and is derived as an offshoot from the Lituolidae. These large, complex forms have become extinct.

In the Textulariidae (Pl. 11), the early stages are planispiral, but the later ones become twisted and elongate, finally settling down to a test with the chambers making a half turn of 180°, the aperture toward the central axis. From this type are derived many diverse forms. These developed early in the Paleozoic, and many of the complex ones developed there did not persist to later times. In the same series are the Valvulinidae (Pl. 12), which have developed very large and complex forms in the late Cretaceous and again in the Middle Eocene.

The Miliolidae (Pl. 14) started their development in the Paleozoic, but did not reach their height until the Upper Cretaceous. They kept the power of adding arenaceous material to the outside of the calcareous test in the early, more primitive forms, but lost it in the more specialized ones. The calcareous test is imperforate. There are characters, such as those found in the pseudopodia, and the color of the protoplasm, which may help to show the distinctions in this group even more plainly than at present. Associated with the Miliolidae are the Ophthalmidiidae (Pl. 15); both of these families have imperforate calcareous tests with a peculiar milky white color. These have some similarities, but many differences, which distinctly separate them from one another.

From the Ammodiscidae have probably developed the planispiral forms, one group of which, the Peneroplidae, Alveolinellidae, and Keramosphaeridae, have an imperforate calcareous test in the adult, but the earliest stages in some forms are apparently perforate. In the other group are the Nonionidae, which are mostly planispiral, but in their later development have trochoid and uncoiled forms. Along a similar line developed the Camerinidae, with their greatest development in the Eocene, but in their more primitive forms perhaps going back to the late Paleozoic.

Also developing from planispiral forms are the Lagenidae (Pl. 16), with their very finely perforate tests and glassy appearance. The planispiral forms uncoiled, and the coiling stage is lost in many forms. The Lagenidae are a primitive group in many ways, reaching their height in some respects in the Jurassic, although there is a considerable development of large forms of *Frondicularia*, *Palmula*, *Kyphopyxa*, etc., in the Cretaceous. They still continue as a very plastic group, very baffling in their characters to the systematist who would divide them into hard and fast species or genera. The radiate aperture with the apertural chamberlet is a characteristic structure, and persists also in the Polymorphinidae (Pl. 18), which developed from the Lagenidae in the Jurassic.

Also from a planispiral ancestry are the Heterohelicidae (Pl. 21), which, in the Cretaceous especially, developed many diverse and special-
great range in the amount of chitinous material used, some forms such

ized forms. There are many parallelisms in this group with the Buliminidae and other families.

The other two great groups developed from elongate spires, more or less cylindrical in the Buliminidae (Pl. 22), and a flaring, conical type in the Rotaliidae (Pl. 24). The Buliminidae have their beginnings as far back as the Jurassic, and have developed along many different lines, usually terminating in a rectilinear uniserial test. These are abundant in the Tertiary, and have many varied forms in the present ocean. From the Buliminidae are derived the Ellipsoidinidae (Pl. 23).

The Rotaliidae and families derived from it are all perforate and calcareous in the adult, but some of them are evidently arenaceous in the early stages of the fossil series, such as *Cymbalopora* in the Cretaceous. Many of them have an inner chitinous layer, representing the early primitive wall in the early chambers. It is very probable that these groups have had more than one source, and probably several. Some of them may have developed directly from *Trochammina*. Although some of the genera of the Rotaliidae are recorded, by Brady particularly, from the Paleozoic, these have not since been confirmed, and there may be some error in these records, which often rest on a single specimen. There are some primitive forms in the Jurassic, and these are evidently developed from conical spiral, undivided forms. Their real development is in the Cretaceous and Tertiary. From the Rotaliidae have developed numerous families, some very interesting in their adaptations. The Globigerinidae and Globorotaliidae (Pl. 27), have become specially adapted to a pelagic life with many modifications. The young of the highly globose, very spinose Globigerinidae, in the miscrospheric form, is a trochoid, smooth, more or less flattened, *Discorbis*-like form, showing the early ancestry of this group.

Some groups become attached, and like the Homotremidae and Rupertiidae have developed peculiar forms, which, except that their early stages are trochoid forms, like others of the Rotaliidae, would hardly be classed with the rest of the foraminifera. Finally, in the Orbitoididae, complex large forms are developed, which in the late Cretaceous and early Tertiary become very much specialized. These are excellent guide fossils for various formations during this particular time.

By taking into consideration the various stages in development, especially in the microspheric forms, and the known geologic history of each group, the following classification of the foraminifera has been developed. That it is perfect is not for a moment claimed, but that it is based upon known truth in regard to the past and present history of the group, and on developmental stages, makes it at least nearer the truth than the more artificial classifications that have preceded it.

CHAPTER X

RESEARCH PROBLEMS CONCERNING THE FORAMINIFERA

MANY PROBLEMS relating to fossil or recent foraminifera are still unsolved. A number of lines of research are indicated in the next few paragraphs, and these represent only a few of the problems that will suggest themselves to workers.

Little is known in detail about the living animal and its functions. Workers with access to marine laboratories have a wide field for research. Little is known of the complete life history, although we think the main facts are on record. The protoplasmic structure, especially the types of pseudopodia in different family groups, needs detailed study. Physiological reactions in different groups also offer possible facts having a far-reaching influence on the relationships in the group. Food habits are practically unknown. Little has been published on the rate of growth of the test and the length of life of the individual. The stages in formation of new chambers have been only touched upon in the literature. The types of test produced under acid and alkaline conditions would be most interesting, as well as the selective ability of specimens with many different things in their environment. What type of test will be developed by an arenaceous form without any source of foreign material?

Very much needs to be known of the developmental stages, especially in microspheric forms. These, in many groups, can only be studied by the use of sections, which are tedious to prepare but nevertheless will reward the conscientious, open-minded worker with many new and valuable facts. It is only in this way that our final classification of groups can be made, coupled with a clearer and more adequate knowledge of fossil forms.

It is known from numerous, well-demonstrated cases that the arenaceous forms develop into calcareous ones. This is seen in the different stages of the same individual in several species, and probably will be found in many others, especially the more primitive genera or species of present calcareous forms. The application of weak acid to supposedly calcareous forms, especially in the older fossil series, has shown that the cement or the fragments may be calcareous, yet the cementing material or the fragmental material may be of other sorts, and the test retain its shape and character without further dissolving. Careful treatment with weak acid in this way will give much needed information as to the real structure of many so-called "calcareous" forms.

Additional research on the early Paleozoic, which has been so fruitful in the discovery of many primitive arenaceous forms, should be continued into unexplored areas. Sections of the early forms should be studied, with the aid of ultra-violet and polarized light, to determine more definitely the structure. Likewise, the chemical and mechanical composition of the test, in recent and fossil forms, should be studied much more fully than has been done.

Faunas, particularly of the Triassic, should be thoroughly studied for the origin of many of the Jurassic and later groups. The migrations of species and genera through various geologic formations should yield valuable information as to changes in ecologic conditions and paleogeography. Careful papers on fossil faunas with good illustrations are needed, particularly in economic work. The very poor or inaccurate figures given with some papers very largely detract from their usefulness.

Recent faunas should be studied with regard to the occurrence of species in definite ecologic conditions, especially of depth and temperature, so that fossil faunas may be interpreted in the light of the conditions under which living related forms now occur.

The method of using cores from the present ocean bottom gives a field for the interpretation, by means of the foraminifera, of various problems concerning changes of sea level, temperature, etc., which have wide application in geologic problems.

In the matter of nomenclature, there is need of careful study and refiguring of the original early types. Studies of genera in a monographic way, by one who is able and willing to consult the early collections, will give a fund of welcome information from many angles other than that of straightening out an already tangled mass of names. The bringing up to date of Sherborn's invaluable "Index to the Genera and Species of the Foraminifera" would place anyone who did it thoroughly under the obligation of workers on the foraminifera for a long while to come. It is understood that the records since Sherborn are to be published when conditions allow.

CHAPTER XI

SYSTEMATIC ARRANGEMENT OF THE FORAMINIFERA

In the following pages, the families, subfamilies, and genera are taken up in systematic order. Descriptions are short and simple, since to give fully all the details of the genera would require more than a single volume. Figures are given of practically all the genera treated here. The type figure is often copied, and where it is not adequate for the full understanding of the genus, a later figure of the species is usually given. Other species are given in many genera for comparison, both fossil and recent species are figured where space allows, and wherever possible the species represented in the key plates are different from those in the text. Many more figures might have been used to advantage, but there are limits to the number of illustrations that can be included. A very careful selection of figures has been made in order to use those which will give the clearest idea of the necessary structural points. Sections are used throughout for a fuller understanding of the structure.

A number of the plates show the relationships of the genera in the families, and the lines of development. The other plates give various structures to illustrate these, as words cannot do as clearly. The Key contains figures of practically all the genera, with references to the text. These plates with black backgrounds make the specimens appear more as they do under the microscope, and will be found perhaps of more value than the plates in the text with white backgrounds. The latter are mostly to show relationships. A table showing the geologic distribution is given, as a sample of what the student may make for his own use from the text itself. The Key includes the keys to the families, and to the genera under each family. Many of the genera in various families assume so many forms that it is difficult to get the full details into a statement of a line or so. As the student will gain much more from working with actual specimens than with the best of books, he will soon learn in which families various forms belong. The student should not depend upon these keys except as aids, and then should carefully check the fuller generic characters in the text.

The genotypes are given under each genus. The terms genoholotype, by designation, monotype, lectotype, etc., are not used, as the genotypes are now nearly all well worked out and accepted by practically all workers on the group. Likewise, synonyms have been greatly reduced. The generic names are given for synonyms, so that an advanced worker may have an idea under what genera the various species may have been placed in the older literature. Many of the older names cannot be definitely placed as synonyms, since their types have not been thoroughly studied,

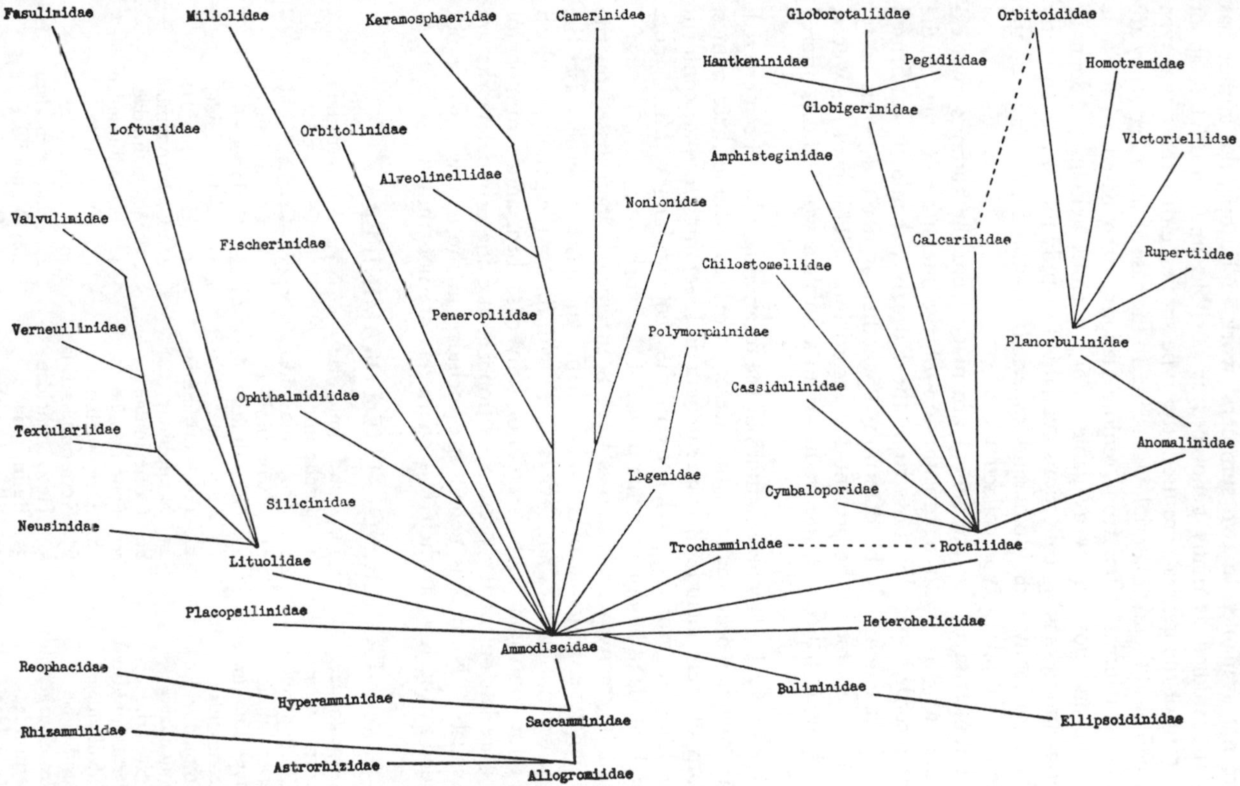

Relationships of the Families of the Foraminifera

FIGURE 8

or are not available. In two summers' work in Europe, the author has seen and studied as many genotypes as possible, and, in all but about 5 per cent of the genera here used, the type species, either the actual type specimen or authentic material, was seen. In the case of the others, they are usually rare forms, often from remote regions, often based on a single specimen, and not readily available. For generous help in the loan of material and in making collections available for study the author is greatly indebted to many individuals and especially to those in charge of the numerous European type collections.

The ranges given for the genera are made from the author's own experience, together with a consultation of the literature. Some of the earlier records, especially those from the Paleozoic, should be confirmed before it can definitely be determined whether or not the genus as recorded is the same as the present-day record for it. This book, however, is a base upon which to build, rather than a finished work, and is offered as such.

In recent years many new families of the foraminifera have been erected. No attempt has been made in the present work to include these names, as many of them include the same genera and seemingly do not conform to the knowledge of the phylogeny or the geologic ranges. Much more research needs to be done on this phase of the work.

Similarly, many generic names have been erected, some based on the same genotype, which at once makes them synonyms. In other cases the generic description does not always agree with that of the genotype, which makes for confusion. Others are not fully described, and a number of others are left out of the present edition because the original works are not available. Numerous genera are included in the text which are not included in the Key, as their full characters are not known.

FAMILIES OF THE FORAMINIFERA

1. Allogromiidae
2. Astrorhizidae
3. Rhizamminidae
4. Saccamminidae
5. Hyperamminidae
6. Reophacidae
7. Ammodiscidae
8. Lituolidae
9. Textulariidae
10. Verneuilinidae
11. Valvulinidae
12. Fusulinidae
13. Neoschwagerinidae
14. Loftusiidae
15. Neusinidae
16. Silicinidae
17. Miliolidae
18. Ophthalmidiidae
19. Fischerinidae
20. Trochamminidae
21. Placopsilinidae
22. Orbitolinidae
23. Lagenidae
24. Polymorphinidae
25. Nonionidae
26. Camerinidae
27. Peneroplidae
28. Alveolinellidae
29. Keramosphaeridae
30. Heterohelicidae
31. Buliminidae
32. Ellipsoidinidae
33. Rotaliidae
34. Pegidiidae
35. Amphisteginidae
36. Calcarinidae
37. Cymbaloporidae
38. Cassidulinidae
39. Chilostomellidae
40. Globigerinidae
41. Hantkeninidae
42. Globorotaliidae
43. Anomalinidae
44. Planorbulinidae
45. Rupertiidae
46. Victoriellidae
47. Homotremidae
48. Orbitoididae
49. Discocyclinidae
50. Myogypsinidae

ORDER FORAMINIFERA

Animals with a single cell, a protoplasmic body usually developing some sort of test, of chitin, agglutinated material, or of calcareous or rarely siliceous material secreted by the animal, typically with one or more definite apertures and the test usually perforate, in some families imperforate, pseudopodia of fine threads freely anastomosing to form a network.

Family 1. ALLOGROMIIDAE

Test either wanting or of chitin, not porous; aperture either single or at each end of the test, the surface of the test sometimes with attached foreign materials; fresh and brackish water, sometimes marine, not known as fossils.

For most of the students who will use this book, the Allogromiidae are of only slight interest. They are important from the point of view of simple primitive types in the foraminifera however, and figures are given of the important genera. No fossil records exist, as it is obvious that such tests as they possess would not leave their records in fossilization.

While the Allogromiidae are undoubtedly primitive, and give an excellent clue to what the earliest types of foraminifera were, they are in reality a very little known group. Most of the genera and species rest upon the original record of their discoverers, and have not been seen since. Most of the workers on the foraminifera have probably seen few if any of them.

The structure of the pseudopodia or of the main protoplasmic body has been little studied, and the whole family needs special study. Only the best characterized genera are given here. Others sometimes included with these may not belong here. For these various reasons, little space is given to this family here, and they are not included in the accompanying key.

Subfamily 1. Myxothecinae

Test when present of thin chitin, not rigid; the aperture not definite, but the pseudopodia capable of being pushed out through any part of the exterior; mostly marine.

Genus SCHULTZELLA Rhumbler, 1903

Plate 8, figure 1

Genotype, *Schultzia diffluens* Gruber

Schultzella Rhumbler, Arch. Prot., vol. 3, 1903, p. 197.
Lieberkühnia (part) Gruber, 1884. *Schultzia* Gruber, 1889 (not Grimm, 1877).

Test hardly distinguishable, of thin chitin which may be broken at any point for emission of the pseudopodia, generally spherical, without foreign material on the surface. — Recent, Mediterranean.

Genus MYXOTHECA Schaudinn, 1893

Plate 8, figure 2

Genotype, *Myxotheca arenilega* Schaudinn

Myxotheca SCHAUDINN, Zeitschr. Wiss. Zool., vol. 57, 1893, p. 18.
Pleurophrys GRUBER, 1884 (not CLAPARÈDE and LACHMANN, 1859).

Test consisting of a thin chitinous layer, the body of the animal generally spherical or flattened on the base, somewhat changeable in shape, surface with occasional attached foreign material.—Recent, Mediterranean and Atlantic.

Genus BODERIA Str. Wright, 1867

Plate 8, figure 3

Genotype, *Boderia turneri* Str. Wright

Boderia STR. WRIGHT, Journ. Anat. and Physiol., vol. 1, 1867, p. 335.

Test extremely thin and delicate, colorless body of changeable form, flattened, more or less angular, from the angles are long, fine, somewhat branching pseudopodia.—Recent, shallow water of North Sea.

Genus PLAGIOPHRYS Claparède and Lachmann, 1859

Plate 8, figure 4

Genotype, *Plagiophrys cylindrica* Claparède and Lachmann

Plagiophrys CLAPARÈDE and LACHMANN, Mém. Inst. Geneva, vol. 6, 1859, p. 453.

No definite test apparent, body short, cylindrical, pseudopodia from one end, the opposite end rounded.—Recent, fresh water.

PLATE 8

FAMILY 1. ALLOGROMIIDAE

FIG.

1. *Schultzella diffluens* (Gruber). (From Rhumbler, after type figure.)
2. *Myxotheca arenilega* Schaudinn. (From Rhumbler, after type figure.)
3. *Boderia turneri* Str. Wright. (From Rhumbler, after type figure.)
4. *Plagiophrys cylindrica* Claparède and Lachmann. × 110. (From Rhumbler, after type figure.)
5. *Dactylosaccus vermiformis* Rhumbler. × 15. (From Rhumbler, after type figure.)
6. *Allogromia mollis* (Gruber). (From Rhumbler, after type figure.)
7. *Allogromia lagenoides* (Gruber). (From Rhumbler, after type figure.)
8. *Lieberkühnia wageneri* Claparède and Lachmann. × 35. (From Rhumbler, after Verworn.)
9. *Shepheardella taeniformis* Siddall. × 8. (From Rhumbler, after type figure.)
10. *Rhynchosaccus immigrans* Rhumbler. × 20. (After type figure.)
11. *Rhynchogromia variabilis* Rhumbler. × 75. (After type figure.)
12. *Diplogromia brunneri* (Blanc). × 75. (From Rhumbler, after type figure.)
13. *Amphitrema wrightianum* Archer. × 200. (After Chapman.)
14. *Diaphoropodon mobile* Archer. × 65. (After Chapman.)

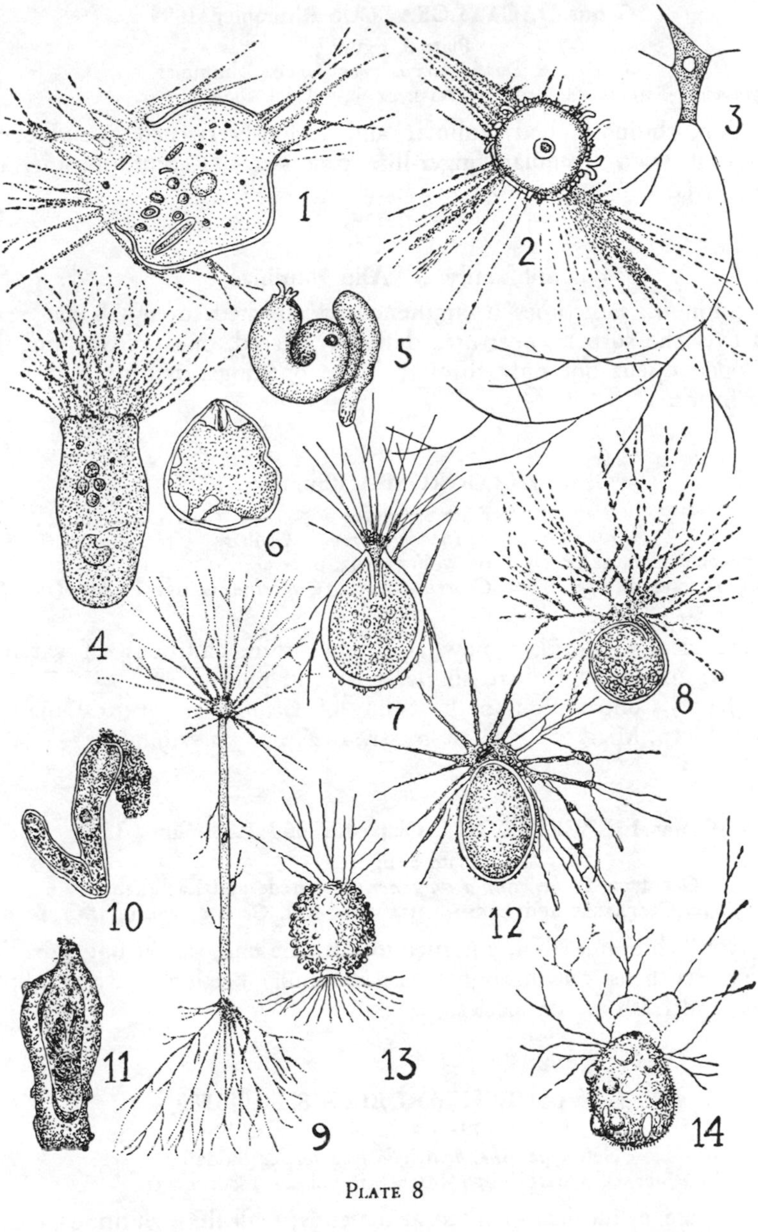

PLATE 8

Genus DACTYLOSACCUS Rhumbler, 1894

Plate 8, figure 5

Genotype, *Dactylosaccus vermiformis* Rhumbler

Dactylosaccus RHUMBLER, Zeitschr. Wiss. Zool., vol. 57, 1894, p. 601.

Test thin, chitinous, body tubular and variously twisted, pseudopodia thrust out from irregular finger-like processes.—Recent, marine, off Scandinavia.

Subfamily 2. Allogromiinae

Test chitinous, sometimes strengthened by attached foreign bodies scattered over the surface; apertures definitely placed, one or two, and the pseudopodia sent out only through these openings; marine and fresh water forms.

Genus ALLOGROMIA Rhumbler, 1903

Plate 8, figures 6, 7

Genotype, *Craterina mollis* Gruber

Allogromia RHUMBLER, Arch. Prot., vol. 3, 1903, p. 203.

Gromia DUJARDIN, 1839 (not *Gromia* 1835). *Craterina* GRUBER, 1884 (not ST. VINCENT, 1824).

Test spherical to ovoid, somewhat flexible, thin, chitinous, of variable thickness; aperture terminal, single.—Recent, marine.

See Jepps, Contribution to the Study of *Gromia oviformis* Dujardin (Quart. Journ. Micr. Sci., vol. 70, n. ser., 1926, pp. 701–719, pls. 37–39).

Genus LIEBERKÜHNIA Claparède and Lachmann, 1859

Plate 8, figure 8

Genotype, *Lieberkühnia wageneri* Claparède and Lachmann

Lieberkühnia CLAPARÈDE and LACHMANN, Mém. Inst. Geneva, vol. 6, 1859, p. 464.

Test ovoid, chitinous, thin, aperture toward the end, but at one side, the pseudopodia thrust out through a sort of tubular portion.—Recent, both marine and fresh water species.

Genus SHEPHEARDELLA Siddall, 1880

Plate 8, figure 9

Genotype, *Shepheardella taeniformis* Siddall

Shepheardella SIDDALL, Quart. Journ. Micr. Sci., vol. 20, 1880, p. 131.

Test elongate, cylindrical, pointed at the ends; wall thin, chitinous, colorless; pseudopodia mostly from the apertures at either end of the test.—Recent, marine.

Genus RHYNCHOSACCUS Rhumbler, 1894

Plate 8, figure 10

Genotype, *Rhynchosaccus immigrans* Rhumbler

Rhynchosaccus RHUMBLER, Zeitschr. Wiss. Zool., vol. 57, 1894, p. 595.

Test tubular, arcuate or with a sharp bend, both ends rounded, the apertural end with a projecting snout, wall thin, chitinous.—Recent, marine.

Genus RHYNCHOGROMIA Rhumbler, 1894

Plate 8, figure 11

Genotype, *Rhynchogromia variabilis* Rhumbler

Rhynchogromia RHUMBLER, Zeitschr. Wiss. Zool., vol. 57, 1894, p. 590.

Test usually elongate, of a single chitinous layer, the protoplasmic contents usually not completely filling the interior, which contains also more or less foreign material; aperture terminal.—Recent, fresh water and marine.

Genus DIPLOGROMIA Rhumbler, 1903

Plate 8, figure 12

Genotype, *Gromia brunneri* Blanc

Diplogromia RHUMBLER, Arch. Prot., vol. 3, 1903, p. 214.
Gromia (part) BLANC, 1886 (not DUJARDIN, 1835).

Test ovoid, with a double wall, the outer layer of flinty grains and other foreign material, the inner layer chitinous, clear; aperture terminal.—Recent, fresh water, Switzerland.

Genus DIAPHOROPODON Archer, 1869

Plate 8, figure 14

Genotype, *Diaphoropodon mobile* Archer

Diaphoropodon ARCHER, Quart. Journ. Micr. Sci., vol. 9, 1869, p. 394.

Test ovate, of chitin, with the exterior of foreign material; aperture terminal.—Recent, fresh water, Ireland.

Genus AMPHITREMA Archer, 1870

Plate 8, figure 13

Genotype, *Amphitrema wrightianum* Archer

Amphitrema ARCHER, Quart. Journ. Micr. Sci., vol. 10, 1870, p. 122.

Test short, fusiform, the ends truncate and open, exterior with foreign material.—Recent, fresh water, Ireland.

Family 2. ASTRORHIZIDAE

Test free, consisting of a central chamber from which radiate tubular channels to the exterior, either simple or branching; wall with a thin chitinous inner layer on all or part of which is agglutinated arenaceous material; apertures formed by the peripheral ends of the arms or by openings in the peripheral wall.

The Astrorhizidae consist of some of the simplest and most primitive of the foraminifera. The test is composed of agglutinated material on the outside showing little selectivity, and a thin chitinous lining within. In its simplest form, that of *Astrorhiza*, there is a central body from which the protoplasm streams in varying directions until the material of the environment is more or less cemented about the stream margins, and a stellate test results. From this simple form, the approach to which may be seen in observing a living foraminifer in active feeding, building up foreign material on the test and along the pseudopodial lines, it is a simple step to the other forms. A great accumulation of material would give rise to such a form as *Crithionina* with channels through the very thick soft walls to the exterior. *Astrammina* and *Pelosphaera* have short arms from a large chamber. *Vanhoeffenella* is the most interesting genus with its clear central portion of chitin showing clearly the protoplasmic body within, and the peripheral portion built of arenaceous material.

The group has usually been recognized as very primitive, not as an ancestral family, but as one which has kept its simple character through geologic time. Authors since Brady's time have placed with this family other groups which are now separated from it.

Although the type of test is difficult to recognize in many types of sedimentary rocks, five of the genera take the early history of the family back at least as far as the Silurian.

Genus ASTRORHIZA Sandahl, 1858

Key, plate 1, figures 1–3

Genotype, *Astrorhiza limicola* Sandahl

Astrorhiza Sandahl, Öfv. Svensk. Vet.-Akad. Förh., vol. 14, 1857 (1858), p. 299.
Arenistella Fischer and deFolin, 1872. *Astrodiscus* F. E. Schulze, 1875. *Haeckelina* Bessels, 1875. *Ammodiscus* Carpenter and Jeffreys, 1870 (not Reuss).

Test free, flattened or tubular, stellate or subcylindrical, composed of a central chamber with tubular portions to the exterior in the compressed stellate species, or an irregular tubular chamber in the subcylindrical ones; wall of loosely cemented mud or sand showing little selection, the interior with a thin chitinous lining; aperture at the outer ends of the tubular portions.—Jurassic to Recent, undoubtedly earlier.

Most of the species are characteristic of cool water conditions, and temperature is more of a control than depth. The proportion of cement is small, and it is not surprising that this very primitive form has not yet been found in the earliest rocks.

Genus PSEUDASTRORHIZA Eisenack, 1932

Key, plate 1, figure 4

Genotype, *Pseudastrorhiza silurica* Eisenack

Pseudastrorhiza EISENACK, Pal. Zeitschr., vol. 14, 1932, p. 259.

Test free, very similar to *Astrorhiza*, with a central chamber and several radiating arms in varying planes; wall thin, of fine sand grains; apertures not apparent. — Silurian of the Baltic region.

There may be as many as seven arms to the test. It probably had a chitinous base to hold together the thin arenaceous wall.

Genus MASONELLA H. B. Brady, 1889

Key, plate 1, figure 13

Genotype, *Masonella planulata* H. B. Brady

Masonella H. B. BRADY, Ann. Mag. Nat. Hist., ser. 6, vol. 3, 1889, p. 295.

Test stellate or circular, much compressed, consisting of a large central chamber with fine branching tubules extending to the periphery; wall finely arenaceous, thin, especially between the tubules; apertures at the open ends of the tubules. — Recent, Indian Ocean.

Genus RHABDAMMINA M. Sars, 1869

Key, plate 1, figures 9–12

Genotype, *Rhabdammina abyssorum* M. Sars

Rhabdammina M. SARS, Forh. Selsk. Christiania, 1868, p. 248 (*nomen nudum*); in W. B. CARPENTER, Ann. Mag. Nat. Hist., ser. 4, vol. 4, 1869, p. 288.

Test free, either radiate, subcylindrical or branching; wall firmly cemented, usually of sand grains (not mud), but occasionally with sponge spicules or other foreign fragments, showing some selection, interior with a thin chitinous lining, cement usually yellowish-brown; open ends of the tubes serving as apertures. — Silurian to Recent.

Temperature is more of a control than depth, and it is often abundant in comparatively shallow depths in Arctic and Subarctic regions.

Genus CRITHIONINA Goës, 1894

Key, plate 1, figures 7, 8

Genotype, *Crithionina mamilla* Goës

Crithionina GOËS, Kongl. Svensk. Vet.-Akad. Handl., vol. 25, No. 9, 1894, p. 14.

Test free, spherical, lenticular or variously shaped, consisting of a single chamber with tubules, simple or branching, connecting with the exterior; wall of sponge spicules and very fine sand, often chalky, soft, with little cement; color white or grayish. — Silurian to Recent.

Moreman's records from the Silurian are the earliest fossil occurrences. It often occurs in great numbers in cold waters of medium depths.

Genus VANHOEFFENELLA Rhumbler, 1905

Key, plate 1, figure 14

Genotype, *Vanhoeffenella gaussi* Rhumbler

Vanhoeffenella RHUMBLER, Verhandl. deutsch. Zool. Gesell., Jahr. 15, 1905, p. 105.

Test free, consisting of a compressed central chamber with thin chitinous wall, the exterior a polygonal, peripheral, tubular chamber of sand grains with apertural tubes at the angles; apertures at the ends of the tubes. — Recent, particularly the Antarctic.

This genus is somewhat similar to an *Astrorhiza* with the outer material removed from the central disc, leaving the chitinous lining.

Genus ASTRAMMINA Rhumbler, 1931

Key, plate 1, figure 6

Genotype, *Astrammina rara* Rhumbler

Astrammina RHUMBLER, in WIESNER, Deutsche Südpolar Exped., vol. XX, Zool., 1931, p. 77.

Test free, consisting of a large central chamber with an arenaceous wall from which radiate tubular projections in one or several planes. — Recent, Antarctic.

Genus PELOSPHAERA Heron-Allen and Earland, 1932

Key, plate 1, figure 5

Genotype, *Pelosphaera cornuta* Heron-Allen and Earland

Pelosphaera HERON-ALLEN and EARLAND, Journ. Roy. Micr. Soc., vol. 52, 1932, p. 255.

Test free, large, roughly spherical, with two or more conical arms; wall soft and friable on the outside, firm and smooth within; outer ends of the arms without definite apertures. — Recent, Antarctic.

Genus ORDOVICINA Eisenack, 1937

Key, plate 42, figure 1

Genotype, *Ordovicina oligostoma* Eisenack

Ordovicina EISENACK, Pal. Zeitschr., vol. 19, 1937, p. 234.

Test free, single-chambered, irregularly polygonal in shape with tubular extensions, wall largely chitinous with finely arenaceous surface; apertures probably at the ends of the tubular extensions. — Silurian.

Genus AMPHITREMOIDEA Eisenack, 1937

Key, plate 42, figure 2

Genotype, *Amphitremoidea citroniforma* Eisenack

Amphitremoidea EISENACK, Pal. Zeitschr., vol. 19, 1937, p. 235.

Test free, single-chambered, fusiform; wall thick, smooth, finely arenaceous; apertures at the tapering ends of the test. — Silurian.

Family 3. RHIZAMMINIDAE

Test consisting of a tubular chamber open at both ends; wall with a chitinous lining, and exterior of agglutinated foreign material, sand, sponge spicules or other foraminifera; apertures formed by the open ends of the tubes.

The Rhizamminidae are all of very simple structure, consisting of a tube open at both ends. The basal inner layer is of chitin, as can be easily demonstrated in *Rhizammina* by dissolving away the calcareous covering. On this chitinous base is built up a wall of various foreign materials. The Rhizamminidae may be derived directly from such a simple tubular form as *Shepheardella* in the Allogromiidae.

The fossil record goes back to the early Paleozoic. In the present oceans, the species are widely distributed, some species of *Marsipella* occurring in great numbers in shallow warm water off Florida, and some in deep cold water.

Genus RHIZAMMINA H. B. Brady, 1879

Key, plate 1, figures 18, 19

Genotype, *Rhizammina algaeformis* H. B. Brady

Rhizammina H. B. Brady, Quart. Journ. Micr. Sci., vol. 19, 1879, p. 39.

Test a simple or dichotomously branching tube; wall flexible, chitinous, with various foreign bodies, usually other foraminiferal tests attached to the exterior; apertures formed by the open ends of the tubes. — Jurassic to Recent.

Genus MARSIPELLA Norman, 1878

Key, plate 1, figure 22

Genotype, *Marsipella elongata* Norman

Marsipella Norman, Ann. Mag. Nat. Hist., ser. 5, vol. 1, 1878, p. 281.
Proteonina W. B. Carpenter, 1869 (not Williamson, 1858).

Test free, tubular, cylindrical or fusiform, sometimes recurved at the ends; wall thin, firmly cemented, composed wholly or in part of sponge spicules, or the middle part of sand grains; apertures formed by the open ends of the tube, one end sometimes closed by a loosely aggregated knob of spicules. — Ordovician to Recent.

Genus BATHYSIPHON M. Sars, 1872

Key, plate 1, figures 15–17

Genotype, *Bathysiphon filiformis* M. Sars

Bathysiphon M. Sars, in G. O. Sars, Forh. Vidensk.-Selsk. Christiania, 1871 (1872), p. 251.

Test free, cylindrical, often tapering slightly, straight or slightly curved, sometimes constricted externally; wall with a base of broken sponge spicules firmly cemented and overlaid with fine-grained, amorphous material, soft or firmly cemented, often with a very thin surface coating; apertures at the ends of the tube. — Silurian to Recent.

Genus ARENOSIPHON Grubbs, 1939

Key, plate 42, figure 3

Genotype, *Arenosiphon gigantea* Grubbs

Arenosiphon Grubbs, Journ. Pal., vol. 13, 1939, p. 544.

Test free, cylindrical, rarely branching, commonly tapering slightly, straight or slightly curved, occasionally constricted at intervals; wall of medium to fine sand grains, firmly cemented, surface rough; apertures at ends of tube. — Silurian.

Genus HIPPOCREPINELLA Heron-Allen and Earland, 1932

Key, plate 1, figures 20, 21

Genotype, *Hippocrepinella hirudinea* Heron-Allen and Earland

Hippocrepinella Heron-Allen and Earland, Journ. Roy. Micr. Soc., vol. 52, 1932, p. 257.

Test elongate, tubular; wall thin, of fine sand and mud, little cement, flexible in life; apertures at the two ends. — Recent, Arctic and Antarctic.

This is a primitive form which retains the flexible character of the wall.

Family 4. SACCAMMINIDAE

Test free or attached, composed typically of a single chamber, or occasionally with chambers of the same sort loosely united; wall lined with chitin, the exterior of agglutinated material of various sorts, sand grains, sponge spicules or other foraminiferal tests; aperture usually single, of various shapes.

The early history of this primitive family goes back to the early Paleozoic where a number of genera are now known. In its simplest form the test is spherical, the inner wall of chitin, and the exterior of sand grains or other agglutinated material without definite apertures. Such forms are known from the early Paleozoic to the present time. In its next higher stage the test develops a definite single aperture as in *Saccammina*. These forms also are known from the early Paleozoic to the present. There is a certain amount of selection in the materials used for the exterior of the test. Certain species use sponge spicules to the exclusion of almost everything else; others use mica flakes; and in *Technitella thompsoni*, the test is entirely made of the ambulacral plates of brittle stars. In some species there are two layers of sponge spicules laid in opposite directions so that a very strong test is produced. In the case of those species which use mica flakes entirely, a very weak test necessarily results. There is a great range in the amount of chitinous material used, some forms such as *Leptodermella* being largely composed of chitin, while other forms may have a very thin inner layer. A few genera have more than one aperture, and in *Thurammina* there are numerous apertures scattered over the entire surface.

There are a considerable number of attached forms in this family some of which have very definite shapes, while others are irregular in form. Very few of the attached forms occur as fossils, partly due to the fact that they are difficult to distinguish from the surface to which they are attached. Most of the genera of this family are most abundant in rather cool deep waters, but a few, such as *Iridia* and *Diffusilina*, occur in shallow warm waters.

It is possible that some of the genera grouped in this family may belong to the Filosa, and are perhaps not true foraminifera. This can only be determined by a study of the living animal.

Subfamily 1. Psammosphaerinae

Test without a definite aperture.

Genus PSAMMOSPHAERA Schulze, 1875

Key, plate 2, figure 1
Genotype, *Psammosphaera fusca* Schulze

Psammosphaera SCHULZE, II Jahr. Comm. Wiss. Unt. deutsch. Meer in Kiel, 1875, p. 113.
Psammella RHUMBLER, 1935.

Test free or attached, globular; wall composed of a thin layer of chitin with an outer layer of sand grains, mica flakes, sponge spicules or other foraminiferal tests, firmly cemented; aperture indefinite. — Silurian to Recent.

Various species of *Psammosphaera* show high selectivity in the choice of materials for the test. The genus *Orbulinaria* Rhumbler, 1911, may belong here if it can be recognized. Its characters are not well defined.

Genus PILALLA Rhumbler, 1935

Key, plate 42, figure 4
Genotype, *Pilalla exigua* Rhumbler

Pilalla RHUMBLER, Schrift. Nat. Ver. Schleswig-Holstein, Bd. XXI, 1935, p. 150.

Test free, single-chambered, globular; wall chitinous, impregnated with fine mineral grains; no definite aperture. — Recent.

Genus CAUSIA Rhumbler, 1938

Key, plate 42, figures 5–7
Genotype, *Causia injudicata* Rhumbler

Causia RHUMBLER, Kieler Meeresforschungen, vol. 2, 1938, p. 171.

Test single-chambered, planoconvex, dorsal side convex, ventral side concave, periphery extending outward in a thin, broad flange; wall chitinous, thin, more or less covered with fine mineral grains; aperture on the concave ventral side, a small rounded opening or wanting.—Recent.

Genus BLASTAMMINA Eisenack, 1932

Key, plate 2, figure 4

Genotype, *Blastammina polymorpha* Eisenack

Blastammina EISENACK, Pal. Zeitschr., vol. 14, 1932, p. 261.

Test free or attached, generally rounded, polyhedral or irregular, of single chambers or rarely in irregular colonies; wall of thin, brownish chitin with a thin outer layer of sand grains; apertures indefinite.—Silurian, Baltic region.

Genus SHIDELERELLA Dunn, 1942

Key, plate 42, figure 8

Genotype, *Shidelerella biscuspidata* Dunn

Shidelerella DUNN, Journ. Pal., vol. 16, 1942, p. 328.

Test free, elongate-fusiform or cylindrical, upper portion tapering to a tubular neck, basal end with projections, possibly with apertures; wall thin, of medium to large sand grains, poorly cemented; aperture at end of neck and sometimes at ends of small projections at base.—Silurian.

Genus THEKAMMINA Dunn, 1942

Key, plate 42, figure 9

Genotype, *Thekammina quadrangularis* Dunn

Thekammina DUNN, Journ. Pal., vol. 16, 1942, p. 326.

Test free, with walls compressed to give an angular outline; varying from a flattened triangular to a box-shaped test; wall thin, composed of fine to coarse sand grains, usually poorly cemented; aperture indefinite.—Silurian.

Genus SOROSPHAERA H. B. Brady, 1879

Key, plate 2, figure 2

Genotype, *Sorosphaera confusa* H. B. Brady

Sorosphaera H. B. BRADY, Quart. Journ. Micr. Sci., vol. 19, 1879, p. 28.

Test consisting of a group of inflated independent chambers loosely joined in an aggregate; wall with a chitinous base and outer layer of fine sand grains; apertures minute, interstitial.—Silurian to Recent.

Genus ARENOSPHAERA Stschedrina, 1939

Key, plate 42, figure 10

Genotype, *Arenosphaera perforata* Stschedrina

Arenosphaera STSCHEDRINA, Comptes Rendus Acad. Sci. URSS, vol. 24, 1939, p. 95.

Test free or attached; chambers one or more, up to three, nearly spherical, with porelike spaces between; wall coarsely arenaceous; aperture represented by numerous openings over the surface between the sand grains, of various size and form.—Recent.

Genus PSAMMOPHAX Rhumbler, 1931

Key, plate 2, figure 7

Genotype, *Psammophax consociata* Rhumbler

Psammophax Rhumbler, in Wiesner, Deutsche Südpolar Exped., vol. XX, Zool., 1931, p. 80.

Test composed of two or more chambers in a series, straight or irregularly coiled; wall of sand grains or spicules; apertures minute, interstitial.—Recent, Antarctic.

Psammophax differs from *Sorosphaera* in the more definite arrangement of chambers, but the two are evidently very closely related.

Genus STORTHOSPHAERA Schulze, 1875

Key, plate 2, figure 3

Genotype, *Storthosphaera albida* Schulze

Storthosphaera Schulze, II Jahr. Comm. Wiss. Unt. deutsch. Meer in Kiel, 1875, p. 113.

Test free, irregularly rounded, single-chambered; wall thick, of whitish fine sand, very loosely cemented; aperture indefinite.—Recent.

Genus STEGNAMMINA Moreman, 1930

Key, plate 2, figure 6

Genotype, *Stegnammina cylindrica* Moreman

Stegnammina Moreman, Journ. Pal., vol. 4, 1930, p. 49.

Test free, a straight cylindrical or subcylindrical chamber; wall thin, composed of small to medium sized sand grains, well cemented; aperture indefinite.—Lower Paleozoic, America.

Genus CERATAMMINA Ireland, 1939

Key, plate 42, figures 11, 12

Genotype, *Ceratammina cornucopia* Ireland

Ceratammina Ireland, Journ. Pal., vol. 13, 1939, p. 194.

"Test free, horn-shaped; wall composed of fine sand grains; surface smooth; aperture not apparent."—Devonian.

Genus RAIBOSAMMINA Moreman, 1930

Key, plate 2, figure 5

Genotype, *Raibosammina mica* Moreman

Raibosammina Moreman, Journ. Pal., vol. 4, 1930, p. 50.

Test free or attached, subcylindrical, straight, crooked, or irregularly branched, interior of chamber not of uniform diameter; wall of unequal thickness, composed of poorly sorted sand grains; aperture not apparent. — Lower Paleozoic, America.

Subfamily 2. Saccammininae

Test free, with a definite aperture; wall of firmly agglutinated sand, sponge spicules or other objects, usually not thick nor with fine white amorphous material.

Genus SACCAMMINA M. Sars, 1869

Key, plate 2, figure 8

Genotype, *Saccammina sphaerica* M. Sars

Saccammina M. SARS, Förh. Vidensk.-Selsk. Christiania, 1868 (1869), p. 248 (nomen nudum); in W. B. CARPENTER, Ann. Mag. Nat. Hist., ser. 4, vol. 4, 1869, p. 289.

Test typically free, sometimes attached, globular, usually single, occasionally several attached; wall with a chitinous base and outer single layer of sand grains, firmly cemented; aperture single, often with a slight neck. — Silurian to Recent.

The genus *Saccamminopsis* Sollas, 1921, is based upon *Saccammina carteri* H. B. Brady which is probably not a foraminifer.

Genus PROTEONINA Williamson, 1858

Key, plate 2, figure 9

Genotype, *Proteonina fusiformis* Williamson

Proteonina WILLIAMSON, Rec. Foram. Great Britain, 1858, p. 1.

Reophax (part) of authors. *Difflugia* EGGER, 1895 (not LECLERC, 1815).

Saccammina (part) EIMER and FICKERT, 1899.

Test free, a fusiform or flask-shaped undivided chamber; wall a thin chitin layer on which are cemented sand grains, mica flakes, other tests, etc.; aperture usually circular, often with a slight neck which may become elongate. — Silurian to Recent.

It is possible that *Brachysiphon* Chapman, 1906, may belong here.

The genus *Archaelagena* Howchin, 1888, is very obscure as to its real characters, and needs further investigation.

Genus LAGENAMMINA Rhumbler, 1911

Key, plate 2, figure 10

Genotype, *Lagenammina laguncula* Rhumbler

Lagenammina RHUMBLER, Foram. Plankton-Exped., pt. 1, 1911, p. 92.

Test free, bottle-shaped; wall consisting of a pseudochitinous layer on which is a thin covering of foreign bodies; aperture circular, neck elongate. — Silurian to Recent.

Genus LAGUNCULINA Rhumbler, 1903

Key, plate 2, figure 11
Genotype, *Ovulina urnula* Gruber

Lagunculina RHUMBLER, Archiv. Prot., vol. 3, 1903, p. 248.
Ovulina GRUBER, 1884 (not EHRENBERG, 1855).

Test free, a flask-shaped simple chamber; wall of fine sand grains; aperture circular, with a short neck and flaring lip.—Recent, Gulf of Mexico, shallow water.

Genus MILLETTELLA Rhumbler, 1903

Key, plate 2, figure 12
Genotype, *Reophax pleurostomelloides* Millett

Millettella RHUMBLER, Archiv. Prot., vol. 3, 1903, p. 250.
Reophax (part) MILLETT, 1899.

Test free, ovate; wall arenaceous, of fine grains; aperture large, semicircular, in a depression near one end of the test, and at one side.—Eocene to Recent.

Genus MARSUPULINA Rhumbler, 1903

Key, plate 2, figure 15
Genotype, *Marsupulina schultzei* Rhumbler

Marsupulina RHUMBLER, Archiv. Prot., vol. 3, 1903, p. 249.
Ovulina SCHULTZE, 1854 (not EHRENBERG, 1854).

Test free, pouch-like, ellipsoid or kidney-shaped; wall with amorphous calcareous material; aperture circular, at one side near the end of the test.—Recent, Ancona, Mediterranean, in shallow water.

Genus PLACENTAMMINA Majzon, 1943

Key, plate 42, figure 13
Genotype, *Reophax placenta* Grzybowski

Placentammina MAJZON, Mitt. Jahrb. K. Ungarn. Geol. Anstalt, vol. 37, 1943, p. 151.
Reophax (part) of authors.

Test single-chambered, round, strongly compressed, with a thickened periphery; wall arenaceous; aperture a small rounded opening with a thickened rim in the face of the test removed from the margin.—Upper Cretaceous and Tertiary.

Genus URNULINA Gruber, 1884

Key, plate 2, figure 16
Genotype, *Urnulina difflugiaeformis* Gruber

Urnulina GRUBER, Nova Acta Acad. Leop., vol. 46, 1884, p. 496.

Test free, obovate, initial end pointed, apertural end broad; wall thin, arenaceous; aperture terminal, large, and rounded.—Miocene to Recent.

Genus LEPTODERMELLA Rhumbler, 1935

Key, plate 2, figure 14

Genotype, *Pseudarcella arenata* Cushman

Leptodermella RHUMBLER, Schrift. Nat. Ver. Schleswig-Holstein, Bd. XXI, 1935, p. 177.

Pseudarcella CUSHMAN, 1930 (not SPANDEL, 1909).

Test single-chambered, plano-convex, rounded in dorsal view, semi-elliptical in side view, dorsal side convex, ventral side flattened or concave; wall thin, of chitin with a layer of fine, arenaceous material on the surface; aperture in the middle of the ventral side. — Eocene to Recent.

According to Rhumbler, the test of the type species of *Pseudarcella* is calcareous, thick-walled, and perforate. The position of the genus seems obscure until further study of the types can be made.

Genus AMMOSPHAEROIDES Cushman, 1910

Key, plate 2, figure 13

Genotype, *Ammosphaeroides distoma* Cushman

Ammosphaeroides CUSHMAN, Bull. 71, U. S. Nat. Mus., pt. 1, 1910, p. 51.

Test free, consisting of an elongate or subspherical chamber; wall finely arenaceous with a large proportion of reddish-brown cement; aperture typically double, at the end of short tubular portions of the test. — Silurian, Recent.

Genus THURAMMINA H. B. Brady, 1879

Key, plate 2, figure 17

Genotype, *Thurammina papillata* H. B. Brady

Thurammina H. B. BRADY, Quart. Journ. Micr. Sci., vol. 19, 1879, p. 45.

Thyrammina RHUMBLER, 1903. *Lituola* (part) W. B. CARPENTER, 1875.

Test typically free, usually nearly spherical, sometimes compressed; chamber, typically simple, occasionally divided; wall thin, chitinous, with fine sand; apertures several to many, at the end of nipple-like protuberances from the surface, occasionally wanting. — Silurian to Recent.

Under the name *Thuramminopsis* Haeusler has described peculiar forms from the Jurassic, apparently arenaceous, and with a network of tubular structures on the inside of the wall, and probably filling the interior. These peculiar forms, if they are true foraminifera, need further study to determine their relationships.

Genus THURAMMINOIDES Plummer, 1945

Key, plate 42, figures 15, 16

Genotype, *Thuramminoides sphaeroidalis* Plummer

Thuramminoides PLUMMER, Univ. Texas Publ. 4401, 1945, p. 218.

Test unicellular, rounded, somewhat compressed; wall finely arenaceous, interior labyrinthic; apertures one or several or wanting. — Pennsylvanian.

Genus ARMORELLA Heron-Allen and Earland, 1932

Key, plate 42, figure 14
Genotype, *Armorella sphaerica* Heron-Allen and Earland
Armorella HERON-ALLEN and EARLAND, Journ. Roy. Micr. Soc., vol. 52, 1932, p. 256.

Test free, spherical, consisting of a single chamber with a number of tubular extensions; wall arenaceous, of fine sand, diatoms and sponge spicules firmly cemented with a large proportion of light colored cement; apertures at the ends of the tubes. — Recent.

Genus GASTROAMMINA Dunn, 1942

Key, plate 42, figure 19
Genotype, *Gastroammina williamsae* Dunn
Gastroammina DUNN, Journ. Pal., vol. 16, 1942, p. 335.

Test free, stomach-shaped, large, composed of a proloculum and strongly inflated second chamber; wall thin, composed of rather large sand grains poorly cemented, surface rough; aperture at end of a broad tubular portion of the test. — Silurian.

Subfamily 3. Pelosininae

Test free; wall typically of matted spicules and fine amorphous material; aperture usually single.

Genus PELOSINA H. B. Brady, 1879

Key, plate 2, figure 18
Genotype, *Pelosina variabilis* H. B. Brady
Pelosina H. B. BRADY, Quart. Journ. Micr. Sci., vol. 19, 1879, p. 30.

Test free, rounded, cylindrical or irregularly elongate; wall with a chitinous inner layer on which is a thick layer of mud or amorphous material except about the apertural neck; aperture terminal, rounded. — Carboniferous (?) to Recent.

Genus TECHNITELLA Norman, 1878

Key, plate 2, figure 20
Genotype, *Technitella legumen* Norman
Technitella NORMAN, Ann. Mag. Nat. Hist., ser. 5, vol. 1, 1878, p. 279.

Test free, composed of a single, simple, elongate, subcylindrical, fusiform or oval chamber; wall thin, of sponge spicules and fine sand; aperture terminal, rounded, with or without a neck. — Tertiary and Recent, typically in cold or deep water.

Genus PILULINA W. B. Carpenter, 1870

Key, plate 2, figure 21
Genotype, *Pilulina jeffreysii* W. B. Carpenter
Pilulina W. B. CARPENTER, Desc. Cat. Objects Deep Sea Dredging, 1870, p. 5.

Test free, globular or ovate, consisting of a single undivided chamber; wall of felted sponge spicules with a minimum of fine sand and cement; aperture elongate, in a somewhat depressed area, or occasionally with a slightly raised lip. — Recent, deep or cold water.

Genus CRONEISELLA Dunn, 1942

Key, plate 42, figure 17
Genotype, *Croneisella typa* Dunn
Croneisella DUNN, Journ. Pal., vol. 16, 1942, p. 334.

Test free, cylindrical, constricted in the central portion and sometimes slightly bent, with both ends sloping toward pointed necks; wall thin, composed of medium to fine sand grains, well cemented, surface rough; apertures at ends of necks. — Silurian.

Subfamily 4. Webbinellinae

Test attached; wall of agglutinated foreign material.

Genus WEBBINELLA Rhumbler, 1903

Key, plate 2, figure 22
Genotype, *Webbina hemisphaerica* Jones, Parker and H. B. Brady
Webbinella RHUMBLER, Archiv. Prot., vol. 3, 1903, p. 228.
Webbina JONES, PARKER and H. B. BRADY, 1865 (not D'ORBIGNY, 1839). *Psammosphaera* (part) EIMER and FICKERT, 1899.

Test attached, plano-convex, circular in outline, central part convex, surrounded by a flattened border; wall of fine sand grains, smoothly finished; no general aperture, pseudopodia thrust out along the basal rim. — Silurian to Recent.

Genus IRIDIA Heron-Allen and Earland, 1914

Key, plate 2, figure 26
Genotype, *Iridia diaphana* Heron-Allen and Earland
Iridia HERON-ALLEN and EARLAND, Trans. Zool. Soc. London, vol. 20, pt. 12, 1914, p. 371.

Test attached, base of clear chitin, upper surface convex, of sand grains or other foreign material; apertures irregular, peripheral. — Eocene (?), Recent, shallow water.

Genus AMPHIFENESTRELLA Rhumbler, 1935

Key, plate 42, figure 18

Genotype, *Amphifenestrella wiesneri* Rhumbler

Amphifenestrella RHUMBLER, Schrift. Nat. Ver. Schleswig-Holstein, Bd. XXI, 1935, p. 169.

Test free, disc shaped, the two flattened sides of transparent chitin with arenaceous material about the periphery; no aperture apparent. — Recent.

Genus COLONAMMINA Moreman, 1930

Key, plate 2, figure 25

Genotype, *Colonammina verruca* Moreman

Colonammina MOREMAN, Journ. Pal., vol. 4, 1930, p. 55.

Test attached, plano-convex, circular or elliptical in outline, attached surface surrounded by a more or less flattened border; wall thin, composed of fine sand, well cemented; aperture single, on the convex surface. — Lower Paleozoic, America.

Genus KERIONAMMINA Moreman, 1933

Key, plate 42, figure 22

Genotype, *Kerionammina favus* Moreman

Kerionammina MOREMAN, Journ. Pal., vol. 7, 1933, p. 397.

Test attached, early portion obscure, later portion spreading in a flattened, irregular manner, interior labyrinthic but tending to become divided into somewhat regularly arranged chambers; wall arenaceous; apertures at ends of the peripheral, tubular arms. — Upper Ordovician.

Genus THOLOSINA Rhumbler, 1895

Key, plate 2, figure 24

Genotype, *Placopsilina bulla* H. B. Brady (part)

Tholosina RHUMBLER, Nachr. Köngl. Ges. Wiss. Göttingen, 1895, p. 82.

Placopsilina (part) H. B. BRADY, 1879. *Pseudoplacopsilina* EIMER and FICKERT, 1899.

Test attached, a hemispherical chamber undivided; wall of sand grains with a large proportion of calcareous cement; apertures small, circular, protuberant or irregular, just above the base of the test. — Silurian, Recent.

Genus VERRUCINA Goës, 1896

Key, plate 2, figure 23

Genotype, *Verrucina rudis* Goës

Verrucina GOËS, Bull. Mus. Comp. Zoöl., vol. 29, 1896, p. 25.

Test attached, irregular, ovoid, interior somewhat labyrinthic; wall coarsely arenaceous; aperture usually double, in a depressed area in the middle of the dorsal side. — Recent, Pacific coast of Mexico, 772 fathoms, attached to *Rhabdammina*.

Genus URNULA Wiesner, 1931

Key, plate 2, figure 27
Genotype, *Urnula quadrupla* Wiesner
Urnula WIESNER, Deutsche Südpolar Exped., vol. XX, Zool., 1931, p. 82.

Test attached, in young stage similar to *Webbinella*, later adding chambers and each becoming polygonal in outline, whole test plano-convex, rounded; base with a thin chitinous layer, dorsal side with a fine arenaceous covering; aperture in adults, rounded, one near the middle of each chamber. — Recent, Arctic and Antarctic.

Genus DIFFUSILINA Heron-Allen and Earland, 1924

Key, plate 2, figure 28
Genotype, *Diffusilina humilis* Heron-Allen and Earland
Diffusilina HERON-ALLEN and EARLAND, Journ. Linn. Soc. Zool., vol. 35, 1924, p. 614.

Test attached, of irregular outline; chamber consisting of a mass of tubules; wall of sand and mud; apertures inconspicuous, at the top of small pustules on the outer surface. — Recent, attached to algae, South Pacific.

Family 5. HYPERAMMINIDAE

Test free or attached, consisting of a globular proloculum and a more or less elongate but not close coiled, sometimes branching portion, not divided into chambers; wall of variously agglutinated materials with a basal layer of chitin; aperture formed by the open end of the tubular portion.

In its simplest terms the test of the Hyperamminidae consists of a globular proloculum which is comparable to the test of the Saccamminidae to which is added a tubular second chamber. There is a chitinous lining on which the agglutinated wall is built, as is usual in arenaceous forms. The tubular chamber may be variously branched and modified, as will be seen on our plate.

The genera *Ophiotuba* and *Dendrotuba* are unusual forms of obscure structure which may prove not to be foraminifera. The genus *Hospitella* Rhumbler may also be related to these.

The early history of the family goes well back into the Paleozoic and will be extended with further research. From the Hyperamminidae were developed early in the Paleozoic the Reophacidae by the division of the tubular portion into definite chambers, the intermediate stages of which may be still seen in such genera as *Kalamopsis* and *Aschemonella*.

Subfamily 1. Hyperammininae

Test free, simple.

Genus HYPERAMMINA H. B. Brady, 1878

Key, plate 3, figures 1, 2

Genotype, *Hyperammina elongata* H. B. Brady

Hyperammina H. B. Brady, Ann. Mag. Nat. Hist., ser. 5, vol. 1, 1878, p. 433.
Rhabdopleura (?) Dawson, 1871. *Bactrammina* Eimer and Fickert, 1899.

Test free, elongate, consisting of a proloculum and long, undivided, tubular, second chamber; wall of sand grains, amount of cement varying greatly in different species, interior chitinous, smooth; aperture formed by the open end of the tube. — Cambrian (?), Silurian to Recent.

Genus NUBECULARIELLA Awerinzew, 1911

Key, plate 3, figure 5

Genotype, *Nubeculariella birulai* Awerinzew

Nubeculariella Awerinzew, Mém. Acad. Sci. St.-Pétersbourg, ser. 8, vol. 20, pt. 3, 1911, p. 8.

Test attached, consisting of an indistinct proloculum and curved cylindrical tube; wall of chitin with sand grains of various sizes attached to the exterior; aperture rounded, with a slightly flaring lip. — Recent, Arctic.

Genus NORMANINA Cushman, 1928

Key, plate 3, figure 13

Genotype, *Haliphysema confertum* Norman

Normanina Cushman, Contr. Cushman Lab. Foram. Res., vol. 4, 1928, p. 7.
Haliphysema (part) Norman, 1878 (not Bowerbank).

Test consisting of a globular proloculum and slender, tubular, second chamber, individuals gathered together in masses, tubular portions toward the center of the mass; wall chitinous with sand grains and other foraminiferal tests on the exterior; aperture at the end of the tubular chamber. — Recent, Davis Strait, 1750 fathoms.

Genus JACULELLA H. B. Brady, 1879

Key, plate 3, figure 4

Genotype, *Jaculella acuta* H. B. Brady

Jaculella H. B. Brady, Quart. Journ. Micr. Sci., vol. 19, 1879, p. 35.

Test free, consisting of a proloculum and elongate, conical, second chamber; wall thick, of sand grains firmly cemented, chitinous within; aperture circular, at the open end of the tubular chamber. — Jurassic (?), Miocene to Recent.

Genus HIPPOCREPINA Parker, 1870

Key, plate 3, figure 9

Genotype, *Hippocrepina indivisa* Parker

Hippocrepina Parker, in Dawson, Canad. Nat., n. ser., vol. 5, 1870, p. 176.

Test free, elongate, tapering, apertural end somewhat contracted; wall thin, of fine sand grains with a yellowish-brown cement, often with fine mica flakes giving a luster to the surface; aperture narrow, curved or irregular, sometimes with a slight lip. — Cretaceous to Recent.

Genus HYPERAMMINOIDES Cushman and Waters, 1928

Key, plate 3, figures 7, 8

Genotype, *Hyperamminella elegans* Cushman and Waters

Hyperamminoides Cushman and Waters, Contr. Cushman Lab. Foram. Res., vol. 4, 1928, p. 112.

Hyperamminella Cushman and Waters, 1928 (not deFolin, 1881).

Test free, with a proloculum and elongate, tapering, second chamber, compressed or rounded, constricted on the exterior; wall finely arenaceous with siliceous cement; aperture terminal, circular or elliptical, often with a slight lip. — Pennsylvanian and Permian.

Genus EARLANDIA Plummer, 1930

Key, plate 3, figure 6

Genotype, *Earlandia perparva* Plummer

Earlandia Plummer, Univ. Texas, Bull. 3019, 1930, p. 12.

Test free, very elongate, composed of a globular or subglobular proloculum and an elongate, nonseptate, second chamber; wall of minute, crystalline, calcareous granules bound by a calcareous cement, imperforate, smoothly finished; aperture a broad, circular opening at the end of the tube. — Pennsylvanian, Texas.

Subfamily 2. Dendrophryinae

Test attached, usually branching.

Genus CHITINODENDRON Eisenack, 1937

Key, plate 42, figure 20

Genotype, *Chitinodendron bacciferum* Eisenack

Chitinodendron Eisenack, Pal. Zeitschr., vol. 19, 1937, p. 236.

Test small, single-chambered, irregular in shape, with a chitinous tubular extension which branches and occasionally has another chamber attached; wall somewhat arenaceous, tubular portion chitinous; aperture at the end of the tubular extension. — Silurian.

Genus SACCORHIZA Eimer and Fickert, 1899

Key, plate 3, figure 3

Genotype, *Hyperammina ramosa* H. B. Brady

Saccorhiza Eimer and Fickert, Zeitschr. Wiss. Zool., vol. 65, 1899, p. 670.

Hyperammina (part) H. B. Brady, 1879.

Test free, with an ovoid proloculum and a branching, tubular, second chamber; wall with a thin, inner, chitinous layer, and an outer layer of sand grains and sponge spicules; apertures at the ends of the tubular branches. — Mississippian to Recent.

Genus DENDROPHRYA Str. Wright, 1861

Key, plate 3, figure 12
Genotype, *Dendrophrya erecta* Str. Wright
Dendrophrya STR. WRIGHT, Ann. Mag. Nat. Hist., ser. 3, vol. 8, 1861, p. 133.

Test attached, with a proloculum and tubular second chamber, simple or branched, erect or with spreading arms; wall with a chitinous lining and outer arenaceous layer; aperture at the end of the tubular portion. — Jurassic (?), Cretaceous to Recent.

Genus SACCODENDRON Rhumbler, 1935

Key, plate 42, figure 21
Genotype, *Saccodendron heronalleni* Rhumbler
Saccodendron RHUMBLER, Schrift. Nat. Ver. Schleswig-Holstein, Bd. XXI, 1935, p. 173.

Test attached, consisting of a central body of irregular shape according to the base of attachment with numerous branching arms; wall arenaceous; apertures at the ends of the tubular arms. — Recent.

Genus DENDRONINA Heron-Allen and Earland, 1922

Key, plate 3, figure 10
Genotype, *Dendronina arborescens* Heron-Allen and Earland
Dendronina HERON-ALLEN and EARLAND, British Antarctic ("*Terra Nova*") Exped., 1910, Zool., vol. 6, No. 2, 1922, p. 78.

Test attached or free, with a basal pad extending from which is a branching nonseptate tube; wall chitinous within, and mud, sand grains and spicules outside; apertures terminal, sometimes with groups of spicules. — Recent, New Zealand and Antarctic.

Genus HALIPHYSEMA Bowerbank, 1862

Key, plate 3, figure 14
Genotype, *Haliphysema tumanowiczii* Bowerbank
Haliphysema BOWERBANK, Philos. Trans., vol. 152, 1862, p. 1105.
Squamulina (part) CARTER, 1870.

Test attached, base expanded, from which is a columnar erect portion, either simple or branched; wall arenaceous with numerous included sponge spicules, especially toward the outer end; aperture at the free end, partially obscured by the irregular clustering of spicules. — Recent.

Genus SAGENINA Chapman, 1900

Key, plate 3, figure 15

Genotype, *Sagenella frondescens* H. B. Brady

Sagenina CHAPMAN, Journ. Linn. Soc., Zool., vol. 28, 1900, p. 4.
Sagenella H. B. BRADY, 1879 (not *Sagenella* HALL).

Test attached, with a proloculum and dichotomously or irregularly branching second chamber; wall arenaceous, usually with calcareous cement; apertures at ends of the branches. — Recent.

Genus PSAMMATODENDRON Norman, 1881

Key, plate 3, figure 11

Genotype, *Psammatodendron arborescens* Norman

Psammatodendron NORMAN, in H. B. BRADY, Denkschr. k. Akad. Wiss. Wien, vol. 43, 1881, p. 98.
Hyperammina (part) H. B. BRADY, 1884.

Test attached, with a bulbous proloculum and tubular, dichotomously branched second chamber; wall arenaceous, with a thin chitinous lining and ferruginous cement; apertures at the open ends of the tubes. — Silurian (?), Recent.

Genus SYRINGAMMINA H. B. Brady, 1883

Key, plate 3, figure 16

Genotype, *Syringammina fragillissima* H. B. Brady

Syringammina H. B. BRADY, Proc. Roy. Soc. London, vol. 35, 1883, p. 155.

Test free or adherent, consisting of a bulbous base and many branching arms, or of anastomosing tubes in a rounded mass; wall finely arenaceous with a small amount of cement; apertures at the ends of the tubular portions. — Recent.

Genus OPHIOTUBA Rhumbler, 1894

Genotype, *Ophiotuba gelatinosa* Rhumbler

Ophiotuba RHUMBLER, Zeitschr. Wiss. Zool., vol. 57, 1894, p. 604.

Test attached to the interior of larger foraminiferal tests, irregularly winding; wall chitinous; aperture at the open end of the tube. — Recent.

Genus DENDROTUBA Rhumbler, 1894

Genotype, *Dendrotuba nodulosa* Rhumbler

Dendrotuba RHUMBLER, Zeitschr. Wiss. Zool., vol. 57, 1894, p. 606.

Test attached to the interior of other foraminifera, irregularly winding and anastomosing; wall chitinous; apertures, the open ends of the tubes. — Recent.

Family 6. REOPHACIDAE

Test consisting of either an irregular or a generally rectilinear series of chambers, typically increasing in size as added, simple or labyrinthic; wall chitinous, with usually an exterior of agglutinated material, sand grains, sponge spicules or the tests of other foraminifera; aperture usually terminal, simple or multiple.

The Reophacidae are for the most part simple and primitive forms which probably had their origin in forms such as the Hyperamminidae. The simpler, more primitive genera of the family, such as *Aschemonella* and *Kalamopsis*, are often imperfectly divided into chambers. The material of the wall is essentially a chitinous one, on which is built up the outer wall of foreign material of various kinds. Some genera, *Nodellum* and *Turriclavula*, retain the primitive chitinous test almost without further covering, and there are gradations to the thick-walled tests of *Haplostiche*. Most of the genera have simple chambers, but they gradually become labyrinthic in *Haplostiche*, comparable to what is seen in the end forms of so many other families.

There is a complete series connecting this group with the other primitive arenaceous families, and no evidence that they have ever come from *Nodosaria* or other strictly perforate, calcareous groups, or are related to them.

The early history of the group goes back to the earliest Paleozoic where Cambrian forms are found. *Reophax* and *Nodosinella* are abundant in the Pennsylvanian, and it is often difficult to distinguish the two, authors usually placing those with coarse arenaceous material with a small amount of cement in *Reophax*, and those species with much cement in *Nodosinella*. *Nodosinella* was evidently derived from *Reophax* by the addition of more and more cementing material, as was *Hormosina*.

In the present oceans the family is represented in many habitats, the type species of *Reophax* being extremely abundant in shallow warm waters of coral reefs.

Subfamily 1. Aschemonellinae

Chambers irregular.

Genus ASCHEMONELLA H. B. Brady, 1879

Key, plate 3, figure 17

Genotype, *Aschemonella scabra* H. B. Brady

Aschemonella H. B. Brady, Quart. Journ. Micr. Sci., vol. 19, 1879, p. 42.
Astrorhiza (part) Norman, 1876.

Test free, with several tubular or inflated chambers in a single or branching series, irregular in size and form; wall thin, chitinous within, arenaceous outside, with much cement; apertures often several, at the end of the chambers. — Cretaceous to Recent.

Somewhat similar, partially divided, tubular forms extend back as far as the Silurian, and may belong here.

Genus KALAMOPSIS deFolin, 1883

Key, plate 3, figure 19

Genotype, *Kalamopsis vaillanti* deFolin

Kalamopsis deFolin, Congrès Scient. Dax, 1882 (1883), p. 320.

Test free, with a globular proloculum and an irregular series of linear chambers making a cylindrical test; wall chitinous, with some arenaceous material; aperture rounded, terminal. — Recent.

Genus HOSPITELLA Rhumbler, 1911

Key, plate 3, figure 18

Genotype, *Hospitella fulva* Rhumbler

Hospitella Rhumbler, Plankton Exped., vol. 3, 1909 (1911), p. 227.

Test a series of connected chambers in the interior of the tests of larger foraminifera; wall chitinous; aperture at the end of the last-formed chamber. — Recent.

These may not be true foraminifera, and are apparently not really parasitic but simply occupy empty foraminiferal tests.

Subfamily 2. Reophacinae

Chambers typically in a regular rectilinear series.

Genus REOPHAX Montfort, 1808

Key, plate 3, figures 27, 28

Genotype, *Reophax scorpiurus* Montfort

Reophax Montfort, Conch. Syst., vol. 1, 1808, p. 331.

Nodosaria (part) of authors (not Lamarck). *Lituola* (part) of authors (not Lamarck). *Haplostiche* Schwager, 1865 (not Reuss). *Nodulina* Rhumbler, 1895. *Protoschista* Eimer and Fickert, 1899.

Test free, elongate, composed of a series of undivided chambers, ranging from overlapping to remotely separated, in a straight or curved linear series; wall typically with a chitinous base and an outer wall of agglutinated material, firmly cemented, sand grains, mica flakes, sponge spicules or other foraminifera; aperture simple, terminal, sometimes with a slight neck. — Cambrian, Pennsylvanian to Recent.

Ammofrondicularia Schubert, 1902, is known only from a section of a portion of a compressed test, perhaps of a *Reophax*.

Genus SULCOPHAX Rhumbler, 1931

Key, plate 3, figure 23
Genotype, *Sulcophax claviformis* Rhumbler

Sulcophax RHUMBLER, in WIESNER, Deutsche Südpolar-Exped., vol. XX, Zool., 1931, p. 93.

Test free, elongate, tapering, chambers rapidly increasing in size; wall coarsely arenaceous; aperture terminal, elongate, in the middle of a terminal furrow. — Recent, Antarctic.

Genus NODOSINELLA H. B. Brady, 1876

Key, plate 3, figures 29, 30
Genotype, *Nodosinella digitata* H. B. Brady

Nodosinella H. B. BRADY, Pal. Soc. Mon., vol. 30, 1876, p. 102.
Dentalina (part) of authors (not D'ORBIGNY, 1826).

Test free, straight or arcuate; chambers usually distinct, typically enlarging in size as added, simple; walls finely to coarsely arenaceous with much cement, often apparently with a double wall; aperture usually simple or terminal. — Carboniferous to Cretaceous.

Genus HORMOSINA H. B. Brady, 1879

Key, plate 3, figures 20–22
Genotype, *Hormosina globulifera* H. B. Brady

Hormosina H. B. BRADY, Quart. Journ. Micr. Sci., vol. 19, 1879, p. 56.

Test free, composed of a straight, curved, or irregular linear series of subglobular, fusiform or pyriform chambers; wall with a chitinous lining, finely arenaceous, with much ferruginous cement; aperture circular, terminal, or at one side, often with a definite neck. — Jurassic to Recent.

The large, single-chambered, megalospheric form, usually abundant with the two or more chambered forms, can with difficulty be distinguished from *Saccammina*.

Genus POLYCHASMINA Loeblich and Tappan, 1946

Key, plate 42, figures 23, 24
Genotype, *Polychasmina pawpawensis* Loeblich and Tappan

Polychasmina LOEBLICH and TAPPAN, Journ. Pal., vol. 20, 1946, p. 242.

Test free, flattened, composed of a linear series of chambers; wall thick, coarsely arenaceous, interior labyrinthic; aperture terminal, consisting of a single row of elongate slits, whose trend parallels the flattened sides of the test. — Lower Cretaceous.

Genus HAPLOSTICHE Reuss, 1861

Key, plate 3, figures 25, 26

Genotype, *Dentalina foedissima* Reuss

Haplostiche Reuss, Sitz. Böhm. Ges. Wiss., 1861, p. 16.

Nodosaria (part) of authors. *Dentalina* Reuss, 1860 (not d'Orbigny). *Lituola* (part) Parker and Jones, 1860. *Cribratina* Sample, 1932.

Test free, cylindrical or tapering, composed of a linear series of chambers, interior labyrinthic; wall thick, coarsely arenaceous; aperture in the young simple, later of several pores or dendritic, terminal, occasionally with a short neck. — Jurassic to Recent.

Genus TURRICLAVULA Rhumbler, 1911

Key, plate 3, figure 24

Genotype, *Turriclavula interjecta* Rhumbler

Turriclavula Rhumbler, Plankton Exped., vol. 3, 1909 (1911), p. 421.

Test free, elongate, of several chambers in a rectilinear series; wall chitinous, with some scattered arenaceous material on the exterior; aperture terminal, elongate, elliptical. — Recent.

Genus NODELLUM Rhumbler, 1913

Key, plate 3, figure 31

Genotype, *Reophax membranacea* H. B. Brady

Nodellum Rhumbler, Plankton Exped., vol. 3, 1913, p. 473.

Reophax (part) H. B. Brady, 1879.

Test free, elongate, made up of a rectilinear series of inflated chambers; wall entirely chitinous; aperture circular, terminal. — Cretaceous to Recent.

Subfamily 3. Sphaeramminінae

Later chambers extending back and enclosing the earlier ones.

Genus SPHAERAMMINA Cushman, 1910

Key, plate 3, figure 33

Genotype, *Sphaerammina ovalis* Cushman

Sphaerammina Cushman, Proc. U. S. Nat. Mus., vol. 38, 1910, p. 439.

Test free, composed of a series of globular chambers, the last-formed one completely enveloping the preceding ones, axis usually straight; wall arenaceous; aperture elliptical or rounded. — Recent, Philippine region.

Genus AMMOSPHAERULINA Cushman, 1912

Key, plate 3, figure 32

Genotype, *Ammosphaerulina adhaerens* Cushman

Ammosphaerulina Cushman, Proc. U. S. Nat. Mus., vol. 42, 1912, p. 229.

Test spherical, adherent, composed of two or more chambers, each included by the one next formed, eccentric; wall arenaceous; aperture small, rounded. — Recent, Philippine region.

Family 7. AMMODISCIDAE

Test composed of a globular proloculum and long, tubular, undivided, second chamber, usually close coiled, at least in the young, planispiral, conical spiral, or irregularly winding, free or attached; wall of fine arenaceous material, usually with an abundance of yellowish- or reddish-brown cement and a chitinous inner layer; aperture formed by the open end of the tubular chamber.

The test in all the genera is composed of arenaceous material with a greatly variable amount of cement. In Recent species there is a great range in the relative amount of cement, some consisting of almost pure cement with only occasional fragments of foreign material, and others very largely arenaceous. The cement is usually yellowish- or reddish-brown similar to that in the Astrorhizidae, Saccamminidae, Hyperamminidae, Lituolidae, Trochamminidae and in the primitive arenaceous forms of the Miliolidae. This characteristic color is often shown in well-preserved Paleozoic material.

It has been stated that there is no chitinous inner layer in this group, but its presence may be very strikingly demonstrated by anyone who will take the trouble, in *Ammolagena*, for instance, by simply dissolving the attachment. An *Ammolagena* attached to *Globorotalia* may be entirely freed by dissolving the latter with weak acid. The base of the *Ammolagena* is then revealed as a thin almost transparent layer of chitin which formed the floor of the proloculum and second chamber, and this inner tubular sac may be carefully teased away from the arenaceous outer wall. The same chitinous inner layer may be seen less strikingly in the other genera. There is no evidence to show that *Ammodiscus* or its relatives were ever derived from calcareous forms, the Paleozoic forms under polarized light showing fragments of varied minerals.

The different genera show the varied forms that may be derived from the simple planispiral coil of *Ammodiscus*. The free forms comprise *Hemidiscus*, similar to *Ammodiscus*, but with the later chambers irregularly placed at one side; *Turritellella* in an elongate spiral; *Ammodiscoides*, conical in the young, then tending to become almost planispiral; *Glomospira*, irregularly close coiled in varying planes, and from it, derived *Lituotuba* with the later portion uncoiling. *Psammonyx* is a large form from off Japan, compressed and growing to large size. In *Discammina*, described from the Mediterranean, is a form in which the peripheral portion of the otherwise undivided second chamber becomes labyrinthic, resembling the labyrinthic forms developed in the late history of so many of the foraminiferal families.

The attached forms may form a conical spiral, as in *Trepeilopsis*, which coils about the spines of *Productus*; a close-coiled planispiral form later winding back and forth as in *Ammovertella*; wandering about irregularly over the attached surface as in *Tolypammina*; or with the proloculum large and the tubular chamber small as in *Ammolagena*. Most of the genera are known from the Paleozoic where they are often very abundant. They have been confused with the imperforate calcareous forms of the later Paleozoic which often assume somewhat similar forms.

Subfamily 1. Ammodiscinae

Test free.

Genus AMMODISCUS Reuss, 1861

Plate 9, figure 1; Key, plate 4, figures 1, 2
Genotype, *Operculina incerta* d'Orbigny

Ammodiscus Reuss, Sitz. Akad. Wiss. Wien, vol. 44, 1861, p. 365.
Operculina (part) d'Orbigny, 1839. *Orbis* Strickland, 1848 (not Philipps). *Spirillina* Williamson, 1858 (not Ehrenberg). *Trochammina* (part) of authors. *Cornuspira* (part) of authors. *Involutina* (part) Terquem.

Test free, planispiral, with a proloculum and long, tubular, undivided, second chamber; wall arenaceous, varying greatly in the size of particles and relative amount of cement; aperture formed by the open end of the tubular chamber. — Silurian to Recent.

Paleozoic records for *Spirillina* are probably based on specimens of *Ammodiscus*.

Genus HEMIDISCUS Schellwien, 1898

Plate 9, figure 2; Key, plate 4, figure 3
Genotype, *Hemidiscus carnicus* Schellwien

Hemidiscus Schellwien, Palaeontographica, vol. 44, 1898, p. 266.

Test free, with a proloculum and long, tubular, undivided, second chamber, early portion planispiral, later portion more or less involute or irregular

---

PLATE 9

Relationships of the Genera of the Ammodiscidae

FIG.
1. *Ammodiscus incertus* (d'Orbigny).
2. *Hemidiscus carnicus* Schellwien. (After Schellwien.)
3. *Turritellella shoneana* (Siddall). (After H. B. Brady.)
4. *Ammodiscoides turbinatus* Cushman.
5. *Psammonyx vulcanicus* Döderlein. (After Rhumbler.)
6. *Discammina fallax* Lacroix. (After Lacroix.)
7. *Glomospira gordialis* (Jones and Parker). (After H. B. Brady.)
8. *Lituotuba lituiformis* (H. B. Brady). (After H. B. Brady.)
9. *Trepeilopsis grandis* (Cushman and Waters). (After Cushman and Waters.)
10. *Ammovertella inversus* (Schellwien). (After Schellwien.)
11. *Tolypammina vagans* (H. B. Brady). (After H. B. Brady.)
12. *Ammolagena clavata* (Jones and Parker).

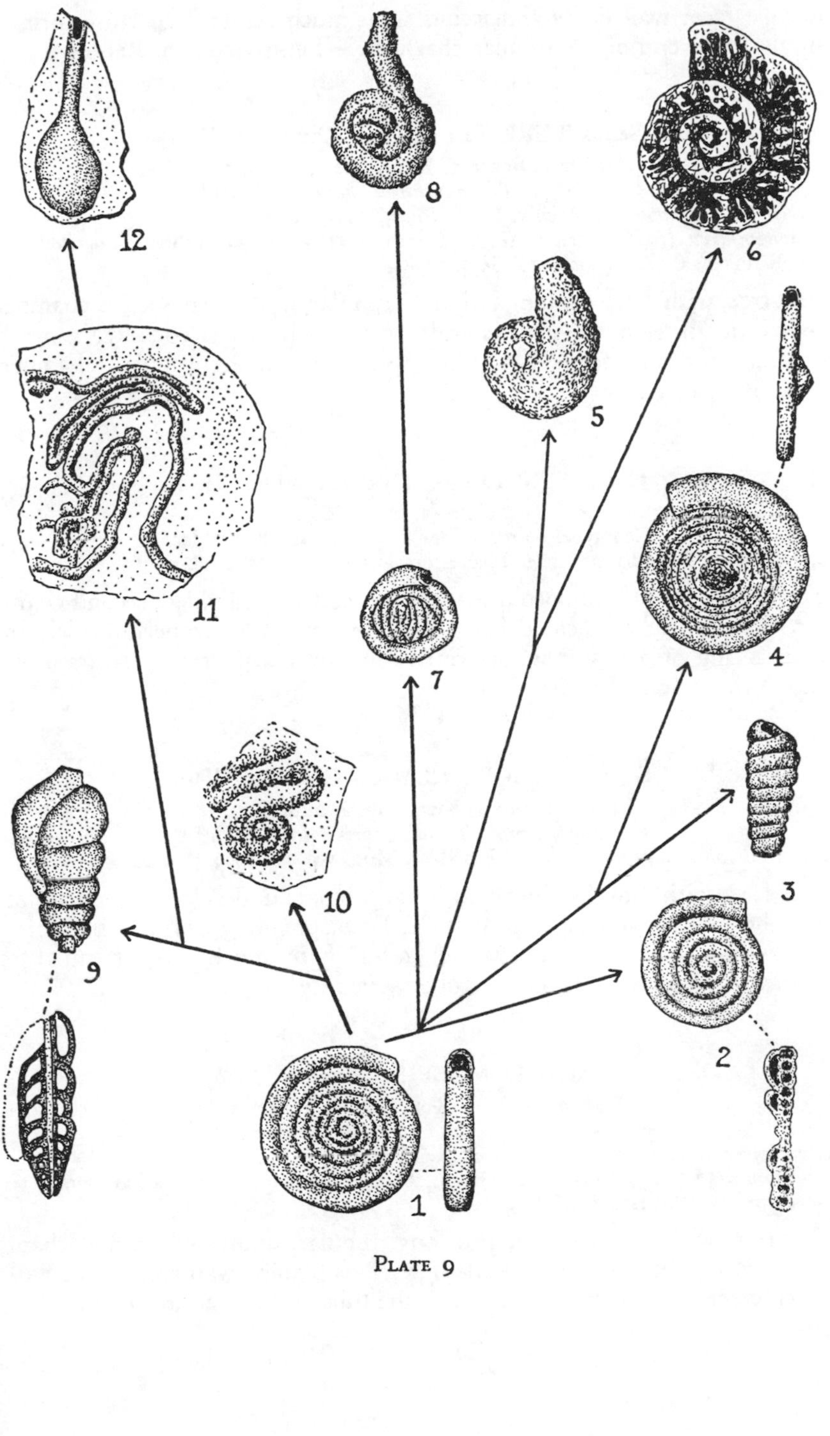

Plate 9

on one face; wall finely arenaceous with much cement; aperture formed by the open end of the tubular chamber. — Pennsylvanian, Recent.

Genus TURRITELLELLA Rhumbler, 1903

Plate 9, figure 3; Key, plate 4, figure 14
Genotype, *Trochammina shoneana* Siddall

Turritellella RHUMBLER, Archiv. Prot., vol. 3, 1903, p. 283.
Trochammina (part) SIDDALL, 1878. *Ammodiscus* (part) of authors. *Turritellopsis* RHUMBLER, 1895 (not G. O. SARS, 1878).

Test free, with a proloculum and long, tubular, undivided, second chamber coiled in an elongate close spiral; wall finely arenaceous with much cement; aperture formed by the open end of the tubular chamber. — Silurian to Recent.

Genus SPIRILLINOIDES Rhumbler, 1938

Key, plate 42, figures 26, 27
Genotype, *Spirillinoides circumcinctus* Rhumbler

Spirillinoides RHUMBLER, Kieler Meeresforschungen, vol. 2, 1938, p. 174.

Test composed of a proloculum and long, coiled, undivided, second chamber, typically in a trochoid coil; wall thin, of clear, imperforate chitin, with a ring of arenaceous material about the periphery; aperture at the end of the tube. — Recent.

Genus AMMODISCOIDES Cushman, 1909

Plate 9, figure 4; Key, plate 4, figures 17, 18
Genotype, *Ammodiscoides turbinatus* Cushman

Ammodiscoides CUSHMAN, Proc. U. S. Nat. Mus., vol. 36, 1909, p. 424.

Test free, with a proloculum and long, tubular, undivided, second chamber, the early coils in a conical spire, later spreading out and becoming nearly planispiral; wall finely arenaceous with much cement; aperture formed by the open end of the tube. — Pennsylvanian to Recent.

Genus GLOMOSPIRA Rzehak, 1888

Plate 9, figure 7; Key, plate 4, figures 4–8
Genotype, *Trochammina gordialis* Jones and Parker

Glomospira RZEHAK, Verh. k. k. Geol. Reichs., 1888, p. 191.
Trochammina (part) JONES and PARKER, 1860. *Ammodiscus* (part) of authors. *Gordiammina* RHUMBLER, 1895.

Test free, with a proloculum and long, tubular, undivided, second chamber winding about its earlier coils in various planes; wall arenaceous with much cement; aperture at the end of the tube. — Silurian to Recent.

Genus GLOMOSPIRELLA Plummer, 1945

Key, plate 42, figure 25

Genotype, *Glomospira umbilicata* Cushman and Waters

Glomospirella PLUMMER, Univ. Texas Publ. 4401, 1945, p. 233.
Glomospira (part) of authors.

Test composed of a proloculum and elongate, tubular second chamber, in the early stages closely wound about the proloculum in varying planes, later becoming nearly planispiral; wall arenaceous; aperture formed by the open end of the tubular chamber. — Pennsylvania.

Genus LITUOTUBA Rhumbler, 1895

Plate 9, figure 8; Key, plate 4, figures 9, 10

Genotype, *Trochammina lituiformis* H. B. Brady

Lituotuba RHUMBLER, Nachr. Köngl. Ges. Wiss. Göttingen, 1895, p. 83.
Trochammina (part) of authors.

Test free, with a proloculum and long, tubular, second chamber, sometimes constricted externally but not divided, early portion irregularly coiled, later uncoiling; wall arenaceous with much cement; aperture at the end of the tube. — Silurian to Recent.

Genus AMMOFLINTINA Earland, 1934

Key, plate 42, figures 28, 29

Genotype, *Ammoflintina trihedra* Earland

Ammoflintina EARLAND, *Discovery* Reports, vol. 10, 1934, p. 98.

Test free, planispiral, bilaterally compressed, with a proloculum and one or two coils about it, periphery subacute; chambers three to a coil, V-shaped in section; wall arenaceous, thin and fragile, of fine mineral grains with much cement; aperture large and simple, at the outer end of the final chamber. — Recent.

Genus PSAMMONYX Döderlein, 1892

Plate 9, figure 5; Key, plate 4, figures 15, 16

Genotype, *Psammonyx vulcanicus* Döderlein

Psammonyx DÖDERLEIN, Verh. Deutsch. Zool. Ges., 1892, p. 145.

Test free, with a proloculum and undivided second chamber, loosely coiled in the early stages, later uncoiled, compressed, and increasing in height toward the open end; wall of fine sand grains; aperture large, elongate, formed by the open end of the test. — Devonian, Recent.

Genus BIFURCAMMINA Ireland, 1939

Key, plate 42, figures 30, 31

Genotype, *Bifurcammina bifurca* Ireland

Bifurcammina IRELAND, Journ. Pal., vol. 13, 1939, p. 201.

"Test free, composed of an ovoid proloculum and a tubular second chamber planispirally coiled in the early stages, the last coil divided into two tubes with various characteristics but always with a definite hump at the point where the added tube starts; wall of fine sand; apertures at the ends of the tubes." — Silurian.

Genus DISCAMMINA Lacroix, 1932

Plate 9, figure 6; Key, plate 4, figures 21–23
Genotype, *Discammina fallax* Lacroix

Discammina LACROIX, Bull. Instit. Océanographique, No. 600, 1932, p. 2.

Test free, with a proloculum and long second chamber, much compressed and gradually widening, interior with peripheral portion labyrinthic, but not divided; wall of fine sand and a few spicules, firmly cemented; aperture elongate, widest at the base, at the end of the tube. — Recent, Mediterranean.

Subfamily 2. Tolypammininae

Test attached.

Genus TOLYPAMMINA Rhumbler, 1895

Plate 9, figure 11; Key, plate 4, figure 11
Genotype, *Hyperammina vagans* H. B. Brady

Tolypammina RHUMBLER, Nachr. Köngl. Ges. Wiss. Göttingen, 1895, p. 83.

Hyperammina (part) H. B. BRADY, 1879. *Serpulella* and *Ammonema* EIMER and FICKERT, 1899. *Girvanella* (part) of authors (not of NICHOLSON and ETHERIDGE, 1878).

Test attached, with proloculum and long, tubular, undivided, second chamber, earliest portion sometimes coiled, later irregular; wall arenaceous with much cement; aperture at the end of the tube. — Silurian to Recent.

If the attachment is to a small object the test may completely envelop it, and appear to be free.

Genus AMMOVERTELLA Cushman, 1928

Plate 9, figure 10; Key, plate 4, figure 12
Genotype, *Psammophis inversus* Schellwien

Ammovertella CUSHMAN, Contr. Cushman Lab. Foram. Res., vol. 4, 1928, p. 8.

Psammophis SCHELLWIEN, 1898 (not BOIE, 1827).

Test attached, with proloculum and long, tubular, second chamber, early portion planispiral, later and larger portion bending back and forth but progressing forward in one general direction; wall clearly arenaceous with much cement; aperture at the end of the tube. — Pennsylvanian to Recent.

Schellwien proved the arenaceous character of the wall by tests with polarized light.

Genus THALMANNINA Majzon, 1943

Key, plate 42, figure 32

Genotype, *Thalmannina nothi* Majzon

Thalmannina MAJZON, Mitt. Jahrb. K. Ungarn. Geol. Anstalt, vol. 37, 1943, p. 154.

Test tubular, irregular in shape, taking on a series of U-shaped forms in the adult; wall arenaceous; aperture at the ends of the tubular chamber. — Upper Cretaceous and Tertiary.

Genus XENOTHEKA Eisenack, 1937

Key, plate 42, figure 33

Genotype, *Xenotheka klinostoma* Eisenack

Xenotheka EISENACK, Pal. Zeitschr., vol. 19, 1937, p. 239.

Test attached, consisting of a single chamber and elongated tubular neck; wall chitinous; aperture an irregular opening in the wall of the chamber and at the end of the tubular neck. — Silurian.

Genus AMMOLAGENA Eimer and Fickert, 1899

Plate 9, figure 12; Key, plate 4, figures 19, 20

Genotype, *Trochammina clavata* Jones and Parker

Ammolagena EIMER and FICKERT, Zeitschr. Wiss. Zool., vol. 65, 1899, p. 673.

Trochammina (part) of authors. *Webbina* (part) H. B. BRADY (not D'ORBIGNY, 1839).

Webbinella (part) RHUMBLER, 1903.

Test attached, with an oval proloculum and long, tubular, second chamber of variable length but of nearly uniform diameter; wall with a thin chitinous inner layer, outer layer finely arenaceous with much cement; aperture at the end of the tube. — Pennsylvanian to Recent.

Genus TREPEILOPSIS Cushman and Waters, 1928

Plate 9, figure 9; Key, plate 4, figure 13

Genotype, *Turritellella grandis* Cushman and Waters

Trepeilopsis CUSHMAN and WATERS, Contr. Cushman Lab. Foram. Res., vol. 4, 1928, p. 38.

Turritellella CUSHMAN and WATERS, 1927 (not RHUMBLER, 1903).

Test attached to *Productus* spines, with a proloculum and long, tubular, second chamber, early portion spirally coiled, later bending back and making nearly a straight tube over the earlier whorls; wall finely arenaceous with much cement; aperture at the end of the tube. — Mississippian to Permian.

Family 8. LITUOLIDAE

Test free, planispiral at least in the young, later portion in some genera uncoiled or even becoming discoid, divided into chambers, either simple or labyrinthic; wall arenaceous with varying proportions of cement in different genera and species, usually with a yellowish- or reddish-brown cement, the last-formed chamber in the adult often white; aperture simple or compound.

This family contains many genera, and shows a great diversity of form in the different types of structure which the various members of the family have adopted. The forms are essentially planispiral, at least in the young, but in the adult the last chambers may become uncoiled or annular, and are often very complex in the internal structure. All the members of the family are definitely arenaceous, although the amount of cement used varies greatly, as it does in other arenaceous groups. The family undoubtedly originated from the Ammodiscidae by the division of the second tubular chamber into definite smaller chambers. The stages in this process may be seen in *Trochamminoides* which goes back to the Carboniferous and probably earlier. From such forms as this, develops *Haplophragmoides*, also known from the Paleozoic and probably to be looked for among the earliest of the foraminifera. These forms have a chitinous inner wall and an outer arenaceous one. From *Haplophragmoides* have developed in the Paleozoic and later a number of genera. In *Cribrostomoides*, the aperture which is typically at the base of the last-formed chamber is broken up into a number of rounded pores. In *Ammobaculites*, the young is coiled, but in the adult becomes uncoiled with a rectilinear series of chambers, usually inflated. *Ammomarginulina* is very similar but very strongly compressed throughout, and the early portion usually evolute. In *Flabellammina* and *Frankeina*, both known from the Cretaceous, the young stages are planispiral. In *Flabellammina* the later chambers are compressed and frondicularian, while in *Frankeina* the uncoiled chambers are triangular in section. *Triplasia* develops from *Frankeina* by the reduction of the coiled stage or its total loss in the megalospheric form.

Also from *Haplophragmoides* there develops a group of forms in which the chambers become much divided. The simplest of these is *Cyclammina* where there is an ingrowth from the walls in a labyrinthic manner partially filling the interior of the test. This develops in the Cretaceous, and continues to the present oceans. From *Cyclammina* is developed *Pseudocyclammina* in which the later chambers are uncoiled. *Lituola* is also planispiral in the young, becomes uncoiled, has the chambers divided by radial partitions, and the aperture is often multiple. From *Cyclammina* is developed *Choffatella*, which is much compressed and has the chambers partially divided. In *Dictyopsella*, the chambers are divided by branching divisions, and in *Yaberinella* they are divided by criss-cross divisions. There are four other genera related to these which have been placed in various families. They are known from the Jurassic and Cretaceous. By some authors they have been erroneously placed with the calcareous imperforate forms, but the wall is finely arenaceous, as shown by a study of the originals of several of these genera. *Spirocyclina* has the early chambers uncoiled, the later ones annular. *Cyclolina* of d'Orbigny has simply divided chambers, and very little is known about this particular form. It needs further study for one to be certain of its true characters. *Cyclopsinella* is almost entirely an unknown form. The original authors say that it is arenaceous, but all that is known of it from figures is a portion of a schematic section. *Orbitopsella* has a very peculiar test which is thin in the center, and greatly thickened toward the exterior. From a study of original material of this genus, it seems that it should be placed here.

Another form derived from *Haplophragmoides*, or similar forms from *Glomospira*, is *Endothyra*. It has been claimed by one or two authors that the test of *Endothyra* is entirely calcareous. Sections of well-preserved material when studied with polarized light will show the presence of angular fragments of various types of minerals, and the cementing material has much chitin. This can be easily demonstrated by dissolving the calcareous material in very weak acid so as not to break up the more delicate chitinous and arenaceous elements of the test as is done by the use of strong acids and the consequent explosive force of the rapid effervescence. The same treatment of *Bradyina* and other Paleozoic forms yields very similar results, and proves conclusively that this entire group belongs to definitely arenaceous types. In *Bradyina* and *Glyphostomella*, the apertures consist of elongate slits at the base of the chambers, and in *Cribrospira* the aperture is cribrate. From such forms as *Endothyra* is developed *Endothyranella*, in which the later chambers are uncoiled and the aperture simple, and the name *Septammina* is applied to similar forms in which the aperture is cribrate.

From such forms as *Endothyra* the Fusulinidae are undoubtedly derived. The Paleozoic forms, particularly those occasionally referred to as calcareous, need much more careful study in section and with polarized light together with treatment with dilute acid, and with patient treatment can be shown to be essentially arenaceous.

Endothyra has sometimes been taken for a simple radicle from which numerous other families are supposed to have been derived. It is, however, probably the ancestor of the Fusulinidae as already noted, but did not give rise to other groups.

Subfamily 1. Haplophragmiinae

Test composed of simple chambers, not labyrinthic, cement not usually dominant.

Genus TROCHAMMINOIDES Cushman, 1910

Plate 10, figure 1; Key, plate 4, figures 24, 25
Genotype, *Trochammina proteus* Karrer

Trochamminoides Cushman, Bull. 71, U. S. Nat. Mus., pt. 1, 1910, p. 97.
Trochammina (part) of authors. *Ammodiscus* (part) Rhumbler, 1903.

Test free, of several coils, not involute, divided more or less irregularly into chambers with large openings between; wall of fine sand with yellowish-brown cement; aperture simple, at the end of the last-formed chamber. — Cretaceous to Recent.

Genus HAPLOPHRAGMOIDES Cushman, 1910

Plate 10, figure 2; Key, plate 4, figures 26–28
Genotype, *Nonionina canariensis* d'Orbigny

Haplophragmoides Cushman, Bull. 71, U. S. Nat. Mus., pt. 1, 1910, p. 99.
Nonionina (part) d'Orbigny, 1839. *Placopsilina* (part) Parker and Jones, 1857.
Lituola (part), *Haplophragmium* (part), and *Trochammina* (part) of authors.
Ammochilostoma (part) Eimer and Fickert, 1899.

Test free, planispiral, of several coils, usually not completely involute; chambers simple; wall single, arenaceous or with sponge spicules, firmly cemented, amount of cement varying greatly in different species; aperture simple, at the base of the apertural face of the chamber. — Carboniferous to Recent.

Genus RECURVOIDES Earland, 1934

Key, plate 42, figures 34–36
Genotype, *Recurvoides contortus* Earland

Recurvoides Earland, *Discovery* Reports, vol. 10, 1934, p. 90.

Test free, spiral, the coils of the early portion planispiral, later ones nearly at right angles to the earlier ones; chambers numerous, undivided; wall arenaceous, of fine sand grains with a large proportion of ferruginous cement; aperture small, on inner edge of final chamber, sometimes with protruding lips. — Recent.

Genus ALVEOLOPHRAGMIUM Stschedrina, 1936

Key, plate 42, figures 37, 38
Genotype, *Alveolophragmium orbiculatum* Stschedrina

Alveolophragmium Stschedrina, Zool. Anzeiger, vol. 114, 1936, p. 314.
Haplophragmium (part) of authors. *Haplophragmoides* (part) of authors.

Test similar to *Haplophragmium* but with the walls alveolar, the chambers not labyrinthic. — Recent.

Genus CRIBROSTOMOIDES Cushman, 1910

Plate 10, figure 3; Key, plate 4, figure 29
Genotype, *Cribrostomoides bradyi* Cushman

Cribrostomoides Cushman, Bull. 71, U. S. Nat. Mus., pt. 1, 1910, p. 108.
Haplophragmium (part) H. B. Brady, 1884.

Test free, planispiral, composed of numerous simple chambers in several coils; wall arenaceous with much cement; aperture in early stage, a simple elongate slit at the base of the apertural face, in later chambers subdivided by tooth-like processes, in the adult with a linear series of distinct, rounded openings. — Cretaceous to Recent.

Genus LABROSPIRA Höglund, 1947

Key, plate 43, figure 1
Genotype, *Haplophragmium crassimargo* Norman

Labrospira Höglund, Zool. Bidrag Uppsala, vol. 26, 1947, p. 141.
Nonionina (part), *Lituola* (part), *Haplophragmium* (part), and *Haplophragmoides* (part) of authors.

Test free, planispiral or nearly so, of several coils, often not completely involute; chambers simple; wall single, arenaceous; aperture simple, oval or crescentiform, slightly above the base of the last-formed chamber.—Recent.

Genus AMMOMARGINULINA Wiesner, 1931

Plate 10, figure 6; Key, plate 5, figures 13, 14
Genotype, *Ammomarginulina ensis* Wiesner

Ammomarginulina Wiesner, Deutsche Südpolar Exped., vol. XX, Zool., 1931, p. 97.

Test similar to *Ammobaculites* but much compressed, aperture narrow, linear, terminal.—Jurassic to Recent.

Genus AMMOSCALARIA Höglund, 1947

Key, plate 43, figure 2
Genotype, *Haplophragmium tenuimargo* H. B. Brady.

Ammoscalaria Höglund, Zool. Bidrag Uppsala, vol. 26, 1947, p. 151.
Haplophragmium (part), *Proteonina* (part), *Ammobaculites* (part), and *Haplophragmoides* (part) of authors.

Test similar to *Ammobaculites* but strongly compressed, chambers less definite and interior divisions very slight, and the chamber divisions very thin and not conforming with the aperture.—Cretaceous to Recent.

Genus AMMOBACULITES Cushman, 1910

Plate 10, figure 5; Key, plate 5, figures 10–12
Genotype, *Spirolina agglutinans* d'Orbigny

Ammobaculites Cushman, Bull. 71, U. S. Nat. Mus., pt. 1, 1910, p. 114.
Spirolina (part) d'Orbigny, 1846. *Haplophragmium* (part) of authors.

Test free, early chambers close coiled, later ones in typically a linear series, simple; wall arenaceous, with a chitinous lining; aperture in the early stages at the base of the apertural face, in the adult rounded, terminal, simple.—Carboniferous to Recent.

Genus FLABELLAMMINA Cushman, 1928

Plate 10, figure 8; Key, plate 5, figure 16
Genotype, *Flabellammina alexanderi* Cushman

Flabellammina Cushman, Contr. Cushman Lab. Foram. Res., vol. 4, 1928, p. 1.

Test free, much compressed, early stages close coiled, later developing low broad chambers in an inverted V-shape, microspheric form broad and fan-shaped, megalospheric form more elongate and narrow; wall coarsely arenaceous with much fine material and cement; aperture in the adult terminal, elliptical. — Jurassic and Cretaceous.

Genus FRANKEINA Cushman and Alexander, 1929

Plate 10, figure 9; Key, plate 5, figure 17

Genotype, *Frankeina goodlandensis* Cushman and Alexander

Frankeina Cushman and Alexander, Contr. Cushman Lab. Foram. Res., vol. 5, 1929, p. 61.

Test free, early stages planispiral, compressed, later uncoiling and becoming triangular in section; chambers simple; wall coarsely arenaceous

---

PLATE 10

Relationships of the Genera of the Lituolidae

FIG.

1. *Trochamminoides proteus* (Karrer). (After H. B. Brady.)
2. *Haplophragmoides scitulum* (H. B. Brady). (After H. B. Brady.) *a*, side view; *b*, apertural view.
3. *Cribrostomoides bradyi* Cushman.
4. *Orbignyna ovata* Hagenow. (After Hagenow.) *a*, side view; *b*, apertural view. (Valvulinidae.)
5. *Ammobaculites calcareum* (H. B. Brady). *a*, side view; *b*, apertural view.
6. *Ammomarginulina foliacea* (H. B. Brady). (After H. B. Brady.) *a*, side view; *b*, apertural view.
7. *Haplophragmium lituolinoideum* (Goës). *a*, side view; *b*, apertural view.
8. *Flabellammina saratogaensis* Cushman. *a*, side view; *b*, apertural view.
9. *Frankeina goodlandensis* Cushman and Alexander. (After Cushman and Alexander.) *a*, side view; *b*, apertural view.
10. *Lituola mexicana* Cushman.
11. *Cyclammina pauciloculata* Cushman. *a*, side view; *b*, apertural view; *c*, *C. cancellata* H. B. Brady. (After H. B. Brady), section showing labyrinthic interior.
12. *Pseudocyclammina lituus* (Yokoyama). (Adapted from Yokoyama.)
13. *Choffatella decipiens* Schlumberger. (After Schlumberger.) Section.
14. *Dictyopsella kiliani* Schlumberger. (After Schlumberger.) Section.
15. *Yaberinella jamaicensis* Vaughan. (After Vaughan.) *a*, side view; *b*, portion of section.
16. *Orbitopsella praecursor* (Gümbel). (After Gümbel).
17. *Endothyra bowmani* Phillips. (After H. B. Brady.) *a*, side view; *b*, apertural view.
18. *Cribrospira panderi* Möller. (Adapted from Möller.) *a*, apertural view; *b*, side view.
19. *Bradyina nautiliformis* Möller. (Adapted from Möller.) *a*, side view; *b*, apertural view.
20. *Glyphostomella triloculina* (Cushman and Waters). (After Cushman and Waters.)
21. *Endothyranella powersi* (Harlton). (After Harlton.) *a*, side view; *b*, apertural view.
22. *Septammina bradyi* Cushman. (After H. B. Brady.) *a*, side view; *b*, apertural view.

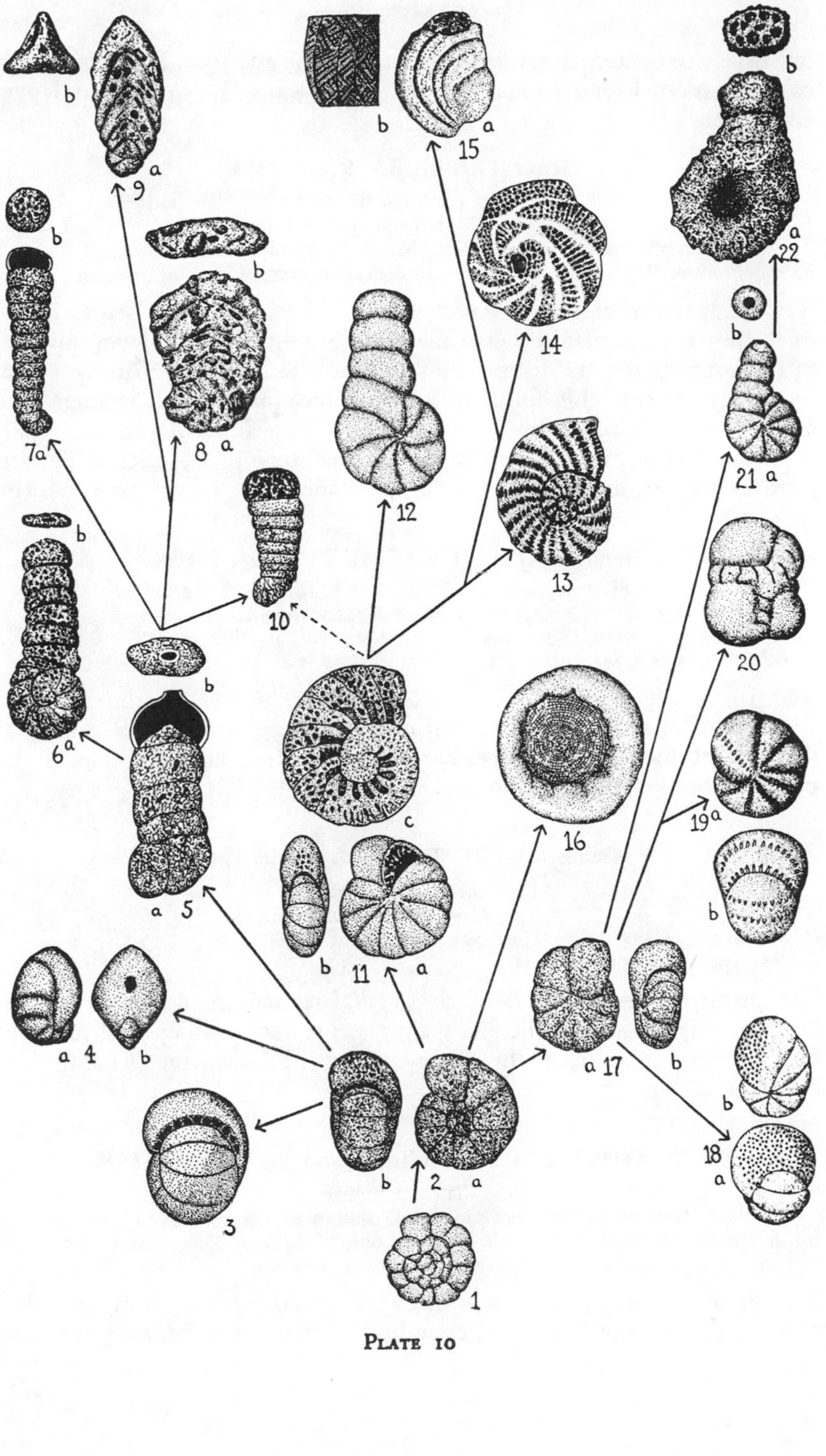

b
b
a
9
a
15
b
a
b
b
22
a
b
14
b
7a
8
a
21
a
b
12
13
10
20
b
6a
c
16
19a
b
a
5
b
11
a
a
4
b
a
17
b
b
18
a
b
2
a
3
1

Plate 10

but firmly cemented; aperture at the base of the apertural face in the early coiled portion, later terminal, simple.—Jurassic (?), Lower and Upper Cretaceous.

Genus TRIPLASIA Reuss, 1854

Key, plate 40, figure 6
Genotype, *Triplasia murchisoni* Reuss
Triplasia Reuss, Denkschr. Akad. Wiss. Wien, vol. 7, 1854, p. 65.
Rhabdogonium Reuss, 1860. *Orthocerina* (part) of authors.

Test elongate, triangular or quadrangular in section, early stages in the microspheric form planispirally coiled, in the megalospheric form uniserial throughout; chambers distinct, not inflated; wall rather coarsely arenaceous, usually smoothly finished; aperture terminal, rounded, often with a short neck.—Cretaceous.

Study of type material in Vienna showed that this species is derived from *Frankeina*, by the loss of the coiled stage in the megalospheric form.

Genus HAPLOPHRAGMIUM Reuss, 1860

Plate 10, figure 7; Key, plate 5, figures 18, 19
Genotype, *Spirolina aequalis* Roemer
Haplophragmium Reuss, Sitz. Akad. Wiss. Wien, vol. 40, 1860, p. 218.
Spirolina Roemer, 1841 (not Lamarck).

Test free, early chambers close coiled, later ones in a rectilinear series, simple; wall arenaceous with a chitinous lining; aperture in the young at the base of the apertural face, becoming terminal and multiple in the uncoiled portion.—Carboniferous to Recent.

Genus CRIBROSPIRELLA Marie, 1941

Key, plate 43, figures 3, 4
Genotype, *Lituola difformis* Lamarck
Cribrospirella Marie, Mém. Mus. Nat. Hist. Nat., n. ser., vol. 12, 1941, p. 28.
Lituola (part) of authors.

Test planispirally coiled in the early stages, later tending to uncoil slightly, bilaterally symmetrical; chambers simple, not labyrinthic; wall arenaceous; aperture cribrate, at the outer end of the last-formed chamber.—Cretaceous.

Genus AMMOASTUTA Cushman and Bronnimann, 1948

Key, plate 43, figure 5
Genotype, *Ammoastuta salsa* Cushman and Bronnimann
Ammoastuta Cushman and Bronnimann, Contr. Cushman Lab. Foram. Res., vol. 24, 1948, p. 17.

Test in the earliest portion planispiral, close coiled, becoming uncoiled very early and in the adult the chambers elongate, curved, each forming more than half the periphery of the oval test; wall arenaceous; aperture a series of very minute pores on the curved lower end of the last-formed chamber.—Recent.

Subfamily 2. Endothyrinae

Test composed of simple chambers, not labyrinthic, cement usually dominant.

Genus ENDOTHYRA Phillips, 1846

Plate 10, figure 17; Key, plate 5, figures 1, 2
Genotype, *Endothyra bowmani* Phillips

Endothyra PHILLIPS, Rept. Proc. Geol. Poly. Soc. West Riding Yorkshire, 1844–45 (1846), p. 277.

Involutina (part) of authors. *Nonionina* EICHWALD, 1860 (not D'ORBIGNY, 1826).

Test free, close coiled, often completely involute, early chambers with the plane of coiling changing; chambers numerous, distinct, simple; wall arenaceous, usually with a large amount of cement, sometimes appearing double, exterior smoothly finished; aperture simple, typically narrow, at the base of the apertural face.—Devonian (?), Mississippian to Triassic.

Thin sections carefully made show the angular irregular fragments of variable size and of various sources.

See Scott, Zeller, and Zeller, The Genus *Endothyra* (Journ. Pal., vol. 21, 1947, pp. 557–562, pls. 83, 84, 2 text figs.).

The generic name *Nanicella* (genotype *Endothyra gallowayi* Thomas) was proposed by Henbest (Journ. Washington Acad. Sci., vol. 25, 1935, pp. 34, 35) for a form from the Devonian related to the fusulinid group. Further details are needed before this genus can be definitely placed.

Genus BRADYINA Möller, 1878

Plate 10, figure 19; Key, plate 5, figure 3
Genotype, *Bradyina nautiliformis* Möller

Bradyina MÖLLER, Mém. Acad. Imp. Sci. St.-Pétersbourg, ser. 7, vol. 25, no. 9, 1878, p. 78.

Nonionina (part) EICHWALD, 1860 (not D'ORBIGNY). *Lituola* (part) H. B. BRADY, 1876 (not LAMARCK).

Test free, mostly involute, close coiled, bilaterally symmetrical in the adult; wall finely arenaceous, with a very large proportion of cement; aperture a single opening or series of openings at the base of the apertural face, high and narrow with a supplementary series in a crescentic line near the peripheral margin.—Carboniferous.

The types of *Bradyina* have not been studied by later students, and more should be known of its detailed structure before its exact relationship to *Glyphostomella* can be accurately determined.

I have been able to study some of the type material of *Bradyina nautiliformis* from the type locality of Miatechkowo identified by Möller. When subjected to strong acid, the fragmentary calcareous material dissolves with much effervescence, leaving the entire skeletal form with the vacuoles and all in a clear, flexible, chitinous material. With high magnification the fragmentary character of the wall is very evident in the specimens, and it is therefore a truly arenaceous test.

Genus GLYPHOSTOMELLA Cushman and Waters, 1928

Plate 10, figure 20; Key, plate 5, figures 4–6

Genotype, *Ammochilostoma* (?) *triloculina* Cushman and Waters

Glyphostomella CUSHMAN and WATERS, Contr. Cushman Lab. Foram. Res., vol. 4, 1928, p. 53.

Ammochilostoma (?) CUSHMAN and WATERS, 1927 (not EIMER and FICKERT).

Test free, involute, early chambers not bilaterally symmetrical, later ones nearly so; wall finely arenaceous, with a large proportion of cement; apertures in the early stages parallel to the margin of the chamber and suture, later developing at right angles to the base of the chamber, several to a chamber and occasionally irregular ones in the apertural face, the apertures connecting with the interior by a funnel-shaped structure. — Pennsylvanian.

While there are similarities between *Glyphostomella* and *Bradyina* the early development of the Russian form is as yet unknown. Our American species of *Glyphostomella* show a distinct residue of angular fragments when dissolved in strong acid.

Genus CRIBROSPIRA Möller, 1878

Plate 10, figure 18; Key, plate 5, figures 7, 8

Genotype, *Cribrospira panderi* Möller

Cribrospira MÖLLER, Mém. Acad. Imp. Sci. St.-Pétersbourg, ser. 7, vol. 25, No. 9, 1878, p. 86.

Test free, mostly involute, bilaterally symmetrical, at least in the adult; wall arenaceous with much cement; aperture cribrate, consisting of numerous, small, rounded openings on the apertural face. — Carboniferous, Cretaceous.

Genus ENDOTHYRANELLA Galloway and Harlton, 1930

Plate 10, figure 21; Key, plate 5, figure 15

Genotype, *Ammobaculites powersi* Harlton

Endothyranella GALLOWAY and HARLTON, in GALLOWAY and RYNIKER, Oklahoma Geol. Surv., Circular 21, 1930, p. 13.

Ammobaculites (part) of authors.

Test free, early chambers close coiled, later ones typically a linear series, simple; wall finely arenaceous with a large proportion of cement; aperture at the base of the apertural face, in the adult rounded, terminal, simple. — Mississippian and Pennsylvanian.

This genus is an uncoiled form derived from *Endothyra*.

Genus SEPTAMMINA Meunier, 1888

Plate 10, figure 22; Key, plate 5, figure 20
Genotype, *Septammina renaulti* Meunier

Septammina MEUNIER, Bull. Soc. Hist. Nat. Autun, vol. 1, 1888, p. 235. — GALLOWAY, Man. Foram., 1933, p. 160.
Endothyra (part) MÖLLER, 1879.

Test similar to *Endothyranella* but with cribrate aperture. — Carboniferous.

The original figure given by Meunier is very obscure and little can be determined from it.

Subfamily 3. Lituolinae

Test composed of labyrinthic chambers.

Genus CYCLAMMINA H. B. Brady, 1876

Plate 10, figure 11; Key, plate 6, figures 2–4
Genotype, *Cyclammina cancellata* H. B. Brady

Cyclammina H. B. BRADY, in NORMAN, Proc. Roy. Soc., vol. 25, 1876, p. 214.
Lituola (part) and *Trochammina* (part) of authors.

Test free, planispiral, partially or wholly involute; wall thick, of arenaceous material, showing sorting, often with a very large proportion of cement, the outer part with fine vacuoles but covered by a thin imperforate coat on the exterior, interior labyrinthic; aperture, a curved fissure at or near the base of the apertural face, supplemented by numerous pores in the central portion of the apertural face. — Cretaceous to Recent.

The detailed finer structure of *Cyclammina* is of special interest in comparison with similar forms in the Paleozoic.

Genus BISERIAMMINA Tchernysheva, 1941

Key, plate 43, figures 6, 7
Genotype, *Biseriammina uralica* Tchernysheva

Biseriammina TCHERNYSHEVA, Comptes Rendus Acad. Sci. URSS, vol. 32, 1941, p. 69.

Test free, polythalamous, close-coiled, consisting of few whorls, completely involute; chambers arranged in two series, alternating on the periphery; wall agglutinated, coarsely granulated; aperture narrow, slit-like, on the inner border of the septal surface. — Carboniferous.

This genus, resembling *Cassidulina* in its structure, is a good example of parallelism.

Genus PSEUDOCYCLAMMINA Yabe and Hanzawa, 1926

Plate 10, figure 12; Key, plate 6, figures 5, 6
Genotype, *Cyclammina lituus* Yokoyama

Pseudocyclammina YABE and HANZAWA, Sci. Rep. Tohoku Imp. Univ., ser. 2 (Geol.), vol. 9, 1926, p. 10.
Cyclammina YOKOYAMA, 1890 (not H. B. BRADY).

Test in the early chambers planispiral and completely involute, later uncoiling in a rectilinear series, peripheral portion of chambers labyrinthic; wall thick, arenaceous, with much cement; aperture in the adult cribrate. —Jurassic, Cretaceous.

Genus CHOFFATELLA Schlumberger, 1904

Plate 10, figure 13; Key, plate 6, figures 9, 10
Genotype, *Choffatella decipiens* Schlumberger

Choffatella SCHLUMBERGER, Bull. Soc. Géol. France, ser. 4, vol. 4, 1904, p. 763.

Test planispiral, much compressed, composed of numerous, narrow, elongate chambers, nearly completely involute, coils increasing rapidly in height; wall arenaceous, with much cement, becoming labyrinthic, especially on the sides and periphery of the chambers; aperture, an elongate series of small pores on the narrow apertural face.—Jurassic and Cretaceous.

Genus DICTYOPSELLA Munier-Chalmas, 1899

Plate 10, figure 14; Key, plate 6, figures 7, 8
Genotype, *Dictyopsella kiliani* Munier-Chalmas

Dictyopsella MUNIER-CHALMAS, in SCHLUMBERGER, Bull. Soc. Géol. France, ser. 3, vol. 27, 1899, p. 462.

Test planispiral, much compressed, partially involute in the young, later becoming complanate; chambers distinct, interior divided into a complex network of small chamberlets; wall finely arenaceous.—Upper Cretaceous.

Genus YABERINELLA Vaughan, 1928

Plate 10, figure 15; Key, plate 6, figures 17, 18
Genotype, *Yaberinella jamaicensis* Vaughan

Yaberinella VAUGHAN, Journ. Pal., vol. 2, 1928, p. 8.

Test planispiral in the young, involute, later becoming complanate, and in microspheric adults annular, much compressed, interior of chambers peculiarly divided by irregularly crossed partitions; wall finely arenaceous; apertures numerous on the apertural face.—Middle Eocene, Jamaica.

Genus LITUOLA Lamarck, 1804

Plate 10, figure 10; Key, plate 6, figure 1
Genotype, *Lituola nautiloidea* Lamarck

Lituola LAMARCK, Ann. Mus., vol. 5, 1804, p. 243.

Test in the early stages planispiral, later portion typically uncoiled and straight, interior labyrinthic; wall arenaceous with much cement; aperture in early stages at the base of the apertural face, single, later becoming multiple and in the terminal face. — Carboniferous to Recent.

Genus SPIROCYCLINA Munier-Chalmas, 1887

Key, plate 6, figures 13–16

Genotype, *Spirocyclina choffati* Munier-Chalmas

Spirocyclina Munier-Chalmas, Compte-rendu som. séan. Soc. Géol. France, 1887, No. 7, p. xxx.

Test in early stages planispiral, later chambers bending back at each end, and in the adult becoming annular, forming a flat disc; chambers labyrinthic and very irregularly divided; wall of calcareous sand; apertures many, on the peripheral face. — Upper Jurassic and Cretaceous.

Genus CYCLOLINA d'Orbigny, 1846

Genotype, *Cyclolina cretacea* d'Orbigny

Cyclolina d'Orbigny, Foram. Foss. Bass. Tert. Vienne, 1846, p. 139.

Test, a thin disc composed of undivided annular chambers; character of wall unknown but probably finely arenaceous; apertures, numerous pores on the peripheral face. — Cretaceous.

Little is known of this form except the very meager description given by d'Orbigny. As *Cyclopsinella* is described as similar to two superimposed Cyclolinas, and the wall of *Cyclopsinella* is arenaceous, and as both forms come from the same locality, it may be inferred that *Cyclolina* also has an arenaceous test, and not a porcellanous one as has been suggested. Both these genera need careful study of the early stages to determine their relationship to other foraminifera of the Upper Cretaceous.

Genus ORBITOPSELLA Munier-Chalmas, 1902

Plate 10, figure 16; Key, plate 6, figures 11, 12

Genotype, *Orbitulites praecursor* Gümbel

Orbitopsella Munier-Chalmas, Bull. Soc. Géol. France, ser. 4, vol. 2, 1902, p. 351.

Test discoidal, the periphery greatly thickened, early stages spiral, later annular; chambers regularly divided into rectilinear chamberlets; wall finely arenaceous, with much cement; apertures small, rounded, peripheral. — Jurassic.

Genus CYCLOPSINELLA Galloway, 1933

Genotype, *Cyclopsina steinmanni* Munier-Chalmas

Cyclopsinella Galloway, Man. Foram., 1933, p. 138.

Cyclopsina Munier-Chalmas, 1887 (not Milne-Edwards).

Test discoid, early stages undescribed, later chambers annular in two parallel planes, those of each plane independent of the other; chambers undivided; wall arenaceous, exterior grainy; apertures numerous, peripheral, in two rows. — Upper Cretaceous.

The only known figure of this is a silhouette of a bit of a vertical section. I did not see the type in Paris. It must remain with *Cyclolina* d'Orbigny as a problematical form until actual material is studied.

Family 9. TEXTULARIIDAE

Test in the earliest stages, at least in primitive forms, planispiral, later in all but the most accelerated forms developing a biserial stage, final development taking various forms, usually becoming uniserial in the more specialized types; wall typically arenaceous, with a varying proportion of cement in different genera and species; aperture typically at the inner margin of the last-formed chamber in the biserial forms, becoming terminal and sometimes multiple in the uniserial forms.

The Textulariidae are directly derived from a planispiral ancestry such as the Lituolidae by the addition of the biserial character, resulting from an elongate spiral settling down to a turn of 180° for each chamber. In the more primitive genera, the early stages at least in the microspheric form show the primitive planispirally coiled stage, but this is gradually lost in the more highly developed genera.

The material of the wall of the test varies greatly. Some species are almost entirely formed of quartz sand with ferruginous cement, and may show a chitinous lining. The large species of shallow warm waters, where there is an abundance of lime, use calcareous sand with a calcareous cement. In the later Paleozoic there is developed a calcareous layer inside of the outer arenaceous wall in some forms, and these need careful study in thin section with modern petrographic methods to determine with scientific accuracy their exact structure. In others, even from the early Paleozoic, there is the typical arenaceous test.

Many of the early figures of Paleozoic forms such as those of Möller are much conventionalized, and based on his interpretation of the form, rather than on complete evidence from the specimens themselves.

From the biserial forms there is a tendency to develop a uniserial form, and finally by acceleration of development the biserial stage becomes progressively shorter, and in the nearly uniserial forms of the Permian, *Geinitzina*, *Spandelina*, and *Monogenerina*, the biserial stage is only seen in the early chambers of the microspheric form, if at all.

Some of the genera which have commonly been associated with the Textulariidae have been proven on detailed study to belong elsewhere. Instead of *Textulariella* and *Cuneolina* being developed from *Textularia*, the early stages show that they are related to *Arenobulimina*, and were developed in the Cretaceous, persisting to the present oceans. The simpler genera, such as *Spiroplectammina*, *Textularia*, and *Bigenerina*, developed early in the Paleozoic, and have continued to the present oceans, while the more specialized forms, such as *Climacammina*, *Deckerella*, *Mono-*

generina, *Geinitzina*, etc., reached their climax at the end of the Paleozoic, and mostly became extinct there.

There is no tangible evidence that the typical arenaceous Textularias have ever arisen from a calcareous ancestry. The only forms which develop calcareous perforate tests, if they are truly so, have been shown not to be true Textularias as the early stages are not biserial. Even in these forms, the test starts with a purely arenaceous test, and develops the calcareous material only in its later stages, adding further evidence on this point. Many species of the Textulariidae, especially the more primitive ones, have the same yellowish-brown cement and chitinous lining characteristic of the arenaceous group from the Astrorhizidae onward, and confirm the development of the Textulariidae through the Ammodiscidae and Lituolidae.

The genus *Nodosaroum* Rhumbler, 1913, has as its type species, *Nodosaria index* Ehrenberg. The type figure, based on a very poor section, is without description and entirely unrecognizable. Möller's later figure referred to the same species is much idealized as are most of his figures, and it is difficult to know what he had until a re-examination of his types is possible. The genus is, therefore, not recognized here.

The genus *Pseudobolivina* Wiesner, 1931, proposed for the minute, slender forms from the Antarctic which are biserial with a large amount of cement may belong here. These are peculiar forms and need further study.

Subfamily 1. Spiroplectammininae

Test with the chambers simple, the early ones distinctly planispiral in both microspheric and megalospheric forms, and forming a considerable portion of the test, later chambers biserial.

Genus SPIROPLECTAMMINA Cushman, 1927

Plate 11, figure 1; Key, plate 7, figures 1–3

Genotype, *Textularia agglutinans*, var. *biformis* Parker and Jones

Spiroplectammina Cushman, Contr. Cushman Lab. Foram. Res., vol. 3, 1927, p. 23.
Textularia (part) Parker and Jones (not Defrance). *Spiroplecta* H. B. Brady, 1884 (not Ehrenberg).

Test free, early chambers planispiral in both microspheric and megalospheric forms, and forming a considerable portion of the test, later chambers biserial; wall arenaceous, amount of cement variable; aperture, a low opening at the base of the inner margin.—Pennsylvanian to Recent.

Genus MORULAEPLECTA Höglund, 1947

Key, plate 43, figure 8

Genotype, *Morulaeplecta bulbosa* Höglund

Morulaeplecta Höglund, Zool. Bidrag Uppsala, vol. 26, 1947, p. 165.

Test with the early chambers irregularly placed about the proloculum, remainder of test biserial; wall arenaceous; aperture a low arched opening at the base of the inner margin of the last-formed chamber.—Recent.

This genus must remain somewhat indefinite until more is known of the early stages.

Genus SEPTIGERINA Keijzer, 1941

Key, plate 43, figures 9, 10
Genotype, *Septigerina dalmatica* Keijzer

Septigerina Keijzer, Proc. Ned. Akad. Wetensch., vol. 44, No. 8, 1941, p. 1006.

Test similar to *Spiroplectammina* but the chambers of the biserial portion subdivided by vertical projections. — Eocene.

Genus AMMOSPIRATA Cushman, 1933

Plate 11, figure 3; Key, plate 7, figure 16
Genotype, *Pavonina mexicana* Cushman

Ammospirata Cushman, Contr. Cushman Lab. Foram. Res., vol. 9, 1933, p. 32.
Pavonina (part) Cushman, 1926 (not d'Orbigny).

Test compressed, early portion planispiral, later biserial, and in the adult uniserial; wall finely arenaceous, with much calcareous cement; apertures in the adult one to several, elongate, at the periphery. — Eocene and Oligocene.

Genus SEMITEXTULARIA Miller and Carmer, 1933

Key, plate 43, figure 11
Genotype, *Semitextularia thomasi* Miller and Carmer

Semitextularia Miller and Carmer, Journ. Pal., vol. 7, 1933, p. 428.

Test free, small, compressed, irregularly rhomboid in outline; early stages planispiral, then biserial, and in the adult uniserial; wall finely arenaceous with a large proportion of cement; aperture multiple, consisting of two rows of small pores in an elongate depression of the outer face of the last-formed chamber. — Devonian.

Genus AMMOBACULOIDES Plummer, 1932

Plate 11, figure 2; Key, plate 7, figure 4
Genotype, *Ammobaculoides navarroensis* Plummer

Ammobaculoides Plummer, Amer. Mid. Nat., vol. 13, 1932, p. 86.

Test with the earliest chambers planispiral, followed by chambers arranged biserially, and the adult with uniserial chambers, simple; wall arenaceous; aperture terminal, rounded. — Cretaceous, Texas.

Genus SPIROPLECTELLA Earland, 1934

Key, plate 43, figures 14, 15
Genotype, *Spiroplectella cylindroides* Earland

Spiroplectella Earland, *Discovery* Reports, vol. 10, 1934, p. 113.

Test free, early stages planispiral, followed by a biserial stage and in the adult uniserial; wall arenaceous, with much cement; aperture simple and terminal. — Recent.

Subfamily 2. Textulariinae

Test typically biserial or becoming uniserial, usually free; chambers simple or labyrinthic; wall arenaceous; aperture simple or cribrate.

Genus TEXTULARIA Defrance, 1824

Plate 11, figure 4; Key, plate 7, figures 5, 6
Genotype, *Textularia sagittula* Defrance

Textularia DEFRANCE, Dict. Sci. Nat., vol. 32, 1824, p. 177.
Textilaria EHRENBERG and later authors. *Plecanium* REUSS, 1862. *Grammostomum* (part) of authors. *Palaeotextularia* SCHUBERT, 1920. *Pseudobolivina* WIESNER, 1931.

Test free, elongate, tapering, typically compressed with the zigzag line between the chambers on the middle of the flattened side, early chambers in the microspheric form usually planispirally coiled, later biserial, chambers simple, not labyrinthic; wall arenaceous, cement of various sorts, the relative amount variable; aperture, typically an arched slit at the inner margin of the chamber, occasionally in the apertural face. — Cambrian (?), Devonian to Recent.

The structure of the early Paleozoic forms is not thoroughly known, and needs study of thin sections by modern petrographic methods. Lacroix has recently shown that the type species, *Textularia sagittula* Defrance, has a coiled stage in both the microspheric and megalospheric forms, but that they are short compared to *Spiroplectammina biformis* and related species.

Genus PSEUDOPALMULA Cushman and Stainbrook, 1943

Key, plate 46, figure 36
Genotype, *Pseudopalmula palmuloides* Cushman and Stainbrook

Pseudopalmula CUSHMAN and STAINBROOK, Contr. Cushman Lab. Foram. Res., vol. 19, 1943, p. 78.

Test in the earliest stages elongate and irregularly biserial, later much compressed and triangular; chambers biserial throughout, the later ones in the adult becoming much elongate, reaching backward nearly to the base of the test; wall finely arenaceous; aperture in the adult slightly elongate at one side of the apertural tip of the test, on the narrow edge. — Devonian.

Genus SILICOTEXTULINA Deflandre, 1934

Genotype, *Silicotextulina diatomitarum* Deflandre

Silicotextulina DEFLANDRE, Comptes Rendus, Acad. Sci. Paris, vol. 198, 1934, p. 1447.

This genus has a siliceous test with a proloculum followed by biserial chambers. Figures are not given, and the apertural characters not defined. It may belong in the Textulariidae or elsewhere. — Miocene.

See Deflandre, Sur un Foraminifère Siliceux Fossile des Diatomitès Miocènes de Californie: *Silicotextulina diatomitarum* n.g., n.sp. (Comptes Rendus. Acad. Sci. Paris, vol. 198, 1934, pp. 1446–1448).

Genus SIPHOTEXTULARIA Finlay, 1939

Key, plate 43, figure 13
Genotype, *Siphotextularia wairoana* Finlay
Siphotextularia Finlay, Trans. Roy. Soc. New Zealand, vol. 68, 1939, p. 510.

Test and development similar to *Textularia*, but usually quadrangular in section; aperture a distinct, short, slit-like tube, in the apertural face somewhat above the basal margin. — Cretaceous to Recent.

Genus TEXTULARIOIDES Cushman, 1911

Plate 11, figure 5; Key, plate 7, figure 9
Genotype, *Textularioides inflata* Cushman
Textularioides Cushman, Bull. 71, U. S. Nat. Mus., pt. 2, 1911, p. 26.
Haddonia Heron-Allen and Earland, 1924 (not Chapman, 1898).

Test similar to *Textularia* but attached, the attached side somewhat flattened. — Recent.

Genus HAEUSLERELLA Parr, 1935

Key, plate 43, figure 12
Genotype, *Haeuslerella pukeuriensis* Parr
Haeuslerella Parr, Trans. Roy. Soc. New Zealand, vol. 65, 1935, p. 82.

"Test in the early stages regularly biserial, becoming loosely biserial; aperture in the early stages textularian, becoming semicircular and subterminal in the adult chambers." — Miocene and Pliocene.

Genus BIGENERINA d'Orbigny, 1826

Plate 11, figure 6; Key, plate 7, figures 10, 11
Genotype, *Bigenerina nodosaria* d'Orbigny
Bigenerina d'Orbigny, Ann. Sci. Nat., vol. 7, 1826, p. 261.
Gemmulina d'Orbigny, 1826. *Palaeobigenerina* Galloway, 1933.

Test free, early chambers biserial, later ones uniserial, in a rectilinear series, not labyrinthic; wall arenaceous, relative amount of cement varying greatly; aperture in the biserial stage as in *Textularia*, in the adult uniserial stage terminal, simple. — Pennsylvanian to Recent.

Typical species occur since the Pennsylvanian, and the peculiar double wall shown by Schellwien needs demonstration.

Genus VULVULINA d'Orbigny, 1826

Plate 11, figure 10; Key, plate 7, figures 13, 14
Genotype, *Vulvulina capreolus* d'Orbigny

Vulvulina D'ORBIGNY, Ann. Sci. Nat., vol. 7, 1826, p. 264.
Nautilus (part) BATSCH, 1791. *Bigenerina* (part) of authors (not D'ORBIGNY). *Schizophora* REUSS, 1861. *Grammostomum* (part) PARKER and JONES, 1863 (not EHRENBERG). *Venilina* GÜMBEL, 1868. *Textilaria* (part) GÜMBEL. *Trigenerina* SCHUBERT, 1902.

Test free, much compressed throughout, early stages biserial, or planispirally coiled in the microspheric form, later chambers uniserial, simple; wall arenaceous, with a large proportion of cement; aperture elongate, elliptical, simple, terminal. — Cretaceous to Recent.

See Cushman, The Genus *Vulvulina* and Its Species (Contr. Cushman Lab. Foram. Res., vol. 8, 1932, pp. 75–85, pl. 10).

Genus SPANDELINA Cushman and Waters, 1928

Plate 11, figure 9; Key, plate 7, figure 18
Genotype, *Spandelina excavata* Cushman and Waters

Spandelina CUSHMAN and WATERS, Journ. Pal., vol. 2, 1928, p. 363.

Test uniserial, the chambers in a generally rectilinear series, the earlier ones at least compressed, especially in the microspheric form; wall calcareous, finely arenaceous, with a thin coating; aperture simple, terminal, elliptical or rounded. — Pennsylvanian (?), Permian.

Without the thin outer covering, the wall of *Spandelina* appears perforate, especially when calcitized as is common.

The genera *Padangia* and *Pachyphloia* Lange, 1925, described from the Permian of Sumatra, are known only from photographic sections which lack full details of structure. These may be allied to *Spandelina*. Both forms need more detailed study for their satisfactory determination as to family relationships.

Genus PLEUROSTOMELLOIDES Majzon, 1943

Genotype, *Pleurostomelloides andreasi* Majzon

Pleurostomelloides MAJZON, Mitt. Jahrb. K. Ungarn. Geol. Anstalt, vol. 37, 1943, p. 157.

It seems difficult to separate this form from the genus *Spandelina*. The aperture seems to be lateral instead of terminal, as in *Spandelina*.

Genus MONOGENERINA Spandel, 1901

Plate 11, figure 7; Key, plate 7, figure 15
Genotype, *Monogenerina atava* Spandel

Monogenerina SPANDEL, Festschr. Nat. Ges. Nürnberg, 1901, p. 181.

Test free, elongate, cylindrical or slightly arcuate, tapering; chambers in the earliest stages of the microspheric form biserial, otherwise uniserial;

wall finely arenaceous, with much cement and a thin outer covering; aperture simple, rounded, terminal.—Mostly Permian.

The early microspheric stage, which is biserial, shows the relationships to the biserial genera of the Textulariidae. Spandel was unable to determine from his types whether the wall was sandy or calcareous.

Genus GEINITZINA Spandel, 1901

Plate 11, figure 8; Key, plate 7, figure 19

Genotype, *Textularia cuneiformis* Jones (not d'Orbigny)

Geinitzina SPANDEL, Festschr. Nat. Ges. Nürnberg, 1901, p. 189.

Textularia (part) JONES, 1850 (not DEFRANCE). *Geinitzella* SPANDEL, 1898 (not WAAGEN and WENTZEL, 1886).

Test free, much compressed, especially along the median line, in the microspheric form showing a trace of the biserial ancestry; wall finely

---

PLATE 11

RELATIONSHIPS OF THE GENERA OF THE TEXTULARIIDAE AND VERNEUILINIDAE

TEXTULARIIDAE

FIG.

1. *Spiroplectammina biformis* (Parker and Jones). (After Heron-Allen and Earland.)
2. *Ammobaculoides navarroensis* Plummer. (After Plummer.) *a*, exterior; *b*, showing section of early coiled chambers.
3. *Ammospirata mexicana* (Cushman).
4. *Textularia gramen* d'Orbigny.
5. *Textularioides inflata* Cushman. *a*, side view; *b*, apertural view.
6. *Bigenerina ciscoensis* Cushman and Waters. (After Cushman and Waters.)
7. *Monogenerina atava* Spandel. (After Spandel.) Section.
8. *Geinitzina postcarbonica* (Spandel). (After Spandel.) Sections: *a*, longitudinal; *b*, transverse.
9. *Spandelina excavata* Cushman and Waters. (After Cushman and Waters.) *a*, side view; *b*, apertural view.
10. *Vulvulina pennatula* (Batsch). (After H. B. Brady.) *a*, front view; *b*, apertural view.
11. *Cribrostomum textulariforme* Möller. (After Möller.)
12. *Deckerella clavata* Cushman and Waters. (After Cushman and Waters.) *a*, side view; *b*, apertural view.
13. *Climacammina antiqua* H. B. Brady. (After H. B. Brady.)
14. *Cribrogenerina*. (Idealized section.)

VERNEUILINIDAE

15. *Eggerella bradyi* (Cushman). (VALVULINIDAE.)
16. *Tritaxia pyramidata* Reuss. (After Reuss.) *a*, front view; *b*, apertural view.
17. *Dorothia subrotundata* (Schwager). (After H. B. Brady.) *a*, front view; *b*, apertural view. (VALVULINIDAE.)
18. *Heterostomella rugosa* (d'Orbigny). (After d'Orbigny.) *a*, front view; *b*, apertural view.
19. *Spiroplectinata annectens* (Parker and Jones).
20. *Gaudryinella delrioensis* Plummer.

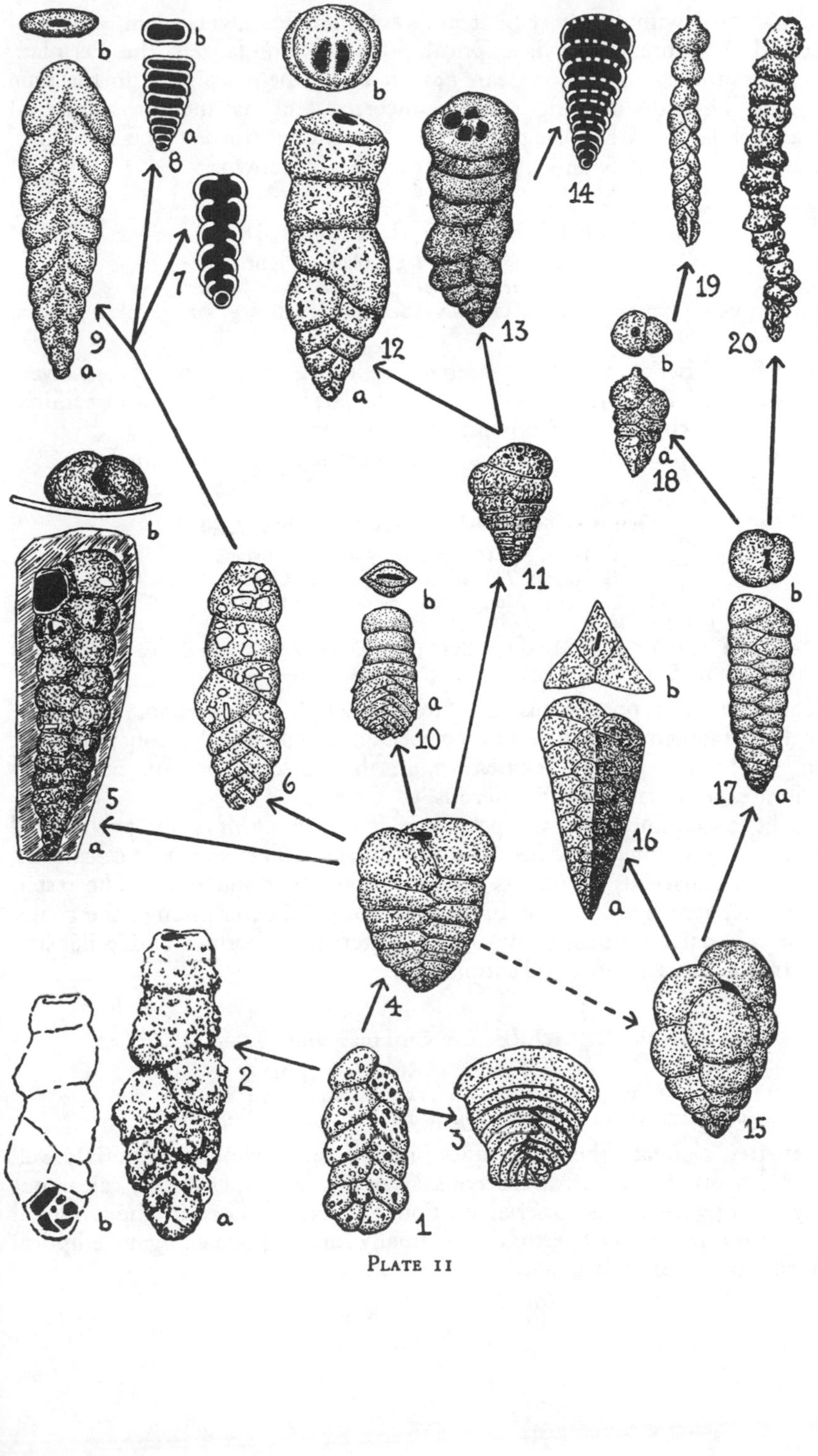

b
b
a
8
7
9
a
b
12
a
13
14
19
20
b
a
18
11
b
5
a
b
6
b
10
a
b
16
a
b
17
a
4
2
3
1
b
a
15

PLATE II

arenaceous, with much cement and a thin outer layer when well preserved; aperture terminal, elliptical. — Often abundant in the Permian.

An examination of European Permian specimens assigned to the type species shows the early figures to be incorrect and that the test is uniserial except for the earliest stages in the microspheric form. The subgenus, *Lunucammina* of Spandel, is very imperfectly known.

Genus CRIBROSTOMUM Möller, 1879

Plate 11, figure 11; Key, plate 7, figure 7

Genotype, *Cribrostomum textulariforme* Möller

Cribrostomum Möller, Mém. Acad. Imp. Sci. St.-Pétersbourg, ser. 7, vol. 27, No. 5, 1879, p. 39.

Test free, biserial; wall finely arenaceous, thick, with an outer thin layer; apertures of the early stages textularian, later cribrate, on the terminal face of the chamber. — Carboniferous to Permian.

These may be only a stage in the development of *Climacammina.*

Genus CLIMACAMMINA H. B. Brady, 1873

Plate 11, figure 13; Key, plate 7, figure 12

Genotype, *Textularia antiqua* H. B. Brady

Climacammina H. B. Brady, Mem. Geol. Survey Scotland, Expl. Sheet 23, 1873, p. 94; Pal. Soc. Mon., vol. 30, 1876, p. 67.

Textularia (part) H. B. Brady, 1871 (not Defrance). *Moellerina* Eimer and Fickert, 1899 (not Ulrich or Schellwien).

Test free, early portion biserial, later uniserial; wall arenaceous, mostly of fine fragments but including coarser ones, cement calcareous; aperture in the biserial portion textularian, in the uniserial portion irregularly cribrate, terminal. — Carboniferous to Permian.

I have examined Brady's specimens of *Climacammina antiqua* as well as duplicates which have been tested with acid. The cement is calcareous, but the arenaceous fragments are of various sorts and sizes. The test is truly arenaceous as stated by Brady in his original description of the genus. The original specimens show the characteristic distortion and collapsing so frequently seen in the Textulariidae.

Genus DECKERELLA Cushman and Waters, 1928

Plate 11, figure 12; Key, plate 7, figure 8

Genotype, *Deckerella clavata* Cushman and Waters

Deckerella Cushman and Waters, Journ. Pal., vol. 2, 1928, p. 128.

Test free, elongate, the early stages biserial, later chambers uniserial; wall with an outer, opaque, arenaceous layer and an inner, clear, calcareous layer; aperture in the biserial portion textularian, later in the uniserial part pushing into the apertural face, finally forming two elongate, elliptical openings. — Pennsylvanian.

Genus CRIBROGENERINA Schubert, 1907

Plate 11, figure 14; Key, plate 7, figure 17
Genotype, *Bigenerina sumatrana* Volz

Cribrogenerina SCHUBERT, Neues Jahrb. für Min., Beil. Bd. 25, 1907, p. 245.
Bigenerina (part) of authors.

Test free, typically uniserial throughout unless the earliest chambers in the microspheric form are biserial; wall finely arenaceous; aperture in the early chambers simple, in later ones in a cribrate terminal plate. — Carboniferous, Permian.

Family 10. VERNEUILINIDAE

Test, at least in the early stages, triserial and typically triangular in transverse section, later biserial in some genera, and some forms probably becoming uniserial; wall arenaceous, the amount of cement varying in different genera and species; aperture simple, typically without a tooth.

The family Verneuilinidae as here restricted includes those forms which are entirely arenaceous. The group early assumed a form with sharp angles to the triserial test. The earliest known species of *Verneuilina* are from the Jurassic. They are very elongate, and resemble *Textularia* in all respects except that they are triserial. The only difference between these Jurassic forms and the type is in the sharp angles of the chambers in the later forms of the Cretaceous. Some specimens of *Gaudryina* from the Triassic also have an intermediate form in the rounded angles of the chambers. Typical biserial Textularias occur with these triserial Verneuilinas, and the resemblance is very close. To derive these very elongate forms from the flat, scale-like *Trochammina fusca* as has been proposed, or to unite this species with those forms called by Haeusler *Valvulina*, seems to be going very far afield when a simple explanation is at hand. It may be noted that there are broad biserial Textularias figured by Hauesler as occurring with the Valvulinas.

There seems to be no definite valvular tooth developed in the aperture of the Verneuilinidae. The triangular forms of *Verneuilina* are very close to *Valvulina*, which develops a very definite valvular tooth in the aperture, as do the various genera derived from it. For that reason the type of *Clavulina* is transferred to that family as derived from *Valvulina* rather than from *Verneuilina*. There is a short neck developed in the biserial genera *Heterostomella* and *Barbourinella*.

Biserial forms developed as early as the late Jurassic where undoubted Gaudryinas occur. Most of the forms usually assigned to *Gaudryina*, which have more than three chambers to a whorl in the early stage, belong in the Valvulinidae. In *Tritaxia* there is a tendency for the last chamber in the adult to take a terminal position. The uniserial forms usually assigned to *Clavulina* need careful study to determine their exact relationships. The genus *Gaudryinella*, in the type species, has the later portion becoming irregularly uniserial; the early portion is triangular in transverse section, in later formations becoming more regularly uniserial. The biserial stage is fairly short and rather irregular. In *Spiroplectinata* the early stage is triserial, followed by a rather long and regular biserial stage, and the last chambers are uniserial.

Similar stages are seen in the uniserial genera *Pseudoclavulina* and *Clavulinoides*. In the genus *Rudigaudryina*, the later part becomes very irregular and assumes many shapes.

For details in regard to the species and genera of this family, see Cushman, A Monograph of the Foraminiferal Family Verneuilinidae (Cushman Lab. Foram. Res. Special Publ. 7, 1937, pp. i–xiii, 1–157, pls. 1–20) and Supplement (Special Publ. 7A, 1946, pp. 1–43, pls. 1–4).

Genus VERNEUILINA d'Orbigny, 1840

Key, plate 7, figures 21, 22

Genotype, *Verneuilina tricarinata* d'Orbigny

Verneuilina d'Orbigny, Mém. Soc. Géol. France, ser. 1, vol. 4, 1840, p. 39.

Test triserial throughout, usually triangular in transverse section, but the earlier species usually with the angles rounded; chambers distinct, three making up each whorl, usually increasing in size toward the apertural end; sutures usually distinct, either depressed or in some species raised and somewhat limbate; wall arenaceous, variable in coarseness and relative amount of cement; aperture textularian, usually a low opening, occasionally somewhat arched, at the base of the inner margin of the last-formed chamber. — Mostly Jurassic and Cretaceous, rare in the Tertiary and Recent.

Genus UVIGERINAMMINA Majzon, 1943

Genotype, *Uvigerinammina jankoi* Majzon

Uvigerinammina Majzon, Mitt. Jahrb. K. Ungarn. Geol. Anstalt, vol. 37, 1943, p. 158.

Test triserial, elongate; wall arenaceous; aperture at the outer end of the last-formed chamber. — Paleocene (?).

Without a more detailed description of the early stages, it is difficult to place this genus with certainty.

Genus TRITAXIA Reuss, 1860

Plate 11, figure 16; Key, plate 7, figure 23

Genotype, *Textularia tricarinata* Reuss

Tritaxia Reuss, Sitz. Akad. Wiss. Wien, vol. 40, 1860, p. 227.
Textularia (part) Reuss, 1845.

Test triserial, triangular in transverse section, the angles acute or somewhat rounded; chambers usually distinct but not inflated, three making up each whorl but the last one or more chambers in the adult terminal, usually somewhat contracted; sutures usually distinct and slightly depressed; wall arenaceous, usually of fine particles, with much cement, and usually the exterior smoothly finished; aperture in the early stages textularian, a low opening at the inner margin of the last-formed chamber, in the adult becoming terminal and usually rounded.—Lower and Upper Cretaceous.

Genus BARBOURINELLA Bermudez, 1940

Key, plate 43, figure 18
Genotype, *Barbourina atlantica* Bermudez

Barbourinella Bermudez, Mem. Soc. Cubana Hist. Nat., vol. 14, 1940, p. 410.
Barbourina Bermudez, l. c., vol. 13, 1939, p. 9.

Test triserial throughout, triangular in section; wall arenaceous; aperture terminal, in terminal face of the last-formed chamber, rounded or elliptical with a short neck and slightly thickened lip.—Tertiary and Recent.

Genus FLOURENSINA Marie, 1938

Key, plate 43, figures 19, 20
Genotype, *Flourensina douvillei Marie*

Flourensina Marie, Bull. Soc. Géol. France, ser. 5, vol. 8, 1938, p. 91.

Test free, triserial throughout, interior not labyrinthic; chambers added at an angle of 120°, outer end with a large spinose projection; wall arenaceous, of siliceous sand with little cement; aperture elongate, rounded, at the inner margin of the last-formed chamber, in a slight depression.—Cretaceous.

Genus GAUDRYINA d'Orbigny, 1839

Key, plate 7, figure 24
Genotype, *Gaudryina rugosa* d'Orbigny

Gaudryina d'Orbigny, in De la Sagra, Hist. Phys. Pol. Nat. Cuba, 1839, "Foraminifères," p. 109.

Test with the chambers triserial, the adult biserial, the early portion typically triangular with distinct angles, but in some species the chambers rounded, and the angles obtuse or broadly rounded, adult portion with the biserial chambers of various shapes, either rounded or more or less quadrate, or with one series having the periphery truncate and the opposite series with a distinct angle; chambers usually distinct, the early ones sometimes obscure; sutures usually distinct except in the early portion, typically depressed, very rarely limbate; wall arenaceous throughout, either coarsely arenaceous with little cement or composed of fine fragments and much cement, the surface ranging from very rough to smooth; aperture in the early stages at the inner margin of the last-formed chamber, usually with a distinct re-entrant, in some species tending to become terminal in the final chamber.—Jurassic to Recent.

Subgenus GAUDRYINA sensu strictu

Subgenotype, *Gaudryina rugosa* d'Orbigny

Test with the early portion triserial, and the later portion biserial with inflated chambers, usually of the same general shape in both series.—Jurassic to Recent.

Subgenus SIPHOGAUDRYINA Cushman, 1935

Key, plate 43, figure 16

Subgenotype, *Gaudryina* (*Siphogaudryina*) *stephensoni* Cushman

Siphogaudryina Cushman, Smithsonian Misc. Coll. vol. 91, No. 21, 1935, p. 3; Special Publ. No. 7, Cushman Lab. Foram. Res., 1937, p. 72.

Test with the early stages triserial, usually triangular with sharp angles, one of the ridges becoming divided and forming a quadrangular test with four distinct angles, usually somewhat compressed with two narrow sides and two broad sides, one of which is somewhat wider than the other, the angles frequently expanded into fistulose processes, which become broken on the exterior, showing a row of openings along the angles of the test, end view usually quadrangular.—Upper Cretaceous to Recent.

Subgenus PSEUDOGAUDRYINA Cushman, 1936

Key, plate 43, figure 17

Subgenotype, *Gaudryina* (*Pseudogaudryina*) *atlantica* (Bailey)

Pseudogaudryina Cushman, Special Publ. No. 6, Cushman Lab. Foram. Res., 1936, p. 12; Special Publ. No. 7, 1937, p. 85.

Test in the early portion triserial and triangular in section with distinct angles, in the adult biserial and triangular in section, the peripheral margin of one series of chambers being squarely truncate, those of the opposite side ending in a distinct angle, usually one of the three flattened faces narrower than the other, and marked by simple horizontal sutures, representing the line of meeting between the series of chambers with truncate periphery, the other two sides usually broader and marked by a series of zigzag lines representing the alternating chambers.—Upper Cretaceous to Recent.

Genus DIGITINA Crespin and Parr, 1941

Key, plate 43, figure 21

Genotype, *Digitina recurvata* Crespin and Parr

Digitina Crespin and Parr, Journ. Proc. Roy. Soc. New South Wales, vol. 74, 1941, p. 306.

Test free, elongate, curved, generally almost circular in cross section, early chambers in the form of a cone, later chambers biserial, interior of chambers undivided; wall coarsely arenaceous, consisting of a single layer;

aperture an arched slit in a re-entrant angle at the base of the last-formed chamber. — Permian.

The systematic position of this genus is in doubt until more studies are made. It is evidently close to *Gaudryina*.

Genus MIGROS Finlay, 1939

Key, plate 43, figure 22

Genotype, *Gaudryina medwayensis* Parr

Migros FINLAY, Trans. Roy. Soc. New Zealand, vol. 69, 1939, p. 312.
Gaudryina (part) of authors.

Test similar to *Gaudryina* but with the aperture in the adult rounded, in the apertural face, but connected with the inner margin by a narrow opening. — Tertiary and Recent.

Genus SPIROPLECTINATA Cushman, 1927

Plate 11, figure 19; Key, plate 7, figure 20

Genotype, *Textularia annectens* Parker and Jones

Spiroplectinata CUSHMAN, Contr. Cushman Lab. Foram. Res., vol. 3, 1927, p. 62.
Textularia (part) of authors (not DEFRANCE). *Spiroplecta* (part) of authors. *Spiroplectina* CUSHMAN, 1927 (not SCHUBERT).

Test with the early chambers triserial, sides flattened or concave, later becoming biserial, the adult uniserial, and the chambers nodose; wall distinctly arenaceous; aperture in the adult rounded, terminal, with a distinct neck. — Lower and Upper Cretaceous.

The early descriptions and figures of this form are incorrect, as a study of typical material has shown.

Genus BERMUDEZINA Cushman, 1937

Key, plate 43, figure 23

Genotype, *Heterostomella* (?) *cubensis* D. K. Palmer and Bermudez

Bermudezina CUSHMAN, Special Publ. No. 7, Cushman Lab. Foram. Res., 1937, p. 102.
Heterostomella (part) of authors.

Test similar to *Gaudryina* in general structural characters, especially the subgenus *Pseudogaudryina*, but differing in the apertural characters which consist of a rounded, tubular neck in a terminal position on the last-formed chamber, the aperture circular. — Upper Eocene to Miocene.

Genus GAUDRYINELLA Plummer, 1931

Plate 11, figure 20; Key, plate 7, figure 26

Genotype, *Gaudryinella delrioensis* Plummer

Gaudryinella PLUMMER, Amer. Midland Nat., vol. 12, 1931, p. 341.

Test with the early stage triserial, triangular in transverse section, later biserial, and in the adult uniserial; wall arenaceous; aperture in the adult terminal, rounded. — Cretaceous, Eocene (?).

Genus PSEUDOCLAVULINA Cushman, 1936

Plate 12, figure 2; Key, plate 43, figure 24

Genotype, *Clavulina clavata* Cushman

Pseudoclavulina CUSHMAN, Special Publ. No. 6, Cushman Lab. Foram. Res., 1936, p. 16; Special Publ. No. 7, 1937, p. 107.

Clavulina (part) of authors.

Reophax W. BERRY (not MONTFORT).

Test with the early chambers triserial and the test usually triangular in transverse section, the angles acute or rounded, later portion uniserial, usually nodose, rounded in transverse section; chambers of the early portion usually indistinct, of the uniserial portion often strongly inflated; wall variable in texture, arenaceous, frequently with the exterior roughly finished; aperture in the adult terminal, rounded, often with a distinct neck. — Upper Cretaceous to Recent.

Genus CLAVULINOIDES Cushman, 1936

Key, plate 8, figure 2

Genotype, *Clavulina trilatera* Cushman

Clavulinoides CUSHMAN, Special Publ. No. 6, Cushman Lab. Foram. Res., 1936, p. 119; Special Publ. No. 7, 1937, p. 119.

Clavulina (part) of authors.

Test elongate, in most species triangular, in some with the latest chambers rounded, earliest portion triserial, triangular, frequently with an early biserial stage following the triserial portion, in which one of the three sides is narrower than the other two, adult uniserial; chambers in many species distinct, in a few somewhat obscure, particularly in the early portion; wall coarsely to finely arenaceous, amount of cement variable, surface either smooth or somewhat roughly finished; aperture in the adult becoming terminal, rounded, or with radiating portions toward the angles of the test, without a distinct tooth. — Upper Cretaceous to Recent.

Genus PSEUDOGAUDRYINELLA Cushman, 1936

Key, plate 43, figure 25

Genotype, *Gaudryinella capitosa* Cushman

Pseudogaudryinella CUSHMAN, Special Publ. No. 6, Cushman Lab. Foram. Res., 1936, p. 23; Special Publ. No. 7, 1937, p. 138.

Gaudryinella (part) CUSHMAN (not PLUMMER).

Test in the earliest stages triserial, triangular in transverse section, angled, later becoming biserial, with the chambers in two distinct series, one with a truncate periphery, the other with an angular periphery, the two broader sides with the sutures showing a zigzag line, the other narrower side with the sutures simple, nearly horizontal, adult with the chambers becoming typically uniserial, inflated, rounded in transverse section; aperture in the biserial stage in a distinct re-entrant at the inner margin of the last-formed chamber, in the adult becoming rounded and terminal. — Upper Cretaceous.

Genus HETEROSTOMELLA Reuss, 1865

Plate 11, figure 18; Key, plate 7, figure 25
Genotype, *Sagrina rugosa* d'Orbigny

Heterostomella Reuss, Sitz. Akad. Wiss. Wien, vol. 52, pt. 1, 1865, p. 448.
Gaudryina (part) of authors. *Tritaxia* Marsson, 1878 (not Reuss).

Test in the early stages triserial, triangular in section, later biserial with one of the angles of the test dividing and forming a somewhat quadrangular test, the ridges formed by the angles becoming fistulose, and occasionally supplementary ridges add to the four primary ones; wall arenaceous, usually with much cement; aperture in the young as in *Gaudryina*, later with a terminal aperture, frequently with a slight neck. — Upper Cretaceous.

Genus RUDIGAUDRYINA Cushman and McCulloch, 1939

Key, plate 43, figures 26, 27
Genotype, *Rudigaudryina inepta* Cushman and McCulloch

Rudigaudryina Cushman and McCulloch, Univ. So. Calif. Publ., Allen Hancock Pacific Expeditions, vol. 6, No. 1, 1939, p. 94.

Test in the early stages triserial, at least in the microspheric form, followed by a biserial series as in *Gaudryina* and in the adult with the chambers in an irregular spreading series; wall finely arenaceous, firmly and smoothly cemented; aperture in the adult chambers generally terminal, rounded, often with a slight lip. — Recent.

Family 11. VALVULINIDAE

Test in the earliest stages of the more primitive forms triserial, later with a secondary spiral development increasing the number of chambers to 4, 5 or more in a whorl, in later development becoming triserial, biserial, uniserial, or spreading, and even annular; chambers simple, or in the higher forms labyrinthic; wall arenaceous with a chitinous inner layer, in the higher forms with an additional thin outer coating, and in a few genera becoming almost entirely calcareous in the adult; aperture simple or cribrate.

The Valvulinidae as here considered contain a very varied group of adult forms, all of which, however, show similar stages in the young, and are connected by intermediate forms showing that the entire group is a compact whole. Much recent work has been done showing the detailed stages of the development of many of these genera. Nearly all the genera show the development of a very distinctive flat tooth in the aperture.

The earliest forms, such as *Valvulina*, are triserial at least in the young, and often triangular in section. Such forms are known before the Cretaceous, but the remaining groups have developed since the beginning of the Cretaceous. *Clavulina* with its uniserial later development is derived directly from *Valvulina*, and shows the distinctive tooth. Many of the forms previously referred to *Clavulina* belong elsewhere as will be noted. *Cribrobulimina*, which developed in the Tertiary of Europe and Australia, has numerous chambers in the adult, and developed a cribrate aperture. Directly from *Valvulina* in the Lower Cretaceous develops *Arenobulimina*, triserial in the young, but in the adult with numerous chambers to a whorl. From this developed several different lines largely through the reduction in number of chambers in a whorl. *Eggerella* has three chambers in the adult, but four or five in the young stages. This has produced *Chrysalidina*, which has a cribrate aperture and a distinctly arenaceous test, but which still keeps the triserial arrangement in the adult. In *Marssonella* the round form is kept, but the number of chambers to a whorl is reduced to two. In *Dorothia* the adult is biserial, often compressed, and the apertural face raised. *Plectina* has the aperture terminal, although the biserial arrangement is kept, and in *Goësella* the uniserial form is developed.

Also from *Arenobulimina* is developed *Martinottiella*, by becoming directly uniserial, and *Schenckiella*, which passes through a triserial and biserial stage before becoming uniserial. These have been known previously as *Clavulina*, but the early stages have four or five chambers to the whorl, and are rounded. Also from *Arenobulimina* in the Tertiary is developed *Valvulammina*, in the Eocene of the Paris-Basin, in which there are five or six chambers in the last-formed coil, and a very prominent tooth in the aperture. Also in the Tertiary develops *Karreriella*, the microspheric form of which has as many as five chambers in the whorl, and the megalospheric form develops to the biserial stage. The aperture is tubular, somewhat above the base of the chamber.

An interesting specialized line developed from *Arenobulimina* is that developed in the Cretaceous. In *Ataxophragmium* the test becomes uniserial with a simple terminal aperture. In *Pernerina*, also in the Upper Cretaceous, is developed a flattened form with a broadly spreading apertural face, the aperture moving in from the border of the test. In the Lower Cretaceous, from such a form developed the series *Lituonella*, *Coskinolina*, *Dictyoconus*, and *Gunteria*. These developed a generally conical test which becomes compressed in *Gunteria*. The apertures are on the ventral side, and are represented by numerous pores. All these forms from *Ataxophragmium* have, in the central portion at least, definite pillars from floor to roof. In the Eocene forms, *Lituonella* continues this structure of simple pillars. In *Coskinolina* there are added simple subdivisions of the chamber at the periphery, and in *Dictyoconus* and *Gunteria* these are again subdivided into cellules. This is an interesting group, very definitely connected at all its stages with the simpler *Arenobulimina*, and showing a definite progressive complexity, coupled with the geologic sequence.

From *Hagenowella*, which in the Upper Cretaceous develops usually four chambers in the adult and divides the peripheral portion of the chambers, are developed such forms as *Textulariella*, which is biserial, and has the chambers subdivided at the periphery. From this developed in the Tertiary *Liebusella*, in which the interior is entirely labyrinthic, and the test becomes uniserial, and *Tritaxilina*, which has a rather long stage with four chambers in each whorl, then gradually becomes uniserial. From *Textulariella*, also in the Cretaceous, is developed *Cuneolina*, in which the test becomes much flattened and fan-shaped, and the interior rather irregularly subdivided, and directly derived from it is *Dicyclina*, which becomes still more compressed, and the chambers annular.

This is one of the best examples in the foraminifera to show how a simple form may become greatly modified in later development, and were it not that the intermediate stages have been well preserved, and the early stages of various genera sectioned, these might be thought to belong to different groups. As it is, they form a very compact whole, now that the development of the various forms is well known.

For details in regard to the species and genera of this family, see Cushman, A Monograph of the Foraminiferal Family Valvulinidae (Cushman Lab. Foram. Res. Special Publ. 8, 1937, pp. i–xiii, 1–210, pls. 1–24) and Supplement (Special Publ. 8A, 1947, pp. 1–69, pls. 1–8).

Subfamily 1. Valvulininae

Genus VALVULINA d'Orbigny, 1826

Plate 12, figure 1; Key, plate 8, figure 1
Genotype, *Valvulina triangularis* d'Orbigny

Valvulina d'Orbigny, Ann. Sci. Nat., vol. 7, 1826, p. 270.
Rotalina (part) Williamson, 1858.

Test free or attached, spiral, conical, triserial, usually umbilicate, typically with three chambers throughout, but the megalospheric form with the adult having more than three chambers to a whorl; wall arenaceous, of coarse or fine material, surface very rough to smooth; aperture with a distinct, valvular tooth.—Jurassic to Recent.

Genus EGGERELLINA Marie, 1941

Key, plate 43, figure 28
Genotype, *Bulimina brevis* d'Orbigny

Eggerellina Marie, Mém. Mus. Nat. Hist. Nat., n. ser., vol. 12, 1941, p. 31.
Bulimina (part) of authors.

Test free, conical or ovoid, triserial throughout, spire trochoid, the vertical axis more or less oblique, composed of three or four whorls, each with three chambers, laterally embracing, spherical or oval, not depressed near the aperture, interior simple; sutures limbate, slightly depressed; aperture simple, narrow, slightly curved, at the end angled and at the base connecting by an internal tube with the outer end of the previously formed chamber; wall imperforate, thin, composed of arenaceous material, mostly calcareous, with a large proportion of cement, surface smooth. Translation. — Upper Cretaceous.

This genus is based upon d'Orbigny's species which, from material I have seen, would appear to be perforate, and thus the genus would be synonymous with *Bulimina*. The genus is recorded here until further studies can be made.

Genus PSEUDOGOËSELLA Keijzer, 1945

Key, plate 46, figures 1–3

Genotype, *Pseudogoësella cubana* Keijzer

Pseudogoësella Keijzer, Outline geol. eastern part Prov. Oriente, Cuba, Utrecht, 1945, p. 190.

"Test arenaceous, starting with 3 (?) chambers in a whorl, and remaining triserial and triangular for a considerable proportion of the test. Aperture in the earlier stages simple at the inner margin, later cribrate in an apertural plate on the front of the chambers. Finally a biserial stage is developed, with a tendency to become uniserial, while pillars connect roofs and floors of the chambers. The aperture in this stage remains of the same type, but shifts to a still more terminal position." — Eocene.

Genus CLAVULINA d'Orbigny, 1826

Key, plate 8, figure 3

Genotype, *Clavulina parisiensis* d'Orbigny

Clavulina d'Orbigny, Ann. Sci. Nat., vol. 7, 1826, p. 268.

Orthocerina d'Orbigny, 1839. *Verneuilina* (part), *Valvulina* (part), and *Tritaxia* (part) of authors.

Test elongate, triserial, early portion triangular in section; chambers simple, later portion uniserial at once, following the triserial stage; wall arenaceous, very rough to smoothly finished; aperture in the adult terminal, rounded or lobed, typically with a tooth. — Middle Eocene to Recent.

The forms with rounded early stages are usually not triserial, see *Goësella*, *Schenckiella*, and *Martinottiella*. I examined the originals of Lamarck's "*Spirolina cylindrica*, var. β" in Caen, and they are Clavulinas.

Genus CRIBROBULIMINA Cushman, 1927

Plate 12, figure 3; Key, plate 9, figure 13

Genotype, *Valvulina mixta* Parker and Jones

Cribrobulimina Cushman, Contr. Cushman Lab. Foram. Res., vol. 2, pt. 4, 1927, p. 80.

Valvulina (part) Parker and Jones, 1862.

Test in the early stages triserial, angled, later in a loose spiral, five or more in a whorl; wall arenaceous, with much cement; aperture in the young as in *Valvulina*, later developing a series of openings to form a cribrate aperture. — Tertiary, Recent.

Genus VALVULAMMINA Cushman, 1933

Plate 12, figure 14; Key, plate 8, figure 17
Genotype, *Valvulina globularis* d'Orbigny

Valvulammina CUSHMAN, Contr. Cushman Lab. Foram. Res., vol. 9, 1933, p. 37.
Valvulina (part) D'ORBIGNY.

Test in early stages conical, with four or five chambers making up the initial whorl, later portion trochoid with inflated chambers, adult usually with five or six chambers in a whorl, ventral side umbilicate; wall finely arenaceous, with a chitinous lining in early portion, later portion becoming almost entirely calcareous, surface with a thin outer layer; aperture large, ventral, with a large rounded tooth. — Paleocene to Oligocene.

Genus MAKARSKIANA van Soest, 1942

Key, plate 46, figures 4, 5
Genotype, *Makarskiana trochoidea* van Soest

Makarskiana VAN SOEST, Thesis Univ. Utrecht, 1942, p. 27.

Test trochoid throughout, with four to five chambers in the early whorls, later rapidly expanding with three to four chambers in each whorl, ventral side slightly concave; wall finely agglutinated and fairly thick; aperture on the ventral side with a short, rounded tooth. Translation. — Eocene.

Subfamily 2. Eggerellinae

Genus ARENOBULIMINA Cushman, 1927

Plate 12, figure 4; Key, plate 8, figures 4, 5
Genotype, *Bulimina preslii* Reuss

Arenobulimina CUSHMAN, Contr. Cushman Lab. Foram. Res., vol. 2, pt. 4, 1927, p. 8c
Bulimina (part) REUSS and later authors.

Test with the earlier chambers triserial, later chambers spirally arranged, more than three to a whorl, close coiled; wall arenaceous, with much cement, typically with a chitinous lining; aperture typically with a broad rounded tooth. — Cretaceous to Eocene.

Genus EGGERELLA Cushman, 1933

Plate 11, figure 15; plate 12, figure 5; Key, plate 8, figure 9
Genotype, *Verneuilina bradyi* Cushman

Eggerella CUSHMAN, Contr. Cushman Lab. Foram. Res., vol. 9, 1933, p. 33.
Verneuilina (part) of authors.

Test a trochoid spire, in the early stages in the microspheric form with five chambers in a whorl, later reduced to four, and in the adult to three; wall finely arenaceous, with calcareous cement or becoming almost entirely calcareous; aperture, a low arched slit at the base of the inner margin. — Cretaceous to Recent.

Genus EGGERINA Toulmin, 1941

Key, plate 46, figures 6, 7

Genotype, *Eggerina cylindrica* Toulmin

Eggerina Toulmin, Journ. Pal., vol. 15, 1941, p. 573.

"Test a trochiform spire, with three chambers to a whorl in the adult, the last three chambers making up nearly the whole exterior; wall very finely arenaceous with much calcareous cement, or almost entirely calcareous, smoothly finished; aperture a low arched opening at the inner margin of the last-formed chamber, adjacent to the upper end of an earlier chamber of the final whorl." — Tertiary.

Genus CHRYSALIDINA d'Orbigny, 1839

Plate 12, figure 6; Key, plate 9, figure 12

Genotype, *Chrysalidina gradata* d'Orbigny

Chrysalidina d'Orbigny, in De la Sagra, Hist. Phys. Pol. Nat. Cuba, vol. 5, 1839, "Foraminifères," p. 109.

Test large, conical, triserial in the adult, early stages unknown, chambers inflated; wall arenaceous, with a thin, outer, epidermal layer; aper-

---

PLATE 12

Relationships of the Genera of the Valvulinidae

FIG.

1. *Valvulina oviedoiana* d'Orbigny. *a*, front view; *b*, apertural view.
2. *Pseudoclavulina mexicana* (Cushman). (Verneuilinidae.)
3. *Cribrobulimina polystoma* (Parker and Jones).
4. *Arenobulimina preslii* (Reuss).
5. *Eggerella bradyi* (Cushman).
6. *Chrysalidina gradata* d'Orbigny.
7. *Marssonella oxycona* (Reuss).
8. *Dorothia bulletta* (Carsey). *a*, side view; *b*, apertural view.
9. *Plectina ruthenica* (Reuss). (After Reuss.) *a*, side view; *b*, apertural view.
10. *Goësella rotundata* (Cushman).
11. *Martinottiella communis* (d'Orbigny).
12. *Karreriella siphonella* (Reuss).
13. *Schenckiella primaeva* (Cushman).
14. *Valvulammina globulosa* (d'Orbigny). *a*, dorsal view; *b*, apertural view.
15. *Textulariella barrettii* (Jones and Parker). (After H. B. Brady.) *a*, side view; *b*, interior.
16. *Cuneolina pavonia* d'Orbigny. (After d'Orbigny.) *a*, side view; *b*, end view.
17. *Liebusella soldanii* (Jones and Parker). (After H. B. Brady.)
18. *Tritaxilina caperata* (H. B. Brady). (After H. B. Brady.) *a*, side view; *b*, apertural view.
19. *Hagenowella gibbosa* (d'Orbigny). *a*, side view; *b*, section.
20. *Pernerina depressa* (Perner). *a*, apertural view; *b*, dorsal view.
21. *Ataxophragmium variabile* (d'Orbigny).
22. *Lituonella liburnica* Schubert. (After Schubert.) *a*, side view; *b*, apertural view.
23. *Coskinolina liburnica* Stache. (After Schubert.) *a*, side view; *b*, apertural view.
24. *Dictyoconus indicus* Davies. (After Davies.) *a*, side view; *b*, apertural view.
25. *Gunteria floridana* Cushman and Ponton. *a*, side view; *b*, apertural view.

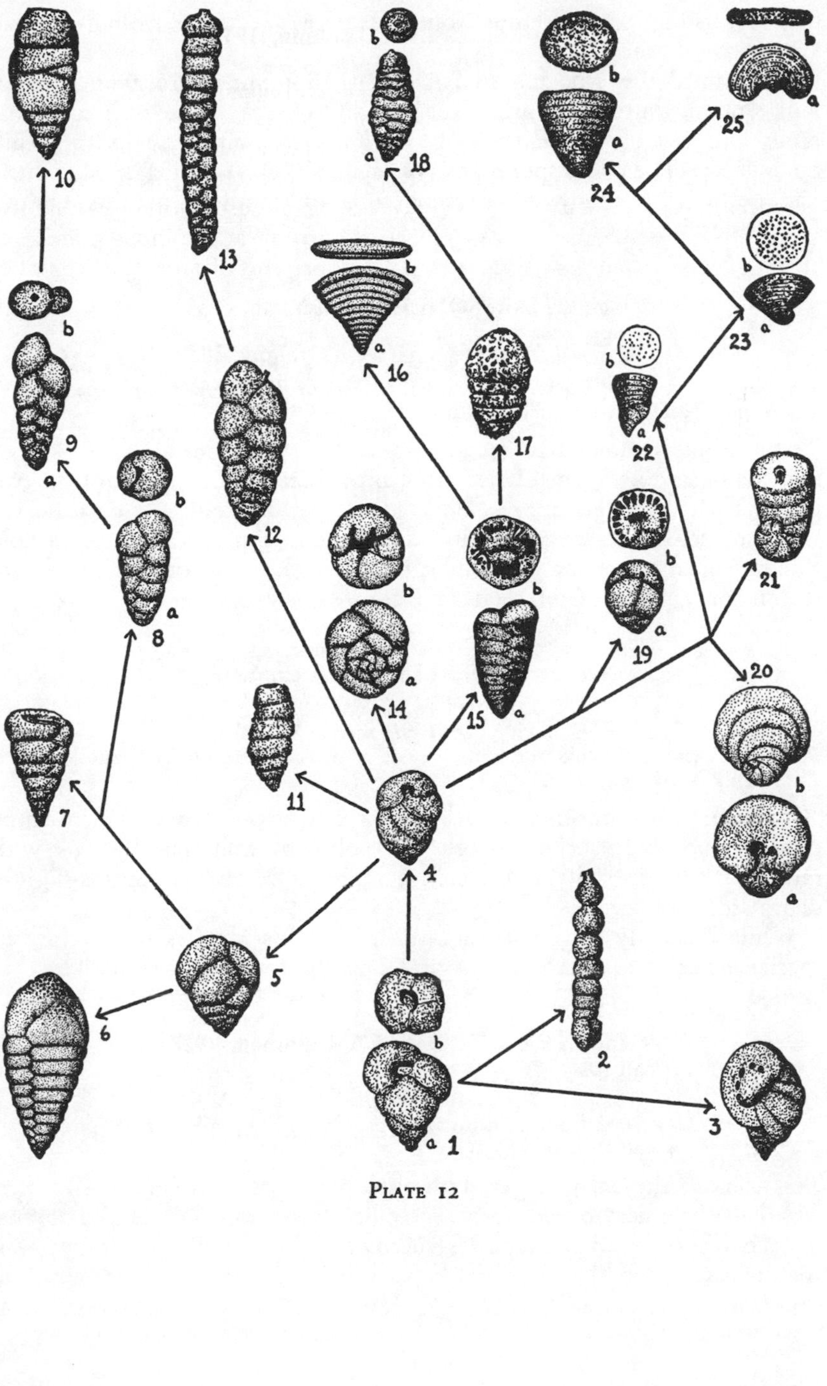
10
13
b
a
18
b
a
24
b
a
25
b
a
16
b
a
9
b
a
17
b
a
23
b
a
22
b
a
8
12
b
a
14
b
a
15
b
a
19
21
20
7
11
4
b
a
5
6
b
a
1
2
3

PLATE 12

ture consisting of numerous rounded pores in the terminal face.—Cretaceous, France.

I examined the type material in Paris. Specimens are about 7 mm. long (no measurements are given by d'Orbigny). The wall is arenaceous like that of *Textulariella*, *Cuneolina*, etc., arenaceous with a thin outer layer. The Paris specimens are labelled "Falaise du l'île Madame," and occur with many other arenaceous foraminifera of this famous Cretaceous locality.

Genus MARSSONELLA Cushman, 1933

Plate 12, figure 7; Key, plate 8, figure 23

Genotype, *Gaudryina oxycona* Reuss

Marssonella CUSHMAN, Contr. Cushman Lab. Foram. Res., vol. 9, 1933, p. 36.
Gaudryina (part) REUSS and later authors.

Test trochoid, rounded in section, in early stages conical with four or five chambers to a whorl, later reduced to three, and in the adult to two in a whorl; chambers simple, undivided, apertural end flat or concave; wall arenaceous, calcareous with a chitinous lining; aperture, a low elongate opening, at the inner margin of the chamber, or extending into the chamber wall.—Cretaceous to Eocene.

Genus REMESELLA Vasicek, 1947

Key, plate 46, figure 8

Genotype, *Remesella mariae* Vasicek

Remesella VASICEK, Vestnik Statniho Geologisceho Ustavu Republiky Ceskoslovenske, Rocnik XXII, 1947, p. 246.

"Chambers with uncomplete secondary transversal septa, which shine through the wall; surface rough but polished; material siliceous with much cement, especially on the later chambers. Secondary sutures slightly depressed."—Eocene.

While the early stages are not described, this appears to be close to species referred to *Textulariella* and probably is closely related to that genus.

Genus TEXTULARIELLA Cushman, 1927

Plate 12, figure 15; Key, plate 8, figures 24, 25

Genotype, *Textularia barrettii* Jones and Parker

Textulariella CUSHMAN, Contr. Cushman Lab. Foram. Res., vol. 3, 1927, p. 24.
Textularia JONES and PARKER, 1863 (not DEFRANCE).

Test conical, the early stages with three or more chambers in the first whorl in the microspheric form, later becoming rapidly reduced to two in each whorl, biserial, chambers subdivided toward the periphery; wall arenaceous, smoothly finished; aperture textularian.—Cretaceous to Recent.

Genus CUNEOLINA d'Orbigny, 1839

Plate 12, figure 16; Key, plate 9, figures 1–4
Genotype, *Cuneolina pavonia* d'Orbigny

Cuneolina D'ORBIGNY, in DE LA SAGRA, Hist. Phys. Pol. Nat. Cuba, vol. 5, 1839, "Foraminifères," p. 150.
Textularia (part) GOËS, 1882 (not DEFRANCE).

Test conical in the young, compressed in the adult so that the zigzag line between the chambers is on the narrow side of the test, early stage with as many as five chambers in the first whorl, becoming rapidly reduced to two, labyrinthic; wall arenaceous, with much cement, exterior with a thin imperforate layer; aperture in the adult an elongate slit or a series of rounded openings at the base of the inner margin of the chamber.—Cretaceous to Recent.

Genus CUNEOLINELLA Cushman and Bermudez, 1941

Key, plate 46, figures 9–11
Genotype, *Cuneolinella lewisi* Cushman and Bermudez

Cuneolinella CUSHMAN and BERMUDEZ, Contr. Cushman Lab. Foram. Res., vol. 17, 1941, p. 101.

Test large, much compressed so that the zigzag lines between the chambers are on the narrow sides of the test, in the earliest stages conical, then compressed, and in the adult with the angles extending backward toward the initial end; chambers in the earliest stages with as many as five chambers in the first whorl, becoming reduced to two, interior labyrinthic; wall arenaceous, with much cement, exterior with a thin imperforate layer; aperture in the adult consisting of a series of elongate openings at the base of the inner margin of the last-formed chamber.—Miocene.

Genus PSEUDORBITOLINA Douvillé, 1910

Genotype, *Pseudorbitolina marthae* Douvillé

Pseudorbitolina DOUVILLÉ, Bull. Soc. Géol. France, ser. 4, vol. 10, 1910, p. 57.

This genus is not well known, and the types have not been restudied nor have I had topotypes. It is placed here until more is definitely known of this genus.

Genus DICYCLINA Munier-Chalmas, 1887

Key, plate 9, figures 5, 6
Genotype, *Dicyclina schlumbergeri* Munier-Chalmas

Dicyclina MUNIER-CHALMAS, Compte-rendu som. séan. Soc. Géol. France, 1887, No. 7, p. xxx.

Test discoidal, early stages forming a globular mass, adult consisting of two series of annular chambers alternating with one another, interior labyrinthic, the interior with generally radiate portion, nearer the surface with a finer meshwork, the surface with a thin coating; wall finely arenaceous; apertures numerous, in a line at the meeting of the two chambers at the periphery. — Upper Cretaceous.

This represents a very specialized development from *Cuneolina* in becoming annular, somewhat parallel to the development of *Gunteria* from *Dictyoconus*. The walls are finely arenaceous, not porcellanous, and it has no relation to *Marginopora* and its relatives.

Genus DOROTHIA Plummer, 1931

Plate 11, figure 17; plate 12, figure 8; Key, plate 8, figure 10
Genotype, *Gaudryina bulletta* Carsey

Dorothia PLUMMER, Univ. Texas Bull. 3101, 1931, p. 130.
Gaudryina (part) of authors.

Test a trochoid spire, in the early stage in the microspheric form with five or six chambers in a whorl, rapidly reducing the number to four, then three, and in the adult with two in each whorl; wall arenaceous, becoming largely calcareous in some living forms; aperture narrow, at the base of the inner margin. — Cretaceous to Recent.

Genus MATANZIA D. K. Palmer, 1936

Key, plate 46, figures 12–14
Genotype, *Matanzia bermudezi* D. K. Palmer

Matanzia D. K. PALMER, Mem. Soc. Cubana Hist. Nat., vol. 10, 1936, p. 125.— CUSHMAN, Special Publ. No. 8, Cushman Lab. Foram. Res., 1937, p. 102.

"Test superficially resembling *Gaudryina* and *Dorothia*; agglutinated, usually smoothly finished; elongate, cylindrical; apex with approximately 5 chambers per whorl; test quickly assuming biserial habit. Aperture a low arch at base of final chamber. Chambers subdivided by narrow, sharply defined partitions radiating from the outer chamber wall but not reaching the inner (apertural) wall; supplementary chamber partitions usually not exceeding 10 in number in mature chambers." — Eocene and Oligocene.

This genus resembles *Dorothia* externally, and its interior in some respects resembles the structure of *Textulariella*.

Genus CUBANINA D. K. Palmer, 1936

Key, plate 46, figures 15–19
Genotype, *Cubanina alavensis* D. K. Palmer

Cubanina D. K. PALMER, Mem. Soc. Cubana Hist. Nat., vol. 10, 1936, p. 123.— CUSHMAN, Special Publ. No. 8, Cushman Lab. Foram. Res., 1937, p. 103.

Test superficially resembling *Clavulina*, elongate, cylindrical, early portion triserial, later uniserial; chambers subdivided by inwardly projecting radial partitions, not all of which reach the center; wall arenaceous; aperture in the adult terminal, central, sometimes slightly denticulate.— Eocene and Oligocene.

Genus PLECTINA Marsson, 1878

Plate 12, figure 9; Key, plate 8, figure 11
Genotype, *Gaudryina ruthenica* Reuss

Plectina Marsson, Mitth. Nat. Ver. Neu-Vorpommern u. Rügen, Jahrg. 10, 1878, p. 160.
Gaudryina (part) Reuss and later authors.

Test in early stages rounded, with as many as five chambers to a whorl, rapidly reducing to two in a whorl in the adult; chambers simple, undivided; wall arenaceous, often with much cement; aperture in early stages textularian, in the adult rounded, in the terminal face, but without a neck.—Cretaceous to Recent.

Genus GOËSELLA Cushman, 1933

Plate 12, figure 10; Key, plate 8, figure 12
Genotype, *Clavulina rotundata* Cushman

Goësella Cushman, Contr. Cushman Lab. Foram. Res., vol. 9, 1933, p. 34.
Clavulina (part) of authors.

Test in the early stages with four or five chambers in a whorl, rapidly reducing to three, then two, and in the adult uniserial; chambers simple, undivided; wall finely to coarsely arenaceous; aperture in the adult rounded, terminal, usually sunken, without a lip.—Cretaceous to Recent.

Genus CRIBROGOËSELLA Cushman, 1935

Key, plate 46, figure 20
Genotype, *Bigenerina robusta* H. B. Brady

Cribrogoësella Cushman, Smithsonian Misc. Coll., vol. 91, No. 21, 1935, p. 4; Special Publ. No. 6, Cushman Lab. Foram. Res., 1936, p. 34; Special Publ. No. 8, 1937, p. 118.
Bigenerina H. B. Brady (not d'Orbigny).

Test elongate, subcylindrical, the early portion tapering, later portion with the sides nearly parallel, rounded in transverse section, earliest whorl with four or five chambers, rapidly reducing to three, and then to a biserial stage which continues for a considerable period, followed in the adult by uniserial chambers, interior undivided; wall arenaceous; aperture in the biserial portion at the inner margin of the last-formed chamber, in the uniserial portion becoming terminal, central, and gradually increasing from one opening in the early stage to many in the adult, occupying the central portion of the terminal face.—Miocene to Recent.

Genus KARRERIELLA Cushman, 1933

Plate 12, figure 12; Key, plate 8, figures 13, 14
Genotype, *Gaudryina siphonella* Reuss

Karreriella CUSHMAN, Contr. Cushman Lab. Foram. Res., vol. 9, 1933, p. 34.
Gaudryina (part) REUSS and later authors.

Test in the young a trochoid spiral with five or more chambers in a whorl, later reducing to four, three, and in the adult two to a whorl, chambers simple; wall finely arenaceous, smooth; aperture just above the base of the inner chamber margin, with a distinct neck.—Eocene to Recent.

Genus MARTINOTTIELLA Cushman, 1933

Plate 12, figure 11; Key, plate 8, figures 6–8
Genotype, *Clavulina communis* d'Orbigny

Martinottiella CUSHMAN, Contr. Cushman Lab. Foram. Res., vol. 9, 1933, p. 37.
Clavulina (part), *Listerella* (part), and *Schenckiella* (part) of authors.

Test in early stages a trochoid spire with several chambers, usually five in the first whorl of the microspheric form, and keeping four or five for several whorls, later becoming uniserial very abruptly; wall arenaceous, with a chitinous lining; aperture in the adult terminal, typically an elongate, narrow, arcuate opening about a rounded tooth, sometimes with a slight lip.—Upper Cretaceous to Recent.

Genus SCHENCKIELLA Thalmann, 1942

Plate 12, figure 13; Key, plate 8, figure 15
Genotype, *Clavulina primaeva* Cushman

Schenckiella THALMANN, Amer. Midland Nat., vol. 28, 1942, p. 458.
Listerella CUSHMAN, 1933 (not JAHN, 1906).
Clavulina (part) of authors.

Test in the early stages a trochoid spire with four or five chambers to a whorl, reducing later to three, then a series of twos, and in the adult uniserial; chambers simple, undivided; wall finely arenaceous; aperture terminal, with a slender neck.—Eocene to Recent.

Genus TRITAXILINA Cushman, 1911

Plate 12, figure 18; Key, plate 8, figure 22
Genotype, *Clavulina caperata* H. B. Brady

Tritaxilina CUSHMAN, Bull. 71, U. S. Nat. Mus., pt. 2, 1911, p. 71.
Clavulina (part) H. B. BRADY, 1881. *Tritaxia* (part) H. B. BRADY, 1884. *Clavulinella* SCHUBERT, 1920.

Test in the early stage with as many as five chambers in the first whorl, then successively reducing to four, three, two, to a rectilinear series in the adult; chambers labyrinthic; wall arenaceous, calcareous, with a distinct chitinous lining in the earliest portion, insoluble in acid; aperture in the adult becoming terminal, rounded, with a slight lip, with a series of teeth projecting in and partially closing the opening. — Eocene to Recent.

Genus CAMAGUEYIA Cole and Bermudez, 1944

Key, plate 46, figure 21
Genotype, *Camagueyia perplexa* Cole and Bermudez

Camagueyia Cole and Bermudez, Bull. Amer. Pal., vol. 28, no. 113, 1944, p. 5.

"Test in the early stages with several chambers to the whorls with some reduction of the number of chambers in the adult portion of the test; wall finely arenaceous with considerable cement; chambers low, compressed, between thick roofs and floors; roofs and floors not extending to the center of the test which has pillar-like structures, particularly near the aperture; aperture located in the center of the truncate apertural face surrounded by inward projecting teeth." — Eocene.

Genus LIEBUSELLA Cushman, 1933

Plate 12, figure 17; Key, plate 8, figures 18–21
Genotype, *Lituola soldanii* Jones and Parker

Liebusella Cushman, Contr. Cushman Lab. Foram. Res., vol. 9, 1933, p. 36.
Lituola (part) Jones and Parker (not Lamarck). *Haplostiche* H. B. Brady (not Reuss).
Arenodosaria Finlay, 1939.

Test in early stages with four or five chambers in the first whorl, rapidly dropping to four, then three, and in the adult uniserial, interior labyrinthic; wall coarsely arenaceous, calcareous; aperture complex, irregularly radiate in adult, terminal. — Eocene to Recent.

Genus HAGENOWELLA Cushman, 1933

Plate 12, figure 19; Key, plate 8, figure 16
Genotype, *Valvulina gibbosa* d'Orbigny

Hagenowella Cushman, Amer. Journ. Sci., vol. 26, 1933, p. 21.
Valvulina (part) d'Orbigny, 1840. *Bulimina* (part) of authors.

Test in a trochoid spiral, early portion with three or more chambers in a whorl, interior simple in the early stages, later with radial partitions growing in from the outer wall, partially dividing the periphery of each chamber; wall arenaceous, of calcareous and siliceous fragments, smoothly cemented; aperture oblique, elongate, from the inner margin, with a tooth. — Cretaceous.

Genus ATAXOPHRAGMIUM Reuss, 1861

Plate 12, figure 21; Key, plate 9, figures 7–9
Genotype, *Bulimina variabilis* d'Orbigny

Ataxophragmium Reuss, Verz. Gypsmodellen Foraminiferen, 1861, Nos. 8, 9.
Bulimina (part) d'Orbigny, 1840, and later authors.
Ataxogyroidina Marie, 1941.

Test with the earliest portion with three or more chambers in each whorl, later several chambers in the adult whorl and usually spreading, in the adult uncoiling tending to become uniserial, interior of the adult chambers with a few simple pillars from floor to roof; wall finely or coarsely arenaceous, often with much calcareous cement; aperture in the adult tending to become terminal by working into the apertural face, and becoming cut off from the margin. — Upper Cretaceous.

Genus ORBIGNYNA Hagenow, 1842

Plate 10, figure 4; Key, plate 5, figure 9
Genotype, *Orbignyna ovata* Hagenow

Orbignyna Hagenow, Neues Jahrb. für Min., 1842, p. 573.
Haplophragmium (part) Reuss, and later authors. *Lituola* (part) of authors (not Lamarck). *Spirolina* Reuss (not Lamarck).

Test in the early stages irregularly trochoid with several chambers to a whorl, later in the adult becoming planispiral and involute; chambers with the periphery subdivided by radiating partitions running in from the outer wall; wall arenaceous, with calcareous sand and fragments of mollusc prisms; aperture at the base of the apertural face in the young stage, with a definite tooth, in the adult rounded or elliptical, in the middle of the apertural face. — Upper Cretaceous.

Genus LITUONELLA Schlumberger, 1905

Plate 12, figure 22; Key, plate 9, figures 14–20
Genotype, *Lituonella roberti* Schlumberger

Lituonella Schlumberger, Bull. Soc. Géol. France, ser. 4, vol. 5, 1905, p. 291.

Test trochoid in the young, low, with several chambers in a whorl, later becoming conical; chambers uniserial, not entirely in a straight line, not truly labyrinthic but with a series of simple, buttress-like pillars from floor to roof, outer face made up of a "central shield" about which is the "marginal ridge," and outside this a distinct "marginal trough"; wall finely arenaceous, outer layer of cone imperforate; apertures small, rounded, confined to the "central shield." — Lower and middle Eocene.

Genus COPROLITHINA Marie, 1941

Key, plate 46, figure 22
Genotype, *Coprolithina subcylindrica* Marie

Coprolithina Marie, Mém. Mus. Nat. Hist. Nat., n. ser., vol. 12, 1941, p. 37.

Test elongate, irregularly reniform, circular in transverse section; chambers somewhat covered, in a trochoid spiral in the early stages, becoming progressively uniserial and rectilinear in the adult, interior of the chambers with a circular series of projections radiating inward from the outer wall; aperture complex, terminal, at the outer end of the last-formed chamber; wall smooth, composed of various foreign materials with an abundance of calcareous cement. Translation. — Upper Cretaceous.

Genus COSKINOLINA Stache, 1875

Plate 12, figure 23; Key, plate 9, figures 21–24
Genotype, *Coskinolina liburnica* Stache

Coskinolina Stache, Verh. k. k. Geol. Reichs., 1875, p. 335.

Test with the early portion asymmetrical, trochoid, with several chambers in a broad trochoid spire, later conical, similar in general structure to *Lituonella*, but in the adult with vertically set radial partitions at the outer portion of each chamber, but the inner portion interrupted only by the pillars. — Lower Cretaceous to middle Eocene.

Genus COSKINOLINOIDES Keijzer, 1942

Key, plate 46, figures 23, 24
Genotype, *Coskinolinoides texanus* Keijzer

Coskinolinoides Keijzer, Proc. Ned. Akad. Wetensch., vol. 45, 1942, p. 1016.

"Test conical, arenaceous, many chambered. Early portion a trochoid spire, starting with 4 chambers in a whorl, later becoming uniserial. Uniserial chambers, and possibly also later chambers of spiral portion subdivided by radial septa. Aperture of the spiral portion at the inner margin of the last-formed chamber. Aperture of uniserial portion consisting of rounded pores in the central area of the roof of the last-formed chamber." — Lower Cretaceous.

Genus DICTYOCONUS Blanckenhorn, 1900

Plate 12, figure 24; Key, plate 9, figures 25–27
Genotype, *Patellina egyptiensis* Chapman

Dictyoconus Blanckenhorn, Zeitschr. Deutsch. geol. Ges., vol. 52, 1900, p. 419.
Patellina (part) of authors (not Williamson). *Cushmania* Silvestri, 1919.

Test similar to *Coskinolina* in general structure but with the peripheral portion of each chamber further subdivided into "cellules," showing at the surface only when eroded. — Lower Cretaceous to middle Eocene.

For details of structure of this and allied genera, see Davies, "The genus *Dictyoconus* and Its Allies" (Trans. Roy. Soc. Edinburgh, vol. 56, pt. 2, 1930, pp. 485–505, pls. I, II).

Genus PERNERINA Cushman, 1933

Plate 12, figure 20; Key, plate 9, figures 10, 11
Genotype, *Bulimina depressa* Perner

Pernerina Cushman, Amer. Journ. Sci., vol. 26, 1933, p. 19.
Bulimina (part) and *Ataxophragmium* (part) of authors.

Test in the earliest stages a trochoid spiral with several chambers to a whorl, later chambers expanding, apertural face becoming very broad and rounded; chambers in the adult with pillars from floor to roof, otherwise undivided; wall arenaceous, with smooth finish; aperture elongate, at base of inner margin, with a flattened tooth. — Cretaceous.

Genus GUNTERIA Cushman and Ponton, 1933

Plate 12, figure 25; Key, plate 9, figures 28–30
Genotype, *Gunteria floridana* Cushman and Ponton

Gunteria CUSHMAN and PONTON, Contr. Cushman Lab. Foram. Res., vol. 9, 1933, p. 25.

Test in general structure similar to *Dictyoconus*, but in the adult becoming much compressed, the ends of the chambers becoming curved back and a saddle-shaped test results, tending toward an annular growth; apertures in a narrow band representing the central shield. — Middle Eocene.

According to Davies (Trans. Roy. Soc. Edinburgh, vol. 59, pt. 3, 1939, pp. 779, 780) the wall is perforate, and therefore the genus belongs elsewhere.

FUSULINE FORAMINIFERA

By Carl O. Dunbar

This group is herewith divided into two families: the Fusulinidae, including the subfamilies Fusulininae and Schwagerininae, and the new family Neoschwagerinidae, including the subfamilies Verbeekininae and Neoschwagerininae. Reasons for this revision are given below.

The fusulines are one of the dominant groups of late Paleozoic foraminifera. They appeared about the beginning of Pennsylvanian time, underwent a meteoric evolution during the Pennsylvanian and Permian periods, and then suffered extinction. They are also the largest and most complexly organized of the Paleozoic foraminifera. Average species resemble grains of wheat or oats in both size and shape, and larger species range up to two inches in length. Only a few are microscopic in size. The "fusuline limestones" that occur at many horizons in the late Paleozoic rocks are veritable coquinas made of fusuline shells. Fusulines are found chiefly in limestones or calcareous shale and in calcareous sandstones.

Up to 1948 a total of over 600 species have been described and 33 genera have been proposed, of which at least 24 now appear valid and useful. There are doubtless many synonyms among the described species, but new forms are still being found in abundance, and the total number may eventually reach 1000.

The organization of a typical fusuline shell is shown in plate 13. It is normally coiled planispirally about an elongate axis (less commonly a short axis), and is bilaterally symmetrical with respect to an equatorial or sagittal plane. The steep face of the last chamber is the antetheca. Each volution has an outer wall, the spirotheca, and is divided into long meridional chambers by septa that were formed by an inflection of the outer wall into the axial plane as each chamber in turn was added.

Taxonomic characters are chiefly internal, and identification requires the study of specially oriented thin sections or polished slices. Three such sections are needed to show all the characters of any species, one axial, one sagittal, and one tangential to the surface and near the inner

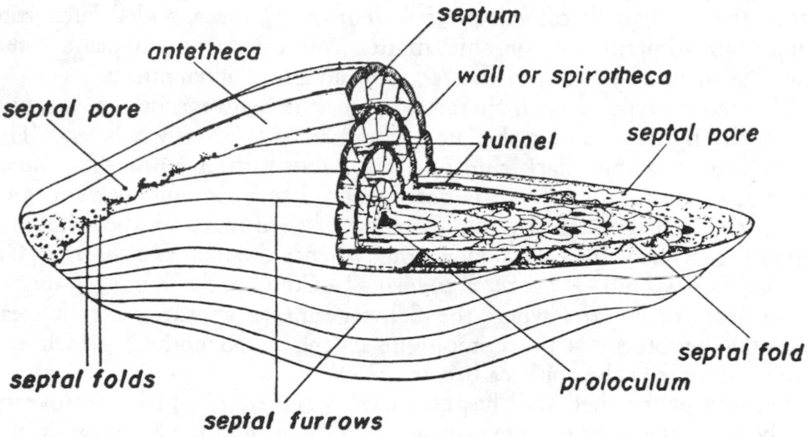

Plate 13. Diagram to show the Structure of a Fusulinid Test (After Dunbar and Condra)

(basal) margin of the penultimate whorl. The tangential slice shows better than any other the form of the septal folds. Unfortunately, the taxonomic importance of the septal folds was only realized in recent years and this character was neither adequately described nor figured by the early workers. As a result, many of the species cannot be accurately placed as to genera from the literature.

The microscopic structure of the shell wall is amazingly complex and has been the subject of much controversy. Two major types may be recognized, (1) the fusulinellid, and (2) the schwagerinid wall. The first, which is the more primitive, and is characteristic of the subfamily Fusulininae, is illustrated by figure 1 on plate 10. It consists of the primary wall or *protheca* (seen alone in the last 4 chambers of the diagram) plus an *epitheca* or secondary deposit added somewhat later. The protheca consists of a layer of clear, transparent calcite covered on the outside by a thin, dark, rind-like film. The clear layer is the *diaphanotheca* and the rind is the *tectum*. The epitheca lines the roof, sides and floor of the earlier formed chambers with a film of translucent, grey calcite (tectorium). In thin section, therefore, the wall of the inner whorls appears to have four layers differing in opacity; the tectum (t) appearing as a dark line, the diaphanotheca (d) as a clear layer, and both inner and outer tectoria (te) as translucent grey layers.

It is probable, as suggested by Gubler, that the opacity of the shell substance varies with the organic impurity and that the distinct layers thus reflect original differences in chemical composition. According to this interpretation, the tectum represents a first-formed film of nitrogenous material (tectine) closely akin to chitin, on which a nearly pure calcite layer, the diaphanotheca, was precipitated; the epitheca, added later, contained an admixture of organic matter and calcite and appears intermediate in opacity between the tectum and the diaphanotheca.

The second type of wall, characteristic of the Schwagerininae, is represented by figures 1 and 5 of plate 12, and consists of only 2 layers. The outer layer is a thin, dark rind and is obviously the tectum. The inner layer is relatively thick but has been reduced to a sieve-like structure of tubular alveoli separated by thin lamellae of shell material, the whole resembling in form a piece of honeycomb, whence the name *keriotheca* [Gr. *kerion*, honeycomb + *theka*, a covering]. Comparative study of species from near the horizon where the Schwagerininae appear makes it clear that the keriotheca is the homologue of the diaphanotheca which has grown thicker as the epitheca became obsolete.

Porosity of the shell wall has also been a subject of much controversy. Early workers appear to have assumed that the wall of the Schwagerininae was perforate, but Girty in 1904 called attention to the tectum which appeared to be imperforate. Hayden in 1909 observed pores to pass through the tectum and correctly described the structure, but his work was not appreciated and the fusuline wall was generally considered to be imperforate until White in 1932 verified Hayden's observations. Dunbar and Skinner demonstrated porosity in numerous species and several genera, and there is no longer reason to doubt that the fusuline wall was normally perforate.

In the primitive type of wall of the subfamily Fusulininae, the pores are extremely fine capillary tubes and are commonly visible only in exceptionally preserved or artificially stained specimens. In the Schwagerininae, on the contrary, the pores have become greatly enlarged in the keriothecal layer, where they appear as alveoli, separated by thin lamellae of shell substance. They remain very fine, however, in the tectum and commonly escape attention. As Dunbar and Skinner have shown, the evolution of the alveolar, schwagerinid type of wall from the primitive, fusulinellid type involved (1) the obsolescence of the tectoria, (2) a thickening of the diaphanotheca, and (3) the great expansion of the tubular mural pores into the alveoli. The Schwagerininae certainly developed out of the Fusulininae through the genus *Triticites*.

The septa are radially disposed with respect to the axis of coiling, and extend from pole to pole, separating short meridional chambers. The final septum forms an abrupt face, the antetheca, closing the last volution. It is formed like the others by a sharp inflection of the tectum of the spiral wall into the radial plane, where it descends to meet the penultimate whorl. The shell is marked externally by shallow, meridional furrows, caused by this inbending of the wall.

In the most primitive genera, such as *Staffella*, the septa are perfectly plane, but in all more advanced types they are ruffled or folded, especially near their inner (basal) margin and near the polar extremities (Pl. 10, figs. 10 and 21, and Pl. 11, figs. 3 and 4). In axial sections these septal folds are cut across so that they appear as loops between the walls of successive volutions (Pl. 10, figs. 11, 20, and 23). Their true form is best studied in tangential slices (Pl. 10, fig. 22). In the less specialized genera, the septa are so widely spaced and the folds so shallow that successive septa do not touch except in the polar regions. In the Schwagerininae there is a marked evolution in the septal folding, leading from *Triticites*, in which the septa are only gently folded across the equatorial region, to *Schwagerina*, in which they are folded from pole to pole with such depth and regularity that the meridional chambers are subdivided basally into a series of chamberlets, each septal fold touching that of the next septum fore or aft (Pl. 10, fig. 22). In *Parafusulina* the specialization is carried further, in that the top of each septal fold is upturned into an open arch where it joins the opposed fold of the next septum (Pl. 11, figs. 3, 6, and 7). As a result, the septa failed to touch the floor of the volution at these points, leaving a series of open arches or foramina extending from one meridional chamber to the next (Pl. 11, fig. 5). In this genus the folds of successive septa are fused edge to edge along the sides of these arches, and their junction with the floor of the volution runs spirally around the shell instead of parallel to the axis, as in more primitive genera. This may be seen in etched specimens (Pl. 11, fig. 6), or in tangential slices cut near the floor of the volution (Pl. 11, figs. 8 and 12).

Successive chambers are in communication by means of a low equatorial tunnel formed by the resorption of the basal margins of the septa near the equator. It has commonly been supposed that the tunnel is a trace of the aperture. However, perfectly preserved shells, with the antetheca exposed, show no such aperture (Pl. 10, figs. 10, 16, 17, and 21 and Pl. 11, fig. 4). Moreover, axial sections of such specimens show the last few septa to be complete. Since this is true for shells at all stages of growth, it is clear that the tunnel is produced by resorption, and is not a primary structure (compare Pl. 11, figs. 4 and 5). In many, and probably all shells, the last few septa are perforated by small septal pores that permitted communication between the chambers and allowed for the extrusion of pseudopodia. Such pores commonly cannot be seen in the earlier volutions, but since they appear in the outer volutions of juvenile specimens, it seems clear that they were present in the antetheca at all stages of growth, and were the means of external communication; furthermore, that they have been filled and closed in the older septa, their function being taken over by the tunnel.

The tunnel is commonly bounded on each side by a ridge of secondary deposit, the chomata (Pl. 10, figs. 4, 19, and 20), which usually shows a concentric laminar structure. The chomata are clearly a part of the secondary deposit that forms the tectorium. In the Schwagerininae, such secondary deposit is usually limited to the zones beside the tunnel, where it forms a levee-like ridge on the floor of the volution, and commonly extends for a short distance along and up the sides of the septa. In many species of *Schwagerina*, and in occasional species of other genera, however, the axial zone of the shell is filled up with a secondary deposit.

Dimorphism has been observed in many of the fusuline genera, and is to be expected in all. Generally the microspheric shells have a juvenarium of 2 or 3 initial whorls coiled askew to the axis of later whorls. These juvenile whorls, furthermore, consist of short globular chambers and strikingly resemble in form the adult shell of *Endothyra*, whence they are described as endothyroid. In some of the genera, as *Fusulinella* and *Fusulina*, among the Fusulininae, there is little difference in size and external appearance between the megalospheric and microspheric individuals, but in the genera *Parafusulina* and *Polydiexodina* the microspheric shells are commonly at least twice as long as the megalospheric. In several of the small, primitive genera, such as *Schubertella* and *Fusiella*, the megalospheric individuals are very rare. The disposition of some earlier students to base genera on the occurrence of an obliquely coiled (microspheric) juvenarium now appears to be unjustified.

Family 12. FUSULINIDAE Möller, 1878 emend Dunbar

Foraminifera of medium to relatively large size; tests normally fusiform to subcylindrical and planispirally coiled about an elongate axis [in some primitive genera the axis is short, and in a few terminal and aberrant genera the coiling becomes evolute (*Codonofusiella*) or irregular (*Rauserella, Nipponitella*)]; surface divided by meridional furrows into numerous melon-like lobes; volutions subdivided internally into meridional chambers by septa that are formed by inflections of the spiral wall; septal folds are common except in the most primitive genera. A low median tunnel is formed through resorption of part of the basal margin of the septa, and is bordered by chomata in early genera. The spiral wall is calcareous and perforate. — Basal Pennsylvanian (possibly late Mississippian) to late Permian.

Subfamily 1. Fusulininae Rhumbler, 1895 emend. Dunbar and Henbest, 1930

Fusulines of medium or small size, ranging from lenticular through spheroidal to fusiform or subcylindrical shape; the spiral wall consists of a protheca (tectum + diaphanotheca), commonly not over 10 to 15 microns thick, which is generally covered both inside and out by epitheca, the total wall thickness being not over 30–35 microns (in a few genera one or more of these layers is lacking); the mural pores are extremely fine tubules, commonly invisible except where specially preserved or artificially stained; the septa are plane in a few primitive genera but in the rest are gently to deeply folded; a median tunnel of elliptical or slit-like form is present and is bordered by chomata, but there is no external aperture. Septal pores are numerous.

Dimorphism is not uncommon but there is little difference in size and external appearance between the megalospheric and microspheric forms. Microspheric individuals have an endothyroid juvenarium, commonly coiled askew to the rest of the shell. — Late Mississippian (?) to late Permian, but largely confined to the lower part of the Pennsylvanian system (Morrow, Lampasas and Des Moines series).

Genus STAFFELLA Ozawa, 1925

Key, plate 10, figures 2, 3; plate 45, figures 5, 6

Genotype, by original designation, *Fusulina sphaerica* Möller = *Fusulina sphaerica* Abich

Staffella Ozawa, Jour. Coll. Sci., Imp. Univ. Tokyo, vol. 45, 1925, art. 4, p. 24. — Likharev, Atlas of Leading Forms of the Fossil Fauna of the U.S.S.R., vol. 6, Permian, 1939, p. 34 [? not *Staffella* of most authors which is *Pseudostaffella*].

The genotype and only certainly recognized species is thickly nautiliform, its axis being only about three fourths the equatorial diameter and its periphery broadly rounded. It consists of about ten low and slowly expanding volutions, and reaches an equatorial diameter commonly more than 3 mm. and exceptionally as much as 4.75 mm. The spiral wall is thin and is said to consist of four layers like that of *Fusulinella.* The septa are plane, the tunnel narrow, and chomata low and broad. — Permian of Armenia.

Many small spheroidal species, commonly referred to *Staffella*, have been assigned to a new genus, *Pseudostaffella*, by Thompson. This appears to be justified, but the genotype of *Staffella* is still too poorly known to permit a positive decision.

In his original description of *Staffella*, Ozawa cited as type "*Fusulina sphaerica* Möller," overlooking the fact that the species was described first by Abich. He later explained (personal communication, January 16, 1929), "When I established the new genus, *Staffella,* I selected at random a spheroidal species in the paper by Möller." The original description by Abich stated that the species is abundant and widespread in the upper part of the Bergkalk in southern Armenia, and that it is recognizable by its round form. Abich added that the usual size is between 1.5 and 3 mm., occasionally rising to 4 mm., but he gave no other characters. Möller's description of 1878 was based on material from Arax Valley in southern Armenia, and implies that he was restudying Abich's collections. He states that all the specimens are silicified. His single figure of an axial section is very inadequate, but he states that this specimen had ten volutions and measured 4.75 mm. in diameter and 3.5 mm. along the axis. Likharev (1935) restudied Abich's material and figured an axial and a sagittal section, but because of the silicified condition they do not adequately show the structures. From his description it is clear that the genotype is relatively large, has at least ten volutions which are low and closely coiled, and that the early whorls are nautiloid rather than spheroidal. The chomata appear low and thin, though he describes them as somewhat massive.

Genus PSEUDOSTAFFELLA Thompson, 1942

Key, plate 10, figure 4

Genotype, by original designation, *Pseudostaffella needhami* Thompson
Pseudostaffella Thompson, Amer. Jour. Sci., vol. 240, 1942, p. 410.
Staffella of most authors [not Likharev 1939 or Thompson 1942].

Test small, subspherical, rounded or slightly umbilicate at the poles; normally comprising less than eight whorls and measuring less than 2 mm. in diameter. The spiral wall is thin and commonly consists of three layers (diaphanotheca, tectum, and thick outer tectorium); an inner tectorium appears locally in adult whorls. The septa are plane, the tunnel is narrow, chomata are massive and wide, grading laterally into the tectorium that lines the basal part of the chambers. — Basal Pennsylvanian to late Permian.

This is the stem-form, the most conservative, and the longest ranging genus of the family.

As here recognized, *Pseudostaffella* is common in the lower part of the Pennsylvanian system but ranges up through the Permian. It is normally less than 2 mm. in diameter and varies in shape from almost spheroidal to broadly nautiliform, with the axis appreciably shorter than the equatorial diameter. The chomata are massive. The epitheca is thick over the floor (i.e., the inner side) of the volution but is thin or lacking on the roof; as a result the outer tectorium is thick and the inner very thin or only recognizable in adult whorls.

Without more knowledge of the wall structure in the genotype of *Staffella*, there is room for doubt whether it differs generically from that in *Pseudostaffella*. The types of these genera differ materially in size and somewhat in shape, but such differences alone are not of generic significance. It appears probable, however, that *Staffella* is an Oriental, Permian genus with little direct relation to *Pseudostaffella*.

Genus NUMMULOSTEGINA Schubert, 1907

Key, plate 45, figures 1, 2

Genolectotype (Cushman, 1928), *Nummulostegina velebitana* Schubert, 1909

Nummulostegina SCHUBERT, Verh. k. k. geol. Reichsanst., Wien, 1907, p. 212; Jahrb. k. k. geol. Reichsanst., vol. 58, 1908, p. 377.—CUSHMAN, Foraminifera, etc., 1928, p. 209; ibid., 2nd ed., 1933, p. 196.—THOMPSON, Jour. Pal., vol. 9, 1935, p. 114.

Fusulinella (part) of STAFF, DEPRAT, COLANI.

Orobias (part) of GALLOWAY.

Small, nautiliform shells with a short axis and rounded periphery. Judging solely by the external features figured by Schubert, it differs from *Staffella* only in having a shorter axis, and it may take priority over *Nankinella*; but the shell of the genotype has never been critically studied or adequately described, and until this is done *Nummulostegina* cannot be referred to the Fusulinidae with certainty.—"*Schwagerina* dolomite" (probably Sakmarian) of Austria.

Genus PISOLINA Lee, 1933

Genotype, by original definition, *Pisolina excessa* Lee

Pisolina LEE, Nat. Research Inst. Geology (Shanghai) Mem. 14, 1933, p. 19.

Genus not well established, being based on a few sections of an inadequately described species. The genotype is spherical, about 4 mm. in diameter, and has seven to eight rather low volutions. It has plane septa, a narrow tunnel, and distinct chomata. The wall is thin and consists of "tectum and ill-defined keriotheca, of which the alveolar structure is by no means clear." The proloculum is large for a shell of this size.

Apparently a descendant of *Staffella*; possibly a synonym of the latter.—Lower Permian of the Yangtze gorge, China.

Genus MILLERELLA Thompson, 1942

Key, plate 45, figures 7, 8

Genotype, by original designation, *Millerella marblensis* Thompson

Millerella THOMPSON, Amer. Jour. Sci., vol. 240, 1942, p. 404.

Minute discoidal tests with a short axis and commonly not over seven or eight whorls. Young shells of three volutions or less are completely involute, slightly umbilicate, and about twice as high as wide; but subsequent whorls become partly evolute until the limbs of the last volution reach only about half the distance to the axis. Early whorls are evenly rounded, but the last adult whorl becomes narrow and subacutely rounded at the periphery. Spirotheca thin and not clearly differentiated into distinct layers; septa thin, arched but not folded; tunnel narrow, elliptical in section, bordered by sharp slender chomata.—Late Mississippian (Kinkaid formation) and early Pennsylvanian (Morrow and Lampasas series) of central and southwestern United States.

There is room for serious doubt whether *Millerella* belongs with the Fusulinidae. It is extremely small, its wall is too thin and unspecialized to be significant, and no one has determined whether the elliptical openings at the base of the septa record an external aperture (as in many of the "smaller" foraminifera) or are due to resorption (as in the fusulines). Chomata are one of the most distinctive features of the early fusulines but, if present in *Millerella*, they are in a very rudimentary state.

The doubts thus expressed are of a negative character and may indicate merely that this is the most primitive genus of the fusulines, as Thompson (1942) argued. In any event, it appeared somewhat earlier than other fusulines. On the contrary, its extreme lateral compression and the evolution of its outer whorl are rather marked specializations, in exactly the opposite direction from that displayed by the fusulines. Furthermore, the ontogeny of the fusuline shell shows no stage resembling the adult *Millerella*. Instead, it clearly suggests *Endothyra*, as indicated by Dunbar and Henbest (1942) and by Scott, Zeller and Zeller (1947). Therefore, if related in any way to the fusulines, *Millerella* would appear to be a very early aberrant specialization from a stem-form more like *Staffella*.

Genus SPHÆRULINA Lee, 1933

Genotype, by original designation, *Sphærulina crassispira*, Lee
Sphærulina Lee, Mem. Nat. Res. Inst. Geol. (Shanghai), no. 14, 1933, p. 16.

Minute spheroidal shells superficially like *Staffella* but differing therefrom in two respects: (1) the inner whorls are narrow and shaped like *Nummulostegina* and the outer ones gradually broaden and become spheroidal; (2) the wall consists of 2 layers, a thin tectum and a "keriotheca."—Permian of China.

This Permian genus, based on a single species, may represent a specialized offshoot of the staffelloid stock in which the mural pores have specialized into alveoli; but inasmuch as Lee described the keriotheca as "finely alveolar," it is possible he observed normal mural pores. However, he describes this layer as "fairly thick" and says the preservation is excellent. The lack of tectoria would probably distinguish this type from *Staffella*.

Genus LEËLLA Dunbar and Skinner, 1937

Key, plate 44, figure 5
Genotype, by original definition, *Leëlla bellula* Dunbar and Skinner, 1937
Leëlla Dunbar and Skinner, Univ. of Texas Bull. 3701, 1937, p. 603.

Minute fusulines developing a staffelloid form during the first 3 or 4 volutions and then elongating rapidly to become fusiform; wall structure as in *Fusulinella*; septa almost plane with low and relatively wide tunnel bordered by well-developed chomata. — High Permian of west Texas.

This genus is most similar to *Sphaerulina* Lee, which has narrow staffelloid inner whorls and later becomes spherical. If they are actually related, *Sphaerulina* must be the more primitive and conservative type, but they are probably independent developments out of *Staffella*.

Genus RAUSERELLA Dunbar, 1944

Key, plate 45, figures 9–11

Genotype, by original designation, *Rauserella erratica* Dunbar

Rauserella Dunbar, Geol. Soc. Am., Special Paper 52, 1944, p. 37.

Minute fusulines, derived out of *Staffella*, in which the first three or four volutions are planispirally coiled and have a short axis and rounded periphery, after which the axis of coiling abruptly shifts to a new orientation and succeeding whorls elongate rapidly to form a thickly fusiform shell. The outer whorls are few in number and irregular in form.

The spiral wall is thin and its structure obscure, appearing to consist of tectum and diaphanotheca in the outer whorls, but to possess a basal or outer tectorium in the nautiloid early whorls. The septa are plane in the early whorls and highly irregular but not distinctly folded in the outer. — High Permian, Guadalupian series, in Mexico, Texas, and the Orient.

This is considered a highly aberrant specialization.

Genus OZAWAINELLA Thompson, 1935

Key, plate 44, figures 6, 7

Genotype, by original designation, *Fusulinella angulata* Colani, 1924

Ozawainella Thompson, Jour. Pal., vol. 9, 1935, p. 114. — Dunbar and Skinner, Univ. of Texas Bull. 3701, 1937, p. 599.

Ozawaina Lee, Pal. Sinica (B), vol. 4, fac. 1, 1927, p. 13.

Fusulinella (part) of Deprat and Colani.

Staffella (part) of Ozawa, Lee, Chen, and Lee.

Orobias of Galloway and Harlton, and Galloway [not Eichwald].

Lenticular fusulinids in which the axis is the shortest diameter and the periphery is subacutely or acutely angular; the coiling is bilaterally symmetrical from the start, and the periphery is angular in all but the earliest whorls; the septa are plane and the tunnel appears as a depressed reniform opening at the base of each septum; the wall is thin, and is comprised of a thin tectum, a thin diaphanotheca and thicker tectoria. — Pennsylvanian and Permian.

In 1928 Galloway and Harlton (p. 348) revived the long-forgotten name *Orobias* Eichwald, designated *Nummulina antiquior* Rouillier and Vosinsky as its genotype, and applied it to these small lenticular fusulines. Lee had previously introduced for these the name *Ozawaina*, but, unfortunately, he did not designate a genotype, and Galloway, in 1933 (p. 396), designated as its genotype *N. antiquior* Rouillier and Vosinsky, thus making *Ozawaina* a straight synonym of *Orobias*. Thompson (1935, p. 114) showed this course to be unjustified and proposed the new name *Ozawainella* to be used in the sense Lee intended for *Ozawaina*. Dunbar and Skinner (1937, p. 600) showed that *Nummulina antiquior* is probably not related to the fusulines.

In its lenticular form this genus mimics *Nankinella*, but it is less than half as large, has fewer and more rapidly expanding whorls, and a subacute instead of well-rounded periphery. The numerous species now known do not indicate a gradation between the two types, and they appear to be a distinct generic line.

Genus NANKINELLA Lee, 1933

Key, plate 44, figures 8, 9; plate 45, figures 3, 4

Genotype, by original designation, *Staffella discoidea* Lee

Nankinella Lee, Nat. Research Inst. Geology (Shanghai), Mem. 14, 1933, p. 14.

Shell large, lenticular, with rounded periphery, consisting of eight to fourteen volutions. Spiral wall thin, consisting of tectum and diaphanotheca in the inner whorls but adding tectoria in the outer whorls, the outer tectorium being well developed but the inner very thin. Septa plane; tunnel narrow and crescentic in section; chomata slender. The microspheric generation possesses an endothyroid juvenarium.—Permian of the Orient.

The genotype of *Nankinella* possesses thirteen to fourteen closely wound volutions, and normally has an equatorial diameter of 5 to 6 mm., and rarely as much as 8 mm.

Genus FUSULINELLA Möller, 1877

Key, plate 10, figures 5–7; plate 44, figure 19

Genotype (monotypical), *Fusulinella bocki* Möller, 1878

Fusulinella Möller, Neues Jahrb., 1877, p. 144; (part) Mem. Acad. Imp. Sci. St. Petersbourg (7), vol. 25, no. 9, 1878, p. 101.—Douvillé, Compte Rend., Acad. Sci., Paris, 1906, p. 258.—Ozawa, Jour. Coll. Sci., Imp. Univ., Tokyo, vol. 45, art. 4, 1925, pp. 7 and 24.—Dunbar and Henbest, Am. Jour. Sci. (5), vol. 20, 1930, p. 357.—Galloway, Manual of Foraminifera, 1933, p. 400.—Dunbar, in Cushman, Foraminifera, etc., 2nd ed., 1933, p. 132.—Dunbar and Skinner, Univ. of Texas, Bull. 3701, 1937, p. 561.

Neofusulinella (part) of Deprat, Mém. Serv. Géol. Indochine, vol. 2, fasc. 1, 1913, pp. 40–44 [not *N. lantenoisi* Deprat, which is the genotype of *Neofusulinella*].—(part) Colani, ibid., vol. 11, fasc. 1, 1924, pp. 101 and 144.—Lee, Pal. Sinica, Ser. B, vol. 4, fasc. 1, 1927, pp. 13 and 16.—Lee and Chen, Mem. Nat. Res. Inst. Geol. (Shanghai), no. 9, 1930, p. 118.

Test small, fusiform; wall composed of tectum, diaphanotheca, and inner and outer tectoria; septa nearly plane across the middle, with slight folds near the poles (and in advance species folds become deeper and spread toward the middle in the last whorl); tunnel rather narrow and bordered by massive chomata.

Dimorphism has been observed by Dunbar and Henbest but the microspheric shells are rare. The dimorphic forms do not differ in size or external appearance, but the microspheric shells have a very minute proloculum and an endothyroid juvenarium commonly coiled askew to the rest of the shell. — Lower and Middle Pennsylvanian (pre-Missourian) of America; Moscovian of Eurasia. (Reported occurrences in the Permian of Indo-China and Japan need stratigraphic confirmation.)

Differs from *Fusulina* in having less deeply and extensively folded septa and more massive chomata.

Genus FUSIELLA Lee and Chen, 1930

Key, plate 10, figure 9

Genotype, by original designation, *Fusiella typica* Lee and Chen, 1930

Fusiella LEE and CHEN, Mem. Nat. Res. Inst. Geol. (Shanghai), no. 9, 1930, p. 107. — SKINNER, Jour. Pal., vol. 5, 1931, p. 253. — GALLOWAY, Manual of Foraminifera, 1933, p. 398. — DUNBAR, in CUSHMAN, Foraminifera, etc., 1933, p. 134. — DUNBAR and SKINNER, Univ. of Texas Bull. 3701, 1937, p. 567.

Profusulinella RAUSER-CHERNOUSSOVA, BELJAEV and REITLINGER, Trans. Polar Commission, Acad. Sci. U. S. S. R., no. 28, 1936, p. 175, pl. 1, figs. 1–6.

Extremely small fusulines resembling *Fusulinella*, from which they differ only in having a thinner wall of three instead of four layers. The wall is formed chiefly of the protheca, which is scarcely differentiated into distinct tectum and keriotheca. A thin layer of epitheca forms on the base (inner side) of the volutions, and in sections appears as an outer tectorium. The wall then appears three-layered. The original types were microspheric and possessed an endothyroid juvenarium, but megalospheric shells predominate in the American species *F. primaeva* Skinner. — Huanglung limestone of China, Lampasas series of Texas, and Moscovian of the U. S. S. R.

In proposing *Profusulinella*, Rauser-Chernoussova and her colleagues did not mention *Fusiella*, but based their genus upon the same type of three-layered wall. Both genotypes occur at about the same horizon.

Genus BOULTONIA Lee, 1927

Genotype, by original designation, *B. willsi* Lee, 1927

Boultonia LEE, Pal. Sinica (Ser. B), vol. 4, fasc. 1, 1927, p. 10.

Minute, fusiform shells resembling *Fusiella*, but having a wall of only two layers, tectum and diaphanotheca, and having deeply plicated septa. The types possessed an endothyroid juvenarium and probably represent the microspheric form; megalospheric form not known. — Sakmarian horizon in North China.

Genus CODONOFUSIELLA Dunbar and Skinner, 1937

Key, plate 44, figures 3, 4

Genotype, by original designation, *C. paradoxica* Dunbar and Skinner, 1937

Codonofusiella DUNBAR and SKINNER, Univ. of Texas Bull. 3701, 1937, p. 606.

Exceedingly small, fusiform shells with a very thin wall, an endothyroid juvenarium, and deeply folded septa; approaching maturity the outer whorl rapidly increases in length and height and becomes evolute, producing a trumpet-like flare. The wall appears to consist of two layers only, a very thin tectum overlying a homogeneous shell layer.

Numerous sectioned specimens all possess an endothyroid juvenarium coiled askew to the rest of the shell. These are probably microspheric individuals. It is not known whether the megalospheric form differs appreciably in external appearance.—Late Permian of Texas and of British Columbia; Middle Productus limestone of the Salt Range, India.

Genus WEDEKINDELLINA Dunbar and Henbest, 1933

Key, plate 10, figures 13–15

Genotype, by original designation, *W. euthysepta* Henbest, 1928

Wedekindella DUNBAR and HENBEST, Am. Jour. Sci. (5), vol. 20, 1930, p. 357.

Wedekindia DUNBAR and HENBEST, ibid., vol. 21, 1931, p. 458.

Wedekindellina DUNBAR and HENBEST, in CUSHMAN, Foraminifera, etc., 1933, p. 134.—THOMPSON, Univ. of Iowa Studies, vol. 16, 1934, p. 278.—RAUSER-CHERNOUSSOVA, Compt. Rendu Acad. Sci., U. S. S. R., IV (IX), 1935, pp. 117–120.

Boultonia (part) LEE, Pal. Sinica, Ser. B., vol. 4, fasc. 1, 1927, p. 11.

Fusulinella (part) HENBEST, Jour. Pal., vol. 2, 1928, p. 80.

Fusulina (part) WHITE, Univ. of Texas Bull. 3211, 1932, p. 24.

Minute, fusiform, commonly very slender and tightly coiled fusulines; wall of four layers as in *Fusulinella*; septa nearly plane or very slightly folded in the end zone of the outer whorls; tunnel narrow and bordered by massive and generally broad chomata; the epitheca is exceptionally developed, taking the form of a solid axial filling of the shell.—Pennsylvanian.

Genus WEARINGELLA Thompson, 1942

Genotype, by original designation, *Wearingella spiveyi* Thompson

Wearingella THOMPSON, Amer. Jour. Sci., vol. 240, 1942, p. 413.

Closely resembling *Wedekindella* but said to differ in the obsolescence of the inner tectorium (so that the wall is essentially three-layered), and in a reduction in the amount of axial filling, and a tendency for septal folding in the end zones and in the last whorl.—Upper Pennsylvanian (Missouri and Virgil series) of the Mid-Continent region of the United States.

Wedekindellina is common and widespread in the middle portion of the Des Moines series, and, there, shows a very distinct, clear diaphanotheca and a well-developed inner tectorium. Its septa are somewhat irregular near the poles but show no real folding. *Wearingella* includes its descendants in the Upper Pennsylvanian. The distinction is one of degree rather than of kind and will be difficult to apply unless preservation is excellent.

Genus SCHUBERTELLA Staff and Wedekind, 1910

Key, plate 10, figure 8

Genotype (monotypical), *S. transitoria* Staff and Wedekind, 1910

Schubertella STAFF and WEDEKIND, Bull. Geol. Inst., Univ. of Upsala, vol. 10, 1910, p. 121. — THOMPSON, Jour. Pal., vol. 11, 1937, pp. 118–123.

? *Depratella* OZAWA, Contr. Cushman Lab. Foram. Res., vol. 4, 1928, p. 9.

? *Eoschubertella* THOMPSON, Jour. Pal., vol. 11, 1937, p. 123.

Small, thickly fusiform shells of 4 to 6 volutions in which the wall is relatively thick and is comprised of tectum and diaphanotheca, without tectoria; septa plane as in *Fusulinella*; chomata well developed.

The predominant form is microspheric, with an endothyroid juvenarium of 1 to 2 volutions coiled askew to the rest of the shell. The megalospheric form does not differ appreciably in size or shape. — Permian (and possibly Pennsylvanian) of North America and Eurasia.

Staff and Wedekind believed the wall of their type species to consist of a single homogeneous layer, but Thompson's study of supposedly topotypical material indicates two layers as in the American species, *S. kingi*, described by Dunbar and Skinner.

Genus EOSCHUBERTELLA Thompson, 1937

Genotype, by original designation, *Schubertella lata* Lee and Chen, 1930

Eoschubertella THOMPSON, Jour. Pal., vol. 11, 1937, p. 123.

Schubertella LEE and CHEN, Mem. Nat. Res. Inst. Geol. (Shanghai), no. 9, 1930, p. 109.

Schubertella (part) of authors.

Minute fusiform shells like *Schubertella* in all respects except the structure of the spiral wall, which in this genus is considered to be differentiated into four layers, as in *Fusulinella*, instead of two, as in *Schubertella.* — Lower Pennsylvanian of Eurasia.

This genus was separated from *Schubertella* by Thompson chiefly on the basis of descriptions of the wall structure found in the literature. The forms embraced here are of Lower Pennsylvanian age and those referred to *Schubertella* s. s. are from the Lower Permian. Until a critical study of oriental material can be undertaken, it cannot be stated that *Eoschubertella* is proven to be different from *Schubertella.*

Genus YANGCHIENIA Lee, 1933

Genotype, by original designation, *Yangchienia iniqua* Lee

Yangchienia LEE, Mem. Nat. Res. Inst. Geol. (Shanghai), no. 14, 1933. — DUNBAR and SKINNER, Univ. of Texas Bull. 3701, 1937, p. 569.

A very small, fusiform fusuline differing from *Schubertella* only in the possession of very massive and broad chomata. — Lower Permian (Chihshia limestone) of China.

This genus, based on only a few sections of one species, is not securely established. In spite of the heavy chomata, these shells should probably be referred to *Schubertella*, which occurs in about the same stratigraphic position.

Genus FUSULINA Fischer, 1829

Key, plate 10, figures 10–12

Genotype (monotypical), also designated by Yabe in 1903, *Fusulina cylindrica* Fischer

Fusulina FISCHER DE WALDHEIM, Bull. Soc. Imp. Nat. Moscou, vol. 1, 1829, p. 330; Oryctographie, Moscou, 1830–1837, p. 126. — DUNBAR and HENBEST, Am. Jour. Sci. (5), vol. 20, 1930, p. 357. — GALLOWAY, Manual of Foraminifera, 1933, p. 401. — DUNBAR, in CUSHMAN, Foraminifera, etc., 1933, p. 134. — THOMPSON, Am. Jour. Sci., vol. 32, 1936, p. 287. — DUNBAR and SKINNER, Univ. of Texas Bull. 3701, 1937, p. 562.

Schellwienia STAFF and WEDEKIND, Bull. Geol. Inst. Univ. Upsala, vol. 10, 1910, p. 113 (proposed as a subgenus to include the *Fusulina* s. s., hence a direct synonym of *Fusulina* according to Articles 9 and 29 of the International Rules).

Girtyina LEE (non STAFF), Pal. Sinica (B), vol. 4, fasc. 1, 1927, p. 22. — LEE and CHEN, Mem. Nat. Res. Inst. Geol. (Shanghai), no. 9, 1930, p. 129.

Beedeina GALLOWAY, Manual of Foraminifera, 1933, p. 401.

Test fusiform to subcylindrical, with wall of four layers as in *Fusulinella*; differing from *Fusulinella* chiefly in having the septa fully and deeply folded even at the center of the shell, and in having the chomata less massive. — Lower and Middle Pennsylvanian of America; Moscovian of Eurasia.

The real nature of *Fusulina* has been a source of much confusion. Fischer described the genus briefly and illustrated it poorly, basing it on the common species from the Moscovian white limestone at Mjatchkova, near Moscow. Möller misinterpreted it and figured a form with alveolar wall, and subsequently the name *Fusulina* came to be widely used for the species now embraced in the subfamily Schwagerininae, while the name *Fusulinella* was extended to embrace the true *Fusulina*. Dunbar and Henbest in 1930 restudied topotype material from Mjatchkova, showed the true nature of *F. cylindrica*, and redefined the genus accordingly.

Subfamily 2. Schwagerininae Dunbar and Henbest, 1930

Shells fusiform, subcylindrical or subglobular; wall alveolar, consisting of two layers, tectum and keriotheca; septa more or less deeply folded; a median tunnel is invariably present and, in the most specialized genera, cuniculi and accessory tunnels provide additional communication between chambers; chomata are well developed in the ancestral genus, *Triticites*, but either obsolete or represented only in the early ontogeny of the other genera.

Although tectoria have disappeared, secondary deposits of shell substance may largely fill certain of the chambers, either along the axis or in a transverse belt some distance on each side of the middle.

Marked dimorphism has been observed in some of the genera, the microspheric shells being much larger than the megalospheric. The former possess a minute proloculum and an endothyroid juvenarium of about two volutions, commonly coiled askew to the axis of the adult whorls. Megalospheric shells are bilaterally symmetrical and completely involute, with the exception of the highly aberrant genus *Nipponitella* which fails to coil after the first two or three whorls. — Mid-Pennsylvanian to late Permian.

Genus TRITICITES Girty, 1904

Key, plate 10, figures 16–20; plate 44, figure 10

Genotype, by original designation, *Miliolites secalica* Say, 1824

Triticites GIRTY, Am. Jour. Sci. (4), vol. 17, 1904, p. 234. — DUNBAR and CONDRA, Nebraska Geol. Surv. (2), Bull. 2, 1927 [1928], pp. 53–60. — GALLOWAY, Manual of Foraminifera, 1933, p. 402. — DUNBAR, in CUSHMAN, Foraminifera, etc., 1933, p. 135. — CHEN, Pal. Sinica, Ser. B, vol. 4, fasc. 2, 1934, p. 21. — DUNBAR and SKINNER, Univ. of Texas Bull. 3701, 1937, p. 613.

Girtyina STAFF, Neues Jahrb., Beil.-Bd. 27, 1909, pp. 490 and 506. (Genotype, by original designation, "*Girtyina ventricosa* Meek" [and HAYDEN]. Although the named genotype is a typical species of *Triticites*, STAFF used the name for species belonging to *Fusulina* s. s.)

Grabauina LEE, Bull. Geol. Soc. China, vol. 3, no. 1, 1924, p. 51 (based on a crushed shell of *Triticites*).

Fusulina (part) of many authors (e.g., SILVESTRI, Boll. Soc. Geol. Italiana, vol. 54, 1935, pp. 203–219).

Schellwienia (part) of authors.

? *Hemifusulina* MÖLLER, Neues Jahrb., 1877, p. 144; Mem. Acad. Imp. Sci. de St. Petersbourg (7), vol. 25, no. 9, 1878, pp. 74–78. [Not definitely recognizable; may be a synonym of *Schwagerina*.]

Fusiform to subcylindrical or subglobular; wall distinctly alveolar, consisting of tectum and keriotheca, with traces of an outer tectorium in some species; septa nearly plane or slightly folded at the middle of the shell but more deeply folded toward the poles; folds of adjacent septa commonly do not touch so as to subdivide the longitudinal chambers into chamberlets; septa pierced by a slit-like median tunnel which is bordered by well-developed chomata; septal pores are commonly conspicuous in the outer whorls. — Mid-Pennsylvanian to early Permian.

Triticites differs from *Schwagerina* [*Pseudofusulina* of authors] in two respects, (1) the less deep and regular septal folds, and (2) the possession of conspicuous chomata.

Genus SCHWAGERINA Möller, 1877

Key, plate 10, figures 21–23; plate 44, figures 2A, 11, 20

Genotype, by original designation, *Borelis princeps* Ehrenberg

Schwagerina MÖLLER, Neues Jahrb., 1877, p. 143. — DUNBAR and SKINNER, Jour. Pal., vol. 10, 1936, pp. 83–91; Univ. of Texas Bull. 3701, 1937, p. 623.

[Not *Schwagerina* of most authors, which = *Pseudoschwagerina* or *Paraschwagerina*.]

Fusulina (part) of authors.

Pseudofusulina DUNBAR and SKINNER, Am. Jour. Sci. (5), vol. 22, 1931, p. 252. — GALLOWAY, Manual of Foraminifera, 1933, p. 404. — DUNBAR, in Cushman, Foraminifera, etc., 1933, p. 136. — CHEN, Pal. Sinica, Ser. B, vol. 4, fasc. 2, 1934, p. 50. — RAUSER-CHERNOUSSOVA, Bull. Acad. Sci. U. S. S. R., 1936, pp. 573–584; Trans. Polar Commission, Acad. Sci. U. S. S. R., no. 28, 1936, p. 224. — BELJAEV and RAUSER-CHERNOUSSOVA, Studies of the Geol. Inst. Acad. Sci. U. S. S. R., vol. 7, 1938, pp. 169–196.

Leeina GALLOWAY, Manual of Foraminifera, 1933, p. 406.

Chusenella LEE, Bull. Geol. Soc. China, vol. 22, 1942, p. 171.

Nagatoella THOMPSON, Jour. Geol. Soc. Japan, vol. 43, 1936, pp. 195–202.

Schellwienia (part) of authors.

Fusiform to subcylindrical or subglobular fusulines having a distinctly alveolar wall consisting of tectum and keriotheca; volutions planispirally coiled and *gradually expanding*; septa regularly and very deeply folded so that the lower part of opposed folds in adjacent septa meet, subdividing the meridional chambers into a series of cell-like chamberlets (plate 10, figure 22); median tunnel low and slit-like; chomata are lacking or, in some early species of the genus, appear in a rudimentary condition in the inner whorls only.

In various species of this genus a conspicuous deposit of epitheca is laid down, largely filling the chambers in certain parts of the shell. This filling may be localized in a narrow zone along the axis or in a belt in each end zone some distance from the tunnel. Various attempts have been made to erect genera on the basis of such axial filling, but it is clearly not a generic character. Such filling assumes a distinctive pattern in many species but it has developed repeatedly in species obviously not closely related, and it occurs in other genera such as *Wedekindellina*, *Parafusulina* and *Polydiexodina*. — Chiefly early Permian, i.e., Wolfcamp and early Leonard in America, and Sakmarian and Artinskian in Eurasia.

Schwagerina has been the subject of unfortunate confusion. Möller defined it in 1877, designating *Borelis princeps* Ehrenberg as genotype, but he did not describe that species. A year later he applied the name *Schwagerina princeps* to a globular shell superficially resembling *B. princeps* but having a very different interior. Möller's *S. "princeps"* of 1878 had a tightly coiled juvenarium followed by abrupt and rapid inflation, and this subsequently came to be regarded as the diagnostic character of *Schwagerina*. Meanwhile, the real *Borelis princeps* was never restudied until the types were sectioned and described by Dunbar and Skinner in 1936. It was then discovered that expansion was gradual and the shell features were those for which Dunbar and Skinner had previously proposed the name *Pseudofusulina*. *Schwagerina* was then rediagnosed in accordance with its genotype, *Pseudofusulina* was suppressed as a synonym, and the form so long mistaken for *Schwagerina* was described as a new genus, *Pseudoschwagerina*.

The true *Schwagerina* expands gradually as does *Triticites*, from which it differs in its deep and regular septal folding and in the absence of chomata.

In December of 1942, Lee published a diagnosis of a new genus, *Chusenella*, based upon a new species, *Chusenella ishanensis* Hsu, described on a subsequent page of the same issue of the Bulletin of the Geological Society of China. In neither was a genotype clearly designated, so that the name may not be legally defensible.

A more serious criticism is that the genus appears to have been founded on a misinterpretation of shell structures. An axial and a sagittal section of the species on which it was based are here reproduced as figures 14 and 15 on plate 45. Lee believes that this species is closely allied to *Pseudodoliolina*, and that what appear to be septal loops are "split parachomata." To me such an interpretation is incredible. The shell appears to have had regularly and intensely folded septa and to fit within the complex of species now embraced in the genus *Schwagerina*.

Subgenus RUGOFUSULINA (Rauser-Chernoussova)

Key, plate 44, figures 2B, C

Genotype, by original designation, *Rugofusulina prisca* Ehrenberg emend. Möller

Rugofusulina Rauser-Chernoussova, Studies in Micropal., Moscow Univ., vol. 1, fasc. 1, 1937, pp. 9–26.

Distinguished from *Schwagerina* s. s. because of the "rugosity" of its spiral wall (plate 44, figs. 2B, C).

Since the wall appears to undulate, regardless of the orientation of the section, it is evident that the inequalities are of the nature of dimples and mounds rather than rugae. In the older species, the entire wall is flexed as in plate 44, figure 2B, but in later species only the tectum is affected and the dimples are on a very small scale.

The significance of this feature is not understood but it is clearly an original shell character, not a modification during fossilization, and it seems to characterize a group of related species in the U. S. S. R. and has been observed in forms from the Carnic Alps and from Texas, all of the same limited stratigraphic range. We tentatively regard this as a subgenus of *Schwagerina*.—Lower Permian (Sakmarian).

Genus DUNBARINELLA Thompson, 1942

Key plate 45, figures 12, 13

Genotype, by original definition, *Dunbarinella ervinensis* Thompson

Dunbarinella Thompson, Amer. Jour. Sci., vol. 240, 1942, p. 416.

Test small, fusiform; proloculus small; early volutions thin-walled and closely coiled, later ones gradually expanding and becoming thicker walled; spiral wall alveolar. Septa rather strongly folded near their basal margin as in *Schwagerina*; tunnel narrow; chomata rather slender but persisting at least into the penultimate whorl. Axial filling conspicuous in the end zones of early whorls but sparse or lacking in the outer whorls.—Upper Pennsylvanian to Early or possibly Mid-Permian.

The genotype is a late Pennsylvanian species structurally intermediate between *Triticites* and a part of the *Schwagerina*-complex. Septal folding is far advanced but chomata are still retained. In these respects it is transitional and does not well fit into either of these genera. Heavy axial filling is never seen in *Triticites* but characterizes a good many species of the *Schwagerina*-complex. The combination of a small proloculus, low, thin-walled, early whorls, and massive axial filling with the distribution shown by *D. ervinensis* serve to distinguish a tribe that kept its identity up into the Permian.

Genus PSEUDOSCHWAGERINA Dunbar and Skinner, 1936

Key, plate 44, figures 15, 16

Genotype, by original designation, *Schwagerina uddeni* Beede and Kniker

Pseudoschwagerina Dunbar and Skinner, Journ. Pal., vol. 10, 1936, p. 89; Univ. of Texas Bull. 3701, 1937, p. 656.

Schwagerina (part) of authors.

Thickly fusiform to subspherical fusulines in which the inner volutions are closely coiled as in *Triticites*, forming a compact juvenarium of 2 to 5 whorls, which is followed by a rapid change to the high, inflated volutions of the adult shell; the wall is alveolar as in *Triticites*, but the keriotheca is commonly relatively thin; the septa are gently and irregularly folded and so widely spaced that opposed folds do not commonly touch; the median tunnel is low and slit-like; chomata are present in the juvenarium but rudimentary or lacking in the inflated whorls.—Wolfcamp and equivalent horizons in America and Sakmarian horizon in Eurasia.

The great height of the inflated whorls in some species is noteworthy, but the actual height varies with the size of the species; the diagnostic character is the abrupt change in the plan of coiling, which leaves the juvenarium sharply marked off from the outer whorls. The ontogeny clearly shows that this genus developed out of *Triticites*.

Subgenus ZELLIA Kahler and Kahler, 1937

Genotype, by original designation, *Pseudoschwagerina* (*Zellia*) *heritschi* Kahler and Kahler

Pseudoschwagerina (*Zellia*) Kahler and Kahler, Paleontographica, Bd. 87, Abt. A, Lief. 1, 1937, pp. 20–21.

Distinguished from *Pseudoschwagerina* s. s. by having a thicker wall and thicker septa and in the great abundance and prominence of the septal pores. The septa are said to have a distinct layer of epitheca on both front and back side.—Lower Permian (Schwagerinakalk) of the Carnic Alps.

Genus PARASCHWAGERINA Dunbar and Skinner, 1936

Key, plate 11, figures 1, 2; plate 44, figures 12–14

Genotype, by original designation, *Schwagerina gigantea* White

Paraschwagerina DUNBAR and SKINNER, Jour. Pal., vol. 10, 1936, p. 89; Univ. of Texas Bull. 3701, 1937, p. 666.

Schwagerina (part) of authors.

Differing from *Pseudoschwagerina* in that the whorls of the tightly coiled juvenarium are slender and elongate and the septa at all stages of growth are deeply and regularly folded as in *Schwagerina* s. s., and that chomata are inconspicuous or lacking even in the juvenarium, and normally are completely absent in the inflated whorls. — Wolfcamp and equivalent formations in America and Sakmarian horizon in Eurasia.

The ontogeny clearly shows that this genus developed from *Schwagerina* s. s., from which it differs only in the abrupt inflation at the end of the juvenarium.

Genus PARAFUSULINA Dunbar and Skinner, 1931

Key, plate 11, figures 3–8

Genotype, by original designation, *Parafusulina wordensis* Dunbar and Skinner

Parafusulina DUNBAR and SKINNER, Am. Jour. Sci. (5), vol. 22, 1931, p. 258. — GALLOWAY, Manual of Foraminifera, 1933, p. 406. — DUNBAR, in CUSHMAN, Foraminifera, etc., 1933, p. 137. — CHEN, Pal. Sinica, Ser. B, vol. 4, fasc. 2, 1934, p. 80. — DUNBAR and SKINNER, Univ. of Texas Bull. 3701, 1937, p. 672.

Fusulina (part) of authors.

Schellwienia (part) of authors.

Elongate, fusiform to subcylindrical fusulines, commonly attaining a large size. Wall as in *Schwagerina*, consisting of tectum and keriotheca; septa regularly and very deeply folded, with the tips of opposed folds in adjacent septa meeting before reaching the floor and joining to form arch-like foramina. The repetition of these basal foramina under successive septa forms a series of spiral galleries or *cuniculi* running around the shell; the basal margins of the septa are joined along the sides of these cuniculi to form a series of wavy basal sutures running around the shell at right angles to the axis (plate 11, figs. 6 and 8).

Very marked dimorphism has been observed in several species of this genus. Microspheric shells are rare but greatly exceed in size the megalospheric. The microspheric form has an endothyroid juvenarium of about two volutions, commonly coiled askew to the axis of later whorls. Moreover, these shells have no tunnel, except in the first three or four volutions succeeding the juvenarium. The giants among the fusulines are microspheric shells of this genus and *Polydiexodina*. — Permian.

This genus was derived from *Schwagerina* by progressive specialization of the septa, particularly the development of cuniculi and the spiral basal sutures of the septa. These distinctive features are best seen in polished surfaces tangential to the surface and near the floor of a volution.

Genus POLYDIEXODINA Dunbar and Skinner, 1931

Key, plate 11, figures 9–12

Genotype, by original designation, *Polydiexodina capitanensis* Dunbar and Skinner

Polydiexodina DUNBAR and SKINNER, Am. Jour. Sci. (5), vol. 22, 1931, p. 263. — GALLOWAY, Manual of Foraminifera, 1933, p. 406. — DUNBAR, in CUSHMAN, Foraminifera, etc., 1933, p. 137. — DUNBAR and SKINNER, Univ. of Texas Bull. 3701, 1937, p. 693.

Fusulina (part) of authors.

Mostly large and very elongate fusulines, resembling *Parafusulina*, but distinguished by the presence of a series of accessory tunnels paired on opposite sides of the median tunnel. Generally the whorls are low and tightly coiled, and considerable secondary shell material is deposited as an axial filling. The proloculum is commonly large, thin-walled and irregular in shape.

Marked dimorphism occurs here precisely as in *Parafusulina*. — High Permian of North America and Central Asia.

Genus NIPPONITELLA Hanzawa, 1938

Key, plate 44, figures 17, 18

Genotype, by original designation, *Nipponitella explicata* Hanzawa, 1938

Nipponitella HANZAWA, Proc. Imp. Acad. Tokyo, vol. 14, no. 7, 1938, p. 256.

An irregularly uncoiled fusuline apparently derived out of *Triticites*. It possesses a juvenarium of two or three normally coiled, fusiform volutions after which the shell grows out in a rectilinear form as an irregularly undulating ribbon. The juvenile whorls have a median tunnel and well-developed chomata and the wall is alveolar, consisting of tectum and keriotheca. In the uncoiled part of the shell the septa are closely folded. — Permian (Maiya group) of Japan.

Genus GALLOWAIINELLA Chen, 1937

Genotype (monotypical), *Gallowaiina meitienensis* Chen, 1934

Gallowaiinella CHEN, in DUNBAR and SKINNER, Univ. of Texas Bull. 3701, 1937, p. 571.

Gallowaiina CHEN [non ELLIS], Bull. Geol. Soc. China, vol. 13, 1934, p. 237.

Based on a single species of subcylindrical form which differs from *Schwagerina* only in wall structure, its spirotheca being very thin (hardly exceeding 20 microns even in the outer whorls) and chiefly composed of a homogeneous clear layer coated both inside and out by a thin dark film. Of these dark films the outer is the more distinct and may be homologized with the tectum.

The septa are regularly and deeply folded as in *Schwagerina*, the tunnel is slit-like, and there are no chomata. — Permian (?), south China.

There is uncertainty as to the taxonomic position and value of this form. Its thin and nearly structureless wall suggests "degeneracy." It does not have the distinct alveolar structure characteristic of the Schwagerininae, but its size, general organization, and association suggest that it may be a specialized or degenerate offshoot of this subfamily rather than a member of the Fusulininae.

Genus QUASIFUSULINA Chen, 1934

Genotype, by original definition, *Fusulina longissima* Möller

Quasifusulina CHEN, Pal. Sinica, Ser. B, vol. 4, fasc. 2, 1934, p. 91. — DUNBAR and SKINNER, Univ. of Texas Bull. 3701, 1937, p. 570.

Habitus like that of a slender, subcylindrical species of *Schwagerina* from which it is distinguished by wall structure alone.

The wall is quite thin, commonly about 30 microns (and rarely as much as 50 microns) thick in the outer whorls. This appears to consist of a single, clear layer in which mural pores can be seen. Chen believed a tectum to be present but this needs confirmation.

The taxonomic position of *Quasifusulina* is not certain. It may be a specialized and "degenerate" offshoot of *Fusulina*. The types figured by Chen are from the basal Permian (Chuanshan) of China, but the named species is common in the *Triticites* zone in Russia. Another species, *Q. tenuissima* Schellwien, occurs in the Permian of the Carnic Alps. Whether *Quasifusulina* is a late derivative from *Fusulina* or a specialized and degenerate branch from *Schwagerina* s. s. needs further investigation.

Genus PALEOFUSULINA Deprat, 1913

Key, plate 10, figure 24

Genotype (monotypical), *Paleofusulina prisca* Deprat

Paleofusulina DEPRAT, Acad. Sci.., Paris, Compte Rendus, vol. 154, 1912, p. 1548; Mém. Serv. Géol. Indochine, vol. 2, fasc. 1, 1913, p. 36. — COLANI, ibid., vol. 11, fasc. 1, 1924, pp. 24, 52, 79, and 133. — LIKHAREV, Bull. Com. Géol. Leningrad, vol. 45, 1926, p. 59.

Test small, thickly fusiform; spiral wall only about 15 to 20 microns thick and either homogeneous or with very obscure structure; septa strongly and regularly folded, the folds having almost equal intensity from bottom to top so that in axial slices they appear as pillar-like rods rather than loops; tunnel narrow and rather high; chomata present but chiefly in the form of secondary deposits coating the septa near the tunnel; septal pores abundant. — Lower Permian of the Orient.

The above description is based on the literature and upon a small collection of beautifully preserved specimens from near Likiang, northwestern Yunnan. These are believed to belong to the type species *Paleofusulina prisca* Deprat.

This is believed to be a specialized and aberrant genus in which the wall has lost all pristine structures. It probably is a descendant of *Schwagerina*.

Family 13. NEOSCHWAGERINIDAE Dunbar nov.

The subfamilies Verbeekininae and Neoschwagerininae are here transferred from the Fusulinidae to a new family, Neoschwagerinidae.

They resemble the fusulines only in their planispiral growth (normally about an elongate axis) and in the alveolar wall structure possessed by many genera of both groups. It is also true that they descended from *Staffella*, which is the stem-form of the Fusulinidae; but the Neoschwagerinidae first appeared in the Permian, after the fusulines were far advanced, and they specialized in quite distinctive ways so that there was no genetic contact with the fusulines after their origin. The neoschwagerines are essentially an oriental tribe, though they made forays westward along the Tethyan geosyncline to Greece and Tunisia, and by way of the northwest into British Columbia.

In the Neoschwagerinidae the septa remained plane and the shell was strengthened by linear pendants (septulae) from the spirotheca which served as I-beams (Key, plate 12, figures 7 and 11). These appeared first as a series directed spirally, crossing the true septa and subdividing the meridional chambers into rectangular chamberlets. Later, another set appeared, running meridionally and alternating with the true septa. Still later, several sets of these axial septulae were introduced.

Communication between chambers was provided by a row of rounded foramina along the base of each septum. These are present in the outer whorl and almost certainly represent external apertures. Secondary deposits take the form of slender, hoop-like ridges on the floor of the volution, alternating in position with the foramina.

To recapitulate, the fusulines strengthened their shells by septal folds; the neoschwagerines by pendants (septulae) from the spirotheca. The antetheca of the fusuline had no aperture but a large number of minute septal pores; the neoschwagerines possessed a row of round foramina along the base of the antetheca; the fusulines secured communication between chambers by resorption of part of the basal margin of the septa to form a tunnel; the neoschwagerines retained the foramina for communication.

Subfamily 1. Verbeekininae Staff and Wedekind, 1910

Melon-shaped or subspherical shells in which the septa are plane and (except in *Eoverbeekina*) do not have a tunnel but are perforated along the basal margin by a single row of round foramina. Parachomata are present in the later ontogeny of primitive forms and in all stages of more advanced genera.

The wall is normally alveolar, consisting of a thin tectum and a keriotheca of fine texture, but one specialized genus (*Pseudodoliolina*) has a thin, compact and homogeneous wall.

Dimorphism is common and the microspheric shells have a juvenarium of endothyroid form coiled askew to the axis of later whorls. There is little difference, however, in size and external appearance between the microspheric and megalospheric shells. — Permian of the Orient and the Tethyan region.

Apparently this stock arose directly from *Staffella* during Permian time, the genus *Eoverbeekina* being the connecting link.

Genus EOVERBEEKINA Lee, 1933

Genotype, by original designation, *Eoverbeekina intermedia* Lee

Eoverbeekina LEE, Mem. Nat. Res. Inst. Geol. (Shanghai), no. 14, 1933, p. 18. — CHEN, Pal. Sinica, Ser. B, vol. 4, fasc. 2, 1934, p. 103. — DUNBAR and SKINNER, Univ. of Texas Bull. 3701, 1937, p. 573.

Subspherical shells of small size in which the inner whorls are nautiliform, gradually changing to spheroidal at maturity. The wall is thin, consisting of tectum and a finely alveolar keriotheca; the septa are plane and are perforated by a slit-like median tunnel and, in addition, in the outer whorls, by numerous round basal foramina on each side of the middle. Chomata are obsolescent but rudimentary parachomata appear in the outer whorls. — Lower Permian (Chihsia limestone of China).

This genus differs from *Verbeekina* in smaller size, in possessing a median tunnel, in the rudimentary character of its parachomata, and in the fact that its early whorls are narrow instead of spheroidal. It forms an almost ideal link between *Staffella* and *Verbeekina*.

Genus VERBEEKINA Staff, 1909

Key, plate 12, figure 2

Genotype (monotypical), *Fusulina verbeeki* Geinitz

Verbeekina STAFF, Neues Jahrb., Beil. Bd. 27, 1909, p. 476. — OZAWA, Jour. Coll. Sci. Imp. Univ. Tokyo, vol. 45, art. 4, 1925, p. 25; ibid., vol. 45, art. 6, 1925, p. 48. — DUNBAR and CONDRA, Nebraska Geol. Surv. Bull. 2, 2nd ser., 1927, p. 74. — DUNBAR, in CUSHMAN, Foraminifera, etc., 1933, p. 138. — TAN SIN HOK, Wetensch. Medeel, Mijnbouw Nederlandisch-Indië, no. 25, 1933, p. 57. — THOMPSON, Jour. Pal., vol. 10, 1936, p. 193. — DUNBAR and SKINNER, Univ. of Texas Bull. 3701, 1937, p. 573.

Doliolina (part) of SCHELLWIEN.

Schwagerina (part) of authors prior to 1909, and of DEPRAT, 1911–14.

Test spheroidal, consisting of numerous volutions; wall composed of tectum and thin keriotheca; septa plane, without a median tunnel, but with a regular row of round foramina along the basal margin; parachomata are lacking or rudimentary in the inner whorls but well developed in the outer ones.

The proloculus is minute and is followed in the microspheric shells by a few obliquely coiled juvenile whorls.—Permian of the Orient and the Tethyan region.

Closely allied to *Misellina* from which it differs in shape and in the fact that parachomata are developed later in the ontogeny.

Genus MISELLINA Schenck and Thompson, 1940

Key, plate 12, figures 3–5

Genotype, by original designation, *Doliolina ovalis* Deprat

Misellina SCHENCK and THOMPSON, Jour. Pal., vol. 14, 1940, p. 584.

Möllerina SCHELLWIEN, *Palaeontographica*, vol. 44, 1898, p. 238 (Name preoccupied.)

Doliolina SCHELLWIEN, Schrift. Phys.-Oekon. Gesell. Königsberg, Jahrg. 43, 1902, p. 67; in FUTTERER, Durch Asien, vol. 3, 1902, p. 125.—DEPRAT, Mém. Serv. Géol. Indochine, vol. 1, fasc. 3, 1912, p. 42, vol. 4, fasc. 1, 1915, p. 27.—YABE and HANZAWA, Proc. Imp. Acad. Japan, vol. 8, no. 2, 1932, p. 41.—GALLOWAY, Manual of Foraminifera, 1933, p. 408.—DUNBAR, in CUSHMAN, Foraminifera, etc., 1933, p. 138.—DUNBAR and SKINNER, Univ. of Texas Bull. 3701, 1937, p. 574. (Name preoccupied by *Doliolina* BORGERT, 1894, a tunicate.)

Test melon-shaped, somewhat elongated and bluntly rounded at the ends; microspheric individuals possessing an endothyroid juvenarium but megalospheric shells are planispiral throughout; walls rather thin, consisting of a tectum and a finely alveolar keriotheca; septa plane, with a basal row of foramina but no median tunnel; parachomata well developed at all stages of growth.—Permian of the Orient.

Subgenus BREVAXINA Schenck and Thompson, 1940

Subgenotype, by original designation, *Doliolina compressa* Deprat

Brevaxina SCHENCK and THOMPSON, Jour. Pal., vol. 14, 1940, p. 587.

Differing from *Misselina* s. s. only in having a subspheroidal form with the axis the shortest diameter.—Permian of the Orient.

Genus PSEUDODOLIOLINA Yabe and Hanzawa, 1932

Key, plate 12, figure 6

Genotype, by original designation, *Pseudodoliolina ozawai* Yabe and Hanzawa

Pseudodoliolina YABE and HANZAWA, Proc. Imp. Acad. Japan, vol. 8, 1932, p. 40.—GALLOWAY, Manual of Foraminifera, 1933, p. 410.—DUNBAR, in CUSHMAN, Foraminifera, etc., 1933, p. 139.—DUNBAR and SKINNER, Univ. of Texas Bull. 3701, 1937, p. 575.

Doliolina (part) of DEPRAT and OZAWA.

Differing from *Doliolina* in wall structure alone, the keriotheca being obsolete and the wall consisting of a single, thin layer.—Permian of the Orient.

Subfamily 2. Neoschwagerininae Dunbar and Condra, 1927

Shells mostly of large size and thickly fusiform to subglobular shape, in which the spiral wall is specialized by the development of pendant lamellae called septula.

There are usually numerous, closely coiled volutions; the septa are plane; there is no median tunnel but, instead, a row of rounded foramina along the base of the septa; parachomata are invariably present. The spiral wall consists of a tectum and finely alveolar keriotheca in the ancestral and more conservative genera, but the keriotheca undergoes extensive specializations in this subfamily. The first specialization took the form of local thickening of the keriotheca into septum-like pendants, the septula, which hung down into the chambers (plate 12, fig. 7). In the most primitive genus (*Cancellina*), these septula formed a meridional series crossing the true septa at right angles; in the next genus (*Neoschwagerina*) a second set of septula appeared, paralleling the true septa and crossing the meridional septula to form a grid of intersecting plates. In later genera the lower edges of these septula became solid instead of alveolar and, eventually, in the most highly specialized genus (*Lepidolina*), the entire keriotheca was reduced to a thin, compact and apparently homogeneous layer from which hung the thin compact septula.

The solidification of the septula began at the free edge and progressed upward toward the spiral wall; it also began late in the ontogeny and was gradually pushed back into the early whorls. Lee has proposed genera based on the stages of this progressive evolution. — Permian of the Orient and the Tethyan region, and British Columbia.

Genus CANCELLINA Hayden, 1909

Genotype (selected by Ozawa, 1925), *Neoschwagerina primigenia* Hayden

Cancellina Hayden, Rec. Geol. Surv. India, vol. 38, 1909, p. 249. — Ozawa, Jour. Coll. Sci. Imp. Univ. Tokyo, vol. 45, art. 4, 1925, pp. 18, 26; in Cushman, Foraminifera, etc., 1928, p. 138. — Galloway, Manual of Foraminifera, 1933, p. 410. — Dunbar, in Cushman, Foraminifera, etc., 1933, p. 139. — Dunbar and Skinner, Univ. of Texas Bull. 3701, 1937, p. 575.

Like *Neoschwagerina* but with a single series of septula which run transverse to the axis and subdivide the longitudinal chambers into rectangular chamberlets. — Permian of Afghanistan and the Orient, and of British Columbia.

Genus NEOSCHWAGERINA Yabe, 1903

Key, plate 12, figure 7

Genotype, by original designation, *Schwagerina craticulifera* Schwager

Neoschwagerina Yabe, Jour. Geol. Soc. Tokyo, vol. 10, no. 113, 1903, p. 5; Jour. Coll. Sci. Imp. Univ. Tokyo, vol. 21, art. 5, 1906, p. 3. — Deprat, Mém. Serv. Géol. Indochine, vol. 1, fasc. 3, 1912, pp. 6, 15; vol. 3, tasc. 1, 1914, p. 24 — Ozawa, Jour. Coll. Sci. Imp. Univ. Tokyo, vol. 45, art. 4, 1925, pp. 18, 24. —

GALLOWAY, Manual of Foraminifera, 1933, p. 410. — DUNBAR, in CUSHMAN, Foraminifera, etc., 1933, p. 140. — DUNBAR and SKINNER, Univ. of Texas Bull. 3701, 1937, p. 576.

Test large, thickly fusiform to subspherical, consisting of numerous, closely wound volutions; wall thin, formed of tectum and keriotheca; septa plane; septula of two series well developed, one set alternating with the true septa and paralleling the axis, while the other set runs at right angles to the septa, the two sets of septula forming a rectangular grid that hangs pendant from the wall; septa perforated by a row of rounded basal foramina; parachomta well developed. — Permian of the Orient and the Tethyan region, and British Columbia.

This genus shows an advance over *Cancellina* in the addition of axial septula.

Genus COLANIA Lee, 1933

Genotype, by original designation, *Colania kwangsiana* Lee.

Colania LEE, Mem. Nat. Res. Inst. Geol. (Shanghai), no. 14, 1933, p. 20. — DUNBAR and SKINNER, Univ. of Texas Bull. 3701, 1937, p. 576.

Intermediate between *Neoschwagerina* and *Yabeina*, the inner whorls having exactly the character of those in the former genus, and the outer whorls following the pattern of *Yabeina*. — Permian of the Orient.

There is room for considerable doubt as to the usefulness of this generic distinction. It is evident that *Yabeina* descended from *Neoschwagerina*, and it is to be expected that early species of the former would show in their ontogeny some recapitulation of this racial history. *Colania kwangsiana* does so precisely. It is, of course, an incompletely developed *Yabeina* and might be admitted to that genus. The unique species is apparently rare since but a single figure was given to supplement the brief specific description of the genotype. We are inclined for the present to regard *Colania* as a synonym of *Yabeina*.

Genus YABEINA Deprat, 1914

Key, plate 12, figures 8, 9, 10

Genotype, by original designation, *Neoschwagerina* (*Yabeina*) *inoueyi* DEPRAT = *Neoschwagerina globosa* YABE = *Yabeina globosa* (Yabe)

Yabeina DEPRAT, Mém. Serv. Géol. Indochine, vol. 3, fasc. 1, 1914, p. 30. — OZAWA, Jour. Coll. Sci. Imp. Univ. Tokyo, vol. 45, art. 4, 1925, pp. 18, 26. — GALLOWAY, Manual of Foraminifera, 1933, p. 411. — DUNBAR, in CUSHMAN, Foraminifera, etc., 1933, p. 140. — DUNBAR and SKINNER, Univ. of Texas Bull. 3701, 1937, p. 577.

Like *Neoschwagerina* except that the distal parts of the pendant septula are solid plates, the alveolar texture being lost here by a thickening and fusing of the lamellae. — Upper Permian of the Orient.

This genus is clearly a specialized descendant of *Neoschwagerina* from which it differs in the modification of the structure of its wall pendants.

There is also a tendency toward the introduction of a greater number of axial septula than in the parent genus, *Neoschwagerina* commonly having 1 to 3 septula between each pair of true septa and *Yabeina* from 3 to 6.

Genus LEPIDOLINA Lee, 1933

Genotype, by original designation, *Neoschwagerina (Sumatrina) multiseptata* Deprat

Lepidolina LEE, Mem. Nat. Res. Inst. Geol. (Shanghai), no. 14, 1933, p. 21.—DUNBAR and SKINNER, Univ. of Texas Bull. 3701, 1937, p. 578.

Like *Yabeina* except that the fusion of the keriothecal lamellae is complete in both the wall and its pendants, the spirotheca consisting of a thin, homogeneous layer and the septula being compact plates. While the general form is precisely that of *Yabeina*, the wall structure is that of *Sumatrina*.—Permian of the Orient.

We are uncertain of the validity of this genus because of discrepancies between the observations of Deprat, Colani, and Lee on the structure of the type species. In the original description Deprat indicated that the wall and septula both have an alveolar structure. Colani also discussed the species at length and stated (1924, p. 124) that the lamellae of the wall are long and thin, resembling somewhat those of *Neoschwagerina craticulifera*. She concluded that the species belongs in *Neoschwagerina*, not *Sumatrina*. Her illustrations (1924, Pl. 25, figs. 5, 6, 12) seem to confirm her observations. Lee, on the contrary, states that both wall and septula are compact as in *Sumatrina* and proposes the separation of the genus from *Yabeina* on this basis.

Until Lee's observations are confirmed or supported by adequate illustrations, *Lepidolina* must be considered a probable synonym of *Yabeina* or *Neoschwagerina*.

Genus SUMATRINA Volz, 1904

Key, plate 12, figure 11

Genotype, by original designation, *Sumatrina annae* Volz

Sumatrina Volz, Geol. Pal. Abh. Koken (Jena), vol. 10, pt. 2, 1904, pp. 24, 98, 177.—DEPRAT, Mém. Serv. Géol. Indochine, vol. 1, fasc. 3, 1912, p. 56; vol. 3, fasc. 1, 1914, p. 34.—OZAWA, Jour. Coll. Sci. Imp. Univ. Tokyo, vol. 45, art. 4, 1925, pp. 19, 26.—GALLOWAY, Manual of Foraminifera, 1933, p. 411.—DUNBAR, in CUSHMAN, Foraminifera, etc., 1933, p. 140.—LEE, Mem. Nat. Res. Inst. Geol. (Shanghai), no. 4, 1933, p. 21.—DUNBAR and SKINNER, Univ. of Texas Bull. 3701, 1937, p. 577.

Elongate fusiform, consisting of several rather closely coiled volutions; wall thin and compact, consisting of a single layer; septa plane; both axial and transverse septula abundant but short, and having the form of compact solid lamellae, thin near the outer wall but much thickened toward their free margin. Three to six axial septula intervene between each pair of true septa.—Upper Permian of the Orient.

Sumatrina appears to be a descendant of some *Neoschwagerina* in which the alveolar texture of the wall and its pendants has been lost. It has been considered to be a further specialization in the direction indicated by *Yabeina.* Indeed, Deprat considered *Yabeina* to be a direct connecting link between *Neoschwagerina* and *Sumatrina.*

It must be noted, however, that the known species of *Sumatrina* are relatively more slender than any of the species of *Neoschwagerina* or *Yabeina.* Furthermore, as Lee (1933, p. 21) has pointed out, the pendant septula are subequal in length in *Sumatrina* and in the other genera appear as two or more series of unequal length in each meridional chamber, the first-formed series being longer than the next.

Family 14. LOFTUSIIDAE

Test large, fusiform, close coiled about an elongate axis, chambers labyrinthic; wall arenaceous; aperture along the base of the apertural face of the chamber.

Genus LOFTUSIA H. B. Brady, 1869

Key, plate 13, figure 1

Genotype, *Loftusia persica* H. B. Brady

Loftusia H. B. Brady, in Carpenter and H. B. Brady, Philos. Trans., 1869, p. 721.

Test fusiform, coiled about the elongate axis, early stages much more nearly rounded, and the adult greatly increasing in length, interior labyrinthic; wall arenaceous; apertures near the base along the apertural face of the chamber. — Upper Cretaceous.

See Cox, The Genus *Loftusia* in southwestern Iran (Eclog. Geol. Helvet., vol. 30, 1937, pp. 431–450, pls. 35–37, 4 text figs.).

This is one of the largest of the foraminifera. Beautifully sectioned specimens studied in the British Museum show that the early stages are much more compressed along the line of the axis, and indicate that this family is the outgrowth of such an arenaceous form as *Cyclammina* by the elongation of the axis similar to the development of the Fusulinidae from the arenaceous *Endothyra,* and fusiform members of the Alveolinellidae from the globular *Borelis.* These three groups show an example of the striking parallelisms that occur in the foraminifera.

Family 15. NEUSINIDAE

Test apparently attached, of many varied external forms, interior of the whole body of the test labyrinthic; wall arenaceous with much chitin, flexible; apertures numerous, small.

The various forms grouped in this family are very different in general form, but they have a number of peculiar features in common. The wall is labyrinthic with large vacuoles composed of arenaceous material with a large amount of chitin, so that the test when wet is flexible. The apertures usually are indistinct in the tubular forms, but *Neusina* has a series of openings about the periphery.

All the members of this family are known only from the present oceans. Other genera, such as *Polyphragma*, *Bdelloidina*, and *Diffusilina*, have nothing in common with the members of this family.

Genus NEUSINA Goës, 1892

Key, plate 13, figures 6–8

Genotype, *Neusina agassizi* Goës

Neusina Goës, Bull. Mus. Comp. Zoöl., vol. 23, 1892, p. 195.

Test in the early stages irregularly planispiral, later uncoiled and very much compressed and of varying shape; wall arenaceous, with much chitin, flexible, labyrinthic, with root-like processes of chitin from portions of the periphery; apertures numerous, along the periphery of the chamber. — Recent, Pacific.

Genus JULLIENELLA Schlumberger, 1890

Key, plate 13, figures 2–5

Genotype, *Jullienella foetida* Schlumberger

Jullienella Schlumberger, Mém. Soc. Zool. France, vol. 3, 1890, p. 213.

Test large, early portion obscure, later spreading in a flattened irregular manner with the periphery produced into simple or branched processes, interior somewhat labyrinthic; wall arenaceous, with chitin, flexible; apertures apparently at the ends of the peripheral tubes. — Recent, coast of Africa (Liberia).

The types in Paris certainly seem to show relationships to those of *Neusina*.

Genus BOTELLINA W. B. Carpenter, 1869

Key, plate 13, figures 13, 14

Genotype, *Botellina labyrinthica* H. B. Brady

Botellina W. B. Carpenter, Proc. Roy. Soc. London, vol. 18, 1869, p. 444.

Test generally cylindrical, sometimes branched, one end rounded and swollen, whole test labyrinthic, with a common tubular chamber through the length of the test; wall arenaceous, with much chitin, flexible; apertures at the ends of the tubular portions, through the interstices of the closing material. — Recent.

Genus PROTOBOTELLINA Heron-Allen and Earland, 1929

Key, plate 13, figures 10–12

Genotype, *Protobotellina cylindrica* Heron-Allen and Earland

Protobotellina Heron-Allen and Earland, Journ. Roy. Micr. Soc., 1929, p. 326.

Test irregularly cylindrical, undivided, non-labyrinthic, open at one end, closed at the other; wall of fine sand grains and broken sponge spicules, firmly agglutinated, but with little visible cement.—Recent, South Atlantic.

This form has many characteristics in common with *Bathysiphon*, but is closed at one end. It also somewhat resembles *Botellina*, but the wall does not seem to be labyrinthic.

Genus SCHIZAMMINA Heron-Allen and Earland, 1929

Key, plate 13, figure 9

Genotype, *Schizammina labyrinthica* Heron-Allen and Earland

Schizammina Heron-Allen and Earland, Journ. Roy. Micr. Soc., 1929, p. 103.

Test free, tubular, branching in one plane, ends typically closed by a cap of sand grains, interior with a common tubular cavity; wall labyrinthic, arenaceous, with much chitin, flexible; apertures at the ends of the branches between the sand grains.—Recent, South Atlantic.

Family 16. SILICINIDAE

Test planispiral, at least in the early stages, consisting of a proloculum and long, closely coiled, tubular, second chamber, either undivided or partially divided into chambers, in some forms with chambers a half coil in length, the sides of the test often with secondary thickening; wall arenaceous, usually siliceous, sometimes partially calcareous; aperture terminal, rounded or contracted, sometimes with a portion thickened or infolded to resemble a tooth.

This family apparently arose from *Ammodiscus* or its relatives by the test developing a siliceous cement, and the division of the tubular chamber into chambers, at first irregular in the Jurassic forms, but later in *Rzehakina* with the chambers each a half coil in length but still planispiral. The plane of coiling changes and forms tests, sigmoid in end view in *Silicosigmoilina*, or irregularly quinqueloculine or triloculine in *Miliammina.*

The suggestion has been made by Heron-Allen and Earland that the three genera with chambers a half coil in length be made a subfamily. These differ in this respect from the earlier Jurassic genera, and form a definite sequence.

The development of the tooth-like structure is interesting as it seems to be a thickening or infolding of the rim of the aperture of the last chamber instead of an outgrowth attached to the previous whorl. In *Miliammina* the apertural end often forms a short neck, and this infolding is then very apparent, and the tubular portion below is circular.

The Miliolidae have so-called arenaceous species, but in the more recent species the arenaceous portion represents the primitive character and is retained on the calcareous inner wall, and in the higher groups this primitive arenaceous character is lost.

Annulina Terquem, 1862, is sometimes placed with the *Silicinidae* but it probably does not belong to the foraminifera.

Subfamily 1. Involutininae

Test planispiral, chambers not definitely a half coil in length.

Genus SILICINA Bornemann, 1874

Key, plate 13, figure 15

Genotype, *Involutina polymorpha* Terquem

Silicina BORNEMANN, Zeitschr. deutsch. geol. Ges., vol. 26, 1874, p. 731.

Test planispiral, evolute, the early portion undivided, last coil divided into chamberlets; wall arenaceous, usually siliceous and imperforate; aperture rounded, terminal.—Jurassic (Lias).

Genus INVOLUTINA Terquem, 1862

Key, plate 13, figures 16, 17

Genotype, *Involutina silicea* Terquem

Involutina TERQUEM, Mém. Acad. Imp. Metz, vol. 42, 1862, p. 450.

Test typically planispiral, the tubular chamber partially divided by incomplete walls; wall arenaceous, mostly siliceous; aperture circular, terminal.—Jurassic.

Genus PROBLEMATINA Bornemann, 1874

Key, plate 13, figure 18

Genotype, *Involutina deslongchampsi* Terquem

Problematina BORNEMANN, Zeitschr. deutsch. geol. Ges., vol. 26, 1874, p. 733.

Involutina (part) TERQUEM, 1863.

Test planispiral, the tubular chamber partially divided, central portion of test secondarily thickened; wall arenaceous, with much cement; aperture circular, at the end of the tubular chamber.—Jurassic (Lias).

Subfamily 2. Rzehakininae

Test planispiral in the young or throughout, in the adult often changing to a sigmoid or milioline plan.

Genus RZEHAKINA Cushman, 1927

Key, plate 13, figures 19, 20

Genotype, *Silicina epigona* Rzehak

Rzehakina CUSHMAN, Contr. Cushman Lab. Foram. Res., vol. 3, 1927, p. 31.

Silicina RZEHAK, 1895 (not BORNEMANN).

Test planispiral, compressed, especially the central portion of each side; chambers each forming a half coil; wall finely arenaceous, siliceous, imperforate; aperture narrow, constricted, terminal. — Upper Cretaceous to Oligocene.

Genus SILICOSIGMOILINA Cushman and Church, 1929

Key, plate 13, figures 21, 22

Genotype, *Silicosigmoilina californica* Cushman and Church

Silicosigmoilina Cushman and Church, Proc. Calif. Acad. Sci., ser. 4, vol. 18, 1929, p. 502.

Test in the early stages nearly planispiral, chambers a half coil in length, later added in varying planes and becoming sigmoid in end view; wall finely arenaceous, with siliceous cement; aperture rounded, terminal, without an apertural tooth. — Upper Cretaceous.

Genus MILIAMMINA Heron-Allen and Earland, 1930

Key, plate 13, figures 23–26

Genotype, *Miliolina oblonga* (Montagu), var. *arenacea* Chapman

Miliammina Heron-Allen and Earland, Journ. Roy. Micr. Soc., 1930, p. 41.

Miliolina (part) of authors.

Test in the earliest stages planispiral, chambers half a coil in length, later added in varying planes and becoming irregularly triloculine or quinqueloculine in end view; wall finely arenaceous, with siliceous cement; aperture rounded, terminal, with the inner edge often infolded to resemble a tooth. — Cretaceous, Recent.

The study of fine sets of the several species kindly furnished by the authors shows that in section the early stages are apparently planispiral, and that the tooth is unlike that of the Miliolidae. It is a part of the aperture due to a thickening or infolding and not based on the previous whorl. A similar contraction of the aperture often occurs in *Silicosigmoilina* and *Rzehakina.*

Genus SPIROLOCAMMINA Earland, 1934

Key, plate 46, figures 25, 26

Genotype, *Spirolocammina tenuis* Earland

Spirolocammina Earland, *Discovery* Reports, vol. 10, 1934, p. 109.

Test free, compressed, planispiral or slightly sigmoid; chambers similar to *Spiroloculina*, one-half coil in length; wall thin, chitinous with very minute mineral particles; aperture terminal, with a neck but no tooth. — Recent.

Family 17. MILIOLIDAE

Test typically coiled about an elongate axis in various planes, at least in the microspheric young of even the specialized genera; chambers typically a half coil in length, simple in most genera, in a few with complex interiors, in the adult of many forms variously arranged, several in a coil or even rectilinear; wall with a chitinous base on which is a calcareous imperforate wall with an outer arenaceous layer persisting in the more primitive genera, under brackish conditions developing a chitinous test; aperture terminal, simple or cribrate or even complex radiate, typically with a tooth.

This family is one of the finest in the whole range of the foraminifera to show the development through the geologic series from the primitive, undivided *Agathammina* to the very complex *Lacazina* of the Upper Cretaceous.

The most primitive member is *Agathammina* which developed in the Paleozoic, probably from a *Glomospira*-like ancestor. The coiling is about an elongate axis, a character which has been held to the present by all the simpler genera. The step from *Agathammina* coiling in various planes to the arenaceous forms of *Quinqueloculina* is an almost imperceptible one. This method of coiling began in *Agathammina*, and, established in *Quinqueloculina*, is the primitive stage in all the Miliolidae in the microspheric form at least, regardless of the complexity and diversity of form in the adult of such different appearing genera as *Periloculina*, *Hauerina*, *Articulina* and others. It is one of the best examples anywhere of the retention of nepionic characters through long geologic time and in great diversity of later form.

From *Quinqueloculina* various modifications take place. *Miliolinella* has a large flat tooth nearly filling the aperture. *Miliola* has a cribrate aperture. The chambers may swing to one plane, but keep the two chambers to a coil and simple chambers. This gives us *Massilina*, which passes without sharp division into *Spiroloculina*, where the microspheric form keeps the quinqueloculine stage even though the megalospheric form loses it by acceleration. A similar development with a cribrate aperture takes place in *Heterillina*, which passes to *Hauerina* where the chambers become progressively shorter. *Nummoloculina* has a somewhat similar form, but with a flat tooth in the aperture. *Riveroina* has the chambers subdivided. In *Sigmoilina* the quinqueloculine stage gives way to a series with more than 180° and a sigmoid appearance in end view. In *Articulina*, *Nubeculina* and *Tubinella*, genera abundant in shallow water tropical conditions, especially of the Indo-Pacific, the chambers after the quinqueloculine stage become uniserial. In *Ptychomiliola*, developed in the Indo-Pacific, the chambers tend to uncoil loosely. *Schlumbergerina* has the chambers finally more than five in a series, developing as many as eight in a series in the adult. From this, by developing in one plane, is *Ammomassilina*, also a tropical genus.

From *Quinqueloculina*, by reduction of the number of chambers in a whorl to three, *Triloculina* is developed. *Triloculina* in turn gave rise to *Flintina* with its highly specialized apertural characters, and to *Austrotrillina* with a labyrinthic interior and cribrate aperture.

Pyrgo develops from *Triloculina* by reduction of the number of chambers in a whorl to two. In *Biloculinella* a large flat tooth is developed, nearly filling the aperture and in *Pyrgoella* a series of small apertures develops. *Fabularia* developed a complex interior and cribrate aperture in the Eocene. By the development of chambers spreading out into an evolute test *Pyrgo* gave rise to *Flintia*.

The series from *Pyrgo* toward a single chamber forming the whole exterior is seen in *Nevillina*, and the more complex Upper Cretaceous genera *Idalina*, *Periloculina*, and *Lacazina*.

It will be seen that the family reached its highest development in the large, complex, Upper Cretaceous genera just mentioned. In the Middle and Upper Eocene there was a great development of forms with cribrate apertures and labyrinthic interiors. Such forms with others from the Milioline limestones of the Paris Basin and in the Upper Eocene have a wide distribution. *Nevillina* still persists today but has been found in a very limited area in the Malay and southern Philippine area. This region is peculiar in having still persisting numerous genera such as *Trigonia*, *Nautilus*, etc., which are characteristic of much older geological periods.

The test in the Miliolidae has a chitinous base that can be easily demonstrated by dissolving the test in weak acid. In *Agathammina* the test is similar to that of *Glomospira* from which it is only separated by the more definite elongate axis. The same yellowish-brown cementing material, seen in the Ammodiscidae and in many other typical arenaceous genera, persists in such species as *Quinqueloculina fusca* and many others. In tropical waters it is usually replaced by entirely calcareous cement. The primitive arenaceous character is kept by many of the simpler genera as *Quinqueloculina*, *Massilina*, *Sigmoilina*, and others, but this character is lost in the more specialized genera such as *Articulina*, *Tubinella*, *Fabularia*, *Lacazina* and others. Its retention in the more primitive genera is another example of the persistence of ancestral characteristics in the less specialized genera of a family.

The development of tooth-like processes in the aperture and the cribrate plate sometimes called a "trematophore" are characteristics of the group not to be found in the Ophthalmidiidae, which also do not have the characteristic milioline basis of development, but a planispiral young which came from *Cornuspira*. While there are a number of parallelisms in the two families, they are not so close as exist in numerous other families of the foraminifera.

The Miliolidae have kept to their free habit, while many of the Ophthalmidiidae adopted an attached method of life in the Paleozoic, and it has been kept to the present in a number of genera.

The Miliolidae have reached their greatest development in shallow warm waters of the Upper Cretaceous and Eocene, and are now particularly characteristic of coral reef conditions in the tropics. Certain genera, particularly *Sigmoilina* and *Pyrgo*, have become adapted to deep cold waters, and a few other genera are found in the polar regions.

The student who wishes to understand the various problems connected with the relationships of the species and genera of the Miliolidae must be prepared to study them in section, as only in that way can these be seen in complete detail.

Genus AGATHAMMINA Neumayr, 1887

Plate 14, figure 1; Key, plate 14, figures 1, 2

Genotype, *Serpula pusilla* Geinitz

Agathammina NEUMAYR, Sitz. Akad. Wiss. Wien, vol. 95, pt. 1, 1887, p. 171.
Serpula (part) GEINITZ, 1846 (not LINNÉ). *Trochammina* (part) of authors.

Test tubular, undivided, winding about an elongate axis; wall imperforate, calcareous, with arenaceous material at the surface; aperture formed by the open end of the tubular chamber. — Carboniferous to Jurassic.

Genus QUINQUELOCULINA d'Orbigny, 1826

Plate 14, figure 2; Key, plate 14, figures 3–5

Genotype, *Serpula seminulum* Linné

Quinqueloculina D'ORBIGNY, Ann. Sci. Nat., vol. 7, 1826, p. 301.
Serpula (part) LINNÉ. *Adelosina* D'ORBIGNY, 1826. *Uniloculina* D'ORBIGNY, 1846. *Miliolina* (part) WILLIAMSON, 1858, and later authors.

Test with the coiling in five planes, the chambers a half coil in length, and added successively in planes 144° apart, five chambers completing a cycle of two turns about the axis in section, but two and a half coils lengthwise, each chamber 72° from its next adjacent one, but 144° from its immediate predecessor; wall with an interior chitinous layer outside of which is a calcareous imperforate layer, in some species with an outer layer of sand grains; aperture at the end of the chamber, rounded, typically with a simple tooth. — Jurassic to Recent.

Adelosina is the megalospheric young of *Quinqueloculina*, and *Retorta* may represent an inadequately drawn figure of a similar stage. All stages in the development of various species in the Miocene of the Vienna Basin may be seen linking the "*Adelosina*" young with their adults. *Uniloculina* also represents the extreme megalospheric young of large Indo-Pacific species, all stages of which to the adult can be found in the same sample.

Genus MILIOLINELLA Wiesner, 1931

Key, plate 15, figure 6

Genotype, *Quinqueloculina lamellidens* Reuss

Miliolinella WIESNER, Deutsche Südpolar-Exped., vol. XX, Zool., 1931, p. 65.
Quinqueloculina (part) of authors.

Test similar to *Quinqueloculina*, but the aperture with a large flat tooth nearly filling the aperture. — Tertiary and Recent.

Genus DENTOSTOMINA Carman, 1933

Key, plate 14, figure 6

Genotype, *Dentostomina bermudiana* Carman

Dentostomina CARMAN, Contr. Cushman Lab. Foram. Res., vol. 9, 1933, p. 31.
Quinqueloculina (part) of authors.

Test in general quinqueloculine; wall with a chitinous lining, outer wall calcareous and imperforate, with an arenaceous covering; aperture loop-shaped to circular, with a long bifid tooth, inner surface of aperture with small, regularly spaced, tooth-like projections. — Recent.

Genus MASSILINA Schlumberger, 1893

Plate 14, figure 9; Key, plate 14, figures 7–9

Genotype, *Quinqueloculina secans* d'Orbigny

Massilina SCHLUMBERGER, Mém. Soc. Zool. France, vol. 6, 1893, p. 218.
Quinqueloculina (part) D'ORBIGNY, 1826, and later authors. *Miliolina* (part) of authors.

Test with the early chambers quinqueloculine, later ones added on opposite sides in a single plane, quinqueloculine stage present in both megalospheric and microspheric forms; wall often with an arenaceous layer; aperture simple, typically with a bifid tooth. — Lower Cretaceous to Recent (probably older).

Genus PROEMASSILINA Lacroix, 1938

Key, plate 46, figure 27

Genotype, *Massilina rugosa* Sidebottom

Proemassilina LACROIX, Bull. Instit. Océanographique, No. 754, 1938, p. 3 (754).

Structure identical with that of *Massilina*, but differs in having the under porcellanous layer covered with a layer of arenaceous material; aperture variable but with a bifid tooth. — Recent.

Genus PSEUDOMASSILINA Lacroix, 1938

Key, plate 46, figure 28

Genotype, *Massilina australis* Cushman

Pseudomassilina LACROIX, Bull. Instit. Océanographique, No. 754, 1938, p. 3 (754).

Test more or less compressed, early quinqueloculine chambers in an obscure mass, last chambers spiroloculine; wall with very fine canals, ending at the outer surface in minute openings, each at the bottom of a shallow pit; aperture elongate, without a tooth. — Recent.

Genus SPIROLOCULINA d'Orbigny, 1826

Plate 14, figure 10; Key, plate 14, figures 10, 11

Genotype, *Spiroloculina depressa* d'Orbigny

Spiroloculina D'ORBIGNY, Ann. Sci. Nat., vol. 7, 1826, p. 298.

Test with the early chambers in the microspheric form quinqueloculine, later ones in a single plane, in the megalospheric form the quinqueloculine stage reduced or wanting; chambers a half coil in length; wall occasionally with an outer arenaceous layer; aperture usually with a neck and lip. simple, with a simple or bifid tooth. — Upper Cretaceous to Recent.

The difference between *Spiroloculina* and *Massilina* is one of degree only. All species in both genera in the microspheric form have a quinqueloculine stage, often greatly reduced in *Spiroloculina*, but in the megalospheric form *Spiroloculina* may entirely skip this stage by acceleration. *Spiroloculina* has developed directly from *Massilina* by acceleration. *Spirophthalmidium*, sometimes confused with *Spiroloculina*, has a very different early development.

See Cushman and Todd, The Genus *Spiroloculina* and Its Species (Special Publl. 11, Cushman Lab. Foram. Res., 1944, pp. 1–82, pls. 1–9).

Genus RIVEROINA Bermudez, 1939

Key, plate 46, figure 31

Genotype, *Riveroina caribaea* Bermudez

Riveroina BERMUDEZ, Mem. Soc. Cubana Hist. Nat., vol. 13, 1939, p. 248.

Test similar in development to *Spiroloculina* but the interior divided into chamberlets by oblique partitions; wall calcareous, imperforate; aperture elongate transversely to the axis of the test. — Recent.

Genus SIGMOILINA Schlumberger, 1887

Plate 14, figure 8; Key, plate 14, figures 14, 15

Genotype, *Planispirina sigmoidea* H. B. Brady

Sigmoilina SCHLUMBERGER, Bull. Soc. Zool. France, vol. 12, 1887, p. 118.

Spiroloculina (part) COSTA, 1857 (not D'ORBIGNY). *Planispirina* (part) H. B. BRADY, 1884 (not SEGUENZA).

Test with the early chambers quinqueloculine, later ones added in planes slightly more than 180° from one another, making a continuously revolving spiral, in transverse section giving a sigmoid appearance; wall often with an outer arenaceous layer; aperture rounded, with a simple tooth. — Tertiary, Recent.

See Cushman, The Genus *Sigmoilina* and Its Species (Contr. Cushman Lab. Foram. Res., vol. 22, 1946, pp. 29–45, pls. 5, 6).

Genus NUMMOLOCULINA Steinmann, 1881

Plate 14, figure 4; Key, plate 14, figures 12, 13

Genotype, *Biloculina contraria* d'Orbigny

Nummoloculina STEINMANN, Neues Jahrb. für Min., 1881, p. 31.

Biloculina (part) D'ORBIGNY, 1846. *Planispirina* (part) H. B. BRADY, 1884 (not SEGUENZA).

Test with the earliest chambers quinqueloculine, later ones in a single plane, several to a coil; wall calcareous; aperture rounded, with a flat semicircular tooth. — Jurassic to Recent.

This is not to be confused with *Planispirina* which does not have a tooth and which has a planispiral early stage.

Genus ARTICULINA d'Orbigny, 1826

Plate 14, figure 6; Key, plate 14, figure 18
Genotype, *Articulina nitida* d'Orbigny

Articulina D'ORBIGNY, Ann. Sci. Nat., vol. 7, 1826, p. 300.
Nautilus BATSCH, 1791 (not LINNÉ). *Ceratospirulina* EHRENBERG, 1858. *Vertebralina* (part) of authors (not D'ORBIGNY).

Test with the early chambers quinqueloculine, later ones in a rectilinear series; wall calcareous; aperture in the adult rounded, often elliptical, with a short neck and phialine lip. — Middle Eocene to Recent.

See Cushman, The Genus *Articulina* and Its Species (Special Publ. 10, Cushman Lab. Foram. Res., 1944, pp. 1–21, pls. 1–4).

Genus POROARTICULINA Cushman, 1944

Key, plate 46, figures 32, 33
Genotype, *Poroarticulina glabra* Cushman

Poroarticulina CUSHMAN, Contr. Cushman Lab. Foram. Res., vol. 20, 1944, p. 52.

Test with the early chambers triloculine, later ones in a uniserial, rectilinear series; aperture in the adult terminal, cribrate. — Miocene.

PLATE 14

RELATIONSHIPS OF THE GENERA OF THE MILIOLIDAE

FIG.

1. *Agathammina pusilla* (Geinitz). (After H. B. Brady.)
2. *Quinqueloculina vulgaris* d'Orbigny. *a*, *b*, opposite sides; *c*, apertural view; *d*, section (After Schlumberger).
3. *Miliola saxorum* Lamarck. (After Terquem.) *a*, side view; *b*, apertural view.
4. *Nummoloculina contraria* (d'Orbigny). *a*, side view; *b*, apertural view (After H. B. Brady); *c*, section (After Steinmann).
5. *Hauerina bradyi* Cushman. (After H. B. Brady.) *a*, side view; *b*, apertural view.
6. *Articulina sagra* d'Orbigny. (After H. B. Brady.)
7. *Tubinella funalis* (H. B. Brady). (After H. B. Brady.)
8. *Sigmoilina herzensteini* Schlumberger. (Section after Schlumberger.)
9. *Massilina secans* (d'Orbigny). (After Schlumberger.) *a*, side view; *b*, section.
10. *Spiroloculina depressa* d'Orbigny. (After Martinotti.) *a*, side view; *b*, apertural view; *c*, section.
11. *Triloculina laevigata* d'Orbigny. (After Schlumberger.) *a*, *c*, opposite sides; *b*, end view; *d*, section.
12. *Austrotrillina howchini* (Schlumberger). (Section after Schlumberger.)
13. *Flintina triquetra* (H. B. Brady). *a*, side view; *b*, apertural view.
14. *Pyrgo bradyi* (Schlumberger). *a*, front view; *b*, apertural view; *c*, section.
15. *Fabularia discolithes* Defrance. (Section after Schlumberger.)
16. *Nevillina coronata* (Millett). (After Sidebottom.) *a*, front view; *b*, end view.
17. *Idalina antiqua* (d'Orbigny). (After Munier-Chalmas and Schlumberger.)
18. *Periloculina zitteli* Munier-Chalmas and Schlumberger. (After Munier-Chalmas and Schlumberger.)
19. *Lacazina compressa* (d'Orbigny). (After Munier-Chalmas and Schlumberger.) *a*, apertural view; *b*, side view.

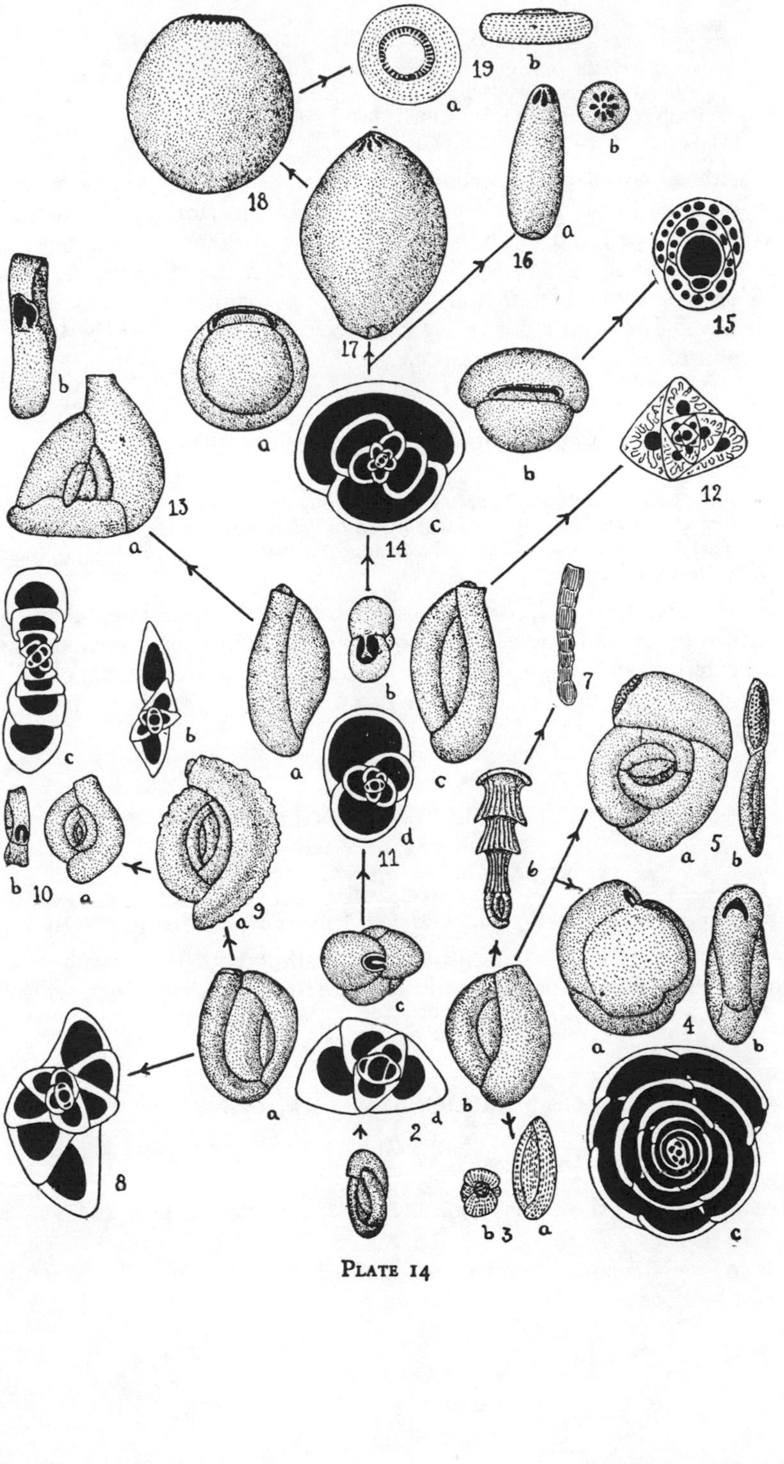
19
a
b
18
a
b
16
15
17
a
b
b
13
a
c
14
12
b
a
c
c
b
7
b
a
c
d
11
6
a
5
b
b
10
a
a
9
c
4
a
b
a
2
d
b
8
b
3
a
c

PLATE 14

Genus TUBINELLA Rhumbler, 1906

Plate 14, figure 7; Key, plate 14, figures 16, 17
Genotype, *Articulina inornata* H. B. Brady
Tubinella RHUMBLER, Zool. Jahrb., Abteil. Syst., vol. 24, 1906, p. 25.
Articulina (part) H. B. BRADY, 1884 (not D'ORBIGNY).

Test with an ovoid early portion, remainder of the test nearly straight, cylindrical, partially divided; wall calcareous, imperforate, bluish-white; aperture, the open end of the tube. — Tertiary, Recent.

In 1931 Wiesner proposed the new generic name *Tubinellina* limiting *Tubinella* Rhumbler to the peculiar species called *perforata*. As the genotype of *Tubinella* has already been fixed, the name must be kept for the group to which it is now applied.

Genus NUBECULINA Cushman, 1924

Key, plate 14, figures 19, 20
Genotype, *Nubecularia divaricata* H. B. Brady
Nubeculina CUSHMAN, Publ. 342, Carnegie Inst. Washington, 1924, p. 52.
Sagrina (part) H. B. BRADY, 1879 (not D'ORBIGNY). *Nubecularia* (part) H. B. BRADY, 1884 (not DEFRANCE).

Test elongate, uniserial, early portion irregularly milioline, chambers distinct, simple; wall imperforate, porcellanous, with sand grains attached to the exterior; aperture at the end of an elongated tubular neck, with an everted phialine lip, the apertural opening with a series of inwardly pointing teeth. — Recent.

Genus PTYCHOMILIOLA Eimer and Fickert, 1899

Key, plate 14, figures 21–23
Genotype, *Miliolina separans* H. B. Brady
Ptychomiliola EIMER and FICKERT, Zeitschr. Wiss. Zool., vol. 65, 1899, p. 687.
Miliolina (part) H. B. BRADY, 1881. *Adelosina* SCHLUMBERGER, 1890 (not D'ORBIGNY).

Test in the early stages triloculine, in the adult with the chambers uncoiling and tending to become uniserial; aperture with a distinct, typically bifid tooth. — Recent, Indo-Pacific.

Genus MILIOLA Lamarck, 1804

Plate 14, figure 3; Key, plate 14, figure 24
Genotype, *Miliola saxorum* Lamarck
Miliola LAMARCK, Ann. Mus., vol. 5, 1804, p. 349.
Quinqueloculina (part) D'ORBIGNY, 1826. *Pentellina* MUNIER-CHALMAS and SCHLUMBERGER, 1905.

Test in its structure similar to *Quinqueloculina* but the aperture cribrate. — Eocene.

Genus HETERILLINA Munier-Chalmas and Schlumberger, 1905

Key, plate 14, figure 25
Genotype, *Heterillina guespellensis* Schlumberger

Heterillina Munier-Chalmas and Schlumberger, Bull. Soc. Géol. France, ser. 4, vol. 5, 1905, p. 131.

Test in the young similar to *Miliola*, the adult with chambers in a single plane, but two making a complete coil; aperture cribrate. — Upper Eocene, Oligocene.

This genus has evidently developed from *Miliola*, as *Massilina* has come from *Quinqueloculina*.

Genus HAUERINA d'Orbigny, 1839

Plate 14, figure 5; Key, plate 14, figure 26
Genotype, *Hauerina compressa* d'Orbigny

Hauerina d'Orbigny, in De la Sagra, Hist. Phys. Pol. Nat. Cuba, "Foraminifères," 1839, p. xxxviii.

Test with the early chambers quinqueloculine, later ones more or less in one plane, making a half coil, later gradually shortening so that more than two make up a coil; aperture cribrate. — Eocene to Recent.

In old age specimens the chambers may greatly lengthen, as divisions are not completed.

See Cushman, The Genus *Hauerina* and Its Species (Contr. Cushman Lab. Foram. Res., vol. 22, 1946, pp. 2–15, pls. 1, 2).

Genus RAADSHOOVENIA van den Bold, 1946

Key, plate 46, figures 34, 35
Genotype, *Raadshoovenia guatemalensis* van den Bold

Raadshoovenia van den Bold, Thesis Univ. Utrecht, 1946, p. 123.

"Test calcareous, imperforate. Early chambers quinqueloculine, later ones close-coiled, 4–6 to a whorl, adult uncoiling. Labyrinthic chambers developed in later stages. Aperture cribrate, terminal." — Eocene.

Genus SPIROSIGMOILINA Parr, 1942

Key, plate 46, figures 29, 30
Genotype, *Spiroloculina tateana* Howchin

Spirosigmoilina Parr, Mining and Geol. Journ., vol. 2, 1942, p. 361.
Spiroloculina (part), *Sigmoilina* (part), and *Hauerina* (part) of authors.

Test with the early chambers arranged as in *Sigmoilina*, later chambers spiroloculine; wall calcareous, imperforate; aperture terminal at the end of the last-formed chamber. — Tertiary.

Genus SCHLUMBERGERINA Munier-Chalmas, 1882

Key, plate 14, figure 27

Genotype, *Schlumbergerina areniphora* Munier-Chalmas

Schlumbergerina Munier-Chalmas, Bull. Soc. Géol. France, ser. 3, vol. 10, 1882, p. 424.

Miliolina (part) of authors.

Test with the chambers in the early stages quinqueloculine, later having as many as eight chambers making up a whorl; wall with the exterior thickly coated with sand grains; aperture cribrate. — Late Tertiary, Recent.

Hofker has recently shown that there may be as many as eight chambers in the whorl. It is apparently a fairly recent development in shallow tropical waters.

Genus AMMOMASSILINA Cushman, 1933

Key, plate 14, figure 28

Genotype, *Massilina alveoliniformis* Millett

Ammomassilina Cushman, Contr. Cushman Lab. Foram. Res., vol. 9, 1933, p. 32.

Massilina (part) Millett, 1898.

Test in the early stages similar to *Schlumbergerina*, but in the adult developing the chambers in a single plane; wall with an arenaceous outer layer; aperture cribrate. — Recent.

Genus TRILOCULINA d'Orbigny, 1826

Plate 14, figure 11; Key, plate 15, figures 1–3

Genotype, *Miliolites trigonula* Lamarck

Triloculina d'Orbigny, Ann. Sci. Nat., vol. 7, 1826, p. 299.

Trillina Munier-Chalmas, 1882.

Miliolites (part) Lamarck, 1804. *Miliolina* (part) Williamson, 1858, and later authors.

Test with the early chambers quinqueloculine, at least in the microspheric form, later ones added in planes 120° from one another, each added in the plane of the third preceding and covering it so that the exterior of the test is composed of but three chambers, chambers not labyrinthic; wall calcareous, porcellanous, only occasionally with an arenaceous outer layer; aperture typically with a bifid tooth. — Triassic to Recent, earlier records somewhat doubtful.

The large forms with the aperture in the form of an "X," called by d'Orbigny *Cruciloculina*, need further study.

Genus CRIBROLINOIDES Cushman and LeRoy, 1939

Key, plate 47, figure 1

Genotype, *Quinqueloculina curta* Cushman

Cribrolinoides Cushman and LeRoy, Contr. Cushman Lab. Foram. Res., vol. 15, 1939, p. 15.

Test with a globular proloculum followed by a second chamber a coil in length and planispiral, next by a series in typical quinqueloculine arrangement, and in the adult sometimes becoming more or less compressed toward a single plane; wall calcareous throughout; aperture in the earliest stages with a simple, linear tooth, later becoming bifid, then one of the tips elongating and joining the opposite side, the other tip curling about and attaching to the first, leaving a rounded opening, after which various teeth may develop from the central area and from the borders of the aperture to form a complex cribrate aperture. — Pliocene to Recent.

Genus AUSTROTRILLINA Parr, 1942

Plate 14, figure 12; Key, plate 15, figure 4
Genotype, *Trillina howchini* Schlumberger

Austrotrillina Parr, Mining and Geol. Journ., vol. 2, 1942, p. 361.
Trillina (part) of authors.

Test with chambers arranged as in *Triloculina*, each chamber completely surrounded by its own wall, chamber wall thick, alveolate in that portion on the outside of the chamber cavity, non-alveolate in the portion overlapping the previous whorl; chamber cavity not divided; wall calcareous, imperforate; aperture doubtfully cribrate. — Miocene.

Genus FLINTINA Cushman, 1921

Plate 14, figure 13; Key, plate 15, figure 5
Genotype, *Flintina bradyana* Cushman

Flintina Cushman, Bull. 100, U. S. Nat. Mus., vol. 4, 1921, p. 465.
Miliolina (part) H. B. Brady, 1884.

Test in the early stages, at least of the microspheric form, quinqueloculine, followed very early by a triloculine stage, and in the adult planispiral, usually three chambers in a complete cycle; aperture large, with thickened border, tooth in the early stage simple, later often becoming very complex. — Tertiary, Recent.

Genus PYRGO Defrance, 1824

Plate 14, figure 14; Key, plate 15, figures 7–10
Genotype, *Pyrgo laevis* Defrance

Pyrgo Defrance, Dict. Sci. Nat., vol. 32, 1824, p. 273.
Miliola (part) Lamarck, 1804. *Biloculina* d'Orbigny, 1826, and later authors.

Test with the chambers, at least in the microspheric form, quinqueloculine, followed by a triloculine series, and in the adult added in planes 180° apart and involute; two chambers making up the exterior, interior simple; aperture typically with a broad bifid tooth. — Jurassic to Recent.

Genus PYRGOELLA Cushman and White, 1936

Key, plate 47, figures 2–4
Genotype, *Biloculina sphaera* d'Orbigny

Pyrgoella Cushman and White, Contr. Cushman Lab. Foram. Res., vol. 12, 1936, p. 90.

Test with the earliest stages in the microspheric form similar to those of *Pyrgo*, first quinqueloculine, then triloculine, and later biloculine; chambers in the adult with the last two making up the whole exterior, the final chamber enclosing all but a small rounded area of the next preceding; wall calcareous, imperforate, thin; aperture in the early stages with a flattened, triangular tooth almost completely filling the opening, in the adult becoming Y-shaped, and in some specimens broken up into a series of 3, 4, or more, elongate, often sinuous openings. — Late Tertiary and Recent.

Genus BILOCULINELLA Wiesner, 1931

Key, plate 15, figure 11
Genotype, *Biloculina labiata* Schlumberger
Biloculinella Wiesner, Deutsche Südpolar-Exped., vol. XX, Zool., 1931, p. 69.
Biloculina (part) of authors.

Test similar to *Pyrgo* but with a broad flat tooth from the base of the exterior of the aperture, nearly filling the aperture. — Tertiary and Recent.

Genus CRIBROPYRGO Cushman and Bermudez, 1946

Key, plate 47, figures 5, 6
Genotype, *Cribropyrgo robusta* Cushman and Bermudez
Cribropyrgo Cushman and Bermudez, Contr. Cushman Lab. Foram. Res., vol. 22, 1946, p. 119.

Test calcareous, imperforate, similar in general structure to *Pyrgo* but differing from that genus in its cribrate aperture and from *Fabularia* in its simple, undivided chambers. — Recent.

Genus FABULARIA Defrance, 1820

Plate 14, figure 15; Key, plate 15, figures 12, 13
Genotype, *Fabularia discolithes* Defrance
Fabularia Defrance, Dict. Sci. Nat., vol. 16, 1820, p. 103.

Test similar to *Pyrgo*, but the chambers labyrinthic and the aperture cribrate. — Eocene to Pliocene.

Genus FLINTIA Schubert, 1911

Key, plate 15, figure 14
Genotype, *Spiroloculina robusta* H. B. Brady
Flintia Schubert, Abhandl. k. k. Geol. Reichs., vol. 20, pt. 4, 1911, p. 124.
Spiroloculina (part) of authors (not d'Orbigny).

Test in the early stages completely involute like *Pyrgo*, only two chambers making up the exterior, in the adult becoming evolute, spiroloculine; aperture with a broad flat tooth with curved ends. — Tertiary, Recent.

In the adult, *Flintia* superficially resembles *Massilina*, but *Massilina* developed directly from *Quinqueloculina*, while *Flintia* has developed from *Pyrgo*.

Genus NEVILLINA Sidebottom, 1905

Plate 14, figure 16; Key, plate 15, figures 15, 16
Genotype, *Nevillina coronata* Sidebottom

Nevillina SIDEBOTTOM, Mem. Proc. Manchester Lit. Philos. Soc., vol. 49, pt. 2, No. 11, 1905, p. 1.

Test similar to *Pyrgo* in the young, but in the adult the last-formed chamber almost completely embracing the earlier ones and forming an elongate egg-shaped test, interior not labyrinthic; aperture radiate, formed by numerous incurved lamellae meeting centrally. — Recent.

Genus IDALINA Munier-Chalmas and Schlumberger, 1884

Plate 14, figure 17; Key, plate 15, figure 17
Genotype, *Biloculina antiqua* d'Orbigny

Idalina MUNIER-CHALMAS and SCHLUMBERGER, Bull. Soc. Géol. France, ser. 3, vol. 12, 1884, p. 629.
Biloculina (part) D'ORBIGNY, 1850.

Test in the microspheric form quinqueloculine, followed by triloculine, then biloculine series, and in the adult with the penultimate chamber showing as a narrow strip at one side near the base, and the final chamber making up the remainder of the surface of the test; aperture radiate, but the chambers not labyrinthic. — Upper Cretaceous.

Genus PERILOCULINA Munier-Chalmas and Schlumberger, 1885

Plate 14, figure 18; Key, plate 15, figures 18, 19
Genotype, *Periloculina zitteli* Munier-Chalmas and Schlumberger

Periloculina MUNIER-CHALMAS and SCHLUMBERGER, Bull. Soc. Géol. France, ser. 3, vol. 13, 1885, p. 308.

Test similar to *Idalina* but the last-formed chamber completely involute, the chambers labyrinthic; aperture irregularly radiate. — Upper Cretaceous.

Genus LACAZINA Munier-Chalmas, 1882

Plate 14, figure 19; Key, plate 15, figure 20
Genotype, *Alveolina compressa* d'Orbigny

Lacazina MUNIER-CHALMAS, Bull. Soc. Géol. France, ser. 3, vol. 10, 1882, p. 472.
Alveolina (part) D'ORBIGNY.

Test in the young similar to *Periloculina* but in the adult compressed into a flattened spheroid, the apertures appearing as a ring of pores near the periphery on the dorsal side. — Upper Cretaceous.

Family 18. OPHTHALMIDIIDAE

Test free or attached, with the early chambers at least planispiral, except in degenerate forms, later chambers adopting many different shapes; wall calcareous, imperforate, without an arenaceous coating; aperture typically open, without a tooth, rarely cribrate.

The radicle from which this family is derived is *Cornuspira*. *Cornuspira* itself is undoubtedly derived from forms such as *Ammodiscus* by the adoption of calcareous cement and the consequent development of an imperforate calcareous test. Such a form is represented in the young of all except the most degenerate forms of this entire family. *Cornuspira* itself and a number of its derivatives occur in the late Pennsylvanian and Permian. Some of them such as *Nubeculinella*, *Ophthalmidium*, and *Spirophthalmidium* were abundant in the Jurassic, while some of the most complex and largest forms such as *Cornuspiroides*, *Cornuspirella*, and *Discospirina* are only known from the present oceans or late Tertiary. The various developments directly from *Cornuspira* are shown in the accompanying plate, and in the other plates showing the genera of this family. A considerable group have added characters to the typical close coiled *Cornuspira*, while in the late Paleozoic, forms developed such as *Orthovertella* and *Calcitornella*, in which the later portion uncoiled and bent back and forth more or less irregularly. This zigzag bending became definite in *Calcivertella*, in which the later portion becomes entirely uncoiled and nearly straight. In *Apterrinella* the uncoiling started much earlier in its development so that in the megalospheric form only a partial coil exists about the proloculum. In *Plummerinella* is seen one of the most peculiar developments in this group, the early stages being close coiled like *Cornuspira*, then the chamber becoming bent in a zigzag manner but coiling about the early chambers in a planispiral manner. In *Nodophthalmidium* and *Vertebralina* there is a distinct uncoiling, and very definite chambers develop, the aperture being without the tooth, and often at one side of the axis. Attached forms of *Cornuspira* gave the sources of *Nubecularia* and the other attached forms derived from it, some of which became degenerate and single chambered. In *Nubeculinella* the later chambers uncoil and in *Sinzowella* there is a piling up of the chambers about the early portion, but the apertures remain distinct, are reniform in shape, and without a tooth. *Calcituba*, *Parrina*, and *Squamulina* form a progressively degenerate series.

From *Ophthalmidium* develops *Discospirina*, which is the largest and most complex member of the group. It is superficially somewhat like the Peneroplidae, but has an entirely different early stage which is exactly like *Ophthalmidium*. Other genera in this same group develop different adult types as shown in the figures, and in *Trisegmentina* and *Polyseg-*

mentina there is developed a definite cribrate aperture, the only genera in this family to develop such a character.

Various members of this family have been at times classed with the Miliolidae. There are definite characters which separate all the members of this family from the Miliolidae. None of the Ophthalmidiidae develop an arenaceous exterior such as is developed in many of the primitive genera of the Miliolidae. None of them develops a definite tooth in the aperture, a feature which is a characteristic of the Miliolidae. The Ophthalmidiidae are based upon a planispiral ancestor, *Cornuspira*, while the Miliolidae have a coiled early stage in which the chambers change the plane of coiling continuously.

Altogether the Ophthalmidiidae show excellently the great range of form that can be developed from a simple ancestry, keeping all the essential characters such as the calcareous imperforate test, the planispiral young, and the open aperture without a tooth, at the same time developing great diversity of adult form.

Subfamily 1. Cornuspirinae

Test made up of a proloculum and elongate second chamber, normally tubular and planispiral in the young, but assuming various forms in the adult, usually free but occasionally attached, not divided into definite chambers.

Genus CORNUSPIRA Schultze, 1854

Plate 15, figure 1; Key, plate 16, figures 1–3
Genotype, *Cornuspira planorbis* Schultze

Cornuspira SCHULTZE, Organismus Polythal., 1854, p. 40.
Orbis (part) PHILIPPI, 1844. *Operculina* (part) CZJZEK, 1844. *Spirillina* (part) WILLIAMSON, 1858.

Test free, with a proloculum and long, planispirally coiled, second chamber, rounded or complanate; wall calcareous, imperforate; aperture formed by the open end of the chamber, sometimes constricted and with a thickened lip.—Carboniferous to Recent.

Genus VIDALINA Schlumberger, 1899

Plate 15, figure 2; Key, plate 16, figures 9, 10
Genotype, *Vidalina hispanica* Schlumberger

Vidalina SCHLUMBERGER, Bull. Soc. Géol. France, ser. 3, vol. 27, 1899, p. 459.

Test similar to *Cornuspira*, but completely involute, the umbonal region thickened.—Cretaceous.

Genus RECTOCORNUSPIRA Warthin, 1930

Plate 15, figure 3; Key, plate 16, figure 4

Genotype, *Rectocornuspira lituiformis* Warthin

Rectocornuspira WARTHIN, Oklahoma Geol. Surv., Bull. 53, 1930, p. 15.

Test similar to *Cornuspira*, but somewhat involute and uncoiling in the adult. — Pennsylvanian.

PLATE 15

RELATIONSHIPS OF THE GENERA OF THE OPHTHALMIDIIDAE AND FISCHERINIDAE

OPHTHALMIDIIDAE

FIG.

1. *Cornuspira involvens* Reuss. *a*, side view; *b*, section.
2. *Vidalina hispanica* Schlumberger. (Adapted from Schlumberger.) Section.
3. *Rectocornuspira lituiformis* Warthin. (After Warthin.)
4. *Cornuspiramia antillarum* (Cushman).
5. *Cornuspirella diffusa* (Heron-Allen and Earland). (After Heron-Allen and Earland.)
6. *Cornuspiroides striolata* (H. B. Brady). (After Heron-Allen and Earland.)
7. *Hemigordius calcareus* Cushman and Waters. (After Cushman and Waters.)
8. *Gordiospira fragilis* Heron-Allen and Earland. (After Heron-Allen and Earland.)
9. *Orthovertella protea* Cushman and Waters. (After Cushman and Waters.)
10. *Plummerinella complexa* Cushman and Waters. (After Cushman and Waters.)
11. *Calcivertella adherens* Cushman and Waters. (After Cushman and Waters.)
12. *Apterrinella grahamensis* (Harlton). (After Cushman and Waters.) *a*, microspheric; *b*, megalospheric.
13. *Calcitornella elongata* Cushman and Waters. (After Cushman and Waters.)
14. *Calcitornella heathi* Cushman and Waters. (After Cushman and Waters.)
15. *Nubecularia lucifuga* Defrance. (After H. B. Brady.)
16. *Nubeculinella bigoti* Cushman.
17. *Sinzowella deformis* (Karrer and Sinzow). (After Karrer and Sinzow.)
18. *Calcituba polymorpha* Roboz. (After Roboz.)
19. *Parrina bradyi* (Millett). (After Millett.)
20. *Squamulina laevis* Schultze. (Adapted from Schultze.)
21. *Wiesnerella auriculata* (Egger). *a*, side view; *b*, apertural view.
22. *Renulina opercularia* Lamarck.
23. *Planispirinella exigua* (H. B. Brady). (Adapted from H. B. Brady.)
24. *Trisegmentina sidebottomi* Cushman. (After Sidebottom.) *a*, side view; *b*, apertural view.
25. *Planispirina communis* Seguenza. (After Seguenza.) *a*, side view; *b*, apertural view.
26. *Ophthalmidium inconstans* H. B. Brady.
27. *Spirophthalmidium acutimargo* (H. B. Brady). (After H. B. Brady.)
28. *Discospirina tenuissima* (W. B. Carpenter). (Adapted from H. B. Brady.)
29. *Vertebralina striata* d'Orbigny. (After Williamson.)
30. *Nodophthalmidium compressum* (Rhumbler). (After Rhumbler.)

FISCHERINIDAE

31. *Fischerina helix* Heron-Allen and Earland. (After Heron-Allen and Earland.) *a*, dorsal view; *b*, ventral view; *c*, side view.

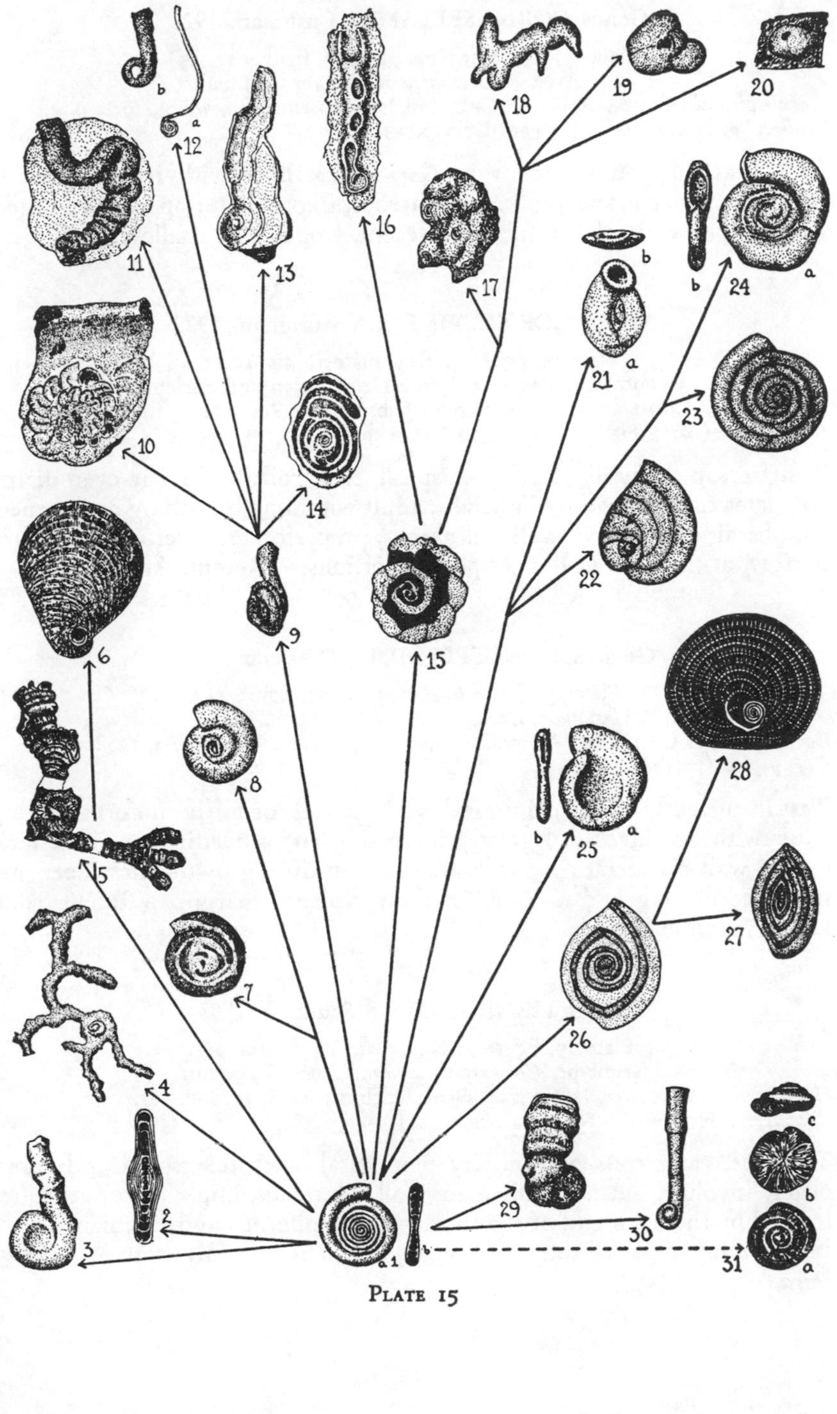

b
a
12
11
13
16
18
19
20
17
b
b
24
a
21
a
10
14
23
22
6
9
15
28
8
b
a
25
5
27
7
26
4
c
b
29
2
30
3
a 1
b
31
a

PLATE 15

Genus CORNUSPIRAMIA Cushman, 1928

Plate 15, figure 4; Key, plate 16, figures 14, 15
Genotype, *Nubecularia antillarum* Cushman
Cornuspiramia CUSHMAN, Contr. Cushman Lab. Foram. Res., vol. 4, 1928, p. 4.
Nubecularia CUSHMAN, 1922 (not DEFRANCE).

Test attached, young similar to *Cornuspira*, later with irregular, compressed, tubular branches; wall calcareous, imperforate; apertures formed by the open ends of the tubes. — Recent, tropical, in shallow water.

Genus CORNUSPIRELLA Cushman, 1928

Plate 15, figure 5; Key, plate 16, figure 13
Genotype, *Cornuspira diffusa* Heron-Allen and Earland
Cornuspirella CUSHMAN, Contr. Cushman Lab. Foram. Res., vol. 4, 1928, p. 4.
Cornuspira (part) HERON-ALLEN and EARLAND.

Test free, in the early stages planispiral, close coiled, of fairly even diameter, later coils expanding in height, adult with long branching or flattened peripheral extensions; wall calcareous, imperforate; aperture elongate, narrow, at the ends of the peripheral portions. — Recent, Atlantic.

Genus CORNUSPIROIDES Cushman, 1928

Plate 15, figure 6; Key, plate 16, figure 11
Genotype, *Cornuspira striolata* H. B. Brady
Cornuspiroides CUSHMAN, Contr. Cushman Lab. Foram. Res., vol. 4, 1928, p. 3.
Cornuspira (part) of authors.

Test in the early stages planispiral, and the coils of fairly uniform height, adult with height of coil greatly increasing and spreading out in a fan-shape; wall calcareous, imperforate, with distinct growth lines; aperture in the adult long and narrow, on the peripheral margin. — Recent, cold water, Atlantic.

Genus HEMIGORDIUS Schubert, 1908

Plate 15, figure 7; Key, plate 16, figures 5–8
Genotype, *Cornuspira schlumbergeri* Howchin
Hemigordius SCHUBERT, Jahrb. k. k. Geol. Reichs., vol. 58, 1908, p. 381.
Cornuspira HOWCHIN, 1895 (not SCHULTZE).

Test free, early coils not entirely planispiral, later planispiral and com pletely involute, but not umbonate; wall calcareous, imperforate; aperture formed by the open end of the tube. — Carboniferous and Permian.

I have topotypes of the type species, and it is certainly close to *Cornuspira.*

Genus MEANDROSPIRA Loeblich and Tappan, 1946

Key, plate 47, figure 7
Genotype, *Meandrospira washitensis* Loeblich and Tappan
Meandrospira LOEBLICH and TAPPAN, Journ. Pal., vol. 20, 1946, p. 248.

Test free, composed of proloculum followed by a tubular second chamber which spirals about the proloculum in sort zigzag bends so that a side view shows numerous loops reaching toward the umbilicus, the loops being formed by the tubular chamber swinging back upon itself frequently; wall calcareous, imperforate; aperture simple, terminal. — Lower Cretaceous.

Genus GORDIOSPIRA Heron-Allen and Earland, 1932

Plate 15, figure 8; Key, plate 16, figure 12
Genotype, *Gordiospira fragilis* Heron-Allen and Earland
Gordiospira HERON-ALLEN and EARLAND, Journ. Roy. Micr. Soc., vol. 52, 1932, p. 254.

Test free, early coils in varying planes, later planispiral and nearly involute; wall calcareous, imperforate; aperture somewhat triangular, terminal. — Recent.

In many respects this is very close to *Hemigordius*.

Genus ORTHOVERTELLA Cushman and Waters, 1928

Plate 15, figure 9; Key, plate 16, figures 17, 18
Genotype, *Orthovertella protea* Cushman and Waters
Orthovertella CUSHMAN and WATERS, Contr. Cushman Lab. Foram. Res., vol. 4, 1928, p. 45.

Test free, early coils in varying planes, later portion uncoiled and more or less straight; wall calcareous, imperforate; aperture rounded, terminal. — Pennsylvanian, Permian.

Orthovertella has been confused with the forms of similar shape but arenaceous walls.

Genus CALCITORNELLA Cushman and Waters, 1928

Plate 15, figures 13, 14; Key, plate 16, figure 19
Genotype, *Calcitornella elongata* Cushman and Waters
Calcitornella CUSHMAN and WATERS, Contr. Cushman Lab. Foram. Res., vol. 4, 1928, p. 45.

Test attached, the early portion close coiled, later uncoiling and bending back and forth in long swings, returning to the early portion, forming an elongate test; wall calcareous, imperforate; aperture terminal. — Pennsylvanian, Permian.

Calcitornella and its more specialized derivatives, *Calcivertella* and *Plummerinella*, should not be confused with the arenaceous *Ammovertella*.

Genus CALCIVERTELLA Cushman and Waters, 1928

Plate 15, figure 11; Key, plate 16, figure 16
Genotype, *Calcivertella adherens* Cushman and Waters

Calcivertella Cushman and Waters, Contr. Cushman Lab. Foram. Res., vol. 4, 1928, p. 48.

Test attached, early portion close coiled, later uncoiling in short zigzag bends, forming a nearly even width to the test, later unbending; wall calcareous, imperforate; aperture rounded, terminal. — Pennsylvanian, Permian.

Genus PLUMMERINELLA Cushman and Waters, 1928

Plate 15, figure 10; Key, plate 16, figure 22
Genotype, *Plummerinella complexa* Cushman and Waters

Plummerinella Cushman and Waters, Contr. Cushman Lab. Foram. Res., vol. 4, 1928, p. 49.

Test attached, early portion in a simple close planispiral, the coils in the adult composed of a zigzag, bent, tubular chamber, in the last portion greatly expanded, on the dorsal side obscured by thickening, all stages of development seen on the attached side; wall calcareous, imperforate; sutures somewhat limbate; aperture in early stages rounded, terminal, in an inflated chamber at the outer angles. — Pennsylvanian.

This is a very specialized development for a single tube without chambers.

Genus APTERRINELLA Cushman and Waters, 1928

Plate 15, figure 12; Key, plate 16, figures 20, 21
Genotype, *Tolypammina grahamensis* Harlton

Apterrinella Cushman and Waters, Contr. Cushman Lab. Foram. Res., vol. 4, 1928, p. 64.

Tolypammina Harlton, 1928 (not Rhumbler).

Test attached, early portion in the microspheric form consisting of proloculum and several close, planispiral coils, in the megalospheric form the adult, irregular, tubular chamber starts abruptly with but a half turn about the large bulbous proloculum, adult chamber increasing rapidly in size; wall calcareous, imperforate, coarsely cancellated; aperture rounded, terminal. — Pennsylvanian to Jurassic.

The test, especially in the thin-walled, early stage, has the characteristic bluish-white appearance so common in the Ophthalmidiidae even in Pennsylvanian specimens.

Genus CARIXIA Macfadyen, 1941

Key, plate 47, figure 8
Genotype, *Carixia langi* Macfadyen

Carixia MACFADYEN, Philos. Trans., Roy. Soc. London, ser. B, Biol. Sci., No. 576, vol. 231, 1941, p. 27.

"An adherent reticulation of unsegmented, imperforate, calcareous tubes, set in a groundwork of calcareous cement; apertures, the simple open ends of the tubes; early development of the test unknown." — Jurassic.

Subfamily 2. Nodophthalmidiinae

Early portion as in *Cornuspira*, followed by chambers in a rectilinear series.

Genus NODOPHTHALMIDIUM Macfadyen, 1939

Plate 15, figure 30; Key, plate 16, figures 25, 26
Genotype, *Nodobacularia compressa* Rhumbler

Nodophthalmidium MACFADYEN, Journ. Roy. Micr. Soc., vol. 59, 1939, p. 167.
Nubecularia (part) JONES and PARKER (not DEFRANCE). *Nodobacularia* (part) of authors.

Test free, consisting of a globular proloculum followed by a planispiral, tubular, second chamber and in the adult a few chambers in a rectilinear series; wall calcareous, imperforate; aperture simple, terminal, with a lip.— Tertiary (?), Recent.

See Cushman and Todd, Species of the Genera *Nodophthalmidium*, *Nodobaculariella*, and *Vertebralina* (Contr. Cushman Lab. Foram. Res., vol. 20, 1944, pp. 64–77, pls. 11, 12).

Genus NODOBACULARIELLA Cushman and Hanzawa, 1937

Key, plate 47, figure 9
Genotype, *Nodobaculariella japonica* Cushman and Hanzawa

Nodobaculariella CUSHMAN and HANZAWA, Contr. Cushman Lab. Foram. Res., vol. 13, 1937, p. 41.

Test free, compressed, early portion planispiral, later becoming uncoiled, nearly or sometimes completely bilaterally symmetrical; chambers consisting of a globular proloculum, immediately followed by a planispiral, tubular chamber ½ coil in length, and then by several, rapidly widening chambers, each normally ½ coil in length, sometimes shorter so that three chambers may make up a coil, the adult stage with somewhat involute chambers, partially concealing the earlier ones, and in the final development a single, uncoiled chamber; wall calcareous, imperforate; aperture long, narrow, in the median portion of the terminal face of the chamber, with an everted lip, but without teeth. — Eocene to Recent.

Genus VERTEBRALINA d'Orbigny, 1826

Plate 15, figure 29; Key, plate 16, figures 23, 24
Genotype, *Vertebralina striata* d'Orbigny
Vertebralina D'ORBIGNY, Ann. Sci. Nat., vol. 7, 1826, p. 283.

Test with the early chambers in a trochoid spiral, often involute, later ones in a rectilinear series; wall calcareous, imperforate; aperture simple, long, narrow, terminal or somewhat lateral, with a lip but no teeth. — Eocene to Recent.

Subfamily 3. Ophthalmidiinae

Test free, planispiral, in the later stage usually two or more chambers making up a coil, later chambers variously arranged in different genera.

Genus OPHTHALMIDIUM Zwingli and Kübler, 1870

Plate 15, figure 26; Key, plate 17, figures 1, 2
Genotype, *Oculina liasica* Kübler and Zwingli
Ophthalmidium ZWINGLI and KÜBLER, Foram. Schweiz. Jura, 1870, p. 46.
Oculina KÜBLER and ZWINGLI, 1866. *Hauerina* (part) H. B. BRADY, 1879 (not D'ORBIGNY).
Hauerinella SCHUBERT, 1920

Test planispiral, compressed, not involute, consisting of a globular proloculum followed by a planispiral tubular chamber of usually two or more coils, following chambers decreasing in length, loose coiled, intermediate area filled by a thin plate; aperture at the open end of the chamber, rounded, without lip or tooth. — Jurassic to Recent.

See Macfadyen, On *Ophthalmidium*, and Two New Names for Recent Foraminifera of the Family *Ophthalmidiidae* (Journ. Roy. Micr. Soc., vol. 59, 1939, pp. 162–169, text figs. 1–3); Wood and Barnard, *Ophthalmidium*, a Study of Nomenclature, Variation, and Evolution in the Foraminifera (Quart. Journ. Geol. Soc. London, vol. 102, July 1946, pp. 77–113, pls. 4–10); and Wood, The Type Specimen of the Genus *Ophthalmidium* (l. c., April 1947, pp. 461–463, pls. 29, 30).

Genus SPIROPHTHALMIDIUM Cushman, 1927

Plate 15, figure 27; Key, plate 17, figure 7
Genotype, *Spiroloculina acutimargo* H. B. Brady (part)
Spirophthalmidium CUSHMAN, Contr. Cushman Lab. Foram. Res., vol. 3, 1927, p. 37.
Spiroloculina (part) H. B. BRADY, and later authors.

Test similar to *Ophthalmidium* but accelerated, the stage having two in a coil quickly reached, plate between chambers usually present; wall calcareous, imperforate; aperture simple, without teeth. — Jurassic to Recent.

This should not be confused with *Spiroloculina* which has teeth in the aperture, and, in the microspheric form at least, a quinqueloculine stage.

Genus OPHTHALMINA Rhumbler, 1936

Key, plate 47, figures 10, 11
Genotype, *Ophthalmina kilianensis* Rhumbler
Ophthalmina RHUMBLER, Kieler Meeresforschungen, vol. 1, 1936, p. 217.

Test attached, mostly planispiral, evolute in the adult, compressed, periphery rounded; chambers irregular, in the early stages undivided and *Cornuspira*-like, later a half coil in length, irregularly coiling, in the adult planispiral; wall calcareous, imperforate; aperture at the open end of the last-formed chamber, without a tooth.—Lower Cretaceous, Recent.

Genus DISCOSPIRINA Munier-Chalmas, 1902

Plate 15, figure 28; Key, plate 17, figure 8
Genotype, *Pavonina italica* Costa
Discospirina MUNIER-CHALMAS, Bull. Soc. Géol. France, ser. 4, vol. 2, 1902, p. 352.
Pavonina COSTA, 1856 (not D'ORBIGNY). *Orbitolites* (part) of authors.
Cyclophthalmidium LISTER, 1903.

Test in the young similar to *Ophthalmidium*, later chambers annular with incomplete divisions into chamberlets at the periphery of the very thin test.—Upper Cretaceous to Recent.

Genus PLANISPIRINA Seguenza, 1880

Plate 15, figure 25; Key, plate 17, figures 5, 6
Genotype, *Planispirina communis* Seguenza
Planispirina SEGUENZA, Atti R. Accad. Lincei, ser. 3, vol. 6, 1880, p. 310.

Test in the early stages like *Cornuspira*, later divided into chambers, several to a coil, somewhat involute; aperture simple, without a tooth. —Cretaceous to Recent.

This genus should not be confused with *Nummoloculina* which has a quinqueloculine early stage and a tooth in the aperture.

Genus POLYSEGMENTINA Cushman, 1946

Key, plate 47, figures 12, 13
Genotype, *Hauerina circinata* H. B. Brady
Polysegmentina CUSHMAN, Contr. Cushman Lab. Foram. Res., vol. 22, 1946, p. 1.
Hauerina (part) of authors.

Test in the early stages similar to *Cornuspira* with proloculum and planispirally coiled second chamber several coils in length, in the adult the final coil divided into as many as six or seven short chambers and the test becoming more or less involute; wall calcareous, imperforate; aperture in the adult an elongate area nearly the whole height of the chamber, cribrate.—Recent.

Genus RENULINA Lamarck, 1804

Plate 15, figure 22; Key, plate 17, figures 11, 12
Genotype, *Renulina opercularia* Lamarck

Renulina LAMARCK, Ann. Mus., vol. 5, 1804, p. 354.

Test in the early stages planispiral, in the adult chambers becoming progressively shorter and broader, one side of the test nearly a straight line, opposite end extending back to the earlier coils, in later stage chambers extending back at both ends and becoming annular; aperture very narrow, peripheral. — Eocene.

Genus PLANISPIRINELLA Wiesner, 1931

Plate 15, figure 23; Key, plate 17, figures 3, 4
Genotype, *Planispirina exigua* H. B. Brady

Planispirinella WIESNER, Deutsche Südpolar-Exped., vol. XX, Zool., 1931, p. 69.
Planispirina (part) H. B. BRADY.

Test in early stages similar to *Cornuspira* with later coils each divided into two or three chambers; wall calcareous, imperforate; aperture simple. — Tertiary, Recent.

Genus WIESNERELLA Cushman, 1933

Plate 15, figure 21; Key, plate 17, figures 9, 10
Genotype, *Planispirina auriculata* Egger

Wiesnerella CUSHMAN, Contr. Cushman Lab. Foram. Res., vol. 9, 1933, p. 33.
Planispirina (part) of authors.

Test nearly planispiral throughout, mostly evolute, later chambers somewhat embracing, a half coil in length in the adult; wall thin, calcareous, imperforate; aperture large, nearly circular, with a broad flaring lip at one side of the end of the chamber. — Recent.

Genus TRISEGMENTINA Wiesner, 1931

Plate 15, figure 24; Key, plate 17, figure 14
Genotype, *Trisegmentina sidebottomi* Cushman = *Hauerina compressa* Sidebottom (not d'Orbigny)

Trisegmentina WIESNER, Deutsche Südpolar-Exped., vol. XX, Zool., 1931, p. 70.
Hauerina (part) SIDEBOTTOM, 1904 (not D'ORBIGNY).

Test in the early stages like *Cornuspira*, later developing chambers usually three to a coil, planispiral; wall calcareous, imperforate; aperture in the adult cribrate. — Recent.

The types of *Hauerina compressa* d'Orbigny have an early quinqueloculine stage, and I have examined a number of them from the type locality. This genus has the young stages like *Cornuspira*, and a new name must be given to the form found by Sidebottom.

Genus MEANDROLOCULINA Bogdanowicz, 1935

Key, plate 47, figures 14–16
Genotype, *Meandroloculina bogatschovi* Bogdanowicz
Meandroloculina Bogdanowicz, Bull. Acad. Sci. de L'URSS, 1935, p. 695.

Test free, composed of a proloculum followed by one or two chambers one coil in length coiled about it, followed in turn by a series of chambers a half coil in length alternating in a zigzag manner, and in the adult with two or three uniserial chambers; wall calcareous, imperforate; aperture rounded, terminal, with a slight lip but without a tooth. — Miocene.

Subfamily 4. Nubeculariinae

Test typically attached, at least in the early stages, coiled in the young, later irregular and degenerate, even consisting of a single attached chamber.

Genus NUBECULARIA Defrance, 1825

Plate 15, figure 15; Key, plate 17, figure 13
Genotype, *Nubecularia lucifuga* Defrance
Nubecularia Defrance, Dict. Sci. Nat., vol. 35, 1825, p. 210.
Amorphina Parker, 1857.

Test typically coiled, at least in the young, free or usually attached, with an oval proloculum and coiled tubular chamber, followed in the adult by irregular chambers varying more or less with the attached surface. — Jurassic to Recent, perhaps earlier.

Genus NUBECULINELLA Cushman, 1929

Plate 15, figure 16; Key, plate 17, figure 15
Genotype, *Nubeculinella bigoti* Cushman
Nubeculinella Cushman, Bull. Soc. Linn. Normandie, ser. 8, vol. 2, 1929, p. 133.

Test attached, early stages with a single chamber coiled about the proloculum, followed by a series of chambers irregularly placed, earlier ones more or less in a linear series; wall calcareous, imperforate; aperture simple, terminal. — Jurassic.

Genus RHIZONUBECULA LeCalvez, 1935

Key, plate 47, figures 17, 18
Genotype, *Rhizonubecula adherens* LeCalvez
Rhizonubecula LeCalvez, Protistologica, vol. LV, 1935, p. 96.

Test attached, early stage composed of a proloculum and second chamber of one coil in length about it, followed by a series of somewhat pyriform chambers in an irregular coil, in the adult becoming irregularly dichotomously branched; wall calcareous, imperforate; aperture at the outer end of the chambers, without a tooth. — Recent.

Genus SINZOWELLA Cushman, 1933

Plate 15, figure 17; Key, plate 17, figures 18–20

Genotype, *Nubecularia novorossica* var. *deformis* Karrer and Sinzow

Sinzowella CUSHMAN, Contr. Cushman Lab. Foram. Res., vol. 9, 1933, p. 33.

Nubecularia KARRER and SINZOW, 1876 (not DEFRANCE).

Test in the early stages like *Cornuspira* but quickly becoming involute, and in the adult with numerous, irregular, simple chambers built up into various shapes; wall calcareous, imperforate; apertures of adult elongate, slightly arcuate, with a lip but no teeth. — Oligocene and Miocene.

I studied the types of the various species of this genus in Vienna, and they represent a very specialized development from *Nubecularia.*

Genus CALCITUBA Roboz, 1884

Plate 15, figure 18; Key, plate 17, figure 21

Genotype, *Calcituba polymorpha* Roboz

Calcituba ROBOZ, Sitz. Akad. Wiss. Wien, vol. 88, pt. 1, 1883 (1884), p. 420.

Test adherent, branched, of irregular chambers, more or less cylindrical; wall calcareous, imperforate; apertures simple, at the ends of the branches. — Recent.

Genus PARRINA Cushman, 1931

Plate 15, figure 19; Key, plate 17, figures 16, 17

Genotype, *Nubecularia inflata* H. B. Brady (not Terquem) = *N. bradyi* Millett

Parrina CUSHMAN, Contr. Cushman Lab. Foram. Res., vol. 7, 1931, p. 20.

Nubecularia (part) H. B. BRADY (not DEFRANCE). *Silvestria* SCHUBERT, 1920 (not VERHOEFF).

Test with the early chambers similar to *Calcituba,* later ones inflated, irregularly coiled; wall calcareous, imperforate; aperture rounded, irregularly placed. — Recent.

Genus GLOMULINA Rhumbler, 1936

Key, plate 47, figures 19, 20

Genotype, *Glomulina fistulescens* Rhumbler

Glomulina RHUMBLER, Kieler Meeresforschungen, vol. 1, 1936, p. 198.

Test free or fixed, consisting of a proloculum and encircling second chamber, the following chambers irregularly coiling about the center, forming a generally spherical mass, later chambers more or less a half coil in length; wall calcareous, imperforate; aperture at the end of the final tubular chamber with fistulose, accessory openings on the chamber wall. — Recent

Genus SQUAMULINA Schultze, 1854

Plate 15, figure 20; Key, plate 17, figure 22

Genotype, *Squamulina laevis* Schultze

Squamulina SCHULTZE, Organismus Polythal., 1854, p. 56.

Test adherent, consisting of a single inflated chamber; wall calcareous, imperforate; aperture simple, on the convex surface. — Recent.

Family 19. FISCHERINIDAE

Test free, coiled, earlier coils somewhat planispiral, later ones trochoid, all coils visible from dorsal side only, the last-formed typically involute on the ventral side; chambers distinct but not inflated, usually four or five in the last-formed whorl; wall calcareous, imperforate; aperture rounded, terminal.

This family, represented by the single genus *Fischerina*, represents the attempt in fairly recent times to develop a trochoid form. There is a considerable range from nearly planispiral forms to a rather high trochoid spire.

Genus FISCHERINA Terquem, 1878

Plate 15, figure 31; Key, plate 17, figure 23

Genotype, *Fischerina rhodiensis* Terquem

Fischerina Terquem, Mém. Soc. Géol. France, ser. 3, vol. 1, 1878, p. 80.

Test free, low trochoid spiral; early chambers long, later ones shortened, and four or five in a whorl, ventral side usually entirely involute, all whorls visible from dorsal side; wall calcareous, imperforate; aperture terminal, rounded, simple. — Pliocene to Recent.

Family 20. TROCHAMMINIDAE

Test in general trochoid, later chambers somewhat irregular in a few genera; wall arenaceous with a distinctly chitinous base, cement usually yellowish- or reddish-brown; aperture typically on the ventral side, at least in the young of all forms.

The Trochamminidae through the most primitive genus *Trochammina*, which is known from as early as the Silurian, have been derived from a chitinous ancestry. The arenaceous material of the test varies from much to very little, so that tests of Recent species are known, made almost entirely of chitin. The flexibility of the test is shown by the distortion that occurs in fossil members of the family. Often specimens are so crushed as to be almost unrecognizable specifically, while thin-walled calcareous forms found with them are unbroken. The group has no relation to *Endothyra* or its relatives except that both groups were derived from the Ammodiscidae, the former through *Glomospira* with the cement becoming somewhat more calcareous, and *Trochammina* from the trochoid spiral forms of the Ammodiscidae.

There is a tendency in *Globivalvulina* and *Tetrataxis* with their derivatives to have a thickened wall, but the chitinous material is often retained as can be seen by dissolving out the calcareous material slowly with very weak acid. The number of chambers becomes much greater in *Polytaxis*.

Specialized genera such as *Nouria*, *Cystammina*, *Rotaliammina* and others are known mainly from Recent material, as is also *Carterina* which secretes specialized spicular bodies. In *Mooreinella* the later growth becomes irregularly biserial and in *Ammocibicides* uncoiled.

The number of chambers varies greatly especially in the more primitive *Trochammina,* where the number in an adult whorl may range from three or possibly two to as many as ten, the number often varying greatly in the microspheric and megalospheric forms of the same species.

There is a great variation in the relative amounts of cement and arenaceous material even in the same species. Throughout the group the characteristic yellowish-brown cementing material is present that is so characteristic of arenaceous forms from the Astrorhizidae through the Ammodiscidae and other related groups.

Subfamily 1. Trochammininae

Test trochoid, chambers in spiral whorls; aperture ventral.

Genus TROCHAMMINA Parker and Jones, 1859

Key, plate 18, figures 1–3

Genotype, *Nautilus inflatus* Montagu

Trochammina PARKER and JONES, Ann. Mag. Nat. Hist., ser. 3, vol. 4, 1859, p. 347.
Nautilus (part) MONTAGU, 1808 (not LINNÉ). *Rotalina* (part) WILLIAMSON, 1858 (not D'ORBIGNY). *Lituola* (part) PARKER and JONES, 1865 (not LAMARCK).
Haplophragmium (part) SIDDALL, 1879 (not REUSS). *Ammoglobigerina* EIMER and FICKERT, 1899. *Tritaxis* SCHUBERT, 1920. *Glomerina* FRANKE, 1928.

Test free or adherent, spiral, trochoid, all chambers visible from dorsal side, only those of the last-formed whorl from the ventral, varying from much compressed to nearly globular; wall arenaceous, with chitinous base, amount of cement very variable; aperture an arched slit at the inner margin of the ventral side of the chamber. — Silurian to Recent.

Genus CONOTROCHAMMINA Finlay, 1940

Key, plate 47, figures 21, 22

Genotype, *Conotrochammina whangaia* Finlay

Conotrochammina FINLAY, Trans. Roy. Soc. New Zealand, vol. 69, 1940, p. 448.

Test trochoid, umbilicate; chambers several to a whorl, obscure; wall arenaceous, of coarse sand grains with little cement; aperture a small circular opening in the terminal face of the last-formed chamber. — Upper Cretaceous.

Genus TROCHAMMINITA Cushman and Bronnimann, 1948

Key, plate 48, figures 4, 5

Genotype, *Trochamminita irregularis* Cushman and Bronnimann

Trochamminita CUSHMAN and BRONNIMANN, Contr. Cushman Lab. Foram. Res., vol. 24, 1948, p. 17.

Test in the early stages trochoid as in *Trochammina,* later with the chambers added in a very irregular manner; wall arenaceous, thin, partially chitinous; aperture in the irregular adult portion consisting of a rounded opening in the chamber wall with a slightly raised border. — Recent.

Genus TROCHAMMINELLA Cushman, 1943

Key, plate 47, figures 23–25

Genotype, *Trochamminella siphonifera* Cushman

Trochamminella CUSHMAN, Contr. Cushman Lab. Foram. Res., vol. 19, 1943, p. 95.

Test trochoid, free in the early stages, sometimes attached in the later stages; wall arenaceous; aperture in the unattached forms a rounded opening near the margin of the ventral face of the last-formed chamber, usually surrounded by a slightly raised ring; the attached adult surrounded by an irregular rim of material similar to that of the wall and extending out in a tubular neck with a rounded aperture. — Recent.

Genus REMANEICA Rhumbler, 1938

Key, plate 48, figure 1

Genotype, *Remaneica helgolandica* Rhumbler

Remaneica RHUMBLER, Kieler Meeresforschungen, vol. 2, 1938, p. 194.
Patellina (part) of authors. *Trochammina* (part) of authors.

Test attached, trochoid, circular, greatly compressed; chambers numerous, all visible from the dorsal side, on the ventral side in the adult the ventral wall absorbed, making the whole base into one large chamber, later chambers with internal plications from the peripheral wall, partially dividing the chamber; wall of thin, imperforate chitin, with some adhering foreign material about the peripheral portion. — Recent.

Genus ROTALIAMMINA Cushman, 1924

Key, plate 18, figures 6, 7

Genotype, *Rotaliammina mayori* Cushman

Rotaliammina CUSHMAN, Publ. 342, Carnegie Inst. Washington, 1924, p. 11.

Test trochoid, attached by the ventral side, all chambers visible from above, only those of the last-formed whorl from below; dorsal wall of matted spicules and much chitin, flexible, ventral wall of thin chitin; aperture ventral, along the edge of the chamber. — Recent, Samoa in shallow water.

Genus ENTZIA Daday, 1883

Key, plate 40, figure 5

Genotype, *Entzia tetrastomella* Daday

Entzia DADAY, Orvos-term. ertes., vol. 8, 1883, p. 209.

Test trochoid, dorsal side convex, ventrally flattened; chambers distinct, not inflated, of rather uniform shape, gradually increasing in size as added; wall chitinous, with included irregular bodies, nearly transparent;

apertures in two pairs near the base of the apertural face, rounded or elliptical.—Recent, salt pools of Hungary.

This form seems to be related in its wall characters to *Rotaliammina* and *Carterina*.

Genus CARTERINA H. B. Brady, 1884

Key, plate 18, figures 4, 5

Genotype, *Rotalia spiculotesta* Carter

Carterina H. B. BRADY, Rep. Voy. *Challenger*, Zoology, vol. 9, 1884, p. 345.
Rotalia CARTER, 1877 (not D'ORBIGNY).

Test trochoid, usually attached, later chambers irregularly spreading; wall of cement in which are thin, translucent, fusiform bodies; aperture ventral, near the umbilicus.—Recent.

Genus AMMOCIBICIDES Earland, 1934

Key, plate 48, figure 2

Genotype, *Ammocibicides proteus* Earland

Ammocibicides EARLAND, *Discovery* Reports, vol. 10, 1934, p. 106.

Test probably attached in the early stages, later becoming free, early whorls trochoid, later irregular, dorsal side flattened, ventral side convex, periphery somewhat thickened and irregular in outline; chambers all visible from the dorsal side, those of the last-formed whorl visible from the ventral side; wall very finely arenaceous, with a large amount of cement; aperture circular or elliptical, on the periphery or ventral surface, occasionally more than one in the earlier chambers.—Eocene to Recent.

Subfamily 2. Globotextulariinae

Test irregularly spiral, the chambers globose; aperture in the open umbilical area or nearly terminal in irregular forms.

Genus GLOBOTEXTULARIA Eimer and Fickert, 1899

Key, plate 18, figure 14

Genotype, *Haplophragmium anceps* H. B. Brady

Globotextularia EIMER and FICKERT, Zeitschr. Wiss. Zool., vol. 65, 1899, p. 679.
Haplophragmium (part) H. B. BRADY, 1884 (not REUSS).

Test irregularly spiral; chambers globose, the last-formed ones increasing rapidly in size; wall arenaceous; aperture in the open umbilical area.—Recent.

Genus MOOREINELLA Cushman and Waters, 1928

Key, plate 18, figures 8–10

Genotype, *Mooreinella biserialis* Cushman and Waters

Mooreinella CUSHMAN and WATERS, Contr. Cushman Lab. Foram. Res., vol. 4, 1928, p. 50.

Test in the early stages trochoid, later developing at one side into a biserial form; chambers alternating along an elongate axis; wall rather coarsely arenaceous; aperture becoming rounded and subterminal.—Pennsylvanian.

Subfamily 3. Ammosphaeroidininae

Test with the early portion trochoid, later chambers few, embracing; aperture arched, at the base of the chamber or becoming terminal.

Genus AMMOSPHAEROIDINA Cushman, 1910

Key, plate 18, figure 11

Genotype, *Haplophragmium sphaeroidiniformis* H. B. Brady

Ammosphaeroidina CUSHMAN, Bull. 71, U. S. Nat. Mus., pt. 1, 1910, p. 128.
Haplophragmium (part) H. B. BRADY, 1884 (not REUSS).

Test in the early stages trochoid, in the adult globose, involute, the last three chambers making up the entire surface; wall coarsely arenaceous; aperture arched, at the umbilical border of the chamber.—Tertiary, Recent.

Genus NOURIA Heron-Allen and Earland, 1914

Key, plate 18, figures 12, 13

Genotype, *Nouria polymorphinoides* Heron-Allen and Earland

Nouria HERON-ALLEN and EARLAND, Trans. Zool. Soc. London, vol. 20, 1914, p. 375.

Test free, of several chambers, irregularly spiral or later biserial; wall arenaceous; aperture simple, terminal.—Eocene (?), Recent.

Genus CYSTAMMINA Neumayr, 1889

Key, plate 18, figure 15

Genotype, *Trochammina pauciloculata* H. B. Brady

Cystammina NEUMAYR, Die Stämme des Thierreichs, vol. 1, 1889, p. 167.
Trochammina (part) of authors. *Ammochilostoma* EIMER and FICKERT, 1899.

Test with the early stages coiled, in the adult subglobose, with but two or three chambers making up the surface of the test; wall finely arenaceous, with much cement; aperture at the base or in the apertural face of the last-formed chamber.—Recent.

Subfamily 4. Tetrataxinae

Test trochoid, whorls becoming more or less definitely four chambered at least in the young, later more chambers in a whorl, in some several; wall finely arenaceous, with much cement, becoming calcareous, thickened; aperture ventral.

The wall structure in this group needs careful study by modern petrographic methods.

Genus GLOBIVALVULINA Schubert, 1920

Key, plate 18, figures 16, 17
Genotype, *Valvulina bulloides* H. B. Brady

Globivalvulina SCHUBERT, Pal. Zeitschr., vol. 3, 1920, p. 153.
Valvulina H. B. BRADY, 1876 (not D'ORBIGNY).

Test trochoid, subglobular or plano-convex, ventral side flattened, dorsal side strongly convex; chambers few, inflated; wall finely arenaceous, with much cement, often appearing to have a thickened inner wall and an outer thinner wall; aperture low, arched, at the umbilical margin of the chamber. — Pennsylvanian, Permian.

See Plummer, Morphology of *Globivalvulina* (Amer. Midland Nat., vol. 39, 1948, pp. 169–173, text figs. 1–5), which indicates that this genus is calcareous and biserial and possibly related to the Cassidulinidae.

Genus TETRATAXIS Ehrenberg, 1843

Key, plate 18, figures 18, 19
Genotype, *Tetrataxis conica* Ehrenberg

Tetrataxis EHRENBERG, Bericht Preuss. Akad. Wiss. Berlin, 1843, p. 106.
Valvulina (part) H. B. BRADY, 1876 (not D'ORBIGNY).

Test conical, consisting of proloculum and elongate second chamber, later broken up into elongate, crescentic chambers; wall often with apparently two layers, an outer finely arenaceous one and the interior one clear and alveolar; aperture elongate, opening into the four-lobed umbilicus. — Pennsylvanian to Permian.

Genus POLYTAXIS Cushman and Waters, 1928

Key, plate 18, figures 21, 22
Genotype, *Polytaxis laheei* Cushman and Waters

Polytaxis CUSHMAN and WATERS, Contr. Cushman Lab. Foram. Res., vol. 4, 1928, p. 51.
Tetrataxis (part) of authors (not EHRENBERG).

Test in the early stages similar to *Tetrataxis*, earliest stage coiled, followed by elongate chambers in series of fours, then in the adult spreading, chambers numerous in each whorl, ventral side concave, irregular; apertures several, on the ventral side. — Pennsylvanian.

Genus RUDITAXIS Schubert, 1920

Key, plate 18, figure 20
Genotype, *Valvulina rudis* H. B. Brady

Ruditaxis SCHUBERT, Pal. Zeitschr., vol. 3, 1920, p. 180.
Valvulina (part) H. B. BRADY, 1876 (not D'ORBIGNY).

Test in general structure like *Tetrataxis*, but the chambers labyrinthic; wall more coarsely and roughly arenaceous. — Pennsylvanian, Permian.

Family 21. PLACOPSILINIDAE

Test attached; chambers numerous and distinct, the early ones often coiled or trochoid, interior simple or labyrinthic; wall arenaceous, usually of calcareous fragments; apertures of various forms.

The forms included in this family are all attached, at least in their early stages. It may be that some of these are more or less degenerate forms, and do not have the same ancestral source, but all have certain characters in common.

Forms referred to *Placopsilina* are known from the Silurian, and continue to the present oceans. The forms included present a variable group, and little is known of some of them. The Paleozoic forms referred to *Bullopora* belong here, as *Bullopora* is a calcareous perforate form particularly abundant in the Cretaceous and Eocene. Dr. Paalzow has examined originals of *Bullopora rostrata*, and they are similar to the Polymorphinidae. There is much question concerning *Bdelloidina* which, from Brady's figures and description, seems to be arenaceous. Carter's original figure shows nothing of the early stage which is coiled. I have recently seen material possibly referable to this genus, which makes it appear that it may be a degenerate form from some of the rotaliform families. It needs further study. *Haddonia* is another form that needs further study. Chapman records the early stages as coiled, but Heron-Allen and Earland have had somewhat similar forms with young stages resembling the Textulariidae. It is quite possible that the two are entirely different in their origin, and that Heron-Allen and Earland's form may be *Textularioides* or a new genus.

Polyphragma has been sometimes referred to the Bryozoa, but the material of both the Bohemian and Saxon Cretaceous, as well as our own, shows that this belongs to the foraminifera. Reuss may possibly have had two distinct things under this name, but his original types of this could not be found. *Stylolina* is a peculiar form developed in the Miocene of Europe which seems similar to *Polyphragma*. The early stages are unknown. *Stacheia* contains many apparently abundant forms which it is difficult to group with other genera.

Subfamily 1. Placopsilininae

Chambers simple, not labyrinthic.

Genus PLACOPSILINA d'Orbigny, 1850

Key, plate 19, figures 1–3

Genotype, *Placopsilina cenomana* d'Orbigny

Placopsilina D'ORBIGNY, Prodr. Pal., vol. 2, 1850, p. 96.

Lituola JONES and PARKER, 1860 (not LAMARCK). *Bullopora* of authors (not QUENSTEDT).

Test attached, composed of numerous chambers, early portion close coiled, later portions uncoiled and spreading out in a generally linear series, last chambers sometimes growing upward from the attachment; wall coarsely or finely arenaceous; aperture rounded, at the end of the last-formed chamber. — Silurian to Recent.

Genus PLACOPSILINELLA Earland, 1934

Key, plate 48, figure 3

Genotype, *Placopsilinella aurantiaca* Earland

Placopsilinella Earland, *Discovery* Reports, vol. 10, 1934, p. 95.

Test sessile; chambers in early stages in a loose spiral line of single chambers, later in an irregularly curving line with at first two, then three or four chambers abreast; wall chitinous with some admixture of ferruginous cement; aperture not visible. — Recent.

Genus BDELLOIDINA Carter, 1877

Key, plate 19, figures 4, 5

Genotype, *Bdelloidina aggregata* Carter

Bdelloidina Carter, Ann. Mag. Nat. Hist., ser. 4, vol. 19, 1877, p. 201.

Test attached, of irregular chambers, broad and low, early ones more or less coiled; wall arenaceous, with sponge spicules; apertures numerous, rounded, on the outer face of the chamber. — Jurassic to Recent.

Genus ACRULIAMMINA Loeblich and Tappan, 1946

Key, plate 48, figures 6–8

Genotype, *Placopsilina longa* Tappan

Acruliammina Loeblich and Tappan, Journ. Pal., vol. 20, 1946, p. 252.

Placopsilina (part) of authors.

Test attached, at least in the early portion; chambers numerous, at first close coiled, later uncoiling, only a few chambers of the coiled portion or all of the coiled portion and much of the uniserial portion may be attached, the later portion usually free, the uniserial portion then becoming cylindrical; wall arenaceous; aperture terminal, a single low slit at the attachment in the early stages, later divided by a median septum and finally cribrate. — Lower Cretaceous.

Subfamily 2. Polyphragminae

Chambers labyrinthic.

Genus HADDONIA Chapman, 1898

Key, plate 19, figures 6, 7

Genotype, *Haddonia torresiensis* Chapman

Haddonia Chapman, Journ. Linn. Soc. London, Zool., vol. 26, 1898, p. 453.

Test attached, early chambers often coiled, later ones broad and low; wall arenaceous, with coarse pores; aperture, a crescent-shaped slit on the upper face of the last-formed chamber. — Eocene to Recent.

Genus POLYPHRAGMA Reuss, 1871

Key, plate 19, figures 8–10
Genotype, *Lichenopora cribrosa* Reuss

Polyphragma Reuss, Sitz. Akad. Wiss. Wien, vol. 64, pt. 1, 1871, p. 277.
Lichenopora (part) Reuss.

Test attached, later growing upward, cylindrical, often branched; chambers short, interior labyrinthic; wall double, outer arenaceous and imperforate, inner hyaline and perforate; aperture cribrate. — Cretaceous.

Genus ADHAERENTIA Plummer, 1938

Key, plate 48, figures 13–15
Genotype, *Adhaerentia midwayensis* Plummer

Adhaerentia Plummer, Amer. Mid. Nat., vol. 19, 1938, p. 242.

Test attached by the proloculum only; chambers of the early stages biserial, last few chambers uniserial, earlier ones simple, later labyrinthic; wall arenaceous, composed of various kinds of material with much cement; aperture in the early stages simple, later terminal and becoming multiple. — Paleocene.

Genus STYLOLINA Karrer, 1877

Key, plate 19, figures 11, 12
Genotype, *Stylolina lapugyensis* Karrer

Stylolina Karrer, Abhandl. k. k. Geol. Reichs., vol. 9, 1877, p. 371.

Test attached, with the early chambers spiral, later ones forming a cylindrical test; wall arenaceous; aperture, a ring of pores near the periphery of the outer face. — Miocene.

Genus STACHEIA H. B. Brady, 1876

Key, plate 19, figures 13–15
Genotype, *Stacheia marginulinoides* H. B. Brady

Stacheia H. B. Brady, Pal. Soc., Mon. 30, 1876, p. 107.

Test attached, early chambers suggesting a spiral arrangement, later ones irregular, labyrinthic; wall arenaceous, with an outer imperforate layer; aperture simple, circular, often with a neck. — Carboniferous to Jurassic.

Family 22. ORBITOLINIDAE

Test usually conical, early chambers spiral, trochoid, later ones discoid, then annular, central portion irregularly divided, outer area of the cone with small cellules below the thin outer coating; wall finely arenaceous with calcareous or iron oxide cement; apertures in radial rows on the ventral surface.

The family Orbitolinidae contains principally the genus *Orbitolina*, a very important and often very abundant genus in the Lower and Upper

Cretaceous, having its beginning perhaps in the Jurassic. This genus has often been confused with the genera *Coskinolina* and *Dictyoconus*, which it superficially resembles and which belong in the Valvulinidae. Those genera are derived from *Arenobulimina*, and are very abundant in the Middle Eocene. Their structure is very different from that of *Orbitolina*. The small conical species of *Orbitolina*, often abundant in the Lower Cretaceous, have been mistaken for *Coskinolina*, which is very different structurally.

The peculiar trochoid genus of the Pennsylvanian, *Howchinia*, has peculiar structures at the periphery reminding one of the markings of the surface of *Valvulinella* also known from the Carboniferous. The latter genus has the chambers divided into cellules just below the surface, and forms intermediate between it and *Orbitolina* may be looked for in the Triassic and Jurassic.

The family reached its climax in the large species of *Orbitolina* in the Cretaceous.

Silvestri has proposed a generic name, *Orbitolinopsis* with the genotype *Orbitolina conulus* Douvillé.

Genus HOWCHINIA Cushman, 1927

Key, plate 19, figures 16–18

Genotype, *Patellina bradyana* Howchin

Howchinia Cushman, Contr. Cushman Lab. Foram. Res., vol. 3, 1927, p. 42.
Patellina Howchin, 1888 (not Williamson).

Test free, conical, trochoid, consisting of an undivided, compressed, spiral chamber; wall characters not given; sutures limbate, externally with a row of pits. — Carboniferous, England.

Genus VALVULINELLA Schubert, 1907

Key, plate 19, figures 19–21

Genotype, *Valvulina youngi* H. B. Brady

Valvulinella Schubert, Verh. k. k. Geol. Reichs., 1907, p. 211.
Valvulina (part) of authors (not Parker and Jones).

Test conical, a trochoid spiral; chambers few to the whorl, divided toward the periphery into usually two horizontal series of chamberlets, the divisions appearing through the thin outer coating; wall finely arenaceous, with much cement; apertures on the ventral side. — Carboniferous.

Genus ORBITOLINA d'Orbigny, 1850

Key, plate 19, figures 22–24

Genotype, *Orbitolites concava* Lamarck

Orbitolina d'Orbigny, Prodr. Pal., vol. 2, 1850, p. 143.
Orbitolites (part) of authors.

Test depressed conical, lower side usually concave, early chambers in a depressed trochoid spire, later chambers covering the entire base of the test, ventral side usually convex, later becoming more or less annular, interior of chambers divided into chamberlets, outer portion with fine cellules, usually two rows to a chamber; wall with an outer, thin, imperforate, epidermal layer, inner wall finely arenaceous, with cement of lime or iron oxide; apertures, fine openings in somewhat radial lines, ventral. — Cretaceous.

The disposition of the triangular elements of the chambers often gives an "engine turned" appearance to slightly worn tests. *Orbitolina* does not have the same structure of pillars, marginal trough, and central shield, seen in *Dictyoconus* and its relatives. The two groups are examples of close parallelisms in structure seen so often in the foraminifera.

For excellent figures and detailed description of *Orbitolina*, see Silvestri, Foraminiferi del Cretaceo della Somalia (Pal. Ital., vol. XXXII (n. ser., vol. II), 1931 (1932), pp. 143–204, pls. IX–XVI), and Davies, An early *Dictyoconus*, and the Genus *Orbitolina*: their Contemporaneity, Structural Distinction, and Respective Natural Allies (Trans. Roy. Soc. Edinburgh, vol. 59, pt. 3, 1939, pp. 773–790, pls. 1, 2, text figs. 1–6).

Genus ORBITOLINOIDES Vaughan, 1945

Genotype, *Orbitolinoides senni* Vaughan

Orbitolinoides VAUGHAN, Geol. Soc. Amer., Mem. 9, 1945, pt. 1, p. 22

Test similar to *Orbitolina* but lacking the peripheral zone crossed by the radiating plates. — Eocene.

Family 23. LAGENIDAE

Test with chambers simple, neither typically biserial, trochoid, nor irregularly spiral, planispiral when coiled, the test uncoiling into straight or arcuate elongate forms, or compressed and in the shape of an inverted "V"; wall calcareous, very finely perforate, with a glassy appearance; aperture typically radiate but in a few genera simple, in the radiate apertured forms with a small chamberlet below the radiate aperture, opening into the main chamber by a simple rounded orifice.

The ancestry of the Lagenidae is unknown. The earliest forms appear in the Triassic, although they have been recorded earlier. The main characters of the family, the glassy appearing test and the radiate aperture, have been kept throughout the history of the group except in the earliest forms which have a rounded opening. Above this primitive rounded opening is developed in the higher forms an apertural chamberlet, the outer opening of which is radiate. In *Robulus* the original rounded opening is seen in the earliest forms, and in the later ones is represented by the ventral slit which is often expanded. The planispiral forms develop early in the history of the group, and except for the genus *Darbyella* are planispiral throughout. In *Darbyella* the later chambers are developed at one side of the axis, giving a somewhat trochoid appearance. In *Planularia* the test becomes greatly flattened. From these coiled forms there are several different genera developed. In *Vaginulina* there are two distinct groups, one to which the name *Citharina* was given by Reuss which is characteristic of the Lower Cretaceous. The two sides are nearly parallel, but in the type species of *Vaginulina* the test is biconvex. In *Lingulina* the test becomes uncoiled, and the aperture is elongate, sometimes with slight tooth-like projections at the border, but not definitely radiate. In *Palmula* the early chambers are coiled, and the later ones are much compressed and have the inverted "V" shape. This is carried further in *Frondicularia*, and often the coiled early stage is completely skipped. *Marginulina* becomes uncoiled, and the aperture is at one side of the test. There are all gradations between *Robulus* and *Lenticulina* and *Marginulina*. The amount of compression of the test varies greatly, not only in different species but often in different stages of the same individual. In *Saracenaria* the test in the adult becomes triangular in section. Of the more completely uncoiled forms, *Dentalina* shows most clearly its derivation from such forms as *Marginulina*, by the arcuate test and the oblique sutures. *Nodosaria* has the sutures nearly at right angles to the elongate axis, and the test is typically straight. *Chrysalogonium* has developed from *Nodosaria* by the regular closing of the slit-like apertures at intervals to form pores. This genus is not related to the Buliminidae as has often been thought. Material of Schwager from the type locality has been studied, and shows the true development of this form. In *Pseudoglandulina* the chambers become involute, and very nearly enclose the early chambers in some species. This should be distinguished from *Glandulina* in which the early stages are biserial.

The very large group included in the genus *Lagena* shows some of the most remarkable forms in the whole group of the foraminifera. There is but a single chamber, but the ornamentation has become exceedingly complex and variable. It is rather evident that the forms included under *Lagena* have probably been derived from various sources, and perhaps but a few of these belong in the family Lagenidae. However, until someone is able to study this group with much patience and with plenty of time, it does not seem wise to divide this group on the basis of superficial form alone.

In many ways the Lagenidae are primitive in their characters. Various species are very variable in form, and the microspheric and megalospheric forms in the same species are often very different in their early stages. Some species, as shown on page 53, have the early stages close coiled and would be placed in *Marginulina*; in the forms with a smaller proloculum there is an arcuate test with somewhat oblique sutures, quite well placed in *Dentalina*; while in the forms with the larger proloculum, the test becomes nearly straight throughout and would be placed naturally in *Nodosaria.* To determine the exact generic position of many of these forms is, therefore, extremely difficult, and the number of generic names proposed for this group is very considerable. No good purpose can be served by the use of too many names, especially in such a variable group, and the number of recognized genera is here limited to a comparatively few.

It is quite possible that a study of Triassic and Jurassic material will show that there are various forms there which may have distinct generic names. Most of our later frondicularian forms have been developed from a coiled ancestry, but it is quite possible, from a study of the Jurassic and Triassic ones, that those forms may never have had a coiled ancestry.

Subfamily 1. Nodosariinae

Test multilocular.

Genus ROBULUS Montfort, 1808

Plate 16, figure 1; Key, plate 20, figures 1–4
Genotype, *Robulus cultratus* Montfort

Robulus Montfort, Conch. Syst., vol. 1, 1808, p. 215, 54th genre.
Phonemus, Pharamum, Patrocles, Spincterules, Herion, Rhinocurus, Lampas, Scortimus, Linthuris, Astacolus (?), and *Periples* Montfort, 1808. *Robulina* d'Orbigny, 1826. *Cristellaria* Lamarck, 1816, and of most later authors.

Test planispiral, bilaterally symmetrical, typically close coiled and involute; chambers numerous, triangular in side view; wall very finely perforate, glassy; aperture rounded in older forms, in modern ones radiate, the median slit enlarged in the middle of the end of the apertural face.—Jurassic to Recent.

In many species it is very difficult to separate *Robulus* from *Lenticulina*, and it may be best as a practical matter to drop *Robulus* and use *Lenticulina* for both. There is no sharp line between *Robulus* and *Saracenaria*, it being difficult to place intermediate forms.

Cribrorobulina (genotype, *Robulina serpens* Seguenza) has been erected by Selli (Giornale di Geologie, Ann. Mus. Geol. Bologna, ser. 2, vol. 14, 1939–40, pp. 1–12, pl. 3, text figs. 1, 2) for forms with the aperture cribrate rather than radiate.

Genus DARBYELLA Howe and Wallace, 1933

Key, plate 20, figure 5
Genotype, *Darbyella danvillensis* Howe and Wallace

Darbyella Howe and Wallace, Louisiana Geol. Surv., Bull. 2, 1933, p. 144.

Test close coiled throughout, early stages planispiral, adult with the chambers at one side of the previous plane of coiling; wall calcareous, very finely perforate; aperture an elongate slit at the peripheral angle.— Cretaceous to Recent.

Genus LENTICULINA Lamarck, 1804

Plate 16, figure 4; Key, plate 20, figure 6

Genotype, *Lenticulina rotulata* Lamarck

Lenticulina LAMARCK, Ann. Mus., vol. 5, 1804, p. 186.

Antenor, *Oreas* (?), and *Clisiphontes* (?) MONTFORT, 1808. *Cristellaria* (part) of authors.

Test similar to *Robulus*, tending to become uncoiled in some species, the aperture radiate, at the peripheral angle, the slits equal.—Permian to Recent, perhaps earlier.

There is no sharp division between this genus and *Marginulina* in the uncoiling forms, nor between *Lenticulina* and the forms sometimes referred to *Astacolus*.

Genus PLANULARIA Defrance, 1824

Key, plate 20, figure 7

Genotype, *Planularia auris* Defrance

Planularia DEFRANCE, Dict. Sci. Nat., vol. 32, 1824, p. 178.

Megathyra EHRENBERG, 1841.

Cristellaria (part) of authors.

Test planispiral, bilaterally symmetrical, very much compressed, the sides nearly parallel, microspheric form more coiled in the young; aperture at the peripheral angle, radiate, sometimes with the ventral slit expanded.— Triassic to Recent.

Typical specimens are easily placed, but less compressed specimens bridge the gap between this and other genera.

Genus MARGINULINA d'Orbigny, 1826

Plate 16, figure 5; Key, plate 21, figures 6, 7

Genotype, *Marginulina glabra* d'Orbigny

Marginulina D'ORBIGNY, Ann. Sci. Nat., vol. 7, 1826, p. 258.

Cristellaria (part) of authors. *Hemicristellaria* (part) and *Hemirobulina* (part) STACHE, 1864.

Test subcylindrical or somewhat compressed, earliest portion close coiled, later uncoiled, final chambers often inflated; aperture radiate, in the early coiled portion as in *Lenticulina*, later becoming central and terminal.— Triassic to Recent, perhaps earlier.

The microspheric form has a close coiled young, whereas the megalospheric form may be much like *Dentalina*. Compressed forms have been referred to *Hemicristellaria*. The type of *Hemicristellaria*, designated by Galloway and Wissler, unfortunately is round in transverse section like *Marginulina*, in the type which I examined in Vienna, and the two become entirely synonymous.

The relationships of *Marginulinopsis* and *Vaginulinopsis* erected by A. Silvestri in 1904 still seem to be doubtful. They are possibly synonymous with *Marginulina*, although some species placed in these genera may belong in *Vaginulina*.

Genus DENTALINA d'Orbigny, 1826

Plate 16, figure 6; Key, plate 21, figures 10, 11
Genotype, *Nodosaria* (*Dentalina*) *obliqua* d'Orbigny

Dentalina d'Orbigny, Ann. Sci. Nat., vol. 7, 1826, p. 254.
Nodosaria (part) of authors. *Svenia* Brotzen, 1936.

Test arcuate, elongate; chambers numerous in a linear series; sutures usually oblique, at least in the early portion; aperture radiate, peripheral in early stages, later nearly central and terminal.—Carboniferous to Recent.

Dentalina passes into *Nodosaria* on the one hand, and into *Marginulina* on the other. It is difficult to place many of the species where microspheric and megalospheric forms show a wide range.

The forms to which the generic name *Svenia* Brotzen has been applied are probably to be included under *Dentalina*.

Genus NODOSARIA Lamarck, 1812

Plate 16, figure 7; Key, plate 21, figures 12–14
Genotype, *Nautilus radicula* Linné

Nodosaria Lamarck, Extrait Cours Zool., 1812, p. 121.
Lagenonodosaria and *Glandulonodosaria* A. Silvestri, 1900.

Test typically with chambers in a straight linear series, curved in the early stages of the microspheric form in many species, not strongly embracing; sutures in the adult at right angles to the axis; aperture terminal, radiate.—Carboniferous to Recent.

The length of neck is entirely a variable feature even at different stages in the same specimen, and the separation of the chambers is also extremely variable even in the same species.

Genus TRISTIX Macfadyen, 1941

Key, plate 48, figure 11
Genotype, *Rhabdogonium liasinum* Berthelin

Tristix Macfadyen, Philos. Trans., Roy. Soc. London, ser. B, Biol. Sci., No. 576, vol. 231, 1941, p. 54.
Triplasia (part), *Rhabdogonium* (part), and *Dentalinopsis* (part) of authors. *Tricarinella* ten Dam and Schijfsma, 1945.

"Test free, hyaline, consisting of a number of chambers, generally triangular in section, joined in a straight series; aperture terminal, simple."—Jurassic to Tertiary.

Van Voorthuysen (Geol. Mijnb., No. 3, March 1947) indicates an internal tube for this form and suggests that it should be placed in the Buliminidae. Further study is necessary to be certain of its relationships.

Genus QUADRATINA ten Dam, 1946

Key, plate 48, figure 12
Genotype, *Quadratina depressula* ten Dam

Quadratina TEN DAM, Bull. Soc. Géol. France, ser. 5, vol. 16, 1946, p. 65.

Test similar to *Tristix* but the test quadrangular in section.—Lower Cretaceous.

Genus CHRYSALOGONIUM Schubert, 1907

Key, plate 21, figure 15
Genotype, *Nodosaria polystoma* Schwager

Chrysalogonium SCHUBERT, Neues Jahrb. für Min., vol. 25, 1907, p. 242.
Nodosaria (part) SCHWAGER.

Test similar to *Nodosaria* but with the slit-like portions of the radiate aperture irregularly divided into pores.—Upper Cretaceous to Recent.

A study of Schwager's material shows that this is hardly to be separated from *Nodosaria*.

Genus PSEUDOGLANDULINA Cushman, 1929

Plate 16, figure 8; Key, plate 21, figures 16, 17
Genotype, *Nautilus comatus* Batsch

Pseudoglandulina CUSHMAN, Contr. Cushman Lab. Foram. Res., vol. 5, 1929, p. 87.
Nodosaria (part) of authors. *Glandulina* (part) of authors (not D'ORBIGNY).

Test similar to *Nodosaria*, but the chambers embracing, the last-formed one making up a large proportion of the test; chambers uniserial throughout; aperture radiate, terminal.—Jurassic to Recent.

Dr. Ozawa studied the type of *Glandulina laevigata* d'Orbigny in Paris, and found it to be biserial, therefore belonging to the Polymorphinidae.

PLATE 16

RELATIONSHIPS OF THE GENERA OF THE LAGENIDAE

FIG.
1. *Robulus crassus* (d'Orbigny). (After H. B. Brady.) *a*, side view; *b*, front view.
2. *Saracenaria italica* Defrance. (After H. B. Brady.) *a*, side view; *b*, end view.
3. *Lingulina carinata* d'Orbigny. *a*, front view; *b*, end view.
4. *Lenticulina convergens* Bornemann. *a*, side view; *b*, apertural view.
5. *Marginulina glabra* d'Orbigny. (After H. B. Brady.)
6. *Dentalina roemeri* Neugeboren.
7. *Nodosaria soluta* Reuss.
8. *Pseudoglandulina*.
9. *Lagena apiculata* (Reuss).
10. *Amphicoryne falx* (Jones and Parker). (After H. B. Brady.)
11. *Vaginulina patens* H. B. Brady. (After H. B. Brady.)
12. *Frondicularia alata* d'Orbigny. (After H. B. Brady.)

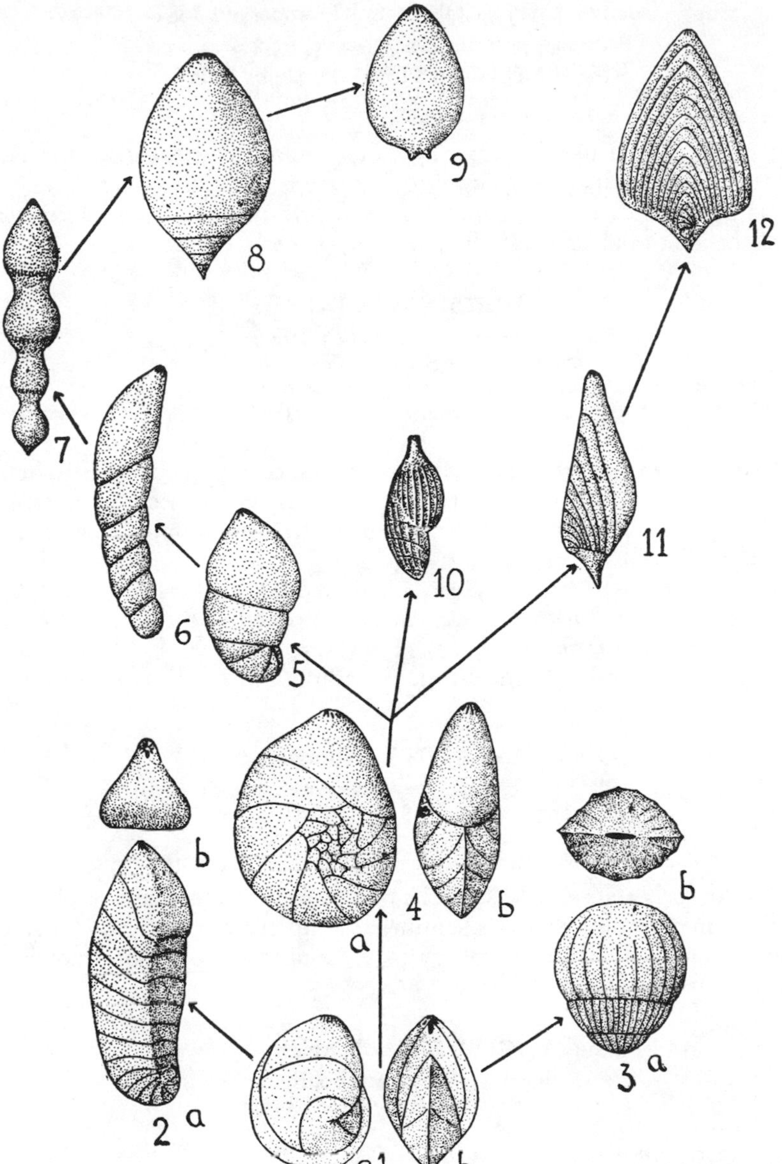

Plate 16

Genus AMPHICORYNE Schlumberger, 1881

Plate 16, figure 10; Key, plate 20, figures 14, 15
Genotype, *Marginulina falx* Jones and Parker

Amphicoryne SCHLUMBERGER, Comptes Rendus Acad. Sci., 1881, p. 881.
Marginulina (part) JONES and PARKER, 1860.

Test in the young like a compressed *Lenticulina* loosely coiled, the last-formed chambers like *Nodosaria.* — Cretaceous, Tertiary, Recent.

It is probable that at least most of the forms assigned to *Amphicoryne* represent abnormal ones rather than truly generic characters.

Genus SARACENARIA Defrance, 1824

Plate 16, figure 2; Key, plate 21, figures 8, 9
Genotype, *Saracenaria italica* Defrance

Saracenaria DEFRANCE, Dict. Sci. Nat., vol. 32, 1824, p. 177.
Cristellaria (part) of authors. *Hemirobulina* (part) STACHE, 1864. *Saracenella* FRANKE, 1936.

Test with the earliest chambers close coiled, especially in the microspheric form, later uncoiling, usually triangular in transverse section; aperture radiate, at the peripheral angle, median ventral slit usually enlarged. — Jurassic to Recent.

Franke (Abhandl. Preuss. Geol. Landes., Heft. 169, 1936, p. 87) has proposed the generic name *Saracenella* for uncoiled forms with the genotype, *Saracenaria trigona* Terquem. However, the type of *Saracenaria* is apparently an uncoiled form also in the adult, and so the two are therefore synonyms.

Genus LINGULINA d'Orbigny, 1826

Plate 16, figure 3; Key, plate 20, figure 13
Genotype, *Lingulina carinata* d'Orbigny

Lingulina D'ORBIGNY, Ann. Sci. Nat., vol. 7, 1826, p. 256.
Nodosaria (*Mucronina*) D'ORBIGNY, 1826. *Lingulinopsis* REUSS, 1860.

Test in the early stages, at least in the microspheric form, planispiral, later chambers in a rectilinear series, compressed; aperture becoming elongate, elliptical, terminal, with a tendency in some species to become radial or with teeth. — Permian to Recent.

Genus VAGINULINA d'Orbigny, 1826

Plate 16, figure 11; Key, plate 20, figures 10–12
Genotype, *Nautilus legumen* Linné

Vaginulina D'ORBIGNY, Ann. Sci. Nat., vol. 7, 1826, p. 257.
Nautilus (part) LINNÉ, 1758. *Citharina* D'ORBIGNY, 1839.

Test compressed, usually with one margin of the test straight, representing the periphery in coiled forms, the other typically convex, early stages often somewhat coiled in the microspheric form, sides flat or convex; aperture at the peripheral angle, radiate. — Triassic to Recent.

The Cretaceous forms named *Citharina* by Reuss are often very distinctive, and perhaps this genus should be recognized. An examination of the literature will show the difficulties authors have had to distinguish *Vaginulina*, *Marginulina*, "*Cristellaria*" and *Dentalina*.

Genus PALMULA Lea, 1833

Key, plate 20, figures 8, 9
Genotype, *Palmula sagittaria* Lea

Palmula LEA, Contrib. to Geology, 1833, p. 219.
Flabellina D'ORBIGNY, 1839. *Frondicularia* (part) of authors. *Frondicularia* MÜNSTER, 1838 (not Lamarck).

Test similar to *Frondicularia*, but the early chambers coiled both in the microspheric and megalospheric forms. — Jurassic to Recent.

There are many aberrant forms of *Palmula* and *Frondicularia*, some of which have been given generic names.

Genus FRONDICULARIA Defrance, 1824

Plate 16, figure 12; Key, plate 21, figures 3–5
Genotype, *Frondicularia complanata* Defrance

Frondicularia DEFRANCE, Dict. Sci. Nat., vol. 32, 1824, p. 178.

Test much compressed, in the early stage in the microspheric form, sometimes partially coiled, megalospheric form not coiled, later chambers extending back on the two sides of the test forming inverted V-shaped (chevron-shaped) chambers; aperture terminal, radiate. — Carboniferous to Recent.

Some of the Cretaceous species have a supplementary angle and have been named *Tribrachia* by Schubert. The name may perhaps be used for these specialized species.

Genus DYOFRONDICULARIA Asano, 1936

Key, plate 48, figure 10
Genotype, *Dyofrondicularia nipponica* Asano

Dyofrondicularia ASANO, Jap. Journ. Geol. Geogr., vol. 13, 1936, p. 330.

"Test much compressed, early chambers inverted V-shaped, later portion biserial; wall calcareous, finely perforate; aperture radiate." — Pliocene.

Genus PARAFRONDICULARIA Asano, 1938

Key, plate 48, figure 9
Genotype, *Parafrondicularia japonica* Asano

Parafrondicularia ASANO, Tohoku Imp. Univ., Sci. Rep'ts, ser. 2 (Geol.), vol. 19, 1938, p. 189.

"Test compressed; early chambers biserial, later portion inverted V-shape; wall calcareous, very finely perforate; aperture terminal, radiate." — Pliocene to Recent.

Genus KYPHOPYXA Cushman, 1929

Key, plate 21, figures 1, 2
Genotype, *Frondicularia christneri* Carsey

Kyphopyxa Cushman, Contr. Cushman Lab. Foram. Res., vol. 5, 1929, p. 1.
Frondicularia (part) of authors.

Test compressed, in the microspheric form with chambers as in the young of *Palmula*, later with a series of alternating biserial chambers, adult as in *Frondicularia* or bending back at the base, and the chambers overlapping; aperture terminal, radiate. — Upper Cretaceous, America.

Genus FLABELLINELLA Schubert, 1900

Key, plate 20, figures 16, 17
Genotype, *Frondicularia tetschensis* Matouschek

Flabellinella Schubert, Zeitschr. Deutsch. Geol. Ges., vol. 52, 1900, p. 551.
Frondicularia (part) of authors.

Test in the early stages similar to *Vaginulina*, the adult similar to *Frondicularia*. — Upper Cretaceous.

Citharinella Marie (Bull. Soc. Geol. France, ser. 5, vol. 8, 1938, p. 99) seems to be very close to this genus. The species included in *Citharinella* are quite distinct from the type of *Flabellinella* in having a basal spine, numerous chambers, and in being strongly ornamented, but these do not seem to be generic differences.

Genus SPORADOGENERINA Cushman, 1927

Plate 22, figure 13; Key, plate 28, figure 16
Genotype, *Sporadogenerina flintii* Cushman

Sporadogenerina Cushman, Contr. Cushman Lab. Foram. Res., vol. 2, pt. 4, 1927, p. 95.
Ramulina (part) Flint.

Test with a globular proloculum followed by several globular chambers usually arranged in a dentaline or marginuline arrangement, later chambers irregularly uniserial and elongate, sometimes branching; wall calcareous, very finely perforate, glassy; apertures in the early stages one to each chamber, terminal, radiate, in the later chambers numerous, protruding, radiate, irregularly placed on the surface of the chamber. — Recent.

See Cushman and Todd, Relationships of the Genus *Sporadogenerina* (Contr. Cushman Lab. Foram. Res., vol. 19, 1943, pp. 93–95, pl. 16, figs. 11–15).

Subfamily 2. Lageninae

Test consisting of a single chamber; aperture variously formed, radiate or rounded.

Genus LAGENA Walker and Jacob, 1798

Plate 16, figure 9; Key, plate 21, figures 18, 24

Genotype, *Serpula (Lagena) sulcata* Walker and Jacob

Lagena WALKER and JACOB, in Kanmacher's ed. of Adams' Essays Micr., 1798, p. 634. *Vermiculum* MONTAGU, 1803. *Lagenula* (?) MONTFORT, 1808. *Oolina* D'ORBIGNY, 1839. *Amphorina* D'ORBIGNY, 1849. *Ovulina* EHRENBERG, 1854. *Phialina* COSTA, 1856. *Tetragonulina, Trigonulina,* and *Obliquina* SEGUENZA, 1862. *Ovolina* TERQUEM, 1866. *Lagenulina* TERQUEM, 1876. *Capitellina* MARSSON, 1876.

Test consisting of a single chamber; wall calcareous, finely to coarsely perforate, often highly ornamented, with or without a neck; aperture radiate (rarely), rounded, elliptical or slit-like, terminal.—Jurassic to Recent, perhaps earlier.

It is very doubtful if many of the forms classed as *Lagena* really belong to this genus or this family. They need much study before their relationships are really known with any certainty. The species are very variable, and there seems little value at the present in trying to distribute them among the various families to which they are probably related. Some of the species are the most extravagantly ornamented of any of the foraminifera. In Vienna I examined specimens of Rzehak's *Balanulina*, and it is an early stage of a barnacle, not a foraminifer at all.

For a discussion of the relationships of this group, see Parr, The Lagenid Foraminifera and their Relationships (Proc. Roy. Soc. Victoria, vol. 58, pt. 1–2 (n. ser.), 1947, pp. 116–133, pls. 6, 7, 1 text fig.).

Family 24. POLYMORPHINIDAE

Test usually free, sometimes attached, in the early stages of most genera spiral, later sigmoid, biserial or uniserial, globular, cylindrical or compressed, in attached forms in a linear series; wall calcareous, very finely perforate, vitreous; aperture terminal, radiate, or in degenerate forms rounded.

The Polymorphinidae although often abundant have only in recent years been carefully studied as a whole from large suites of material from Jurassic to Recent. By a study of the development as seen in section, and from the basal view, a definite idea of the development that has taken place in this family may be clearly seen.

The earliest identifiable genera appear in the Jurassic, although specimens are figured from the Triassic. The simplest form is *Eoguttulina*, which is irregularly coiled about an elongate axis. This was undoubtedly derived from some coiled form of the Lagenidae, such as *Marginulina* or *Vaginulina* by introducing a spiral arrangement of the chambers. From *Eoguttulina* in the Jurassic were developed *Quadrulina*, which had the chambers arranged in a tetraloculine series, and which became extinct in the Jurassic; *Paleopolymorphina*, which became elongate, biserial with inflated chambers, and persisted until the Upper Cretaceous; and *Guttulina*, which probably developed in the late Jurassic, continues to the present ocean, and has its chambers arranged on a quinqueloculine plan.

Of these primitive genera, *Guttulina* gave rise to the genera which have dominated the Cretaceous, and which in turn developed into those specialized forms of the Tertiary and Recent. *Guttulina* gave rise to *Globulina* which is common in the Cretaceous and early Tertiary, but rare after the Miocene. This gradually develops into a triserial globular form which in the Upper Cretaceous became *Pyrulina* by adding elongate biserial

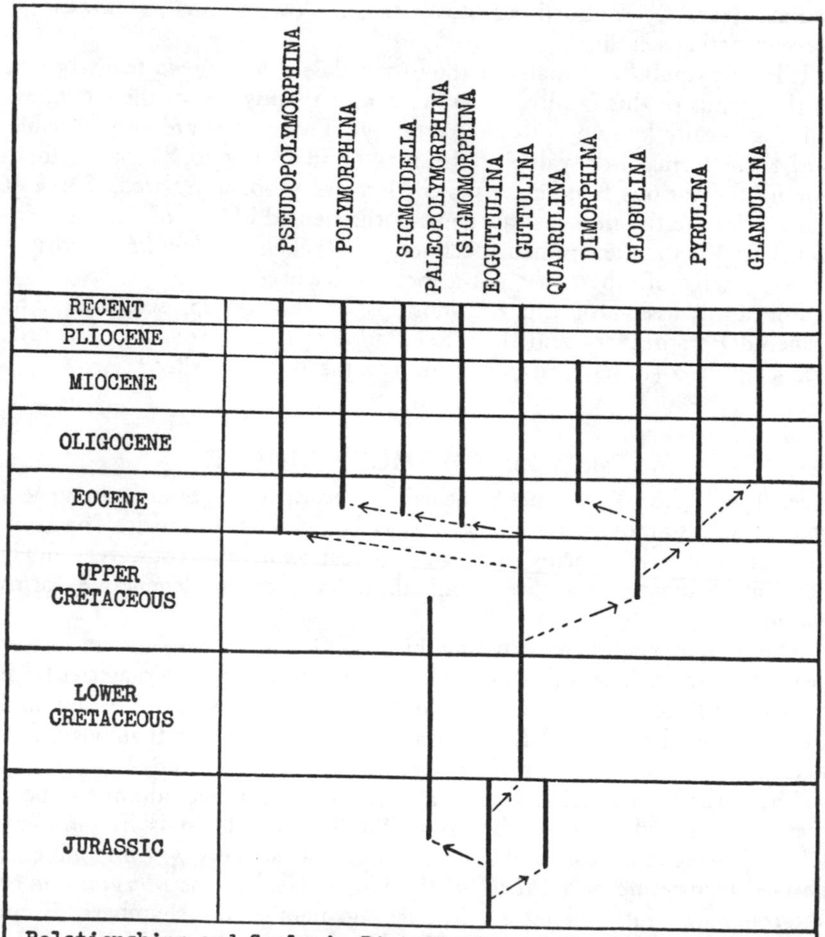

Relationships and Geologic Distribution of the Polymorphinidae

FIGURE 9

chambers. From *Pyrulina* in the Tertiary came *Glandulina* with its biserial stage still persisting in the microspheric form, but often entirely uniserial in the megalospheric form. The uniserial form representing an involute *Nodosaria* is *Pseudoglandulina* of the Lagenidae.

From *Guttulina* in the Upper Cretaceous was developed *Pseudopolymorphina* with biserial chambers, also tending to become uniserial in later development, but not strongly compressed. In the Eocene, *Guttulina* gave rise to forms which became sigmoid and gradually compressed, *Sigmomorphina*, and this in turn to *Polymorphina* where greater compression and a more definite biserial character are taken on. Also in the Eocene from *Guttulina* came *Sigmoidella*, also sigmoid but the chambers extending to the base and embracing.

Dimorphina is a doubtful genus, and as the type is lost, it is difficult to satisfactorily place. Abundant specimens from the Vienna Basin show that most of the specimens are similar to *Marginulina*. Fistulose forms with a great expansion of tubular outgrowths occur in many genera, and represent an adherent condition.

In the Ramulininae there are many doubtful forms, but in the Upper Cretaceous and early Eocene, *Bullopora* is often abundant, and has the same type of wall as the other Polymorphinidae. It should not be confused with *Placopsilina* as has been done in the case of Paleozoic material.

Several generic names have been used for forms that probably belong to this family, as *Misilus*, *Arethusa* and *Cantharus* Montfort, and *Apiopterina* and *Raphanulina* Zborzewski, but the species cannot be recognized, and the genera therefore cannot be used.

The Polymorphinidae when studied carefully in basal view have many parallelisms with the Miliolidae in passing from irregular spiral, through quinqueloculine, triloculine, biloculine and uniserial stages. This parallelism is also seen in the development of the Valvulinidae, where after a maximum of chambers is reached, there is a reduction from five to four, three, two and one in a regular series.

In the Polymorphinidae the geologic history of the group follows excellently the stages in development as shown by the accompanying charts.

For further discussion, and particularly for figures of the species, see Cushman and Ozawa, A Monograph of the Foraminiferal Family Polymorphinidae, Recent and Fossil (Proc. U. S. Nat. Mus., vol. 77, Art. 6, 1930, pp. 1–185, pls. 1–40).

Paalzow has given the generic name *Ramulinella* to a single specimen of a peculiar form from the Jurassic but a further study of the species in a larger series is needed to validate the genus.

Subfamily 1. Polymorphininae

Test with the chambers in a closed spiral or sigmoid series at least in the early stages, later becoming in some genera biserial or uniserial.

Genus EOGUTTULINA Cushman and Ozawa, 1930

Plate 18, figures 1, 2; Key, plate 22, figures 1, 2
Genotype, *Eoguttulina anglica* Cushman and Ozawa
Eoguttulina CUSHMAN and OZAWA, Proc. U. S. Nat. Mus., vol. 77, Art. 6, 1930, p. 16.

Test with the chambers arranged in a spiral series, added in planes less than 90° apart from one another, each succeeding chamber removed farther from the base. — Jurassic, Lower Cretaceous.

Genus QUADRULINA Cushman and Ozawa, 1930

Plate 18, figure 3; Key, plate 22, figures 3, 4
Genotype, *Polymorphina rhabdogonoides* Chapman
Quadrulina CUSHMAN and OZAWA, Proc. U. S. Nat. Mus., vol. 77, Art. 6, 1930, p. 18.
Polymorphina (part) of authors (not D'ORBIGNY).

Test with the chambers added in planes 90° apart from one another, that is, arranged in a tetraloculine series, at least in the later stages. — Jurassic, Cretaceous.

Genus PALEOPOLYMORPHINA Cushman and Ozawa, 1930

Plate 18, figure 4; Key, plate 22, figure 7
Genotype, *Polymorphina pleurostomelloides* Franke
Paleopolymorphina CUSHMAN and OZAWA, Proc. U. S. Nat. Mus., vol. 77, Art. 6, 1930, p. 112.
Polymorphina (part) of authors (not D'ORBIGNY).

Test elongate, early chambers spiral, later ones becoming biserial. — Lower and Upper Cretaceous.

Genus GUTTULINA d'Orbigny, 1839

Plate 18, figures 5, 6; Key, plate 22, figures 5, 6
Genotype, *Polymorphina* (*Guttulina*) *communis* d'Orbigny
"Guttulines, Les," D'ORBIGNY, Ann. Sci. Nat., vol. 7, 1826, p. 266. — *Guttulina* D'ORBIGNY, in DE LA SAGRA, Hist. Phys. Pol. Nat. Cuba, 1839, "Foraminifères," p. 132.
Sigmoidina CUSHMAN and OZAWA, 1928. *Sigmomorpha* CUSHMAN and OZAWA, 1928.

Test with the chambers more or less elongated, added in planes 144° apart from one another, that is, in a quinqueloculine series, each chamber as added removed farther from the base. — Jurassic to Recent.

Genus PSEUDOPOLYMORPHINOIDES van Bellen, 1946

Key, plate 48, figure 16
Genotype, *Pseudopolymorphinoides limburgensis* van Bellen
Pseudopolymorphinoides VAN BELLEN, Med. Geol. Stichting, ser. C-V, No. 4, 1946, p. 41.

"Test short, consisting of an inflated lower part of quinqueloculinely ar-

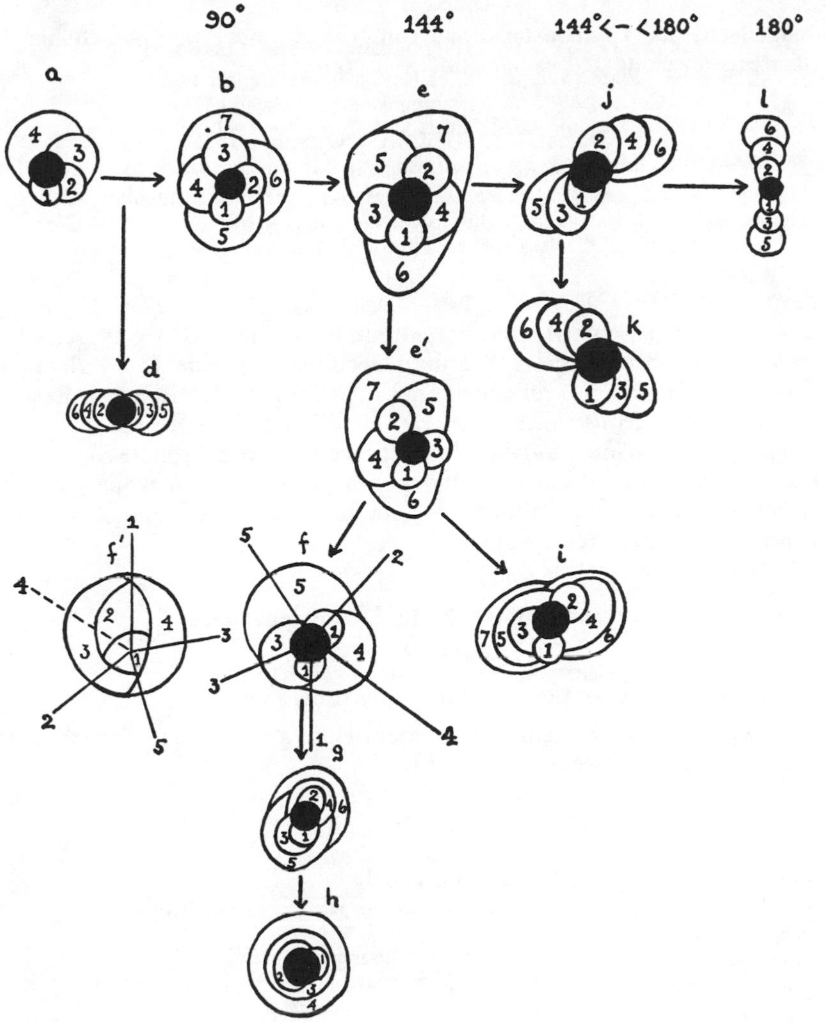

PLATE 17

Idealized basal views of various genera of the Polymorphinidae to show the arrangement of chambers (after Cushman and Ozawa). Proloculum in black. *a. Eoguttulina* (spiral). *b. Quadrulina* (tetraloculine). *d. Paleopolymorphina* (spiral-biserial). *e, e′. Guttulina*; *e*, clockwise quinqueloculine; *e′*, contraclockwise quinqueloculine. *f, f′. Globulina*; *f*, microspheric form; *f′*, megalospheric form. *g. Pyrulina* (quinqueloculine-biserial). *h. Glandulina* (biserial-uniserial). *i. Pseudopolymorphina* (quinqueloculine-biserial). *j, k. Sigmomorphina, Sigmoidella* (sigmoidal); *j*, clockwise; *k*, contraclockwise. *l. Polymorphina* (biserial).

ranged chambers and a terminal, compressed upper part which has an aperture, an elongated radiate slit." — Eocene.

Genus GLOBULINA d'Orbigny, 1839

Plate 18, figure 12; Key, plate 22, figures 8–10

Genotype, *Polymorphina* (*Globulina*) *gibba* d'Orbigny

"Globulines, Les," D'ORBIGNY, Ann. Sci. Nat., vol. 7, 1826, p. 266. — *Globulina* D'ORBIGNY, in DE LA SAGRA, Hist. Phys. Pol. Nat. Cuba, 1839, "Foraminifères," p. 134.

Aulostomella ALTH, 1850.

Test globular or somewhat elongate, rounded or somewhat compressed in section; chambers somewhat quinqueloculine, but due to overlapping, appearing triserial; sutures usually not depressed. — Cretaceous to Recent; abundant Cretaceous to Miocene, rare in Pliocene and Recent.

The generic name, *Raphanulina* Zborzewski, has been used recently instead of *Globulina*, but the only species, *R. humboldti*, while globular, is not identifiable from either figure or description. *Globulina* does not appear before the Cretaceous.

Genus DIMORPHINA d'Orbigny, 1826

Key, plate 22, figure 19

Genotype, *Dimorphina tuberosa* d'Orbigny

Dimorphina D'ORBIGNY, Ann. Sci. Nat., vol. 7, 1826, p. 264.

Test with the early chambers apparently in a globuline triserial form, later uniserial. — Eocene to Miocene.

PLATE 18

RELATIONSHIPS OF THE GENERA OF THE POLYMORPHINIDAE

FIG.

1, 2. *Eoguttulina polygona* (Terquem). Jurassic, England.
3. *Quadrulina rhabdogonioides* (Chapman). (After Chapman.) Cretaceous, England.
4. *Paleopolymorphina pleurostomelloides* (Franke). Cretaceous, Germany.
5. *Guttulina problema* d'Orbigny. Miocene, France.
6. *Guttulina* (*Sigmoidina*) *pacifica* (Cushman and Ozawa). Recent, Philippines.
7. *Sigmoidella kagaensis* Cushman and Ozawa. Recent, Japan.
8. *Sigmomorphina frondiculariformis* (Galloway and Wissler). Pliocene, California.
9. *Polymorphina complanata* d'Orbigny. Miocene, Austria.
10. *Pseudopolymorphina hanzawai* Cushman and Ozawa. Pliocene, Japan.
11. *Pseudopolymorphina jonesi* Cushman and Ozawa. Miocene, France.
12. *Globulina gibba* d'Orbigny. Miocene, Austria.
13. *Pyrulina fusiformis* (Roemer). Oligocene, Germany.
14. *Glandulina laevigata* d'Orbigny. Miocene, Austria.

(Figures after Cushman and Ozawa)

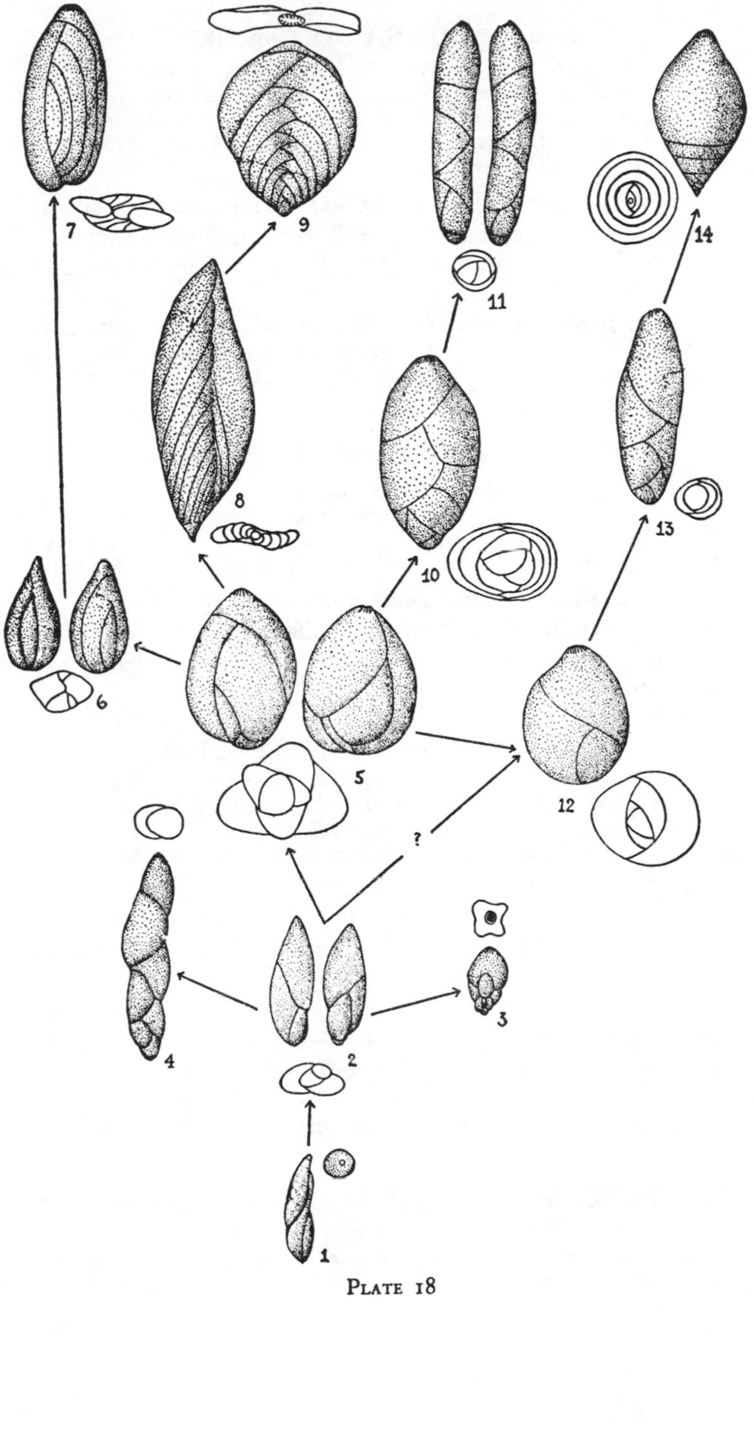

PLATE 18

Genus PYRULINA d'Orbigny, 1839

Plate 18, figure 13; Key, plate 22, figures 11, 12, 14
Genotype, *Polymorphina* (*Pyrulina*) *gutta* d'Orbigny
"Pyrulines, Les," D'ORBIGNY, Ann. Sci. Nat., vol. 7, 1826, p. 267. — *Pyrulina* D'ORBIGNY, in DE LA SAGRA, Hist. Phys. Pol. Nat. Cuba, 1839, "Foraminifères," p. 107.
Pyrulinella CUSHMAN and OZAWA, 1928.

Test elongate or fusiform, with the early chambers triloculine, later biserial, in the early stage of microspheric form quinqueloculine. — Upper Cretaceous to Recent; common Upper Cretaceous to Miocene, rare in Pliocene and Recent.

The generic name, *Apiopterina* Zborzewski, has been used recently instead of *Pyrulina*, but the only species of *Apiopterina*, *A. orbignyi*, is not identifiable from either figure or description.

Genus PYRULINOIDES Marie, 1941

Key, plate 48, figure 18
Genotype, *Pyrulina acuminata* d'Orbigny
Pyrulinoides MARIE, Mém. Mus. Nat. Hist. Nat., n. ser., vol. 12, 1941, p. 169.
Pyrulina (part), *Paleopolymorphina* (part), and *Pyrulinella* (part) of authors.

Test elongate, composed of a regular spiral series of chambers, two to a whorl, 180° apart; chambers elongate, oblique, strongly embracing; all sutures visible on both sides; aperture radiate at the end of the last chamber. — Cretaceous.

Genus GLANDULINA d'Orbigny, 1826

Plate 18, figure 14; Key, plate 22, figure 13
Genotype, *Nodosaria* (*Glandulina*) *laevigata* d'Orbigny
Glandulina D'ORBIGNY, Ann. Sci. Nat., vol. 7, 1826, p. 251.
Guttulina and *Polymorphina* (part) of authors. *Psecadium* NEUGEBOREN, 1856.
Atractolina v. SCHLICHT, 1870.

Test with the early chambers, at least in the microspheric form, biserial, later uniserial, rectilinear, with septa almost horizontal and parallel. — Tertiary, Recent.

The type species is biserial in the microspheric form, and there are other species that show this character. It should be distinguished from *Pseudoglandulina* derived from *Nodosaria*. Authors have been misled by the poor figures of *Psecadium* and *Atractolina* into thinking that the early chambers were coiled, but type material shows them to be biserial.

Genus PSEUDOPOLYMORPHINA Cushman and Ozawa, 1928

Plate 18, figures 10, 11; Key, plate 22, figure 15
Genotype, *Pseudopolymorphina hanzawai* Cushman and Ozawa
Pseudopolymorphina CUSHMAN and OZAWA, Contr. Cushman Lab. Foram. Res., vol. 4, 1928, p. 15.

Polymorphina (part) and *Guttulina* (part) of authors. *Bulimina* (part) BAGG, 1912 (not D'ORBIGNY). *Cristellaria* (part) KARRER, 1868 (not LAMARCK).

Test elongate, usually somewhat compressed, early chambers quinqueloculine, later ones becoming biserial, slightly overlapping. — Upper Cretaceous to Recent.

Genus SIGMOMORPHINA Cushman and Ozawa, 1928

Plate 18, figure 8; Key, plate 22, figures 16–18

Genotype, *Sigmomorphina yokoyamai* Cushman and Ozawa

Sigmomorphina CUSHMAN and OZAWA, Contr. Cushman Lab. Foram. Res., vol. 4, 1928, p. 17.

Polymorphina (part) of authors.

Test compressed, in the adult at least, with chambers added in planes slightly less than 180° and more than 144° from one another, each succeeding chamber farther removed from the base. — Cretaceous to Recent.

Genus SIGMOIDELLA Cushman and Ozawa, 1928

Plate 18, figure 7; Key, plate 22, figure 20

Genotype, *Sigmoidella kagaensis* Cushman and Ozawa

Sigmoidella CUSHMAN and OZAWA, Contr. Cushman Lab. Foram. Res., vol. 4, 1928, p. 18.

Polymorphina (part) of authors.

Test compressed, with the chambers in sigmoid series, but each reaching to the base and embracing the preceding ones of the same series on one side. — Tertiary, Recent.

Genus POLYMORPHINA d'Orbigny, 1826

Plate 18, figure 9; Key, plate 22, figures 21, 22

Genotype, *Polymorphina burdigalensis* d'Orbigny

Polymorphina D'ORBIGNY, Ann. Sci. Nat., vol. 7, 1826, p. 265.

Test usually broad and compressed, early chambers arranged in a sigmoid series, becoming biserial or entirely biserial from the start, the sigmoid character usually seen slightly throughout the biserial test. — Tertiary, Recent.

This is the most advanced group of the family, and only developed from the Eocene onward. It is easily distinguished from *Sigmomorphina* from which it is derived by acceleration.

Genus POLYMORPHINELLA Cushman and Hanzawa, 1936

Key, plate 48, figure 17

Genotype, *Polymorphinella vaginulinaeformis* Cushman and Hanzawa

Polymorphinella CUSHMAN and HANZAWA, Contr. Cushman Lab. Foram. Res., vol. 12, 1936, p. 46.

Test free, compressed, elongate, with one margin straight or slightly convex, the other strongly convex, in the young stage biserial, later becoming more nearly uniserial; wall calcareous, very finely perforate; aperture radiate, with an apertural chamberlet at the dorsal angle. — Upper Cretaceous to Recent.

Genus POLYMORPHINOIDES Cushman and Hanzawa, 1936

Key, plate 48, figure 19

Genotype, *Polymorphinoides spiralis* Cushman and Hanzawa

Polymorphinoides Cushman and Hanzawa, Contr. Cushman Lab. Foram. Res., vol. 12, 1936, p. 48.

Test free, compressed, elongate, early portion coiled, evolute, but bilaterally asymmetrical in the sagittal plane, convex on one side, slightly umbilicate on the other, later portion becoming uncoiled and biserial as in *Polymorphinella*; wall calcareous, finely perforate; aperture radiate at the peripheral angle. — Pleistocene.

Subfamily 2. Ramulininae

Test free or attached, chambers typically separated by stoloniferous connections.

Genus RAMULINA Rupert Jones, 1875

Key, plate 22, figure 23

Genotype, *Ramulina laevis* Rupert Jones

Ramulina Rupert Jones, in J. Wright, Rep't Proc. Belfast Nat. Field Club, 1873–74, App. III, 1875, p. 88 (90).

Test usually free, branching, typically consisting of more or less rounded chambers connected by long stoloniferous tubes; wall calcareous, finely perforated; apertures rounded, at the ends of the tubes. — Jurassic to Recent.

Genus BULLOPORA Quenstedt, 1856

Key, plate 22, figure 24

Genotype, *Bullopora rostrata* Quenstedt

Bullopora Quenstedt, Der Jura, 1856, p. 292.

Webbina (part) d'Orbigny, 1850. *Vitriwebbina* Chapman, 1892. *Nodobacularia* Rhumbler, 1895.

Test attached, consisting of a series of inflated chambers, in a generally linear series, the chambers usually connected by definite tubular necks; wall calcareous, finely perforate, vitreous; aperture the end of the tubular neck, rounded. — Jurassic to Paleocene.

Bullopora is a degenerate attached form, often abundant, particularly in the Cretaceous. The Paleozoic species referred to this genus are entirely different, and belong to *Placopsilina*.

The following genera recently erected by Marie are placed in this family with some question until the early stages can be studied further.

Genus ENANTIOMORPHINA Marie, 1941

Key, plate 48, figure 20
Genotype, *Enantiomorphina lemoinei* Marie
Enantiomorphina Marie, Mém. Mus. Nat. Hist. Nat., n. ser., vol. 12, 1941, p. 144.
Marginulina (part) and *Dentalina* (part) of authors.

Test subcylindrical, circular in transverse section, composed of alternating chambers in a uniserial arrangement; sutures oblique; wall calcareous, finely perforate; aperture terminal, radiate, slightly projecting. — Cretaceous.

Genus ENANTIODENTALINA Marie, 1941

Key, plate 48, figure 21
Genotype, *Dentalina communis* d'Orbigny
Enantiodentalina Marie, Mém. Mus. Nat. Hist. Nat., n. ser., vol. 12, 1941, p. 149.
Dentalina (part), *Nodosaria* (part), and *Svenia* (part) of authors.

Test elongate, subcylindrical, straight or slightly arcuate, chambers alternating in the early portion, a single series in the adult and less embracing, sutures very oblique; aperture terminal, radiate, slightly projecting. — Jurassic (?), Cretaceous to Recent.

Genus ENANTIOVAGINULINA Marie, 1941

Key, plate 48, figure 23
Genotype, *Cristellaria recta* d'Orbigny
Enantiovaginulina Marie, Mém. Mus. Nat. Hist. Nat., n. ser., vol. 12, 1941, p. 160.
Cristellaria (part) of authors.

Test elongate, laterally compressed, slightly triangular in section; chambers slightly curved, alternating throughout; aperture terminal, radiate. — Upper Cretaceous.

Genus ENANTIOCRISTELLARIA Marie, 1941

Key, plate 48, figure 24
Genotype, *Cristellaria navicula* d'Orbigny
Enantiocristellaria Marie, Mém. Mus. Nat. Hist. Nat., n. ser., vol. 12, 1941, p. 162.
Cristellaria (part) and *Lenticulina* (part) of authors.

Test close coiled, composed of alternating chambers; wall calcareous, finely perforate; aperture radiate, at the peripheral angle of the last-formed chamber. — Upper Cretaceous.

Genus ENANTIOMARGINULINA Marie, 1941

Key, plate 48, figure 22
Genotype, *Enantiomarginulina d'Orbignyi* Marie, 1941
Enantiomarginulina Marie, Mém. Mus. Nat. Hist. Nat., n. ser., vol. 12, 1941, p. 163.

Test composed of an early spiral stage and the adult uncoiled; chambers of the earliest portion alternating, later uniserial; wall calcareous, finely perforate; aperture radiate, at the upper end of the peripheral margin of the last-formed chamber. — Upper Cretaceous.

Family 25. NONIONIDAE

Test typically planispiral, more or less involute, in the adult trochoid in a few genera, and even uncoiled; wall calcareous, finely perforate; aperture simple or cribrate, if simple at or near the base of the apertural face.

The Nonionidae were probably derived from a planispirally coiled form as were the Lagenidae. In the Nonionidae the simple aperture is at the base of the apertural face. Such forms may have easily been derived from such forms as *Trochamminoides* and *Glomospira* where the originally arenaceous test has become largely one of pure cementing material. In *Nonion* the test is typically planispiral, but in *Nonionella* trochoid forms are developed to varying degrees. The division between the simple forms of *Nonion* and *Elphidium* is often so slight as to make the generic determination very difficult. The sutural pores and retral processes of *Elphidium* may be developed slightly in the last portion of the test. In the more specialized species of *Elphidium*, a complex test is developed, often of large size. In *Faujasina*, *Polystomellina*, *Notorotalia*, and *Elphidioides* trochoid forms are developed, and in *Ozawaia* the test becomes uncoiled and the aperture terminal and cribrate. *Astrononion* develops peculiar supplementary chambers.

The genera *Bradyina* and *Cribrospira* are placed in the Lituolidae, as these are truly arenaceous forms allied to *Haplophragmoides* and *Endothyra*. *Pullenia* has a similar form to *Nonion*, but its development places it in the Chilostomellidae. *Hantkenina* has occasionally been placed with the Nonionidae, but its early development shows that it developed from the Globigerinidae. *Laticarinina*, placed by its authors first in the Lagenidae and later in the Nonionidae, is truly trochoid in the young, and belongs in the Anomalinidae.

Genus NONION Montfort, 1808

Plate 19, figure 1; Key, plate 23, figures 1, 2

Genotype, *Nautilus incrassatus* Fichtel and Moll

Nonion Montfort, Conch. Syst., vol. 1, 1808, p. 211.

Melonis and *Florilus* Montfort, 1808. *Pulvinulus* (part) Lamarck, 1816. *Placentula* (part) and *Cristellaria* (part) Lamarck, 1822. *Lenticulina* (part) Defrance, 1824 (not Lamarck). *Polystomella* (part) of authors. *Nonionina* d'Orbigny, 1826.

Test free, planispiral, more or less involute, bilaterally symmetrical, periphery, broadly rounded to acute; chambers numerous; wall finely or coarsely perforate, calcareous; aperture median, an arched, usually low opening between the base of the apertural face and the preceding coil.—Jurassic to Recent.

Genus PARANONION Logue and Haas, 1943

Key, plate 48, figure 25
Genotype, *Paranonion venezuelanum* Logue and Haas
Paranonion LOGUE and HAAS, Journ. Pal., vol. 17, 1943, p. 177.

Test similar to *Nonion* but with the aperture in the early stages extending into the apertural face and in the adult entirely separated from the margin. —Tertiary.

Genus HYDROMYLINA de Witt Puyt, 1941

Key, plate 49, figure 1
Genotype, *Hydromylina rutteni* de Witt Puyt
Hydromylina DE WITT PUYT, Geol. Pal. Beschr. Umgebung von Ljubuski, Hercegovina, Utrecht, 1941, p. 54.

Test free, planispiral, bilaterally symmetrical, close coiled; spiral sutures and radial sutures strongly raised into plate-like projections on the exterior, making a complex surface pattern; wall calcareous, perforate; aperture a small, rounded opening in the apertural face.—Eocene.

Genus ASTRONONION Cushman and Edwards, 1937

Key, plate 49, figure 3
Genotype, *Nonionina stelligera* d'Orbigny
Astrononion CUSHMAN and EDWARDS, Contr. Cushman Lab. Foram. Res., vol. 13, 1937, p. 30.

Test free, planispiral, coiled, bilaterally symmetrical or nearly so, periphery broadly rounded; chambers numerous, distinct, usually somewhat inflated, with supplementary tubular or rhomboid chambers on both sides, alternating with the primary chambers; wall calcareous, perforate; aperture of the main series of chambers at the base of the last-formed chamber in the median line, a low arched opening, which in some species may be somewhat subdivided, the supplementary chambers with either rounded openings at the peripheral end, or, in those species with distinctly rhomboid supplementary chambers, with the aperture elongate along the peripheral posterior margin.—Eocene (?), Oligocene to Recent.

Genus NONIONELLA Cushman, 1926

Plate 19, figure 2; Key, plate 23, figures 3, 4
Genotype, *Nonionella miocenica* Cushman
Nonionella CUSHMAN, Contr. Cushman Lab. Foram. Res., vol. 2, 1926, p. 64.
Nonionina (part) of authors. *Pseudononion* ASANO, 1936.

Test free, subtrochoid, dorsal side only partially involute, ventral side completely so, close coiled; chambers in adult inequilateral, ventral side developing a distinct, elongate lobe at the umbilical end, covering the umbilicus; wall calcareous, finely perforate; aperture at the base of the apertural face, low and elongate, extending from the periphery toward the ventral side. — Cretaceous to Recent.

Genus ELPHIDIUM Montfort, 1808

Plate 19, figure 3; Key, plate 23, figure 5
Genotype, *Nautilus macellus* Fichtel and Moll

Elphidium MONTFORT, Conch. Syst., vol. 1, 1808, p. 15.
Geophonus, Pelorus, Andromedes, Sporilus, Themeon, and *Cellanthus* MONTFORT, 1808. *Vorticialis* LAMARCK, 1812. *Polystomella* LAMARCK, 1822. *Robulina* (part) MÜNSTER, 1838. *Geoponus* and *Polystomatium* EHRENBERG, 1839. *Nonionina* (part) BOLL, 1846. *Helicoza* MOEBIUS, 1880.

Test planispiral, bilaterally symmetrical, mostly involute; chambers numerous, with distinct sutures, either depressed, or raised and limbate, with septal bridges ("retral processes"); wall calcareous, perforate; apertures one or more at the base of the apertural face. — Eocene to Recent.

Genus CRIBROELPHIDIUM Cushman and Bronnimann, 1948

Key, plate 48, figure 26
Genotype, *Cribroelphidium kugleri* Cushman and Bronnimann

Cribroelphidium CUSHMAN and BRONNIMANN, Contr. Cushman Lab. Foram. Res., vol. 24, 1948, p. 18.

Test similar to *Elphidium* but with the apertural face with a series of supplementary apertures consisting of small rounded openings, with or without a raised border, in the apertural face. — Tertiary and Recent.

Genus ELPHIDIOIDES Cushman, 1945

Key, plate 49, figure 2
Genotype, *Elphidioides americanus* Cushman

Elphidioides CUSHMAN, Contr. Cushman Lab. Foram. Res., vol. 21, 1945, p. 7.

Test trochoid, dorsal side showing the early coils, ventral side involute; sutures with semicircular openings and retral processes; wall calcareous, perforate; aperture consisting of a low opening on the ventral side at the

---

PLATE 19

RELATIONSHIPS OF THE GENERA OF THE NONIONIDAE

FIG.

1. *Nonion incisum* Cushman. *a,* side view; *b,* apertural view.
2. *Nonionella auris* (d'Orbigny). *a,* dorsal view; *b,* ventral view; *c,* apertural view.
3. *Elphidium macellum* (Fichtel and Moll). *a,* side view; *b,* apertural view.
4. *Faujasina carinata* d'Orbigny. (After d'Orbigny.) *a,* dorsal view; *b,* ventral view; *c,* peripheral view.
5. *Ozawaia tongaensis* Cushman. *a,* side view; *b,* apertural view.
6. *Polystomellina discorbinoides* Yabe and Hanzawa. (After Yabe and Hanzawa.) *a,* dorsal view; *b,* ventral view; *c,* peripheral view.

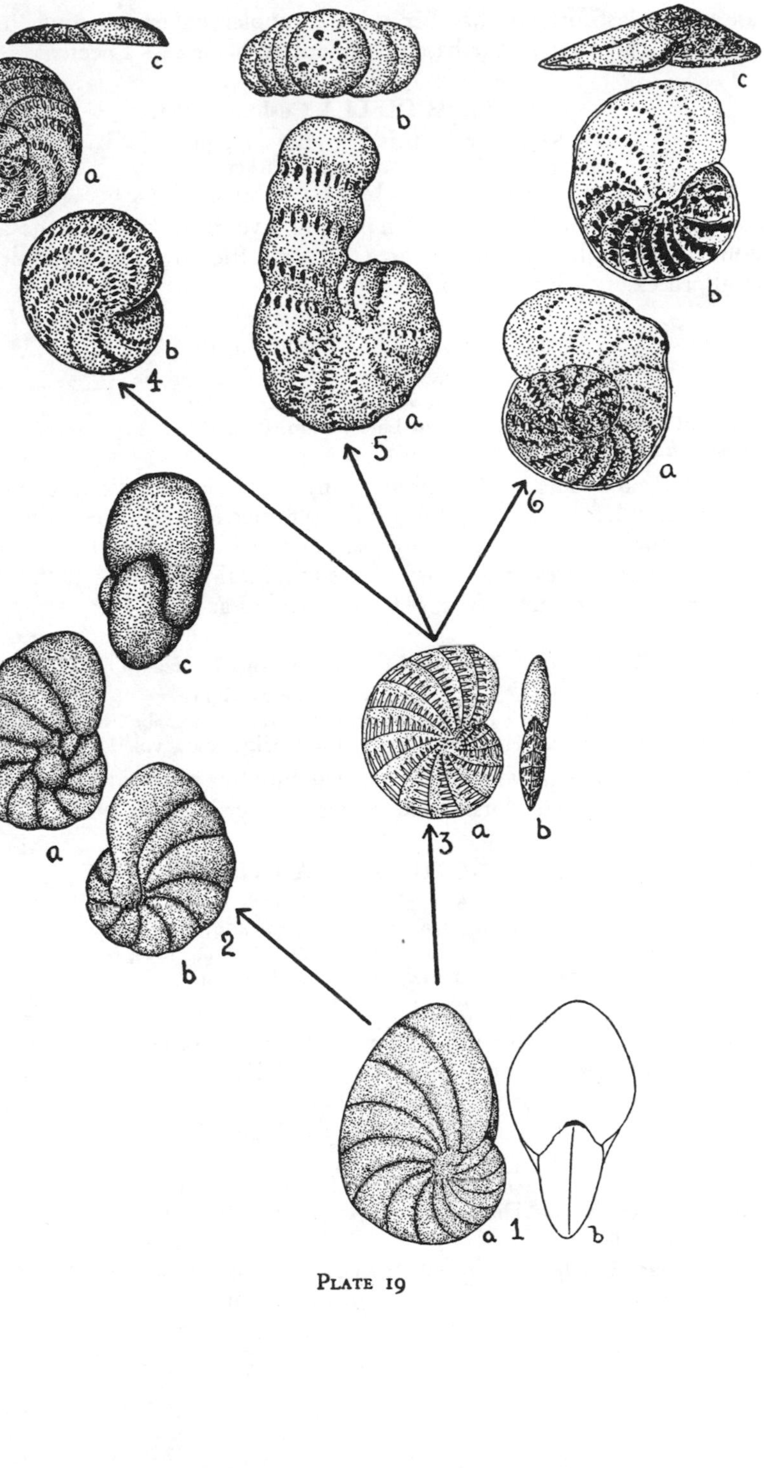

PLATE 19

base of the last-formed chamber with a supplementary opening, irregularly V-shaped, toward the base of the apertural face.—Eocene.

Genus ELPHIDIELLA Cushman, 1936

Key, plate 23, figure 6; plate 49, figure 4
Genotype, *Polystomella arctica* Parker and Jones

Elphidiella Cushman, Contr. Cushman Lab. Foram. Res., vol. 12, 1936, p. 89.

Test differing from *Elphidium* in having two rows of openings at the sutures, and with a thickened area between the rows, without definite retral processes.—Pliocene to Recent.

Genus OZAWAIA Cushman, 1931

Plate 19, figure 5; Key, plate 23, figures 9, 10
Genotype, *Ozawaia tongaensis* Cushman

Ozawaia Cushman, Contr. Cushman Lab. Foram. Res., vol. 7, 1931, p. 80.
Polystomella (part) of authors.

Test free, early stages planispiral, compressed, later uncoiling and becoming circular in section, bilaterally symmetrical; sutures with retral processes throughout; wall calcareous, finely perforate; aperture in young with numerous pores at the base of the apertural face, adult with a series of pores in the terminal face.—Recent, Indo-Pacific.

Genus POLYSTOMELLINA Yabe and Hanzawa, 1923

Plate 19, figure 6; Key, plate 23, figure 7
Genotype, *Polystomella (Polystomellina) discorbinoides* Yabe and Hanzawa

Polystomellina Yabe and Hanzawa, Jap. Journ. Geol. Geogr., vol. 2, 1923, p. 99.

Test similar to *Elphidium* in general structure but trochoid, plano-convex, ventral side flattened, dorsal side convex.—Tertiary, Recent.

Genus NOTOROTALIA Finlay, 1939

Key, plate 49, figure 5
Genotype, *Notorotalia zelandica* Finlay

Notorotalia Finlay, Trans. Roy. Soc. New Zealand, vol. 68, 1939, p. 517.
Rotalia (part) of authors. *Polystomellina* (part) of authors.

Test similar to *Polystomellina* but differing in the heavy reticulation of both dorsal and ventral sides and without a definite aperture at the base of the last-formed chamber.—Tertiary and Recent.

Genus FAUJASINA d'Orbigny, 1839

Plate 19, figure 4; Key, plate 23, figure 8
Genotype, *Faujasina carinata* d'Orbigny

Faujasina d'Orbigny, in De la Sagra, Hist. Phys. Pol. Nat. Cuba, 1839, "Foraminifères," p. 109.

Test similar to *Elphidium* but trochoid, plano-convex, dorsal side flattened, ventral side convex.—Pliocene to Recent.

Family 26. CAMERINIDAE

Test generally planispiral and bilaterally symmetrical, in the early stages involute, in the later stages often evolute; wall calcareous, perforate; in the higher forms with a secondary skeleton and complex canal system.

The simpler forms of the Paleozoic have been placed in this family. Their structure is not yet fully known, but they have been taken by many authors as the source from which the higher forms of later geologic periods developed. There is a progressive development in complexity from the older and simpler forms of *Camerina* to the very complex forms of the Eocene which reached large size and great abundance. From *Camerina* the development is into large complanate compressed forms like *Calcarina*, from which by division into chamberlets *Heterostegina* is derived. From *Heterostegina* it is a simple step to forms like *Cycloclypeus* with annular chambers. The family is a dominant one in the Eocene of Eurasia and Africa, continuing into the late Tertiary in the East Indian region.

Subfamily 1. Archaediscinae

Test not broken up into chambers.

Genus ARCHAEDISCUS H. B. Brady, 1873

Plate 20, figure 1; Key, plate 23, figures 11, 12
Genotype, *Archaediscus karreri* H. B. Brady
Archaediscus H. B. Brady, Ann. Mag. Nat. Hist., ser. 4, vol. 12, 1873, p. 286.

Test lenticular, consisting of a proloculum and long, undivided, second chamber, close coiled; wall thick, calcareous, finely perforate, upper and lower surfaces thickened; aperture at the open end of the chamber.—Carboniferous.

Subfamily 2. Camerininae

Test with numerous chambers.

Genus NUMMULOSTEGINA Schubert, 1907

Plate 20, figure 2; Key, plate 23, figure 13
Genotype, *Nummulostegina velibitana* Schubert
Nummulostegina Schubert, Verhandl. k. k. Geol. Reichs., 1907, p. 212.

Test lenticular, planispiral, bilaterally symmetrical, divided into chambers, without complex secondary skeleton or canal system; wall calcareous, perforate; aperture narrow, at the base of the apertural face.—Carboniferous.

The genus *Orobias* Eichwald has for its genotype *Nummulina antiquior* Rouillier and Vosinsky, from the Carboniferous of Russia. This is a lenticular form which is much more convex on one side than the other. The interior structure given by Möller is like many of his figures, very much conventionalized. As the types have not been carefully restudied, the genus must be treated as very uncertain until this can be done. This genus has also been noted under the Fusulinidae.

Genus CAMERINA Bruguière, 1792

Plate 20, figure 3; Key, plate 23, figures 14, 15
Genotype, *Camerina laevigata* Bruguière

Camerina BRUGUIÈRE, Encyc. Method., "Vers," pt. 1, 1792, p. 395.
Nautilus (part) of authors. *Phacites* BLUMENBACH, 1799. *Nummulites* LAMARCK, 1801. *Lenticulina* (part) LAMARCK, 1804. *Lycophris* and *Egeon* MONTFORT, 1808. *Nummulina* D'ORBIGNY, 1826. *Nummularia* SOWERBY, 1826.

Test lenticular, planispiral, typically bilaterally symmetrical, involute; wall perforate, calcareous, with a secondary skeleton and complicated canal system; aperture simple, at the base of the apertural face, median. — Carboniferous (?) to Oligocene, perhaps earlier.

There are many subgeneric names for *Camerina* and *Assilina*. For a discussion of the apertural characters, see Grimsdale and Smout, Note on the Aperture in *Nummulites* Lamarck (Abst. Proc. Geol. Soc. London, No. 1436, 1947, pp. 14, 15).

Genus OPERCULINOIDES Hanzawa, 1935

Key, plate 49, figure 6
Genotype, *Operculina wilcoxi* (Heilprin)

Operculinoides HANZAWA, Tohoku Imp. Univ., Sci. Rep'ts, ser. 2, vol. 18, 1935, p. 18.

Similar in structure to *Operculina* but involute in the adult coils. — Tertiary.

Genus PARASPIROCLYPEUS Hanzawa, 1937

Genotype, *Camerina chawneri* D. K. Palmer

Paraspiroclypeus HANZAWA, Journ. Pal., vol. 11, 1937, p. 116.
Camerina (part) of authors.

This form is allied to *Operculina* and *Spiroclypeus* and is interpreted by the author as "an advanced type of *Operculinoides*." No complete description is given.

Genus MISCELLANEA Pfender, 1934

Key, plate 49, figures 7–9
Genotype, *Nummulites miscella* d'Archiac and Haime

Miscellanea PFENDER, Soc. geol. France, Comptes rendus, 1935, p. 80.

Test similar to *Camerina* but the supplementary skeleton degenerating into pillars only and appearing at the surface as large granules, and the marginal cord becoming obsolete. — Upper Cretaceous and Eocene.

Miscellanea represents a somewhat degenerate group very close to *Camerina*.

Genus ASSILINA d'Orbigny, 1826

Plate 20, figure 5; Key, plate 23, figure 16
Genotype, *Assilina discoidalis* d'Orbigny

Assilina d'Orbigny, Ann. Sci. Nat., vol. 7, 1826, p. 296 (as a subgenus of *Nummulina*).
Nummulites (part) of authors.

Test similar to *Camerina* but the test flattened, the chambers usually not completely involute so that the earlier coils are not covered, or with the wall very thin so that the earlier coils are visible from the exterior.—Eocene.

Genus OPERCULINELLA Yabe, 1918

Plate 20, figure 4; Key, plate 23, figures 18, 19
Genotype, *Amphistegina cumingii* W. B. Carpenter

Operculinella Yabe, Sci. Rep. Tohoku Imp. Univ., ser. 2 (Geol.), vol. 4, 1918, p. 126.
Amphistegina W. B. Carpenter, 1859 (not d'Orbigny). *Nummulites* H. B. Brady, 1884 (not Lamarck).

Test lenticular and involute in the young, bilaterally symmetrical, in the adult with a broadly flaring, complanate border, chambers simple; wall calcareous, perforate; aperture at the base of the apertural face, median.—Tertiary, Recent.

Genus OPERCULINA d'Orbigny, 1826

Plate 20, figure 6; Key, plate 23, figure 17
Genotype, *Lenticulites complanata* Defrance

Operculina d'Orbigny, Ann. Sci. Nat., vol. 7, 1826, p. 281.
Nautilus (part) of authors. *Lenticulites* (part) Defrance, 1822. *Amphistegina* (part) d'Orbigny, 1826. *Nonionina* (part) Williamson, 1852. *Nummulina* (part) Parker and Jones, 1865.

Test bilaterally symmetrical, planispiral, complanate, usually all coils visible from the exterior, earlier coils sometimes involute, chambers undivided, periphery with a thickened "marginal cord"; wall calcareous, perforate, smooth or ornamented with bosses; aperture single, at the base of the apertural face, median.—Lower Cretaceous to Recent.

Subgenus SULCOPERCULINA Thalmann, 1938

Key, plate 49, figure 10
Subgenotype, *Camerina* (?) *dickersoni* D. K. Palmer

Sulcoperculina Thalmann, Eclogae geologicae Helvetiae, vol. 31, 1938, p. 330.

Test differing from typical *Operculina* in having a peripheral canal.—Upper Cretaceous.

Genus HETEROSTEGINA d'Orbigny, 1826

Plate 20, figure 7; Key, plate 23, figure 20
Genotype, *Heterostegina depressa* d'Orbigny
Heterostegina D'ORBIGNY, Ann. Sci. Nat., vol. 7, 1826, p. 305.

Test in general similar to *Operculina*; early chambers simple, later ones divided into chamberlets; aperture a row of rounded openings on the narrow apertural face. — Eocene to Recent.

Genus SPIROCLYPEUS H. Douvillé, 1905

Key, plate 23, figure 22
Genotype, *Spiroclypeus orbitoideus* H. Douvillé
Spiroclypeus H. DOUVILLÉ, Bull. Soc. Géol. France, ser. 4, vol. 5, 1905, p. 458.

Test somewhat similar to *Heterostegina* but more accelerated, the curved chambers divided into chamberlets beginning almost immediately after the proloculum, lateral chambers and pillars developed at each side of the test. — Eocene to Miocene.

Genus HETEROCLYPEUS Schubert, 1906

Genotype, *Heterostegina cycloclypeus* A. Silvestri
Heteroclypeus SCHUBERT, Centralbl. für Min., 1906, p. 640.
Heterostegina (part) of authors.

Test similar to *Heterostegina* in the young, but the chambers becoming annular in the later development. — Tertiary.

This genus has not been well figured, nor is there any complete description based on authentic specimens.

PLATE 20

RELATIONSHIPS OF THE GENERA OF THE CAMERINIDAE

FIG.

1. *Archaediscus karreri* H. B. Brady. (After H. B. Brady.) *a*, side view; *b*, edge view; *c*, section.
2. *Nummulostegina velibitana* Schubert. (After Schubert.) *a*, side view; *b*, edge view.
3. *Camerina budensis* (Hantken). (After Hantken.) *a*, side view; *b*, edge view; *c*, section.
4. *Operculinella cumingii* (Carpenter). (After H. B. Brady.) *a*, side view; *b*, edge view.
5. *Assilina undata* d'Orbigny. (After d'Orbigny.) *a*, side view; *b*, edge view.
6. *Operculina bartschi* Cushman.
7. *Heterostegina depressa* d'Orbigny. (After H. B. Brady.) *a*, side view; *b*, edge view.
8. *Cycloclypeus guembelianus* H. B. Brady. (After H. B. Brady.) *a*, side view; *b*, edge view.

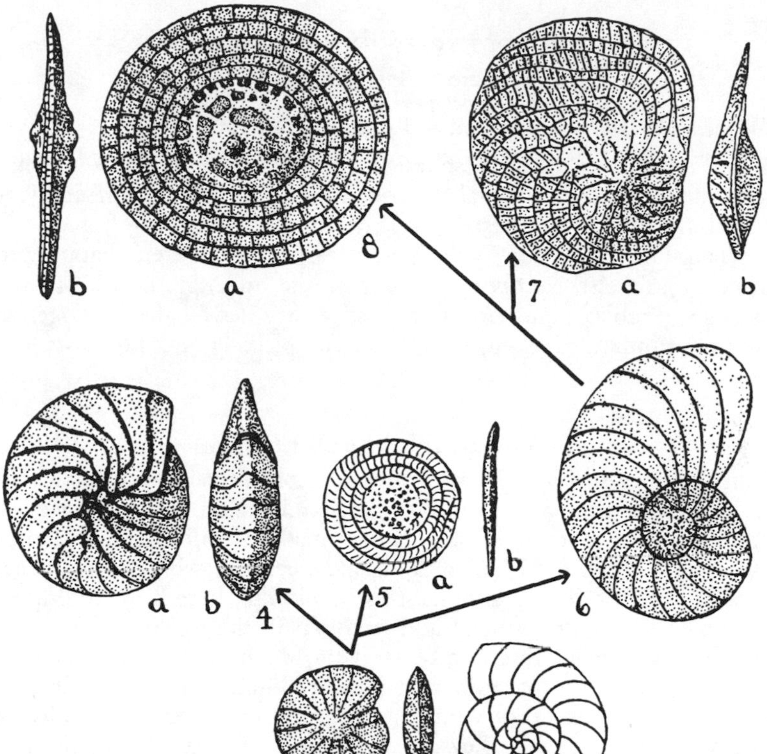

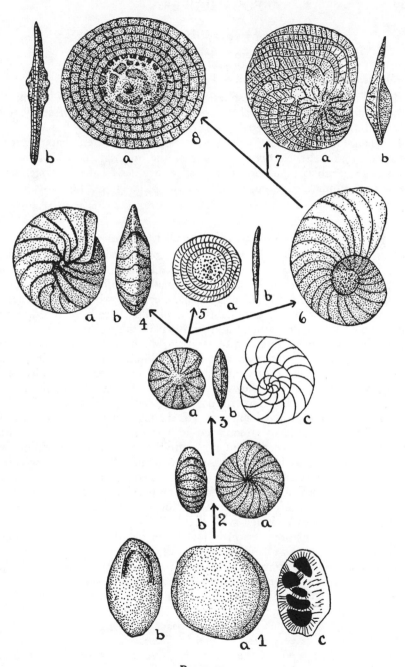

PLATE 20

Genus CYCLOCLYPEUS W. B. Carpenter, 1856

Plate 20, figure 8; Key, plate 23, figure 21
Genotype, *Cycloclypeus carpenteri* H. B. Brady
Cycloclypeus W. B. Carpenter, Philos. Trans., vol. 146, 1856, p. 555.

Test in the young of the microspheric form like *Heterostegina*, later chambers annular, divided by radial portions into rectangular chamberlets, test discoidal and much compressed. — Tertiary, Recent.

See Tan, On the Genus *Cycloclypeus* Carpenter (Wetenschapp. Mededeel. No. 19, Dienst. Mijnbouw. Nederl. Indie, 1932, pp. 1–194, pls. 1–24, 4 text figs., 7 tables), and Cosijn, Statistical Studies on the Phylogeny of Some Foraminifera (Leiden, 1938, pp. 1–70, pls. 1–5, text figs. 1–12).

Family 27. PENEROPLIDAE

Test in the early stages, at least of the more primitive genera, planispiral, then becoming annular or uncoiling, chambers typically divided into chamberlets in all but the most primitive genera; wall calcareous, imperforate, except the proloculum and second chamber which are distinctly perforate; aperture in the simpler forms, slit-like, becoming multiple in the complex forms, or rounded and terminal in the uncoiled forms.

The relationships of this family are very obscure. The early stages of the more primitive forms show that they are bilaterally symmetrical, and from the fact that the proloculum and second chamber are usually perforate, would indicate that they are closely related to the Nonionidae and Camerinidae. Like the Camerinidae they develop large tests in the Eocene and have a similar habitat, warm shallow waters of the coral reef type. The imperforate calcareous test is a secondary development, and probably is entirely a parallelism not definitely related to the Miliolidae and probably not to the Ophthalmidiidae which they more closely resemble. The history of the family is in the Tertiary and Recent oceans.

It seems very probable that the family as here constituted has only a superficial resemblance to those complex forms of the Upper Cretaceous that have often been included with them, *Fallotia*, *Broeckina*, and *Praesorites*. These forms are very little known in detail as to their development in the young, or as to their actual wall characters. The same is true of *Rhapydionina* and *Rhipidionina*, two forms from the Eocene of Italy, which also are very imperfectly known. To place any of these two groups of genera with any certainty is based largely upon guesswork until they are studied in full detail.

Dicyclina of the Upper Cretaceous does not belong here, but in the Valvulinidae, and is very closely allied to *Cuneolina*.

The genus *Peneroplis* is the most primitive, and is very unstable in form. It gave rise to *Dendritina* and *Spirolina*, and probably to *Mona-*

lysidium. The genera with annular chambers and division into chamberlets form a very definite series from *Archaias*, through *Sorites*, *Amphisorus*, *Orbitolites*, and *Marginopora* to *Opertorbitolites*.

Craterites is very doubtfully placed with this family.

Silvestri has given the name *Somalina* (Pal. Ital., vol. 32, suppl. 4, 1939, p. 51) to some very peculiar Eocene forms, but it is difficult to determine whether these belong to the Orbitoids or are complex forms belonging to the Peneroplidae.

Subfamily 1. Spirolininae

Test close coiled in the young, often becoming uncoiled in the adult.

Genus PENEROPLIS Montfort, 1808

Key, plate 24, figures 1, 2

Genotype, *Nautilus planatus* Fichtel and Moll

Peneroplis MONTFORT, Conch. Syst., vol. 1, 1808, p. 259.
Nautilus (part) of authors. *Coscinospira* EHRENBERG, 1840.

Test free, planispiral, young close coiled, usually involute, adult variously shaped, close coiled, flaring, annular, or partially uncoiled; chambers undivided; wall calcareous, imperforate, except proloculum and following chamber; aperture simple, at base of apertural face, or long and slit-like, occasionally divided into pores. — Eocene to Recent.

Genus DENDRITINA d'Orbigny, 1826

Key, plate 24, figure 3

Genotype, *Dendritina arbuscula* d'Orbigny

Dendritina D'ORBIGNY, Ann. Sci. Nat., vol. 7, 1826, p. 285.
Peneroplis (part) of authors.

Test similar to *Peneroplis*, the test usually thick and showing a tendency to uncoil; aperture dendritic, in the apertural face. — Eocene to Recent.

Genus SPIROLINA Lamarck, 1804

Key, plate 24, figure 4

Genotype, *Spirolina cylindracea* Lamarck

Spirolina LAMARCK, Ann. Mus., vol. 5, 1804, p. 244.
Peneroplis (part) of authors.

Test similar to *Peneroplis*, thick; early chambers close coiled, usually not completely involute, later ones uncoiled; aperture rounded or irregular. — Eocene to Recent.

Genus MONALYSIDIUM Chapman, 1900

Key, plate 24, figures 5, 6

Genotype, *Peneroplis* (*Monalysidium*) *sollasi* Chapman

Monalysidium CHAPMAN, Journ. Linn. Soc. Zool., vol. 28, 1900, p. 3.

Test with the early chambers close coiled, later uncoiled in a rectilinear series; wall imperforate, smooth or with vertical rows of minute tubercles; aperture circular, terminal, sometimes with a short neck and lip. — Recent.

Genus TABERINA Keijzer, 1945

Key, plate 49, figures 11, 12

Genotype, *Taberina cubana* Keijzer

Taberina KEIJZER, Outline geol. eastern part Prov. Oriente, Cuba, Utrecht, 1945, p. 200.

Test in the early stages composed of planispirally arranged chambers, subdivided by short, radial septa, centrally provided with pillars, uncoiling in the later portion; wall calcareous, imperforate; aperture multiple, consisting of pores in the face of the chambers. — Upper Cretaceous to Paleocene.

Genus PRAERHAPYDIONINA van Wessem, 1943

Genotype, *Praerhapydionina cubana* van Wessem

Praerhapydionina VAN WESSEM, Geol. Pal. Central Camaguey, Cuba, Utrecht, 1943, p. 43.

"Test elongate, round in transverse section, conical; chambers in the early stages planispirally coiled, later in a rectilinear series, uniserial, entirely divided into chamberlets by septa radiating inward from the outer wall. Wall porcellaneous. One central aperture, terminal." — Upper Cretaceous.

Only sections were figured as no free specimens were found.

Subfamily 2. Archaiasinae

Test discoid, early chambers spiral and simple, later ones divided into chamberlets, later stages variously involute.

Genus ARCHAIAS Montfort, 1808

Key, plate 24, figures 7, 8

Genotype, *Nautilus angulatus* Fichtel and Moll

Archaias MONTFORT, Conch. Syst., vol. 1, 1808, p. 191.

Helenis and *Ilotes* MONTFORT, 1808. *Orbiculina* LAMARCK, 1816.

Test in early stages planispiral and lenticular, bilaterally symmetrical, later stages becoming flaring, even annular; chambers divided into chamberlets; wall imperforate except proloculum and next chamber which are perforate; apertures in several rows, on the narrow apertural face. — Eocene to Recent.

Subfamily 3. Orbitolitinae

Test with early stages planispiral, at least in the microspheric form, later annular, chambers divided into chamberlets; apertures on the peripheral face.

Genus SORITES Ehrenberg, 1840

Key, plate 24, figures 9–11
Genotype, *Sorites dominicensis* Ehrenberg

Sorites Ehrenberg, Abhandl. k. Akad. Wiss. Berlin, 1838 (1840), p. 134.
Orbitolites (part) of authors (not Lamarck). *Taramellina* Munier-Chalmas, 1902.

Test discoid, planispiral in early stages, at least of microspheric form, later annular, completely divided into chamberlets; chambers typically in a single layer but sometimes with lateral divisions, those of each annular chamber communicating with adjacent ones as well as with those of the preceding and second chambers; apertures consisting of small pores in a narrow band in the middle of the periphery, either a single or multiple line. — Oligocene to Recent.

Genus AMPHISORUS Ehrenberg, 1840

Key, plate 24, figures 12–14
Genotype, *Amphisorus hemprichii* Ehrenberg

Amphisorus Ehrenberg, Abhandl. k. Akad. Wiss. Berlin, 1838 (1840), p. 130.
Orbitolites (part) of authors (not Lamarck). *Bradyella* Munier-Chalmas, 1902.

Test discoid, planispiral in early stages, at least of microspheric form, later annular, completely divided into chamberlets, typically in two layers, those of each annular chamber communicating with adjacent ones of the preceding and succeeding annular chambers, and adjacent chambers of the two layers also communicating; wall imperforate except proloculum and second chamber; apertures in a double alternating line along the periphery. — Oligocene to Recent.

Genus MARGINOPORA Blainville, 1830

Key, plate 24, figures 15–17
Genotype, *Marginopora vertebralis* Blainville

Marginopora Blainville, Dict. Sci. Nat., vol. 60, 1830, p. 377 (Quoy and Gaimard, MS.). — Quoy and Gaimard, Voyage de l'*Astrolabe*, 1833, *fide* Blainville, Man. Actin., 1834, p. 412.
Orbitolites (part) of authors.

Test in the early stages similar to *Sorites* with one or two rows of apertures, later with two outer layers and an inner series of chamberlets gradually expanding toward the periphery, the chamberlets in each annular chamber connecting, peripheral wall in the center projecting beyond the lateral series of chamberlets; wall calcareous, imperforate; apertures in more or less vertical rows on the periphery with a horizontal row above and below. — Late Tertiary, Recent.

Genus ORBITOLITES Lamarck, 1801

Key, plate 25, figure 2
Genotype, *Orbitolites complanata* Lamarck
Orbitolites LAMARCK, Syst. Anim. sans Vert., 1801, p. 376.
Discolites (?) MONTFORT, 1808.

Test discoidal; earliest chambers in the microspheric form coiled, later annular, divided into chamberlets, those of the same annular chamber not connecting with each other but with those of the adjacent preceding and succeeding chambers; wall calcareous, imperforate, unless perforate in the earliest chambers; apertures numerous, rounded, peripheral. — Eocene.

Genus OPERTORBITOLITES Nuttall, 1925

Key, plate 25, figure 5
Genotype, *Opertorbitolites douvilléi* Nuttall
Opertorbitolites NUTTALL, Quart. Journ. Geol. Soc., vol. 81, 1925, p. 447.

Test circular, lenticular, consisting of a median chamber-layer resembling that of *Orbitolites* with a thick imperforate lamina of shelly material on each side of the median layer. — Eocene, India and Somalia.

Genera of doubtful relationships.

Genus CRATERITES Heron-Allen and Earland, 1924

Key, plate 25, figure 15
Genotype, *Craterites rectus* Heron-Allen and Earland
Craterites HERON-ALLEN and EARLAND, Journ. Linn. Soc. Zool., vol. 35, 1924, p. 611.

Test probably attached in life, the whole composed of numerous layers of chambers, the basal layer without trace of spiral development, in side view contracted above the base, outer end broadening and convex; chambers very numerous; wall calcareous; outer surface with numerous, small, rounded openings. — Recent, Lord Howe Id., So. Pacific.

Through the courtesy of the authors, I was enabled to study the holotype now in the Heron-Allen and Earland Collection in the British Museum. More specimens are needed to make clear the full relationships. In some respects it resembles *Gypsina* and its allies, but it is here left where the authors placed it, near *Orbitolites*.

Genus FALLOTIA H. Douvillé, 1902

Key, plate 25, figure 1
Genotype, *Fallotia jacquoti* H. Douvillé
Fallotia H. DOUVILLÉ, Bull. Soc. Géol. France, ser. 4, vol. 2, 1902, p. 298.

Test discoid or lenticular, nummulitoid throughout the growth, growing edge always peripheral; chambers involute, V-shaped, divided into chamberlets. — Upper Cretaceous and Eocene.

Genus BROECKINA Munier-Chalmas, 1882

Genotype, *Cyclolina dufresnoyi* d'Archiac

Broeckina Munier-Chalmas, Bull. Soc. Géol. France, ser. 3, vol. 10, 1882, p. 471.
Cyclolina (part) of authors.

Test discoid, consisting of a single layer of chambers, early ones planispiral, later annular, partially divided; wall character not definitely known; apertures peripheral. — Upper Cretaceous.

This genus is imperfectly known but has resemblances to the large Upper Cretaceous forms which are finely arenaceous.

Genus MEANDROPSINA Munier-Chalmas, 1899

Key, plate 25, figures 3, 4

Genotype, *Meandropsina vidali* Schlumberger

Meandropsina Munier-Chalmas, in Schlumberger, Bull. Soc. Géol. France, ser. 3, vol. 27, 1899, p. 336.

Test discoid, the growing edge variously meandering over the flattened faces of the test; chambers with many chamberlets; apertures rounded, in linear rows on the periphery. — Upper Cretaceous.

This genus is as yet imperfectly known as to its early development and definite wall characters.

Genus FASCISPIRA A. Silvestri, 1939

Key, plate 49, figure 13

Genotype, *Fascispira colomi* A. Silvestri

Fascispira A. Silvestri, Boll. Soc. Geol. Ital., vol. 58, 1939, p. 230.

Test differing from *Meandropsina* in becoming involute, but not flaring or annular in later development as in *Archaias*. — Upper Cretaceous.

In external appearance this genus is very similar to *Fallotia*.

Genus PRAESORITES H. Douvillé, 1902

Key, plate 25, figures 6, 7

Genotype, *Praesorites moureti* H. Douvillé

Praesorites H. Douvillé, Bull. Soc. Géol. France, ser. 4, vol. 2, 1902, p. 291.

Test in the early stages planispiral, at least in the microspheric form, later annular, chambers in a single plane, incompletely divided; apertures peripheral. — Upper Cretaceous and Recent.

The wall characters and the early development of this genus are not fully known.

Genus RHAPYDIONINA Stache, 1912

Key, plate 40, figures 7–10
Genotype, *Peneroplis liburnica* Stache
Rhapydionina Stache, Jahrb. k. k. Geol. Reichs., vol. 62, 1912, p. 659.
Peneroplis (part) Stache (not Montfort).

Test elongate, conical, rounded in section, chambers in a rectilinear series, entirely or partially divided into chamberlets; apertures numerous, terminal. — Lower Eocene.

This genus and the following are very imperfectly known as to their wall characters and early development.

Genus RHIPIDIONINA Stache, 1912

Key, plate 40, figure 11
Genotype, *Pavonina liburnica* Stache
Rhipidionina Stache, Jahrb. k. k. Geol. Reichs., vol. 62, 1912, p. 658.
Pavonina (part) Stache.

Test compressed, fan-shaped; chambers divided radially into chamberlets; apertures consisting of peripheral pores. — Lower Eocene.

Family 28. ALVEOLINELLIDAE

Test coiled about an elongate axis, usually fusiform but some forms compressed along the axis of coiling, involute; chambers divided into chamberlets; wall calcareous, imperforate, except the earliest chambers which may apparently be perforate; apertures numerous, in one or more rows on the face of the last-formed chamber.

The Alveolinellidae were evidently developed in the Cretaceous, and reach their climax at present in the Indo-Pacific. They are characteristic of shallow warm waters. From living specimens there is seen a commensal relationship with marine algae, and as marine algae are very limited in their depth by light penetration, this may help to account for the narrow limits of depth at which members of this family live.

Some of the compressed forms are very similar indeed to *Peneroplis* and *Dendritina*, and can be distinguished only by the chamberlets. The idea that the Alveolinellidae were derived from the Miliolidae is based on sections which seem to show irregular chambers following the proloculum. In section the early chamberlets are not in the same plane, and when cut are easily interpreted in this way. If the section is not exactly through the middle of the proloculum, the chamberlets extending outward are cut to show different sizes, and thus present this appearance. I have made many sections without finding any trace of a true milioline young.

There is a regular increase in complexity of structure from the simple genera to the more complex ones.

For discussion and illustrations of this family see Altpeter, Beiträge zur Anatomie und Physiologie von Alveolina (Marburg, 1911), and later papers by Silvestri, Nuttall, Yabe and Hanzawa, Vaughan and others. Silvestri has proposed a new grouping of these forms: Sul modo di presentarsi di alveoline eoceniche, etc. (Mem. Pont. Accad. Sci. Nuovi Lincei, vol. XV, 1931, pp. 203–231, pl. 1, text figs. 1–3), and in later papers.

Recently Reichel has published detailed studies of this group with many figures, and the grouping there given has been largely adopted here. (See Reichel, M. Etude sur les Alveolines. — Mem. Soc. pal. Suisse, vols. 57 (pt.), 59 (pt.), 1936–1937, pp. 1–147, pls. I–XI, 29 text figs.)

The use of the generic name *Neoalveolina* A. Silvestri is in some doubt according to the rules of nomenclature but as used includes those Tertiary forms with a single row of apertures and no post septal canal.

Genus OVALVEOLINA Reichel, 1936

Key, plate 49, figure 14
Genotype, *Alveolina ovum* d'Orbigny
Ovalveolina Reichel, Mem. Soc. pal. Suisse, vols. 57 (pt.), 59 (pt.), 1936–37, p. 69.

Test generally globular or ovoid, planispiral; chambers numerous, short, extending from one pole to the other, not sinuous, divided into high chamberlets, usually undivided and pyriform in section; wall calcareous, imperforate; apertures in a single row on the apertural face, no accessory apertures. — Cretaceous.

Genus PRAEALVEOLINA Reichel, 1933

Key, plate 49, figure 15
Genotype, *Praealveolina tenuis* Reichel
Praealveolina Reichel, Eclog. Geol. Helv., vol. 26, 1933, p. 270.

Test globular or fusiform, planispiral, completely involute; chambers toward the poles divided into several rows of chamberlets corresponding to the incurved apertures which are in a single row near the middle of the test, but in several rows toward the poles, post septal canal lacking. — Cretaceous.

Genus SUBALVEOLINA Reichel, 1936

Key, plate 49, figure 17
Genotype, *Subalveolina dordonica* Reichel
Subalveolina Reichel, Mem. Soc. pal. Suisse, vols. 57 (pt.), 59 (pt.), 1936–37, p. 73.

Test similar to *Praealveolina* but with a series of small alveolae along the inner wall connecting with the outside by a line of pores, not alternating, on the outer chamber wall. — Cretaceous.

Genus BORELIS Montfort, 1808

Key, plate 25, figures 8–10; plate 50, figure 1

Genotype, *Borelis melonoides* Montfort = *Nautilus melo* Fichtel and Moll (part)

Borelis MONTFORT, Conch. Syst., vol. 1, 1808, p. 171.

Clausulus MONTFORT, 1808. *Melonites* LAMARCK, 1812. *Oryzaria* DEFRANCE, 1820. *Melonia* DEFRANCE, 1824. *Alveolina* D'ORBIGNY, 1826. *Fasciolites* GALLOWAY, 1933 (PARKINSON, 1811 (?)). *Neoalveolina* SILVESTRI, 1928.

Test globular or fusiform; chambers divided into a single series of chamberlets, alternating in adjacent chambers, a post septal canal running along the anterior face of each vertical wall of the chamber cutting the posterior extremity of the chamberlets; apertures of the two or more rows alternating.—Eocene to Recent (?).

Reichel recognizes the following subgenera: *Glomalveolina* Reichel, 1936, including small, globular species; *Fasciolites* Parkinson, 1811, including large forms, ovoid, cylindrical or fusiform, with a thickened basal layer to the chambers; and *Eoalveolinella* A. Silvestri, 1928, including very elongate forms with supplementary chamberlets.

Genus FLOSCULINA Stache, 1883

Key, plate 25, figures 11, 12; plate 50, figure 2

Genotype, *Flosculina decipiens* Schwager

Flosculina STACHE, in SCHWAGER, Palaeontographica, vol. 30, 1883, p. 102.

Alveolina (part) of authors.

Test globular or fusiform, planispiral, completely involute, early coils high with few chambers, later ones low and many chambered, divided into chamberlets; wall calcareous, imperforate in adult, apparently perforate in young; apertures numerous, in a single row on the long apertural face.—Eocene.

Genus BORELOIDES Cole and Bermudez, 1947

Key, plate 50, figures 5, 6

Genotype, *Boreloides cubensis* Cole and Bermudez

Boreloides COLE and BERMUDEZ, Bull. Amer. Pal., vol. 31, No. 125, 1947, p. 9.

Test subspherical to fusiform, planispiral, involute, very slightly increasing in height in the coils; chambers divided into a single series of chamberlets by revolving partitions; basal wall thick and with low, conical projections on the outer side; embryonic apparatus bilocular; apertural face developed but apertures not observed.—Eocene.

Genus BULLALVEOLINA Reichel, 1936

Key, plate 50, figure 4

Genotype, *Alveolina bulloides* d'Orbigny

Bullalveolina REICHEL, Eclog. Geol. Helv., vol. 29, 1936, p. 136.

Test differing from *Subalveolina* in the globular shape and in having several rows of secondary apertures.—Oligocene.

Genus FLOSCULINELLA Schubert, 1910

Key, plate 50, figure 3
Genotype, *Flosculina bontangensis* Rutten (?)
Flosculinella Schubert, Neues Jahrb., Beilage-Band 29, 1910, p. 533.
Alveolina (part) and *Flosculina* (part) of authors.

Test differing from *Flosculina* in having more than one row of chambers in the last-formed whorls, the outer added chambers smaller than those proximal to them. — Oligocene, Miocene.

Genus ALVEOLINELLA H. Douvillé, 1906

Key, plate 25, figures 13, 14; plate 49, figure 16
Genotype, *Alveolina quoyi* d'Orbigny
Alveolinella H. Douvillé, Bull. Soc. Géol. France, ser. 4, vol. 6, 1906, p. 585.
Alveolina (part) d'Orbigny and later authors.

Test elongate fusiform, shape varying greatly in megalospheric and microspheric forms; chambers sinuous toward the poles, chamberlets in several tiers, those of the upper rows smaller and alternating, basal wall of chamber very thin, a preseptal canal present at the base of the chamber with a secondary canal sometimes present; apertures in several longitudinal rows along the apertural face, increasing in number toward the poles. — Miocene to Recent. Indo-Pacific.

Family 29. KERAMOSPHAERIDAE

Test globular, the wall imperforate, of generally concentric chambers divided into chamberlets, generally flattened, parallel to the surface in concentric layers, communicating with adjacent chambers both in the same layer and in the layers above and below; apertures rounded, one at the margin of each chamberlet.

Genus KERAMOSPHAERA H. B. Brady, 1882

Key, plate 25, figures 16, 17
Genotype, *Keramosphaera murrayi* H. B. Brady
Keramosphaera H. B. Brady, Ann. Mag. Nat. Hist., ser. 5, vol. 10, 1882, p. 245.

Test globular; chambers arranged in a generally concentric manner, irregularly divided into chamberlets, communicating with adjacent chambers both in the same layer and in the layers above and below; wall calcareous, imperforate; apertures rounded, at the margins of the chamberlets. — Recent.

Family 30. HETEROHELICIDAE

Test in the more primitive forms planispiral in the young, later becoming biserial, in the more specialized genera the spiral and biserial stages reduced or wanting, and the relationships shown only by other characters or intermediate forms; wall calcareous, perforate, ornamentation in specialized genera bilaterally symmetrical; aperture simple, usually large for the size of the test, without teeth, in some forms with an apertural neck and phialine lip.

The Heterohelicidae form a most interesting group. Derived from a planispiral ancestry, this is held in the early stages of all the most primitive genera, at least in the microspheric forms. There is no evidence that the group was derived from the specialized pelagic Globigerinidae, although the genus *Gümbelina* may have been pelagic. The family began its development in the Cretaceous where it developed numerous specialized genera such as *Pseudotextularia* and its derivative *Planoglobulina* with the young at least biserial, and then spiral about an axis. From *Gümbelina* also developed directly with fan-shaped chambers in one plane, *Ventilabrella*, and with uniserial chambers, *Rectogümbelina*. In the Cretaceous, triserial and multiserial forms developed from the biserial ones as is characteristic in other groups, and the genera *Gümbelitria* and *Gümbelitriella* resulted. Directly from *Heterohelix* comes the elongate *Bolivinopsis*, often very numerous in the Upper Cretaceous, but rarely found with the early coiled portion intact. *Spiroplectinata*, formerly placed here, is really triserial in the young, is distinctly arenaceous, and belongs in the Verneuilinidae.

Also developed in the Upper Cretaceous is the group with a terminal aperture and phialine lip: *Eouvigerina*, planispiral in the young of the microspheric form then biserial and tending to become loosely triserial in the adult; *Pseudouvigerina*, normally triserial; and two genera becoming uniserial, *Siphogenerinoides*, with the early stages biserial or triserial and the adult uniserial, and *Nodogenerina* with the chambers uniserial throughout.

In the Tertiary was developed *Plectofrondicularia* with the early stage of the microspheric form at least coiled, later biserial, and the adult compressed, uniserial. From this developed *Amphimorphina* with the early stages as in *Plectofrondicularia* but more accelerated, while in *Nodomorphina* the test is easily mistaken for *Nodosaria*, but for the bilateral character of the ornamentation and the lack of the radiate aperture.

Bolivinoides has a planispiral young in the microspheric form and a thickening of the apertural end. It should not be confused with *Bolivina* of the Buliminidae which it somewhat resembles. *Bolivina* is derived directly from *Virgulina*, and both have elongate spiral ancestry shown in the spirally twisted test of many forms, and an elongate spiral grouping of the early chambers in the microspheric form, a character not seen in the Heterohelicidae. Both *Bolivinita* and *Bolivinitella* were developed in the Cretaceous, and *Bolivinita* persists to the present oceans. *Bolivinella*, with its very compressed form, developed in the Tertiary, and is found today principally in the Indo-Pacific.

Pavonina has been grouped with the Heterohelicidae, but it has been shown by Parr that the young is triserial and it belongs close to *Chrysalidinella* in the Buliminidae.

Subfamily 1. Heterohelicinae

Test in the early stages distinctly planispiral, later chambers biserial; aperture large, at the inner margin of the chamber.

Genus HETEROHELIX Ehrenberg, 1843

Plate 21, figure 1; Key, plate 26, figure 1
Genotype, *Spiroplecta americana* Ehrenberg

Heterohelix Ehrenberg, Abhandl. k. Akad. Wiss. Berlin, 1841 (1843), p. 429.
Spiroplecta Ehrenberg, 1844.

Test with early stages planispiral, forming a considerable portion of the test, the few adult chambers biserial; wall calcareous, perforate, thin; aperture large, on the inner margin of the last-formed chamber, median.— Cretaceous.

Genus BOLIVINOPSIS Yakovlev, 1891

Plate 21, figure 12; Key, plate 26, figures 2–5
Genotype, *Bolivinopsis capitata* Yakovlev

Bolivinopsis Yakovlev, Trav. Soc. nat. Univ. Imp. Kharkov, vol. 24, 1890 (1891), p. 349.—Macfadyen, Journ. Roy. Micr. Soc., vol. 53, 1933, p. 139.
Spiroplectoides Cushman, 1927. *Spiroplecta* (part) of authors.

Test elongate, sides nearly parallel; early chambers planispiral in both microspheric and megalospheric forms, later ones biserial, numerous; wall calcareous, perforate; aperture elliptical, terminal or nearly so in adult.—Cretaceous to Recent.

Genus NODOPLANULIS Hussey, 1943

Key, plate 50, figure 7
Genotype, *Nodoplanulis elongata* Hussey

Nodoplanulis Hussey, Journ. Pal., vol. 17, 1943, p. 166.

Test indistinctly planispiral in the very early stage, remainder of test uniserial; wall calcareous, perforate; aperture terminal, round, with a short neck and flaring lip.—Eocene.

Subfamily 2. Gümbelininae

Test in the early stages of the microspheric form planispiral, often skipped in the megalospheric form, followed by a biserial stage which in specialized genera may be followed by globular chambers, variously arranged.

Genus GÜMBELINA Egger, 1899

Plate 21, figure 2; Key, plate 26, figures 6–8

Genotype, *Textularia globulosa* Ehrenberg

Gümbelina Egger, Abhandl. kön. bay. Akad. Wiss. München, Cl. II, vol. 21, 1899, p. 31.

Textularia (part) of authors (not Defrance).

Test with the early chambers planispiral, at least in the microspheric form, later chambers biserial; wall calcareous, perforate; aperture large and open, arched, at base of inner margin of last-formed chamber.—Cretaceous to Oligocene.

PLATE 21

Relationships of the Genera of the Heterohelicidae

FIG.

1. *Heterohelix americana* (Ehrenberg). (Adapted from Ehrenberg.) *a*, exterior; *b*, section.
2. *Gümbelina globulosa* (Ehrenberg). *a*, side view; *b*, showing aperture.
3. *Gümbelitria cretacea* Cushman. *a*, side view; *b*, showing aperture.
4. *Rectogümbelina cretacea* Cushman.
5. *Tubitextularia bohemica* (Sulc). (After Sulc.)
6. *Ventilabrella carseyi* Plummer. (After Plummer.) *a*. side view; *b*, showing apertures.
7. *Pseudotextularia varians* Rzehak.
8. *Planoglobulina acervulinoides* (Egger).
9. *Bolivinoides rhomboidalis* (Cushman).
10. *Bolivinitella eleyi* (Cushman). *a*, side view; *b*, showing aperture.
11. *Bolivinella folia* (Parker and Jones).
12. *Bolivinopsis rosula* (Ehrenberg). (Adapted from Ehrenberg.) *a*, exterior; *b*, section of another specimen.
13. *Plectofrondicularia mexicana* Cushman. *a*, side view; *b*, apertural view; *c*, section.
14. *Amphimorphina haueriana* Neugeboren. (After Neugeboren.) *a*, side view; *b*, section of compressed portion; *c*, section of inflated portion.
15. *Nodomorphina compressiuscula* (Neugeboren). (After Neugeboren.)
16. *Eouvigerina gracilis* Cushman. *a*, front view; *b*, side view; *c*, apertural view.
17. *Pseudouvigerina cristata* (Marsson). (After Marsson.) *a*, front view; *b*, section.
18. *Siphogenerinoides plummeri* (Cushman).
19. *Nodogenerina bradyi* Cushman.

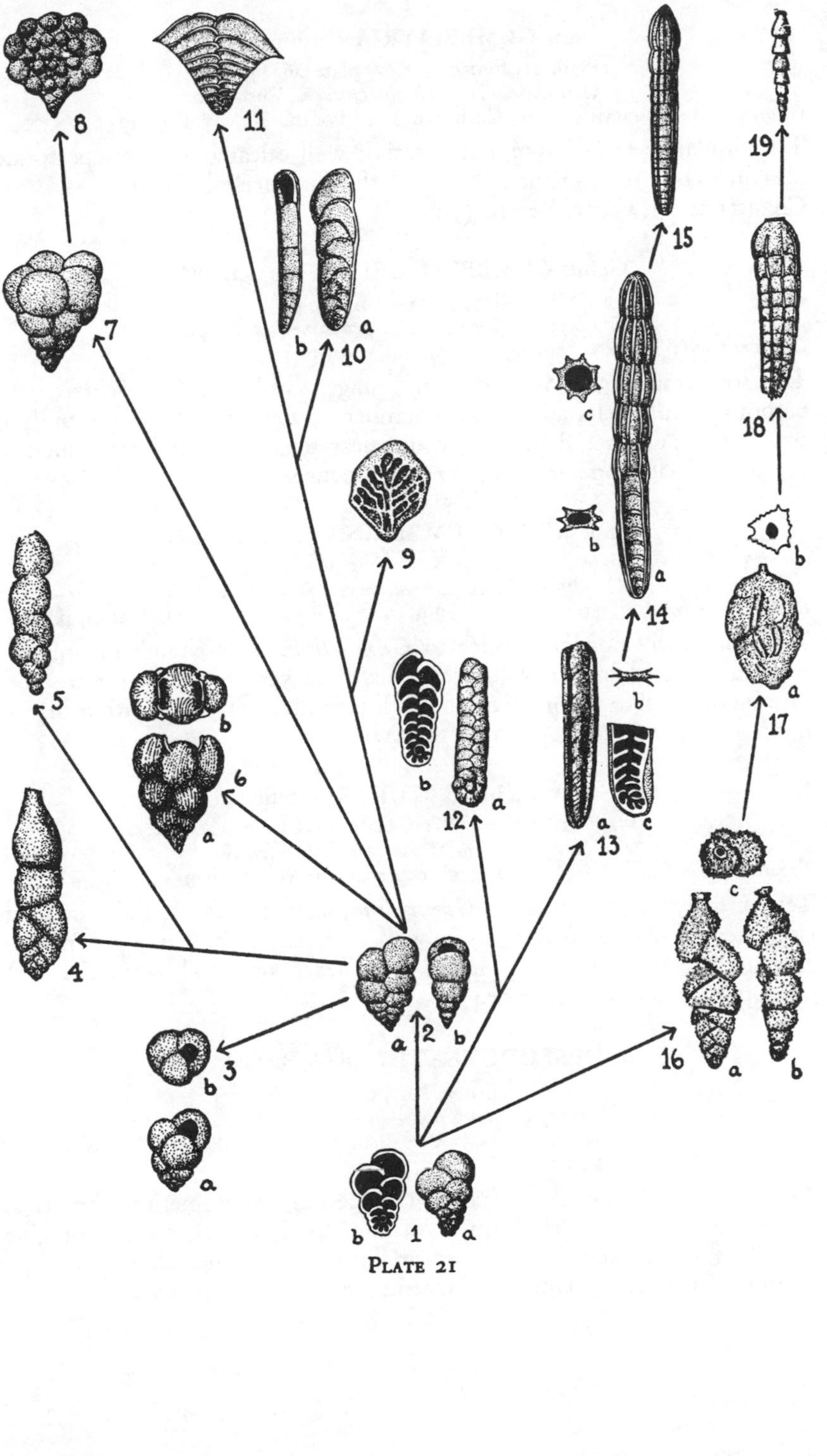

8
11
19
15
7
b
a
10
c
18
9
b
a
b
14
5
b
17
b
b
a
6
b
a
12
a
c
a
13
c
4
a
2
b
a
b
3
16
a
b
a
b
1
a

Plate 21

Genus GÜMBELITRIA Cushman, 1933

Plate 21, figure 3; Key, plate 26, figure 9
Genotype, *Gümbelitria cretacea* Cushman

Gümbelitria Cushman, Contr. Cushman Lab. Foram. Res., vol. 9, 1933, p. 37.

Test similar to *Gümbelina*, but triserial; wall calcareous, finely perforate; aperture large, at the inner edge of the last-formed chamber. — Upper Cretaceous to Eocene, Recent (?).

Genus GÜMBELITRIELLA Tappan, 1940

Key, plate 50, figure 8
Genotype, *Gümbelitriella graysonensis* Tappan

Gümbelitriella Tappan, Journ. Pal., vol. 14, 1940, p. 115.

Test free, small, triserial in the early stage, similar to *Gümbelitria*, later becoming multiserial at the top; chambers globular, increasing rapidly in size; sutures distinct, depressed; wall calcareous, finely perforate; aperture at base of final chamber. — Lower Cretaceous.

Genus RECTOGÜMBELINA Cushman, 1932

Plate 21, figure 4; Key, plate 26, figures 10, 11
Genotype, *Rectogümbelina cretacea* Cushman

Rectogümbelina Cushman, Contr. Cushman Lab. Foram. Res., vol. 8, 1932, p. 6.

Test with early portion similar to *Gümbelina*; later chambers uniserial, rectilinear, subglobular; wall calcareous, thin, very finely perforate; aperture in young like *Gümbelina*, in adult terminal, rounded, with a distinct neck. — Upper Cretaceous to Oligocene.

Genus TUBITEXTULARIA Sulc, 1929

Plate 21, figure 5; Key, plate 26, figures 12, 13
Genotype, *Pseudotextularia bohemica* Sulc

Tubitextularia Sulc, Vestnik stát. geol. ceskosl. rep., vol. 5, 1929, p. 148.

Test in early stages similar to *Gümbelina*, later loosely biserial; chambers elongate and becoming uniserial; wall calcareous, thin, finely perforate; aperture in young as in *Gümbelina*, later nearly terminal, without a neck. — Upper Cretaceous, Europe.

Genus PSEUDOTEXTULARIA Rzehak, 1886

Plate 21, figure 7; Key, plate 26, figure 16
Genotype, *Pseudotextularia varians* Rzehak

Pseudotextularia Rzehak, Verh. Nat. Ver. Brünn, vol. 24, 1885 (1886), Sitz., p. 8.
Gümbelina (part) of authors.

Test with the early chambers as in *Gümbelina*, and sometimes nearly all chambers in microspheric form so arranged, adult with a series of globular chambers arranged in a more or less spiral manner about the upper portion of the test. — Upper Cretaceous, Europe and America.

Genus PLANOGLOBULINA Cushman, 1927

Plate 21, figure 8; Key, plate 26, figure 17
Genotype, *Gümbelina acervulinoides* Egger

Planoglobulina CUSHMAN, Contr. Cushman Lab. Foram. Res., vol. 2, pt. 4, 1927, p. 77.
Gümbelina (part) of authors.

Test in young similar to *Pseudotextularia*, later chambers tending to be added in a single plane and somewhat spreading. — Upper Cretaceous.

Genus VENTILABRELLA Cushman, 1928

Plate 21, figure 6; Key, plate 26, figures 14, 15
Genotype, *Ventilabrella eggeri* Cushman

Ventilabrella CUSHMAN, Contr. Cushman Lab. Foram. Res., vol. 4, 1928, p. 2.
Gümbelina (part) of authors.

Test in the early stages as in *Gümbelina*, in the adult with chambers alternating and spreading out fan-shaped in a single plane; wall calcareous, finely or coarsely perforate; aperture in biserial stage single, in adult chambers one aperture at each side near the base of the median line. — Upper Cretaceous.

Subfamily 3. Bolivinitinae

Test in the adult biserial, compressed; aperture in the median line, at the base of the inner margin, microspheric form with the young planispiral in earliest stages.

Genus BOLIVINOIDES Cushman, 1927

Plate 21, figure 9; Key, plate 26, figures 18–20
Genotype, *Bolivina draco* Marsson

Bolivinoides CUSHMAN, Contr. Cushman Lab. Foram. Res., vol. 2, pt. 4, 1927, p. 89.
Bolivina (part) of authors.

Test compressed, rhomboid, thickest portion near apertural end, usually appearing like a thickened lip without ornamentation, which is in general at right angles to the sutures, earliest stage in microspheric form planispiral, quickly biserial; wall calcareous, finely perforate; aperture fairly large, at the inner margin. — Upper Cretaceous.

Genus BOLIVINITA Cushman, 1927

Key, plate 26, figure 22; plate 50, figure 12
Genotype, *Textularia quadrilatera* Schwager

Bolivinita CUSHMAN, Contr. Cushman Lab. Foram. Res., vol. 2, pt. 4, 1927, p. 90.
Textularia (part) and *Bolivina* (part) of authors.

Test with chambers biserial, periphery and broader sides all concave with strongly developed angles giving a quadrate end view to the test; wall calcareous, perforate; aperture large, at base of inner margin in the median line. — Cretaceous to Recent.

The Cretaceous forms with a terminal aperture have been placed by Marie in a new genus *Bolivinitella.* The genotype of *Bolivinita* has the aperture extending into the apertural face from the inner margin.

Genus BOLIVINITELLA Marie, 1941

Plate 21, figure 10; Key, plate 26, figure 21; plate 50, figure 9
Genotype, *Bolivinita eleyi* Cushman

Bolivinitella Marie, Mém. Mus. Nat. Hist. Nat., n. ser., vol. 12, 1941, p. 189.
Bolivina (part), *Textularia* (part), and *Bolivinita* (part) of authors.

Test biserial, strongly compressed, sides flat or concave, generally carinate, periphery truncate, quadrangular biconcave in section; wall calcareous, perforate; aperture elongate, terminal, with a slightly raised border. — Upper Cretaceous.

Genus BOLIVINELLA Cushman, 1927

Plate 21, figure 11; Key, plate 26, figure 23
Genotype, *Textularia folium* Parker and Jones

Bolivinella Cushman, Contr. Cushman Lab. Foram. Res., vol. 2, pt. 4, 1927, p. 79.
Textularia (part) of authors.

Test much compressed, proloculum in megalospheric form rectangular, in microspheric form the young apparently planispiral; later chambers biserial, long and recurved, not overlapping; wall calcareous, perforate; aperture transverse to compression of test, with numerous papillae at base of opening. — Eocene (?), Lower Oligocene to Recent.

See Cushman, The Genus *Bolivinella* and Its Species (Contr. Cushman Lab. Foram. Res., vol. 5, 1929, pp. 28–34, pl. 5).

Subfamily 4. Plectofrondiculariinae

Test in microspheric form planispiral in the early stages, later biserial and then uniserial, or in the higher forms starting as compressed uniserial tests, ornamentation bilaterally symmetrical; aperture rounded or elliptical, terminal.

Genus PLECTOFRONDICULARIA Liebus, 1903

Plate 21, figure 13; Key, plate 26, figures 24–27
Genotype, *Plectofrondicularia concava* Liebus

Plectofrondicularia Liebus, Jahrb. k. k. Geol. Reichs., vol. 52, 1902 (1903), p. 76.
Frondicularia (part) of authors.

Test elongate, compressed, microspheric form planispiral, then biserial, then uniserial, rectilinear, much compressed; wall calcareous, finely perforate; aperture in adult terminal, elliptical. — Eocene to Recent.

The subgenus, *Mucronina* d'Orbigny, "Les Mucronines" 1826, is an indeterminable form never having been found since d'Orbigny's time, and may possibly be a *Plectofrondicularia.* The early stages are unknown.

Genus AMPHIMORPHINA Neugeboren, 1850

Plate 21, figure 14; Key, plate 26, figures 28–30
Genotype, *Amphimorphina haueriana* Neugeboren
Amphimorphina NEUGEBOREN, Verh. Mitth. siebenbürg. Ver. Nat., vol. 1, 1850, p. 125.
Staffia SCHUBERT, 1911.

Test elongate, uniserial except in the microspheric form which may show traces of the biserial stage, earlier portion of test flattened as in *Plectofrondicularia,* later portion with inflated chambers; aperture in young elliptical, later circular. — Eocene to Pliocene.

Genus NODOMORPHINA Cushman, 1927

Plate 21, figure 15; Key, plate 26, figure 31
Genotype, *Nodosaria compressiuscula* Neugeboren
Nodomorphina CUSHMAN, Contr. Cushman Lab. Foram. Res., vol. 2, pt. 4, 1927, p. 80.
Nodosaria (part) of authors.

Test elongate, early portion compressed; chambers uniserial, in a straight series, early ones quadrilateral in section, later ones inflated and nearly circular, ornamentation bilaterally symmetrical; aperture circular or elliptical, terminal, without teeth or radiate fissures. — Miocene, Pliocene.

Subfamily 5. Eouvigerininae

Test in the earliest stages biserial, later triserial, or in the most specialized genera, uniserial after the triserial stage, or uniserial throughout.

Genus EOUVIGERINA Cushman, 1926

Plate 21, figure 16; Key, plate 26, figure 32
Genotype, *Eouvigerina americana* Cushman
Eouvigerina CUSHMAN, Contr. Cushman Lab. Foram. Res., vol. 2, pt. 1, 1926, p. 4.
Sagrina (part) of authors.

Test elongate, the early chambers at least in the microspheric form planispiral, later chambers biserial, final ones irregularly triserial; wall calcareous, perforate; aperture circular or rhomboid, with a definite neck and usually a phialine lip. — Upper Cretaceous to Eocene.

Genus ZEAUVIGERINA Finlay, 1939

Key, plate 50, figure 10
Genotype, *Zeauvigerina zelandica* Finlay
Zeauvigerina FINLAY, Trans. Roy. Soc. New Zealand, vol. 68, 1939, p. 541.

Test in its general characters similar to *Eouvigerina* but the early chambers without trace of coiling and final chambers completely biserial. — Eocene.

Genus PSEUDOUVIGERINA Cushman, 1927

Plate 21, figure 17; Key, plate 26, figure 34
Genotype, *Uvigerina cristata* Marsson

Pseudouvigerina Cushman, Contr. Cushman Lab. Foram. Res., vol. 2, pt. 4, 1927, p. 81.
Uvigerina (part) of authors.

Test in early stages of microspheric form biserial, later triserial; adult chambers triangular in section, with the outer angle usually truncated; wall calcareous, usually coarsely perforate; aperture terminal, usually with a tubular neck and phialine lip. — Upper Cretaceous, Eocene, perhaps later.

Genus SIPHOGENERINOIDES Cushman, 1927

Plate 21, figure 18; Key, plate 26, figure 35
Genotype, *Siphogenerina plummeri* Cushman

Siphogenerinoides Cushman, Contr. Cushman Lab. Foram. Res., vol. 3, 1927, p. 63.
Siphogenerina (part) of authors.

Test elongate, slightly tapering, early stages biserial, later uniserial; wall calcareous, perforate, usually with longitudinal costae; aperture with a neck and phialine lip. — Upper Cretaceous, Paleocene.

Genus NODOGENERINA Cushman, 1927

Plate 21, figure 19; Key, plate 26, figures 36–38
Genotype, *Nodogenerina bradyi* Cushman

Nodogenerina Cushman, Contr. Cushman Lab. Foram. Res., vol. 2, pt. 4, 1927, p. 79.
Sagrina (part) of authors.

Test elongate, uniserial, straight; chambers increasing in size as added, distinct, inflated; wall calcareous, finely perforate; aperture terminal, central, rounded, with a cylindrical neck and phialine lip. — Cretaceous to Recent.

Family 31. BULIMINIDAE

Test typically an elongate spiral, divided into chambers, in the specialized genera biserial, uniserial or even single chambered; wall calcareous, perforate, often highly ornamented; aperture in the simplest form loop-shaped, in some forms developing a neck and lip, and in some becoming cribrate, the opening often showing a tooth or spiral connected with the interior siphons.

The early forms of the Jurassic show the development of the group from an elongate, spiral, undivided form, *Terebralina*, probably derived from the similar arenaceous *Turritellella*, which develops species of almost pure cement as do many of the Ammodiscidae. From this to *Turrilina* of the Jurassic by the division into chambers is a simple step, then to *Buliminella* and *Bulimina*, and so to the more complex forms. In the earlier genera the aperture has a rounded area somewhat away from the base, connecting with the base by a narrower opening. This "loop-shaped" or "comma-shaped" aperture is characteristic of a number of the genera, and persists even when the adult becomes biserial as in *Virgulina* and *Bolivina*, or even in triserial forms such as *Reussella*. The simpler, more primitive genera have the spiral suture of the ancestral *Terebralina* distinctly more marked than the sutures between the chambers. Later in *Bulimina*, the chambers become more inflated and distinctive, and this primitive character is lost.

From *Turrilina* is derived *Buliminella*, with numerous chambers to a coil, and *Robertina* where the chambers are developed in two series, but the spiral suture and loop-shaped aperture still retained. *Buliminoides* and *Ungulatella* become uniserial, but still retain the spiral twisted shape characteristic of their ancestral form.

From *Bulimina* is developed the sharply triserial *Reussella*, from which is developed *Trimosina*, and finally the biserial *Mimosina*. Both the latter genera have a peculiar vesicular wall structure, and, it is to be suspected, gave rise to some of the species referred to *Lagena* which have similar walls. From *Reussella* also developed *Pavonina*, with its chambers triserial in the young and spreading out in one plane in the adult, and *Chrysalidinella*, with its uniserial development and cribrate aperture. *Chysalidina* was formerly placed here, but is a very large arenaceous form belonging to the Valvulinidae, as our study of type material in Paris has shown. *Chysalogonium* has also been placed here on account of its cribrate aperture, but a study of Schwager's material shows it to belong to the Lagenidae, derived from *Nodosaria*.

From *Bulimina* also developed *Virgulina*, gradually becoming biserial in the adult but usually twisted, and *Bolivina* with the *Virgulina* characters only in the young. *Loxostoma* becomes uniserial in its later development, a character more marked in *Bifarina, Bifarinella, Rectobolivina, Schubertia, Geminaricta*, and *Tubulogenerina*. *Entosolenia* with its *Bulimina*-like aperture and internal tube probably developed into numerous species now usually called *Lagena*. *Globobulimina* is directly derived from *Bulimina* by having the chambers involute at the base.

From *Bulimina* also developed *Uvigerinella* with its aperture with a collar open at one side in primitive forms, and from this came *Uvigerina*. *Uvigerina* with its neck and phialine lip developed into *Hopkinsina* with its biserial adult, and *Siphogenerina* and *Unicosiphonia* uniserial in the adult. *Angulogerina* with its sharply triangular test is triserial, and from it came *Trifarina* which is uniserial in the adult, and *Dentalinopsis* uniserial throughout. From this were probably derived those angled "Lagenas" sometimes called *Trigonulina* and *Tetragonulina*. From *Uvigerina* also developed the uniserial *Siphonodosaria*.

The Buliminidae form a closely related group with a spiral ancestral development, but have developed into many highly specialized and diverse forms.

The genus *Buliminopsis* Rzehak is too poorly described and figured to be of use, but may belong in this family.

Subfamily 1. Terebralininae

Test in an elongate close spiral, not divided into chambers; wall calcareous, perforate; aperture rounded, subterminal.

Genus TEREBRALINA Terquem, 1866

Plate 22, figure 1; Key, plate 27, figure 1
Genotype, *Terebralina regularis* Terquem

Terebralina TERQUEM, Sixième Mém. Foram. Lias, 1866, p. 473.

Test consisting of proloculum and elongate, undivided, tubular, second chamber in an elongate close spiral; wall calcareous, perforate; aperture rounded, terminal. — Jurassic.

PLATE 22

RELATIONSHIPS OF THE GENERA OF THE BULIMINIDAE

FIG.

1. *Terebralina regularis* Terquem. (After Terquem.)
2. *Turrilina andreaei* Cushman. (After Andreae.)
3. *Buliminella elegantissima* (d'Orbigny). (After d'Orbigny.)
4. *Robertina arctica* d'Orbigny. (After d'Orbigny.)
5. *Buliminoides williamsoniana* (H. B. Brady). (After H. B. Brady.)
6. *Ungulatella pacifica* Cushman.
7. *Bulimina.*
8. *Uvigerinella californica* Cushman.
9. *Uvigerina canariensis* d'Orbigny.
10. *Siphogenerina costata* Schlumberger.
11. *Hopkinsina danvillensis* Howe and Wallace. (After Howe and Wallace.)
12. *Siphonodosaria abyssorum* (H. B. Brady). (After H. B. Brady.) *a,* side view; *b,* apertural view.
13. *Sporadogenerina flintii* Cushman. (LAGENIDAE.)
14. *Angulogerina angulosa* (Williamson).
15. *Trifarina bradyi* Cushman.
16. *Dentalinopsis subtriquetra* Reuss.
17. *Reussella spinulosa* (Reuss).
18. *Chrysalidinella dimorpha* (H. B. Brady). (After H. B. Brady.)
19. *Pavonina flabelliformis* d'Orbigny.
20. *Trimosina milletti* Cushman. (After Millett.)
21. *Mimosina hystrix* Millett. (Adapted from Millett.)
22. *Globobulimina pacifica* Cushman. *a,* side view; *b,* apertural view.
23. *Entosolenia globosa* Williamson. (After Williamson.) *a,* section; *b,* apertural view.
24. *Neobulimina canadensis* Cushman and Wickenden. *a,* side view; *b,* apertural view.
25. *Virgulina subsquamosa* Egger.
26. *Bolivina incrassata* Reuss.
27. *Loxostomum plaitum* (Carsey).
28. *Bifarina fimbriata* (Millett).
29. *Tubulogenerina mooraboolensis* (Cushman).
30. *Rectobolivina bifrons* (H. B. Brady). *a,* front view; *b,* apertural view.
31. *Schubertia tessellata* (H. B. Brady).

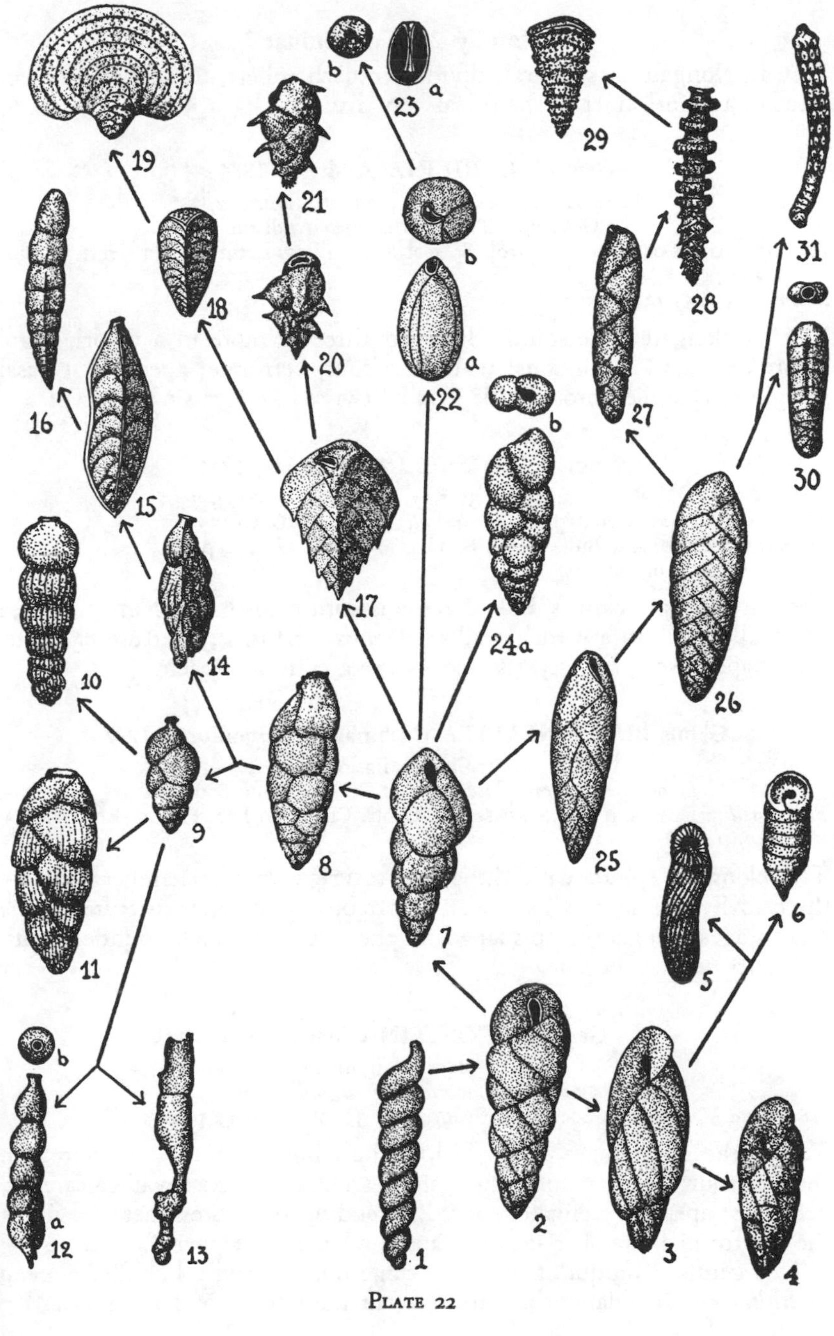
1
2
3
4
5
6
7
8
9
10
11
12
a
b
13
14
15
16
17
18
19
20
21
22
a
b
23
a
b
24a
b
25
26
27
28
29
30
31

PLATE 22

Subfamily 2. Turrilininae

Test an elongate close spiral, divided into chambers, usually more than three to a whorl, lines of the spiral very distinct.

Genus TURRILINA Andreae, 1884

Plate 22, figure 2; Key, plate 27, figures 2, 3
Genotype, *Turrilina alsatica* Andreae

Turrilina Andreae, Abhandl. Geol. Special Karte Elsass-Lothr., vol. 2, Heft 3, 1884, p. 120.
Bulimina (part) of authors.

Test an elongate close spiral; chambers three or more in a whorl, spiral suture deep and continuous; wall calcareous, perforate; aperture at basal margin of chamber, broad, little if at all twisted. — Jurassic to Recent.

Genus BULIMINELLA Cushman, 1911

Plate 22, figure 3; Key, plate 27, figures 4, 5
Genotype, *Bulimina elegantissima* d'Orbigny

Buliminella Cushman, Bull. 71, U. S. Nat. Mus., pt. 2, 1911, p. 88.
Bulimina (part) of authors.

Test an elongate close spiral, the spiral suture distinct; chambers three or usually more in a whorl; wall calcareous, perforate; aperture elongate, loop-shaped, very slightly twisted. — Cretaceous to Recent.

Genus BULIMINELLITA Cushman and Stainforth, 1947

Key, plate 50, figure 11
Genotype, *Buliminellita mirifica* Cushman and Stainforth

Buliminellita Cushman and Stainforth, Contr. Cushman Lab. Foram. Res., vol. 23, 1947, p. 78.

Test elongate, spiral, with the spiral suture distinct; chambers usually three to five in each whorl; wall calcareous, perforate; aperture in the early stages elongate, loop-shaped, in the adult terminal, rounded, with a distinct neck. — Eocene.

Genus LACOSTEINA Marie, 1943

Key, plate 50, figure 18
Genotype, *Lacosteina gouskovi* Marie

Lacosteina Marie, Bull. Soc. Géol. France, ser. 5, vol. 13, 1943, p. 295.

Test in the early stages planispiral, in the adult becoming elongate with three or four chambers to a whorl about an elongate axis; wall calcareous, perforate; aperture elongate, curved, extending from the inner margin of the last-formed chamber into the apertural face. — Upper Cretaceous.

This genus is unique in its early stage but otherwise like *Buliminella* or *Bulimina*. Its relationships must await the finding of other species.

Genus BULIMINOIDES Cushman, 1911

Plate 22, figure 5; Key, plate 27, figures 6, 7
Genotype, *Bulimina williamsoniana* H. B. Brady
Buliminoides CUSHMAN, Bull. 71, U. S. Nat. Mus., pt. 2, 1911, p. 90.
Bulimina (part) of authors.

Test subcylindrical, elongate, spirally twisted; chambers in a spiral, several chambers in a whorl, largely obscured by the heavy longitudinal costae; wall calcareous, perforate; aperture terminal, central, circular, in a depression at the end of the test. — Pliocene to Recent.

Genus UNGULATELLA Cushman, 1931

Plate 22, figure 6; Key, plate 27, figure 8
Genotype, *Ungulatella pacifica* Cushman
Ungulatella CUSHMAN, Contr. Cushman Lab. Foram. Res., vol. 7, 1931, p. 81.

Test with the early portion conical and probably consisting of a series of spirally coiled chambers, later chambers uniserial, forming an elongate subcylindrical test, or somewhat compressed toward the apertural end; wall calcareous, rather coarsely perforate; aperture a loop-shaped opening in the flattened or somewhat concave terminal face. — Recent, Pacific.

Genus COLOMIA Cushman and Bermudez, 1948

Key, plate 50, figures 15–17
Genotype, *Colomia cretacea* Cushman and Bermudez
Colomia CUSHMAN and BERMUDEZ, Contr. Cushman Lab. Foram. Res., vol. 24, 1948, p. 12.

Test conical; earliest chambers indistinct, later ones uniserial, circular in transverse section, interior with vertical columns or tubular structures connecting walls of the adjacent chambers; wall calcareous, perforate; aperture in the adult terminal, a slightly arcuate, narrow opening in the middle of the apertural face. — Upper Cretaceous.

This genus apparently represents a form related to *Ungulatella*, but as the earliest chambers are obscure it is placed here with some question.

Genus ROBERTINA d'Orbigny, 1846

Plate 22, figure 4; Key, plate 27, figures 9, 10
Genotype, *Robertina arctica d'Orbigny*
Robertina D'ORBIGNY, Foram. Foss. Bass. Tert. Vienne, 1846, p. 202.
Bulimina (part) and *Cassidulina* (part) of authors.

Test an elongate, close spiral, the spiral suture distinct; chambers several in each whorl, in microspheric young like *Buliminella*, later forming a double series; wall calcareous, finely perforate; apertures two in number, the primary one elongate, loop-shaped, at basal margin of the chamber, extending into the apertural face, the secondary one at the basal margin extending between the last-formed chambers of the upper and lower series, usually smaller than the primary one. — Eocene to Recent.

The genus *Robertinoides* has been erected by Höglund (Zool. Bidrag Uppsala, vol. 26, 1947, p. 222) for forms differing from *Robertina* in having the supplementary aperture at the base of the apertural face which is supposed not to be present in typical *Robertina.* The only way of settling this problem would be a reëxamination of d'Orbigny's type specimens of *Robertina arctica* to determine the actual apertural characters. Höglund's figure of *R. normani* (chosen as genotype of *Robertinoides*), showing the two ventral and one dorsal aperture, is reproduced on plate 50, figure 13, in the Key.

Genus ELONGOBULA Finlay, 1939

Key, plate 50, figures 19, 20
Genotype, *Elongobula chattonensis* Finlay

Elongobula FINLAY, Trans. Roy. Soc. New Zealand, vol. 69, 1939, p. 321.

Test somewhat similar to *Buliminoides* but the surface without heavy ornamentation and the aperture subterminal and rounded. — Cretaceous and Tertiary.

Genus PSEUDOBULIMINA Earland, 1934

Key, plate 50, figures 21–24
Genotype, *Bulimina chapmani* Heron-Allen and Earland

Pseudobulimina EARLAND, *Discovery* Reports, vol. 10, 1934, p. 133.

Test free, consisting of two series of chambers of very different dimensions, rapidly increasing in size and arranged side by side in a helicoid spiral of more than one convolution; wall calcareous, perforate; aperture a narrow opening at the inner edge of the chamber. — Eocene to Recent.

This is a peculiar form which may possibly be related to the Cassidulinidae.

Subfamily 3. Bulimininae

Test spiral, usually triserial, becoming involute and finally in *Entosolenia* single chambered; wall calcareous, finely perforate; aperture loop-shaped, the larger end away from the inner margin (or rounded in *Entosolenia*), usually with a distinct tooth and internal tube connecting the chambers (or in *Entosolenia* free at the inner end).

Genus BULIMINA d'Orbigny, 1826

Plate 22, figure 7; Key, plate 27, figures 11–14
Genotype, *Bulimina marginata* d'Orbigny

Bulimina D'ORBIGNY, Ann. Sci. Nat., vol. 7, 1826, p. 269.
Pleurites EHRENBERG, 1854. *Cucurbitina* COSTA, 1856.

Test an elongate spiral, generally triserial; chambers inflated, spiral suture more or less obsolete; wall calcareous, perforate; aperture loop-shaped, with a tooth or plate at one side and an internal spiral tube connecting through the chambers between the apertures.—Jurassic to Recent.

Genus NEOBULIMINA Cushman and Wickenden, 1928

Plate 22, figure 24; Key, plate 27, figure 15
Genotype, *Neobulimina canadensis* Cushman and Wickenden

Neobulimina CUSHMAN and WICKENDEN, Contr. Cushman Lab. Foram. Res., vol. 4, 1928, p. 12.

Test in early stages triserial as in *Bulimina*, adult biserial, not compressed; chambers inflated, simple; wall calcareous, perforate; aperture in triserial stage as in *Bulimina*, in adult broader, tending to become subterminal. —Cretaceous.

Genus VIRGULOPSIS Finlay, 1939

Key, plate 50, figure 14
Genotype, *Virgulopsis pustulata* Finlay

Virgulopsis FINLAY, Trans. Roy. Soc. New Zealand, vol. 69, 1939, p. 321.

Test in the early stages triserial, in the adult stage biserial; wall calcareous, perforate, heavily sculptured; aperture elongate oval, terminal, without a tooth.—Tertiary.

Genus GLOBOBULIMINA Cushman, 1927

Plate 22, figure 22; Key, plate 27, figure 16
Genotype, *Globobulimina pacifica* Cushman

Globobulimina CUSHMAN, Contr. Cushman Lab. Foram. Res., vol. 3, 1927, p. 67.
Bulimina (part) of authors.

Test spiral, triserial, early chambers tending to elongate, later ones extending backward, and in the adult becoming involute or nearly so, the last three chambers often making up the whole exterior; wall calcareous, finely perforate; aperture loop-shaped, with a tooth or plate and internal tube.—Tertiary, Recent.

Genus ENTOSOLENIA Ehrenberg, 1848

Plate 22, figure 23; Key, plate 21, figures 19–23, 25; plate 27, figures 17–20
Genotype, *Entosolenia lineata* Williamson

Entosolenia EHRENBERG, in WILLIAMSON, Ann. Mag. Nat. Hist., ser. 2, vol. 1, 1848, p. 5.
Fissurina REUSS, 1850. *Hyaleina* COSTA, 1856. *Ellipsolagena* A. SILVESTRI, 1923.
Lagena (part) of authors.

Test a single chamber with an internal tube, free at the inner end; wall calcareous, finely perforate; aperture elliptical or circular.—Upper Cretaceous to Recent.

Many species referred to *Lagena* evidently belong here. For a discussion of the relationships of this group, see Parr, "The Lagenid Foraminifera and their Relationships" (Proc. Roy. Soc. Victoria, vol. 58, pts. 1–2 (n. ser.), 1947, pp. 116–133, pls. 6, 7, 1 text fig.).

Subfamily 4. Virgulininae

Test usually showing traces of its spiral origin in the twisted early stages, later biserial, and in the end forms uniserial.

For a full discussion of the relationships and the genera and species of this subfamily, see Cushman, A Monograph of the Subfamily Virgulininae of the Foraminiferal Family Buliminidae (Special Publ. 9, Cushman Lab. Foram. Res., 1937, pp. i–xv, 1–228, pls. 1–24).

Genus VIRGULINA d'Orbigny, 1826

Plate 22, figure 25; Key, plate 27, figures 21–24
Genotype, *Virgulina squammosa* d'Orbigny

Virgulina D'ORBIGNY, Ann. Sci. Nat., vol. 7, 1826, p. 267.
Strophoconus EHRENBERG, 1843. *Grammobotrys* EHRENBERG, 1845. *Bulimina* (part) and *Polymorphina* (part) of authors.

Test elongate, more or less compressed, fusiform; early chambers spiral about the elongate axis, especially in the microspheric form, triserial, later becoming irregularly biserial, whole test usually twisted; wall calcareous, finely perforate; aperture elongate, loop-shaped, with an apertural tooth or plate and internal spiral tube. — Lower Cretaceous to Recent.

Subgenus VIRGULINELLA Cushman, 1932

Key, plate 27, figures 23, 24
Subgenotype, *Virgulina pertusa* Reuss

Virgulinella CUSHMAN, Contr. Cushman Lab. Foram. Res., vol. 8, 1932, p. 9.

Test similar to that of typical *Virgulina*, but with the sutures marked by finger-like processes extending backward onto the surface of the preceding chamber, and the intermediate spaces between the processes deeply excavated. — Miocene.

Genus BOLIVINA d'Orbigny, 1839

Plate 22, figure 26; Key, plate 27, figures 25–28
Genotype, *Bolivina plicata* d'Orbigny

Bolivina D'ORBIGNY, Voy. Amér. Mérid., vol. 5, pt. 5, 1839, p. 61.
Sagrina D'ORBIGNY, 1839. *Grammostomum* EHRENBERG, 1840. *Proroporus* EHRENBERG, 1844. *Clidostomum* EHRENBERG, 1845. *Brizalina* COSTA, 1856. *Vulvulina* (part) and *Virgulina* (part) of authors.

Test elongate, usually compressed, tapering, initial end and often whole test twisted; chambers typically biserial; wall calcareous, finely or coarsely perforate; aperture elongate, usually oblique, somewhat loop-shaped, often with a plate-like tooth connecting with an internal tube. — Cretaceous to Recent.

Genus SUGGRUNDA W. Hoffmeister and C. Berry, 1937

Key, plate 50, figure 25

Genotype, *Suggrunda porosa* W. Hoffmeister and C. Berry

Suggrunda W. Hoffmeister and C. Berry, Journ. Pal., vol. 11, 1937, p. 29.

"Test very minute, elongate, tapering, tending to be twisted, biserial throughout; chambers somewhat angular, closely appressed, tending to be spinose; walls calcareous, hyaline, rather coarsely perforated; aperture semi-lunate, at base of apertural face." — Miocene.

Genus LOXOSTOMUM Ehrenberg, 1854

Plate 22, figure 27; Key, plate 27, figures 30–32

Genotype, *Loxostomum subrostratum* Ehrenberg

Loxostomum Ehrenberg, Mikrogeologie, 1854, pl. xvii, fig. 19.
Proroporus of authors (not Ehrenberg). *Bolivina* (part) of authors.

Test in the early stages similar to *Bolivina*, adult tending to become uniserial; aperture terminal. — Cretaceous to Recent.

Genus BIFARINA Parker and Jones, 1872

Plate 22, figure 28; Key, plate 28, figures 1, 2

Genotype, *Dimorphina saxipara* Ehrenberg

Bifarina Parker and Jones, Ann. Mag. Nat. Hist., ser. 4, vol. 10, 1872, p. 198.
Dimorphina Ehrenberg, 1854 (not d'Orbigny).

Test with the earlier chambers biserial, later ones uniserial, uniserial portion making up most of the test in most species; wall calcareous, perforate; aperture in the young as in *Bolivina*, later terminal and rounded. — Jurassic (?), Cretaceous to Recent.

Genus BIFARINELLA Cushman and Hanzawa, 1936

Key, plate 50, figure 26

Genotype, *Bifarinella ryukyuensis* Cushman and Hanzawa

Bifarinella Cushman and Hanzawa, Contr. Cushman Lab. Foram. Res., vol. 12, 1936, p. 45.

Test free, elongate, compressed, tapering, with the greatest width at the apertural end, in the young stage biserial, becoming uniserial in the adult; the early chambers generally triangular in front view with strongly oblique sutures, later chambers with the sutures nearly at right angles to the elongate axis of the test; wall calcareous, finely perforate; aperture in the adult narrow, elongate, with a distinct, everted lip at each side. — Pleistocene.

Genus RECTOBOLIVINA Cushman, 1927

Plate 22, figure 30; Key, plate 27, figure 29
Genotype, *Sagrina bifrons* H. B. Brady

Rectobolivina CUSHMAN, Contr. Cushman Lab. Foram. Res., vol. 3, 1927, p. 68.
Sagrina (part) and *Siphogenerina* (part) of authors.

Test elongate, somewhat compressed, young biserial, adult uniserial; wall calcareous, perforate; aperture in adult terminal, rounded, with a slight lip and internal tube. — Tertiary, Recent.

Genus SCHUBERTIA A. Silvestri, 1911

Plate 22, figure 31; Key, plate 28, figure 3
Genotype, *Sagrina* (?) *tessellata* H. B. Brady

Schubertia A. SILVESTRI, Riv. Ital. Pal., 1911, p. 67, footnote.
Sagrina (part) H. B. BRADY and later authors. *Millettia* SCHUBERT, 1911.

Test free, elongate; early chambers irregularly biserial, later ones uniserial, undivided, later ones much elongated and partially divided into chamberlets; wall calcareous, finely perforate; aperture in adult terminal, with a short neck and phialine lip. — Tertiary, Recent.

See Cushman, The Development and Generic Position of *Sagrina* (?) *tessellata* H. B. Brady (Journ. Washington Acad. Sci., vol. 19, 1929, pp. 337–339, text figs. 1–5).

Genus GEMINARICTA Cushman, 1936

Key, plate 50, figure 27
Genotype, *Bolivinella virgata* Cushman

Geminaricta CUSHMAN, Special Publ. No. 6, Cushman Lab. Foram. Res., 1936, p. 61; Special Publ. No. 9, 1937, p. 208.
Bolivinella CUSHMAN (part), 1929.

Test free, in the early stages biserial, compressed, in the adult uniserial; wall calcareous, perforate; aperture in the adult consisting of a pair of small, rounded openings well separated from one another toward the ends of an elongate, elliptical depression in the terminal wall. — Miocene, Recent.

Genus BITUBULOGENERINA Howe, 1934

Key, plate 50, figure 28
Genotype, *Bitubulogenerina vicksburgense* Howe

Bitubulogenerina HOWE, Journ. Pal., vol. 8, 1934, p. 420.

Test free, early stages triserial, adult biserial; wall calcareous, perforate, with hollow protuberances extending outward, sometimes open and tubular; aperture terminal, with distinct lip and internal tube. — Eocene to Miocene. Recent (?).

Genus TRITUBULOGENERINA Howe, 1939

Key, plate 50, figure 30
Genotype, *Tritubulogenerina mauricensis* Howe
Tritubulogenerina Howe, State of Louisiana, Dept. of Conservation, Geol. Bull. 14, Jan., 1939, p. 69.

"Test free, triserial throughout; wall calcareous, hyaline, perforate, with numerous minute granulations on the surface which appear to be minute tubuli; aperture terminal, ovate or angular, with a distinct lip." — Eocene.

Genus TUBULOGENERINA Cushman, 1927

Plate 22, figure 29; Key, plate 27, figures 33, 34
Genotype, *Textularia (Bigenerina) tubulifera* Parker and Jones
Tubulogenerina Cushman, Contr. Cushman Lab. Foram. Res., vol. 2, pt. 4, 1927, p. 78.
Textularia (part), *Bigenerina* (part), and *Clavulina* (part) of authors.

Test with early stages biserial, sometimes triserial in the microspheric form, adult uniserial, compressed or rounded in section; wall calcareous, perforate, with numerous tubuli extending out from the test, either open or closed, forming lobular connections with the chamber; aperture elongate, narrow, or in the adult numerous rounded openings in the terminal face, interior in some species with a curved structure from roof to floor in the biserial portion. — Eocene to Miocene.

Subfamily 5. Reussellinae

Test distinctly triserial, at least in the young of most forms, in specialized forms becoming uniserial; aperture in simpler forms and young elongate, in some uniserial forms cribrate.

See Cushman, The Species of the Subfamily Reussellinae of the Foraminiferal Family Buliminidae (Contr. Cushman Lab. Foram. Res., vol. 21, 1945, pp. 23–54, pls. 5–8).

Genus REUSSELLA Galloway, 1933

Plate 22, figure 17; Key, plate 28, figure 4
Genotype, *Verneuilina spinulosa* Reuss
Reussella Galloway, Man. Foram., 1933, p. 360.
Reussia Schwager (not McCoy). *Verneuilina* (part) of authors.

Test distinctly triserial, triangular in transverse section, broadest at the apertural end; wall calcareous, finely or coarsely perforate; aperture elongate, oblique, from the base of the chamber in the apertural face. — Cretaceous to Recent.

Genus TRIMOSINA Cushman, 1927

Plate 22, figure 20; Key, plate 28, figure 6
Genotype, *Mimosina spinulosa* Millett, var.
Trimosina CUSHMAN, Contr. Cushman Lab. Foram. Res., vol. 3, 1927, p. 64.
Mimosina (part) MILLETT.

Test triserial; chambers with a single acicular spine, or these may become obsolete; wall calcareous, vesicular; aperture elongate, removed from the edge, sometimes with an added series of rounded pores along the base of the apertural face. — Recent, Indo-Pacific.

See Cushman, The Genus *Trimosina* and Its Relationships to Other Genera of the Foraminifera (Journ. Washington Acad. Sci., vol. 19, 1929, pp. 155–159, text figs. 1–3).

Genus PSEUDOCHRYSALIDINA Cole, 1941

Key, plate 51, figures 1, 2
Genotype, *Pseudochrysalidina floridana* Cole
Pseudochrysalidina COLE, Florida Dept. Conservation, Geol. Bull. 19, 1941, p. 35.

Test conical, triserial, labyrinthic axis through center of test; wall calcareous, perforate; aperture consisting of numerous rounded pores on the last-formed chamber. — Eocene.

Genus MIMOSINA Millett, 1900

Plate 22, figure 21; Key, plate 28, figure 7
Genotype, *Mimosina hystrix* Millett
Mimosina MILLETT, Journ. Roy. Micr. Soc., 1900, p. 547.

Test triserial in the young, later biserial; chambers with a single acicular spine at the outer angle; wall calcareous, vesicular; aperture of two parts, one rounded and nearly terminal, the other below near the inner rim of the chamber, more elongate, arched. — Recent, Indo-Pacific.

Genus PAVONINA d'Orbigny, 1826

Plate 22, figure 19; Key, plate 28, figure 8
Genotype, *Pavonina flabelliformis* d'Orbigny
Pavonina D'ORBIGNY, Ann. Sci. Nat., vol. 7, 1826, p. 260.

Test in the early stages triserial and like *Reussella*, later uniserial, fan-shaped or annular; wall calcareous, coarsely perforate; apertures in the adult many rounded openings on the peripheral margin. — Oligocene to Recent.

See Parr, The Genus *Pavonina* and its Relationships (Proc. Roy. Soc. Victoria, vol. 45, pt. 1, 1933, pp. 28–31, pl. VII).

Genus CHRYSALIDINELLA Schubert, 1907

Plate 22, figure 18; Key, plate 28, figure 5
Genotype, *Chrysalidina dimorpha* H. B. Brady
Chrysalidinella SCHUBERT, Neues Jahrb. für Min., vol. 25, 1907, p. 242.
Chrysalidina of authors (not D'ORBIGNY).

Test tapering, triangular in transverse section, early stage triserial, adult uniserial; wall calcareous, perforate; aperture in adult cribrate, of numerous rounded openings scattered over the triangular apertural face.— Eocene to Recent.

Subfamily 6. Uvigerininae

Test generally triserial, at least in the early stages, later in some forms uniserial or irregular; wall calcareous, perforate; aperture typically terminal, with neck and phialine lip, in some genera with a tooth and internal tube.

Genus UVIGERINELLA Cushman, 1926

Plate 22, figure 8; Key, plate 28, figure 11
Genotype, *Uvigerina (Uvigerinella) californica* Cushman
Uvigerinella CUSHMAN, Contr. Cushman Lab. Foram. Res., vol. 2, pt. 3, 1926, p. 58.
Uvigerina (part) of authors.

Test generally triserial, tapering or fusiform, in some species tending to become biserial; wall calcareous, perforate; aperture loop-shaped similar to *Bulimina* but with a raised, collar-like rim and a tooth, internal tube often present. — Eocene to Recent.

Genus UVIGERINA d'Orbigny, 1826

Plate 22, figure 9; Key, plate 28, figures 9, 10
Genotype, *Uvigerina pigmea* d'Orbigny
Uvigerina D'ORBIGNY, Ann. Sci. Nat., vol. 7, 1826, p. 268.

Test generally triserial, elongate, fusiform, rounded in transverse section; chambers inflated, rounded; wall calcareous, perforate; aperture terminal, rounded, with neck and lip, often with a spiral tooth and internal twisted tube. — Eocene to Recent.

Genus RECTUVIGERINA Mathews, 1945

Key, plate 50, figure 31
Genotype, *Siphogenerina multicostata* Cushman and Jarvis
Rectuvigerina MATHEWS, Journ. Pal., vol. 19, 1945, p. 590.
Siphogenerina (part), *Sagrina* (part), *Uvigerina* (part), *Dimorphina* (part), and *Nodosaria* (part) of authors.

Test free, elongate, round in cross section, early portion triserial, later uniserial, with or without an intermediate biserial stage; wall calcareous, perforate; aperture terminal with a phialine lip, connecting with an internal tube developed in the later chambers. — Tertiary and Recent.

Genus HOPKINSINA Howe and Wallace, 1933

Plate 22, figure 11; Key, plate 28, figure 12
Genotype, *Hopkinsina danvillensis* Howe and Wallace
Hopkinsina Howe and Wallace, Louisiana Geol. Surv. Bull. 2, 1933, p. 182.

Test in the early stages similar to *Uvigerina*, in the adult becoming biserial but somewhat twisted; wall calcareous, perforate; aperture terminal, oval or circular, usually with a short neck and lip. — Eocene to Recent.

Genus SIPHOGENERINA Schlumberger, 1883

Plate 22, figure 10; Key, plate 28, figures 18, 19
Genotype, *Siphogenerina costata* Schlumberger
Siphogenerina Schlumberger, Feuille Jeun. Nat., 1883, p. 117.
Sagrina (part) of authors.

Test elongate, cylindrical, early stages typically triserial, rounded in section, later uniserial; wall calcareous, perforate; aperture in the adult terminal, with a distinct neck, phialine lip and internal tube. — Eocene to Recent.

See Cushman, Foraminifera of the Genera *Siphogenerina* and *Pavonina* (Proc. U. S. Nat. Mus., vol. 67, Art. 25, 1926, pp. 1–24, pls. 1–6).

Genus UNICOSIPHONIA Cushman, 1935

Key, plate 50, figure 29
Genotype, *Unicosiphonia crenulata* Cushman
Unicosiphonia Cushman, Contr. Cushman Lab. Foram. Res., vol. 11, 1935, p. 81.

Test elongate, more or less tapering, uniserial throughout; chambers crenulate at the margin with backwardly projecting, somewhat overlapping processes; sutures distinct, depressed, at right angles to the elongate axis; wall calcareous, finely perforate; aperture rounded, terminal, with a distinct neck and slight lip connecting with an internal, tubular structure connecting the earlier apertures. — Tertiary.

Genus SIPHONODOSARIA A. Silvestri, 1924

Plate 22, figure 12; Key, plate 28, figures 20, 21
Genotype, *Nodosaria abyssorum* H. B. Brady
Siphonodosaria A. Silvestri, Boll. Soc. Geol. Ital., vol. 42, 1923 (1924), p. 18.
Sagrinnodosaria Jedlitschka, 1931.
Nodosaria (part) and *Sagrina* (part) of authors.

Test elongate; chambers in a rectilinear uniserial arrangement, inflated, proloculum often larger than following chambers; wall calcareous, perforate; aperture large, rounded, with neck and lip. — Tertiary, Recent.

Genus ANGULOGERINA Cushman, 1927

Plate 22, figure 14; Key, plate 26, figure 33; plate 28, figures 13, 14
Genotype, *Uvigerina angulosa* Williamson

Angulogerina CUSHMAN, Contr. Cushman Lab. Foram. Res., vol. 3, 1927, p. 69.
Uvigerina (part) of authors.

Test triserial, elongate, whole test angled, with three flattened sides and distinct angles, later chambers often loosely arranged; wall calcareous, perforate; aperture terminal, with a short neck and lip. — Eocene to Recent.

Genus TRIFARINA Cushman, 1923

Plate 22, figure 15; Key, plate 28, figure 15
Genotype, *Trifarina bradyi* Cushman

Trifarina CUSHMAN, Bull. 104, U. S. Nat. Mus., pt. 4, 1923, p. 99.
Rhabdogonium of authors (not REUSS). *Triplasia* of authors (not REUSS).

Test elongate, triangular in transverse section; early chambers in an irregular spire or triserial, later uniserial; wall calcareous, perforate; aperture terminal in adult, rounded, with short neck and lip. — Eocene to Recent.

See Cushman, *Trifarina* in the American Eocene and Elsewhere (Contr. Cushman Lab. Foram. Res., vol. 1, pt. 4, 1926, pp. 86–88).

Genus DENTALINOPSIS Reuss, 1860

Plate 22, figure 16; Key, plate 28, figure 17
Genotype, *Dentalinopsis semitriquetra* Reuss

Dentalinopsis REUSS, Sitz. k. böhm. Ges. Wiss., Jahrg. 1860, p. 91.

Test uniserial; chambers in a generally straight or slightly curved, linear series, early chambers angled and triangular in section, later ones rounded; wall calcareous, perforate; aperture rounded, terminal. — Jurassic, Cretaceous.

Genus WASHITELLA Tappan, 1943

Key, plate 51, figures 3, 4
Genotype, *Washitella typica* Tappan

Washitella TAPPAN, Journ. Pal., vol. 17, 1943, p. 515.

Test free, consisting of well-defined but very irregularly arranged chambers; may be in a linear or slightly coiled series or branch in various ways; wall calcareous, finely perforate; apertures at the end of the series of chambers, occasionally more than one. — Lower Cretaceous.

Until more is known of the developmental stages of this genus its position must be very uncertain. It may be related to *Sporadogenerina* in which case it would belong in the Lagenidae.

Genus DELOSINA Wiesner, 1931

Key, plate 51, figures 5–7
Genotype, *Polymorphina* (?) *complexa* Sidebottom
Delosina WIESNER, Deutsche Südpolar Exped., vol. 20, Zool., 1931, p. 123.

Test fusiform, largely triserial; chambers distinct, greatly increasing in size as added, the final whorl making up a large part of the surface in the adult; wall calcareous, finely perforate; sutures with a series of pores opening into a canal running under the sutures; no general aperture. — Recent.

The position of this genus is in much doubt. Earland, who has made a careful study of its structure, is inclined to place it in the Buliminidae, and we have followed his suggestion.

Family 32. ELLIPSOIDINIDAE

Test with the early stages biserial, becoming uniserial in most of the genera; wall calcareous, finely perforate; aperture usually narrow, elongate, curved, often with a sort of hood-like portion overhanging the aperture itself, a hollow tube or rod-like structure, sometimes in the form of a curved plate connecting the various chambers, similar in general to that found in the Buliminidae.

This family is particularly characterized by the aperture which has a distinct hooded form, which in end view is usually semicircular or crescentic. The internal connecting tubular structure is closely similar to that of the Buliminidae, from which this family is derived, probably through *Virgulina*. The adult biserial character of *Virgulina*, through acceleration, becomes the character of the early stages of most of the Ellipsoidinidae. In *Pleurostomella* this character continues, but in the other more specialized genera it is seen in the early stages or is entirely skipped through acceleration, and the uniserial forms are largely distinguished by the character of the aperture.

There is a closely-knit, continuous series as shown in the accompanying plate. Numerous other generic names have been proposed, but the main groups are provided for by the generic names used here.

The early history of the family starts in the Cretaceous, or possibly in the Jurassic, and the greatest development is in the Upper Cretaceous and Eocene. In the faunas of Trinidad the family is particularly well represented, and Dr. A. Silvestri has described many genera and species from Italy.

In the development of the group there is the usual tendency to produce uniserial forms from biserial ones, and also have the tests become almost completely involute in some genera, two tendencies very common in the foraminifera.

The genera *Stilostomella* and *Daucina*, although sometimes placed here, are difficult to determine with full accuracy until type material shall have been more carefully studied than has yet been done.

Genus PLEUROSTOMELLA Reuss, 1860

Plate 23, figure 1; Key, plate 28, figures 22, 23
Genotype, *Pleurostomella subnodosa* Reuss

Pleurostomella Reuss, Sitz. Akad. Wiss. Wien, vol. 40, 1860, p. 203.
Nodosaria (part) and *Dentalina* (part) Reuss, 1860.

Test usually elongate, biserial; wall calcareous, perforate; aperture an arched opening on the inner side of the chamber, partially closed by two broad teeth at either side at the base with a narrow vertical slit between. — Cretaceous to Recent.

See Cushman and Harris, Notes on the Genus *Pleurostomella* (Contr. Cushman Lab. Foram. Res., vol. 3, 1927, pp. 128–135, 156, pls. 25, 28).

Genus PLEUROSTOMELLINA Schubert, 1911

Key, plate 28, figure 24
Genotype, *Pleurostomella barroisi* Berthelin

Pleurostomellina Schubert, Abhandl. k. k. Geol. Reichs., vol. 20, pt. 4, 1911, p. 59.
Pleurostomella (part) Berthelin (not d'Orbigny).

Test with the general characters of *Pleurostomella* in the young, but quickly becoming uniserial in the adult instead of continuing the biserial character. — Upper Cretaceous.

Genus DENTALINOIDES Marie, 1941

Key, plate 51, figures 8, 9
Genotype, *Dentalinoides canulina* Marie

Dentalinoides Marie, Mém. Mus. Nat. Hist. Nat., n. ser., vol. 12, 1941, p. 207.

Test free, uniserial, elongate, not compressed, circular in transverse section, composed of a series of chambers in a straight or slightly curved form; wall calcareous, perforate; aperture a large circular or oval opening, placed laterally toward the end of the last-formed chamber. — Upper Cretaceous.

This genus may be the same as *Pleurostomellina* but is uniserial throughout and apparently does not have the flat tooth of that genus.

Genus ELLIPSOPLEUROSTOMELLA A. Silvestri, 1903

Plate 23, figure 2; Key, plate 28, figure 25
Genotype, *Ellipsopleurostomella schlichti* A. Silvestri

Ellipsopleurostomella A. Silvestri, Atti R. Accad. Sci. Torino, vol. 39, 1903, p. 216.
Rostrolina (part) von Schlicht, 1870.

Test with the early chambers biserial, later ones uniserial, more or less involute; wall calcareous, perforate; aperture narrow, subelliptical, subterminal. — Cretaceous, Tertiary.

Genus ELLIPSOBULIMINA A. Silvestri, 1903

Plate 23, figure 3; Key, plate 28, figure 26

Genotype, *Ellipsobulimina seguenzai* A. Silvestri

Ellipsobulimina A. Silvestri, Atti R. Accad. Sci. Torino, vol. 39, 1903, p. 11.

Test with the early chambers somewhat biserial, but entirely involute, the last-formed chamber making up nearly the entire exterior of the test; wall calcareous, perforate; aperture narrow, semicircular.—Cretaceous (?), Miocene.

Genus NODOSARELLA Rzehak, 1895

Plate 23, figure 4; Key, plate 28, figure 27

Genotype, *Lingulina tuberosa* Gümbel

Nodosarella Rzehak, Ann. k. k. Nat. Hofmuseums, vol. 10, 1895, p. 220.

Nodosaria (part) and *Lingulina* (part) of authors. *Ellipsoidella* Heron-Allen and Earland, 1910.

Test elongate, with the early chambers showing traces of the biserial ancestry, but the later ones in a rectilinear series, very slightly involute; wall calcareous, perforate; aperture narrow, subterminal, semi-elliptical. —Cretaceous, Tertiary.

Genus ELLIPSONODOSARIA A. Silvestri, 1900

Plate 23, figure 5

Genotype, *Lingulina rotundata* d'Orbigny

Ellipsonodosaria A. Silvestri, Atti Rend. Accad. Sci. Let. Art. Zelanti Stud. Acireale, vol. 10, 1899–1900 (1900), p. 4.

Lingulina (part) d'Orbigny, 1846.

Test elongate, with all the chambers in a rectilinear series, rounded; wall

---

PLATE 23

Relationships of the Genera of the Ellipsoidinidae

FIG.

1. *Pleurostomella pleurostomella* (A. Silvestri). (After A. Silvestri.) *a*, front view; *b*, side view; *c*, apertural view; *d*, *e*, sections.
2. *Ellipsopleurostomella schlichti* A. Silvestri. (After A. Silvestri.)
3. *Ellipsobulimina seguenzai* A. Silvestri. (After A. Silvestri.) *a*, side view; *b*, apertural view; *c*, section.
4. *Nodosarella salmojraghii* Martinotti. (After A. Silvestri.)
5. *Ellipsonodosaria rotundata* (d'Orbigny). (After A. Silvestri.) *a*, microspheric; *b*, megalospheric; *E. chapmani* A. Silvestri, *c*, end view; *d*, section. (After A. Silvestri.)
6. *Ellipsolingulina impressa* (Terquem). (After A. Silvestri.) *a*, front view; *b*, apertural view; *c*, section.
7. *Ellipsoglandulina labiata* (Schwager). (After A. Silvestri.)
8. *Gonatosphaera prolata* Guppy. (After Guppy.) *a*, front view; *b*, apertural view; *c*, section.
9. *Ellipsoidina ellipsoides* Seguenza. (After A. Silvestri.) *a*, front view; *b*, apertural view; *c*, section.
10. *Parafissurina ventricosa* (A. Silvestri). (After A. Silvestri.) *a*, front view; *b*, apertural view; *c*, section.

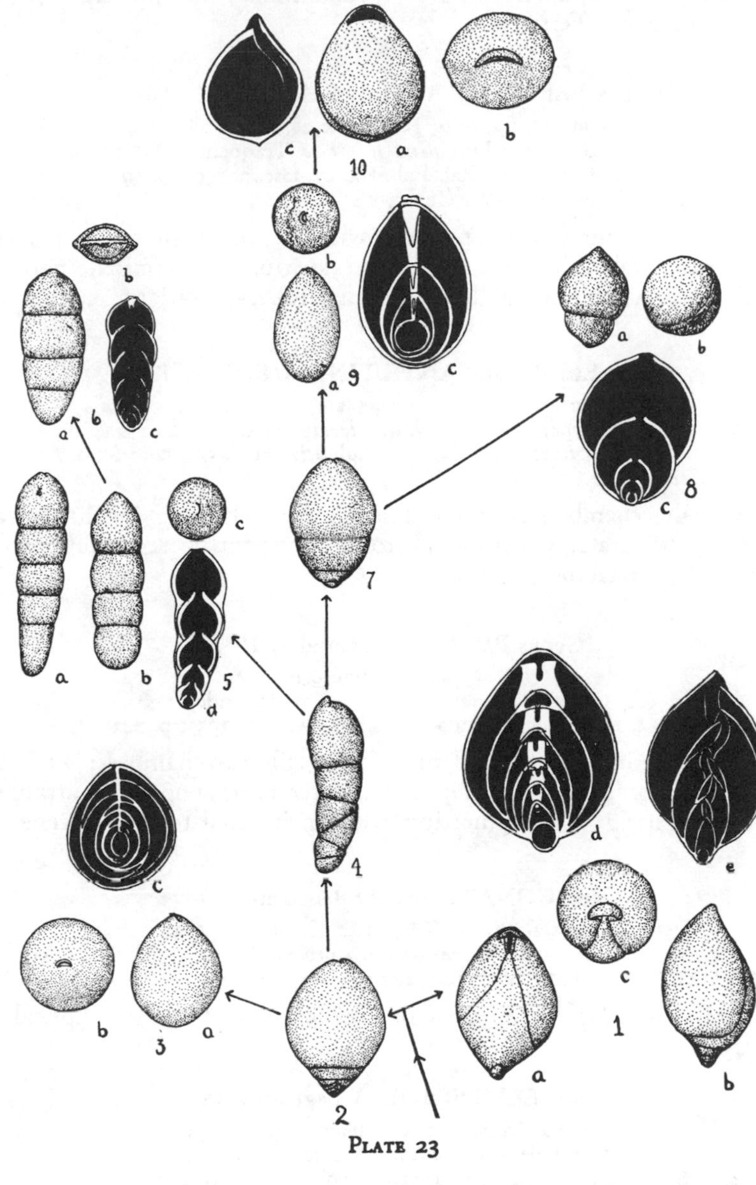

PLATE 23

calcareous, perforate; aperture narrow, subterminal, subelliptical.—Jurassic (?), Cretaceous, Tertiary.

Genus ELLIPSOLINGULINA A. Silvestri, 1907

Plate 23, figure 6; Key, plate 28, figure 28
Genotype, *Lingulina impressa* Terquem

Ellipsolingulina A. Silvestri, Riv. Ital. Pal., vol. 13, fasc. 2, 1907, p. 69.
Lingulina (part) of authors (not d'Orbigny).

Test elongate, somewhat compressed, with all the chambers in a rectilinear series; wall calcareous, perforate; aperture subterminal, narrow, semi-elliptical, sometimes with an internal tube.—Tertiary.

Genus ELLIPSOGLANDULINA A. Silvestri, 1900

Plate 23, figure 7; Key, plate 28, figure 29
Genotype, *Ellipsoglandulina laevigata* A. Silvestri

Ellipsoglandulina A. Silvestri, Atti Rend. Accad. Sci. Let. Art. Zelanti Stud. Acireale, vol. 10, 1899–1900 (1900), p. 9.

Test with the chambers in a rectilinear series becoming involute; wall calcareous, perforate; aperture narrow, subterminal, semi-elliptical.—Jurassic (?), Cretaceous, Tertiary.

Genus PINARIA Bermudez, 1937

Key, plate 51, figure 10
Genotype, *Pinaria heterosculpta* Bermudez

Pinaria Bermudez, Mem. Soc. Cubana Hist. Nat., vol. 11, 1937, p. 242.

Test free, cylindrical, composed of a few rectilinear chambers; wall calcareous, perforate; aperture composed of three narrow openings arranged in a triangular position, connecting with an internal tube.—Eocene.

Genus GONATOSPHAERA Guppy, 1894

Plate 23, figure 8; Key, plate 28, figures 30–32
Genotype, *Gonatosphaera prolata* Guppy

Gonatosphaera Guppy, Proc. Zool. Soc. London, 1894, p. 651.

Test similar to *Ellipsoglandulina* but compressed; aperture elliptical.—Tertiary.

Genus ELLIPSOIDINA Seguenza, 1859

Plate 23, figure 9; Key, plate 28, figure 33
Genotype, *Ellipsoidina ellipsoides* Seguenza

Ellipsoidina Seguenza, Eco Peloritano, ser. 2, vol. 5, 1859, fasc. 9 (10).

Test with the chambers in a rectilinear series, but completely involute, a hollow tubular structure between the apertures of successive chambers; aperture semicircular, narrow, subterminal.—Tertiary.

Genus PARAFISSURINA Parr, 1947

Plate 23, figure 10; Key, plate 28, figures 34, 35
Genotype, *Lagena ventricosa* A. Silvestri

Parafissurina PARR, Proc. Roy. Soc. Victoria, vol. 58, pts. 1–2 (n. ser.), 1947, p. 123.
Lagena (part) of authors.
Ellipsolagena (part) of authors (not A. SILVESTRI, 1923).

"Test calcareous, perforate, consisting of a single, usually compressed, chamber, with an internal tube directed backwards from the sub-terminal aperture, which is an arched or crescentic opening facing the front under a hood-like extension of the ventral wall of the test." — Eocene to Recent.

This genus was erected for the forms previously placed in *Ellipsolagena.* As the genotype of that genus should have been *Lagena acutissima* instead of *Lagena ventricosa*, which would make the genus synonymous with *Entosolenia*, the name *Ellipsolagena* is not valid.

Family 33. ROTALIIDAE

Test generally trochoid except in *Spirillina,* all the chambers visible from the dorsal side except in a very few genera which become partially involute, only those of the last-formed whorl usually visible from the ventral side; wall calcareous, usually rather coarsely perforate, but the early stages in many genera with a chitinous inner wall, and in some of the more primitive genera in the fossil series with an outer arenaceous layer in the young; aperture typically on the ventral side of the test.

The group of the Rotaliidae contains many genera alike in the generally trochoid form. It is very probable that further careful studies of the fossil series and of the early stages will show at least two different groups, and there will be a consequent further refining of the generic descriptions which it is impossible to attempt at present, due to our incomplete knowledge of numerous forms. The early forms of the Mesozoic Trochamminas are so similar to some of the Rotaliidae that the early history should be carefully checked. One group has evidently arisen directly from the arenaceous *Trochammina*, and still retains the chitinous inner wall in the young, and in numerous instances the early stages show traces of the typical ancestral arenaceous wall. Another group has arisen in the Jurassic from the conical *Spirillina*-like forms. These have an umbilicus, either open or filled with a plug or a series of pillars. No inner chitinous wall seems to be present in these forms nor trace of the arenaceous stage.

The second of these groups will be discussed first, as the line of development is well known. From the planispiral *Spirillina,* which early developed from the Ammodiscidae by the calcareous cement becoming the dominant character of the test, arose conical forms with an open umbilical area, *Turrispirillina.* In the Jurassic also develops *Trocholina* in which the umbilical area is filled with pillars appearing as pustules on the exterior, and the coils all on the outer conical surface. It is very similar to *Rotalia* with its central plug or pillars, but is not divided into chambers. *Conicospirillina* has the hollow umbilical area covered, and passes to *Paalzowella* which is partially divided on the ventral side. From this to *Epistomina*, also highly developed in the Jurassic, is but a step.

Patellina and its related forms are primitive in their long undivided early stage, and have usually but two chambers to a whorl. *Annulopatellina* with its annular chambers is the end form of this series.

The series from *Discorbis* includes *Valvulineria* with a valvular lip over the umbilicus, *Lamarckina* with its large umbilicus and highly polished ventral side, *Gyroidina* with an open umbilicus, and *Rotaliatina* with its high spire.

Related to *Epistomina* with its nearly peripheral aperture, and to *Mississippina* with supplementary apertures on both sides of the periphery, are *Siphonina* with a definite neck and lip, the more involute *Siphoninoides*, the uncoiled *Siphoninella*, and the biserial *Siphonides*.

Cancris and *Baggina* have a thin rounded or oval plate on the inner end of the ventral side of the chamber, and probably gave rise to *Ceratobulimina* and the Cassidulinidae. *Eponides*, one of the best marked and most common genera of the family, has the umbilicus typically covered and the aperture ventral. From it directly came *Pegidia* and the Pegidiidae.

From *Rotalia* by very slight steps came *Dictyoconoides*, which is simply a *Rotalia* greatly spread out, and the number of umbilical pillars greatly increased. There are all stages in this development in the Eocene, and the border line between the two genera is often difficult to determine.

From *Rotalia* also developed the Amphisteginidae, Calcarinidae and other families. From forms similar to *Eponides* or *Discorbis* came the Globigerinidae, Planorbulinidae, Anomalinidae and related families.

The other groups previously mentioned possibly are to be derived from *Trochammina*. Some of the forms usually referred to *Discorbis* have a thin inner chitinous wall, and in some forms a thin outer finely arenaceous layer, especially in the fossil forms. Such forms developed into *Cymbalopora* in the Cretaceous, and may probably be related to those Planorbulinidae and Anomalinidae which have the same characters in their early development. Much careful detailed study is necessary to fix the relationships of these different groups. It is probable that offshoots have taken place at different times and places, and that the family Rotaliidae as here used must be still further subdivided as our knowledge of these forms grows.

A number of specialized genera have recently been described which are of interest in showing the parallelisms that have developed in this group. A number of genera show uncoiled adults. These, as is the case with most specialized genera, are relatively short lived.

The idea that *Endothyra* is the ancestor of the Rotaliidae is a fanciful one, or that *Globorotalia* is the ancestor of the Rotaliidae. *Endothyra* is a more or less specialized arenaceous form mostly confined to the Paleozoic. *Globorotalia* was directly derived from *Globotruncana*, and is a specialized genus of a pelagic habit as is *Globotruncana*, appearing in the Cretaceous, while the early forms of the Rotaliidae had already appeared in the Jurassic or earlier.

Hardly any other group of the foraminifera will repay the student with more interesting results than the Rotaliidae.

Nuttallides Finlay, 1939 (Trans. Roy. Soc. New Zealand, vol. 68, 1939, p. 520), seems to be somewhat uncertain, as the type species selected does not appear to possess the apertural characters of the generic description from a study of the actual types.

Subfamily 1. Spirillininae

Test simple, consisting of a proloculum and a planispiral, undivided, tubular, second chamber, open end of the tubular chamber serving as the aperture.

Genus SPIRILLINA Ehrenberg, 1843

Key, plate 29, figures 1–4
Genotype, *Spirillina vivipara* Ehrenberg

Spirillina Ehrenberg, Abhandl. k. Akad. Wiss. Berlin, 1841 (1843), p. 422.
Operculina (part) Reuss, 1849. *Cornuspira* (part) Schultze, 1854. *Cyclolina* Egger, 1857 (not d'Orbigny). *Mychostomina* Berthelin, 1881.

Test typically free, occasionally attached, planispiral, composed of a subglobular or ovoid proloculum and a long, undivided, tubular, second chamber, close coiled, usually in one plane; wall calcareous, finely or coarsely perforate; aperture formed by the open end of the tube. — Carboniferous to Recent.

Subfamily 2. Turrispirillininae

Test simple, conical in shape, consisting of a proloculum and high spired, undivided, second chamber.

Genus TURRISPIRILLINA Cushman, 1927

Plate 24, figure 1; Key, plate 29, figure 5
Genotype, *Spirillina conoidea* Paalzow

Turrispirillina Cushman, Contr. Cushman Lab. Foram. Res., vol. 3, 1927, p. 73.
Spirillina (part) of authors.

Test composed of a proloculum and elongate, tubular, undivided, second chamber in a hollow conical spire, coils not appreciably involute; aperture a semicircular opening, at the end of the tube. — Jurassic to Recent.

Genus CONICOSPIRILLINA Cushman, 1927

Plate 24, figure 2; Key, plate 29, figure 6
Genotype, *Spirillina trochoides* Berthelin
Conicospirillina CUSHMAN, Contr. Cushman Lab. Foram. Res., vol. 3, 1927, p. 73.
Spirillina BERTHELIN, 1879 (not EHRENBERG).

Test coiled in a conical, spiral chamber completely involute on the ventral side; wall calcareous, perforate; aperture a narrow slit on the ventral face of the revolving chamber, from the periphery toward the umbilicus.— Jurassic, Recent.

Genus COSCINOCONUS Leupold, 1935

Key, plate 51, figures 11–13
Genotype, *Coscinoconus alpinus* Leupold
Coscinoconus LEUPOLD, in LEUPOLD and BIGLER, Eclog. geol. Helvetiae, vol. 28, 1935, p. 618.

Test conical, coiled; chamber of several whorls, apparently undivided, on the ventral side with a thin, lattice-like plate across the central cavity; wall calcareous, perforate.— Jurassic.

Genus TROCHOLINA Paalzow, 1922

Key, plate 29, figures 7–9
Genotype, *Involutina conica* Schlumberger
Trocholina PAALZOW, Abhandl. Nat. Geol. Nürnberg, vol. 22, 1922, p. 10.
Involutina (part) of authors.

Test conical, consisting of a coiled, undivided tube at the periphery, the ventral, interior, conical area filled with secondary shell material in the form of pillars, the outer ends ending ventrally in pustules; wall arenaceous, with a calcareous cement or entirely calcareous, coarsely perforate; aperture the open end of the tubular chamber.—Jurassic, Lower Cretaceous.

Genus PAALZOWELLA Cushman, 1933

Key, plate 29, figure 10
Genotype, *Discorbis scalariformis* Paalzow
Paalzowella CUSHMAN, Spec. Publ. 4, Cushman Lab. Foram. Res., 1933, p. 234.
Discorbis (part) of authors.

Test trochoid, conical, consisting of a proloculum and undivided, tubular, second chamber lobed at the base but not completely chambered, ventrally with a pillar or plug filling the central region; wall calcareous, perforate; aperture ventral, narrow.—Jurassic.

The dorsal view shows no divisions but the ventral view has distinct lobes, simulating chambers. Somewhat similar forms occur in the Recent Indo-Pacific faunas, and need careful investigation.

Subfamily 3. Discorbinae

Test chambered, trochoid, umbilical region generally open, dorsal side with all chambers visible, only those of the last-formed whorl visible from the ventral side; aperture ventral, not extending out to the periphery.

Genus PATELLINA Williamson, 1858

Plate 24, figure 3; Key, plate 29, figure 13

Genotype, *Patellina corrugata* Williamson

Patellina WILLIAMSON, Rec. Foram. Gt. Britain, 1858, p. 46.

Test conical or plano-convex, early whorls undivided, microspheric specimens sometimes entirely undivided, later ones usually divided into long chambers, often with internal sinuous septa partially dividing the chamber; wall calcareous, perforate, thin; aperture elongate, at the base of the ventral side of the chamber. — Permian to Recent.

See Cushman, Some Notes on the Genus *Patellina* (Contr. Cushman Lab. Foram. Res., vol. 6, 1930, pp. 11–17, pl. 3).

Genus PATELLINOIDES Heron-Allen and Earland, 1932

Key, plate 29, figures 11, 12

Genotype, *Patellinoides conica* Heron-Allen and Earland

Patellinoides HERON-ALLEN and EARLAND, *Discovery* Reports, vol. IV, 1932, p. 407.

Test similar to *Patellina*, the early stages an undivided trochoid coil, later with two chambers to a whorl, but with the adult chambers not subdivided as in *Patellina*. — Eocene (?), Recent.

Genus PATELLINELLA Cushman, 1928

Key, plate 29, figure 14

Genotype, *Textularia inconspicua* H. B. Brady

Patellinella CUSHMAN, Contr. Cushman Lab. Foram. Res., vol. 4, 1928, p. 5.
Textularia (part) of authors. *Discorbis* (part) CUSHMAN (not LAMARCK).

Test compressed, conical, trochoid, two adult chambers making up each whorl; wall calcareous, perforate; aperture ventral, umbilical. — Tertiary, Recent.

Genus ANNULOPATELLINA Parr and Collins, 1930

Key, plate 29, figure 15

Genotype, *Orbitolina annularis* Parker and Jones

Annulopatellina PARR and COLLINS, Proc. Roy. Soc. Victoria, vol. 43, 1930, p. 92.
Orbitolina PARKER and JONES (not D'ORBIGNY).

Test a low cone, consisting of a proloculum followed by annular chambers in the megalospheric form, in the microspheric form with several semi-annular chambers as in *Patellina*, divided into chamberlets; wall calcareous, perforate; aperture indefinite. — Oligocene, Recent.

Genus CONORBINA Brotzen, 1936

Key, plate 51, figure 14
Genotype, *Conorbina marginata* Brotzen
Conorbina Brotzen, Sver. Geol. Under., ser. C, No. 396, 1936, p. 141.

Test free or attached, trochoid, conical, with an elevated spire, ventral side flattened or concave, umbilicate; wall calcareous, perforate; aperture a narrow opening at the middle of the base of the apertural face.—Cretaceous to Recent.

Genus DISCORBIS Lamarck, 1804

Plate 24, figure 4; Key, plate 29, figures 16, 17
Genotype, *Discorbis vesicularis* Lamarck
Discorbis Lamarck, Ann. Mus., vol. 5, 1804, p. 183.
Rosalina d'Orbigny, 1826. *Turbinulina* (part) d'Orbigny, 1826. *Allotheca* and *Phanerostomum* Ehrenberg, 1843. *Platyoecus* Ehrenberg, 1854. *Aristerospira* (part) Ehrenberg, 1858. *Discorbina* Parker and Jones, 1862.

Test typically plano-convex, the ventral side flattened, microspheric form sometimes showing a long, *Spirillina*-like, second chamber of several coils before division; chambers often produced to partially cover the umbilical area; wall calcareous, perforate; aperture at the base of the umbilical margin on the ventral side of the chamber.—Jurassic to Recent.

PLATE 24

Relationships of the Genera of the Rotaliidae

FIG.

1. *Turrispirillina conoidea* (Paalzow). (After Paalzow.)
2. *Conicospirillina trochoides* (Berthelin). (After Berthelin.)
3. *Patellina corrugata* Williamson. (After Williamson.)
4. *Discorbis rosacea* (d'Orbigny). (After H. B. Brady.)
5. *Valvulineria californica* Cushman.
6. *Lamarckina glabrata* Cushman.
7. *Gyroidina soldanii* d'Orbigny. (After d'Orbigny.)
8. *Rotaliatina mexicana* Cushman. *b*, apertural view; *c*, side view.
9. *Eponides repandus* (Fichtel and Moll). (After H. B. Brady.)
10. *Epistomina elegans* (d'Orbigny). (After H. R. Brady.) *b*, dorsal view; *c*, peripheral view.
11. *Siphonina reticulata* (Czjzek). (After H. B. Brady.)
12. *Siphoninoides echinata* (H. B. Brady). (After H. B. Brady.)
13. *Siphoninella soluta* (H. B. Brady). (After H. B. Brady.)
14. *Cancris auriculus* (Fichtel and Moll). (After H. B. Brady.)
15. *Baggina californica* Cushman.
16. *Rotalia trochidiformis* Lamarck. (After Terquem.)

(In all figures: *a*, ventral view; *b*, peripheral view; *c*, dorsal view, except as otherwise noted.)

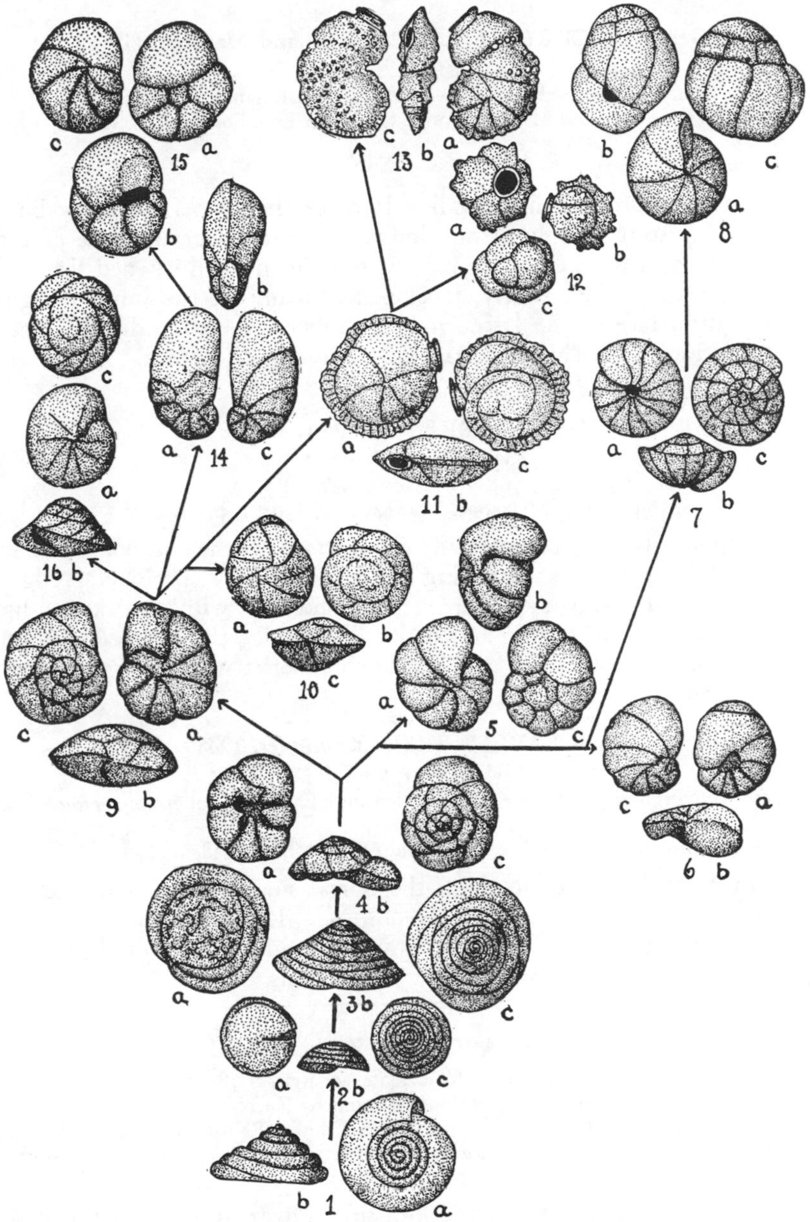
c 15 a
c 13 b a
b c
a 8
b
a b
12
c
b
c
a 14 c
a
a c
11 b
a c
7 b
16 b
b
a b
10 c
a 5 c
c a
c a
9 b
c a
6 b
a c
4 b
a c
3 b
a c
2 b
b 1 a

PLATE 24

Genus DISCORBINELLA Cushman and Martin, 1935

Key, plate 51, figure 15

Genotype, *Discorbinella montereyensis* Cushman and Martin

Discorbinella Cushman and Martin, Contr. Cushman Lab. Foram. Res., vol. 11, 1935, p. 89.

Planulinoides Parr, 1941.

Test trochoid, *Discorbis*-like; wall calcareous, finely perforate, hyaline; aperture consisting mainly of an elongate opening near the base of the apertural face in the axis of coiling, with a distinctly thickened lip, and a supplementary, often poorly developed aperture, at the inner margin of the ventral face of the last-formed chamber, beneath a distinct, flap-like projection of the chamber margin. — Miocene to Recent.

Genus TORRESINA Parr, 1947

Key, plate 51, figure 16

Genotype, *Torresina haddoni* Parr

Torresina Parr, Journ. Roy. Micr. Soc., vol. 64, 1944 (1947), p. 130.

Test trochoid, strongly compressed; chambers with the interior partially subdivided by partitions extending inward from the peripheral edge; apertures two in number, one peripheral, a short slit with a thickened lip nearly in the plane of coiling and inclined ventrally, the other ventral on the inner margin, sometimes as a lunate chamberlet with a narrow opening. — Recent.

Genus EARLMYERSIA Rhumbler, 1938

Key, plate 51, figure 17

Genotype, *Pulvinulina punctulata* Heron-Allen and Earland (not *Rotalia punctulata* d'Orbigny)

Earlmyersia Rhumbler, Kieler Meeresforschungen, vol. 2, 1938, p. 209.

Test similar to *Discorbis* but with the dorsal side carrying a projecting decoration and the walls very finely porous. — Recent.

It would seem that the characters given for this genus would perhaps represent specific rather than generic distinctions.

Genus LAMARCKINA Berthelin, 1881

Plate 24, figure 6; Key, plate 29, figures 18–20

Genotype, *Pulvinulina erinacea* Karrer

Lamarckina Berthelin, Comptes Rendus Assoc. Franc. (Reims, 1880) 1881, p. 555.

Megalostomina Rzehak, 1895. *Pulvinulina* (part), *Rotalina* (part), *Discorbina* (part), and *Valvulina* (part) of authors.

Test trochoid, evidently attached, dorsal side convex, usually ornamented, ventral side usually flattened or concave, very smooth and highly polished; chambers dorsally distinct, ventrally less so, but often with an umbilical projection; wall calcareous, finely perforate; aperture at the umbilical end of the chamber, often enlarged by resorption.—Jurassic (?), Upper Cretaceous to Recent.

See Cushman, The Genus *Lamarckina* and Its American Species (Contr. Cushman Lab. Foram. Res., vol. 2, pt. 1, 1926, pp. 7–14, pl. 4).

Genus HERONALLENIA Chapman and Parr, 1931

Key, plate 29, figure 21

Genotype, *Discorbina wilsoni* Heron-Allen and Earland

Heronallenia Chapman and Parr, Proc. Roy. Soc. Victoria, vol. 43, 1931, p. 236.
Discorbina (part) and *Discorbis* (part) of authors.

"Test compressed, ovate, plano-convex; superior face gently rounded, inferior, flat to concave. Periphery rounded. Sutures and margin on superior face usually limbate. Chambers comparatively few, concave to slightly inflated on upper surface. Aperture a strongly arched slit situated in a depression on the inner face of the last chamber. Shell surface very finely perforate and polished; exogenous beads sometimes developed on the superior face, or single ones near the sutures, when they appear as vesicles. Inferior surface often radially striate."—Eocene to Recent.

Genus VALVULINERIA Cushman, 1926

Plate 24, figure 5; Key, plate 30, figures 1, 2

Genotype, *Valvulineria californica* Cushman

Valvulineria Cushman, Contr. Cushman Lab. Foram. Res., vol. 2, pt. 3, 1926, p. 59.
Rosalina (part) of authors. *Cibicorbis* Hadley, 1934.

Test usually trochoid, close coiled, ventrally umbilicate; chambers numerous; wall calcareous, finely perforate; aperture ventral, large, extending from the umbilicus toward the periphery, often with a thin plate filling the umbilical area, sometimes extending into the face of the chamber, becoming tripartite.—Cretaceous to Recent.

Genus GAVELINELLA Brotzen, 1942

Key, plate 51, figure 18

Genotype, *Discorbina pertusa* Marsson

Gavelinella Brotzen, Sver. Geol. Under., ser. C, No. 451, 1942, p. 7.
Discorbina (part) and *Anomalina* (part) of authors.

Test trochoid in the early stages, nearly planispiral in the adult, biconvex, periphery rounded, umbilicus open; wall calcareous, perforate; aperture a narrow opening along the ventral face of the last-formed chamber extending from the periphery to the umbilicus, with a secondary opening into the umbilical area.—Cretaceous to Recent.

Genus GYROIDINOIDES Brotzen, 1942

Key, plate 51, figure 20
Genotype, *Rotalina nitida* Reuss

Gyroidinoides Brotzen, Sver. Geol. Under., ser. C, No. 451, 1942, p. 19.
Rotalina (part) and *Gyroidina* (part) of authors.

Test differing from *Gyroidina* in having a distinct open umbilicus and the aperture an elongate slit-like opening along the margin of the last-formed chamber, and from *Gavelinella* by the less open umbilicus and more convex ventral side. — Cretaceous to Recent.

Genus PSEUDOVALVULINERIA Brotzen, 1942

Key, plate 51, figure 24
Genotype, *Rosalina lorneiana* d'Orbigny

Pseudovalvulineria Brotzen, Sver. Geol. Under., ser. C, No. 451, 1942, p. 20.
Rosalina (part) and *Anomalina* (part) of authors.

Test differing from *Valvulineria* in not having the ventral lobe over the umbilicus and in having the aperture in the middle of the ventral margin and connecting by a narrow opening with the umbilicus. — Cretaceous.

Genus GYROIDINA d'Orbigny, 1826

Plate 24, figure 7; Key, plate 30, figures 3, 4
Genotype, *Gyroidina orbicularis* d'Orbigny

Gyroidina d'Orbigny, Ann. Sci. Nat., vol. 7, 1826, p. 278.
Rotalina (part) of authors.

Test trochoid, ventral side usually convex, umbilicus small and deep, spiral suture with a depressed channel; wall calcareous, finely perforate; aperture, a low arched opening on the ventral side toward the umbilical area. — Lower Cretaceous to Recent.

Genus STENSIÖINA Brotzen, 1936

Key, plate 52, figure 1
Genotype, *Rotalia exsculpta* Reuss

Stensiöina Brotzen, Sver. Geol. Under., ser. C, No. 396, 1936, p. 164.
Rotalia (part); *Truncatulina* (part); *Cibicides* (part); *Gyroidina* (part) of authors.

Test trochoid, plano-convex, dorsal side flattened, ventral side convex, umbilicate, all chambers visible from the dorsal side, only those of the last-formed whorl from the ventral side; chambers distinct, simple; sutures dorsally raised and ornate, ventrally simple and depressed; wall calcareous, perforate, often coarsely so on the ventral side; aperture an elongate slit at the base of the last-formed chamber on the ventral side. — Cretaceous.

See Cushman and Dorsey, The Genus *Stensiöina* and Its Species (Contr. Cushman Lab. Foram. Res., vol. 16, 1940, pp. 1–6, pl. 1).

The genus *Stensiöina* Brotzen includes peculiar forms from the Upper Cretaceous that seem to belong to *Anomalina* or *Cibicides* but the apertural characters especially in the type species are more like *Gyroidina*.

Genus ROTALIATINA Cushman, 1925

Plate 24, figure 8; Key, plate 30, figure 5
Genotype, *Rotaliatina mexicana* Cushman

Rotaliatina Cushman, Contr. Cushman Lab. Foram. Res., vol. 1, pt. 1, 1925, p. 4.
Rotalina (part) of authors (not d'Orbigny).

Test trochoid, in a high spire, with several chambers in each whorl, ventrally umbilicate; wall calcareous, finely perforate; aperture, an arched slit between the base of the apertural face and the previous coil. — Eocene, Oligocene.

Subfamily 4. Rotaliinae

Test trochoid, umbilical region typically closed, sometimes with a definite conical plug of clear shell material; wall of the test often double, and a tubular canal system developed; aperture ventral, along the margin of the chamber between the periphery and the umbilical area.

Genus EPONIDES Montfort, 1808

Plate 24, figure 9; Key, plate 30, figures 6, 7
Genotype, *Nautilus repandus* Fichtel and Moll

Eponides Montfort, Conch. Syst., vol. 1, 1808, p. 127.
Nautilus (part), *Rotalia* (part), and *Rotalina* (part) of authors. *Pulvinulina* Parker and Jones, 1862. *Placentula* (part) Berthelin (not Lamarck). *Cyclospira* Eimer and Fickert, 1899.

Test trochoid, usually biconvex, umbilical area closed but not typically with a plug; wall calcareous, perforate; aperture, a low opening between the periphery and umbilical area, usually well away from the peripheral margin. — Jurassic to Recent.

Genus PARRELLA Finlay, 1939

Key, plate 51, figure 19; plate 52, figure 11
Genotype, *Anomalina bengalensis* Schwager

Parrella Finlay, Trans. Roy. Soc. New Zealand, vol. 68, 1939, p. 523.
Osangularia Brotzen, 1940.
Planorbulina (part) Parker and Jones, 1865 (not d'Orbigny).
Anomalina (part) Schwager, 1866 (not d'Orbigny).
Pulvinulina (part) and *Pulvinulinella* (part) of authors.

Test trochoid, close coiled; all chambers visible dorsally, only those of the last-formed whorl from the ventral side, umbilical area with a distinct solid mass; sutures on both dorsal and ventral sides strongly oblique; wall calcareous, perforate; aperture on the ventral side, a narrow opening extending from the margin into the ventral face at a distinct angle from the axis of coiling with a short slit-like opening at the margin of the chamber extending toward the umbilicus. — Cretaceous to Recent.

Genus CRESPINELLA Parr, 1942

Key, plate 51, figures 21–23

Genotype, *Operculina* (?) *umbonifera* Howchin and Parr

Crespinella PARR, Mining and Geol. Journ., vol. 2, 1942, p. 361.

Operculina (part) of authors.

Test biconvex, slightly asymmetrical, trochoid in the early stages, adult nearly planispiral; chambers almost entirely involute; wall calcareous, very thick, laminated, closely and distinctly tabulated; aperture a slightly curved slit at the base of the last-formed chamber in the median line with a projecting, thick, hood-like upper margin. — Tertiary.

Genus EPONIDELLA Cushman and Hedberg, 1935

Key, plate 52, figure 2

Genotype, *Eponidella libertadensis* Cushman and Hedberg

Eponidella CUSHMAN and HEDBERG, Contr. Cushman Lab. Foram. Res., vol. 11, 1935, p. 13.

Test trochoid, chambers all visible from the dorsal side, only those of the last-formed whorl visible from the ventral side; chambers of the ventral side of the test in two series, a small supplementary chamber being added over the umbilical portion of each main chamber; wall calcareous, coarsely perforate, that of the main portion very thick, in the supplementary ventral portion much thinner, entire test with a very thin but distinct chitinous inner wall, light yellowish-brown in color; aperture in the earlier stages running in from the ventral side, later becoming semicircular, oval, or nearly circular, in the middle of the apertural face, and sometimes divided by a very thin horizontal partition. — Miocene.

Genus ROTALIDIUM Asano, 1936

Key, plate 52, figure 3

Genotype, *Rotalidium pacificum* Asano

Rotalidium ASANO, Proc. Imp. Acad. Tokyo, vol. 12, 1936, p. 350.

"Test trochoid, all the whorls visible on the dorsal side, only the last whorl visible on the ventral side; chambers numerous, closely appressed, those of the last whorl superposed by supplementary chambers in a series on the ventral side of the test; wall very finely perforate; surface smooth or ornamented with exogenous deposits; aperture a slit at the base of the last chamber." — Recent.

Genus RECTOEPONIDES Cushman and Bermudez, 1936

Key, plate 52, figure 4
Genotype, *Rectoeponides cubensis* Cushman and Bermudez

Rectoeponides Cushman and Bermudez, Contr. Cushman Lab. Foram. Res., vol. 12, 1936, p. 31.

Test in the early portion trochoid, *Eponides*-like, later portion becoming uniserial and rectilinear; wall calcareous, finely perforate; aperture, an elongate, narrow opening, slightly on the ventral side of the terminal face of the last-formed chamber in a distinct depression, without a distinct lip. — Eocene.

Genus COLEITES Plummer, 1934

Key, plate 52, figures 5–7
Genotype, *Pulvinulina reticulosa* Plummer

Coleites Plummer, Amer. Midland Nat., vol. 15, 1934, p. 605.
Pulvinulina Plummer 1927 (part).

Test trochoid in the young, in the adult uncoiling and broad; chambers in a single series; wall calcareous, coarsely perforate; aperture in the young ventral, in the adult subterminal, elliptical, with a slight tooth on the ventral side. — Midway and Wilcox Eocene.

See Plummer, *Epistominoides* and *Coleites*, New Genera of Foraminifera (Amer. Midland Nat., vol. 15, 1934, pp. 601–607, pl. XXIV).

Genus PLANOPULVINULINA Schubert, 1920

Key, plate 30, figure 8
Genotype, *Pulvinulina dispansa* H. B. Brady

Planopulvinulina Schubert, Pal. Zeitschr., vol. 3, 1920, p. 153.
Pulvinulina (part) H. B. Brady, 1884.

Test trochoid in early stages, later very irregular in form but chambers not annular, ventral face with large pores apparently serving as apertures. — Late Tertiary, Recent.

This form should not be confused with *Cibicidella* and *Dyocibicides*, genera which are closely allied to *Cibicides*, but not to *Eponides*, with which *Planopulvinulina* is closely allied.

Genus ROTALIA Lamarck, 1804

Plate 24, figure 16; Key, plate 30, figures 9, 10
Genotype, *Rotalia trochidiformis* Lamarck

Rotalia Lamarck, Ann. Mus., vol. 5, 1804, p. 184.
Nautilus (part) of authors. *Streblus* Fischer, 1817. *Turbinulina* d'Orbigny, 1826. *Rosalina* (part) and *Truncatulina* (part) of authors.

Test trochoid, usually biconvex, umbilical area closed, usually with a conical plug of clear shell material; sutures usually limbate dorsally, ventrally usually deeply depressed, often ornamented along the sides; wall calcareous, perforate, often double, with a canal system; aperture on the ventral side between the periphery and umbilical area. — Cretaceous to Recent.

For structure and type figures see Davies 1932. (Under *Dictyoconoides.*)

In recent publications the genus *Rotalia* has been divided and other names used for certain of the species previously included under this name. There seems to be some question as to which name should be used.

Genus ROTORBINELLA Bandy, 1944

Key, plate 52, figure 8

Genotype, *Rotorbinella calliculus* Bandy

Rotorbinella Bandy, Journ. Pal., vol. 18, 1944, p. 372.
Discorbina (part) and *Rotalia* (part) of authors.

Test free, trochoid, close coiled, ventrally with a simple umbilical plug; ventral sutures sometimes with re-entrants; aperture a slit at the inner margin of the ventral face of the last-formed chamber, not reaching to the periphery. — Tertiary and Recent.

Genus ASANOINA Finlay, 1939

Key, plate 52, figure 12

Genotype, *Rotaliatina globosa* Yabe and Asano

Asanoina Finlay, Trans. Roy. Soc. New Zealand, vol. 68, 1939, p. 541.
Rotaliatina (part) of authors.

Test similar to *Rotalia* in structure but the axis much prolonged, non-umbilicate; aperture a narrow opening near the middle of the basal border of the last-formed chamber. — Pliocene.

Genus LOCKHARTIA Davies, 1932

Key, plate 30, figures 11, 12

Genotype, *Dictyoconoides haimei* Davies

Lockhartia Davies, Trans. Roy. Soc. Edinburgh, vol. 57, 1932, p. 406.
Dictyoconoides (part) Davies, 1927.

Test trochoid, plano-convex or lenticular, similar to *Dictyoconoides* but not developing intercalary whorls in the spire, ventral side with continuous umbilical pillars and transverse successive layers of shell substance; wall calcareous, perforate; aperture ventral. — Upper Cretaceous, Eocene.

Genus SAKESARIA Davies, 1937

Key, plate 52, figures 9, 10

Genotype, *Sakesaria cotteri* Davies

Sakesaria Davies, in Davies and Pinfold, Mem. Geol. Surv. India, n. ser., vol. 24, Mem. No. 1, 1937, p. 49.

"Structure similar to that of *Lockhartia*, from which this genus differs by reason of extreme elongation of the axis, representing an excessive degree of specialization to one pole. The umbilical space is greatly diminished, and the number of umbilical pillars much reduced." — Eocene.

Genus DICTYOCONOIDES Nuttall, 1925

Key, plate 30, figures 13–15
Genotype, *Conulites cooki* Carter

Dictyoconoides Nuttall, Ann. Mag. Nat. Hist., ser. 9, vol. 16, 1925, p. 384.
Conulites of authors (not Fischer, 1832). *Patellina* (part) of authors.

Test in early stages like *Rotalia*, later greatly expanding laterally, and forming a flattened test, spire with intercalary whorls, the ventral umbilical region filled with a secondary mass of calcareous material in the form of pillars and intermediate cavities, outer ends of the pillars forming papillae; wall calcareous, perforate; apertures apparently formed by the openings between the ventral pillars. — Eocene.

For excellent figures and discussion, see Davies, The Genera *Dictyoconoides* Nuttall, *Lockhartia* nov., and *Rotalia* Lamarck: Their Type Species, Generic Differences, and Fundamental Distinction from the Dictyoconus Group of Forms (Trans. Roy. Soc. Edinburgh, vol. 57, 1932, pp. 397–428, 4 pls.).

Subfamily 5. Siphonininae

Test trochoid, at least in the early stages, umbilical area filled, supplementary apertures near the periphery and just below it on the ventral side, sometimes with a neck and lip.

Genus EPISTOMINA Terquem, 1883

Plate 24, figure 10; Key, plate 30, figure 16
Genotype, *Epistomina regularis* Terquem

Epistomina Terquem, Bull. Soc. Géol. France, ser. 3, vol. 11, 1883, p. 37.
Rotalia (*Turbinulina*) (part) d'Orbigny, 1826. *Pulvinulina* (part) of authors. *Placentula* Berthelin, 1882 (not Lamarck).

Test trochoid, biconvex, umbilical area filled, sutures usually limbate; wall calcareous, perforate; apertures of two sorts, one at the inner margin of the ventral side of the chamber or in the face itself, the other elongate, just below the periphery in the axis of coiling, usually filled with clear shell material as growth progresses. — Jurassic to Recent.

Genus POROEPONIDES Cushman, 1944

Key, plate 52, figure 14
Genotype, *Rosalina lateralis* Terquem

Poroeponides Cushman, Special Publ. 12, Cushman Lab. Foram. Res., 1944, p. 34.
Rosalina (part), *Pulvinulina* (part), and *Eponides* (part) of authors.

Test trochoid, biconvex, umbilical area open; wall calcareous, perforate; aperture in the early stages a low opening on the ventral side of the last-formed chamber near the umbilicus, with a slight lip, and in the adult chambers with numerous rounded openings scattered over the ventral face. — Tertiary and Recent.

Genus DISCORINOPSIS Cole, 1941

Key, plate 52, figures 18–20
Genotype, *Discorinopsis gunteri* Cole

Discorinopsis Cole, Florida Dept. Conservation, Geol. Bull. 19, 1941, p. 36.
Discorbis (part) of authors.

Test trochoid, typically plano-convex, ventral side flattened, all whorls visible on dorsal side, only last-formed whorl on ventral side, umbilical area closed by a spongy mass of shell material; wall calcareous, perforate; aperture a series of openings through the mass of shell material filling the umbilicus. — Tertiary and Recent.

Genus EPISTOMARIA Galloway, 1933

Key, plate 30, figure 17
Genotype, *Discorbina rimosa* Parker and Jones

Epistomaria Galloway, Man. Foram., 1933, p. 286.
Discorbina (part) Parker and Jones, 1865. *Epistomella* Cushman, 1928 (not Zittel).

Test trochoid, dorsal side with regular chambers, ventral side with supplementary chambers or alar projections toward the umbilicus, which is covered; wall calcareous, finely perforate; apertures ventrally at periphery of the secondary chambers, with supplementary apertures dorsally at inner edge of chamber along the suture between it and the preceding chamber, narrow, elongate. — Eocene to Recent.

Genus MISSISSIPPINA Howe, 1930

Key, plate 31, figure 1
Genotype, *Mississippina monsouri* Howe

Mississippina Howe, Journ. Pal., vol. 4, 1930, p. 329.
Pulvinulina (part) of authors.

Test trochoid in the young, later almost planispiral, biconvex; wall calcareous, finely perforate, thick, with thinner areas on each side near the periphery; aperture in the young ventral, extending to the periphery in the adult, with supplementary apertures or clear spaces near the periphery on both sides. — Lower Oligocene to Recent.

Genus SIPHONINA Reuss, 1850

Plate 24, figure 11; Key, plate 31, figures 2, 3
Genotype, *Siphonina fimbriata* Reuss

Siphonina Reuss, Denkschr. k. Akad. Wiss. Wien, vol. 1, 1850, p. 372.
Rotalia (part) and *Rotalina* (part) of authors. *Planorbulina* (part) Parker and Jones, 1865 (not d'Orbigny). *Truncatulina* (part) of authors.

Test trochoid, biconvex, umbilical region typically closed; wall calcareous, coarsely perforate; aperture elliptical, just ventral to the periphery, its long axis parallel to it, in fully developed species with a short neck and phialine lip. — Cretaceous to Recent.

See Cushman, Foraminifera of the Genus *Siphonina* and Related Genera (Proc. U. S. Nat. Mus., vol. 72, Art. 20, 1927, pp. 1–15, pls. 1–4).

Genus SIPHONIDES Feray, 1941

Key, plate 52, figure 13

Genotype, *Siphonides biserialis* Feray

Siphonides FERAY, Journ. Pal., vol. 15, 1941, p. 174.

Similar in the young to *Siphonina* but in the adult biserial, the aperture toward the inner margin of the last-formed chamber. — Eocene.

Genus SIPHONINOIDES Cushman, 1927

Plate 24, figure 12; Key, plate 31, figure 6

Genotype, *Planorbulina echinata* H. B. Brady

Siphoninoides CUSHMAN, Contr. Cushman Lab. Foram. Res., vol. 3, 1927, p. 77.

Planorbulina H. B. BRADY, 1879 (not D'ORBIGNY). *Truncatulina* of authors (not D'ORBIGNY). *Siphonina* (part) of authors (not REUSS).

Test in the adult generally globular; chambers irregularly trochoid, involute in the adult; wall calcareous, perforate; aperture circular, with a short neck and flaring lip. — Tertiary, Recent.

Genus SIPHONINELLA Cushman, 1927

Plate 24, figure 13; Key, plate 31, figures 4, 5

Genotype, *Truncatulina soluta* H. B. Brady

Siphoninella CUSHMAN, Contr. Cushman Lab. Foram. Res., vol. 3, 1927, p. 77.

Truncatulina H. B. BRADY, 1881 (not D'ORBIGNY).

Test in early stages similar to *Siphonina*, later becoming uncoiled; wall calcareous, perforate; aperture in adult terminal, elliptical, with a neck and lip. — Eocene to Recent.

Subfamily 6. Bagginінae

Test generally biconvex, the umbilical area closed, the area adjacent to it on each chamber with a thinner, rounded, clear area, usually without perforations; aperture at the base of the ventral margin of the chamber.

Genus CANCRIS Montfort, 1808

Plate 24, figure 14; Key, plate 31, figure 8

Genotype, *Nautilus auriculus* var. β Fichtel and Moll

Cancris MONTFORT, Conch. Syst., vol. 1, 1808, p. 267.

Pulvinulinella EIMER and FICKERT, 1899.

Nautilus (part) FICHTEL and MOLL, 1798 (not LINNÉ). *Rotalina* (part) WILLIAMSON, 1858 (not D'ORBIGNY). *Pulvinulina* (part) of authors.

Test trochoid, nearly equally biconvex, compressed; chambers few, rapidly enlarging; wall calcareous, perforate, umbilical area with a clear plate of rather large dimensions for the size of the test; aperture narrow, on the inner border of the ventral side of the last-formed chamber. — Tertiary, Recent.

See Cushman and Todd, The Genus *Cancris* and Its Species (Contr. Cushman Lab. Foram. Res., vol. 18, 1942, pp. 72–94, pls. 17–24).

Genus BAGGINA Cushman, 1926

Plate 24, figure 15; Key, plate 31, figure 7

Genotype, *Baggina californica* Cushman

Baggina Cushman, Contr. Cushman Lab. Foram. Res., vol. 2, pt. 3, 1926, p. 63.
Pulvinulina (part) of authors.

Test subglobular, trochoid; chambers large and inflated, few, dorsally more or less involute, ventrally completely so; wall calcareous, perforate, with a small, clear, lunate space above the aperture; aperture broadly oval, on the ventral side, without a lip. — Eocene to Recent.

See Cushman and Todd, The Genera *Baggina* and *Neocribrella* and Their Species (Contr. Cushman Lab. Foram. Res., vol. 20, 1944, pp. 97–107, pls. 15–17).

Genus BAGGATELLA Howe, 1939

Key, plate 52, figure 15

Genotype, *Baggatella inconspicua* Howe

Baggatella Howe, Geol. Bull. 14, Louisiana Geol. Survey, 1939, p. 79.

Test trochoid, with a distinct spire, several chambers in each whorl, ventrally involute; wall calcareous, perforate; aperture a high arched slit extending from the base of the last-formed chamber well into the apertural face. — Eocene.

Genus NEOCRIBRELLA Cushman, 1928

Key, plate 31, figure 9

Genotype, *Discorbina globigerinoides* Parker and Jones

Neocribrella Cushman, Contr. Cushman Lab. Foram. Res., vol. 4, 1928, p. 6.
Discorbina (part) Parker and Jones, 1865.

Test trochoid, somewhat involute in later stages; chambers comparatively few, inflated; wall calcareous, perforate; aperture in adult, several small rounded pores, in a depression of the ventral face of the chamber. — Eocene, Europe.

Family 34. PEGIDIIDAE

"Test free, calcareous, perforate, thick-walled, lenticular or sub-spherical in form. Chambers turgid but few in number, rarely more than 3 or 4 in the adult shell, arranged so that each successive chamber is opposed to or partly enveloping its predecessor. Initial chambers either arranged spirally or in opposition, and resorbed in course of growth. Aperture tubular or a series of tubes, either free or perforating a solid mass of shell substance filling up the depression between the final chambers. No canal system."

This family has four genera, all of which are represented by species which occur in comparatively shallow water, frequently where there are strong currents. The wall is usually very thick, and there is a tendency to develop more or less globular tests. The oldest known genus and the simplest of the four is *Pegidia*, known from the Miocene and living in the Indo-Pacific and the West Indies at the present time. The early chambers in *Pegidia* in the microspheric form are coiled, and are hardly distinguishable from some of the thick-walled tropical forms of *Eponides*. The later chambers become somewhat enlarged and more or less involute, although in most of our specimens the early coiled chambers can be seen. Later chambers form a half coil and become greatly thickened, in some species heavily ornamented on the dorsal side. From *Pegidia*, *Sphaeridia* is directly derived by the chambers becoming more involute and the spiral stage becoming reduced. Nearest to these two genera is *Rugidia*, where the aperture has some of the same characteristics, being tubular. The genus *Physalidia* has all the chambers subglobular and opposed, the spiral stage having been completely eliminated. There is nothing to substantiate the theory that this family came from the Globigerinidae which are pelagic for the most part, and have an entirely different wall structure. The genus *Sphaeroidinella* also has no relationships to this group, it being a pelagic form which in the young is like *Globigerina*, and has very delicate fine spines similar to other genera of the Globigerinidae. It may be noted that the early stages, particularly of *Pegidia dubia*, are very finely perforate, and have all the characters of the thick-walled species of *Eponides* found in similar situations. These forms have a wide distribution in the Indo-Pacific, and some of them also in the Atlantic, but all in similar habitats.

Genus PEGIDIA Heron-Allen and Earland, 1928

Key, plate 31, figure 10
Genotype, *Rotalia dubia* d'Orbigny

Pegidia Heron-Allen and Earland, Journ. Roy. Micr. Soc., vol. 48, 1928, p. 290.
Rotalia (part) d'Orbigny.

Test somewhat oval, compressed, typically with a thickened rim; early chambers particularly in the microspheric form trochoid, later two to a whorl, dorsally ornamented or rough, ventrally smooth; wall calcareous, thick, very finely perforate; aperture on the ventral side, consisting of numerous pores connecting with the interior by branching tubes.—Miocene to Recent.

Genus SPHAERIDIA Heron-Allen and Earland, 1928

Key, plate 31, figure 11

Genotype, *Sphaeridia papillata* Heron-Allen and Earland

Sphaeridia Heron-Allen and Earland, Journ. Roy. Micr. Soc., vol. 48, 1928, p. 294.

Test generally spherical; chambers few, rapidly increasing in size, nearly involute; wall calcareous, finely perforate; aperture a series of tubes.—Recent.

Genus PHYSALIDIA Heron-Allen and Earland, 1928

Key, plate 31, figure 12

Genotype, *Physalidia simplex* Heron-Allen and Earland

Physalidia Heron-Allen and Earland, Journ. Roy. Micr. Soc., vol. 48, 1928, p. 288.

Test consisting of a few subglobular chambers, arranged in opposition; wall calcareous, coarsely perforate, smooth; aperture one or more rudimentary tubes where the chambers meet.—Recent.

Genus RUGIDIA Heron-Allen and Earland, 1928

Key, plate 31, figure 13

Genotype, *Sphaeroidina corticata* Heron-Allen and Earland

Rugidia Heron-Allen and Earland, Journ. Roy. Micr. Soc., vol. 48, 1928, p. 289.

Sphaeroidina (part) Heron-Allen and Earland, 1915 (not d'Orbigny).

Test composed of a few subglobular chambers arranged in opposed pairs; wall calcareous, perforate, exterior roughened; aperture consisting of a few tubular openings between the pairs of chambers.—Recent.

Family 35. AMPHISTEGINIDAE

Test trochoid, all chambers visible from the dorsal side except in involute forms of *Amphistegina*, those of the last-formed whorl only, visible on the ventral side, the ventral side with angular supplementary chambers coming in between the regular series, roughly rhomboid in shape as seen from the surface; aperture typically ventral, a slightly arched opening, the area adjacent to the aperture papillate.

This family contains but four genera, of which *Asterigerina* is the simpler. It is very close to *Eponides*, and differs from that genus by the addition of the supplementary chambers on the ventral side. The aperture is typically that of the Rotaliidae except for the papillate area about the opening. In *Amphistegina* the test becomes much more complex, and often more or less involute on the dorsal side, some of the species becoming rather large.

This family is largely limited to the Tertiary and the Recent oceans. *Asterigerina* is often abundant on coral reefs and in warm shallow waters of the West Indian region, but less abundant elsewhere. *Amphistegina* is very abundant in similar conditions and often is present in enormous numbers in shoal water, particularly of the Indo-Pacific. It is frequent in such conditions through the later Tertiary. It is very probable that the genus is limited to about 30 fathoms in its living condition as, like other large foraminifera, it seems to have commensal algae, the limits of which on account of the penetration of sunlight in the ocean are limited to this same depth.

Genus ASTERIGERINA d'Orbigny, 1839

Plate 25, figure 1; Key, plate 31, figures 14, 15

Genotype, *Asterigerina carinata* d'Orbigny

Asterigerina d'Orbigny, in De la Sagra, Hist. Phys. Pol. Nat. Cuba, 1839, "Foraminifères," p. 117.

Test trochoid, biconvex, ventral side usually more strongly so than the dorsal; chambers on the dorsal side regularly coiled, ventrally with angular supplementary chambers between the regular series, large, regularly rhomboid, dorsal sutures simply curved; wall calcareous, finely perforate; aperture ventral, at the base of the chamber margin. — Jurassic to Recent.

See ten Dam, Structure of *Asterigerina* and a New Species (Journ. Pal., vol. 21, 1947, pp. 584–586, text figs. 1–6).

Genus HELICOSTEGINA Barker and Grimsdale, 1936

Key, plate 52, figures 16, 17

Genotype, *Helicostegina dimorpha* Barker and Grimsdale

Helicostegina Barker and Grimsdale, Journ. Pal., vol. 10, 1936, p. 233.

Helicolepidinoides Tan, 1936.

Amphistegina (part) of authors.

Test multichambered, the earliest chambers coiled in an involute trochoid spire, the chambers of the later coils subdividing ventrally into two or more subsidiary chambers or chamberlets, later chambers forming a distinct peripheral flange. — Eocene.

Genus EOCONULOIDES Cole and Bermudez, 1944

Key, plate 52, figures 21, 22

Genotype, *Eoconuloides wellsi* Cole and Bermudez

Eoconuloides Cole and Bermudez, Bull. Amer. Pal., vol. 28, No. 113, 1944, p. 10.

Test conical with an involute, trochoid, multichambered spire, with a subspherical proloculum and smaller second chamber, the final chambers subdivided on the peripheral side into small chamberlets; spiral wall thick, initially with irregularly developed pillars, final spiral wall thinner and without pillars. — Eocene.

Genus AMPHISTEGINA d'Orbigny, 1826

Plate 25, figure 2; Key, plate 31, figures 16, 17
Genotype, *Amphistegina vulgaris* d'Orbigny

Amphistegina d'Orbigny, Ann. Sci. Nat., vol. 7, 1826, p. 304.
Omphalophacus Ehrenberg, 1838.

Test usually lenticular, trochoid, often involute on the dorsal side in the adult, ventral side with supplementary chambers more or less irregularly rhomboid; sutures with a pronounced angle; wall calcareous, finely perforate, without a true secondary canal system; aperture small, ventral, the wall granular about the opening. — Eocene to Recent.

Family 36. CALCARINIDAE

Test typically trochoid in early stages, soon adding a supplementary mass of shell material, over which the new chambers are added, in higher genera chambers extending to the dorsal side and finally covering the whole test in a globular series, test developing surface bosses of shell material which are the outer ends of pillars, also large spines independent of the individual chambers, supplementary canal system well developed; wall calcareous, coarsely perforate; aperture in early trochoid stages as in *Rotalia*, later consisting of numerous smaller openings.

This family in its simplest form is close to *Rotalia*, in fact "*Calcarina calcar*" may be a *Rotalia* as indicated by some authors. The more complex forms have developed secondary shell material between the layers of chambers, and this character is seen in the forms without spines, *Arnaudiella* and *Siderina* of the Upper Cretaceous, and *Pellatispira* in the Tertiary. These genera in some respects resemble the Camerinidae, but lack the marginal cord and other typical structures of that family.

Several of the genera develop large blunt spines independent of the chambers, in the higher forms of *Calcarina*, *Siderolites*, *Baculogypsina*, and *Baculogypsinoides*. A secondary canal system is also developed.

The family has its beginnings in the Upper Cretaceous in *Calcarina*, *Siderolites* with large spines, and *Arnaudiella* and *Siderina* developing the secondary shell material without spines. The spinose group continues to the present in the Indo-Pacific, and the group without spines is represented in the same region by *Pellatispira*, but only in Tertiary sediments.

The genus *Tinoporus* Montfort includes two genera, both in the composite figure and the description, and as such is entirely invalid. The genotype of *Calcarina* has been referred to on the basis of Parker and Jones as "*Calcarina calcar*." Their whole sentence quoted in full is "This [*Rotalia bisaculeata*] is rather a subvariety of the Rotaline genus *Calcarina*, of which Model No. 34 may be taken as the type." This is very ambiguous and may have two interpretations, and the species as already noted may be a *Rotalia*.

Genus CALCARINA d'Orbigny, 1826

Plate 25, figure 3; Key, plate 32, figure 1
Genotype, *Nautilus spengleri* Linné

Calcarina d'Orbigny, Ann. Sci. Nat., vol. 7, 1826, p. 276.
Nautilus (part) of authors.

Test trochoid, biconvex, with radial spines independent of the individual chambers, usually in the plane of coiling, the early stages with the test simple, later with a supplementary mass of shell material over which the new chambers are laid on the ventral side; wall calcareous, perforate, with pillars and canal system; aperture in the adult typically a row of small openings along the inner ventral margin of the chamber. — Cretaceous to Recent.

Calcarina merges in its simplest forms with *Rotalia*. The reference of Parker and Jones, as to the type, is very obscure when their whole sentence is considered, and *Nautilus spengleri* is here considered as the genotype. *Tinoporus* Montfort cannot be used, as it is based on a composite figure and description involving two genera. See Cushman, The Relationships of the Genera *Calcarina*, *Tinoporus*, and *Baculogypsina*, as Indicated by Recent Philippine Material (Bull. 100, U. S. Nat. Mus., vol. 1, pt. 6, 1919, pp. 363–368, pls. 44, 45).

Genus SIDEROLITES Lamarck, 1801

Plate 25, figure 6; Key, plate 32, figures 2, 3
Genotype, *Siderolites calcitrapoides* Lamarck

Siderolites Lamarck, Syst. Anim. sans Vert., 1801, p. 376.
Siderolina Defrance, 1824. *Calcarina* (part) of authors.

Test with the earliest stages trochoid especially in the microspheric form, not planispiral, later becoming loosely planispiral with secondary shell material between whorls, periphery with spinose projections mostly in a single plane; wall calcareous, perforate, with pillars in some species ending at the surface in raised bosses; apertures in the adult consisting of rounded openings in the chamber wall. — Upper Cretaceous to Recent.

Genus BACULOGYPSINOIDES Yabe and Hanzawa, 1930

Plate 25, figure 4; Key, plate 32, figure 4
Genotype, *Baculogypsinoides spinosus* Yabe and Hanzawa = *Siderolites tetraedra* Cushman (not Gümbel)

Baculogypsinoides Yabe and Hanzawa, Sci. Rept. Tohoku Imp. Univ., ser. 2 (Geol.), vol. 14, 1930, p. 43.
Calcarina (part) and *Siderolites* (part) of authors.

Test with early chambers similar to *Calcarina*, later generally globular with several (usually four) large, stout spines not in one plane, independent of the chambers, surface with chambers in generally concentric layers with secondary shell material; wall calcareous, perforate; apertures formed by rounded pores in the outer wall. — Eocene to Recent.

Genus BACULOGYPSINA Sacco, 1893

Plate 25, figure 5; Key, plate 32, figures 5, 6
Genotype, *Orbitolina sphaerulata* Parker and Jones

Baculogypsina SACCO, Bull. Soc. Belg. Géol., vol. 7, 1893, p. 206.
Orbitolina PARKER and JONES (not D'ORBIGNY). *Tinoporus* (part) of authors.

Test in early stages like *Calcarina*, early developing four or more large spines increasing in size independent of the chambers; chambers quickly covering whole surface, supplementary skeleton well developed, consisting of pillars at the chamber angles ending at the surface in rounded bosses, and connected with surrounding ones by radial connecting rods, giving a reticulate appearance to the test; wall calcareous, perforate. — Tertiary, Recent.

Genus ARNAUDIELLA H. Douvillé, 1907

Key, plate 32, figures 7, 8
Genotype, *Arnaudiella grossouvrei* H. Douvillé

Arnaudiella H. DOUVILLÉ, Bull. Soc. Géol. France, ser. 4, vol. 6, 1907, p. 599.

Test lenticular; early chambers involute, later ones compressed and somewhat evolute, supplementary thin-walled chambers developed along the

---

PLATE 25

RELATIONSHIPS OF THE GENERA OF THE AMPHISTEGINIDAE, CALCARINIDAE AND CYMBALOPORIDAE

AMPHISTEGINIDAE

FIG.

1. *Asterigerina carinata* d'Orbigny. (After d'Orbigny.) *a*, dorsal view; *b*, ventral view; *c*, side view.
2. *Amphistegina lessonii* d'Orbigny. *a*, dorsal view; *b*, ventral view; *c*, side view.

CALCARINIDAE

3. *Calcarina spengleri* (Linné). (After W. B. Carpenter.) *a*, dorsal view; *b*, ventral view; *c*, side view.
4. *Baculogypsinoides spinosus* Yabe and Hanzawa.
5. *Baculogypsina sphaerulatus* (Parker and Jones). (After W. B. Carpenter.) *a*, exterior; *b*, section showing developmental stages.
6. *Siderolites calcitrapoides* Lamarck. (After Hofker.) *a*, exterior; *b*, section.

CYMBALOPORIDAE

7. *Cymbalopora radiata* Hagenow. (After Hagenow.) *a*, side view; *b*, ventral view.
8. *Cymbaloporetta squammosa* (d'Orbigny). *a*, side view; *b*, ventral view.
9. *Cymbaloporella tabellaeformis* (H. B. Brady). (After H. B. Brady.) *a*, dorsal view; *b*, side view.
10. *Tretomphalus bulloides* (d'Orbigny). (After H. B. Brady.) *a*, dorsal view; *b*, side view.
11. *Halkyardia minima* (Liebus). (After Heron-Allen and Earland.) *a*, dorsal view; *b*, ventral view; *c*, vertical section.

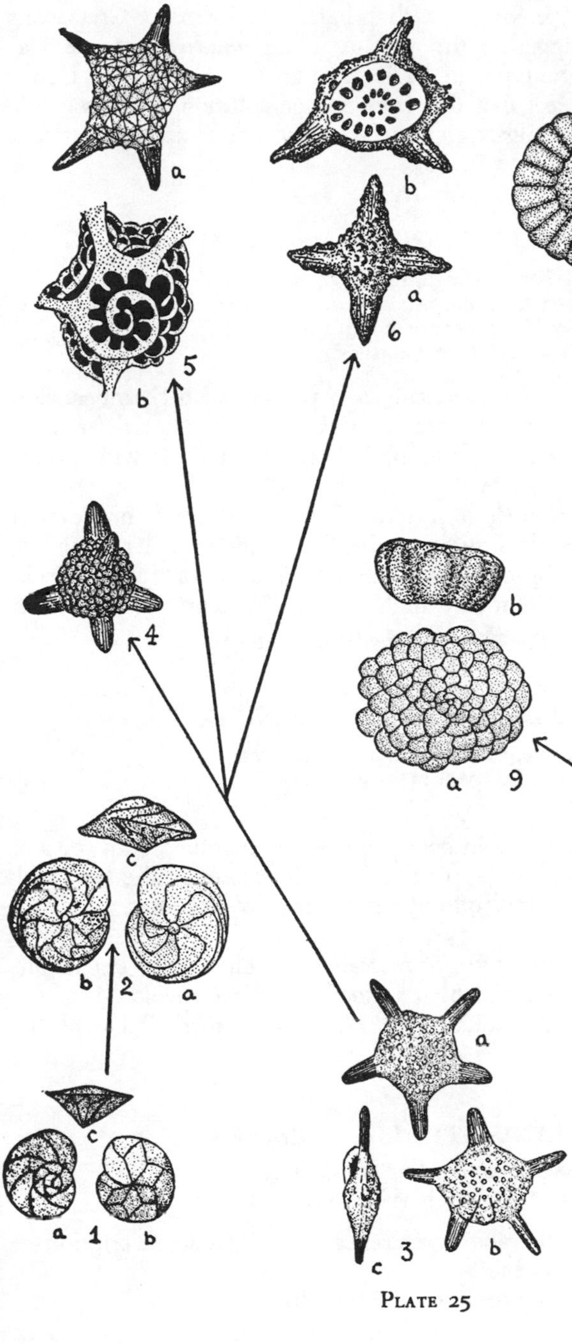

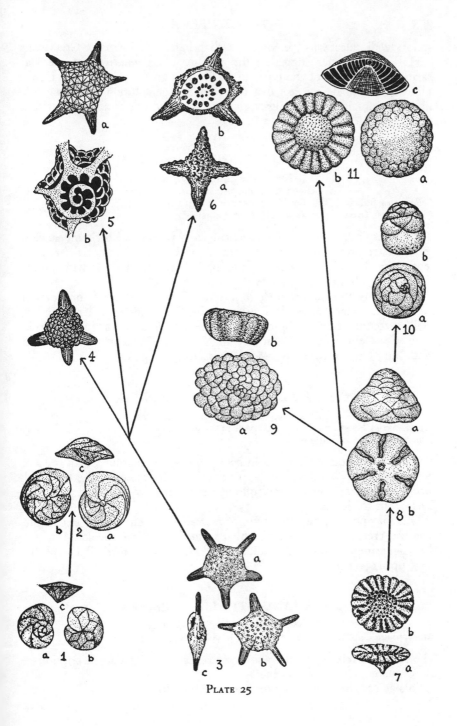

PLATE 25

spiral; wall calcareous, perforate, with pillars. — Uppermost Cretaceous.

Hofker, 1927, has suggested the placing of *Arnaudiella* in the Calcarinidae. A study of the types in Paris through the kindness of Doctor Douvillé has made me feel that this may be one solution of the vacuole-like openings in the test, although otherwise the genus seems related to the Camerinidae.

Genus PELLATISPIRA Boussac, 1906

Key, plate 32, figures 9, 10

Genotype, *Pellatispira douvilléi* Boussac

Pellatispira BOUSSAC, Bull. Soc. Géol. France, ser. 4, vol. 6, 1906, p. 599.
Nummulites (part) HANTKEN, 1876 (not LAMARCK).

Test planispiral, bilaterally symmetrical, early portion with chambers close coiled, later ones loosely coiled with a mass of shell material between, separating the coils; wall calcareous, perforate, lateral walls with pillars, appearing as bosses at the surface. — Eocene.

This genus has some characters like the Camerinidae, but the development of the later portion is similar to the Calcarinidae in its secondary shell material. See Umbgrove, Het Genus *Pellatispira* in het Indo-Pacifische Gebied (with Summary in English) (Wetensch. Med., Mijn. Ned.-Ind., No. 10, 1928, pp. 1–60, 16 pls., 10 text figs.).

Genus PELLATISPIRELLA Hanzawa, 1937

Genotype, *Camerina matleyi* Vaughan

Pellatispirella HANZAWA, Journ. Pal., vol. 11, 1937, p. 114.
Camerina (part) of authors.

Test differing from *Pellatispira* in being completely involute, with the alar prolongation of chambers overlapping toward the center of the test; wall in double layers; aperture multiple along the base of the septa. — Upper Cretaceous to Eocene.

This may possibly be a synonym of *Miscellanea*. The family Pellatispiridae was erected for this genus which seemed to be intermediate between the Camerinidae and Calcarinidae (see Hanzawa, Journ. Pal., vol. 11, 1937, pp. 113–114).

Genus RANIKOTHALIA Caudri, 1944

Genotype, *Nummulites nuttalli* Davies

Ranikothalia CAUDRI, Bull. Amer. Pal., vol. 28, No. 114, 1944, p. 367.

The reader is referred to the above reference as no concise description of the genus is given. — Paleocene.

This genus also may be a synonym of *Miscellanea*.

Genus BIPLANISPIRA Umbgrove, 1937

Genotype, *Biplanispira mirabilis* Umbgrove

Biplanispira Umbgrove, Leidsche Geologische Mededeelingen, vol. 8, 1937, p. 155.
Heterospira Umbgrove, 1936 (not Koken, 1896).

Test discoidal or lenticular, nearly bilaterally symmetrical, central portion with a single layer of chambers, peripheral portion with chambers in two planes on either side of the equatorial plane, marginal cord present; wall calcareous, perforate. — Eocene.

Genus SIDERINA Abrard, 1926

Key, plate 32, figures 11, 12
Genotype, *Siderina douvilléi* Abrard

Siderina Abrard, Comptes Rendus, Somm. Séan. Soc. Géol. France, fasc. 4, Feb. 5, 1926, p. 31.

Test lenticular, thin, central region thickened, more so on one side than on the other; chambers in a spiral, apparently with secondary shell material between the coils; wall calcareous, perforate, surface pustulose. — Uppermost Cretaceous.

Little is known of this genus, either of the material of the wall, of the aperture, or of the details of the structure.

Family 37. CYMBALOPORIDAE

Test in the early stages trochoid, close to *Discorbis*; in the later development the chambers generally in annular series about the periphery; wall calcareous, perforate; apertures numerous, circular pores in the adult variously arranged; in *Tretomphalus* pelagic in the adult.

This family developed in the uppermost Cretaceous with *Cymbalopora*, known from the Maestrichtian of Maestricht, Holland. The test is peculiar in growing dorsally over the proloculum and concealing the early chambers under the piled up walls of later chambers. This same character is seen in the other genera, particularly *Cymbaloporetta* and *Halkyardia*. In *Cymbalopora* the chambers are horizontal and parallel with the base of the test, but in the other genera they become oblique or even vertical. In the Cretaceous species the wall is agglutinated, and the early stages in all the genera are chitinous. The young stage can be cleared by acid of the later chambers, and then is brown and chitinous, easily mistaken for a *Trochammina*.

In *Halkyardia*, in the Eocene, the open umbilical area becomes filled with secondary shell material, but otherwise is like *Cymbaloporetta*. The early chambers are covered in the same way, and the slope of the chambers also similar. In *Cymbaloporella* the chambers surround the umbilicus obliquely in the young, and then become nearly vertical with the whole test spreading. In *Tretomphalus*, at least in the microspheric form, there is developed a peculiar large final chamber with an inner thin globular chamber which can be filled with gas by the animal, and it then floats to the surface. There may be more than one genus represented by these forms with globular chambers, as the young of various species would be placed in *Discorbis*, *Cymbaloporetta*, and *Cymbaloporella* respectively. It may be that all these three genera have this adult stage, in which case the nomenclature becomes much complicated.

These genera form a compact group with similar characters throughout, and should not be confused with such forms as *Planorbulina* or *Dictyoconoides*, genera with which they have no relationship nor true resemblance when the development is studied.

The genus *Pyropilus* is a very peculiar, specialized form in which the early chambers are similar to others of the family, but the later ones become very numerous and irregular.

Genus CYMBALOPORA Hagenow, 1851

Plate 25, figure 7; Key, plate 32, figures 13, 14
Genotype, *Cymbalopora radiata* Hagenow

Cymbalopora Hagenow, Die Bryozoen der Maestrichter Kreide-Bildung, 1851, p. 104, pl. 12, fig. 18.

Test low, conical, in the early stages trochoid; dorsal side of chambers extending over and covering the early ones, later chambers arranged in alternating horizontal series, on the ventral side not meeting at the center, leaving an open umbilicus; wall of the early portion agglutinated with quartz grains and calcareous fragments and with a calcareous cement over a coarsely perforate, thick, chitinous base; apertures in the young ventral, single, in adult chambers along the sides and the inner end often open into the umbilicus. — Upper Cretaceous to Recent.

For details of this genus, see Hofker, Die Foraminiferen aus dem Senon Limburgens. XII. *Cymbalopora radiata* Hagenow (Nat. Hist. Maandblad, Jaarg. 20, 1931, pp. 125–130, 7 text figs.).

Genus CYMBALOPORETTA Cushman, 1928

Plate 25, figure 8; Key, plate 32, figure 15
Genotype, *Rosalina squammosa* d'Orbigny

Cymbaloporetta Cushman, Contr. Cushman Lab. Foram. Res., vol. 4, 1928, p. 7.
Rotalia (part) d'Orbigny. *Rosalina* (part) of authors (not d'Orbigny). *Cymbalopora* of authors (not Hagenow).

Test conical; early chambers trochoid, dorsal wall of chambers extending over and covering the early ones, later chambers in annular series separated somewhat from one another along the periphery with depressions between, the next series of chambers filling these depressions, chambers rapidly becoming oblique to the base and the umbilicus covered by a central shield;

wall of the earliest chambers thick, chitinous, coarsely perforate, later calcareous; aperture in the adult along the sides of the chamber and at the umbilical end, communicating with the exterior by large rounded openings beneath the central shield.—Late Tertiary, Recent.

Genus CYMBALOPORELLA Cushman, 1927

Plate 25, figure 9; Key, plate 32, figures 16, 17
Genotype, *Cymbalopora tabellaeformis* H. B. Brady

Cymbaloporella Cushman, Contr. Cushman Lab. Foram. Res., vol. 3, 1927, p. 81.
Cymbalopora (part) of authors. *Halkyardia* (part) Heron-Allen and Earland.

Test compressed; early chambers trochoid, often somewhat covered on the dorsal side, later ones in alternating annular series about the periphery, quickly assuming an oblique position, then vertical to the base; wall in the young chitinous, brown, later calcareous, coarsely perforate; apertures in the adult a series of openings at the sides of the chamber.—Eocene to Recent.

Genus HALKYARDIA Heron-Allen and Earland, 1919

Plate 25, figure 11; Key, plate 32, figures 18, 19
Genotype, *Cymbalopora radiata*, var. *minima* Liebus

Halkyardia Heron-Allen and Earland, Mem. Proc. Manchester Lit. Phil. Soc., vol. 62, 1917–18 (1919), p. 107.
Cymbalopora Liebus, 1911 (not Hagenow).

Test biconvex; early chambers trochoid, dorsal wall of succeeding chambers extending over and covering the early ones, later ones in annular series, oblique to the base, ventrally umbilicate, later filled with porous, secondary, shell material; wall of early portion probably chitinous, later calcareous, coarsely perforate; apertures connecting with the exterior by the pores of the umbilical filling.—Eocene.

Genus PYROPILUS Cushman, 1934

Key, plate 52, figure 23
Genotype, *Pyropilus rotundatus* Cushman

Pyropilus Cushman, Contr. Cushman Lab. Foram. Res., vol. 10, 1934, p. 100.

Test with the early stages in a trochoid spiral, later building the chambers in an elongate, oval series with an elongate, umbilical depression on the ventral side, the last-formed chambers more or less involute, and the last few covering the growing end as well as a considerable portion of the previous development, particularly on the ventral side; early chambers of a dark, yellowish-brown color, due to the inner, thin, chitinous layer, remainder of the test calcareous, coarsely perforate; last portion of the test without any general aperture, the coarse perforations of the wall evidently serving as apertures.—Recent.

Genus TRETOMPHALUS Moebius, 1880

Plate 25, figure 10; Key, plate 32, figures 20–22
Genotype, *Rosalina bulloides* d'Orbigny

Tretomphalus Moebius, Foram. Insel Mauritius, 1880, p. 98.
Rosalina (part) d'Orbigny. *Cymbalopora* (part) of authors (not Hagenow).

Test in early stages *Discorbis*-like, in other forms with the young like *Cymbaloporetta* or *Cymbaloporella*, the adult with a large globular "float chamber" with an interior thin-walled chamber with a valve-like opening; wall in the young chitinous, later calcareous, perforate; apertures in adult a series of pores on the outer face of the globular chamber. — Tertiary, Recent.

See Earland, On *Cymbalopora bulloides* (d'Orbigny) and Its Internal Structure (Journ. Quekett Micr. Club, ser. 2, vol. 8, 1902, pp. 309–322, pl. 16). *Tretomphalus* has several species, some of which seem to have different origins, and the group needs more study.

Genus CHAPMANINA A. Silvestri, 1931

Key, plate 40, figures 12, 13
Genotype, *Chapmania gassinensis* A. Silvestri

Chapmanina A. Silvestri, Boll. Soc. Geol. Ital., vol. 50, 1931, p. 65.
Chapmania of authors (not Monticelli, 1893).

It is impossible at the present time to give a complete and accurate description of this genus. There has been much confusion in regard to it. There seems to be no doubt as to its calcareous test, removing it from any relation to *Dictyoconus* in the Valvulinidae or to *Orbitolina* in the Orbitolinidae. It belongs either in the Rotaliidae with *Dictyoconoides* and *Lockhartia*, or with the Cymbaloporidae. In both groups there is a tendency for the early chambers on the dorsal side to be covered by the extensions of the dorsal wall in later chambers. In the former group the whorls are either completely spiral or have other series intercalated with them, while in the Cymbaloporidae the chambers are in annular series in the adult. The internal structure seems to be more similar to the Cymbaloporidae than to the Rotaliidae, as no ventral pillars are shown in any of the figures, and the chambers as shown seem to be in annular series. From the indications given it should probably be placed in the Cymbaloporidae. The types are from the Eocene of Italy.

Genus FABIANIA A. Silvestri, 1926

Key, plate 52, figures 24, 25
Genotype, *Patella cassis* (Oppenheim)

Fabiania A. Silvestri, Riv. Ital. Pal., vol. 32, 1926, pp. 15–22, pl. 1, figs. 1–6.
Pseudorbitolina (part) Cushman and Bermudez.
Eodictyoconus Cole and Bermudez, 1944. *Tschoppina* Keijzer, 1945.

Test regularly or irregularly conical, early stage composed of three subspherical chambers; some specimens with a deeply excavated umbilicus are composed of a single layer of chambers, many with a single horizontal plate; others, with the umbilical area filled with chamberlets, have a plate with several large perforations covering the umbilical area on the base of the test. — Eocene.

Our thanks are due to Cole for information on this genus which he found was apparently the same as that of Silvestri's originally described as a coral.

Family 38. CASSIDULINIDAE

Test at least in the early stages trochoid, later chambers in the more specialized genera alternating as they coil, and in the highest genera becoming uncoiled; wall calcareous, perforate; aperture generally elongate, in the plane of coiling, usually with a tooth, in the uncoiled forms becoming more rounded.

There is a very definite development in this family from the Rotaliidae. *Ceratobulimina* is developed as early as the Upper Cretaceous. It has an elongate aperture in the plane of coiling, and it is but a simple step from this to *Pulvinulinella,* also appearing in the Upper Cretaceous where the aperture is elongate, similar to that of *Cassidulina,* and the later chambers on the ventral side in the adult already show a tendency alternately not to reach to the umbilicus. *Cassidulina,* also appearing in the Upper Cretaceous in rather simple forms, has the chambers biserially arranged, the outer end of each added chamber appearing alternately on the opposite side near the periphery. Uncoiling, so common in many groups of the foraminifera, takes place in *Cassidulinoides,* and the aperture becomes more open and nearly terminal. This first appears in simple form in the Upper Eocene, and reaches its greatest development in Recent species. *Orthoplecta* becomes irregularly elongate, and is known only as a Recent form. *Ehrenbergina* beginning in the Eocene uncoils and develops a definite dorsal and ventral side.

In *Cushmanella,* the test becomes involute and bilaterally symmetrical. *Epistominoides* is a very peculiar form, resembling *Saracenaria* in its external appearance, but not in structure.

The Cassidulinidae are most highly developed in the Pacific, and are abundant in the Tertiary of western America, eastern Asia, and in the Pacific Islands. There are some species of *Cassidulina* abundant in the Arctic, and in general the family seems to prefer rather deep or cold waters.

On the basis of the biserial form some of the genera were formerly grouped with the Textulariidae.

Subfamily 1. Ceratobulimininae

Test rotaliform throughout.

Genus CERATOBULIMINA Toula, 1915

Plate 26, figure 1; Key, plate 33, figure 1
Genotype, *Rotalina contraria* Reuss

Ceratobulimina TOULA, Jahrb. k. k. Geol. Reichs., vol. 64, 1914 (1915), p. 665.
Rotalina (part) REUSS, 1851 (not D'ORBIGNY). *Cassidulina* (part) H. B. BRADY, 1881 (not D'ORBIGNY). *Bulimina* (part) of authors (not D'ORBIGNY). *Bulimi-nella* (part) CUSHMAN, 1911. *Pulvinulina* (part) RZEHAK, 1888 (not PARKER and JONES). *Rotalia* (part) PLUMMER, 1927 (not LAMARCK).

Test rotaliform; all chambers visible dorsally, ventrally only those of the last-formed whorl, close coiled, ventrally umbilicate; wall calcareous, finely perforate, added to as growth progresses, laminate, entire exterior polished; aperture elongate, extending into the ventral side of the chamber, and in perfect adult specimens aperture covered by a thin convex plate, merged with the chamber wall above the aperture in a semicircular line, lower end thin, lip-like.—Upper Cretaceous to Recent.

See Cushman, The Genus *Ceratobulimina* and Its Species (Contr. Cushman Lab. Foram. Res., vol. 22, 1946, pp. 107–117, pls. 17–19).

Genus ROGLICIA van Bellen, 1941

Key, plate 53, figure 1
Genotype, *Roglicia sphaerica* van Bellen

Roglicia VAN BELLEN, Proc. Ned. Akad. Wetensch., vol. 44, 1941, p. 1000.

"Shape of the test trochoid, like *Ceratobulimina*. All chambers visible from the dorsal side. Only those of the last formed whorl visible on the ventral side. Chambers distinct. Wall thick. Greatest part of the surface covered with short spines; only a region, surrounding the aperture smooth. The aperture, on the ventral side of the last formed chamber, circular, surrounded by a thickened ring, covered with a thin plate."—Eocene.

Specimens of this genus have not been available for study. In some characters it resembles *Lamarckina*.

Genus PSEUDOPARRELLA Cushman and ten Dam, 1948

Plate 26, figure 2; Key, plate 33, figure 2
Genotype, *Pulvinulinella subperuviana* Cushman

Pseudoparrella CUSHMAN and TEN DAM, Contr. Cushman Lab. Foram. Res., vol. 24, 1948, p. 49.
Pulvinulinella CUSHMAN, 1926 (not EIMER and FICKERT).
Rosalina (part), *Rotalia* (part), *Truncatulina* (part), *Discorbina* (part), *Pulvinulina* (part) of authors.

Test trochoid, close coiled; all chambers visible dorsally, only those of the last-formed whorl from the ventral side, very slightly if at all umbilicate; sutures on the dorsal side oblique, ventrally nearly radial; wall calcareous, perforate; aperture on the ventral side of the peripheral face, elongate, somewhat loop-shaped, nearly parallel to the plane of coiling. — Cretaceous to Recent.

Genus ALABAMINA Toulmin, 1941

Key, plate 53, figure 2
Genotype, *Alabamina wilcoxensis* Toulmin

Alabamina TOULMIN, Journ. Pal., vol. 15, 1941, p. 602.
Pulvinulinella (part) of authors.

"Test trochiform, usually biconvex, umbilical area closed, periphery bluntly acute or narrowly rounded; all chambers visible from the dorsal side only; dorsal sutures oblique, straight or very gently curved, ventral sutures radiate, straight or slightly curved; wall calcareous, finely perforate; aperture a long narrow opening on the ventral side along the base of the septal face, with supplementary false aperture, consisting of a deep indentation of the wall of the septal face, which is parallel to the periphery on the ventral side and carries no opening into the interior of the chamber." — Tertiary.

Subfamily 2. Cassidulininae

Test with the chambers alternating on the two sides of the plane of coiling.

Genus CASSIDULINA d'Orbigny, 1826

Plate 26, figure 3; Key, plate 33, figures 3, 4
Genotype, *Cassidulina laevigata* d'Orbigny

Cassidulina D'ORBIGNY, Ann. Sci. Nat., vol. 7, 1826, p. 282.
Entrochus EHRENBERG, 1841. *Selenostomum* EHRENBERG, 1858. *Burseolina* SEGUENZA (*fide* H. B. BRADY), 1880.

Test close coiled, lenticular or subglobular, usually involute; chambers alternating on the two sides of the periphery, usually smooth, sometimes highly ornate; wall calcareous, perforate; aperture elongate, close to the peripheral plane, often with a plate-like tooth. — Upper Cretaceous to Recent.

See Cushman, Notes on the Genus *Cassidulina* (Contr. Cushman Lab. Foram. Res., vol. 1, pt. 3, 1925, pp. 51–60, pls. 8, 9).

Genus CASSIDULINELLA Natland, 1940

Key, plate 53, figure 3
Genotype, *Cassidulinella pliocenica* Natland

Cassidulinella NATLAND, Journ. Pal., vol. 14, 1940, p. 570.

Early chambers coiled and biserial similar to *Cassidulina*, later chambers biserial but rapidly increasing in length to nearly encircle the test; wall calcareous, perforate; aperture extending full length near the periphery between the last two chambers. — Pliocene and Miocene.

Genus STICHOCASSIDULINA Stone, 1946

Key, plate 53, figure 4
Genotype, *Stichocassidulina thalmanni* Stone
Stichocassidulina STONE, Journ. Pal., vol. 20, 1946, p. 59.

Test in structure similar to *Cassidulina* but differing in having numerous, small, supplementary apertures along the sutures. — Eocene.

Genus EPISTOMINOIDES Plummer, 1934

Key, plate 53, figures 7, 8
Genotype, *Saracenaria wilcoxensis* Cushman and Ponton
Epistominoides PLUMMER, Amer. Midland Nat., vol. 15, 1934, p. 602.
Saracenaria CUSHMAN and PONTON, 1934 (not DEFRANCE 1824).

Test close coiled, trochoid, involute; chambers irregularly biserial on the ventral side; wall calcareous, finely perforate; aperture elongate, at the peripheral angle of the last-formed chamber. — Midway and Wilcox Eocene.

See Plummer, *Epistominoides* and *Coleites*, New Genera of Foraminifera (Amer. Midland Nat., vol. 15, 1934, pp. 601–607, pl. XXIV).

Genus CEROBERTINA Finlay, 1939

Key, plate 53, figure 5, 6
Genotype, *Cerobertina bartrumi* Finlay
Cerobertina FINLAY, Trans. Roy. Soc. New Zealand, vol. 69, 1939, p. 118.

Test similar to *Ceratobulimina* but with a series of supplementary chambers along the umbilical edge of the main chambers, more or less distinctly marked off by a groove or internal line. — Tertiary.

Genus CUSHMANELLA D. K. Palmer and Bermudez, 1936

Key, plate 53, figures 9–12
Genotype, *Nonionina brownii* d'Orbigny
Cushmanella D. K. PALMER and BERMUDEZ, Mem. Soc. Cubana Hist. Nat., vol. 9, 1936, p. 252.

Test free, coiled, involute, bilaterally symmetrical; chambers in two series, one of large conspicuous chambers, and a smaller series alternating with the larger ones at the umbilical margin; wall calcareous, finely perforate; apertures of two kinds, one at the base and the other in the middle of the apertural face, from the latter an internal tube connects with the aperture at the base of the preceding chamber. — Late Tertiary to Recent.

Genus CASSIDULINOIDES Cushman, 1927

Plate 26, figure 4; Key, plate 33, figure 5
Genotype, *Cassidulina parkeriana* H. B. Brady

Cassidulinoides Cushman, Contr. Cushman Lab. Foram. Res., vol. 3, 1927, p. 84.
Cassidulina (part) of authors.

Test in the early stages like *Cassidulina*, but in the adult becoming uncoiled in a series of alternating chambers; aperture in the young like *Cassidulina*, later becoming terminal. — Upper Eocene to Recent.

Genus ORTHOPLECTA H. B. Brady, 1884

Plate 26, figure 5; Key, plate 33, figure 6
Genotype, *Cassidulina (Orthoplecta) clavata* H. B. Brady

Orthoplecta H. B. Brady, Rep. Voy. *Challenger*, Zoology, vol. 9, 1884, p. 432.

Test very elongate, subcylindrical; chambers in an irregular biserial arrangement in the adult, those of the young appearing more as in *Cassidulina*; wall calcareous, perforate; aperture nearly terminal. — Recent.

Subfamily 3. Ehrenbergininae

Test in the early stages as in *Cassidulina*, but the chambers soon becoming compressed in a plane at right angles to that of the early coiling and becoming uncoiled; aperture elongate, on the ventral side near the periphery.

Genus EHRENBERGINA Reuss, 1850

Plate 26, figure 6; Key, plate 33, figure 7
Genotype, *Ehrenbergina serrata* Reuss

Ehrenbergina Reuss, Denkschr. k. Akad. Wiss. Wien, vol. 1, 1850, p. 377.
Cassidulina (part) d'Orbigny, 1839.

Test with microspheric young compressed, planispiral, adult test compressed at right angles to that of the early coiling and becoming uncoiled, developing a dorsal side which is flattened or slightly convex, and a ventral which is thickest near the median line; wall calcareous, perforate; aperture elongate, on the ventral side near the periphery. — Eocene to Recent.

See Cushman, Foraminifera of the Genus *Ehrenbergina* and Its Species (Proc. U. S. Nat. Mus., vol. 70, Art. 16, 1927, pp. 1–8, pls. 1, 2).

Family 39. CHILOSTOMELLIDAE

Test in the early stages of the simpler genera typically trochoid, the chambers all visible from the dorsal side, only those of the last-formed chamber visible from the ventral side, the chambers in later development variously arranged, typically planispiral and involute so that the early stages are completely covered; wall calcareous, perforate; aperture typically on the ventral side, at least in the early stages, in the planispiral forms becoming median.

This family, the early history of which begins in the Upper Cretaceous, was derived from a trochoid rotaliform ancestry, the exact genus of which is unknown although one or more might be guessed at, but with no degree of certainty. In *Allomorphina* the early stages are definitely triserial as is the adult. From this developed *Chilostomella* in which the early stages of the microspheric form are triserial, but the adult has but two chambers to a whorl, and these become distinctly involute. From this developed *Chilostomelloides*, in which the aperture, instead of being low and elongate, develops a rounded opening which may be even separated off from the preceding chamber, and has a definite lip. In the Upper Cretaceous also is developed *Allomorphinella* which becomes very much involute and planispiral. From such a form in the Recent Pacific is developed *Chilostomellina* in which the chambers become still more involute, and the aperture forms a row of small openings at the base of the chamber. These genera have sometimes been placed with the Nonionidae, but a study of their microspheric forms in the early stages will show

---

PLATE 26

RELATIONSHIPS OF THE GENERA OF THE CASSIDULINIDAE AND CHILOSTOMELLIDAE

CASSIDULINIDAE

FIG.

1. *Ceratobulimina pacifica* Cushman and Harris. (After H. B. Brady.) *a*, dorsal view; *b*, ventral view; *c*, edge view.
2. *Pseudoparrella subperuviana* (Cushman). *a*, dorsal view; *b*, ventral view; *c*, edge view.
3. *Cassidulina californica* Cushman and Hughes. *a*, dorsal view; *b*, ventral view; *c*, edge view.
4. *Cassidulinoides parkeriana* (H. B. Brady). (After H. B. Brady.) *a*, dorsal view; *b*, ventral view; *c*, end view.
5. *Orthoplecta clavata* H. B. Brady. (After H. B. Brady.)
6. *Ehrenbergina bradyi* Cushman. (After H. B. Brady.) *a*, dorsal view; *b*, ventral view; *c*, end view.

CHILOSTOMELLIDAE

7. *Allomorphina trigona* Reuss. (After H. B. Brady.) *a*, dorsal view; *b*, ventral view.
8. *Chilostomella oolina* Schwager. (After H. B. Brady.) *a*, dorsal view; *b*, ventral view.
9. *Chilostomelloides oviformis* (Sherborn and Chapman). *a*, dorsal view; *b*, ventral view; *c*, side view; *d*, end view.
10. *Seabrookia pellucida* H. B. Brady. (After H. B. Brady.) *a*, dorsal view; *b*, end view.
11. *Allomorphinella contraria* (Reuss). (After Reuss.) *a*, side view; *b*, edge view.
12. *Chilostomellina fimbriata* Cushman. *a*, dorsal view; *b*, ventral view; *c*, end view.
13. *Pullenia sphaeroides* d'Orbigny. *a*, side view; *b*, apertural view.
14. *Sphaeroidina bulloides* d'Orbigny. *a*, dorsal view; *b*, ventral view.

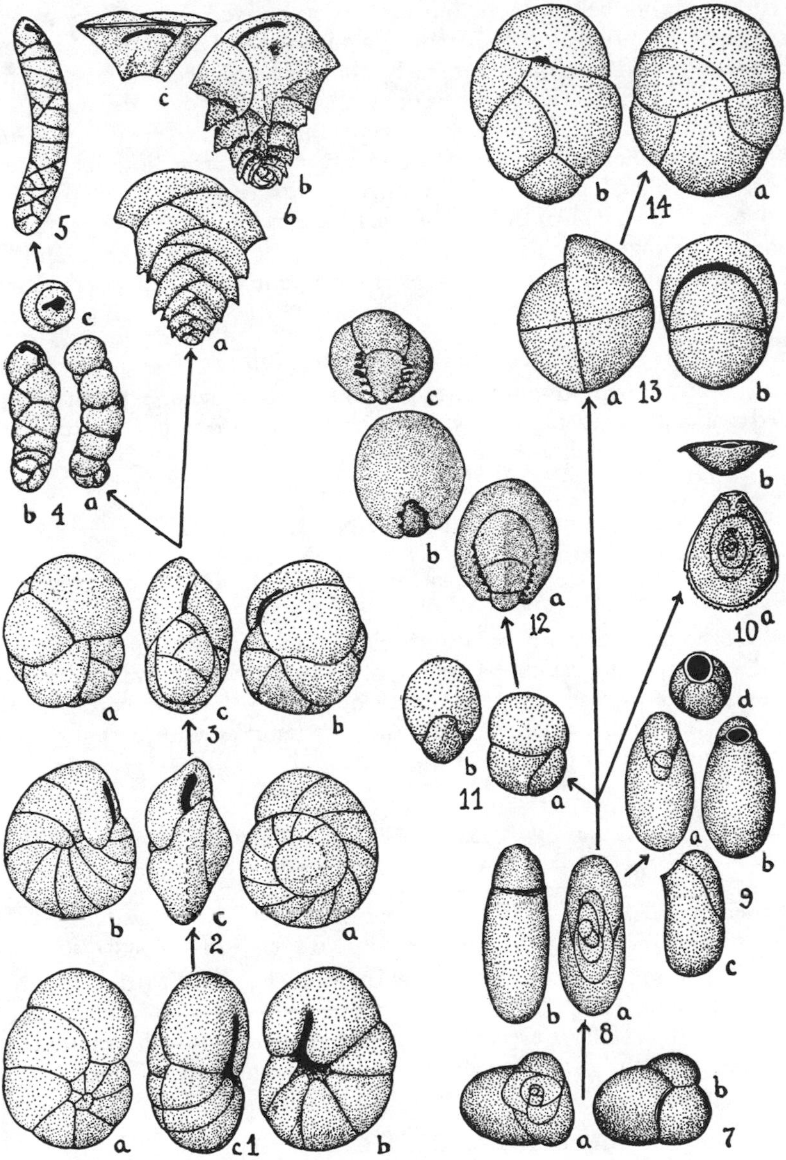

Plate 26

that they do not belong there, but have been derived from *Chilostomella*. In Recent material also is *Seabrookia* in which the involute character becomes very marked, and the early chambers are seen only in section or through the nearly transparent wall. *Pullenia* has developed the bilaterally symmetrical form, but keeps the typical aperture of *Chilostomella*, and the early stages of the microspheric form show its relationships. As an end form in this family is *Sphaeroidina* which develops a more or less globular form, the early stages of which are similar to *Pullenia*. Altogether this group forms a very complete whole, and the geologic history of the various forms is consistent with their development.

Subfamily 1. Allomorphininae

Test in adult with usually three chambers to a whorl; chambers inflated and enlarging rapidly as added; aperture an elongate curved slit at ventral border of last-formed chamber.

Genus ALLOMORPHINA Reuss, 1850

Plate 26, figure 7; Key, plate 33, figure 8
Genotype, *Allomorphina trigona* Reuss
Allomorphina Reuss, Denkschr. k. Akad. Wiss. Wien, vol. 1, 1850, p. 380.

Test trochoid, adult with usually three chambers to a coil; chambers inflated and enlarging rapidly as added, often very involute; wall calcareous, perforate; aperture an elongate arched opening, below the border of the last-formed chamber on the ventral side, sometimes with a slight lip.—Upper Cretaceous to Recent.

Genus QUADRIMORPHINA Finlay, 1939

Key, plate 53, figure 13
Genotype, *Valvulina allomorphinoides* Reuss
Quadrimorphina Finlay, Trans. Roy. Soc. New Zealand, vol. 69, 1939, p. 325.
Valvulina (part), *Allomorphina* (part), and *Valvulineria* (part) of authors.

Test similar to *Allomorphina* but with four chambers to the whorl.—Cretaceous and Tertiary.

Later, in 1941, Marie erected *Gyromorphina* based on the same genotype, but apparently from his description it is somewhat different.

Genus ROTAMORPHINA Finlay, 1939

Key, plate 53, figure 14
Genotype, *Rotamorphina cushmani* Finlay
Rotamorphina Finlay, Trans. Roy. Soc. New Zealand, vol. 69, 1939, p. 325.

Test similar to *Quadrimorphina* but with more than four chambers in a whorl.—Cretaceous and Tertiary.

Subfamily 2. Chilostomellinae

Test in the adult with two chambers making up a coil, the chambers inflated and enlarging rapidly as added; aperture variously modified, lateral or terminal.

Genus CHILOSTOMELLA Reuss, 1850

Plate 26, figure 8; Key, plate 33, figures 9–11
Genotype, *Chilostomella ovoidea* Reuss

Chilostomella Reuss, Denkschr. k. Akad. Wiss. Wien, vol. 1, 1850, p. 379.

Test fusiform, ovoid, or subcylindrical, in the early stages, especially of the microspheric form, with the chambers as in *Allomorphina*, adult with two chambers completing a coil, and embracing so that but a small part of the base of the preceding chamber is visible from the ventral side; wall calcareous, perforate; aperture a narrow curved opening, at the inner margin of the ventral face of the last-formed chamber, often with a slightly upturned lip. — Upper Cretaceous to Recent.

See Cushman, The Genus *Chilostomella* and Related Genera (Contr. Cushman Lab. Foram. Res., vol. 1, pt. 4, 1926, pp. 73–80, pl. 11).

Genus CHILOSTOMELLOIDES Cushman, 1926

Plate 26, figure 9; Key, plate 33, figures 12, 13
Genotype, *Lagena* (*Obliquina*) *oviformis* Sherborn and Chapman

Chilostomelloides Cushman, Contr. Cushman Lab. Foram. Res., vol. 1, pt. 4, 1926, p. 77.

Lagena (*Obliquina*) Sherborn and Chapman, 1886 (not Seguenza). *Chilostomella* (part) of authors.

Test similar in general structure to *Chilostomella*, but the aperture rounded and somewhat offset from the general contour of the test, with a slight neck in some species and a slightly developed lip. — Upper Cretaceous to Oligocene.

Subfamily 3. Seabrookiinae

Test with two chambers forming a coil, but entirely embracing on the ventral side; aperture elliptical and terminal.

Genus SEABROOKIA H. B. Brady, 1890

Plate 26, figure 10; Key, plate 33, figures 14, 15
Genotype, *Seabrookia pellucida* H. B. Brady

Seabrookia H. B. Brady, Journ. Roy. Micr. Soc., 1890, p. 570.
Cerviciferina Goddard and Jensen, 1907.

Test essentially trochoid, the earliest stages as in *Allomorphina*, three chambers making up a whorl, in the adult two chambers involute, the last-formed one making up nearly the whole surface of the test on the ventral side; wall calcareous, thin, perforate; aperture elliptical, at the periphery of the more acute end of the chamber. — Cretaceous to Recent.

Subfamily 4. Allomorphinellinae

Test with the later chambers planispiral, involute; chambers increasing rapidly in size as added; apertures becoming median.

Genus ALLOMORPHINELLA Cushman, 1927

Plate 26, figure 11; Key, plate 33, figure 16
Genotype, *Allomorphina contraria* Reuss

Allomorphinella Cushman, Contr. Cushman Lab. Foram. Res., vol. 3, 1927, p. 86.
Allomorphina (part) of authors.

Test with the adult chambers in a planispiral coil; the chambers involute, rapidly increasing in size as added, and embracing; wall calcareous, perforate; aperture elongate, narrow, at the periphery of the chamber at the median line. — Upper Cretaceous.

Genus CHILOSTOMELLINA Cushman, 1926

Plate 26, figure 12; Key, plate 33, figure 17
Genotype, *Chilostomellina fimbriata* Cushman

Chilostomellina Cushman, Contr. Cushman Lab. Foram. Res., vol. 1, pt. 4, 1926, p. 78.

Test composed of a few inflated chambers, the last-formed one almost completely enveloping the preceding ones, and the chambers rapidly increasing in size as added; wall calcareous, thin, finely perforate; aperture small, crescentiform, the sides of the chamber with a series of reëntrants at each side. — Recent.

Genus PULLENIA Parker and Jones, 1862

Plate 26, figure 13; Key, plate 33, figures 18–20
Genotype, *Nonionina bulloides* d'Orbigny

Pullenia Parker and Jones, in Carpenter, Parker and Jones, Introd. Foram., 1862, p. 184.
Nonionina (part) d'Orbigny, 1826.

Test in the adult planispiral, close coiled; chambers completely involute, a few making up the coil; wall calcareous, perforate; aperture an elongate crescentic opening, at the inner margin of the last-formed chamber. — Cretaceous to Recent.

The type as originally designated by Parker and Jones was *Nonionina sphaeroides* d'Orbigny, but, as that species is indeterminable, it seems better to use *Nonionina bulloides* d'Orbigny.

See Cushman and Todd, The Genus *Pullenia* and Its Species (Contr. Cushman Lab. Foram. Res., vol. 19, 1943, pp. 1–23, pls. 1–4).

Genus CRIBROPULLENIA Thalmann, 1937

Key, plate 53, figures 15–17
Genotype, *Nonion* (?) *marielensis* D. K. Palmer

Cribropullenia THALMANN, Pal. Zeitschr., vol. 10, Review No. 1349, 1937, p. 351.
Antillesina GALLOWAY and HEMINWAY, 1941.
Nonion (part) of authors.

Test similar to *Pullenia* but with a number of small circular apertures along the base of and in the apertural face. — Oligocene.

Subfamily 5. Sphaeroidininae

Test in the earliest stages generally planispiral, later chambers irregularly involute; aperture in the early stages a crescent-shaped slit, in the adult rounded, with a flat, rounded, tooth-like projection.

Genus SPHAEROIDINA d'Orbigny, 1826

Plate 26, figure 14; Key, plate 33, figures 21, 22
Genotype, *Sphaeroidina bulloides* d'Orbigny

Sphaeroidina D'ORBIGNY, Ann. Sci. Nat., vol. 7, 1826, p. 267.
Sexloculina CZJZEK, 1848. *Bolbodium* EHRENBERG, 1872.

Test in the earliest stages generally planispiral, later chambers irregularly embracing; wall calcareous, perforate; aperture in the young a crescent-shaped opening, in the adult rounded, with a flat, rounded, tooth-like projection. — Cretaceous to Recent.

Family 40. GLOBIGERINIDAE

Test, at least in the early stages, trochoid, umbilicate, in later forms becoming planispiral, and in some of the higher forms globular and more or less involute; wall calcareous, usually coarsely perforate, often with a cancellated surface, in well-preserved specimens of most of the genera are fine spines; aperture typically large, but in the higher genera consisting of numerous small openings either at the edge of the chamber or scattered over the entire surface.

This family, with the Globorotaliidae which are directly derived from it, form the pelagic foraminifera of the present oceans. They have held this character since early Cretaceous times at least. They are one of the most highly specialized groups of the foraminifera, and these two families have not given rise to many other groups as is natural from their specialized character. The earliest forms are trochoid with a definite umbilical opening. The early stages of the microspheric form are rare, but show a flattened, *Discorbis*-like early stage, thus showing their derivation from the Rotaliidae in this general character. Two structures have been particularly developed in their adaptation to pelagic life. One of these is the development of spines on the surface, which are very long and slender, and which apparently serve as supports for the protoplasm which is sent out into a vacuolar mass about the test. These spines are a unique feature in this family. They may occur all over the surface as in most of the genera, or may be confined to the outer end of the chamber as in *Hastigerina* and *Hastigerinella*. The other structure is developed through the need of getting the protoplasm to the exterior. This is accomplished either by the large umbilical opening of *Globigerina*, *Globigerinella*, *Hastigerina*, *Hastigerinella*, and *Sphaeroidinella*, or through numerous openings along the border of the chamber as in *Globigerinoides* and *Candeina*, or covering the entire surface as in *Orbulina*. The large opening often seen in *Orbulina* as shown by Earland is probably accidental. In *Orbulina* a complete adult chamber envelops the entire *Globigerina*-like young.

A new name *Candorbulina* has been proposed for peculiar globular forms with a ring of pores in the adult, perhaps representing a globular end form from *Candeina*.

Subfamily 1. Globigerininae

Wall clothed with fine spines, typically trochoid, but in some genera becoming planispiral; wall often cancellated, coarsely perforate.

Genus GLOBIGERINA d'Orbigny, 1826

Plate 27, figure 1; Key, plate 34, figures 1–5
Genotype, *Globigerina bulloides* d'Orbigny

Globigerina D'ORBIGNY, Ann. Sci. Nat., vol. 7, 1826, p. 277.
Rotalia (part) of authors. *Rhynchospira* EHRENBERG, 1845. *Phanerostomum* (part) and *Ptygostomum* (part) EHRENBERG, 1854. *Planulina* (part) EHRENBERG, 1854 (not D'ORBIGNY). *Pylodexia* EHRENBERG, 1858. *Aristerospira* (part) EHRENBERG, 1858.

Test trochoid throughout, umbilicate; chambers in the young especially of the microspheric form in a flattened trochoid form like *Discorbis*, usually smooth and the wall thin, later chambers globular; wall calcareous, thick and cancellated, in well-preserved, especially pelagic specimens, clothed with long slender spines coming from the angles of the cancellated surface areas, the base of such areas with the pores of the wall; aperture large, opening into the umbilicus. — Cretaceous to Recent.

Genus GLOBIGERINOIDES Cushman, 1927

Plate 27, figure 6; Key, plate 34, figures 6–9
Genotype, *Globigerina rubra* d'Orbigny

Globigerinoides CUSHMAN, Contr. Cushman Lab. Foram. Res., vol. 3, 1927, p. 87.
Globigerina (part) of authors.

Test usually trochoid throughout; aperture as in *Globigerina* with numerous, large, supplementary apertures around the margin of the chamber, opening into the chamber, and some of them into the umbilical area, surface in well-preserved specimens clothed with fine spines. — Tertiary, Recent.

Genus GLOBIGERINATELLA Cushman and Stainforth

Key, plate 53, figures 18–20

Genotype, *Globigerinatella insueta* Cushman and Stainforth

Globigerinatella Cushman and Stainforth, Special Publ. 14, Cushman Lab. Foram. Res., 1945, p. 68.

Test trochoid, adult generally spherical in shape, in the early adult stages with rounded supplementary chambers developed along the sutural lines, and later with elongate chambers covering the earlier sutures and with numerous, low, rounded openings on both sides along the margins. — Tertiary.

Genus GLOBIGERINELLA Cushman, 1927

Plate 27, figure 7; Key, plate 34, figures 10–12

Genotype, *Globigerina aequilateralis* H. B. Brady

Globigerinoides Cushman, Contr. Cushman Lab. Foram. Res., vol. 3, 1927, p. 87.
Globigerina (part) of authors.

Test trochoid in the young, at least in the microspheric form, later planispiral; aperture single, large, opening into the umbilicus in the young, in the adult median; fine spines covering the test in well-preserved specimens. — Cretaceous to Recent.

Genus GLOBIGERINELLOIDES Cushman and ten Dam, 1948

Genotype, *Globigerinelloides algeriana* Cushman and ten Dam

Globigerinelloides Cushman and ten Dam, Contr. Cushman Lab. Foram. Res., vol. 24, 1948, p. 42.

Test planispiral, in the early stages similar to *Globigerinella*, in the adult becoming somewhat loosely coiled, the earlier coils all visible in side view; chambers globular in the earlier stages, in the adult with lateral prolongations reaching to and slightly over the preceding coil; wall calcareous, perforate; aperture at the base of the last-formed chamber in the median line. — Upper Cretaceous.

Genus HASTIGERINA Wyville Thomson, 1876

Plate 27, figure 8; Key, plate 34, figure 13

Genotype, *Hastigerina murrayi* Wyville Thomson = *Nonionina pelagica* d'Orbigny, 1839

Hastigerina Wyville Thomson, Proc. Roy. Soc. London, vol. 24, 1876, p. 534.
Nonionina (part) of authors. *Lituola* (part) Jones and Parker, 1860 (not Lamarck).
Globigerina (part) Parker and Jones, 1865 (not d'Orbigny).

Test with the early chambers trochoid, later ones planispiral, involute; wall calcareous, perforate, with comparatively coarse spines, flattened, the edges parallel and toothed, each spine on a base projecting from the surface; aperture large, at the umbilical margin of the chamber. — Miocene to Recent.

Genus HASTIGERINELLA Cushman, 1927

Plate 27, figure 9; Key, plate 34, figures 14–16
Genotype, *Hastigerina digitata* Rhumbler

Hastigerinella Cushman, Contr. Cushman Lab. Foram. Res., vol. 3, 1927, p. 87.
Hastigerina (part) Rhumbler, 1911.

Test similar to *Hastigerina* in the young, in the adult the chambers elongate, club-shaped, the spines limited to the outer ends of the chambers. — Upper Cretaceous to Recent.

Subfamily 2. Orbulininae

Test in the early stages trochoid like *Globigerina*, later developing a globular chamber entirely enclosing the earlier ones which may be later resorbed; wall often of several layers with perforations of various sizes, occasionally large openings which are apparently accidental; exterior with fine spines.

PLATE 27

Relationships of the Genera of the Globigerinidae and Globorotaliidae

Globigerinidae

FIG.

1. *Globigerina bulloides* d'Orbigny. (After H. B. Brady.) *a*, dorsal view; *b*, ventral view; *c*, side view.
2. *Sphaeroidinella dehiscens* (Parker and Jones). (After Parker and Jones.) *a*, dorsal view; *b*, ventral view.
3. *Candeina nitida* d'Orbigny. (After H. B. Brady.) *a*, dorsal view; *b*, side view.
4. *Orbulina universa* d'Orbigny. (After H. B. Brady.)
5. *Pulleniatina obliquiloculata* (Parker and Jones.) (After H. B. Brady.) *a*, side view; *b*, apertural view.
6. *Globigerinoides rubra* (d'Orbigny). (After H. B. Brady.) *a*, dorsal view; *b*, side view; *c*, ventral view.
7. *Globigerinella aequilateralis* (H. B. Brady). (After H. B. Brady.) *a*, side view; *b*, apertural view.
8. *Hastigerina pelagica* (d'Orbigny). (After H. B. Brady.)
9. *Hastigerinella digitata* (Rhumbler). (After Rhumbler — but spines not complete.)

Globorotaliidae

10. *Cycloloculina annulata* Heron-Allen and Earland. (After Heron-Allen and Earland.) *a*, adult; *b*, young.
11. *Globotruncana arca* (Cushman). *a*, dorsal view; *b*, ventral view; *c*, edge view.
12. *Globorotalia tumida* (H. B. Brady). (After H. B. Brady.) *a*, dorsal view; *b*, ventral view; *c*, edge view.

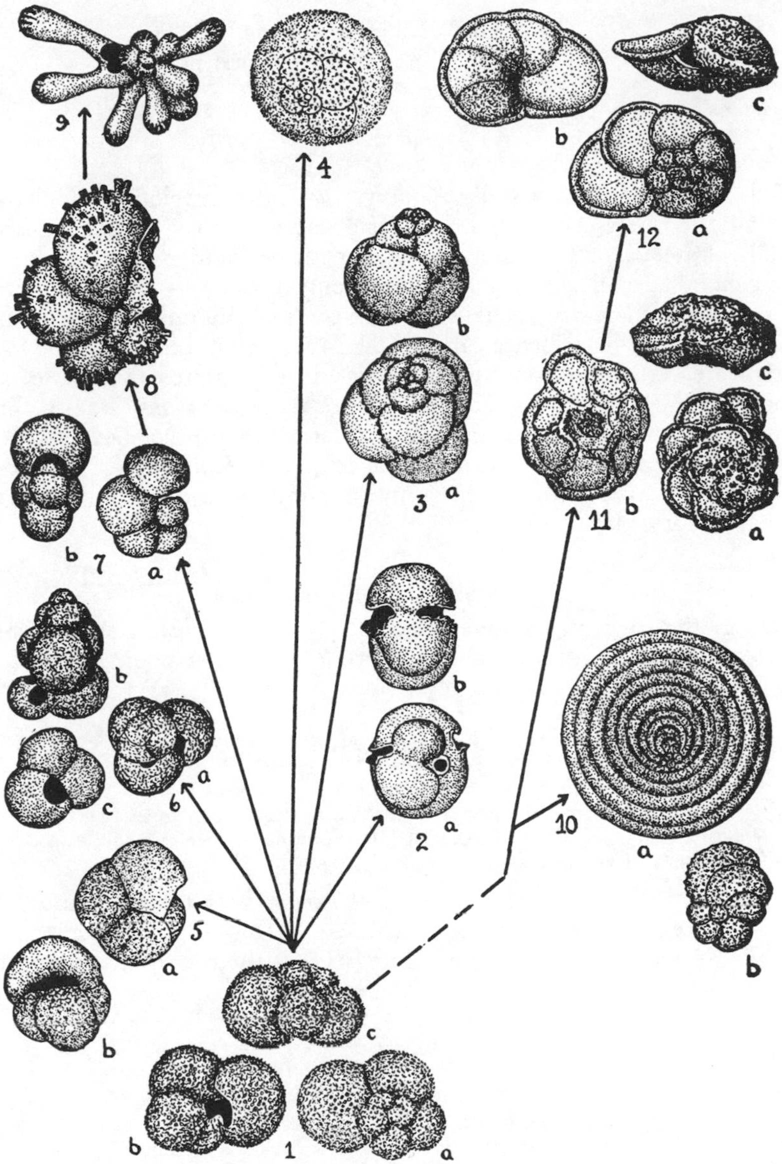

Plate 27

Genus ORBULINA d'Orbigny, 1839

Plate 27, figure 4; Key, plate 34, figures 17, 18
Genotype, *Orbulina universa* d'Orbigny

Orbulina D'ORBIGNY, in DE LA SAGRA, Hist. Phys. Pol. Nat. Cuba, 1839, "Foraminifères," p. 3.
Globigerina (part) of authors.

Test in the early stages like *Globigerina*, later developing a globular chamber entirely enclosing the earlier ones which may be later resorbed; wall calcareous, often of several layers, with perforations of various sizes, no general aperture, exterior with fine elongate spines. — Tertiary, Recent.

Although there are references to "*Orbulina*" in the literature which would indicate its presence early in the fossil series, those from the Cambrian are certainly erroneous, and it is to be suspected that those from the formations before the Tertiary are not truly *Orbulina*. I am inclined to agree with Earland that the so-called larger aperture of *Orbulina* is an accidental opening and not a true aperture. *Orbulina* is an end form, and represents probably the complete attainment of a spherical test adapted for pelagic life.

Subfamily 3. Pulleniatininae

Test in the early stages trochoid and like *Globigerina*, later becoming involute and the later chambers covering the earlier ones, test without spines in the adult; wall coarsely porous.

Genus PULLENIATINA Cushman, 1927

Plate 27, figure 5; Key, plate 35, figures 1–3
Genotype, *Pullenia obliquiloculata* Parker and Jones

Pulleniatina CUSHMAN, Contr. Cushman Lab. Foram. Res., vol. 3, 1927, p. 90.
Pullenia (part) of authors.

Test with the early chambers as in *Globigerina*, with the wall cancellated and apparently with spines, later coarsely perforate but smooth except about the aperture; chambers involute, the last three or four forming the outer surface of the test, without spines in the adult; aperture elongate, arched, at the base of the chamber. — Late Tertiary, Recent.

Genus SPHAEROIDINELLA Cushman, 1927

Plate 27, figure 2; Key, plate 35, figures 4–7
Genotype, *Sphaeroidina dehiscens* Parker and Jones

Sphaeroidinella CUSHMAN, Contr. Cushman Lab. Foram. Res., vol. 3, 1927, p. 90.
Sphaeroidina (part) of authors. *Globigerina* (part) SCHWAGER, 1866 (not D'ORBIGNY).

Test in the early stages trochoid and like *Globigerina* with coarsely cancellated surface and with spines, in the adult with the chambers embracing, two or three forming the exterior; the chambers slightly separated, the edges with somewhat crenulated carinae and without spines; aperture in the deep cavity between the chambers.—Miocene to Recent.

This genus has no real relationships to the Pegidiidae which are derived directly from the Rotaliidae, probably from *Eponides.* The young of *Sphaeroidinella* is easily mistaken for a coarsely cancellated *Globigerina.*

Subfamily 4. Candeininae

Test trochoid, the young with the chambers roughened, spinose, and a single aperture as in *Globigerina,* in the adult the chambers smooth, without spines, and the apertures formed by rows of circular or elliptical openings along the sutures.

Genus CANDEINA d'Orbigny, 1839

Plate 27, figure 3; Key, plate 35, figure 8
Genotype, *Candeina nitida* d'Orbigny

Candeina d'Orbigny, in De la Sagra, Hist. Phys. Pol. Nat. Cuba, 1839, "Foraminifères," p. 107.

Test trochoid, in the young with the chambers somewhat roughened and spinose, and with the aperture as in *Globigerina,* in the adult the chambers smooth, without spines, and the apertures consisting of rows of circular openings along the sutures.—Pliocene to Recent.

Genus CANDORBULINA Jedlitschka, 1933

Key, plate 53, figures 21–23
Genotype, *Candorbulina universa* Jedlitschka

Candorbulina Jedlitschka, Verhandl. Nat. Ver. Brünn, Jahrg. 65, 1934, p. 20.

Test in the early stages similar to *Candeina* with small apertures along the sutures, later developing a final chamber entirely enveloping the earlier ones as in *Orbulina,* with the apertures in a circle.—Miocene.

See Jedlitschka, Ueber Candorbulina, eine neue Foraminiferen-Gattung, und zwei neue Candeina-Arten (Verhandl. Nat. Ver. Brünn, Jahrg. 65, 1934, pp. 17–26, text figs. 1–23).

Family 41. HANTKENINIDAE

Test in the early stages trochoid or planispiral, in later stages usually planispiral and involute; chambers developing one or more large hollow spines from the periphery; wall calcareous, perforate; aperture at the base of the chamber, either a simple arch or with a large opening in the apertural face with lateral lobes toward the base, or a series of pores in the apertural face.

This family now includes three genera, *Schackoina* known from the

Upper Cretaceous, and *Hantkenina* and *Cribohantkenina* from the Eocene. Further studies of these forms seem to show that they were derived from the Globigerinidae and were probably pelagic, at least during part of their life history.

The early stages of *Schackoina* are apparently trochoid, although there is a tendency for it to become planispiral and involute later. In *Hantkenina* the test is mostly planispiral throughout, but in section the microspheric form seems to be somewhat trochoid. Some species of *Hantkenina* in the earliest stages very closely resemble *Globigerinella* in form and surface characters. It may be that the Upper Cretaceous and Eocene forms are not as closely related as seems from their external characters.

In the later Tertiary a recently discovered form may also belong to this family, but its description must await further material.

There is no indication that these genera are related to the Nonionidae.

Genus SCHACKOINA Thalmann, 1932

Key, plate 35, figures 9, 10

Genotype, *Siderolina cenomana* Schacko

Schackoina THALMANN, Eclogae geol. Helv., vol. 29, 1932, p. 289.

Siderolina (part) SCHACKO (not DEFRANCE). *Hantkenina* (part) of authors.

Test in earliest stages slightly trochoid, later nearly planispiral and involute; chambers with a single, peripheral, tubular spine in the young, or several in the adult; wall calcareous, finely perforate; aperture a low arch, at the chamber base. — Cretaceous.

Genus HANTKENINA Cushman, 1924

Key, plate 35, figures 11, 12

Genotype, *Hantkenina alabamensis* Cushman

Hantkenina CUSHMAN, Proc. U. S. Nat. Mus., vol. 66, Art. 30, 1924, p. 1.

Siderolina HANTKEN (not DEFRANCE). *Nonionina* (part) of authors.

Sporohantkenina (part) BERMUDEZ, 1937.

Test planispiral throughout, close coiled; chambers distinct, usually involute, each with an acicular spine at the anterior angle or becoming obsolete in the adult; wall calcareous, perforate; aperture median, at the base of the chamber, typically arched with a basal lobe at either side. — Middle and Upper Eocene, lowest Oligocene.

See Hans E. Thalmann, Die Foraminiferen Gattung *Hantkenina* Cushman, 1924, und ihre regional-stratigraphische Verbreitung (Eclogae geol. Helv., vol. 25, 1932, pp. 287–292); *Hantkenina* in the Eocene of East Borneo (Stanford Univ. Publ., Univ. Ser., Geol. Sci., vol. 3, No. 1, 1942, pp. 1–24, text figs. 1, 2 [maps]); and Foraminiferal Genus *Hantkenina* and its Subgenera (Amer. Journ. Sci., vol. 240, 1942, pp. 809–820, pl. 1, tables 1, 2).

Genus CRIBROHANTKENINA Thalmann, 1942

Key, plate 54, figures 1, 2

Genotype, *Hantkenina* (*Sporohantkenina*) *brevispina* Bermudez (not *H. brevispina* Cushman) = *Cribrohantkenina bermudezi* Thalmann

Cribrohantkenina THALMANN, Amer. Journ. Sci., vol. 240, 1942, p. 812.

Hantkenina (part) and *Sporohantkenina* (part) of authors.

Test similar to *Hantkenina* but with the apertural face having an area of rounded supplementary apertures. — Eocene.

Family 42. GLOBOROTALIIDAE

Test in the early stages trochoid; the chambers with a rough cancellated exterior and often spinose, in the adult resuming the ancestral rotalid form or becoming annular, but often retaining the rough spinose surface; aperture typically opening into the umbilical area, the older species often retaining the protecting covering above the umbilical area, and traces of it appear in the living forms; largely pelagic.

This family is directly derived from the Globigerinidae, and some of the genera of the two families have been associated since the Cretaceous. Both are in a large measure pelagic. *Globotruncana* is directly derived from the compressed Globigerinas in the Cretaceous. *Globorotalia* is directly derived from *Globotruncana* by the greater development of the upper keel and the loss of the lower one. All intermediate stages in this process may be seen. *Globorotalia* is a very specialized genus, and has no relationships with *Endothyra* whatever, as has sometimes been claimed. In *Cycloloculina* and *Sherbornina* are developed annular forms apparently closely allied in their early stages to the Globorotalias.

Genus GLOBOTRUNCANA Cushman, 1927

Plate 27, figure 11; Key, plate 35, figures 14, 17

Genotype, *Pulvinulina arca* Cushman

Globotruncana CUSHMAN, Contr. Cushman Lab. Foram. Res., vol. 3, 1927, p. 91.

Rosalinella MARIE, 1941.

Rosalina (part), *Discorbina* (part), *Globigerina* (part), *Rotalia* (part), and *Pulvinulina* (part) of authors.

Test trochoid in the young; chambers usually globose, rough and cancellated, adult usually much compressed, dorsal and ventral sides either flat or convex, ventral side sometimes slightly concave, periphery truncate, usually with a double keel on dorsal and ventral sides; aperture ventral, in well-preserved specimens with a thin plate-like structure over the umbilical area; apparently pelagic in part. — Upper Cretaceous, Tertiary (?).

See Bolli, Zur Stratigraphie der Oberen Kreide in den höheren helvetischen Decken (Eclog. Geol. Helvet., vol. 37, No. 2, 1944 (1945), pp. 218–328, pl. 9, text figs. 1–6).

Genus GLOBOROTALIA Cushman, 1927

Plate 27, figure 12; Key, plate 35, figures 15, 16
Genotype, *Pulvinulina menardii*, var. *tumida* H. B. Brady
Globorotalia Cushman, Contr. Cushman Lab. Foram. Res., vol. 3, 1927, p. 91.
Rotalia (part), *Rotalina* (part), *Planulina* (part), and *Pulvinulina* (part) of authors.

Test trochoid; earliest chambers often like *Globigerina*, with a rough cancellated exterior, biconvex, dorsal side more or less flattened, ventral side strongly convex; wall calcareous, perforate, frequently spinose in whole or in restricted areas; aperture large, opening into the umbilicus which is either open or partially covered by a lip. — Upper Cretaceous to Recent.

This genus is directly derived from *Globotruncana* by the suppression of one of the keels. It is largely pelagic, a specialized genus associated with the Globigerinidae.

Genus GLOBOROTALITES Brotzen, 1942

Key, plate 35, figure 13
Genotype, *Globorotalia multisepta* Brotzen
Globorotalites Brotzen, Sver. Geol. Under., ser. C, No. 451, 1942, p. 31.
Globorotalia (part) of authors.

Test trochoid, dorsal side flattened or slightly convex, ventral side conical, umbilicate, periphery acute or keeled; chambers all visible on dorsal side, only the last whorl on the ventral side; wall calcareous, perforate; aperture ventral, an elongate opening at the inner margin of the last-formed chamber, extending to the umbilicus. — Cretaceous.

Genus ROTALIPORA Brotzen, 1942

Key, plate 54, figure 3
Genotype, *Rotalipora turonica* Brotzen
Rotalipora Brotzen, Sver. Geol. Under, ser. C, No. 451, 1942, p. 32.

Test similar to *Globorotalia* but with one aperture near the umbilical end of the ventral face at the outer margin, the other rounded, on the sutural side with a slightly projecting neck and thickened lip. — Cretaceous.

Genus CRIBROGLOBOROTALIA Cushman and Bermudez, 1936

Key, plate 54, figures 6–8
Genotype, *Cribrogloborotalia marielina* Cushman and Bermudez
Cribrogloborotalia Cushman and Bermudez, Contr. Cushman Lab. Foram. Res., vol. 12, 1936, p. 63.

Test trochoid, all the chambers visible from the dorsal side, only those of the last-formed whorl from the ventral side; wall calcareous, finely perforate; apertures numerous, forming a cribrate plate over the inner portion of the ventral face of the last-formed chamber. — Eocene.

Genus CYCLOLOCULINA Heron-Allen and Earland, 1908

Plate 27, figure 10; Key, plate 35, figures 18–22
Genotype, *Cycloloculina annulata* Heron-Allen and Earland
Cycloloculina Heron-Allen and Earland, Journ. Roy. Micr. Soc., 1908, p. 533.

Test with the early chambers in a low trochoid spire; chambers globular, then becoming elongate, periphery somewhat spinose, with short conical spines, later chambers still more elongate, finally becoming annular; wall calcareous, coarsely perforate; no general aperture, the large coarse perforations serving as apertures. — Tertiary.

Genus SHERBORNINA Chapman, 1922

Key, plate 35, figures 23, 24
Genotype, *Sherbornina atkinsoni* Chapman
Sherbornina Chapman, Journ. Linn. Soc. Zool., vol. 34, 1922, p. 501.

"Test discoidal, moderately thin, median arch concave. Shell built up of a median annular series of chamberlets with a discorbine commencement; the loculi of the annuli widely spaced. External layer formed of small, overlapping, spatulate chamberlets. The primordial series of about 7 globular to reniform segments, lying in the median system, is discorbine — that is, depressed rotaline. Shell-wall perforated with coarse tubuli." (Chapman). — Miocene, Tasmania and Australia.

Family 43. ANOMALINIDAE

Test free, or attached by the dorsal surface which is typically flattened or concave; chambers arranged in a trochoid manner, at least in the early stages, only those of the last-formed chamber visible from the ventral side; wall calcareous, coarsely perforate; aperture in the adult either at the periphery with an extension on the dorsal side or represented by numerous pores at the outer margin, sometimes with tubular necks.

From the fact that the young in this family is trochoid, it has undoubtedly been derived from the Rotaliidae through such forms as *Discorbis*. The fact that in the early stages at least there is a thin, inner, chitinous layer, and in a few genera traces of an arenaceous outer layer as well, seems to indicate that the group came from those forms of the Rotaliidae already mentioned which have these characters. In none of the group is there an early stage with an undivided coil. The members of the family as a rule are attached or become free. This attachment is by the dorsal side which becomes flattened. From the close coiled forms which are more or less involute as in *Anomalina*, there are derived evolute forms as in

Planulina where on both sides the early chambers are visible. In *Anomalinella* the later development becomes planispiral, and there is an additional aperture at the peripheral angle. In *Ruttenia* developed in the Eocene, the peripheries of the chambers become raised into a peculiar angled form. In *Laticarinina* the early stage is trochoid and related to *Planulina*. The aperture is at one side along the dorsal margin of the chamber as in these other two genera, and a wide keel is developed about the periphery. *Laticarinina*, originally placed by its authors in the Lagenidae and later in the Nonionidae, has only a superficial resemblance to those families.

From the trochoid, attached *Cibicides*, with its aperture largely on the dorsal side, are developed several different modifications. *Dyocibicides* in adult becomes biserial. *Cibicidella* has the later chambers irregularly spreading, each with a definite large opening with a lip but very irregularly arranged. *Cyclocibicides* has irregularly annular chambers in the adult with the apertures peripheral but without tubular necks, while *Annulocibicides* has a somewhat similar general form, but the periphery has numerous apertures with short tubular necks sometimes with a slight lip. In *Rectocibicides* the early portion is similar to *Cibicides*, but the later development is along one axis, and the periphery has tubular apertures. *Stichocibicides* also becomes rectilinear in the adult and *Palmerinella* becomes evolute and planispiral. *Webbina* is a somewhat degenerate form, attached, the later chambers forming a chain-like growth with the aperture tubular, and a slight neck at the end of the last-formed chamber.

This group is an interesting one and very closely knit together although the later development has assumed various diverse forms. While these are derived from the Rotaliidae, they are very distinct from the members of that family.

Subfamily 1. Anomalininae

Test compressed, nearly symmetrical on the two sides in the adult; aperture peripheral.

Genus ANOMALINA d'Orbigny, 1826

Plate 28, figure 1; Key, plate 36, figures 1, 2
Genotype, *Anomalina punctulata* d'Orbigny

Anomalina D'ORBIGNY, Ann. Sci. Nat., vol. 7, 1826, p. 282.
Aspidospira and *Porospira* EHRENBERG, 1844. *Rosalina* (part), *Rotalia* (part), *Discorbina* (part), *Planorbulina* (part), and *Truncatulina* (part) of authors.

Test in the young trochoid, adult often nearly involute dorsally as well as ventrally; chambers added nearly in a planispiral manner, the inner coils of the dorsal side often appearing as a central raised boss; wall calcareous, perforate; aperture in the young ventral, in the adult becoming peripheral, at the base of the last-formed chamber in the median line, sometimes with a boss of clear material over the umbilical region.—Jurassic (?), Lower Cretaceous to Recent.

Genus PLANOMALINA Loeblich and Tappan, 1946

Key, plate 54, figures 4, 5
Genotype, *Planomalina apsidostroba* Loeblich and Tappan
Planomalina LOEBLICH and TAPPAN, Journ. Pal., vol. 20, 1946, p. 257.

Test free, planispiral, partially evolute, so that earlier coils may be seen on both sides of the biumbilicate test; chambers numerous; sutures may be limbate; wall calcareous, coarsely perforate; aperture peripheral, an arch at the base of the apertural face of the last-formed chamber, with a distinct lip. — Lower Cretaceous.

Genus ANOMALINOIDES Brotzen, 1942

Key, plate 54, figure 9
Genotype, *Anomalinoides plummerae* Brotzen = *Anomalina grosserugosa* Plummer (not Gümbel)
Anomalinoides BROTZEN, Sver. Geol. Under., ser. C, No. 451, 1942, p. 23.

Test differing from *Anomalina* in the aperture which extends over onto the dorsal side along the inner margin of the chamber, and from *Cibicides* by its nearly bilaterally symmetrical test. — Cretaceous to Recent.

Genus PALMERINELLA Bermudez, 1934

Key, plate 54, figure 11
Genotype, *Palmerinella palmerae* Bermudez
Palmerinella BERMUDEZ, Mem. Soc. Cubana Hist. Nat., vol. 8, 1934, p. 83.

Test trochoid, compressed, becoming evolute in the adult and planispiral; chambers numerous in the early stages all visible from the dorsal side, only those of the last-formed whorl from the ventral, later all visible from both sides; wall calcareous, distinctly perforate; aperture in the adult elongate, narrow, extending nearly the entire length of the apertural face, with a distinct lip. — Miocene to Recent.

Genus BOLDIA van Bellen, 1946

Key, plate 54, figure 10
Genotype, *Rotalina lobata* Terquem
Boldia VAN BELLEN, Contr. Cushman Lab. Foram. Res., vol. 22, 1946, p. 122.
Terquemia VAN BELLEN, 1946 (not TATE).
Rotalina (part) of authors.

Test free, planoconcave or biconcave, spiral; all whorls visible on the dorsal side, ventral side only somewhat involute, dorsal side mostly flat, ventral side concave, periphery broadly truncate; wall calcareous, perforate; aperture ventral, extending onto the periphery. — Eocene.

Genus RUTTENIA Pijpers, 1933

Plate 28, figure 2; Key, plate 36, figures 5–7
Genotype, *Bonairia coronaeformis* Pijpers

Ruttenia PIJPERS, Contr. Cushman Lab. Foram. Res., vol. 9, 1933, p. 30.
Bonairia PIJPERS, 1933 (not BURRINGTON BAKER).

Test compressed, low, trochoid, dorsal side flat or concave, ventral side involute, somewhat convex, umbilicate; chambers distinct, peripheral portion angled and rising above the early coils; wall calcareous, finely perforate; aperture peripheral. — Eocene.

Genus PLANULINA d'Orbigny, 1826

Plate 28, figure 3; Key, plate 36, figures 3, 4
Genotype, *Planulina ariminensis* d'Orbigny

Planulina D'ORBIGNY, Ann. Sci. Nat., vol. 7, 1826, p. 280.
Anomalina (part) and *Truncatulina* (part) of authors.

Test in the young trochoid, adult much compressed, evolute; earlier chambers visible from both sides in the megalospheric form, microspheric form with the central area raised on the dorsal side; wall calcareous, coarsely perforate; aperture at the base of the chamber at the median line. — Cretaceous to Recent.

Genus KELYPHISTOMA Keijzer, 1945

Key, plate 54, figure 15
Genotype, *Kelyphistoma ampulloloculata* Keijzer

Kelyphistoma KEIJZER, Outline geol. eastern part Prov. Oriente, Cuba, Utrecht, 1945, p. 207.

Test almost planispiral, only very slightly trochoid and much compressed; wall calcareous, perforate; apertures of two kinds, one small, at base of last-formed chamber, with a distinct lip, the other elongate, in the plane of coiling at the peripheral margin of the last whorl. — Tertiary.

Genus LATICARININA Galloway and Wissler, 1927

Plate 28, figure 4; Key, plate 36, figure 9
Genotype, *Pulvinulina repanda*, var. *menardii*, subvar. *pauperata* Parker and Jones

Laticarinina GALLOWAY and WISSLER, Journ. Pal., vol. 1, No. 3, 1927, p. 193.
Pulvinulina (part) of authors. *Pellatispira* CUSHMAN, 1927 (not BOUSSAC). *Carinina* GALLOWAY and WISSLER, 1927 (not HUBRECHT).

Test typically plano-convex, the dorsal side flattened, ventral side convex, in the early stages especially of the microspheric form trochoid, the aperture on the periphery or even on the ventral side as in *Cibicides*, later on the margin of the dorsal side, a low elongate opening at the base of the chamber, a wide flange of clear material separating the later coils and forming a carina about the periphery. — Eocene to Recent.

The early stages of this genus are trochoid, not planispiral.

See Cushman and Todd, The Recent and Fossil Species of *Laticarinina* (Contr. Cushman Lab. Foram. Res., vol. 18, 1942, pp. 14–20, pl. 4).

Genus ANOMALINELLA Cushman, 1927

Plate 28, figure 6; Key, plate 36, figure 8
Genotype, *Truncatulina rostrata* H. B. Brady

Anomalinella Cushman, Contr. Cushman Lab. Foram. Res., vol. 3, 1927, p. 93.
Truncatulina (part) of authors.

Test in the early stages trochoid, adult nearly planispiral; chambers almost entirely involute; wall calcareous, coarsely perforate; aperture ventral, between the periphery and the umbilical area with a supplementary aperture just below the peripheral margin, elongate and parallel to the axis of coiling. — Miocene to Recent.

Subfamily 2. Cibicidinae

Test with the dorsal side flattened or concave, the aperture extending over onto the dorsal side along the inner margin of the chamber or entirely on the dorsal side, test typically attached by the dorsal side.

Genus CIBICIDES Montfort, 1808

Plate 28, figure 5; Key, plate 36, figures 10, 11
Genotype, *Cibicides refulgens* Montfort

Cibicides Montfort, Conch. Syst., vol. 1, 1808, p. 123.
Storilus and *Polyxenes* Montfort (?), 1808. *Nautilus* (part) of authors. *Truncatulina* d'Orbigny, 1826. *Lobatula* Fleming, 1828. *Rosalina* and *Rotalina* (part) of authors. *Aristeropora* Ehrenberg, 1858. *Heterolepa* Franzenau, 1884. *Pseudotruncatulina* Andreae, 1884.

Test plano-convex, trochoid, usually attached by the flattened dorsal side; wall calcareous, coarsely perforate; aperture peripheral, at the base of the chamber, sometimes extending ventrally, but typically with a long slit-like extension dorsally between the inner margin of the chamber and the previous whorl, nearly the length of the chamber. — Jurassic (?), Cretaceous to Recent.

Adherentina Spandel is not completely described, but may belong here.

Somewhat indefinite forms have been called *Karreria* by Rzehak, and these may be close to *Cibicides* or may be specimens somewhat deformed by attachment.

Genus CIBICIDOIDES Brotzen, 1936

Key, plate 54, figure 16
Genotype, *Cibicidoides eriksdalensis* Brotzen

Cibicidoides Brotzen, Sver. Geol. Under., ser. C, No. 396, 1936, p. 186.

Test trochoid, biconvex, both dorsal and ventral sides nearly involute, aperture near the peripheral margin of the last-formed chamber and extending along the inner border of the chamber on the dorsal side. — Cretaceous to Recent.

Genus BUNINGIA Finlay, 1939

Key, plate 54, figures 12–14
Genotype, *Buningia creeki* Finlay
Buningia Finlay, Trans. Roy. Soc. New Zealand, vol. 69, 1939, p. 122.

Test trochoid in the early stages, nearly planispiral in the adult, deeply umbilicate; chambers globose; wall calcareous, perforate; aperture in the adult entirely dorsal, a horizontal opening at the side of the umbilicus. — Tertiary.

Genus STICHOCIBICIDES Cushman and Bermudez, 1936

Key, plate 54, figure 20
Genotype, *Stichocibicides cubensis* Cushman and Bermudez
Stichocibicides Cushman and Bermudez, Contr. Cushman Lab. Foram. Res., vol. 12, 1936, p. 33.

Test attached by the dorsal side, in the early stages close coiled, trochoid, the dorsal side flattened, ventral side convex, later followed by a series of uniserial chambers; wall calcareous, finely perforate; aperture, a generally rounded or elliptical opening on the middle of the ventral side of the last-formed chamber, near the periphery, without a distinct lip. — Lower Cretaceous to Eocene.

PLATE 28

Relationships of the Genera of the Anomalinidae

FIG.
1. *Anomalina punctulata* d'Orbigny. (After d'Orbigny.) *a*, dorsal view; *b*, ventral view; *c*, side view.
2. *Ruttenia coronaeformis* (Pijpers). (After Pijpers.) *a*, dorsal view; *b*, ventral view; *c*, side view.
3. *Planulina ariminensis* d'Orbigny. (After d'Orbigny.) *a*, *b*, opposite sides; *c*, side view.
4. *Laticarinina pauperata* (Parker and Jones). (After H. B. Brady.) *a*, side view; *b*, apertural view.
5. *Cibicides lobatulus* (Walker and Jacob). (After H. B. Brady.) *a*, dorsal view; *b*, ventral view; *c*, side view.
6. *Anomalinella rostrata* (H. B. Brady). (After H. B. Brady.) *a*, side view; *b*, apertural view.
7. *Dyocibicides biserialis* Cushman and Valentine. (After Cushman and Valentine.)
8. *Annulocibicides projectus* Cushman and Ponton. (After Cushman and Ponton.)
9. *Rectocibicides miocenicus* Cushman and Ponton. (After Cushman and Ponton.)
10. *Cibicidella variabilis* (H. B. Brady). (After H. B. Brady.)
11. *Cyclocibicides vermiculatus* (d'Orbigny). (After H. B. Brady.)
12. *Webbina rugosa* d'Orbigny. (After d'Orbigny.) *a*, from outer side; *b*, from edge.

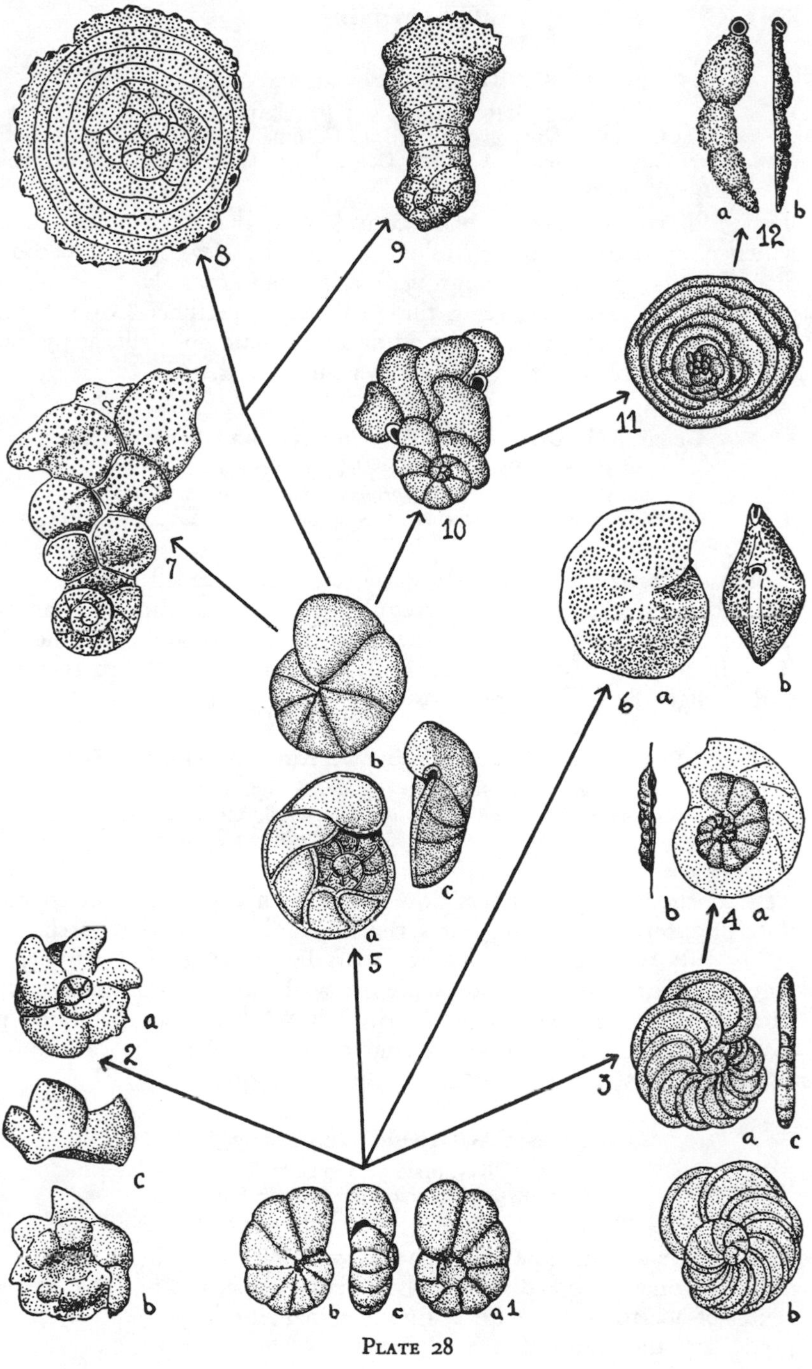

PLATE 28

Genus DYOCIBICIDES Cushman and Valentine, 1930

Plate 28, figure 7; Key, plate 36, figure 12

Genotype, *Dyocibicides biserialis* Cushman and Valentine

Dyocibicides Cushman and Valentine, Contr. Dept. Geol. Stanford Univ., vol. 1, No. 1, 1930, p. 30.

Test in the young, trochoid, plano-convex, ventral side convex, dorsal side flattened, close coiled; the chambers in the later development becoming biserial and rapidly enlarging; wall calcareous, coarsely perforate, test probably attached; aperture in the early stages peripheral or extending slightly to the dorsal side, in the adult an elongate open slit, at the outer end of the chamber, with a lip. — Eocene to Recent.

Genus ANNULOCIBICIDES Cushman and Ponton, 1932

Plate 28, figure 8; Key, plate 36, figures 13, 14

Genotype, *Annulocibicides projectus* Cushman and Ponton

Annulocibicides Cushman and Ponton, Contr. Cushman Lab. Foram. Res., vol. 8, 1932, p. 1.

Test attached by the dorsal side, early stages close coiled similar to *Cibicides*, later chambers becoming irregular, those of the adult annular; wall calcareous, distinctly perforate; apertures in early stages similar to *Cibicides*, in adult consisting of numerous, short, peripheral tubes, sometimes with a slight lip. — Miocene, Florida.

Genus RECTOCIBICIDES Cushman and Ponton, 1932

Plate 28, figure 9; Key, plate 36, figures 15–17

Genotype, *Rectocibicides miocenicus* Cushman and Ponton

Rectocibicides Cushman and Ponton, Contr. Cushman Lab. Foram. Res., vol. 8, 1932, p. 2.

Test attached; early chambers close coiled as in *Cibicides*, later chambers uncoiling, forming a rectilinear series, expanding somewhat as chambers are added; wall calcareous, rather coarsely perforate, especially on the outer side, inner attached side with the wall thin and transparent; apertures in early stages similar to *Cibicides*, in adult a series of tubular projections along the outer growing edge of the chamber, open at the outer end, sometimes with a slight lip. — Miocene, Florida.

Genus VAGOCIBICIDES Finlay, 1939

Key, plate 54, figure 22

Genotype, *Vagocibicides maoria* Finlay

Vagocibicides Finlay, Trans. Roy. Soc. New Zealand, vol. 69, 1939, p. 326.

Early stages similar to *Cibicides*, followed by several biserial chambers which become staggered and finally uniserial, aperture in early stages in center of apertural face with a slightly raised rim, later rounded, in a depression of the apertural face. — Tertiary.

Genus CYCLOCIBICIDES Cushman, 1927

Plate 28, figure 11; Key, plate 36, figures 18, 19
Genotype, *Planorbulina vermiculata* d'Orbigny

Cyclocibicides Cushman, Contr. Cushman Lab. Foram. Res., vol. 3, 1927, p. 93.
Planorbulina (part) of authors. *Rotalia* Jones and Parker, 1860 (not Lamarck).
Pulvinulina (part) of authors. *Planopulvinulina* (part) Schubert.

Test attached, early stages similar to *Cibicides*; chambers elongating in later growth and becoming nearly or completely annular; wall calcareous, coarsely perforate; aperture in the early stages as in *Cibicides*, in the adult formed by the numerous large pores scattered over the surface. — Recent.

Genus CIBICIDELLA Cushman, 1927

Plate 28, figure 10; Key, plate 36, figures 20, 21
Genotype, *Truncatulina variabilis* d'Orbigny

Cibicidella Cushman, Contr. Cushman Lab. Foram. Res., vol. 3, 1927, p. 93.
Truncatulina (part) of authors. *Planorbulina* (part) Egger, 1857 (not d'Orbigny).

Test attached by the flattened dorsal side, early stages similar to *Cibicides*, later chambers irregularly disposed; wall calcareous, coarsely perforate; aperture in the early stages as in *Cibicides*, in the adult rounded, on the dorsal side, with a short neck and distinct lip. — Tertiary, Recent.

Genus WEBBINA d'Orbigny, 1839

Plate 28, figure 12; Key, plate 36, figure 22
Genotype, *Webbina rugosa* d'Orbigny

Webbina d'Orbigny, in Barker-Webb and Berthelot, Hist. Nat. Iles Canaries, 1839, vol. 2, pt. 2, "Foraminifères," p. 125.
Webbum Rhumbler, 1913.

Test attached, consisting of a few chambers, with a neck and circular aperture with a slight lip, the wall calcareous and perforate. — Tertiary, Recent.

Family 44. PLANORBULINIDAE

Test in the early stages coiled, attached by the dorsal surface; chambers at least in the early stages in a spiral arrangement, later spreading in annular series, irregular, or piled up into a solid mass; wall calcareous, usually coarsely perforate, in the early stages with a thin chitinous inner layer; apertures one or two to each chamber, peripheral, or in the massive forms with the apertures formed by the coarse pores of the wall.

This family is derived directly from the Anomalinidae, and the early stages have the same chitinous inner layer. The early stages are usually attached, but the individuals may later become free. The simplest form is *Planorbulina*, in which the chambers are usually in one plane, and the spiral development is continued through the long period. These may become irregularly spreading as in *Planorbulinoides*, where the later chambers often form a network, the apertures of the later chambers tubular similar to some of the genera of the Anomalinidae, but the early stages are definitely like *Planorbulina*. In *Planorbulinella* the chambers become definitely developed in annular series, alternating as they are added. *Linderina* has a secondary development of calcareous material on the central portion of both the dorsal and ventral sides somewhat simulating the structure seen in the orbitoids. In *Acervulina* the early stages are very similar to *Planorbulina*, but the later chambers spread over the attached surface and pile up somewhat upon one another. In *Gypsina* the test which is usually attached in the early stages becomes free, and the chambers pile up irregularly, forming irregular or even definitely spherical masses.

The greatest development of this family is in the Tertiary and Recent oceans, although some of the genera are found in the Upper Cretaceous, at least as far as the records indicate.

Genus PLANORBULINA d'Orbigny, 1826

Plate 29, figure 1; Key, plate 37, figures 1, 2

Genotype, *Planorbulina mediterranensis* d'Orbigny

Planorbulina D'ORBIGNY, Ann. Sci. Nat., vol. 7, 1826, p. 280.

Asterodiscus EHRENBERG, 1838. *Spirobotrys* EHRENBERG, 1844. *Soldanina* COSTA, 1856.

Test in the young coiled, attached by the dorsal surface; very earliest chambers slightly trochoid, closely spiral, later in an irregular series of a single layer about the periphery; wall of early chambers chitinous, later calcareous, coarsely perforate; apertures in the early stages one to each chamber near the periphery, or in the irregular chambers sometimes multiple. — Tertiary, Recent.

PLATE 29

RELATIONSHIPS OF THE GENERA OF THE PLANORBULINIDAE

FIG.

1. *Planorbulina mediterranensis* d'Orbigny. (After d'Orbigny.) *a*, dorsal view; *b*, ventral view; *c*, edge view.
2. *Planorbulinoides retinaculata* (Parker and Jones). (After Parker and Jones.)
3. *Acervulina inhaerens* Schultze. *
4. *Gypsina vesicularis* (Parker and Jones). (After H. B. Brady.)
5. *Planorbulinella larvata* (Parker and Jones). (After H. B. Brady.) *a*, dorsal view; *b*, edge view.
6. *Linderina brugesi* Schlumberger. (Adapted from Schlumberger.) *a*, dorsal view; *b*, section.

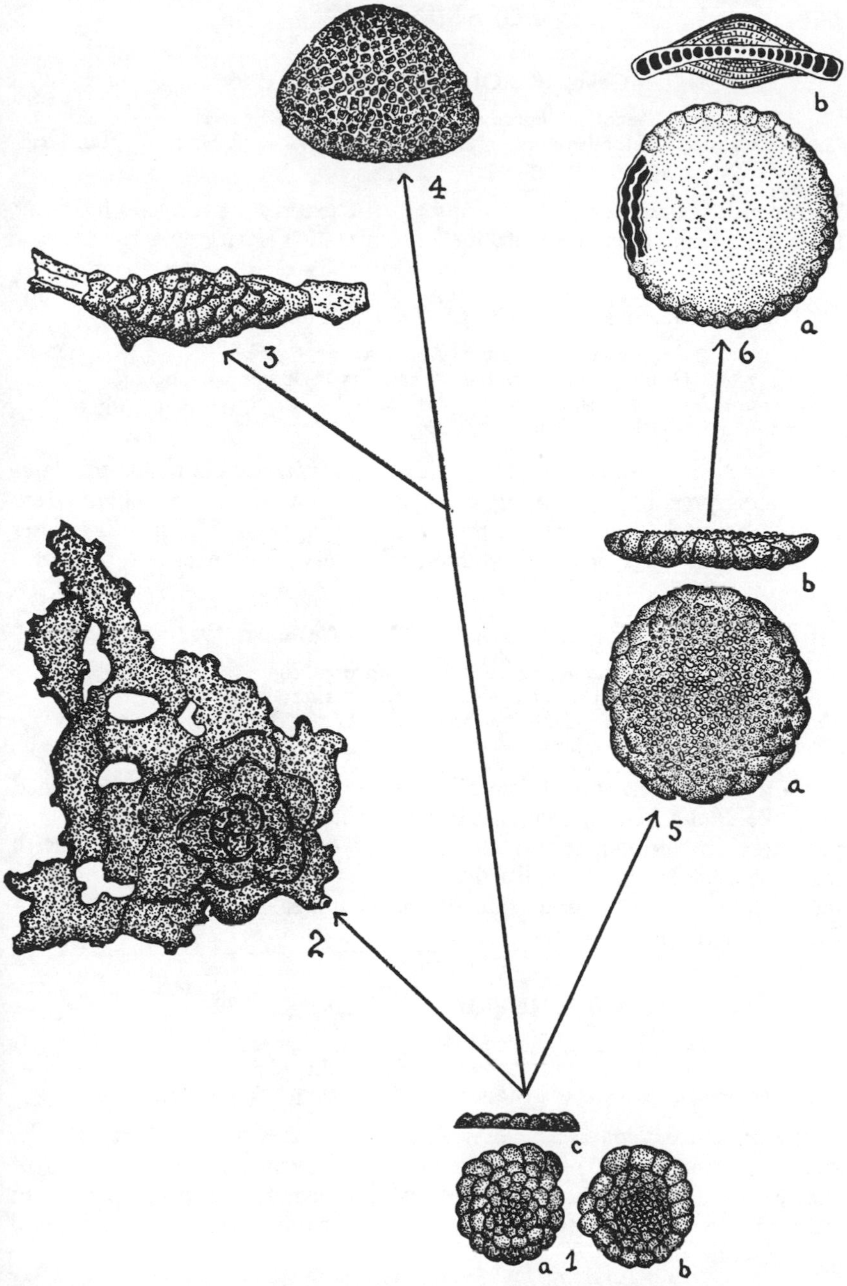

Plate 29

Genus BORODINIA Hanzawa, 1940

Genotype, *Borodinia septentrionalis* Hanzawa

Borodinia HANZAWA, Jubilee Publ. in Commemoration of Prof. H. Yabe's 60th Birthday, 1940, p. 790, pl. 42, figs. 10–12.

This genus is based on thin sections but the exterior of the tests have not been described. They are from well samples of Tertiary age.

Genus PLANORBULINOIDES Cushman, 1928

Plate 29, figure 2; Key, plate 37, figure 4

Genotype, *Planorbulina retinaculata* Parker and Jones

Planorbulinoides CUSHMAN, Contr. Cushman Lab. Foram. Res., vol. 4, 1928, p. 6.

Planorbulina (part) PARKER and JONES, 1865.

Test attached, in the early stages similar to *Planorbulina*, but the later chambers spreading, becoming elongate and more or less separated to form a network; apertures in the early stages as in *Planorbulina*, later several on the sides of the chambers, with very short necks. — Recent.

Genus PLANORBULINELLA Cushman, 1927

Plate 29, figure 5; Key, plate 37, figure 3

Genotype, *Planorbulina larvata* Parker and Jones

Planorbulinella CUSHMAN, Contr. Cushman Lab. Foram. Res., vol. 3, 1927, p. 96.

Planorbulina (part) of authors.

Test in the adult nearly bilaterally symmetrical, in the young attached and like *Planorbulina*, soon having the chambers developed about the periphery in annular series; chambers of each series alternating with those of adjacent ones; wall calcareous, coarsely perforate; apertures in the adult two, one at each side of the chamber in the median line. — Tertiary, Recent.

Genus LINDERINA Schlumberger, 1893

Plate 29, figure 6; Key, plate 37, figures 5–7

Genotype, *Linderina brugesi* Schlumberger

Linderina SCHLUMBERGER, Bull. Soc. Géol. France, ser. 3, vol. 21, 1893, p. 120.

Test similar to *Planorbulinella* with annular series of chambers, but developing a thick layer of clear shell material over the central portion of the test on the two flattened sides; wall calcareous, coarsely perforate; apertures in adult formed by the coarse perforations of the peripheral border. — Eocene.

The genus *Monolepidorbis* Astre, described from the Cretaceous of Southwestern Europe, needs further study to place it accurately. It seems to be closely allied to *Linderina*.

Genus EOANNULARIA Cole and Bermudez, 1944

Key, plate 54, figure 21
Genotype, *Eoannularia eocenica* Cole and Bermudez
Eoannularia Cole and Bermudez, Bull. Amer. Pal., vol. 28, No. 113, 1944, p. 12.

"Test small, fragile, flat or concavo-convex, with or without a small umbo on the convex side. Embryonic apparatus bilocular; either a small, circular initial chamber slightly embraced by a larger chamber, or an ovoid initial chamber completely embraced by the second chamber. Equatorial chambers of two types, those of the annuli adjacent to the embryonic apparatus have curved outer walls and truncated inner ends, those in the final annuli are regularly rectangular with the radial walls in adjacent annuli alternating in position. The entire thickness of the test is composed of the equatorial layer, except a slight deposit of clear shell material over the embryonic apparatus." — Eocene.

Genus ACERVULINA Schultze, 1854

Plate 29, figure 3; Key, plate 37, figures 8–10
Genotype, *Acervulina inhaerens* Schultze
Acervulina Schultze, Organismus Polythal., 1854, p. 67.

Test attached by the dorsal side, at least in the early stages, if attached to a small object often entirely covering it, and then covering its own early chambers, earliest chambers coiled; wall calcareous, coarsely perforate; apertures formed by the coarse perforations of the test. — Jurassic (?), Cretaceous (?), Late Tertiary, Recent.

Genus GYPSINA Carter, 1877

Plate 29, figure 4; Key, plate 37, figure 11
Genotype, *Orbitolina vesicularis* Parker and Jones
Gypsina Carter, Ann. Mag. Nat. Hist., ser. 4, vol. 20, 1877, p. 173.
Orbitolina (part), *Tinoporus* (part), and *Ceriopora* (part) of authors. *Sphaerogypsina* Galloway, 1933.

Test a generally spherical mass of compressed chambers, sometimes arranged in more or less radial rows, early stages attached, on a small object completely covering it and forming a spherical mass, but becoming free early and becoming spherical; apertures formed by the coarse pores of the wall. — Cretaceous (?) to Recent.

Family 45. RUPERTIIDAE

Test in the early stages, trochoid, attached by the dorsal side as in *Cibicides*, later extending upward from the base of attachment still keeping a loose spiral; wall calcareous, coarsely perforate; aperture either at the inner margin of the chamber or becoming terminal and rounded, often with a neck and lip.

The three genera now comprising this family show distinct stages in the development from a *Cibicides*-like ancestor. They are spiral, attached by the dorsal side, and in the adult with the aperture becoming terminal. In *Rupertia* the entire development continues a loose spiral in an erect form. In *Carpenteria* the later development is of various shapes, typically a loose spiral, but may become biserial or even uniserial in the adult. *Neocarpenteria* seems to be a primitive, ancestral form.

These are most characteristic of the Tertiary, and are found in the present oceans, but there are records as far back as the Upper Cretaceous.

Genus RUPERTIA Wallich, 1877

Key, plate 37, figures 12–15
Genotype, *Rupertia stabilis* Wallich
Rupertia WALLICH, Ann. Mag. Nat. Hist., ser. 4, vol. 19, 1877, p. 502.

Test attached, in the young trochoid; the chambers later extending upward from the base of attachment still keeping a loose spiral; wall calcareous, thick, coarsely perforate; aperture in the early stages narrow, at the base of the chamber, becoming more open, rounded, with a thickened lip.— Eocene to Recent.

Genus CARPENTERIA Gray, 1858

Key, plate 37, figures 16–20
Genotype, *Carpenteria balaniformis* Gray
Carpenteria GRAY, Proc. Zool. Soc. London, vol. 26, 1858, p. 269.
Dujardinia GRAY, 1858. *Rhaphidodendron* MOEBIUS, 1876. *Polytrema* (part) CARTER, 1876 (not RISSO).

Test attached, early chambers trochoid, later spreading out over the surface of attachment, the inner ends piled up in a loose spire or the whole test becoming subcylindrical; chambers loosely spiral or even uniserial; wall calcareous, coarsely perforate; aperture in the young, narrow, later stages somewhat rounded, at the end of a tubular projection, uniserial forms sometimes with a tubular neck.— Tertiary, Recent.

Forms described under the name *Semseya* by Franzenau are peculiar, and it is very difficult to place them. They are coarsely perforate, have a single, elongate, somewhat curved aperture with a lip, appear to be single chambered, and have a peculiar depression at the side. Rzehak has suggested that it be placed near *Carpenteria*. I have not seen the types, and the placing of this form must await further work by someone on the original material.

Genus NEOCARPENTERIA Cushman and Bermudez, 1936

Key, plate 54, figures 17–19

Genotype, *Neocarpenteria cubana* Cushman and Bermudez

Neocarpenteria Cushman and Bermudez, Contr. Cushman Lab. Foram. Res., vol. 12, 1936, p. 34.

Test attached by the dorsal side, close coiled, trochoid, dorsal side flattened, ventral side convex; wall calcareous, perforate; aperture large, semicircular, on the margin of the ventral side of the chamber, with a distinct lip. — Eocene.

Family 46. VICTORIELLIDAE

Test free, typically somewhat adherent in the young; early chambers in an irregular trochoid spiral, later forming an irregular rounded mass; wall calcareous, usually double in the young or becoming single in the adult, coarsely perforate; aperture either simple, sublunate to triangular with a distinct lip, or a series of pores.

The three genera comprising this family are not well known. *Eorupertia* is found in the Eocene as an elongate spiral about a hollow center. *Victoriella* is somewhat similar with a solid central axis, and *Hofkerina* has in the adult a generally globular form. These three genera need much further study in detail to show their true relationships. The early stages, however, seem to show that they are rather closely related to *Carpenteria*.

Genus EORUPERTIA Yabe and Hanzawa, 1925

Key, plate 37, figures 23–25

Genotype, *Uhligina boninensis* Yabe and Hanzawa

Eorupertia Yabe and Hanzawa, Sci. Rept. Tohoku Imp. Univ., ser. 2, vol. 7, 1925, p. 77 (footnote).

Uhligina Yabe and Hanzawa, 1922 (not Schubert).

Test perhaps attached in early stages, typically free, early stages coiled, later subcylindrical; chambers in an elongate spiral about a hollow center; wall calcareous, coarsely perforate, with canals and pillars, the latter forming pustules at the surface; aperture typically a long slit at the inner chamber margin. — Eocene.

Genus VICTORIELLA Chapman and Crespin, 1930

Key, plate 37, figures 21, 22

Genotype, *Carpenteria proteiformis*, var. *plecte* Chapman

Victoriella Chapman and Crespin, Proc. Roy. Soc. Victoria, vol. 42, 1930, p. 111.

Carpenteria (part) Chapman (not Gray).

"Test free, consisting of a more or less conoidal aggregate of inflated chambers, either alternating or spirally coiled, chambers not numerous;

surface granulated, the tubercles surrounded by the coarsely tubulated shell-wall; aperture sublunate and limbate. The wall of the test is apparently simple in the later portion, but in the earliest part it consists of two layers as in *Carpenteria*, and is also thicker than in that genus. The surface tubercles are more strongly papillate than those of *Eorupertia*." — Miocene, Australia.

Genus HOFKERINA Chapman and Parr, 1931

Key, plate 37, figure 26
Genotype, *Pulvinulina semiornata* Howchin

Hofkerina CHAPMAN and PARR, Proc. Roy. Soc. Victoria, vol. 43, 1931, p. 237.
Pulvinulina (part) HOWCHIN (not PARKER and JONES).

"Test free, trochoid with a rotaline plan of growth, strongly biconvex, margin rounded, chambers comparatively few, strongly inflated; wall calcareous, thick, laminated, fairly coarsely tubulate, closely papillate above in central portion, inferior face smooth. Aperture cribrate, occupying the umbilical depression." — Oligocene, Miocene, Australia.

Family 47. HOMOTREMIDAE

Test in the early stages trochoid, attached by the dorsal surface, later becoming irregular and growing upward from the area of attachment into a more or less branched mass, all trace of the early arrangement being lost; wall calcareous, coarsely perforate; apertures large, open or covered by a perforated plate; a reddish or orange color strongly developed.

This family has been derived from attached trochoid forms such as *Cibicides*, and the various genera have become highly specialized. This is one of the few groups of the foraminifera that has developed color in any great degree. They form a characteristic element of the foraminiferal fauna of shallow tropical waters especially about coral reefs, and form a considerable factor in the building up of the so-called coral sands. For details of structure of the various genera, the reader is referred to the excellent paper of Hickson — On *Polytrema* and Some Allied Genera (Trans. Linn. Soc. London, Zool., vol. 14, No. 20, 1911).

Genus HOMOTREMA Hickson, 1911

Key, plate 37, figures 27–29
Genotype, *Millepora rubra* Lamarck

Homotrema HICKSON, Trans. Linn. Soc. Zool., vol. 14, 1911, p. 445.
Millepora LAMARCK, 1816 (not LINNÉ). *Polytrema* (part) of authors (not RISSO).

Test attached, the early stages coiled, later extending up in an irregular mass with short, stout, projecting portions; wall calcareous, the surface solid with large scattered foramina covered by a finely perforated plate; color dark red. — **Tertiary, Recent.**

Genus SPORADOTREMA Hickson, 1911

Key, plate 37, figures 30–32
Genotype, *Polytrema cylindricum* Carter

Sporadotrema HICKSON, Trans. Linn. Soc. Zool., vol. 14, 1911, p. 447.
Polytrema CARTER, 1880 (not RISSO).

Test attached, the early stages coiled, later extending outward and upward into short stout branches; numerous chambers apparent about the outer end; wall calcareous, the surface solid with scattered foramina, open, not covered by a plate; color orange or red. — Tertiary, Recent.

Genus MINIACINA Galloway, 1933

Key, plate 37, figures 33–36
Genotype, *Millepora miniacea* Pallas

Miniacina GALLOWAY, Man. Foram., 1933, p. 305.
Polytrema of authors (not RISSO). *Millepora* (part) of authors. *Pustularia* GRAY, 1858 (not SWAINSON).

Test attached, early stages coiled, later with the small chambers piled up in an irregular branching mass with slender projections; wall calcareous, perforate, surface finely perforate with larger open foramina; color light red. — Tertiary, Recent.

Family 48. ORBITOIDIDAE Schubert, 1920

By T. Wayland Vaughan and W. Storrs Cole

Revised by W. Storrs Cole

Test thin or inflated, biconvex or biconcave, discoidal, round lenticular, selliform, or stellate, with a layer of equatorial chambers which may become multiple near the periphery. The equatorial zone is covered on both sides by layers of lateral chambers, either arranged in regular tiers or irregularly overlapping. Initial chambers of the microspheric generation spiral; in the megalospheric generation growth begins with a multilocular embryonic apparatus which may or may not be spiral. Chamber walls perforate, usually also communication between chambers through stoloniferous apertures. *There is no canal system.* — Upper Cretaceous to lower Miocene.

Although a large amount of research has been bestowed upon this family, there are many features which have not been studied adequately. Sweeping conclusions have been drawn from features little known or inadequately described. Any classification, therefore, which may be proposed is admittedly imperfect and unsatisfactory.

Recently, there have been proposed a considerable number of new generic and subgeneric names, many of which are based upon small and often variable features. As the validity and usefulness of certain of these are extremely questionable, many of them are placed in synonymy.

The importance of the arrangement of the stolons connecting the equatorial chambers in phylogenetic and taxonomic studies has been recently emphasized. Van de Geyn and van der Vlerk place so much importance on the stolons that they propose a new classification based largely on this feature. The names proposed by these authors to replace earlier names assigned to several genera by H. Douvillé and Vaughan are invalid.

Stolon systems as now recognized fall into five groups which may be enumerated as follows: (1) Four-stolon system: stolons at two places on each side of each chamber. (2) Stolon system rather indefinite: three, four, or more stolons between chambers, some of them through the lateral walls of the chambers, others radial in position. (3) Five- or six-stolon system: stolons as in number 1, but with a radial stolon through the proximal or the distal wall of a chamber or through both walls. This system when complete is a six-stolon system. Eventually, it may be that all the genera which are now thought to possess four stolons will be classified under the five- or six-stolon system. (4) Six-stolon system: two oblique

---

PLATE 30

FAMILY 48. ORLITOIDIDAE. FAMILY 49. DISCOCYCLINIDAE

FIG.

1–5. *Orbitoides gensacicus* (Leymerie). × 16. (After H. Douvillé.) Early stages of megalospheric individuals showing various divisions of the chambers.

6. *Orbitoides media* (d'Archiac). × 24. (After H. Douvillé.) Portion of vertical section showing median chambers and openings.

7. *Orbitoides gensacicus* (Leymerie). × 16. (After H. Douvillé.) Horizontal section of the embryonic chambers of a teratologic individual.

8–10. *Lepidorbitoides socialis* (Leymerie). 8, Central chambers of horizontal section of microspheric form, × 40. 9, Central chambers of horizontal section of megalospheric form, × 16. 10, Horizontal section of outer median chambers, × 16.

11–15. *Clypeorbis mamillata* (Schlumberger). × 16. (After H. Douvillé.) 11–13, Horizontal sections of the early chambers at different levels. 14, Vertical section. 15, Horizontal section showing pillars.

16, 17. *Pseudorbitoides trechmanni* H. Douvillé. (After H. Douvillé.) 16, Horizontal section of early chambers, × 32. 17, Portion of vertical section, × 16.

18. *Discocyclina.* Diagrammatic horizontal section. (After van der Vlerk and Umbgrove.)

19. *Asterocyclina stella* Gümbel. × 24. (After H. Douvillé.) Portion of horizontal section.

20. *Discocyclina (Discocyclina) nummulitica* Gümbel. × 16. (After H. Douvillé.) Section showing pillars and lateral chambers.

21. *Discocyclina (Discocyclina) archiaci* (Schlumberger). × 16. (After H. Douvillé.) Horizontal section of early chambers.

22. *Discocyclina (Discocyclina) sennesi* H. Douvillé. × 16. (After H. Douvillé.) Horizontal section of early chambers.

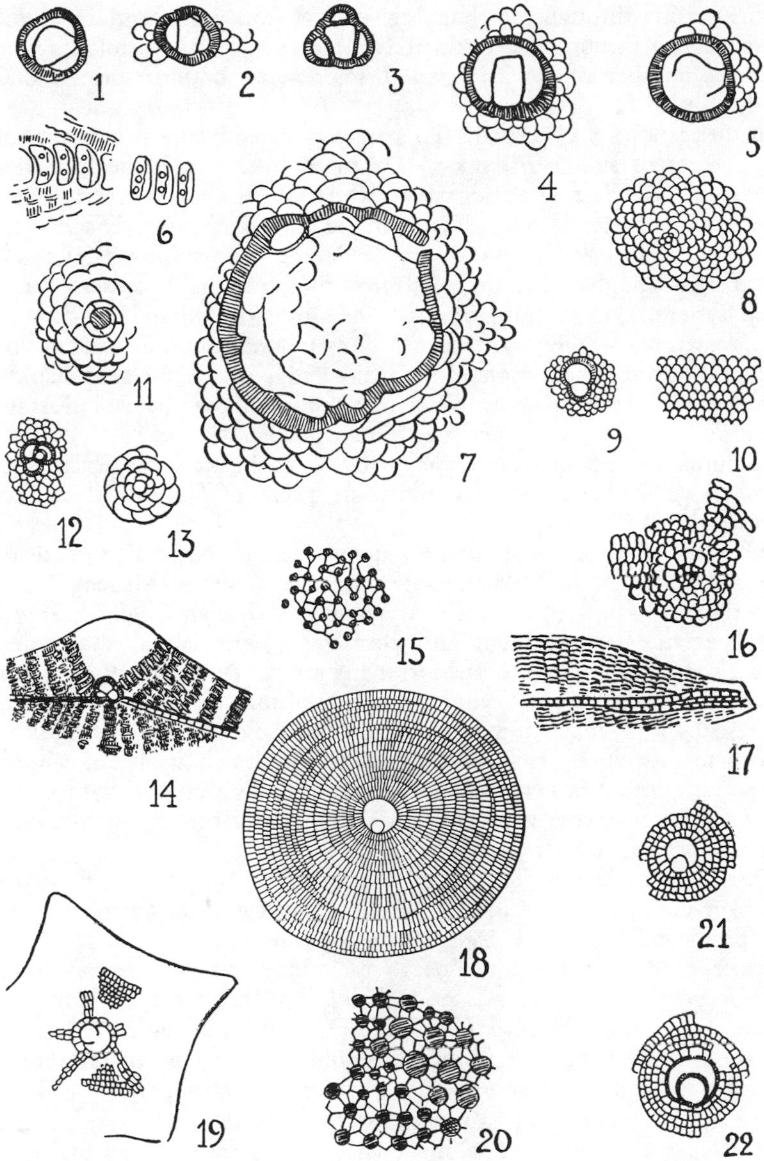

PLATE 30

stolons passing through the chamber walls at four places and an annular stolon passing through the walls at two places. (5) Eight-stolon system: similar to number 4 but with a radial stolon across both the proximal and the distal walls.

Further studies are needed. To base any classification on the incomplete studies of stolon systems so far undertaken would be hazardous. There is no complete agreement as to which of the above numerated categories certain forms belong. For example, *Lepidorbitoides* is specifically mentioned by van de Geyn and van der Vlerk as possessing a four-stolon system, but Vaughan has demonstrated that it actually possesses a six-stolon system. Moreover, Brönnimann has discovered in *Lepidocyclina* (*Lepidocyclina*) *mauretanica* that a four-stolon system is present normally, but toward the periphery six stolons may occur. Great difficulty is encountered in preparing thin sections which show the stolon system unless perfectly preserved specimens are available.

Stoloniferous apertures are occasionally found in the walls of the lateral chambers. They have been reported from species of *Helicolepidina*, *Lepidocyclina*, and *Triplalepidina*.

Other classifications based largely upon a single morphological feature have not succeeded in their purpose. Scheffen (Weten. Meded., No. 21, 1932, pp. 7–16) devised a classification of East Indian *Lepidocyclina* on the appearance, arrangement and character of the lateral chambers of various species of this genus. Tan in many articles evaluates the embryonic apparatus and the peri-embryonic chambers of many genera. He attempts to present the number and type of chambers by means of formulae as well as to introduce an elaborate terminology. Such studies are needed and add important facts to the knowledge of this group of organisms, but caution must interpose in overemphasis of one feature in such a plastic group of organisms.

The initial chambers (embryonic stage) of the megalospheric generation of many of the orbitoids are bilocular, consisting of an initial chamber (protoconch) followed by a second chamber (deuteroconch). These chambers have size relationships to each other varying from equality (*Lepidocyclina* s.s.) to the second chamber so large that it completely embraces the initial chamber (*Eulepidina*). Within a single species intergradation of types may be found, and difficulty may be encountered in placing a specimen in a genus, or, particularly, in a subgenus.

The initial chambers are surrounded partially or completely in many cases by peri-embryonic (nepionic) chambers before the normal equatorial (ephibic) chambers are developed. An attempt has been made to classify the peri-embryonic chambers into various series depending on their development. Three series are recognized: (1) uniserial, in which there is a large peri-embryonic chamber (primary auxiliary) connected by a stoloniferous passage with the second embryonic chamber, and from which there is a single row of smaller peri-embryonic (auxiliary) chambers which encircle the embryonic chambers in one direction; (2) biserial, in which from each side of the large peri-embryonic chamber there develops smaller peri-embryonic chambers which proceed to encircle the initial chambers in two directions; and (3) quadriserial, in which two large peri-embryonic chambers develop, one on each side of the initial chambers, and each of which is connected by a stoloniferous passage with the second embryonic chamber. As each of these large peri-embryonic chambers develop two encircling rows (biserial type) of smaller peri-embryonic chambers, the initial chambers are encircled by four rows of peri-embryonic chambers in four directions.

The usefulness of this arrangement of the peri-embryonic chambers in the definition of a genus, or even a subgenus, may be questioned when it can be proven definitely that individuals of a single species from a given population possess the three types.

Until recently there were two hypotheses regarding the origin of the Orbitoididae. According to H. Douvillé the family is derived from *Arnaudiella* or some similar parent. The Calcarinidae to which *Arnaudiella* is referred have a well-developed supplementary canal system, but the orbitoids are without a canal system. Hofker is of the opinion that the family is derived from his Tinoporidae (*The Foraminifera of the Siboga Expedition*, pt. 1, 1927, pp. 4–6), because of the general similarity of structure, including the absence of any canal system. Formerly, Vaughan was inclined to accept Hofker's suggestion.

Barker and Grimsdale have studied a series of specimens from the Eocene of Mexico which form the most convincing phylogenetic series in linking the advanced *Helicolepidina* with *Amphistegina.* This series would be from *Amphistegina* to *Helicostegina* of the family Amphisteginidae to *Eulinderina* and *Helicolepidina,* both of which are Orbitoididae.

The Cretaceous genus *Lepidorbitoides* shows great similarity to the Lower Eocene or Paleocene genus *Actinosiphon* which in turn is directly connected with *Lepidocyclina* through *Polylepidina.* Barker and Grimsdale consider that *Helicostegina* is the generating point for two series, one ending with *Helicolepidina,* the other advancing from *Helicostegina* to *Eulinderina* to *Polylepidina.* It appears, however, that the progression from *Lepidorbitoides* to *Actinosiphon* to *Polylepidina* and *Pliolepidina* more nearly agrees with the known stratigraphic occurrences of these genera.

If the assumption that an *Amphistegina*-like form was the ancestor of the Helicolepidinae, it may well be that the other subfamilies of the Orbitoididae had a similar ancestry. However, the stock terminating in *Helicolepidina* may represent a separate and distinct one, in which case the subfamily Helicolepidinae should be elevated to family rank. In this case, the remainder of the Orbitoididae may have been initiated from another stock. The final solution must await more analysis of Lower and Upper Cretaceous faunae.

Abnormality of the initial chambers is of rather common occurrence in some genera of the Orbitoididae. It is the multiplication of the initial chambers which produces the phenomenon designated "gigantisme" by H. Douvillé. Vaughan and Cole have demonstrated that the peculiar nucleoconch of *Lepidocyclina tobleri* is due to this fact. *Multilepidina* may be the teratological form of *Eulepidina*, whereas *Simplorbites* is certainly the teratological form of *Orbitoides*.

The orbitoids lived in shallow water at depths from tide level to perhaps one hundred meters in a belt around the earth which included the tropical belt and the equatorial margins of the temperate belts. Vaughan (Proc. Nat. Acad. Sci., vol. 19, 1933, pp. 922–938) summarizes the ecological conditions and probable routes of migration of these organisms. Little can be added to this summary.

Subfamily Pseudorbitoidinae M. G. Rutten

Genus VAUGHANINA D. K. Palmer, 1934

Key, plate 54, figures 23, 24

Genotype, *Vaughanina cubensis* D. K. Palmer

Vaughanina D. K. Palmer, Soc. Cubana Hist. Nat. Mem., 1934, vol. 8, pp. 240–243, pl. 12, fig. 5; pl. 13, figs. 2, 4; text figs. 2, 3; Vaughan and Cole, in Cushman, Foraminifera, 1940, p. 318, Key, pl. 48, figs. 8, 9; Jour. Pal., 1943, vol. 17, pp. 98–100, pl. 17, figs. 3, 4; pl. 18, figs. 1–10; Glaessner, Principles of Micropal., 1945, p. 166.

Test with central umbonate, lenticular, papillate portion, surrounded by an outer thin flange, crossed by radiating plates which pass over the distal edge to form a crenulate margin. Embryonic chambers bilocular, an initial spherical chamber partially embraced by a second chamber. Embryonic chambers followed by a complete or incomplete spiral of six to eight rudely quadrate peri-embryonic chambers. Beyond the peri-embryonic chambers is a zone of equatorial chambers with curved outer walls and more or less truncate inner ends. In the peripheral zone there are successive circles of rectangular chambers produced by the inward extension of the radial, flange plates and the outer walls of the equatorial chambers. Equatorial layer single in the lenticular portion of the test, dividing into three layers in the peripheral portion. Lateral chambers, open distinct, with relatively thin roofs and floors. Flange not covered with lateral chambers. Pillars irregularly present.—Upper Cretaceous, Cuba, Florida.

Vaughanina has many features in common with *Pseudorbitoides*, but the peripheral replacement of normal equatorial chambers by rectangular chamberlets in *Vaughanina* is not known in *Pseudorbitoides*, where the normal equatorial chambers extend to the periphery of the test.

Genus PSEUDORBITOIDES H. Douvillé, 1922

Plate 30, figures 16, 17; Key, plate 38, figures 4–7
Genotype, *Pseudorbitoides trechmanni* H. Douvillé

Pseudorbitoides H. DOUVILLÉ, C. R. Soc. Géol. France, 1922, p. 203; Bull. Soc. Géol. France, ser. 4, vol. 23, 1923, p. 369, figs. 1, 2; VAUGHAN, in CUSHMAN, Foraminifera, 1928, p. 340, pl. 58, figs. 16, 17; Jour. Pal., vol. 3, 1929, p. 168, pl. 21, figs. 4–6; VAUGHAN and COLE, Nat. Acad. Sci. Proc., vol. 18, 1932, p. 614, pl. 2, figs. 1–7; VAUGHAN, in CUSHMAN, Foraminifera, 1933, p. 289, pl. 30, figs. 16, 17; Key, pl. 38, figs. 4–7; GALLOWAY, Man. Foram., 1933, p. 428, pl. 40, fig. 4; M. G. RUTTEN, Jour. Pal., vol. 9, 1935, pp. 543, 544, pl. 62, figs. 2, 5; text figs. 4 E–H, K-O, Q; VAUGHAN and COLE, in CUSHMAN, Foraminifera, 1940, pp. 317, 318, pl. 30, figs. 16, 17; Key, pl. 38, figs. 4–7; M. G. RUTTEN, Geol. en Mijnbouw, n. ser., jaarg. 3, No. 2, 1941, p. 38, pl. 1, figs. 1, 2; GLAESSNER, Principles of Micropal., 1945, p. 166.

Test lenticular, rather compressed or robust, size rather small, from less than 1, to 4 or 5 mm. in diameter. Surface papillate. Embryonic chambers in microspheric specimens form a distinct spiral; in megalospheric specimens there is a large subspherical initial chamber, following and connected with which, through an aperture, is a slightly smaller chamber. Following these two chambers, there are two or more smaller chambers, the entire group forming an irregular spiral. Equatorial chambers in horizontal section irregular polygonal or hexagonal, transverse usually greater than the radial diameter; they tend to form radiating rows; communication by stoloniferous apertures; a single layer in the central part of the test, two or even three layers near the periphery; radiating lines present above the chambers. Lateral chambers form definite tiers with open relatively large chamber cavities. — Upper Cretaceous, Jamaica, Cuba, Louisiana, Texas, Guatemala.

M. G. Rutten (1935) states that the equatorial layer is composed of long, channel-like radial chambers which double or triple near the periphery. The walls of these chambers produce the effect of radiating lines seen in equatorial sections. Vaughan and Cole do not agree that the equatorial layer is composed of channel-like chambers. Recrystallization may have so altered the specimens which Rutten studied that the channel-like chambers were produced by destroying the distal margin of the individual equatorial chambers. As these chambers tend to form radiating rows, the channel-like effect would be readily accomplished.

Subfamily Orbitoidinae Prever

Genus OMPHALOCYCLUS Bronn, 1852

Plate 31, figures 15, 16; Key, plate 39, figures 13–15
Genoholotype, *Orbulites macroporus* Lamarck

Omphalocyclus BRONN, Lethaea geognost., ed. 3, vol. 2, 1851–52, pl. 95; H. DOUVILLÉ, Bull. Soc. Géol. France, ser. 4, vol. 20, 1921, p. 228, pl. 8, figs. 5–14, text figs. 35–37; VAUGHAN, in CUSHMAN, Foraminifera, 1928, p. 355, pl. 56, figs. 4–6; pl. 59, figs. 15, 16; *ibid.*, 1933, pl. 31, figs. 15, 16; Key, pl. 39, figs. 13–15; Galloway, Man. Foram., 1933, p. 439, pl. 41, figs. 14, 16; Vaughan and Cole, in Cushman, Foraminifera, 1940, pp. 314, 315, pl. 31, figs. 15, 16; Key, pl. 39, figs. 13–15; M. G. Rutten, Geol. en Mijnbouw, n. ser., jaarg. 3, No. 2, 1941, p. 41, pl. 1, fig. 6; Glaessner, Principles of Micropal., 1945, p. 166.

Test biconcave lenticular. Embryonic chambers of microspheric test planispiral, of megalospheric test quadrilocular. In the central part of the test there is either a double layer of chambers, or a single layer, which as growth progresses, becomes double, and then a median layer may be intercalated between lateral layers. Although the peripheral part of the test in adult specimens is more than one layer of chambers thick, the lateral chambers are not differentiated from the median, as in the other subfamilies of the orbitoids. The median layer is composed of concentric rings of chambers, whose outer walls are only slightly curved, or are nearly straight. The median chambers communicate with each other by means of stolons, and with the lateral chambers through apertures in the floors of the latter; those on the periphery open to the outside through marginal apertures, which rather regularly alternate in position, and are arranged in one, two, three, or four rows. Both the roofs and floors of the chambers are pierced by very fine perforations, similar to those of the more typical orbitoids. The superficial chambers are arranged in concentric rings, those in one ring alternating in position with those in the adjacent rings. Each chamber communicates with two chambers of both the preceding and the succeeding ring. There are also cribriform perforations. In old specimens, a secondary deposit is laid down in the interior of the chambers, and sometimes obliterates the original structure of the test, producing in its median part a network of canals and irregular cavities. On the outer surface, a network of lozenge-shaped meshes is produced by the immediately underlying layer. Pillars, which may be lamellate, may be formed, and produce wavy, radiating costae.—Upper Cretaceous, Maestrichtian, Europe, India, Cuba.

Genus TORREINA D. K. Palmer, 1934

Key, plate 41, figure 8

Genotype, *Torreina torrei* D. K. Palmer, 1934

Torreina D. K. Palmer, Soc. Cubana Hist. Nat. Mem., vol. 8, 1934, pp. 237, 238, pl. 12, figs. 1, 4; Vaughan and Cole, in Cushman, Foraminifera, 1940, p. 315, Key, pl. 41, fig. 8; M. G. Rutten, Geol. en Mijnbouw, n. ser., jaarg. 3, No. 2, 1941, p. 41, pl. 1, fig. 7.

Test small, nearly spherical in shape. Embryonic chambers of megalospheric individuals composed of 4 to 7 unequal chambers surrounded by a thick fibrous, finely perforate wall. The remainder of the test composed of low arcuate chambers which are arranged in an irregularly concentric manner about the nucleoconch. In tangential sections, these chambers appear nearly circular in shape. The chamber walls are very thin and communicate with adjacent chambers by means of oval, stoloniferous apertures. Pillars are not present. — Upper Cretaceous, Cuba.

Torreina closely resembles *Omphalocyclus*, appearing to differ from that genus mainly in its spherical shape. It is, therefore, probable that it will be reduced to subgeneric rank or even placed in the synonomy of *Omphalocyclus*.

Genus CLYPEORBIS H. Douvillé, 1915

Plate 30, figures 11–15

Genotype, *Orbitoides mamillata* Schlumberger

Clypeorbis H. Douvillé, C. R. Acad. Sci., vol. 161, 1915, p. 669, figs. 18–20; Bull. Soc. Géol. France, ser. 4, vol. 20, 1921, p. 227, figs. 29–34; Vaughan, in Cushman, Foraminifera, 1928, p. 340 (misprinted *Pseudorbitoides*), pl. 58, figs. 11–15; *ibid.*, 1933, p. 285, pl. 30, figs. 11–15; Galloway, Man. Foram., 1933, p. 430, figs. 6, 7; Vaughan, Sci., vol. 83, 1936, p. 485; Vaughan and Cole, in Cushman, Foraminifera, 1940, p. 317, pl. 30, figs. 11–15; M. G. Rutten, Geol. en Mijnbouw, n. ser., jaarg. 3, No. 2, 1941, p. 48, pl. 2, fig. 1.

Test asymmetrically lenticular or low conical with a rounded apex. Surface papillate. Nucleoconch composed of an initial spheroidal chamber, below which there are developed 3 or 4 chambers arranged as a rosette. The embryonic chambers actually form a conical spiral which appears in the arrangement of the early lateral chambers. The equatorial chambers have convex outer walls, and form a network of hexagonal pattern. The chambers are radially shortened, are arranged along radial lines, and increase in number by bifurcation of, or intercalation between, older linear series. Equatorial chambers communicate by means of three, four, or more stolons between chambers, some of which are through the lateral walls of the chambers, others are radial in position. Roofs and floors of the equatorial chambers, and the roofs of the lateral chambers, are cribriform perforate. The anterior chamber wall appears to be a direct continuation of the chamber roof with the maintenance of the same porous texture. — Maestrichtian, Europe.

Genus ORBITOIDES d'Orbigny, 1847

Plate 30, figures 1–7; Key, plate 38, figure 1; plate 40, figures 1–4

Genotype, *Orbitoides media* (d'Archiac)

Orbitoides d'Orbigny, Quart. Journ. Geol. Soc. London, vol. 4, 1847, p. 11; Cours Élément. Pal., vol. 2, 1850, p. 193, fig. 316; Vaughan, in Cushman, Foraminifera, 1928, p. 337, pl. 57, fig. 4; pl. 58, figs. 1–6; *ibid.*, 1933, pp. 286, 288, pl. 30, figs. 1–6; Key, pl. 38, fig. 1; pl. 40, figs. 1–4; Galloway, Man. Foram., 1933, p. 430, pl. 40, figs. 8, 9; Vaughan and Cole, in Cushman, Foraminifera, 1940, p. 316, pl. 30, figs. 1–7; Key, pl. 38, fig. 1; pl. 40, figs. 1–4; M. G. Rutten, Geol. en Mijnbouw, n. ser., jaarg. 3, No. 2, 1941, p. 41, pl. 1, fig. 5; Glaessner, Principles of Micropal., 1945, p. 166.

Hymenocyclus Bronn, Lethaea geognost., vol. 2, 1853, p. 94, pl. 29-1, fig. 29.

Simplorbites DE GREGORIO, Fossili dei Dintorno di Pachino, 1882, p. 10, pl. 6, figs. 21–28, 30; H. DOUVILLÉ, C. R. Acad. Sci., vol. 161, 1915, p. 667, figs. 13–15; Bull. Soc. Géol. France, ser. 4, vol. 20, 1921, p. 217, figs. 19–21; VAUGHAN, in CUSHMAN, Foraminifera, 1928, p. 337, pl. 58, fig. 7; *ibid.*, 1933, p. 288, pl. 30, fig. 7; GALLOWAY, Man. Foram., 1933, p. 431, pl. 40, figs. 10, 11; VAUGHAN and COLE, in CUSHMAN, Foraminifera, 1940, p. 316, pl. 30, fig. 7.
Silvestrina PREVER, Riv. Ital. Pal., vol. 10, 1904, p. 122, pl. 6, figs. 2, 3.
Orbitella H. DOUVILLÉ, C. R. Acad. Sci., vol. 161, 1915, p. 666, figs. 5, 6; Bull. Soc. Géol. France, ser. 4, vol. 20, 1921, p. 214.
Gallowayina ELLIS, Amer. Mus. Novit. No. 568, 1932, p. 1, figs. 1–7.

Test lenticular, more or less compressed, symmetrical or asymmetrical, surface ornamented with vermicular pillars or radiating costae. Embryonic chambers enveloped in a thick shell, at first quadrilocular, later they may become bilocular, by atrophy and fusion of three of the initial chambers, and the production of a smaller, embraced by a larger chamber. Equatorial chambers with a curved outer wall and inwardly converging lateral walls, radial diameter shorter than the transverse. Communication between the chambers by a few, round, lateral apertures.—Upper Cretaceous, Europe, India, West Indies.

In *Simplorbites* (pl. 30, fig. 7) the embryonic apparatus is large, exceeding 2 mm. in diameter, roughly ovoid in form, and enclosed in a thick, porous shell, within which there are many chambers, piled one on another without any definite form. H. Douvillé states that these peculiar embryonic chambers are due to "gigantisme." As *Simplorbites* is similar in its other features to *Orbitoides*, it may be considered a teratologic form.

Genus PSEUDOLEPIDINA Barker and Grimsdale, 1937

Key, plate 41, figures 4, 5

Genotype, *Pseudolepidina trimera* Barker and Grimsdale

Pseudolepidina BARKER and GRIMSDALE, Ann. and Mag. Nat. Hist., ser. 10, vol. 19, 1937, pp. 169–172, pl. 5, figs. 1–3; pl. 8, figs. 1–5; text figs. 1, 2; VAUGHAN and COLE, in CUSHMAN, Foraminifera, 1940, p. 316, Key, pl. 41, figs. 4, 5; M. G. RUTTEN, Geol. en Mijnbouw, n. ser., jaarg. 3, No. 2, 1941, p. 53, text figs. 16, 17.

Test small, usually asymmetrical lenticular, papillate. Nucleoconch composed of two subequal chambers which are surrounded by a common wall and partially embraced by a third chamber lying to one side of the equatorial plane. The third chamber apparently occurs on the less convex side. One or more supplementary chambers may occur between the embryonic and the equatorial chambers. The equatorial zone is composed of a double row of chambers except at the center of the test. The equatorial chambers are irregular arcuate in form and variable in size. Communication is by means of stoloniferous passages to four adjacent chambers and with the chambers of the lateral layers. The lateral chambers are arranged in layers rather than tiers. The equatorial zone is not sharply differentiated from the lateral chambers, and, in some transverse sections, the equatorial zone appears to be formed by interlocking of the extremities of the layers of lateral chambers. Pillars are present. — Lower part of the middle Eocene, Veracruz, Mexico.

Subfamily Lepidocyclinae Tan

Genus LEPIDORBITOIDES A. Silvestri, 1907

Plate 30, figures 8–10; Key, plate 38, figures 2, 3, 8–10
Genotype, *Orbitolites socialis* Leymerie

Lepidorbitoides A. Silvestri, Atti. Accad. Nuovi Lincei, an. 61, 1908, p. 23 (15 Dec., 1907); H. Douvillé, Bull. Soc. Géol. France, ser. 4, vol. 20, 1921, p. 220, figs. 22–24; Vaughan, in Cushman, Foraminifera, 1928, p. 338, pl. 56, figs. 9, 10; pl. 58, figs. 8–10; *ibid.*, 1933, p. 285, pl. 30, figs. 8–10; Key, pl. 38, figs. 2, 3; Galloway, Man. Foram., 1933, p. 432, pl. 40, figs. 15, 16; M. G. Rutten, Proc. Kon. Akad. Wetens., vol. 38, 1935, pp. 186, 187, 1 pl.; Jour. Pal., vol. 9, 1935, p. 533; Thiadens, Jour. Pal., vol. 11, 1937, p. 99; Vaughan, Sci., vol. 83, 1936, p. 485; M. G. Rutten, Geol. en Mijnbouw, n. ser., jaarg. 3, No. 2, 1941, p. 50, pl. 2, fig. 3; Glaessner, Principles of Micropal., 1945, p. 167.

Orbitocyclina Vaughan, Nat. Acad. Sci. Proc., vol. 15, 1929, p. 291; Jour. Pal., vol. 3, 1929, p. 171, pl. 22, figs. 1–6; in Cushman, Foraminifera, 1933, pp. 289, 290, Key, pl. 38, figs. 8–10; Galloway, Man. Foram., 1933, p. 427, pl. 40, fig. 2; Tan, De Mijningenieur, jaarg. 6, 1939, pp. 69–78; Vaughan and Cole, in Cushman, Foraminifera, 1940, pp. 318, 319, pl. 30, figs. 8–10; Key, pl. 38, figs. 2, 3, 8–10; M. G. Rutten, Geol. en Mijnbouw, n. ser., jaarg. 3, No. 2, 1941, p. 50, pl. 2, fig. 3; Glaessner, Principles of Micropal., 1945, p. 167.

Orbitocyclinoides Brönnimann, Schweizerischen Pal. Abhandl., vol. 64, 1944, pp. 5–20, pl. 1, figs. 1–7; pl. 2, figs. 10–12, 14, 15.

Test small to medium size, flat to inflated, lenticular, circular, or stellate. Surface papillate, with or without raised ridges. Nucleoconch bilocular, the first chamber is subspherical, second chamber somewhat larger, slightly reniform, partially embracing the initial one. In some specimens, there is a series of chambers slightly larger than the equatorial chambers partly embracing the embryonic chambers; in others, there may be only one such chamber, or there may be none. The equatorial chambers have curved outer walls and truncate or pointed inner ends, depending on whether the sides of chambers in the same circle do or do not meet, but in some specimens they may become spatuliform or even hexagonal. Gradation may occur in the same specimen. Communication between the equatorial chambers is by means of six stoloniferous apertures. Roofs and floors of the equatorial chambers, and the roofs of the lateral chambers, are cribriform perforate. Pillars present. — Maestrichtian, Europe, India; Upper Cretaceous, Eastern Mexico, Louisiana, Mississippi, Florida, and Cuba.

Orbitocyclina was created to separate American specimens with stoloniferous apertures from European specimens assigned to *Lepidorbitoides* which appeared to be without stoloniferous apertures. Although M. G. Rutten demonstrated that typical *Lepidorbitoides* did possess stoloniferous apertures, Tan retains the genus *Orbitocyclina* because of a slight difference in the arrangement of the peri-embryonic chambers. As this feature is variable within the same species in many genera, it would appear that *Orbitocyclina* is a synonym of *Lepidorbitoides*.

Subgenus ASTERORBIS Vaughan and Cole, 1932

Key, plate 38, figures 14–16

Genotype, *Asterorbis rooki* Vaughan and Cole

Asterorbis VAUGHAN and COLE, Nat. Acad. Sci. Proc., vol. 18, 1932, p. 611, pl. 1, figs. 1–6; VAUGHAN, in CUSHMAN, Foraminifera, 1933, p. 290, Key, pl. 38, figs. 14–16; M. G. RUTTEN, Jour. Pal., vol. 9, 1935, pp. 533–536; TAN, De Mijningenieur, jaarg. 6, 1939, p. 73; VAUGHAN and COLE, in CUSHMAN, Foraminifera, 1940, p. 319, Key, pl. 38, figs. 14–16; M. G. RUTTEN, Geol. en Mijnbouw, n. ser., jaarg. 3, No. 2, 1941, p. 51.

Cryptasterorbis M. G. RUTTEN, Jour. Pal., vol. 9, 1935, pp. 533–536, pl. 60, fig. 5; pl. 61, figs. 3, 8, 9; text figs. 4A, B, I, J; M. G. RUTTEN, Geol. en Mijnbouw, n. ser., jaarg. 3, No. 2, 1941, p. 5, pl. 2, fig. 4.

Asterorbis is distinguished from *Lepidorbitoides* either in possessing a stellate test with elevated, radiating ribs or in the development of a stellate test without raised ribs. The arrangement of the equatorial chambers always shows the rayed structure distinctly.

The subgenus *Cryptasterorbis* was created for those forms which possess a stellate pattern in the equatorial layer which is not reflected on the surface of the test.

Genus ACTINOSIPHON Vaughan, 1929

Key, plate 38, figures 11–13

Genotype, *Actinosiphon semmesi* Vaughan

Actinosiphon VAUGHAN, Jour. Pal., vol. 3, 1929, p. 163, pl. 21, figs. 1–3; in CUSHMAN, Foraminifera, 1933, pp. 288, 289, Key, pl. 38, figs. 11–13; GALLOWAY, Man. Foram., 1933, p. 427, pl. 40, fig. 3; VAUGHAN and COLE, in CUSHMAN, Foraminifera, 1940, pp. 320, 321, Key, pl. 38, figs. 11–13; M. G. RUTTEN, Geol. en Mijnbouw, n. ser., jaarg. 3, No. 2, 1941, pp. 61, 62, text figs. 7, 20.

Orbitosiphon RAO, Current Sci., vol. 9, 1940, pp. 414, 415, fig. 1.

Test lenticular, with a single layer of equatorial chambers and well developed lateral chambers. Pillars and terminal small papillae present. Embryonic apparatus of the megalospheric form consists of a rather large subspherical initial chamber, followed by a smaller chamber whose longer diameter is parallel to the circumference of the first chamber. The latter chamber is followed by about eleven other chambers which entirely encircle the first two chambers, and outside this circle there are several other chambers opposite chamber no. 3. The equatorial chambers tend to form radial rows; roofs and floors perforate. Each chamber communicates by a medianly placed stolon with the adjacent proximal and distal chamber of the same row. Communication with chambers of adjacent rows by means of stolons at the sides of the chambers. — Lower Eocene, State of Veracruz, Mexico; Paleocene, India.

Orbitosiphon, a generic name for specimens first described as *Lepidocyclina* (*Polylepidina*) *punjabensis* Davies, is considered to be a synonym of *Actinosiphon*. The arrangement and structure of the embryonic apparatus and the equatorial chambers are similar to that of *Actinosiphon*. However, the stolon system has not been described completely.

Lepidocyclina (*Polylepidina*) *zeijlmansi* Tan (De Ingen. Nederl. Ind. IV, Mijnb. en Geol., 3 de jaarg., No. 1, 1936, pp. 7–14, 1 pl.) has many features which suggest *Actinosiphon* and probably belongs in this genus.

Genus LEPIDOCYCLINA Gümbel, 1870

Plate 31, figures 1–11; Key, plate 39, figures 3–9; plate 41, figures 9–12; plate 55, figure 7

Genotype, *Nummulites mantelli* Morton

Lepidocyclina Gümbel, Abhandl. k. bay. Akad. Wiss., vol. 10, 1868 (1870), p. 689; Lemoine and R. Douvillé, Mém. Soc. Géol. France, Mém. No. 32, 1904, 41 pp., 3 pls.; Cushman, U. S. Geol. Surv., Prof. Paper 125, 1920, pp. 55–105, pls. 12–35; H. Douvillé, Mém. Soc. Géol. France, n. ser., vol. 1, Mém. No. 2, 1924, pp. 1–49, pls. 5, 6, 48 text figs.; *ibid*., vol. 2, Mém. No. 2, 1925, pp. 51–123, pls. 3, 4, text figs. 49–83; Vaughan, Bull. Geol. Soc. Amer., vol. 35, 1924, pp. 794–802, 807–812, pl. 30, figs. 1–3; pls. 31–35; pl. 36, figs. 1–3; Proc. Acad. Nat. Sci. Phila., vol. 79, 1928, p. 299, pl. 23, figs. 1*a*, *b*, 2; in Cushman, Foraminifera, 1928, p. 345, pl. 56, fig. 11; pl. 57, figs. 1, 3, 5–7; pl. 59, figs. 1–11; *ibid*., 1933, pp. 290–292, pl. 31, figs. 1–11; Key, pl. 39, figs. 3–9; Vaughan and Cole, in Cushman, Foraminifera, 1940, pp. 319, 320, pl. 31, figs. 1–11; Key, pl. 38, figs. 11–13; pl. 39, figs. 3–9; pl. 41, figs. 9–12; M. G. Rutten, Geol. en Mijnbouw, n. ser., jaarg. 3, No. 2, 1941, pp. 54–58; Glaessner, Principles of Micropal., 1945, p. 167, pl. 11, fig. 5, text figs. 37–41.

Cyclosiphon Galloway, Man. Foram., 1933, p. 433, pl. 41, figs. 1–10.

Astrolepidina Silvestri, Riv. Ital. di Pal., vol. 37, 1931, p. 35.

Test flat to inflated, lenticular, circular, selliform or stellate. Surface glabrous or papillate, with or without costae or radial ridges. Embryonic chambers of megalospheric forms normally of one of six types: (1) several large chambers, as many as eight, two or more of which are subequal in size (polylepidine); (2) two large, subequal chambers separated by a straight wall, with variable peri-embryonic chambers, or teratologically, one large chamber with smaller chambers around its periphery (pliolepidine); (3) two equal or unequal chambers, separated by a straight wall (lepidocycline); (4) two unequal embryonic chambers, the larger of which partly embraces the smaller (nephrolepidine); (5) a large central chamber, virtually surrounded by a ring of five to ten smaller chambers (multilepidine); (6) two unequal chambers, the larger of which extends entirely around the smaller, except at the place of attachment of the smaller to the inside wall of the larger (eulepidine). Chamber walls pierced by cribriform perforations. Equatorial chambers in concentric rings, which are modified in species with stellate tests, or radially arranged. In those subgenera with the equatorial chambers arranged in concentric rings, the chambers in one ring usually alternate in position with those in adjacent rings in such a way as to produce intersecting, outwardly convex curves. According to their shape as seen in plan, the following types are recognized: (*a*) outer wall convex, side walls converge to a point, or the inner boundary may be formed by the outer wall of a chamber of an inner ring, the transverse or radial diameter may be the longer; (*b*) type (*a*) grades into rhomboid or diamond shaped chambers, the radial or transverse diagonal may be the longer, or into polygonal shaped chambers; (*c*) spatulate chambers, which have an elongate pointed inner end, radially elongate or short; (*d*) hexagonal chambers, elongate or short. Although for species or even for subgenera, the types of chambers may be characteristic, there is variation, and there is intergradation between the different types. The equatorial chambers communicate with the chambers of the same ring, and with those of the adjacent rings by means of stoloniferous apertures. The roofs of the chambers are minutely cribriform. Lateral chambers are usually well developed, their roofs minutely cribriform, communication also through apertures at the chamber ends. Pillars and papillae or pustules variable in development. They may have their origin at the surface of the equatorial layer or between the ends of the lateral chambers. — Middle Eocene to lower Miocene. Circummundane in tropical and near tropical latitudes in shallow water sediments.

Subgenus POLYLEPIDINA Vaughan, 1924

Plate 31, figure 1; Key, plate 39, figures 6, 9

Genotype, *Lepidocyclina* (*Polylepidina*) *chiapasensis* Vaughan

Polylepidina VAUGHAN, Bull. Geol. Soc. Amer., vol. 35, 1924, p. 807, pl. 30, figs. 1–3, text figs. 4*a-e*; in CUSHMAN, Foraminifera, 1928, p. 350, pl. 59, fig. 1; Nat. Acad. Sci. Proc., vol. 15, 1929, p. 289, pl., figs. 3–8; in CUSHMAN, Foraminifera, 1933, pp. 292, 294, pl. 31, fig. 1; Key, pl. 39, figs. 6, 9; GALLOWAY, Man. Foram., 1933, p. 431, pl. 40, figs. 12, 13; TAN, De Mijningenieur, jaarg. 6, 1939, pp. 59–60, 76–80, pl. 1, fig. 2; pl. 2, figs. 5, 6; VAUGHAN, and COLE, in CUSHMAN, Foraminifera, 1940, p. 321, pl. 31, fig. 1; Key, pl. 39, figs. 6, 9; M. G. RUTTEN, Geol. en Mijnbouw, n. ser., jaarg. 3, No. 2, 1941, p. 58, text figs. 10–12; GLAESSNER, Principles of Micropal., 1945, pp. 167, 168.

Polyorbitoina VAN DE GEYN and VAN DER VLERK, Leidsche Geol. Meded., deel 7, 1935, p. 227.

Embryonic chambers 4 to 10 in number, of which one or two may be somewhat larger than the others. There may be two subequal chambers somewhat larger than two other subequal chambers and the four chambers so arranged as to form a cross. The chambers may be arranged also in a distinct spiral. The chambers tend to grade on one hand in *Pliolepidina*, and on the other into *Lepidocyclina*. Equatorial chambers with curved outer walls, sides converging to a truncate or to a pointed inner end, radial diameter shorter than the transverse. A four-stolon system is present. Each chamber in the type species communicates by openings with two chambers of the next inner and two of the next outer rings. The number of the apertures in a chamber wall ranges from 1 to 3. They occur in the corners of the chambers, and are from 23 to 38 μ in diameter. The equatorial chambers in their form and the apertures resemble rather closely those of *Orbitoides*. Pillars and surface papillae are present. — Middle to upper Eocene of southern Mexico, middle Eocene of Florida, Alabama, Mississippi, Louisiana, and Texas, Eocene of West Indies.

Polylepidina in its embryonic features appears to be very primitive.

Subgenus PLIOLEPIDINA H. Douvillé, 1917

Plate 31, figures 2–4; Key, plate 41, figure 10; plate 55, figure 7

Genotype, *Lepidocyclina* (*Pliolepidina*) *pustulosa* forma *tobleri* H. Douvillé forma *teratologica*

Pliolepidina H. Douvillé, C. R. Paris Acad. Sci., vol. 164, 1917, p. 844, text figs. 1–6; *ibid.*, vol. 161, 1915, p. 727; Mém. Soc. Géol. France, ser. 2, vol. 1, Mém. 2, 1924, pp. 41–44, pl. 1, figs. 2, 3, text figs. 27–32b; text figs. 34, 35; Vaughan, Bull. Geol. Soc. Amer., vol. 35, p. 796, pl. 33, figs. 1, 2; in Cushman, Foraminifera, 1928, p. 350, pl. 59, fig. 2; *ibid.*, 1933, pp. 294, 295, pl. 31, figs. 2–4; Galloway, Man. Foram., 1933, p. 432, pl. 40, fig. 14; Vaughan and Cole, in Cushman, Foraminifera, 1940, p. 322, pl. 31, figs. 2–4; Key, pl. 41, fig. 10; Geol. Soc. Amer. Sp. Paper 30, 1941, pp. 64, 65; M. G. Rutten, Geol. en Mijnbouw, n. ser., jaarg. 3, 1941, pp. 58, 59, text fig. 20; Glaessner, Principles of Micropal., 1945, p. 168.

Multicyclina Cushman, U. S. Nat. Mus. Bull. 103, 1919, p. 96, pl. 41, figs. 2–4.

Orbitoina van de Geyn and van der Vlerk, Leidsche Geol. Meded., deel 7, 1935, p. 227.

Isorbitoina van de Geyn and van der Vlerk, *ibid.*, p. 227.

Pliorbitoina van de Geyn and van der Vlerk, *ibid.*, p. 227.

Neolepidina Brönnimann, Eclogae geol. Helvetiae, vol. 39, 1946, pp. 373–379.

Test with well-developed lateral chambers on each side of the equatorial layer. Embryonic chambers, two, large, subequal, separated by a straight wall, with variable periembryonic chambers; or teratologically, one large chamber with smaller chambers around its periphery. Equatorial chambers with curved outer walls, inner ends pointed or truncate, or lozenge-shaped; each connects by stolons that pierce the chamber walls at four places with four adjacent chambers. Stolons present in the side walls of the lateral chambers. — Upper Eocene of Island of Trinidad, Panama, southern Mexico; middle Eocene, Florida.

Recent studies by Vaughan and Cole have proven that *L.* (*Pliolepidina*) *tobleri*, *L.* (*Isolepidina*) *pustulosa*, and *L.* (*Isolepidina*) *trinitatis*, as well as *L.* (*Multicyclina*) *duplicata* and *L. panamensis*, all belong to one species, of which *L. tobleri* and *L. panamensis* represent a teratologic form and the others represent the normal. This study confirms Douvillé's suggestion that the specimens placed in this subgenus might be teratologic.

Although it was necessary to enlarge the original definition to include the normal generation, this does not invalidate the subgeneric name *Pliolepidina*, as the Law of Priority states that the oldest available name is to be retained "when an animal represents a regular succession of dissimilar generations which have been considered as belonging to different species or even to different genera." Therefore, the name *Neolepidina* Brönnimann is clearly a synonym of *Pliolepidina*.

L. tobleri possesses a four-stolon system which is characteristic of the normal form of the species. Brönnimann has discovered recently (1946), however, that at the outermost periphery of the equatorial layer there is a change from the four-stolon system to a six-stolon system.

Subgenus LEPIDOCYCLINA Gümbel, 1870

Plate 31, figure 5; Key, plate 39, figures 3, 8; plate 41, figures 11, 12

Genotype, *Lepidocyclina mantelli* (Morton) Gümbel

Isolepidina H. Douvillé, C. R. Acad. Sci., vol. 161, 1915, p. 724; Mém. Soc. Géol. France, n. ser., vol. 1, Mém. No. 2, 1924, pp. 11, 34 *et seq.*

Lepidocyclina Vaughan, in Cushman, Foraminifera, 1928, p. 351, pl. 59, fig. 5; Jour. Pal., vol. 3, 1929, p. 28; in Cushman, Foraminifera, 1933, p. 295, pl. 31, fig. 5; Vaughan and Cole, in Cushman, Foraminifera, 1940, pp. 322, 323, pl. 31, fig. 5; Key, pl. 41, figs. 11, 12; M. G. Rutten, Geol. en Mijnbouw, n. ser., jaarg. 3, No. 2, 1941, p. 59, text fig. 13; Glaessner, Principles of Micropal., 1945, p. 168.

Cyclosiphon Galloway, Man. Foram., 1933, p. 433, pl. 41, figs. 1–3.

Nucleoconch bilocular, chambers equal or subequal in size, separated by a straight wall. Equatorial chambers of all the kinds enumerated in stating the characters of the genus. Lateral chambers and pillars of different kinds. Six or eight-stolon systems present. — Upper Eocene to upper Oligocene and perhaps lower Miocene, North, Central and South America, West Indies, East Indies.

A number of American species of this subgenus possess stellate outlines. In some species, there is a tendency for the embryonic chambers to approach the nephrolepidine form.

Subgenus NEPHROLEPIDINA H. Douvillé, 1911

Plate 31, figures 6, 7

Genotype, *Lepidocyclina marginata* (Michelotti)

Nephrolepidina H. Douvillé, Philippine Jour. Sci., vol. 6, 1911, p. 59; Samml. Geol. Reichsmus., Leiden, ser. 2, vol. 8, 1912, pp. 269–270; C. R. Acad. Sci., vol. 161, 1915, p. 727; Mém. Soc. Géol. France, n. ser., vol. 1, Mém. No. 2, 1924, pp. 11, 46 *et seq.*; *ibid.*, vol. 2, Mém. No. 2, pp. 73, *et seq.*, 113, *et seq.*; Vaughan, in Cushman, Foraminifera, 1928, p. 351, pl. 59, fig. 7; *Ibid*, 1933, pp. 295–296, pl. 31, figs. 6, 7; Galloway, Man. Foram., 1933, p. 436, pl. 41, figs. 6, 7; Vaughan and Cole, in Cushman, Foraminifera, 1940, pp. 323, 324, pl. 31, figs. 6, 7; M. G. Rutten, Geol. en Mijnbouw, n. ser., jaarg. 3, No. 2, 1941, p. 60, text fig. 14; Glaessner, Principles of Micropal., 1945, p. 168.

Amphilepidina H. Douvillé, C. R. Acad. Sci., vol. 175, 1922, p. 553; Mém. Soc. Géol. France, n. ser., vol. 1, Mém. No. 2, 1924, pp. 11, 44 *et seq.*; *ibid.*, vol. 2, Mém. No. 2, 1925, p. 100 *et seq.*; Galloway, Man. Foram., 1933, p. 436, pl. 41, figs. 4, 5.

Trybliolepidina VAN DER VLERK, Wetenskap. Med. No. 8, 1928, p. 13, figs. 11–13; Eclog. geol. Helv., vol. 21, 1928, p. 186, pl. 10, figs. 11–13; GALLOWAY, Man. Foram., 1933, p. 437, pl. 41, fig. 8.

Embryonic chambers bilocular, reniform, a smaller, partly embraced by a larger chamber. Equatorial chambers spatulate, with curved outer wall, or lozenge-shaped with the distal end of the chamber angular. Pillars and papillae present. Six-stolon system when complete, but some preparations show only a five-stolon system. — Upper Eocene to lower Miocene, Europe, America, East Indies, Japan.

After establishing the subgeneric name *Nephrolepidina*, Douvillé proposed the subgeneric designation *Amphilepidina* for those species of *Lepidocyclina* that have reniform embryonic chambers, and spatulate equatorial chambers, and in 1925 stated that *L. sumatrensis* may be considered the type. The application of a similar procedure to *Lepidocyclina* s.s. would divide it into two or more subgenera. Furthermore, according to Douvillé's own figures (see 1924, p. 46, figs. 40–41) equatorial chambers with both curved and angular outer walls occur in the same specimen. Vaughan is convinced for these reasons that there is no need for *Amphilepidina*.

In *Trybliolepidina* the initial, embraced chamber has its side-walls perpendicular to its attached side. Caudri (Tertiary Deposits of Soemba, Amsterdam, 1934, pp. 117–121, text fig. 21) has demonstrated that there is complete intergradation in a single species from nephrolepidine to trybliolepidine type of embryonic apparatus. Since this intergradation may be demonstrated in a single species, it appears that two subgenera should not be recognized.

There are several stellate species of *Nephrolepidina*. These include such species as *L. martini* Schlumberger in the lower Miocene of Java and *L. dartoni* Vaughan from the Oligocene of Cuba.

Subgenus EULEPIDINA H. Douvillé, 1911

Plate 31, figures 8–11; Key, plate 39, figures 5, 7
Genotype, *Lepidocyclina dilatata* (Michelotti)

Eulepidina H. DOUVILLÉ, Philippine Journ. Sci., vol. 6, 1911, p. 59; Samml. Geol. Reichsmus., Leiden, vol. 2, 1912, pp. 268, 269; C. R. Acad. Sci., vol. 161, 1915, p. 726; Mém. Soc. Géol. France, vol. 1, Mém. No. 2, 1924, pp. 11, 48, 49; *ibid.*, vol. 2, Mém. No. 2, 1925, pp. 66 *et seq.*, 96 *et seq.*; VAUGHAN, in CUSHMAN, Foraminifera, 1928, p. 352, pl. 57, figs. 5–7; pl. 59, figs. 8–11; *ibid.*, 1933, p. 296, pl. 31, figs. 8–11; Key, pl. 39, figs. 5, 7; GALLOWAY, Man. Foram., 1933, p. 437, pl. 41, figs. 9, 10; VAUGHAN and COLE, in CUSHMAN, Foraminifera, 1940, p. 324, pl. 31, figs. 8–11; Key, pl. 39, figs. 5, 7; M. G. RUTTEN, Geol. en Mijnbouw, n. ser., jaarg. 3, No. 2, 1941, pp. 60, 61, text fig. 15; GLAESSNER, Principles of Micropal., 1945, p. 168.

Embryonic chambers bilocular, a larger chamber completely surrounds a smaller one, except at the place where the smaller is attached to the inside of the wall of the larger. The equatorial chambers are spatulate, the outer wall curved or angular. The radial diameter may be relatively long or rather short.—Middle Oligocene to Miocene, Europe, America, East Indies.

Nephrolepidina and *Eulepidina* intergrade. This has been demonstrated particularly in *L. undosa* Cushman. *Eulepidina* also intergrades with *Trybliolepidina.*

Subgenus MULTILEPIDINA Hanzawa, 1932

Key, plate 41, figure 9

Genotype, *Lepidocyclina* (*Multilepidina*) *irregularis* HANZAWA

Multilepidina HANZAWA, Proc. Imp. Acad. Japan, vol. 8, 1932, pp. 447, 448, figs. 1–6; VAUGHAN and COLE, in CUSHMAN, Foraminifera, 1940, pp. 324, 325, Key, pl. 41, fig. 9; M. G. RUTTEN, Geol. en Mijnbouw, n. ser., jaarg. 3, No. 2, 1941, p. 61, text fig. 21; COLE, B. P. Bishop Mus., Bull. 181, 1945, pp. 283–286, pl. 21, figs. A–H; pl. 22, figs. A–D; GLAESSNER, Principles of Micropal., 1945, p. 168.

Cyclolepidina WHIPPLE, B. P. Bishop Mus., Bull. 119, 1934, pp. 143–145, pl. 20, figs. 1–8; text fig. *b*.

Embryonic apparatus large, multilocular, irregular in outline, composed of a large central chamber, virtually surrounded by a ring of 5 to 10 smaller chambers. Outer wall of the embryonic apparatus thick, finely tubulated; walls separating the component chambers from one another, thin, compact. Outer wall and partitions of the nucleoconch traversed by stoloniferous passages of communication. Equatorial chambers spatulate. Communication between equatorial chambers is by means of a six-stolon system.—Lower Miocene, Formosa, and Vitilevu, Fiji.

Species of *Multilepidina* have commonly radiating surface ridges.

The embryonic chambers of this subgenus resemble the teratological form of *Pliolepidina*, but there are six stolons for each equatorial chamber in *Multilepidina*, whereas in *Pliolepidina* there is a four-stolon system, except in the outermost chambers.

Study of additional preparations of the genoholotype furnished by Professor Hanzawa as well as the type slides of *Cyclolepidina* has demonstrated that *Cyclolepidina* is a synonym of *Multilepidina.*

Genus TRIPLALEPIDINA Vaughan and Cole, 1938

Key, plate 41, figures 1–3

Genotype, *Triplalepidina veracruziana* Vaughan and Cole

Triplalepidina VAUGHAN and COLE, Jour. Pal., vol. 12, 1938, pp. 167–169, pl. 27; VAUGHAN and COLE, in CUSHMAN, Foraminifera, 1940, p. 325, Key, pl. 41, figs. 1–3; M. G. RUTTEN, Geol. en Mijnbouw, n. ser., jaarg. 3, No. 2, 1941, p. 62, text figs. 6, 7.

Test rather small, maximum known diameter about 3.1 mm., lenticular to slightly asymmetrical lenticular, sloping regularly from the central inflated part to the margin, with or without a very narrow marginal flange. Surface ornamentation consists of a reticulate mesh, with or without apical papillae. Embryonic chambers bilocular, one somewhat larger than the other, separated by a straight wall; two or more accessory chambers partially surround the embryonic apparatus and arranged so that the larger accessory chambers are on opposite sides of the embryonic chambers and adjacent to the ends of the dividing wall. Equatorial chambers have curved outer and converging inner walls, which in many chambers do not meet, producing a truncate end on the proximal side of the chamber. Chambers near the center are commonly larger than those near the periphery of the test. There are oblique stolons with at least four openings in each chamber. There are no indications of radial stolons. The presence or absence of annular stolons is left in doubt. Outer portion of the equatorial layer is divided into three parts by a wedge-shaped layer of clear shell material which separates layers of small equatorial chambers. Lateral chambers occur in regular tiers except in the central portion of the test where certain chambers extend over two tiers. Normally, there is a large pillar present on each side of the equatorial layer at the center of the test. Smaller pillars are occasionally noted between the tiers of lateral chambers. Two stoloniferous apertures occur in the walls of numerous lateral chambers. — Upper Eocene of Mexico.

Subfamily Helicolepidinae Tan

Genus EULINDERINA Barker and Grimsdale, 1936

Key, plate 41, figures 6, 7

Genotype, *Eulinderina guayabalensis* (Nuttall)

Eulinderina Barker and Grimsdale, Jour. Pal., vol. 10, 1936, pp. 237, 238, pl. 32, figs. 8, 9; pl. 34, figs. 8, 10, 11; pl. 37, fig. 4; Vaughan and Cole, in Cushman, Foraminifera, 1940, pp. 325, 326, Key, pl. 41, figs. 6, 7; M. G. Rutten, Geol. en Mijnbouw, n. ser., jaarg. 3, No. 2, 1941, pp. 52, 53, text fig. 9; Glaessner, Principles of Micropal., 1945, p. 169.

Eolepidina Tan, De Mijningenieur, jaarg. 6, 1939, p. 69.

Test compressed lenticular to discoidal, surface papillate. Embryonic apparatus consists of two subequal chambers followed by one to six chambers forming a trochoid spiral, which never exceeds one whorl. The spiral has a thick outer wall, and the individual chambers are connected by apertures with counter-septa and anteriorly directed inner lips. The equatorial chambers are similar to those of *Polylepidina* with curved outer walls and truncate or pointed inner ends. The equatorial chambers are arranged in annular rings and communicate by means of stoloniferous apertures; four to each chamber, arranged at first in a single plane, occasionally in two planes toward the periphery. The equatorial layer is overlain on each side by variably thick laminae of perforate shell substance which may show irregularly developed lateral chambers. Pillars are present. — Middle Eocene of eastern Mexico.

Genus HELICOLEPIDINA Tobler, 1922

Plate 31, figures 13, 14

Genotype, *Helicolepidina spiralis* Tobler

Helicolepidina TOBLER, Eclog. geol. Helv., vol. 17, 1922, pp. 380–384, 3 text figs.; H. DOUVILLÉ, *ibid.*, vol. 17, 1923, pp. 566–569, 2 text-figs.; VAUGHAN, in CUSHMAN, Foraminifera, 1928, p. 354, pl. 59, figs. 13, 14; *ibid.*, 1933, pp. 298, 299, pl. 31, figs. 13, 14; GALLOWAY, Man. Foram., 1933, p. 427, pl. 40, fig. 1; BARKER, Jour. Pal., vol. 8, 1934, pp. 344–351, pl. 47, figs. 1–11; 5 text figs.; VAUGHAN, *ibid.*, vol. 10, 1936, pp. 248–252, pl. 39, figs. 1–5; pl. 40, figs. 1–8; VAUGHAN and COLE, in CUSHMAN, Foraminifera, 1940, p. 326, pl. 31, figs. 13, 14; M. G. RUTTEN, Geol. en Mijnbouw, n. ser., jaarg. 3, No. 2, 1941, p. 54, text fig. 18; BRÖNNIMANN, Schweizerischen Pal. Abhandl., vol. 64, 1944, pp. 22–36, pl. 3, figs. 16–21, 23; GLAESSNER, Principles of Micropal., 1945, pp. 169, 170, fig. 44.

Helicocyclina TAN, Proc. Kon. Akad. Wet., Amsterdam, vol. 39, 1936, p. 995.

Test rather small, maximum known diameter of megalospheric specimens about 4 mm., lenticular, sometimes slightly asymmetrical with reference to the equatorial plane; central part arched, sloping to a rather sharp edge. Surface papillate. Embryonic chambers bilocular, one somewhat larger than the other, separated by a straight or slightly curved wall; the smaller chamber may be deformed to the contour of the larger, walls pierced by minute cribriform perforations. The embryonic chambers are partially surrounded by a girdle of 7–9 chambers, some of which are slightly larger than the adjacent equatorial chambers. Equatorial chambers of three types: (*a*) A spiral row of larger chambers which extend through a little more than a complete whorl. The walls of these chambers slope in a curve from their anterior inner ends backward to their posterior outer ends. (*b*) Outside this row there is a row of smaller spiral chambers. (*c*) Smaller chambers which occupy most of the equatorial plane. In larger specimens the growth may become cyclical. The arrangement of these chambers is somewhat irregular, but with a tendency toward occurrence along radial lines. The shape of the chambers varies from arcuate to hexagonal; communication by six stolons, two on each side of a chamber, and a proximal and a distal radial stolon. Roofs cribriform-perforate. Lateral chambers well developed, floors and roofs with cribriform perforations, communication also by passages for protoplasmic stolons. Pillars present. — Upper Eocene of the Caribbean region, Louisiana, Florida, and Peru.

Family 49. DISCOCYCLINIDAE Vaughan and Cole

By T. Wayland Vaughan and W. Storrs Cole

Revised by W. Storrs Cole

Test circular or stellate in plan, thin or inflated in cross section; composed of an equatorial layer with lateral chambers on each side. Embryonic chambers consist of a subspherical inner chamber partly or completely embraced by an outer chamber. Equatorial chambers in annuli connected by stolons with adjacent chambers in the same annulus and with adjacent chambers in the next inner and the next outer annulus. When the radial chamber walls are developed, the chambers are rectangular to faintly hexagonal in plan. An intraseptal and intramural canal system is present.

Although the presence of intraseptal and intramural canals were long suspected in representatives of this family, it was not until recently that Vaughan demonstrated beyond a doubt that these canals are present. This corroboration of the long-held opinion that there is a canal system in *Discocyclina* confirms the dissociation which was made in the last edition of this book of this group from the Orbitoididae.

The Discocyclinidae with canals and marginal cord are related to the Camerinidae and probably are derived from a *Camerina*-like ancestor, whereas the Orbitoididae, which lack canals, are possible mutants from an *Amphistegina*-like ancestor.

In a review of the American Discocyclinidae published in 1945 (Geol. Soc. Amer. Memoir 9, pp. 67–69), Vaughan uses essentially the same classification as that proposed at the time of the introduction of the family name Discocyclinidae. The major difference is the introduction of the subgenus *Asterophragmina* Rao, 1942, for certain stellate specimens from the upper Eocene Yaw shales of Burma which externally resemble *Asterocyclina*, but which have defective or absent radial walls with the annular stolon on the distal side of the chambers.

During the time the memoir by Vaughan was in the process of publication, Brönnimann (Eclogae geol. Helvetiae, vol. 38, 1945, pp. 579–619, pls. 21, 22) suggested that the family Discocyclinidae be divided into two subfamilies, the Discocyclininae in which the equatorial layer is composed of chambers and chamberlets, and the Orbitoclypeinae in which the equatorial layer is composed only of chambers.

Under the Discocyclininae he proposes two groups, those genera with the annular stolon in a proximal position (*Discocyclina* and *Aktinocyclina*), and those genera with the annular stolon in a distal position (*Proporocyclina, Pseudophragmina* and *Athecocyclina*). Under the Orbitoclypeinae there are two genera (*Orbitoclypeus* and *Asterocyclina*).

The classification of Brönniman is based largely upon a detailed analysis of the microspheric generations of certain species of *Discocyclina* and

Asterocyclina. As there is considerable variation in the shape of the chambers in most of the larger Foraminifera, it does not seem advisable at the present time to abandon the older classification of this family.

Genus DISCOCYCLINA Gümbel, 1870

Plate 30, figures 18, 20–22; Key, plate 38, figure 17; plate 41, figure 13; plate 55, figures 5, 10

Genotype, *Orbitulites pratti* Michelin

Discocyclina Gümbel, Abhandl. k. bay. Akad. Wiss., vol. 10, 1868 (1870), p. 109; H. Douvillé, Bull. Soc. Géol. France, ser. 4, vol. 22, 1922, p. 64, numerous figs. in text and on plates; Vaughan, in Cushman, Foraminifera, 1928, p. 341, pl. 56, figs. 1, 7; pl. 58, figs. 18–22; U. S. Nat. Mus. Proc., vol. 76, art. 3, 1929, pp. 1–18, pls. 1–7 (part); Schenck, San Diego Soc. Nat. Hist., Trans., vol. 5, 1929, pp. 211–220, pl. 28, fig. 1; text figs. 3–6; Vaughan, in Cushman, Foraminifera, 1933, p. 299, pl. 30, figs. 18–22; Key, pl. 38, fig. 17; pl. 39, fig. 2; Galloway, Man. Foram., 1933, p. 448, pl. 42, figs. 7, 8; van der Weijden, Diss. Leiden Univ., N. V. Leidsche Courant, 1940 (part), pp. 21–71, pls. 1–12; Vaughan and Cole, in Cushman, Foraminifera, 1940, p. 328, pl. 30, figs. 18–22; Key, pl. 38, fig. 17; pl. 39, fig. 2; pl. 41, fig. 13; Brönnimann, Eclogae geol. Helvetiae, vol. 33, 1940, pp. 252–274, pl. 14, pl. 15; *ibid.*, vol. 38, 1945, pp. 579–592, pl. 21, fig. 3; Glaessner, Principles of Micropal., 1945, p. 170, pl. 12, fig. 1.

Rhipidocyclina Gümbel, Abhandl. k. bay. Akad. Wiss., vol. 10, 1868 (1870), p. 110.

Orthophragmina Munier-Chalmas, Étude du Tithon. du Crét. et du Tert. du Vicentin, 1891, p. 18.

Orbitoclypeus Silvestri, Atti Pont. Accad., vol. 60, 1907, p. 106.

Exagonocyclina Checchia-Rispoli, Giorn. Sci. Nat. Econ., Palermo, vol. 26, 1908, pp. 159, 187.

Eudiscodina van der Weijden, Diss. Leiden Univ., N. V. Leidsche Courant, 1940, pp. 8, 22, 34, 35, 46.

Umbilicodiscodina van der Weijden, *ibid.*, pp. 8, 31, 39.

Trybliodiscodina van der Weijden, *ibid.*, pp. 8, 27, 48.

Hexagonocyclina Caudri, Bull. Amer. Pal., vol. 28, 1944, pp. 362–364, pl. 2, figs. 7, 9.

Test discoidal, lenticular, thin or inflated, selliform or with raised radiating ribs. Surface granulate. Embryonic apparatus consists of a smaller, subglobular chamber, partly embraced by a larger chamber or initial chamber attached to the inside wall of the second chamber either tangentially or by a short peduncle. There may be a third type in which the initial chamber is completely surrounded by the second chamber and extends from the roof to the floor of the equatorial layer. Equatorial chambers of rectangular shape or faintly hexagonal, usually radially elongate, arranged in more or less concentric rings, the radial walls of chambers of contiguous rings usually alternating in position. Chambers of one ring communicate with the chambers of adjacent rings by a row of lateral small apertures 5 to 7 μ in diameter, usually 2 to 4 in each row; chambers in the same ring communicate by annular stolons, situated at the proximal end of the radial wall. Small, cribriform perforations occur in the chamber roofs. Lateral chambers greatly developed, usually composing most of the test. Pillars usually well-developed, terminating on the surface in papillae or granulations of distinctive character, of much value in recognizing species.—Danian to uppermost Eocene; Europe, India, New Zealand, Tonga Islands, East Indies, West Indies, North, Central, and South America.

In certain species of *Discocyclina* and *Asterocyclina* there is a tendency for the equatorial chambers to develop a hexagonal shape. Vaughan (Geol. Soc. Amer. Mem. 9, 1945, p. 66) has stated, "the chamber walls may show a tendency toward faintly hexagonal outlines, a condition not surprising because the radial walls in adjacent annuli alternate in position." For such species the generic designations *Orbitoclypeus, Exagonocyclina* and *Hexagonocyclina* have been proposed by various authors. It would appear that this feature is a specific rather than a generic character. Until more information is available, these species are referred to *Discocyclina.*

Subgenus AKTINOCYCLINA Gümbel, 1870

Key, plate 55, figure 5
Genotype, *Orbitulites radians* d'Archiac

Aktinocyclina GÜMBEL, Abhandl. k. bay. Akad. Wiss., vol. 10, 1868 (1870), p. 110, many figs.; VAUGHAN, in CUSHMAN, Foraminifera, 1928, p. 344; *ibid.*, 1933, p. 300; VAUGHAN and COLE, *ibid.*, 1940, p. 328; BRÖNNIMANN, Eclogae geol. Helvetiae, vol. 38, 1945, pp. 560–578, pl. 20, 14 text figs.; VAUGHAN, Geol. Soc. Amer. Mem. 9, 1945, pp. 85, 86, pl. 32.

Orthophragmina MUNIER-CHALMAS (part), (*op. sup. cit.*), 1891.

Actinocyclina H. DOUVILLÉ, Bull. Soc. Géol. France, ser. 4, vol. 22, 1922, p. 79, pl. 5, figs. 3–8; GALLOWAY, Man. Foram., 1933, p. 449, pl. 42, fig. 9.

Similar to *Discocyclina* except that there are on the surface of the test elevated rays, which, however, do not terminate in protuberant angles on the margin of the test, as is the case in *Asterocyclina.* The elevated rays result from an increase in the number of lateral chambers, whereas the rays in *Asterocyclina* are produced by the equatorial chambers.—Middle and upper Eocene, Europe.

As the annular stolons are regularly on the proximal side of the chambers and the radial walls are perfect, these forms are considered to be a subgenus under *Discocyclina.* The radial walls of the equatorial chambers in adjacent annuli either are in alignment or alternate, whereas in *Discocyclina* they usually alternate. The equatorial layer, however, is clearly defined from the center of the test to the periphery, which is another similarity with *Discocyclina.*

Genus ASTEROCYCLINA Gümbel, 1870

Plate 30, figure 19; Key, plate 39, figure 2
Genotype, *Calcarina* (?) *stella* d'Archiac

Asterocyclina GÜMBEL, Abhandl. k. bay. Akad. Wiss., vol. 10, 1868 (1870), p. 689, many figs.; VAUGHAN, Jour. Pal., vol. 1, 1927, p. 285; in CUSHMAN, Foraminifera, 1928, p. 344, pl. 56, fig. 1; *ibid.*, 1933, p. 301, pl. 30, fig. 19; Key, pl. 39, fig. 2;

Galloway, Man. Foram., 1933, p. 449, pl. 42, figs. 11, 12; Vaughan and Cole, in Cushman, Foraminifera, 1940, p. 329, pl. 30, fig. 19; Key, pl. 39, fig. 2; Brönnimann, Abh. Schweiz. Pal. Gesell., vol. 63, 1942, pp. 13–19; Eclogae geol. Helvetiae, vol. 38, 1945, pp. 592–605, pl. 21, figs. 1, 2, 4–9; Glaessner, Principles of Micropal., 1945, p. 170.

Asteriacites Schlotheim, Die Petrefactenkunde, Nachtrag, 1822, p. 71, pl. 12, fig. 6 (not of Schlotheim of earlier date, *teste* Hodson); Vaughan, Bull. Geol. Soc. Amer., vol. 35, 1924, pp. 790, 793.

Cisseis R. J. L. Guppy, Quart. Jour. Geol. Soc. London, vol. 22, 1866, p. 584, pl. 26, figs. 19*a*, *b* (not Laporte and Gory, 1839).

Asterodiscus Schafhäutl, Sud-Bayern Leth. Geognos., 1863, p. 130 (not Ehrenberg, 1838); H. Douvillé, Bull. Géol. Soc. France, ser. 4, vol. 22, pp. 60 (figs. 13, 14), 76.

Asterodiscocyclina Berry, Eclogae geol. Helvetiae, vol. 21, 1928, pp. 405–407, pl. 1, figs. 1–7.

Orthocyclina van der Vlerk, Verh. Geol.-Mijn. Gen. Ned. Koloniën., Geol. Ser., deel 7, 1933, pp. 91–98, plate with figs. 1, 2; Galloway, Man. Foram., 1933, p. 449, pl. 42, fig. 10.

Isodiscodina van der Weijden, Diss. Leiden Univ., N. V. Leidsche Courant, 1940, p. 38.

Test flat to lenticular, stellate, the rays either prominent, extending from the center to beyond the periphery, or faint to absent on the surface of the test, but distinct in median sections. The number of arms ranges from four to more than twelve. In many species the ends of the rays determine the shape of the test, as seen in plan, and project beyond the margin of the interradial portions of the test. The initial chamber of the embryonic apparatus typically is partly embraced by the second chamber. However, the embryonic chambers may be nearly equal in size and separated by a straight wall, or the initial chamber may be surrounded completely by the second chamber in which case the initial chamber apparently extends from the roof to the floor of the equatorial layer. Peri-embryonic chambers completely surround the embryonic chambers in nearly all species. The rays are produced by the equatorial chambers. — Middle and upper Eocene, Europe, America, Pacific Islands, Morocco.

Nodocyclina Heim (Abh. Schweiz. pal. Gesell., vol. 35, 1908, p. 271, figs. 25, 26) represents either *Asterocyclina* or *Aktinocyclina*.

Asterocyclina has been considered to be a subgenus of *Discocyclina*. Brönnimann (1945) has demonstrated in one species that in the microspheric generation there appears to be a slightly different development in the initial chambers from that of *Discocyclina*. In *Asterocyclina* the initial chambers are first arcuate, then spatulate, and finally rectangular, whereas in *Discocyclina* they are arcuate, followed by segment-like chambers, and finally annular chambers. This development in connection with the distinctive external shape, the expansion of the equatorial chambers to form the rays, and the stellate pattern of the equatorial layer seems to warrant generic rank for *Asterocyclina*.

Genus PSEUDOPHRAGMINA H. Douvillé, 1923

Key, plate 39, figure 1; plate 41, figure 14; plate 55, figures 3, 4, 8, 9
Genotype, *Orthophragmina floridana* Cushman

Pseudophragmina H. Douvillé, Bull. Soc. Géol. France, ser. 4, vol. 23, 1923, p. 373; pl. 12, figs. 1–3; Vaughan, in Cushman, Foraminifera, 1933, pp. 299, 300; Galloway, Man. Foram., 1933, p. 446, pl. 42, figs. 4, 5; Vaughan, Jour. Pal., vol. 10, 1936, pp. 258, 259, pl. 43, figs. 3–7; Vaughan and Cole, in Cushman, Foraminifera, 1940, pp. 329, 330, Key, pl. 39, fig. 1; pl. 41, fig. 14.

Test circular in plan, discoidal, lenticular, thin or inflated, occasionally undulate. Embryonic apparatus consists of a smaller, subglobular chamber, partly embraced by a larger chamber. The radial walls of the equatorial chambers may be complete, incomplete, absent or indistinct. Annular stolons connecting the equatorial chambers are usually on the distal side of the equatorial chambers, but not necessarily confined to this position. Lateral chambers are present. — Paleocene, lower, middle and upper Eocene, America. (The Paleocene determination is not absolutely established and may represent lower Eocene.)

The basic differences between *Discocyclina* and *Pseudophragmina* are: (1) the radial chamber walls in adjacent annuli alternate in position in *Discocyclina*, but are in alignment in *Pseudophragmina*; (2) annular stolons in *Discocyclina* are on the proximal side of the chambers, but they are usually distally situated in *Pseudophragmina*; (3) the equatorial layer of *Discocyclina* is thicker than that of *Pseudophragmina*.

In *Pseudophragmina* s.s. the distal part of the radial chamber wall is degenerate, in places represented by rows of granules.

Subgenus PROPOROCYCLINA Vaughan and Cole, 1940

Key, plate 55, figure 8
Genotype, *Discocyclina perpusilla* Vaughan

Proporocyclina Vaughan and Cole, in Cushman, Foraminifera, 1940, p. 330; Vaughan and Cole, Geol. Soc. Amer., Sp. Paper 30, 1941, pp. 57, 61; Brönnimann, Abh. Schweiz. Pal. Gesell., vol. 63, 1942, pp. 1–18, pls. 1, 2; Vaughan, Geol. Soc. Amer. Mem. 9, 1945, pp. 67–69; Glaessner, Principles of Micropal., 1945, p. 170; Brönnimann, Eclogae geol. Helvetiae, vol. 38, 1945, p. 612.

Test circular in plan. Equatorial chambers rectangular; annular stolons usually on the distal sides of the chambers; chamber walls complete, those in adjacent chambers of adjacent annuli usually in alignment. — Lower, middle Eocene, Tampico embayment, Texas, California; upper Eocene, southeastern United States, Haiti, Trinidad, Venezuela.

Subgenus ATHECOCYCLINA Vaughan and Cole, 1940

Key, plate 55, figures 3, 4
Genotype, *Discocyclina cookei* Vaughan

Athecocyclina Vaughan and Cole, in Cushman, Foraminifera, 1940, p. 330; Vaughan and Cole, Geol. Soc. Amer., Sp. Paper 30, 1941, pp. 57, 62; Vaughan,

Geol. Soc. Amer. Mem. 9, 1945, pp. 67–69, 100, pl. 45, figs. 1, 2; GLAESSNER, Principles of Micropal., 1945, p. 170; BRÖNNIMANN, Eclogae geol. Helvetiae, vol. 38, 1945, p. 612.

Test circular in plan, it may be undulate. Radial chamber walls absent or indistinct. — Lower Eocene, Mexico, Alabama, Florida, and Trinidad; middle Eocene, Barbados.

Subgenus ASTEROPHRAGMINA Rao, 1942

Key, plate 55, figure 9

Subgenotype, *Pseudophragmina* (*Asterophragmina*) *pagoda* Rao

Asterophragmina RAO, Geol. Sur. India Records, Prof. Paper 12, vol. 77, 1942, pp. 8–10, pl. 1, fig. 3; pl. 2, figs. 1, 2, 4, 6; VAUGHAN, Geol. Soc. Amer. Mem. 9, 1945, pp. 67–69.

Test stellate with six to eight rays radiating from the center to the margin, resembling *Asterocyclina*. Only the microspheric form is known. In equatorial sections the radial chamber walls are either absent or defective, and the annular stolon is on the distal side of the chambers. In the type description it is stated that "radial septa of adjacent annuli alternate." — Upper Eocene, Burma.

Specimens of this subgenus need more study to substantiate completely their position. In the genus *Pseudophragmina* the radial chamber walls are in alignment, whereas in *Asterophragmina* these walls are described as alternating, a feature of *Asterocyclina*.

Family 50. MIOGYPSINIDAE Tan

By T. Wayland Vaughan and W. Storrs Cole

Revised by W. Storrs Cole

Test small, trigonal to suborbicular in form; embryonic chambers with periembryonic chambers or with an initial spiral, a spiral canal and interseptal canals present, but *there is no marginal cord*; well-developed, lozenge-shaped, equatorial chambers, a net of canals in the chamber walls; lateral chambers well-developed or represented by appressed laminae over the equatorial layer.

Barker and Grimsdale (Ann. and Mag. Nat. Hist., ser. 10, vol. 19, 1937, pp. 161–168, pl. 5, fig. 6; pl. 6, figs. 1–6, 8; pl. 7, figs. 2–4; pl. 9, fig. 6) have found interseptal canals in each of the subgenera of *Miogypsina* as recognized by them. They suggest the subfamily Miogypsininae be transferred from the Orbitoididae, where it was placed originally by Vaughan, to the Rotaliidae. Previously (1936), Tan (see De Ingen. Nederl. Ind. IV, Mijnb. en Geol. 3 de jaarg., 1936, No. 3, pp. 45–61, pls. 1, 2; No. 5, pp. 84–98; No. 7, pp. 109–123; 4 de jaarg., 1937, No. 3, pp. 35–45, pls. 1–3; No. 6, pp. 87–111, pls. 4–7) independently discovered a canal system in *Mio-*

gypsinoides and *Miogypsina*. He elevates the subfamily to family rank, but also suggests a rotalid ancestor. Élisabeth David-Sylvain (Géol. Soc. France, n. ser., vol. 16, Mem. 33, 1937, pp. 1–44, pls. 1–4) suggests the derivation of *Miogypsina* from the Calcarinidae, whereas Cole placed the subfamily in the Camerinidae.

There are sufficient differences to warrant the creation of a family for this group which is composed of two genera and one subgenus. The present evidence indicates that the Miogypsinidae are specialized, short-ranged descendants of some advanced member of the Rotaliidae.

Genus MIOGYPSINA Sacco, 1893

Plate 31, figure 12; Key, plate 39, figures 10–12; plate 55, figures 1, 2
Genotype, *Nummulites globulina* Michelotti

Miogypsina Sacco, Bull. Soc. belge Géol., vol. 7, 1893, p. 205; Vaughan, in Cushman, Foraminifera, 1928, p. 354, pl. 56, figs. 3, 8; pl. 57, fig. 2; pl. 59, fig. 12; *ibid.*, 1933, pp. 297, 298, pl. 31, fig. 12; Key, pl. 39, figs. 10–12; Galloway, Man. Foram., 1933, pp. 438, 439, pl. 41, fig. 12; Barker and Grimsdale, Ann. and Mag. Nat. Hist., ser. 10, vol. 19, 1937, pp. 161–168, pl. 5, figs. 4–10; pl. 6, figs. 1–6, 8; pl. 7, figs. 1–4; pl. 8, fig. 6; pl. 9, fig. 6; Vaughan and Cole, in Cushman, Foraminifera, 1940, p. 331, pl. 31, fig. 12; Key, pl. 39, figs. 10–12; Hanzawa, Yabe's Jubilee Publ., 1940, pp. 773, 774, 777–778; Brönnimann, Abh. Schweiz. Pal. Gesell., vol. 63, 1942, pp. 60–77; Glaessner, Principles of Micropal.,1947, pp. 159–160, pl. 11, fig. 4, text fig. 36.

Flabelliporus Dervieux, Atti R. Accad. Sci. Torino, vol. 29, 1893, p. 59; Riv. Ital. Pal., 1900, vol. 6, p. 146.

Lepidosemicyclina Rutten, Konin. Akad. van Weten., Amsterdam, 1911, vol. 13, pp. 1135, 1136.

Miogypsinopsis Hanzawa, Yabe's Jubilee Publ., 1940, pp. 773, 776.

Test compressed, small, trigonal or suborbicular in plan; surface papillate. Embryonic apparatus excentric in position, in some forms apical or subapical, in others nearer the center than the periphery, composed of two equal or subequal chambers, around which there are several chambers whose features are intermediate in character between the embryonic and the usual equatorial chambers. The early chambers may be distinctly spiral in arrangement, later growth mostly on only a segment of the periphery, and thereby producing a triangular outline, or cyclical, but greater on one than on the other side of the nucleoconch. A spiral canal and interseptal canals are present. Equatorial chambers rhomboid or elongate hexagonal; a net of canals in the chamber walls. Lateral chambers well-developed. Pillars present; terminating in surface papillae.—Oligocene, Miocene, Europe, America, East Indies, Morocco.

In *Miogypsina* s.s. the nucleoconch is apical or subapical and lateral chambers are present.

Miogypsinopsis was created for specimens which show a well-developed coil of quadrangular peri-embryonic chambers around the initial chambers. As all the other features are identical with *Miogypsina* s.s., and as there is considerable variation in the development of the peri-embryonic chambers, such specimens should be referred to *Miogypsina* s.s.

Subgenus MIOLEPIDOCYCLINA A. Silvestri, 1907

Key, plate 55, figures 1, 2

Genotype, *Orbitoides burdigalensis* Gümbel

Miolepidocyclina A. Silvestri, Bull. Naturalista, Siena, vol. 27, 1907, p. 11; Riv. Ital. Pal., vol. 13, 1907, p. 80; Pont. Accad. R. Nuov. Lincei, Mem., vol. 28, 1910, pp. 119, 137; Vaughan, in Cushman, Foraminifera, 1933, p. 298; Galloway, Man. Foram., 1933, p. 438, pl. 41, fig. 11; Barker and Grimsdale, Ann. and Mag. Nat. Hist., ser. 10, vol. 19, 1937, pp. 162, 166, 167, pl. 7, fig. 4; pl. 9, fig. 6; Vaughan and Cole, in Cushman, Foraminifera, 1940, p. 331; Hanzawa, Yabe's Jubilee Publ., 1940, pp. 775, 778.

Heterosteginoides Cushman, U. S. Nat. Mus. Bull. 103, 1919, p. 97, pl. 43, figs. 1–8; Hanzawa, Jour. Pal., vol. 21, 1947, pp. 260–263, pl. 41, figs. 1–13.

Miolepidocyclina is similar to *Miogypsina* s.s. except that the initial chambers form a portion of a subcentral spiral. In *Miogypsina* s.s. the initial chambers are near the periphery of the test.

PLATE 31

Family 48. Orbitoididae. Family 50. Miogypsinidae

FIG.

1. *Lepidocyclina* (*Polylepidina*) *chiapasensis* Vaughan. × 16. (After Vaughan.) Early chambers in oblique section.

2–4. *Lepidocyclina* (*Pliolepidina*) *pustulosa* (H. Douvillé) forma *tobleri* (H. Douvillé), forma *teratologica.* × 24. (After H. Douvillé.) 2, Horizontal section of embryonic chambers of megalospheric form of a teratological individual. 3, 4, Vertical sections of embryonic chambers of megalospheric form of teratological individuals.

5. *Lepidocyclina* (*Lepidocyclina*) *ocalana* Cushman. × 16. (After H. Douvillé.) Horizontal section of early chambers of megalospheric form.

6. *Lepidocyclina* (*Nephrolepidina*). Diagrammatic horizontal section. (After van der Vlerk and Umbgrove.)

7. *Lepidocyclina* (*Nephrolepidina*) *chaperi* Lemoine and R. Douvillé. × 16. (After H. Douvillé.) Early chambers of horizontal section of megalospheric form.

8–10. *Lepidocyclina* (*Eulepidina*) sp(?). × 16. (After H. Douvillé.) 8, Horizontal section of early chambers of megalospheric form. 9, 10, Vertical sections.

11. *Lepidocyclina* (*Eulepidina*) *dilatata* (Michelotti). × 16. (After H. Douvillé.) Horizontal section of early chambers of megalospheric form.

12. *Miogypsina.* Diagrammatic horizontal section. (After van der Vlerk and Umbgrove.)

13, 14. *Helicolepidina spiralis* Tobler. × 24. (After H. Douvillé.) 13, Horizontal section of early chambers of megalospheric form. 14, Vertical section.

15, 16. *Omphalocyclus macroporus* (Lamarck). (After H. Douvillé.) 15, Vertical section, × 8. 16, Horizontal section, × 16.

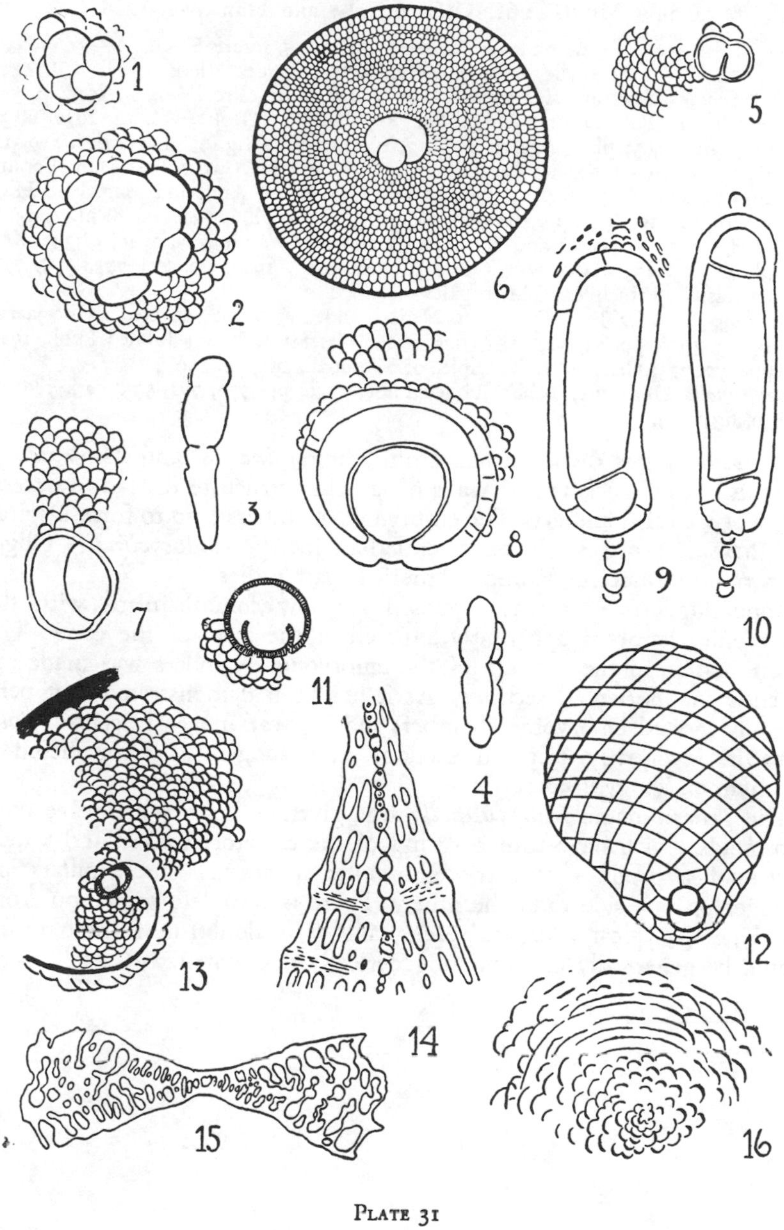

PLATE 31

Genus MIOGYPSINOIDES Yabe and Hanzawa, 1928

Key, plate 41, figure 15; plate 55, figure 6
Genotype, *Miogypsina dehaarti* van der Vlerk

Miogypsinoides YABE and HANZAWA, Imp. Acad. Japan, Proc., vol. 4, 1928, p. 535, text fig. 1; Tohoku Imp. Univ., Sci. Rep't, ser. 2 (Geol.), vol. 14, 1930, p. 32, pl. 3, figs. 4, 5; pl. 4, figs. 3, 4; pl. 7, fig. 12; pl. 9, fig. 9; pl. 11, figs. 1–6, 12; VAUGHAN, in CUSHMAN, Foraminifera, 1933, p. 298; GALLOWAY, Man. Foram., 1933, p. 439, pl. 41, fig. 13; BARKER and GRIMSDALE, Ann. and Mag. Nat. Hist., ser. 10, vol. 19, 1937, pp. 162–163, 168, pl. 5, fig. 6; pl. 6, figs. 1–6, 8; pl. 7, fig. 1; pl. 8, fig. 6; VAUGHAN and COLE, in CUSHMAN, Foraminifera, 1940, p. 332, Key, pl. 41, fig. 15; HANZAWA, Yabe's Jubilee Publ., 1940, pp. 771–773, 775, 776; GLAESSNER, Principles of Micropal., 1947, p. 159.

Conomiogypsinoides TAN, De Ingen. Nederl. Ind. IV, Mijnb. en Geol. 3 de jaarg., No. 3, 1936, pp. 51, 52, pl. 1, figs. 8–12; HANZAWA, Yabe's Jubilee Publ., 1940, pp. 772, 773; GLAESSNER, Principles of Micropal., 1947, p. 159.

Miogypsinella HANZAWA, Yabe's Jubilee Publ., 1940, pp. 770, 771, 775, 779, 780, pl. 39, figs. 1–9.

Miogypsinoides is distinguished from *Miogypsina* s.s. and *Miolepidocyclina* in having thick lateral walls of lamellar structure instead of several layers of lateral chambers. The embryonic chambers tend to form a spiral, but they are more nearly peripheral than in *Miolepidocyclina.*—Oligocene, middle Miocene, Europe, America, East Indies.

Conomiogypsinoides was proposed for low conical forms with the embryonic chambers apparently situated at the apex of the cone. The determination of the position of the embryonic chambers was made apparently in unoriented sections. As it has been demonstrated that peripherally located embryonic chambers may appear in nearly central positions in incorrectly oriented sections, this subgenus is considered a synonym of *Miogypsinoides*.

The generic name *Miogypsinella* was given to certain primitive miogypsinoids which have thin covering of the equatorial layer and which develop heavy pillars over the embryonic apparatus. These pillars are stronger on one side than the other. There is complete gradation from this type to typical *Miogypsinoides*, and it is doubtful if these forms should be generically or even subgenerically designated.

CHAPTER XII

BIBLIOGRAPHY

THE LITERATURE dealing with the foraminifera has increased greatly in these last years, and it is difficult to present an adequate bibliography in the space that can be given to it here. Papers without illustrations have mostly been left out unless they have subject matter of especial interest. As will be seen by a glance at the following pages, many of the important papers are in languages other than English. Some reading knowledge of French, German, and Italian is essential to one who is to do any serious research work on the foraminifera. The distribution of many forms is wide, and the intelligent consultation of foreign literature is very necessary for an understanding of the American faunas.

The works will be found under various headings, the Recent, Tertiary, Cretaceous, Jurassic, Triassic, and Paleozoic, with the papers grouped roughly by geographic divisions where there are many. Papers are arranged alphabetically by authors under these headings. The later papers on Classification, Structure, etc., are arranged by authors. It has been necessary to place each paper under one heading only, whereas often it would have been useful to have it under several headings, but space has had to be considered, and each paper is placed under the heading which seems most appropriate. References are given to the best-known bibliographies which, with the *Zoological Record*, will give the titles of practically all papers published on the foraminifera. It is hoped that this condensed list will furnish the worker with at least a clue to the more important papers.

I. RECENT

GENERAL

BRADY, H. B. Report on the foraminifera dredged by H. M. S. "Challenger" during the years 1873–1876. — Rep. Voy. *Challenger*, Zoology, vol. 9, 1884, 1 vol. text, 814 pp.; 1 vol. pls. 1–115.

CUSHMAN, J. A. A Monograph of the Foraminifera of the North Pacific Ocean. — Bull. 71, U. S. Nat. Mus., pts. 1–6, 1910–1916, 596 pp., 473 text figs., 135 pls.

The Foraminifera of the Atlantic Ocean. — Bull. 104, U. S. Nat. Mus., pts. 1–8, 1918–1931, 1064 pp., 200 pls.

EARLAND, A. Foraminifera. Part II. South Georgia. — *Discovery* Reports, vol. 7, 1933, pp. 27–138, pls. I–VII. Part III. The Falklands Sector of the Antarctic (excluding South Georgia). — l. c., vol. 10, 1934, pp. 1–208, pls. I–X. Part IV. Additional Records from the Weddell Sea Sector from Material Obtained by the S. Y. "Scotia." — l. c., vol. 13, 1936, pp. 1–59, pl. I.

EGGER, J. G. Foraminiferen aus Meeresgrundproben, gelothet von 1874 bis 1876 von S. M. Sch. "Gazelle." — Abhandl. kön. bay. Akad. Wiss., München, Cl. II, vol. 18, Abt. 2, 1893, pp. 195–458, 21 pls., 1 map.

FLINT, J. M. Recent Foraminifera. — Ann. Rep't U. S. Nat. Mus., 1897 (1899), pp. 249–349, 80 pls.

GOËS, A. A Synopsis of the Arctic and Scandinavian Recent Marine Foraminifera Hitherto Discovered. — Kongl. Svensk. Vet.-Akad. Handl., vol. 25, No. 9, 1894, pp. 1–127, pls. 1–25, text figs.

The Foraminifera (Galapagos, etc.). — Bull. Mus. Comp. Zoöl., vol. 29, 1896, pp. 1–103, pls. 1–9.

HERON-ALLEN, E., and A. EARLAND. Some New Foraminifera from the South Atlantic. I.— Journ. Roy. Micr. Soc., vol. 49, 1929, pp. 102–108, 3 pls.; II. — l. c., pp. 324–334, 4 pls.; III. *Miliammina*, A New Siliceous Genus. — l. c., vol. 50, 1930, pp. 38–45, 1 pl.; IV. Four New Genera from South Georgia. — l. c., vol. 52, 1932, pp. 253–261, 2 pls.

Foraminifera. Part I. The Ice-free Area of the Falkland Islands and Adjacent Seas. — *Discovery* Reports, vol. IV, 1932, pp. 291–460, pls. VI–XVII.

NUTTALL, W. L. F. The localities whence the Foraminifera figured in the report of H. M. S. "Challenger" by Brady were derived. — Ann. Mag. Nat. Hist., ser. 9, vol. 19, 1927, pp. 209–241.

ORBIGNY, A. D. D'. Voyage dans l'Amérique Méridionale. Foraminifères. — 4to, *Paris* and *Strasbourg*, 1839, vol. 5, pt. 5, pp. 1–86, pls. 1–9.

RHUMBLER, L. Systematische Zusammenstellung der rezenten Reticulosa (Nuda & Foraminifera). — Arch. Prot., vol. 3, 1903 (1904), pp. 181–294, 142 text figs.

Die Foraminiferen (Thalamophoren) der Plankton-Expedition, etc., Pt. 1. Systematik. — Ergeb. Plankton-Exped. Humboldt Stiftung, Bd. 3, 1911, pp. 1–331, pls. 1–39. Pt. 2, l. c., — 1911, pp. 333–476, text figs. 1–65.

SCHLUMBERGER, C. Revision des Biloculines des Grands Fonds. — Mém. Soc. Zool. France, vol. 4, 1891, pp. 155–191, pls. 10–12, 46 text figs.

WESTERN ATLANTIC

BERMUDEZ, P. J. Foraminíferos de la Costa Norte de Cuba. — Mem. Soc. Cubana Hist. Nat., vol. 9, 1935, pp. 129–224, pls. 10–17, 3 text figs., map.

Foraminíferos recientes colectados por el Dr. Luis Howell Rivero en Jamaica. — l. c., vol. 11, 1937, pp. 249–252.

Aguayoina asterostomata, un foraminifero nuevo del mar caribe. — l. c., vol. 12, 1938, pp. 385–388, pl. 29.

Resultados de la Primera Expedicion en las Antillas del Ketch Atlantis bajo los Auspicios de las Universidades de Harvard y Habana. Nuevo Género y Especies Nuevas de Foraminíferos. — l. c., vol. 13, 1939, pp. 9–12, pls. 1, 2.

BRADY, H. B., W. K. PARKER and T. R. JONES. On some Foraminifera from the Abrohlos Bank. — Trans. Zool. Soc., vol. 12, 1888, pp. 211–239, pls. 40–46, with chart.

BROECK, E. VANDEN. Étude sur les Foraminifères de la Barbade (Antilles) recueillis par L. Agassiz, précédée de quelques considérations sur la Classification et la Nomenclature des Foraminifères. — Ann. Soc. Belg. Micr., vol. 2, 1876, pp. 55–152, pls. 2, 3.

CUSHMAN, J. A. Foraminifera of the Woods Hole Region. — Proc. Boston Soc. Nat. Hist., vol. 34, 1908, pp. 21–34, pl. 5.

Foraminifera from the North Coast of Jamaica. — Proc. U. S. Nat. Mus., vol. 59, 1921, pp. 47–82, pls. 11–19, 16 text figs.

Shallow-water Foraminifera of the Tortugas Region.—Publ. 311, Carnegie Inst. Washington, vol. 17, 1922, pp. 1–85, pls. 1–14.

Recent Foraminifera from Porto Rico.—Publ. 344, Carnegie Inst. Washington, 1926, pp. 73–84, pl. 1.

Fourteen New Species of Foraminifera.—Smithsonian Misc. Coll., vol. 91, No. 21, 1935, pp. 1–9, pls. 1–3.

Foraminifera from the Shallow Water of the New England Coast.—Special Publ. 12, Cushman Lab. Foram. Res., 1944, pp. 1–37, pls. 1–4.

CUSHMAN, J. A., and F. L. PARKER. Recent Foraminifera from the Atlantic Coast of South America.—Proc. U. S. Nat. Mus., vol. 80, Art. 3, 1931, pp. 1–24, pls. 1–4.

DAWSON, G. M. On Foraminifera from the Gulf and River St. Lawrence.—Canad. Nat., vol. 5, 1870, pp. 172–177, text figs.

GOËS, A. On the Reticularian Rhizopoda of the Caribbean Sea.—Kongl. Svensk. Vet.-Akad. Handl., vol. 19, No. 4, 1882, pp. 1–151, 12 pls.

HADLEY, W. H. Recent Foraminifera from near Beaufort, North Carolina.—Journ. Elisha Mitchell Sci. Soc., vol. 52, pp. 35–37, text fig. 1.

LALICKER, C. G., and P. J. BERMUDEZ. Some Foraminifera of the Family Textulariidae Collected by the first "Atlantis" Expedition.—Torreia, No. 8, 1941, pp. 1–19, pls. 1–4.

ORBIGNY, A. D. D'. Foraminifères in De la Sagra, Histoire physique, politique et naturelle de l'Île de Cuba.—French ed., 8vo, *Paris*, 1839, pp. i–xlviii, 1–224, 12 pls.

SHUPACK, B. Some Foraminifera from Western Long Island and New York Harbor.—Amer. Mus. Novitates, No. 737, 1934, pp. 1–12, 1 pl., 2 text figs.

EASTERN ATLANTIC

BALKWILL, F. P., and F. W. MILLETT. The Foraminifera of Galway.—Journ. Micr., vol. 3, 1884, pp. 19–28, 78–90, 4 pls.

BALKWILL, F. P., and J. WRIGHT. Report on some Recent Foraminifera found off the Coast of Dublin and in the Irish Sea.—Trans. Roy. Irish Acad., vol. 28, 1885, Sci., pp. 317–372, 3 pls., text figs.

BRADY, H. B. Contributions to our Knowledge of the Foraminifera. On the Rhizopodal Fauna of the Shetlands.—Trans. Linn. Soc., vol. 24, 1864, pp. 463–475, pl. 48.

A Catalogue of the Recent Foraminifera of Northumberland and Durham.—Nat. Hist. Trans. Northumberland and Durham, vol. 1, 1865 (1867), pp. 83–107, pl. 12.

In Brady, G. S., D. Robertson and H. B. Brady. The Ostracoda and Foraminifera of Tidal Rivers.—Ann. Mag. Nat. Hist., ser. 4, vol. 6, 1870, pp. 273–306, pls. 11, 12.

A Synopsis of the British Recent Foraminifera.—Journ. Roy. Micr. Soc., ser. 2, vol. 7, 1887, pp. 872–927.

CHASTER, G. W. Report upon the Foraminifera of the Southport Society of Natural Science District.—Southport Soc. Nat. Sci., vol. 1, 1892, pp. 54–71, pl. 1.

EARLAND, A. The Foraminifera of the Shore-sand at Bognor, Sussex.—Journ. Quekett Micr. Club, ser. 2, vol. 9, 1905, pp. 187–232, pls. 11–14.

HALKYARD, E. Recent Foraminifera of Jersey.—Trans. Manchester Micr. Soc., 1896, pp. 55–72, pls. 1, 2.

HERON-ALLEN, E., and A. EARLAND. Foraminifera, Clare Island Survey.—Proc. Roy. Irish Acad., vol. 31, pt. 64, 1913, pp. 1–188, 13 pls.

Foraminifera of South Cornwall. — Journ. Roy. Micr. Soc., 1916, pp. 29–55, pls. 5–9.
The Foraminifera of the West of Scotland. — Trans. Linn. Soc. Zool., vol. 11, pt. 13, 1916, pp. 197–299, pls. 39–43, maps.
The Foraminifera of the Plymouth District. I. — Journ. Roy. Micr. Soc., vol. 50, 1930, pp. 46–84, pls. I–III.

Hofker, J. De Protozoën. From Flora en Fauna der Zuiderzee, 1922, pp. 127–183, 91 text figs.
Zoology of the Faroes. IIa. Foraminifera. — Copenhagen, 1930, 21 pp., 31 text figs.

Höglund, H. Foraminifera in the Gullmar Fjord and the Skagerak. — Zoologiska Bidrag från Uppsala, vol. 26, 1947, pp. 1–328, 32 pls., 312 text figs., 2 maps, 7 tables.

Kiaer, H. Synopsis of the Norwegian Marine Thalamophora. — Rep't Norwegian Fish. Mar. Invest., vol. 1, 1900, No. 7, pp. 1–58, 1 pl.

Norman, A. M. On the Genus Haliphysema, with a description of several forms apparently allied to it. — Ann. Mag. Nat. Hist., ser. 5, vol. 1, 1878, pp. 265–284, pl. 16.

Orbigny, A. D. d'. "Foraminifères" in Barker-Webb and Berthelot, Histoire naturelle des Îles Canaries. — 4to, *Paris*, 1839, vol. 2, pt. 2, pp. 119–146, 3 pls.

Rhumbler, L. Nordische Plankton — Foraminiferen. — Nordische Plankton, vol. 14, 1900, 32 pp., 33 text figs.
Rhizopoden der Kieler Bucht, gesammelt durch A. Remane. I. Teil. — Schrift. Nat. Ver. Schleswig-Holstein, Bd. 21, Heft 2, 1935, pp. 143–194, pls. 1–9.
Foraminiferen der Kieler Bucht, gesammelt durch A. Remane, II Teil. (Ammodisculinidae bis einschl. Textulinidae.) — Kieler Meeresforschungen, vol. 1, 1936, pp. 179–242, text figs. 127–246.
Foraminiferen aus dem Meeressand von Helgoland, gesammelt von A. Remane (Kiel). — l. c., vol. 2, 1938, pp. 157–222, text figs. 1–64.

Schlumberger, C. Note sur quelques Foraminifères nouveaux, ou peu connus du Golfe de Gascogne. Campagne du Travailleur, 1880. — Feuille Jeunes Nat., Ann. 13, 1883, pp. 105–108, 117–120, pls. 2, 3, text figs.

Schott, W. Die jüngste Vergangenheit des äquatorialen Atlantischen Ozeans auf Grund von Untersuchungen an Boden proben der "Meteor"-Expedition. — Sitz. Abhandl. Nat. Gesell. Rostock, Dritte Folge, Bd. 4, 1933 (June 7, 1934), pp. 48–59, 3 text figs.
Die Foraminiferen in dem äquatorialen Teil des Atlantischen Ozeans. — Wissenschaftliche Ergebnisse der deutschen Atlantischen Expedition auf dem Forschungsund Vermessungsschiff "Meteor" 1925–1927, Bd. 3, Pt. 3, Section B, 1935, pp. 43–134, 3 charts, text figs. 18–57, maps.
Stratigraphie rezentere Tief-seesedimente auf Grund der Foraminiferen fauna. — Geologischen Rundschau, Bd. 29, 1938, pp. 330–333.

Terquem, O. Essai sur le Classement des Animaux qui vivent sur la Plage et dans les Environs de Dunkerque. — 3 pts., 8vo, *Paris*, fasc. 1, 1875, pp. 1–54, pls. 1–6; fasc. 2, 1876, pp. 55–100, pls. 7–12; fasc. 3, 1881, pp. 101–152, pls. 13–17.

Wallich, G. C. On Rupertia stabilis, a new Sessile Foraminifer from the North Atlantic. — Ann. Mag. Nat. Hist., ser. 4, vol. 19, 1877, pp. 501–504, pl. 20.

Williamson, W. C. On the Recent Foraminifera of Great Britain. — Ray Soc., 4to, *London*, 1858, pls.

WRIGHT, J. Report on the Foraminifera obtained off the South-west of Ireland during the Cruise of the "Flying Falcon," 1888.—Proc. Roy. Irish Acad., ser. 3, vol. 1, 1891, pp. 460–502, pl. 20.

MEDITERRANEAN

BATSCH, A. J. G. K. Sechs Kupfertafeln mit Conchylien des Seesandes, gezeichnet und gestochen von A. J. G. K. Batsch.—4to, *Jena*, 1791.

COLOM, G. Una Contribucion al Conocimiento de los Foraminiferos de la Bahia de Palma de Mallorca.—Instit. Español Oceanografía, Notas y Resúmenes, Ser. 2, No. 108, 1942, pp. 1–49, pls. 1–11.

DOLGOPOLSKAJA, M., and W. PAULI. On the Foraminifera of the Black Sea near the Biological Station Karadagh.—Travaux Sta. Biol. Karadagh, vol. 4, 1931, pp. 23–47, pls. I–III, 5 text figs.

FORNASINI, C. Globigerine Adriatiche.—Mem. Accad. Sci. Istit. Bologna, ser. 5, vol. 7, 1899, pp. 575–586, 4 pls.

Contributo a la Conoscenza de le Testilarine Adriatiche.—l. c., vol. 10, 1903, pp. 299–316, 1 pl.

GAUTHIER-LIÈVRE, L. Sur une des singularités de l'Oued Rhir: des Foraminifères thalassoides vivant dans les eaux sahariennes.—Bull. Soc. Hist. Nat. de l'Afrique du Nord, vol. 26, 1935, pp. 142–147, pl. XII, text figs.

HERON-ALLEN, E., and A. EARLAND. Les Foraminifères des "Sables Rouges" du Golfe d'Ajaccio (Côte Nord).—Bull. Soc. Sci. Hist. Nat. Corse, 1922, pp. 109–149, pls. 1, 2.

HOFKER, J. Notizen über die Foraminiferen des Golfes von Neapel.—Pubbl. Staz. Zool. Napoli, vol. 10, 1930, pp. 365–406, pls. 12, 13, text figs. 1–22; III. Die Foraminiferenfauna der Ammontatura.—l. c., vol. 12, 1932, pp. 61–144, text figs. 1–45.

LACROIX, E. Les Astrorhizides du littoral méditerranéen entre Saint-Raphaël et Monaco.—Bull. Inst. Océanographique, No. 545, 1929, p. 1–22, text figs. 1–32.

Les Lituolides du plateau continental méditerranéen entre Saint-Raphaël et Monaco.—l. c., No. 549, 1930, pp. 1–16, text figs. 1–21.

Textularidae du plateau continental méditerranéen entre Saint-Raphaël et Monaco.—l. c., No. 591, 1932, pp. 1–28, text figs. 1–33.

Discammina: nouveau genre méditerranéen de Foraminifères arénacés.—l. c., No. 600, 1932, pp. 1–4, text figs. *a–e*.

Nouvelles recherches sur les spécimens méditerranéens de *Textularia sagittula* (Defrance).—l. c., No. 612, 1933, pp. 1–23, 9 text figs., tables.

Discammina fallax et *Haplophragmium emaciatum*.—l. c., No. 667, 1935, pp. 1–16, 9 text figs.

LE CALVEZ, J. Sur quelques Foraminifères de Villefranche et de Banyuls.—Protistologica, vol. 55, 1935, pp. 79–98, figs. I–XI (in text).

MARTINOTTI, A. Foraminiferi della Spiaggia di Tripoli.—Atti Soc. Ital. Sci. Nat., vol. 59, 1920, pp. 249–334, pls. 10–13, 176 text figs.

SCHLUMBERGER, C. Monographie des Miliolidées du Golfe de Marseille.—Mém. Soc. Zool. France, vol. 6, 1893, pp. 199–228, pls. 1–4, 37 text figs.

SIDEBOTTOM, H. Report on the Recent Foraminifera from the Coast of the Island of Delos (Grecian Archipelago).—Mem. Proc. Manchester Lit. Philos. Soc., 1904–1909, 22 pls., text figs.

Report on the Recent Foraminifera from the Bay of Palermo, Sicily, 14–20 fms. (Off the Harbour).—l. c., vol. 54, pt. 3, 1910, pp. 1–36, 3 pls., text figs.

SILVESTRI, A. Contribuzione allo studio dei Foraminiferi adriatici. Nota Prima. — Atti Accad. Sci. Acireale, vol. 7, 1895–96, pp. 27–63; Nota Seconda. — l. c., 1896, pp. 1–114; App. I. — l. c., vol. 9, 1897–98, pp. 1–46, 1 pl.; Lageninae del Mare Tirreno. — Mem. Accad. Pont. Lincei, vol. 19, 1902, pp. 133–172, 74 text figs.

WIESNER, H. Zur Systematik adriatischer Nubecularien, Spiroloculinen, Miliolinen und Biloculinen. — Arch. Prot., vol. 25, 1912, pp. 201–239, text figs.

INDO-PACIFIC

ASANO, K. Foraminifera from Siogama Bay, Miyagi Prefecture, Japan. — Saito Ho-on Kai Museum Research Bull. No. 13, 1937, pp. 109–119, pls. XV, XVI.

On the Japanese Species of Cassidulina. — Jap. Journ. Geol. Geog., vol. 14, 1937, pp. 143–153, pls. XIII, XIV.

CARTER, H. J. Description of a new species of Foraminifera (Rotalia spiculotesta). — Ann. Mag. Nat. Hist., ser. 4, vol. 20, 1877, pp. 470–473, pl. 16.

CHAPMAN, F. On some Foraminifera obtained by the Royal Indian Marine Survey's S. S. "Investigator," from the Arabian Sea, near the Laccadive Islands. — Proc. Zool. Soc., 1895, pp. 1–55, pl. 1.

Foraminifera from the Lagoon at Funafuti. — Journ. Linn. Soc., Zool., vol. 28, 1900, pp. 161–210, pls. 19, 20.

On the Foraminifera Collected round the Funafuti Atoll from Shallow and Moderately Deep Water. — l. c., 1901, pp. 379–417, pls. 35, 36.

On some Foraminifera and Ostracoda obtained off Great Barrier Island, New Zealand. — Trans. N. Zealand Inst., vol. 38, 1905, pp. 77–112, pl. 3.

Recent Foraminifera of Victoria; some Littoral Gatherings. — Journ. Quekett Micr. Club, ser. 2, vol. 10, 1907, pp. 117–146, pls. 9, 10.

Report on the Foraminifera from the Subantarctic Islands of New Zealand. — From Subantarctic Islands of New Zealand, art. 15, 1909, pp. 312–371, pls. 13–17.

On the Foraminifera and Ostracoda from Soundings (Chiefly Deep-water) collected round Funafuti by H. M. S. "Penguin." — Journ. Linn. Soc. Zool., vol. 30, 1910, pp. 387–444, pls. 54–57.

Report on the Foraminifera and Ostracoda obtained by F. I. S. "Endeavour" from the East Coast of Tasmania, and off Cape Wiles, South Australia. — Biol. Results "Endeavour," 1909–14 (1915), pp. 1–51, pls. 1–3.

Report on Foraminiferal Soundings and Dredgings of the F.I.S. "Endeavour" along the Continental Shelf of the South-east Coast of Australia. — Trans. Roy. Soc. South Australia, vol. 65, pt. 2, 1941, pp. 145–211, pls. 7–9.

CHAPMAN, F., and W. J. PARR. Foraminifera and Ostracoda from Soundings Made by the Trawler "Bonthorpe" in the Great Australian Bight. — Journ. Roy. Soc. Western Australia, vol. 21, 1934–35 (March 25, 1935), pp. 1–7, pl. I.

CUSHMAN, J. A. Foraminifera of the Philippine and Adjacent Seas. — Bull. 100, U. S. Nat. Mus., vol. 4, 1921, 608 pp., 52 text figs., 100 pls.

Samoan Foraminifera. — Publ. 342, Carnegie Inst. Washington, 1924, pp. 1–75, pls. 1–25.

Foraminifera of the Tropical Central Pacific. — Bernice P. Bishop Mus., Bull. 27, 1925, pp. 121–144.

The Foraminifera of the Tropical Pacific Collections of the "Albatross," 1899–1900. Part 1, Astrorhizidae to Trochamminidae. — Bull. 161, U. S. Nat. Mus., pt. 1, 1932, pp. 1–84, pls. 1–17; Part 2. Lagenidae to Alveolinellidae. — l. c., pt. 2, 1933, pp. i–vi, 1–79, pls. 1–19; Part 3. Heterohelicidae and Buliminidae. — l. c., pt. 3, 1942, pp. i–v, 1–67, pls. 1–15.

Some New Recent Foraminifera from the Tropical Pacific. — Contr. Cushman Lab. Foram. Res., vol. 9, 1933, pp. 77–95, pls. 8–10.

A Recent *Gümbelitria* (?) from the Pacific. — l. c., vol. 10, 1934, p. 105, pl. 13 (pt.).

Hada, Y. A Contribution to the Study on the Plankton and the Allied Protozoa in the Northern Waters of Japan. I. A list of the Littoral Foraminifera of Hokkaido. — Trans. Sapporo Nat. Hist. Soc., vol. XI, 1929, pp. 9–15, text figs. *a–d*.

Report of the Biological Survey of Mutsu Bay. 19. Notes on the Recent Foraminifera from Mutsu Bay. — Sci. Rep't Tohoku Imp. Univ., ser. 4 (Biol.), vol. VI, 1931, pp. 45–148, text figs. 1–95.

Studies on the Foraminifera of Brackish Waters. I. Hijirippu and Mochirippu Lakes. — Zool. Mag., vol. 48, 1936, pp. 847–860, text figs. 1–12. II. Hachiro-Gata. III. Koyama-Ike. — l. c., vol. 49, 1937, pp. 341–347, text figs. 1–7.

Some new monothalamous Foraminifera from northern Japanese waters. — Trans. Sapporo Nat. Hist. Soc., vol. 14, pt. 4, 1936, pp. 242–245, text figs. 1–5.

Heron-Allen, E., and A. Earland. The Foraminifera of the Kerimba Archipelago. — Trans. Zool. Soc. London, vol. 20, pt. 1, 1914, pp. 363–390, pls. 35–37; pt. 2, 1915, pp. 543–794, pls. 40–53.

The Foraminifera of Lord Howe Island, South Pacific. — Journ. Linn. Soc. Zool., vol. 35, 1924, pp. 599–647, pls. 35–37.

Hofker, J. The Foraminifera of the Siboga Expedition. Part 1. — Monograph IV, Siboga Exped., pt. 1, 1927, pp. 1–78, pls. 1–38, text figs. 1–11; Part. 2. — Monograph IVa, pt. 2, 1930, pp. 79–170, pls. 39–64, text figs. 12–33.

Resultats Scientifiques du Voyage aux Indes Orientales Neerlandaises. — Vol. II, fasc. 1, Sur Quelques Foraminifères, 1930, pp. 1–12, pls. I–III.

Foraminifera of the Malay Archipelago. — Vidensk. Medd. fra Dansk naturh. Foren., Bd. 93, 1932–33 (1933), pp. 71–167, pls. II–VI, 35 text figs.

Sur Quelques Foraminifères. — Résultats Scientifiques Voy. Ind. Orientales Néerlandaises, vol. 2, fasc. I, 1930, pp. 1–12, pls. I–III.

Le Roy, L. W. A Preliminary Study of the Microfaunal Facies Along a Traverse Across Peper Bay, West Coast of Java. — "De Ingenieur in Nederlandsch-Indië," IV, Mijnbouw Geol., Jaargang V, 1938, pp. 130–133, 2 text figs. (maps).

Millett, F. W. Report on the Recent Foraminifera of the Malay Archipelago collected by Mr. A. Durrand, F.R.M.S. — Journ. Roy. Micr. Soc., 1898–1904, 17 parts.

Moebius, K. Foraminiferen von Mauritius, In Moebius, F. Richter, and E. von Martens, Beiträge zur Meeresfauna der Insel Mauritius und der Seychellen. — 4to, *Berlin*, 1880, pp. 63–110, pls. 1–14.

Parr, W. J. Victorian and South Australian Shallow-Water Foraminifera. Part I. — Proc. Roy. Soc. Victoria, vol. 44, pt. 1 (n. ser.), 1932, pp. 1–14, pl. 1, text figs. A–E. Part II. — l. c., pt. 2, 1932, pp. 218–234, pls. 21, 22.

Recent Foraminifera from Barwon Heads, Victoria. — l. c., vol. 56, pt. 2, 1945, pp. 189–227, pls. 8–12, figs. 1, 2 (maps).

Rhumbler, L. Foraminiferen von Laysan und den Chatham-Inseln. — Zool. Jahrb., Abteil. Syst., vol. 24, 1906, pp. 21–80, pls. 2–5.

Sidebottom, H. Lagenae of the South-West Pacific Ocean. — Journ. Queckett Micr. Club, vol. 11, 1912, pp. 375–434, pls. 14–21; vol. 12, 1913, pp. 161–210, pls. 15–18.

Report on the Recent Foraminifera dredged off the East Coast of Australia. — Journ. Roy. Micr. Soc., 1918, pp. 1–25, 121–153, 249–264, pls. 1–6.

PACIFIC COAST OF AMERICA

CHURCH, C. C. Some Recent Shallow Water Foraminifera Dredged near Santa Catalina Island, California. — Journ. Pal., vol. 3, 1929, pp. 302–305, text figs. 1–3.

CUSHMAN, J. A. Recent Foraminifera from British Columbia. — Contr. Cushman Lab. Foram. Res., vol. 1, 1925, pp. 38–47, pls. 6, 7.

Recent Foraminifera from off the West Coast of America. — Bull. Scripps Inst. Oceanography, Tech. Ser., vol. 1, 1927, pp. 119–188, pls. 1–6.

Some Recent Angulogerinas from the Eastern Pacific. — Contr. Cushman Lab. Foram. Res., vol. 8, 1932, pp. 44–48, pl. 6.

CUSHMAN, J. A., and B. KELLETT. Recent Foraminifera from the West Coast of South America. — Proc. U. S. Nat. Mus., vol. 75, Art. 25, 1929, pp. 1–16, pls. 1–5.

CUSHMAN, J. A., and L. T. MARTIN. A New Genus of Foraminifera, *Discorbinella*, from Monterey Bay, California. — Contr. Cushman Lab. Foram. Res., vol. 11, 1935, pp. 89, 90, pl. 14, figs. 13*a–c*.

CUSHMAN, J. A. and I. MCCULLOCH. A Report on Some Arenaceous Foraminifera. — Allan Hancock Pacific Exped., vol. 6, No. 1, 1939, pp. 1–113, pls. 1–12.

Some Nonionidae in the Collections of the Allan Hancock Foundation. — l. c., No. 3, 1940, pp. 145–178, pls. 17–20.

Some Virgulininae in the Collections of the Allan Hancock Foundation. — l. c., No. 4, 1942, pp. 179–230, pls. 21–28.

The Species of *Bulimina* and Related Genera in the Collections of the Allan Hancock Foundation. — l. c., No. 5, 1948, pp. 231–294, pls. 29–36.

CUSHMAN, J. A., and D. A. MOYER. Some Recent Foraminifera from off San Pedro, California. — Contr. Cushman Lab. Foram. Res., vol. 6, 1930, pp. 49–62, pls. 7, 8.

CUSHMAN, J. A. and R. TODD. Foraminifera from the Coast of Washington. — Special Publ. 21, Cushman Lab. Foram. Res., 1947, pp. 1–23, pls. 1–4.

CUSHMAN, J. A., and W. W. VALENTINE. Shallow-water Foraminifera from the Channel Islands of Southern California. — Contr. Dept. Geol. Stanford Univ., vol. 1, No. 1, 1930, pp. 1–51, pls. 1–10, map.

CUSHMAN, J. A., and R. T. D. Wickenden. Recent Foraminifera from off Juan Fernandez Islands. — Proc. U. S. Nat. Mus., vol. 75, Art. 9, 1929, pp. 1–16, pls. 1–6.

HANNA, G. D., and C. C. CHURCH. A Collection of Recent Foraminifera taken off San Francisco Bay, California. — Journ. Pal., vol. 1, 1928, pp. 195–202.

LALICKER, C. G., and I. MCCULLOCH. Some *Textulariidae* of the Pacific Ocean. — Allan Hancock Pacific Exped., vol. 6, No. 2, 1940, pp. 115–143, pls. 13–16.

NATLAND, M. L. New Species of Foraminifera from off the West Coast of North America and from the Later Tertiary of the Los Angeles Basin. — Bull. Scripps Inst. Oceanography, Tech. Ser., vol. 4, 1938, pp. 137–164, pls. 3–7.

PALMER, D. K. A Note on the Occurrence of *Patellina corrugata* Williamson in the San Juan Archipelago, Washington. — l. c., vol. 3, 1929, pp. 306, 307.

ARCTIC

AWERINZEW, S. Zur Foraminiferen-Fauna des Sibirischen Eismeeres. — Mém. Acad. Imp. Sci. St.-Pétersbourg, ser. 8, vol. 29, No. 3, 1911, pp. 1–27, 1 pl.

BRADY, H. B. On the Reticularian and Radiolarian Rhizopoda (Foraminifera and Polycystina) of the North Polar Expedition of 1875–76. — Ann. Mag. Nat. Hist., ser. 5, vol. 1, 1878, pp. 425–440, pls. 20, 21.

CUSHMAN, J. A. The Foraminifera of the Canadian Arctic Expedition, 1913–18.—Rep't Canad. Arctic Exped., 1913–18, vol. 9, pt. M, 1920, pp. 1–13, pl. 1.

Results of the Hudson Bay Expedition, 1920. I. The Foraminifera.—Contr. Canadian Biol., 1921 (1922), pp. 135–147.

New Arctic Foraminifera Collected by Capt. R. A. Bartlett from Fox Basin and off the Northeast Coast of Greenland.—Smithsonian Misc. Coll., vol. 89, No. 9, 1933, pp. 1–8, pls. 1, 2.

Arctic Foraminifera.—Special Publ. 23, Cushman Lab. Foram. Res., 1948, pp. 1–79, pls. 1–8.

NORVANG, A. Foraminifera.—The Zoology of Iceland, vol. 2, pt. 2, Foraminifera, Copenhagen and Reykjavik, 1945, pp. 1–79, text figs. 1–14.

STSCHEDRINA, Z. Zur Kenntnis der Foraminiferenfauna der Arktischen Meere der U. S. S. R.—Trans. Arctic Inst., vol. 33, 1936, pp. 51–64.

ANTARCTIC

CHAPMAN, F. Report on the Foraminifera and Ostracoda out of Marine Muds from Soundings in the Ross Sea.—British Antarctic Exped., Geol., vol. 2, 1916, pp. 57–78, pls. 1–6.

CHAPMAN, F., and W. J. PARR. Foraminifera.—Australasian Antarctic Expedition, ser. C, vol. 1, pt. 2, 1937, pp. 1–190, pls. VII–X.

CUSHMAN, J. A. Foraminifera of the United States Antarctic Service Expedition 1939–1941.—Proc. Amer. Philos. Soc., vol. 89, No. 1, 1945, pp. 285–288, 1 pl.

EARLAND, A. Foraminifera, Part II. South Georgia.—*Discovery* Rep'ts, vol. VII, 1933, pp. 27–138, pls. I–VII.

HERON-ALLEN, E., and A. EARLAND. Foraminifera.—British Antarctic ("Terra Nova") Exped., vol. 6, No. 2, 1922, pp. 25–268, pls. 1–8.

PEARCEY, F. G. Foraminifera of the Scottish National Antarctic Expedition.—Trans. Roy. Soc. Edinburgh, vol. 49, pt. 4, 1914, pp. 991–1044, pls. 1, 2.

WARTHIN, A. S., JR. Foraminifera from the Ross Sea.—Amer. Museum Novitates, No. 71, 1934, pp. 1–4, text figs. 1–5.

WIESNER, H. Die Foraminiferen der Deutsche Südpolar-Expedition 1901–1903.—Deutsche Südpolar-Exped., vol. XX, Zool., 1931, pp. 53–165, pls. I–XXIV.

II. FOSSILS—GENERAL

ARCHIAC, LE VICOMTE D', and J. HAIME. Description des Animaux fossiles du groups nummulitique de l'Inde, précédé d'un résumé géologique et d'une Monographie des Nummulites.—2 vols. 4to, *Paris*, i, 1853, ii, 1854, 373 pp., 36 pls.

BRONN, H. G. Lethaea Geognostica.—2 vols. 8vo, *Stuttgart*, 1837, 1838, with 4to, Atlas. Ed. 3, 3 vols. 8vo, *Stuttgart*, 1851–56.

CUSHMAN, J. A. Paleoecology as Shown by the Foraminifera.—Contr. Cushman Lab. Foram. Res., vol. 15, 1939, pp. 40–43.

DOUVILLÉ, R. Lépidocyclines et *Cycloclypeus* malgaches.—Ann. Soc. Roy. Malac. Belg., vol. 44, 1909, pp. 125–139, pls. 5, 6.

GERTH, H. The Distribution and Evolution of the Larger Foraminifera in the Tertiary Sediments.—Proc. Kon. Akad. Wet. Amsterdam, vol. 38, No. 4, 1935, pp. 1–8, table.

GLAESSNER, M. F. Planktonforaminiferen aus der Kreide und dem Eozän und ihre Stratigraphische Bedeutung.—Studies in Micropaleontology, Moscow Univ., vol. 1, fasc. 1, 1937, pp. 27–52, 2 pls., 6 text figs.

HARPE, P. DE LA. Étude des Nummulites de la Suisse et revision des espèces éocènes des Genres Nummulites et Assilina. Pt. 1. — Mém. Soc. Pal. Suisse, vol. 7, 1881, pp. 1–104, pls. 1, 2; Pt. 2. — l. c., pp. 105–140; Pt. 3. — l. c., vol. 10, 1883, pp. 141–180, 5 pls.

HERON-ALLEN, E., and A. EARLAND. The Recent and Fossil Foraminifera of the Shoresands at Selsey Bill, Sussex. — Journ. Roy. Micr. Soc., 1908, pp. 529–543, pl. 12; 1909, pp. 306–336, pls. 15, 16; pp. 442–446, pls. 17, 18; pp. 677–698, pls. 20, 21; 1910, pp. 401–426, pls. 6–11; pp. 693–695; 1911, pp. 298–343, pls. 9–13; pp. 436–448.

JONES, T. R. Catalogue of the Fossil Foraminifera in the British Museum (Natural History). — 8vo, *London*, 1882, 100 pp.

LEMOINE, P., and R. DOUVILLÉ. Sur le genre *Lepidocyclina* Gümbel. — Mém. Soc. Géol. France, Pal., No. 32, 1904, pp. 1–41, 3 pls.

LIEBUS, A. Die Fossilen Foraminiferen eine Einführung in die Kenntnis Ihrer Gattungen. — Knihovna Stat. Geol. Ustavu Cesk. Rep. Svazek 14B, 1931, 159 pp., 349 text figs.

ORBIGNY, A. D. D'. Prodrome de Paléontologie Stratigraphique universelle des Animaux mollusques et rayonnés. — 3 vols. 8vo, *Paris*. Foraminifères, vol. 1, 1849, pp. 161, 242, 293, 324; vol. 2, 1850, pp. 41, 95, 110, 143, 185, 210, 279, 334, 407, 427; vol. 3, 1852, pp. 151, 190.

SCHLUMBERGER, C. Deuxième Note sur les Miliolidées Trematophorées. — Bull. Soc. Géol. France, ser. 4, vol. 5, 1905, pp. 115–134, pls. 2, 3, 29 text figs.

THALMANN, H. E. Regional Distribution of the Genus *Globotruncana* Cushman, 1927 in Upper Cretaceous Sediments. — Proc. Geol. Soc. Amer. for 1933 (June, 1934), p. 111.

Über geographische Rassenkreise bei fossilen Foraminiferes. — Pal. Zeitschr., vol. 16, 1934, pp. 115–121.

VAUGHAN, T. W. American and European Tertiary Larger Foraminifera. — Bull. Geol. Soc. Amer., vol. 35, 1924, pp. 785–822, pls. 30–36.

VAUGHAN, T. W., and W. S. COLE. Preliminary Report on the Cretaceous and Tertiary Larger Foraminifera of Trinidad, British West Indies (with an appendix on new species of *Helicostegina* from Soldado Rock by T. F. Grimsdale). — Geol. Soc. Amer., Spec. Papers, No. 30, 1941, pp. 1–137, pls. 1–36.

III. TERTIARY

BRITISH ISLES

BURROWS, H. W., and R. HOLLAND. The Foraminifera of the Thanet Beds of Pegwell Bay. — Proc. Geol. Assoc., vol. 15, 1897, pp. 19–52, pls. 1, 2.

JONES, T. R., W. K. PARKER and H. B. BRADY. A Monograph of the Foraminifera of the Crag. — Pal. Soc., Mon., Pt. 1, 1866, pp. i–vi, 1–72, App. I and II, pls. 1–4; Pt. 2, 1895, pp. 73–210, pls. 5–7; Pt. 3, 1896, pp. 211–314; Pt. 4, 1897, pp. vii–xv, 315–402 (Pts. 2–4, by T. R. Jones assisted by numerous other authors).

MACFADYEN, W. A. Foraminifera from some Late Pliocene and Glacial Deposits of East Anglia. — Geol. Mag., vol. 69, 1932, pp. 481–497, pls. 34, 35.

Report on the Foraminifera (in Baden-Powell, D. F. W. On the Holocene Marine Fauna from the Implementiferous Deposits of Island Magee, County Antrim). — Journ. Animal Ecology, vol. 6, 1937, pp. 87–91.

On a Marine Holocene Fauna in North-Western Scotland. — l. c., pp. 273–283 (Foraminifera, pp. 274–276).

Post-glacial Foraminifera from the English Fenlands. — Geol. Mag., vol. 75, 1938, pp. 409–17.

SHERBORN, C. D., and F. CHAPMAN. On some Microzoa from the London Clay exposed in the Drainage Works, Piccadilly, London, 1865. — Journ. Roy. Micr. Soc., ser. 2, vol. 6, 1886, pp. 737–764, 3 pls.

WRIGHT, J. Boulder-clays from the North of Ireland with Lists of Foraminifera. — Proc. Belfast Nat. Field Club, vol. 3, 1910–11, App. 1, pp. 1–8, pl. 1.

Foraminifera from the Estuarine Clays of Magheramorne, Co. Antrim, and Limavady Station, Co. Derry. — l. c., App. 2, pp. 9–11, pl. 2.

EASTERN AND CENTRAL EUROPE

ANDREAE, A. Ein Beitrag zur Kenntniss des Elsässer Tertiärs. — Abhandl. Geol. Special-Karte Elsass-Lothringen, vol. 2, Heft 3, 1884, pp. 1–239, pls. 4–12.

VAN BELLEN, R. C. Some Eocene Foraminifera from the Neighbourhood of Ricice near Imotski, E. Dalmatia, Yugoslavia. — Proc. Ned. Akad. Wetensch., Amsterdam, vol. 44, No. 8, 1941, pp. 1–12, pl., table.

Foraminifera from the Middle Eocene in the southern part of the Netherlands Province of Limburg. — Mededeel. Geol. Stichting, ser. C–V, No. 4, 1946, pp. 1–145, pls. 1–13, text figs. 1–11, chart.

BEUTLER, K. Ueber Foraminiferen aus dem jungtertiaren Globigerinenmergel von Bahna in Distrikt Mehediuti (rumänische Karpathen). — Neues Jahrb. für Min., pt. 2, 1909 (1910), pp. 140–162, pl. 18.

BOGDANOWICZ, A. Über *Meandroloculina bogatschovi* nov. gen., nov. sp. ein neues Foraminifer aus den Miocänschichten Transkaukasiens. — Bull. Accad. Sci. de l'URSS, 1935, pp. 691–696, text figs. 1–5.

BONTE, A. Observations sur le Foraminifères du Tuffean landénien de Lille (Port de Gand). — Ann. Soc. Géol. du Nord, vol. 59, 1934, pp. 67–82, pl. III, text figs. 1–8.

BORNEMANN, J. G. Die mikroskopische Fauna des Septarienthones von Hermsdorf bei Berlin. — Zeitschr. Deutsch. geol. Ges., vol. 7, 1855, pp. 307–371, pls. 12–21.

Bemerkungen über einige Foraminifera aus den Tertiärbildungen der Umgegend von Magdeburg. — l. c., vol. 12, 1860, pp. 156–160, pl. 6.

BROTZEN, F. The Swedish Paleocene and its Foraminiferal Fauna. — Sveriges Geologiska Undersökning, ser. C, No. 493, 1948, pp. 1–140, pls. 1–19, text figs. 1–41, table 1.

CAUDRI, C. M. Beitrag zur Alter-bestimmung des Flysches der Niesen-Decke. — Eclogae geologicae Helvetiae, vol. 30, 1937, pp. 403–418, pls. XXX–XXXII.

CUSHMAN, J. A. Foraminifères du Stampien du Bassin de Paris. — Bull. Soc. Sci. Seine-et-Oise, ser. 2, vol. 9, 1928, pp. 47–57, pls. 1–3.

A Fossil Member of the Family Pegididae. — Journ. Washington Acad. Sci., vol. 19, 1929, pp. 125–127, text fig.

CUSHMAN, J. A., and F. L. PARKER. Notes on Some European Miocene Species of Bulimina. — Contr. Cushman Lab. Foram. Res., vol. 13, 1937, pp. 46–54, pls. 6, 7.

CZJZEK, J. Beitrag zur Kenntniss der fossilen Foraminiferen des Wiener Beckens. — Haidinger's Nat. Abhandl., vol. 2, 1848, pp. 137–150, pls. 12, 13.

TEN DAM, A. Die Stratigraphische Gliederung des Niederländischen Paläozäns und Eozäns nach Foraminiferen (mit Ausnahme von Süd-Limburg). — Mededeel. Geol. Stichting, ser. C–V, No. 3, 1944, pp. 1–142, pls. 1–6, diagrams, charts, maps.

TEN DAM, A., and T. REINHOLD. Die Stratigraphische Gliederung des Niederländischen Plio-Plistozäns nach Foraminiferen. — l. c., No. 1, 1941, pp. 1–66, pls. 1–6, diagrams, charts, maps.

Die Stratigraphische Gliederung des Niederländischen Oligo-Miozäns nach Foraminiferen (mit Ausnahme von S. Limburg). — l. c., No. 2, 1942, pp. 1–106, pls. 1–10, diagrams, charts, map.

EGGER, J. G. Die Foraminiferen des Miocän-Schichten bei Ortenburg in Nieder-Bayern. — Neues Jahrb. für Min., 1857, pp. 266–311, pls. 5–15.

FRANKE, A. Die Foraminiferen des Unter-Eocäntones der Ziegelei Schwarzenbeck. — Jahrb. Kön. Preuss. Geol. Landes, vol. 32, pt. 2, 1911, pp. 106–111, pl. 3.

Die Foraminiferen des norddeutschen Unter-Oligocäns mit besonderen Berücksichtigung der Funde an der Fritz-Ebert-Brücke in Magdeburg. — Abhandl. Ber. Mus. Natur. und Heimatkunde und Nat. Ver., vol. 4, 1925, pp. 146–190, pls. 5, 6.

FRANZENAU, A. Die Foraminiferen-Fauna des Mergels neben dem Buda-Eörser-Weg. — Math. Nat. Ber. ungarn, vol. 7, 1889, pp. 61–90, pls. 3, 4.

Bujtur Fossil Foraminiferai. — Termesz. Füzetek, vol. 13, 1890, pp. 95–109 (in Hungarian); pp. 161–172, pl. 2, 6 text figs. (in German).

GLAESSNER, M. F. Studien ueber Foraminiferen aus der Kreide und dem Tertiaer des Kaukasus. I. Die Foraminiferen de aeltesten Tertiaerschichten des Nordwestkaukasus. — Problems of Paleontology, Moscow Univ., vols. 2–3, 1937, pp. 349–410, pls. I–IV.

Das Vorkommen von *Siderolites vidali* Douv. und *Arnaudiella grossouvrei* Douv. im Kaukasus. — Studies in Micropaleontology, Moscow Univ., vol. 1, fasc. 1, 1937, pp. 53–56.

GOCEV, P. Paläontologische und stratigraphische Untersuchungen über das Eocän von Varna. — Zeitschr. Bulgarischen Geologischen Gesellschaft, Jahrg. V, Heft 1, 1933, 82 pp., 7 pls., 14 text figs.

GÜMBEL, C. W. Beiträge zur Foraminiferenfauna der nordalpinen Eocängebilde. — Abhandl. kön. bay. Akad. Wiss., München, vol. 10, 1868 (1870), pp. 581–730, pls. 1–4.

HANTKEN, M. A Clavulina Szabói rétegek Faunája. I. Foraminiferak. — Magyar kir. földt. int. évkönyve, vol. 4, 1875 (1876), pp. 1–82, pls. 1–16 (in Hungarian).

Die Fauna der Clavulina Szaboi Schichten. I. Foraminiferen. — Mitth. Ung. Geol. Anstalt, vol. 4, 1875 (1881), pp. 1–93, pls. 1–16 (in German).

HECHT, F. E. Die Verwertbarkeit der Mikropaläontologie bei Erdöl-Aufschlussarbeiten im norddeutschen Tertiär und Mesozoikum. — Senckenbergiana, vol. 19, 1937, pp. 200–225, text figs.

KARRER, F. Ueber das Auftreten der Foraminiferen in dem marinen Tegel des Wiener Beckens. — Sitz. Akad. Wiss. Wien, vol. 44, 1861, pp. 427–458, pls. 1, 2.

Die Miocene Foraminiferen-Fauna von Kostej im Banat. — Sitz. Akad. Wiss. Wien, vol. 58, 1868, pp. 111–193, pls. 1–4.

KLÄHN, H. Die Fossilien des Tertiars Zwischen Lauch und Fecht. — Mitth. Nat. Ges. Colmar, vol. 14, 1916–17 (1920), pp. 1–73, pls. 4–14.

LAMARCK, J. B. P. A. M. Suite des Mémoires sur les Fossiles des Environs de Paris (Application des Planches relatives aux coquilles fossiles des Environs de Paris). — Annales du Museum, vol. 5, 1804, pp. 179–180, 237–245, 349–357; vol. 8, 1806, pp. 383–387, pl. 62; vol. 9, 1807, pp. 236–240, pl. 17.

LIEBUS, A. Zur Altersfrage der Flyschbildungen in nordöstlichen Mähren. — Deutsch. nat. Ver. "Lotos," vol. 70, 1922, pp. 23–66, 1 pl.

Beitrag zur Kenntnis der Neogenablagerungen aus der Umgebung von Olmütz. — Deutsch. nat. Ver. "Lotos," vol. 72, 1924, pp. 81–135, pls. 3, 4.

Neue Beiträge zur Kenntnis der Eozänfauna des Krappfeldes in Kärnten. — Jahrb. Geol. Bund., vol. 77, 1927, pp. 333–392, pls. 12–14.
Neue Foraminiferen-Funde aus dem Wienerwald-Flysch. — Verhandl. Geol. Bundesanstalt, 1934, pp. 65–70, 6 text figs.
Orbitella apiculata im Wienerwaldflysch. — l. c., 1938, pp. 143–147, 6 text figs.

MADSEN, V. Istidens Foraminiferer I Danmark og Holstein. — Meddel. fra Dansk Geol. Forening, No. 2, 1895, pp. 1–229, 1 pl.

MAJZON, L. Adatok Egyes Karpataljai Flis-Retegekhez, Tekintettel a Globotruncanakra (Beiträge zur Kenntnis Einiger Flysch-Schichten des Karpatenvorlandes mit Besonderer Rücksicht auf die Globotruncanen). — Mitt. Jahrb. K. Ungar. Geol. Anstalt, vol. 37, 1943, pp. 1–169, pls. 1, 2.

NEUGEBOREN, J. L. Foraminiferen von Felsö-Lapugy unweit Dobra im Carlsburger District; ehemals Hunyader Comitat. Artikel I. Glandulina. — Verh. Mitth. siebenbürg. Ver. Nat., 1850, pp. 45–48, 50–53, pl. 1; II. Frondicularia und Amphimorphina. — l. c., pp. 118–127, pls. 3, 4; III. Marginulina. — l. c., 1851, pp. 118–135, 140–145, pls. 4, 5; IV. Nodosaria. — l. c., 1852, pp. 34–42, 50–59, pl. 1.
Die Foraminiferen aus der Ordnung der Stichostegier von Ober-Lapugy in Siebenbürgen. — Denkschr. Akad. Wiss. Wien, vol. 12, 1856, pp. 65–108, pls. 1–5.
Die Cristellarien und Robulien aus der Thierklasse der Foraminiferen aus dem marinen Miocän bei Ober-Lapugy in Siebenbürgen. — Arch. Vereins Siebenbürgischen Landeskunde, vol. 10, 1872, pp. 273–290, pls. 1–3.

ORBIGNY, A. D. D'. Foraminifères fossiles du Bassin tertiarie de Vienne. — 4to, *Paris*, 1846, 312 pp., pls. 1–21.

PAALZOW, R. Die Foraminiferen des Cyrenmergels und des Hydrobientones des Mainzer Beckens. — Ber. Offenbacher Ver. Nat., 1912, pp. 59–74, pls. 1, 2.
Foraminiferen aus den Cerithiensanden von Offenbach a. M. — l. c., 1912–24 (1924), pp. 8–28, pls. 1, 2.

PROTESCU, O. La Microfaune des marnes à glauconie de la region Tintea (Distr. Prahova) et l'importance stratigraphique de l'espèce Clavulina szaboi Hantk. — Publ. Soc. Nat., Romania, No. 11, 1932, pp. 1–28, pls. 1–4.

REUSS, A. E. Neue Foraminiferen aus den Schichten des österreichischen Tertiärbeckens. — Denkschr. Akad. Wiss. Wien, vol. 1, 1850, pp. 365–390, pls. 46–51.
Ueber die fossilen Foraminiferen und Entomostraceen der Septarienthone der Umgegend von Berlin. — Zeitschr. Deutsch. Geol. Ges., vol. 3, 1851, pp. 49–92, pls. 3–7.
Beiträge zur Charakteristik der Tertiärschichten des nördlichen und mittleren Deutschlands. — Sitz. Akad. Wiss. Wien, vol. 18, 1855, pp. 197–272, pls. 1–12.
Beiträge zur tertiären Foraminiferen-Fauna. — l. c., vol. 42, 1860, pp. 355–370, pls. 1, 2; vol. 48, 1863 (1864), pp. 36–69, pls. 1–8.
Les Foraminifères du Crag d'Anvers. — Bull. Acad. Roy. Belge, ser. 2, vol. 15, 1863, pp. 137–162, pls. 1–3.
Zur Fauna des deutschen Ober-Oligocäns. — Sitz. Akad. Wiss. Wien, vol. 50, 1864 (1865), pp. 435–482, pls. 1–5.
Zur fossilen Fauna der Oligocänschichten von Gaas. — l. c., vol. 59, 1869, pp. 446–486, pls. 1–6.
Die Foraminiferen des Septarien-Thones von Pietzpuhl. — l. c., vol. 62, 1870, pp. 455–493.

RZEHAK, A. Die Foraminiferenfauna der Neogenformation der Umgebung von Mähr-Ostrau. — Verh. Nat. Ver. Brünn, vol. 24, 1885 (1886), pp. 77–123, 1 pl.
Ueber einige merkwürdige Foraminiferen aus dem österreichischen Tertiär. — Ann. k. k. Nat. Hofmuseums, vol. 10, 1895, pp. 213–230, pls. 6, 7.

SCHLICHT, E. VON. Die Foraminiferen des Septarienthones von Pietzpuhl. — 4to, *Berlin*, 1870, 38 pls. See Reuss, 1870.

SCHLUMBERGER, C. Note sur le genre *Miogypsina.* — Bull. Soc. Géol. France, ser. 3, vol. 28, 1900, pp. 327–333, pls. 2, 3.

SCHUBERT, R. J. Die Ergebnisse der mikroskopischen Untersuchung der bei die äranschen Tiefbohrung zu Wels durchteuften Schichten. — Ver. k. k. Geol. Reichs., vol. 53, 1903 (1904), pp. 385–422, pl. 19.

SPANDEL, E. Der Rupelton des Mainzer Beckens, seine Abteilungen und deren Foraminiferenfauna, sowie einige weitere geologisch-palaeontologischen Mitteilungen über das Mainzer Becken. — Offenbach Ber. Ver. Nat., vols. 1901–1909 (1909), pp. 57–230, 2 pls.

TERQUEM, O. Les Foraminifères de l'Eocene des Environs de Paris. — Mém. Soc. Géol. France, ser. 3, vol. 2, Mém. 3, 1882, pp. 1–193, pls. 9–28.

TOULA, F. Ueber eine kleine Mikrofauna der Ottnanger (Schlier) Schichten. — Verh. k. k. geol. Reichsanst., Jahrg., 1914, pp. 203–218, text figs. 1–7.

TOUTKOWSKY, P. Foraminifera is gretechnich e melonvich otlojaney Kieva. Pt. 1. Foraminifera kievskago mailovago mergelya. — Zapiski Kiev. Obsch. Estest., vol. 8, pt. 2, 1887, pp. 345–360, pls. 3–7; Pt. 2. Foraminifera goloubovatoi glini is bourovoi skvajini na Pololai. — l. c., vol. 9, pt. 1, 1888, pp. 1–62, pls. 1–9.

UHLIG, V. Ueber eine Mikrofauna aus dem Alttertiär der westgalizschen Karpathen. — Jahrb. k. k. geol. Reichsanst., vol. 36, Heft 1, 1886, pp. 141–214, pls. 2–5.

SOUTHERN EUROPE

AMICIS, G. A. DE. Contribuzione alla Conoscenza dei Foraminiferi Pliocenici I Foraminiferi del Pliocene Inferiore di Trinité-Victor (Nizzardo). — Boll. Soc. Geol. Ital., vol. 12, 1893, pp. 1–188, pl. 3.

I Foraminiferi del Pliocene Inferiore di Bonfornello Presso Termini Imerese (Sicilia). — Nat. Siciliano, vol. 14, 1894–95 (1895), Nos. 4, 5, pp. 7–74; Nos. 6, 7, pp. 12–127, pl. 1.

BOGDANOWICZ, A., and A. FEDOROV. On Some Representatives of the Genus Elphidium of the Sarmatian Deposits of the Lower Kuban River Course. — 50 pp., 1 pl., 34 text figs. Published in Russian with English summary, 1932.

CHECCHIA-RISPOLI, G. I Foraminiferi eocenici del gruppo del M. Judica e dei dintorni di Catenannova in provincia di Catania. — Boll. Soc. Geol. Ital., vol. 23, 1904, pp. 25–66, pl. 2.

Sopra alcune Alveoline eoceniche della Sicilia. — Pal. Ital., vol. 11, 1905, pp. 147–167, pls. 12, 13.

Nuova Contribuzione all Conoscenza delle Alveoline eoceniche della Sicilia. — l. c., vol. 15, 1909, pp. 59–70, pl. 3.

La Serie nummulitica dei dintorni di Bagheria in provincia di Palermo. — Palermo Giorn. Sci. Nat. Econ., vol. 28, 1911, pp. 107–200, 7 pls.

I Foraminiferi dell' Eocene dei dintorni di S. Marco la Catola in Capitanata. — Pal. Ital., vol. 19, 1913, pp. 103–120, pls. 5, 6.

COLOM, G. Notas sobre foraminíferos. — Bull. Inst. Catalana d'Hist. Nat., vol. 33, 1933, pp. 205–207, 4 text figs.

Notes sobre Foraminifers. — l. c., vol. 35, 1935, pp. 1–12, pls. 6, 7, text figs. 1–8.

Los foraminiferos de las margas azules de Enguera (prov. de Valencia). — l. c., vol. 36, 1936, pp. 205–226, pls. XXV–XXIX, text figs. 1–9.

Los Foraminiferos de "concha arenacea" de las margas burdigalienses de Mallorca. — Instit. Invest. Geol., Estudios Geológicos, Num. 2, 1945, pp. 1–33, pls. 1–12.

Estudio preliminar de las microfaunas de Foraminiferos de las margas eocenas y oligocenas de Navarra. — l. c., pp. 35–84, pls. 1–7.

Los sedimentos burdigalienses de las Baleares. — l. c., Num. 3, 1946, pp. 21–112, pls. 1–16.

Los Foraminiferos de las margas vindobonienses de Mallorca. — l. c., pp. 113–180, pls. 1–14, map.

Costa, O. G. Foraminiferi Fossili delle Marne Terziarie di Messina. — Mem. Accad. Sci. Napoli, vol. 2, 1855 (1857), pp. 127–147, pls. 1, 2; pp. 367–373, pl. 3.

Foraminiferi Fossili della Marna Blu del Vaticano. — l. c., pp. 113–126, pl. 1.

Paleontologia del Regno di Napoli, Pt. 2. — Atti Accad. Pontaniana, vol. 7, fasc. 1, 1853, pp. 105–112; fasc. 2, 1856, pp. 115–378, pls. 9–27.

Dervieux, E. Osservazioni Sopra le Tinoporinae e descrizione del nuovo genre Flabelliporus. — Atti R. Accad. Sci. Torino, vol. 29, 1893–94, pp. 57–61, 1 pl.

Douvillé, R. Observations sur les faunes à Foraminifères du sommet du Nummulitique italien. — Bull. Soc. Géol. France, ser. 4, vol. 8, 1908, pp. 88–91, pl. 2.

Egger, J. G. Fossile Foraminiferen von Monte Bartolomeo am Gardasee. — Jahrb. XVI, Nat. Hist. Ver. Passau, 1895, pp. 1–49, pls. 1–5.

Fornasini, C. Nota preliminare sui Foraminiferi della marna pliocenica del Ponticello di Savena nel Bolognese. — Boll. Soc. Geol. Ital., vol. 2, fasc. 2, 1883, pp. 176–190, pl. 2.

Textularina e altri Foraminiferi fossili nella marna miocenica di San Rufillo presso Bologna. — l. c., vol. 4, 1885 (1886), pp. 102–116, pl. 6.

Di alcuni Biloculine fossili negli strati a Pecten hystrix del Bolognese. — l. c., vol. 5, 1886, pp. 255–263, pls. 4, 5.

Varieta di Lagena fossile negli strati a Pecten hystrix del Bolognese. — l. c., 1886 (1887), pp. 350–353, pl. 8.

Indice delle Textularie italiane appunti per una monografia. — l. c., vol. 6, 1887, pp. 379–398, pl. 10.

Contributo alla Conoscenza della Microfauna Terziaria Italiana. Numerous papers under this heading in Mem. Real. Accad. Sci. Istit. Bologna, follow:

Lagenidi Pliocenici del Catanzarese, — ser. 4, vol. 10, 1890, pp. 461–472, 1 pl. Di alcune forme Plioceniche della *Frondicularia complanata*, — ser. 5, vol. 1, 1891, pp. 477–483, 1 pl. Di alcune forme Plioceniche della *Nodosaria obliqua*, — ser. 5, vol. 2, 1892, pp. 561–569, 1 pl. Foraminiferi delle marne Messinesi, collezioni O. G. Costa e G. Seguenza (Museo di Napoli), — ser. 5, vol. 4, 1893, pp. 201–230, 3 pls. Foraminiferi delle marne Messinesi che fanno parte della Collezione O. G. Costa, etc. — ser. 5, vol. 5, 1894, pp. 1–18, 2 pls. Di alcune forme Plioceniche della *Textilaria candeiana* e della *T. concava*, — ser. 5, vol. 6, 1896, pp. 1–6, 1 pl. Foraminiferi del Pliocene superiore di San Pietro in Lama presso Lecce, — ser. 5, vol. 7, 1898, pp. 205–212, 1 pl. Revisione delle Lagene reticolate fossili in Italia. — Rend. Accad. Sci. Istit. Bologna, vol. 13, 1909, pp. 63–69, 1 pl. Revisione delle Lagene scabre fossili in Italia. — l. c., vol. 14, 1910, pp. 65–70, 1 pl.

Gandolfi, R. Ricerche Micropaleontologiche e Stratigrafiche sulla scaglia e sul Flysch Cretacici dei Dintorni di Balerna (Canton Ticino). — Riv. Ital. Pal., vol. 48, 1942, Suppl., 1942, pp. 1–160, pls. 1–14, text figs. 1–49.

Keijzer, F. Mitteleozäne Foraminiferen aus dem Flysch der Umgegend von Omis, Dalmatien. — Proc. Kon. Ned. Akad. Wetenschappen, vol. 41, No. 9, 1938, pp. 1–7, text figs. 1–20.

Liebus, A. Die Tertiärformation in Albanien. Die Foraminiferen. — Palaeontographica, vol. 70, 1928, pp. 41–114, pl. 5, 49 figs.

MALAGOLI, M. Fauna miocenica a Foraminiferi del Vecchio Castello di Baiso.—Boll. Soc. Geol. Ital., vol. 6, 1888, pp. 517–524, pl. 13.

MARIANI, E. Foraminiferi delle marne plioceniche di Savona.—Atti Soc. Ital. Sci., vol. 31, 1888, pp. 91–128, 1 pl.

MARTINOTTI, A. Foraminiferi delle Molassa di Varano (Varesotto). Atti Soc. Ital. Sci. Nat., vol. 62, 1923, pp. 317–354, pl. 7, 34 text figs.
Alcune Forme Notevoli della Microfauna di Gorbio (Alpi Marittime).—l. c., vol. 64, 1925, pp. 175–180, pl. 4.

OVEY, C. D. Some Tertiary Foraminifera from Cyprus.—Journ. Roy. Micr. Soc., vol. 57, 1937, pp. 106–134, 29 text figs.

SCHLUMBERGER, C., and H. DOUVILLÉ. Sur Deux Foraminifères Eocenes, *Dictyoconus egyptiensis* Chapm. et *Lituonella Roberti*, nov. gen. et sp.—Bull. Soc. Géol. France, ser. 4, vol. 5, 1905, pp. 291–304, pl. 9, 7 text figs.

SCHUBERT, R. J. Neue und interessante Foraminiferen aus dem Südtiroler Altertiär.—Beitr. Pal. Oesterr.-Ung., vol. 14, 1902, pp. 9–26, pl. 1.
Ueber *Lituonella* und *Coskinolina liburnica* Stache sowie deren Beziehungen zu den anderen Dictyoconinen.—Jahrb. k. k. Geol. Reichs., vol. 62, 1912, pp. 195–208, pl. 10.

SEGUENZA, G. Descrizione dei Foraminiferi Monotalamici delle Marne Mioceniche del Distretto di Messina.—4to, *Messina*, 1862, 84 pp., 2 pls.
Prime Ricerche intorno ai Rhizopodi fossili delle Argille Pleistoceniche dei dintorni di Catania.—Atti Accad. Gioenia Sci. Nat., ser. 2, vol. 18, 1862, pp. 85–126, pls. 1–3.
Le Formazioni Terziarie nella Provinciadi Reggio (Calabria).—Atti Accad. Lincei, ser. 3, vol. 6, 1880, pp. 1–446, pls. 1–17.

SELLI, R. Una Microfauna Eocenica Inclusa nelle Argille Scagliose del Passo dell' Abbadessa (Ozzano-Bologna).—Giornale di Geologie, Ann. Mus. Geol. Bologna, ser. 2, vol. 17, 1943–44 (1944), pp. 33–91, pls. 1, 2.

SILVESTRI, A. Foraminiferi Pliocenici della provincia di Siena. Parte I.—Mem. Pont. Accad. Nuovi Lincei, vol. 12, 1896, pp. 1–204, pls. 1–5; Parte II.—l. c., vol. 15, 1899, pp. 155–381, pls. 1–16.
Ricerche strutturali su alcune forme dei trubi di Bonfornello (Palermo).—l. c., vol. 22, 1904, pp. 235–276, 14 text figs.
Miliolidi trematoforate nell'Eocene della terra d'Otranto.—Riv. Ital. Pal., vol. 14, 1908, pp. 117–148, pl. 9.
Lagenine terziarie Italiane.—Boll. Soc. Geol. Ital., vol. 31, 1912, pp. 131–180.
Fossili rari o nuovi in formazioni del paleogene.—l. c., vol. 39, 1920, pp. 57–80, pl. 4, text figs.
Singolari Nodosarine dell'Eocene piemontese.—Riv. Ital. Pal., anno 29, 1923, pp. 11–24, pl. 2.
Fauna paleogenica di Vasciano presso Todi.—Boll. Soc. Geol. Ital., vol. 42, 1923, pp. 7–29, pl. 1.
Sulle Ellissonodosarine della Molassa di Varano in Lombardia.—Atti Soc. Ital. Sci. Nat., vol. 64, 1925, pp. 49–60, text figs. 1–8.
Sulla Diffusione Stratigrafira del Genera "Chapmania" Silv.—Mem. Pont. Accad. Sci. Nuovi Lincei, vol. 8, 1925, pp. 31–60, pl. 1, text figs. 1–10.
Su di alcuni foraminiferi terziarii della Sirtica.—Missione della R. Accademia d'Italia a Cufra. (Reale Accademia d'Italia) 1934, pp. 1–28, pls. I–III.
Foraminiferi dell 'Eocene della Somalia.—Palaeontographia Italica, vol. 32, Suppl. 4, 1939, pp. 79–180 (1–102), pls. XI–XXII (I–XII).

SILVESTRI, O. Saggio di studj sulla Fauna microscopica fossile appartenente al terreno subappenino Italiano. Memoria Prima Monographia delle Nodosarie.—Atti Accad. Giocenia Sci. Nat., ser. 3, vol. 7, 1872, pp. 1–108, pls. 1–11.

TELLINI, A. Le Nummulitidi della Majella delle Isole Tremiti e del Promontorio Garganico.—Boll. Soc. Geol. Ital., vol. 9, fasc. 2, 1890, pp. 1–69, pls. 11–14.

TERQUEM, O. Les Foraminifères et les Entomostracés—Ostracodes du Pliocène supérieur de l'Île de Rhodes.—Mém. Soc. Géol. France, ser. 3, vol. 1, 1878, pp. 1–133, pls. 1–14.

TERRIGI, G. Fauna Vaticana a Foraminiferi delle Sabbie Gialle nel Plioceno subappenino superiore.—Atti Accad. Pont. Nuovi Lincei, vol. 33, 1880, pp. 127–219, pls. 1–4.

I depositi lacustri e Marini riscontrati nella trivellazione presso la Via Appia Antica.—Mem. Reg. Com. Geol. Ital., vol. 4, 1891, pp. 53–131, pls. 1–4.

VAN DER WEIJDEN, W. J. M. Het Genus *Discocyclina* in Europa. Een Monografie naar Aanleiding van een Heronderzoek van het Tertiair-profiel van Biarritz.—Leiden, Feb. 29, 1940, pp. 1–116, pls. 1–12, text figs. and charts.

DE WITT PUYT, J. F. C. Geologische und Paläontologische Beschreibung der Umgebung von Ljubuski, Hercegovina.—Diss. Utrecht, 1941, pp. 1–99, pls. 1–5.

AFRICA AND ASIA

ASANO, K. New Foraminifera from the Kakegawa District, Totomi, Japan.—Jap. Journ. Geol. Geogr., vol. 13, 1936, pp. 327–331, pls. 36, 37.

Pseudononion, a New Genus of Foraminifera found in Muraoka-mura, Kamakura-gori, Kanagawa Prefecture.—Journ. Geol. Soc. Japan, vol. 43, No. 512, 1936, pp. 50, 51, text figs. A–C.

Foraminifera from Muraoika-mura, Kamakura-gori, Kanagawa Prefecture. (Studies on the Fossil Foraminifera from the Neogene of Japan. Part I.)—l. c., pp. 603–614, pls. 30, 31.

Foraminifera from Kuromatunai-mura, Suttu-gun, Hokkaido. (Studies on the Fossil Foraminifera from the Neogene of Japan. Part II.)—l. c., pp. 615–622, pls. 32, 33.

A Pliocene Species of *Elphidium* from Japan.—l. c., vol. 44, 1937, pp. 787–790, text figs. 1, 2.

On Some Pliocene Foraminifera from the Setana Beds, Hokkaido.—Jap. Journ. Geol. Geogr., vol. 15, 1938, pp. 87–103, pls. IX–XI, text figs. 1, 2.

BOURCART, J., and E. DAVID. Étude stratigraphique et paléontologique des grès à foraminifères d'Ouezzan au Maroc (Oligocène et Miocène inférieur).—Mém. Soc. Sci. Nat. Maroc., vol. 37, 1933, pp. 1–57, 14 pls.

CHAPMAN, F. On a *Patellina*-Limestone from Egypt.—Geol. Mag., dec. 4, vol. 7, 1900, pp. 3–17, pl. 2.

On an *Alveolina*-Limestone and Nummulitic Limestone from Egypt.—l. c., vol. 9, 1902, pp. 106–114, pls. 4, 5.

On a Foraminiferal Limestone of Upper Eocene Age from the Alexandria Formation, South Africa.—Ann. So. African Mus., vol. 28, 1930, pp. 291–296, pl. 37.

CUSHMAN, J. A., and Y. OZAWA. Some Species of Fossil and Recent Polymorphinidae Found in Japan.—Jap. Journ. Geol. Geogr., vol. 6, 1929, pp. 63–78, pls. 13–16.

DAVIES, L. M. The Ranikot beds at Thal (North-West Frontier Provinces of India).—Quart. Journ. Geol. Soc., vol. 83, 1927, pp. 260–290, 5 pls., 8 text figs.

The Genus *Dictyoconus* and Its Allies: A Review of the Group, Together with a Description of Three New Species from the Lower Eocene Beds of Northern Baluchistan.—Trans. Roy. Soc. Edinburgh, vol. 56, 1930, pp. 485–505, pls. 1, 2.

DAVIES, L. M., and E. S. PINFOLD. The Eocene Beds of the Punjab Salt Range. — Mem. Geol. Surv. India, n. ser., vol. 24, Mem. No. 1, 1937, pp. i–iii, 1–79, pls. I–VII.

FLANDRIN, J. La faune de Tizi Renif, près Dra el Mizan (Algérie). — Bull. Soc. Géol. France, ser. 5, vol. 4, 1934, pp. 251–272, pls. XIV–XVI.

Contribution à l'Étude Paléontologique du Numulitique Algérien. — Matériaux pour la Carte Géologique de l'Algérie, ser. 1, Paléontologie, No. 8, 1938, pp. 1–158, Atlas, pls. 1–15.

FLANDRIN, J., and F. JACQUET. Les Nummulites de l'Eocene moyen du Sénégal. — Bull. Soc. Géol. France, ser. 5, vol. 6, 1936, pp. 363–373, pl. XXIV.

GREIG, D. A. *Rotalia viennoti*, an Important Foraminiferal Species from Asia Minor and Western Asia. — Journ. Pal., vol. 9, 1935, pp. 523–526, pl. 58.

HANZAWA, S. Notes on Tertiary Foraminiferous Rocks from the Kwanto Mountainland, Japan. — Sci. Rep't Tohoku Imp. Univ., ser. 2 (Geol.), vol. 12, 1931, pp. 141–157 (1–17), pls. 24–26 (1–3).

On Some Miocene Rocks with *Lepidocyclina* from the Izu and Boso Peninsulas. — l. c., pp. 159–170 (1–12), pls. 27, 28 (1, 2), text figs. 1, 2.

Notes on Some Eocene Foraminifera Found in Taiwan (Formosa), with Remarks on the Age of the Hori Slate Formation and Crystalline Schists. — l. c., pp. 171–194 (1–24), pl. 29 (1).

Some Fossil Operculina and Miogypsina from Japan, and their Stratigraphical Significance. — l. c., vol. 18, 1935, pp. 1–29, pls. I–III.

Studies on the Foraminifera Fauna Found in the Bore Cores from the Deep Well in Kita-Daito-Sima (North Borodino Island). — Proc. Imp. Acad. Tokyo, vol. 14, 1938, pp. 384–390.

HARPE, P. DE LA. Monographie der in Aegypten und der libyschen Wüste vorkommenden Nummuliten. In Zittel, Beiträge zur Geologie und Paläontologie der libyschen Wüste und der angrenzenden Gebiete. — Palaeontographica, vol. 30, 1883, pp. 157–216, pls. 30–35.

HENSON, F. R. S. Larger Foraminifera from Aintab, Turkish Syria. — Eclogae geologicae Helvetiae, vol. 30, 1937, pp. 45–57, pls. II–VI, 5 text figs.

HERON-ALLEN, E., and A. EARLAND. Foraminifera from the Eocene Clay of Nigeria. — Bull. 3, Geol. Surv. Nigeria, 1922, pp. 138–148, pl. 12.

HODSON, H. K. Lower Miocene Fossils from Portuguese East Africa. — Journ. Pal., vol. 2, 1928, pp. 1–6, pls. 1–3.

JORDAN, L. No. 1, Basbirin Kuyusundaki Kucuk Foraminiferanin bir Mutalaasi. A Study of the Small Foraminifera in the Basbirin Well No. 1. — Publications of M. T. A. Institute, Ankara, 1937, pp. 1–8 (Turkish), 1–7 (English), pls. I–IV.

MACFADYEN, W. A. Miocene Foraminifera from the Clysmic Area of Egypt and Sinai. With an Account of the Stratigraphy and a Correlation of the Local Miocene Succession. — Geol. Surv. Egypt, 1930 (1931), pp. 1–149, pls. 1–4, text figs. 1, 2.

NUTTALL, W. L. F. Indian Reticulate Nummulites. — Ann. Mag. Nat. Hist., ser. 9, vol. 15, 1925, pp. 661–667, pls. 37, 38.

The Stratigraphy of the Laki Series (Lower Eocene) of parts of Sind and Baluchistan (India); with a Description of the Larger Foraminifera contained in those Beds. — Quart. Journ. Geol. Soc., vol. 81, 1925, pp. 417–453, pls. 23–27, 5 text figs.

The Zonal Distribution and Description of the Larger Foraminifera of the Middle and Lower Kirthar Series (Middle Eocene) of Parts of Western India. — Rec. Geol. Surv. India, vol. 59, 1926, pp. 115–164, pls. 1–8.

NUTTALL, W. L. F., and A. G. BRIGHTON. Larger Foraminifera from the Tertiary of Somaliland. — Geol. Mag., vol. 68, 1931, pp. 49–65, pls. 1–4, text figs. 1–3.

SCHWAGER, C. Die Foraminiferen aus den Eocaenablagerungen der libyschen Wüste und Aegyptens. In Zittel, Beiträge zur Geologie und Paläontologie der libyschen Wüste und der angrenzenden Gebiete. — Palaeontographica, vol. 30, 1883, pp. 81–153, pls. 24–29.

SILVESTRI, A. I Foraminiferi. — Res. Sci. Miss. Oasi Giarabub, 1926–27 (1928), pp. 171–199, pls. 27–31, 1 text fig.

Di Alcune *Facies* Lito-paleontologogiche del Terziario di Derna, nella Cirenaica. — Boll. Soc. Geol. Ital., vol. 47, 1928, pp. 109–113, pl. 6.

Nummuliti, operculina e planorbulina di Derna nella Cirenaica. — Mem. Pont. Accad. Sci. Nuovi Lincei, vol. 11, 1928, pp. 263–278, pl. 1, text figs. A–C.

Fossile eocenico singolare della Tripolitania. — Boll. Soc. Geol. Ital., vol. 56, pt. 2, 1937, pp. 203–208, pl. IX.

Foraminiferi dell' Oligocene e del Miocene della Somalia. — Pal. Ital., vol. 32, Suppl. 2, 1937, pp. 43–264, pls. IV–XXII (I–XIX).

Foraminiferi dell' Eocene della Somalia, Parte I. — l. c., Suppl. 3, 1938, pp. 48–89 (37–77), pls. III–XII (I–X).

YABE, H., and S. HANZAWA. Tertiary Foraminiferous Rocks of Taiwan (Formosa). — Proc. Imp. Acad., vol. 4, 1928, pp. 533–536, text figs. 1–3.

Tertiary Foraminiferous Rocks of the Philippines. — Sci. Rep't Tohoku Imp. Univ., ser. 2 (Geol.), vol. 11, 1929, pp. 137–190 (1–54), pls. 15–27 (1–13).

Tertiary Foraminiferous Rocks of Taiwan (Formosa). — l. c., vol. 14, 1930, pp. 1–46, pls. 1–16, 1 text fig., 2 tables.

INDO-PACIFIC

ASANO, K. Japanese Fossil Nodosariidae, with Notes on the Frondiculariidae. — Tohoku Imp. Univ., Sci. Rep'ts, ser. 2 (Geol.), vol. 19, 1938, pp. 179–220 (1–42), pls. XXIV–XXXI (I–VIII).

BOOMGAART, L., and J. VROMAN. Smaller Foraminifera from the Marl Zone between Sonde and Modjokerto (Java). Proc. Roy. Acad. Amsterdam, vol. 39, 1936, pp. 421–425.

BRADY, H. B. On some Fossil Foraminifera from the West-coast district of Sumatra. — Geol. Mag., dec. 2, vol. 2, 1875, pp. 532–539, pls. 13, 14.

Note on the so-called "Soapstone" of Fiji. — Quart. Journ. Geol. Soc., vol. 44, 1888, pp. 1–10, pl. 1.

CAUDRI, C. M. B. Tertiary Deposits of Soemba. — Amsterdam, 1934, pp. i–xiii, 1–224, pls. I–V, 3 maps, 21 text figs.

Lepidocyclinen von Java. — Verhandl. Geol. Mijn. Gen. Ned. Kol., Geol. ser., vol. 12, 1939, pp. 135–257, pls. 1–10, text figs.

CHAPMAN, F. Notes on the Older Tertiary Foraminiferal Rocks on the West Coast of Santo, New Hebrides. — Proc. Linn. Soc. N. S. Wales, 1905, pp. 261–274, pls. 5–8.

On the Tertiary Limestones and Foraminiferal Tuffs of Malekula, New Hebrides. — l. c., 1907, pp. 745–760, pls. 37–48.

A Study of the Batesford Limestone. — Proc. Roy. Soc. Victoria, vol. 22, pt. 2, 1909 (1910), pp. 263–314, pls. 52–55.

Description of a Limestone of Lower Miocene Age from Bootless Inlet, Papua. — Proc. Roy. Soc. N. S. Wales, vol. 48, 1914, pp. 281–301, pls. 7–9.

Report on the Foraminifera and Ostracoda from Elevated Deposits on the Shores of the Ross Sea. — British Antarctic Exped., Geol., vol. 2, 1916, pp. 27–51, pls. 1–6.

On the Occurrence of the Foraminiferal Genus *Miogypsinoides* in New Zealand. — Rec. Cant. Mus., vol. 3, 1932, pp. 491–493, pl. 63.

CHAPMAN, F., and I. CRESPIN. Rare Foraminifera from Deep Borings in the Victorian Tertiaries — Victoriella, gen. nov., Cycloclypeus communis Martin, and Lepidocyclina borneënsis Provale. — Proc. Roy. Soc. Victoria, vol. 42, 1930, pp. 110–115, pls. 7, 8.

The Sequence and Age of the Tertiaries of Southern Australia. — Rep't Melbourne (1935) Meeting of Australian and New Zealand Assoc. Adv. Sci., 1935, pp. 118–126.

CHAPMAN, F., and W. J. PARR. Tertiary Foraminifera of Victoria, Australia. The Balcombian deposits of Port Phillip. Part II. — Journ. Linn. Soc. Zool., vol. 36, 1926, pp. 373–399, 5 pls.

Notes on New and Aberrant Types of Foraminifera. — Proc. Roy. Soc. Victoria, vol. 43, 1931, pp. 236–238, pl. 9 (part).

Australian and New Zealand Species of the Foraminiferal Genera *Operculina* and *Operculinella*. — Proc. Roy. Soc. Victoria, vol. 50, 1938, pp. 279–299, pls. XVI, XVII, 1 text fig.

CHAPMAN, F., W. J. PARR and A. C. COLLINS. Tertiary Foraminifera of Victoria, Australia. — The Balcombian Deposits of Port Phillip. Part III. — Linn. Soc. Journ. Zool., vol. 38 (No. 262), 1934, pp. 553–577, pls. 8–11.

COLE, W. S. Large Foraminifera from Guam. — Journ. Pal., vol. 13, 1939, pp. 183–189, pls. 23, 24, 1 text fig. (map).

Larger Foraminifera of Lau, Fiji (in LADD and HOFFMEISTER). — Bull. 181, Bernice P. Bishop Mus., 1945, pp. 272–297, pls. 12–30.

CRESPIN, I. The Larger Foraminifera of the Lower Miocene of Victoria. — Pal. Bull., Bull. 2, 1936, pp. 1–15, pls. 1, 2 (map).

The Occurrence of *Lacazina* and *Biplanispira* in the Mandated Territory of New Guinea. — l. c., No. 3, 1938, pp. 1–8, pls. 1–2.

CUSHMAN, J. A. New Late Tertiary Foraminifera from Vitilevu, Fiji. — Contr. Cushman Lab. Foram. Res., vol. 7, 1931, pp. 25–32, pl. 4.

Smaller Foraminifera from Vitilevu, Fiji. — Bernice P. Bishop Museum, Bull. 119, 1934, pp. 102–142, pls. 10–18.

Notes on Some Foraminifera Described by Schwager from the Pliocene of Kar Nicobar. — Journ. Geol. Soc. Japan, vol. 46, No. 546, 1939, pp. 149–154, pl. 10 (6).

CUSHMAN, J. A., and S. HANZAWA. New Genera and Species of Foraminifera of the Late Tertiary of the Pacific. — Contr. Cushman Lab. Foram. Res., vol. 12, 1936, pp. 45–48, pl. 8 (pt.).

DAVIES, A. M. Lower Miocene Foraminifera from Pemba Island. — Rep. Pal. Zanzibar Protectorate, 1927, pp. 7–12, pls. 1, 2.

DEPRAT, J. Les dépôts éocènes néocalédoniens; leur analogie avec ceux de la région de la sonde. Description de deux espèces nouvelles d'Orbitoids. — Bull. Soc. Géol. France, ser. 4, vol. 5, 1905, pp. 485–516, pls. 16–19.

DOUVILLÉ, H. Les Foraminifères dans le Tertiaire des Philippines. — Philippine Journ. Sci., ser. D, vol. 6, 1911, pp. 53–80, pls. A–D.

Quelques Foraminifères de Java. — Samml. Geol. Reichs. Mus., ser. 1, vol. 8, 1912, pp. 279–294, pls. 22–24.

Les Foraminifères de l'Île de Nias. — l. c., pp. 253–278, pls. 19–21.

Les Foraminifères des Couches de Rembang. — l. c., ser. 1, vol. 10, 1916, pp. 19–35, pls. 3–6.

van Eek, D. Foraminifera from the Telisa — and the Lower Palembang-Beds of South Sumatra. — "De Ingenieur in Nederlandsch-Indië," IV, Mijnbouw Geol., Jaargang IV, 1937, pp. 47–55, pl. I.

Finlay, H. J. New Zealand Foraminifera: the Occurrence of *Rzehakina, Hantkenina, Rotaliatina,* and *Zeauvigerina.* — Trans. Roy. Soc. New Zealand, vol. 68, 1939, pp. 534–543.

New Zealand Foraminifera: Key Species in Stratigraphy. — No. 1, l. c., pp. 504–533, pls. 68, 69; No. 2, l. c., vol. 69, 1939, pp. 89–128, pls. 11–14; No. 3, l. c., pp. 309–329, pls. 24–29; No. 4, l. c., vol. 69, 1940, pp. 448–472, pls. 62–67.

Fritsch, K. Einige eocän Foraminiferen von Borneo. — Palaeontographica, Suppl. 3, 1878, pp. 139–146, pls. 18, 19.

Germeraad, J. H. Geology of Central Seran, in L. Rutten and W. Hotz, Geological, Petrographical and Palaeontological Results of Explorations, carried out from September 1917 till June 1919 in the Island of Ceram, 1946, pp. 1–135, pls. 1–12, 5 tables, map.

Hanzawa, S. Note on Foraminifera Found in the Lepidocyclina-Limestone from Pabeasan, Java. — Sci. Rep't Tohoku Imp. Univ., ser. 2 (Geol.), vol. 14, 1930, pp. 85–96 (1–12), pls. 26–28 (1–3).

A New Type of *Lepidocyclina* with a Multilocular Nucleoconch from the Taito Mountains, Taiwan (Formosa). — Proc. Imp. Acad., vol. 8, 1932, pp. 446–449, 6 text figs.

Heron-Allen, E., and A. Earland. The Miocene Foraminifera of the "Filter Quarry," Moorabool River, Victoria, Australia. — Journ. Roy. Micr. Soc., 1924, pp. 121–186, pls. 7–14.

Howchin, W. Notes on the Geological Sections Obtained by Several Borings Situated on the Plains between Adelaide and Gulf St. Vincent. Part II. Cowandilla (Government) Bore. — Trans. Roy. Soc. So. Australia, vol. 60, 1936, pp. 1–34, pl. 1.

Howchin, W., and W. J. Parr. Notes on the Geological Features and Foraminiferal Fauna of the Metropolitan Abattoirs Bore, Adelaide. — Trans. Roy. Soc. So. Australia, vol. 62, 1938, pp. 287–317, pls. XV–XIX.

Karrer, F. Die Foraminiferen der Tertiären Thone von Luzon. In von Drasche, Fragmente zu einer Geologie der Insel Luzon. — 4to, *Wien,* 1878, pp. 75–99, pl. 5.

Koch, R. Die jungtertiäre Foraminiferenfauna von Kabu (Res. Surabaja, Java). — Eclog. geol. Helv., vol. 18, No. 2, 1923, pp. 342–361, 1 pl.

Mitteltertiäre Foraminiferen aus Bulongan, Ost-Borneo. — l. c., vol. 19, No. 3, 1926, pp. 722–751, 26 figs.

Krijnen, W. F. Het genus *Spiroclypeus* in het Indo-Pacifische gebied (with summary in English). — Verhandl. Geol.-Mijn. Gen. Nederland en Kolonien, Geol. Ser., vol. 9, 1931, pp. 77–112, pls. 1–3.

LeRoy, L. W. Some Small Foraminifera, Ostracoda and Otoliths from the Neogene ("Miocene") of the Rokan-Tapanoeli Area, Central Sumatra. — Nat. Tijd. Ned.-Indie, vol. 99, 1939, pp. 215–296, pls. 1–14, map, chart.

Small Foraminifera from the Late Tertiary of the Nederlands East Indies. — Colorado School of Mines Quarterly, vol. 36, No. 1, 1941, pp. 1–132, 13 pls., 12 text figs.

Miocene Foraminifera from Sumatra and Java, Netherlands East Indies. — l. c., vol. 39, No. 3, 1944, pp. 1–113, 15 pls., 2 text figs.

NUTTALL, W. L. F. A Revision of the Orbitoids of Christmas Island. — Quart. Journ. Geol. Soc., vol. 82, 1926, pp. 22–42, pls. 4, 5, text figs. 1–3.

PARR, W. J. Notes on Australian and New Zealand Foraminifera. No. 2: The Genus *Pavonina* and Its Relationships. — Proc. Roy. Soc. Victoria, vol. 45, pt. 1 (n. ser.), 1933, pp. 28–31, pl. VII.

Tertiary Foraminifera from Chalky Island, S. W. New Zealand. — Trans. Roy. Soc. New Zealand, vol. 64, 1934, pp. 140–146, pl. 20, 3 text figs.

Some Foraminifera from the Awamoan of the Medway River District, Awatere, Marlborough, New Zealand. — l. c., vol. 65, 1935, pp. 75–87, pls. 19, 20.

Tertiary Foraminifera from near Bombay, Franklin County, Auckland, New Zealand. — l. c., vol. 67, 1937, pp. 71–77, pl. 15.

Upper Eocene Foraminifera from Deep Borings in King's Park, Perth, Western Australia. — Journ. Roy. Soc. W. Australia, vol. 24, 1937–38, pp. 69–101, pls. I–III, 1 text fig.

Foraminifera of the Pliocene of South-Eastern Australia. — Mining and Geological Journal, vol. 1, 1939, pp. 65–71, 1 pl.

PARR, W. J., and A. C. COLLINS. Notes on Australian and New Zealand Foraminifera. No. 1 — The Species of Patellina and Patellinella, with a Description of a New Genus, Annulopatellina. — Proc. Roy. Soc. Victoria, vol. 43, pt. I (n. ser.), 1930, pp. 89–95, pl. 4.

Notes on Australian and New Zealand Foraminifera. No. 3, Some Species of the Family Polymorphinidae. — Proc. Roy. Soc. Victoria, vol. 50 (n. ser.), 1937, pp. 190–211, pls. XII–XV, 7 text figs.

RUTTEN, L. Studien über Foraminiferen aus Ost-Asien. — Leiden Samml. Geol. Reichsmus., ser. 1, vol. 9, 1912, pp. 201–217, pls. 12, 13; 1914, pp. 281–325, pls. 21–27.

Some Notes on Foraminifera from the Dutch Indies. — Proc. Kon. Akad. Wetens., vol. 27, 1924, pp. 1–6, text figs. 1–9.

Roches et Fossiles de l'Île Pisang et de la Nouvelle-Guinée. — Bull. Mus. Roy. Hist. nat. Belgique, vol. 12, No. 10, 1936, pp. 1–14, pls. 1–4.

SCHEFFEN, W. Ostindische Lepidocyclinen. I. Theil. — Wetenschappelijke Mededeelingen, No. 21, 1932, pp. 1–76, pls. 1–14, text figs. 1–6.

SCHUBERT, R. J. Die fossilen Foraminiferen des Bismarksarchipels und einiger angrenzender Inseln. — Abhandl. Geol. Reichs., vol. 20, pt. 4, 1911, pp. 1–130, 6 pls.

SCHWAGER, C. Fossile Foraminiferen von Kar-Nicobar. — *Novara*-Exped., Geol. Theil, vol. 2, 1866, pp. 187–268, pls. 4–7.

STACHE, G. Die Foraminiferen des Tertiaren Mergel des Whaingaroa-Hafens (Provinz Auckland). — l. c., vol. 1, 1864, Pal., pp. 161–304, pls. 21–24.

TAN, S. H. Over Spiroclypeus, met opmerkingen over zijn stratigrafische verspreiding. (Preliminary Communication, with English summary). — De Mijningenieur, Jaargang 11, 1930, pp. 180–184, text figs. *a–d*.

Über mikrosphäre Lepidocyclinen von Ngampel (Rembang, Mitteljava). — "De Ingenieur in Nederlandsch-Indië," IV. Mijnbouw en Geologie, "De Mijningenieur," Jaarg. I, 1934, pp. 203–211, pls. 1, 2, 4 text figs.

Über Lepidocyclina gigantea K. Martin von Süd-Priangan (Westjava), Tegal (Mitteljava) und Benkoelen (Südsumatra). — l. c., Jaarg. II, 1935, pp. 1–8, pls. 1–3.

Zwei neue mikrosphäre Lepidocyclinen von Java. — l. c., pp. 9–18, pls. (I) IV–(IV) VII.

Lepidocyclina zeijlmansi nov. sp., eine polylepidine Orbitoidide von Zentral-Borneo, nebst Bemerkungen über die verschiedenen Einteilunsweisen der Lepidocyclinen. — l. c., Jaarg. III, 1936, pp. 7–14, pl. 1.

Zur Kenntnis der Miogypsiniden. — l. c., pp. 45–61, pls. I, II; pp. 84–98, text figs. 1–13; pp. 109–123.
Note on Miogypsina kotôi Hanzawa. — l. c., Jaarg. IV, 1937, pp. 31, 32, 1 plate.
Weitere Untersuchungen über die Miogypsiniden I. — l. c., pp. 35–45, pls. I–III.
On the Genus Spiroclypeus H. Douvillé, with a Description of the Eocene *Spiroclypeus vermicularis* nov. sp. from Koetai in East Borneo. — l. c., pp. 177–193, pls. I–IV, 1 text fig.

TOBLER, A. Notiz über einige foraminiferenführende Gesteine von der Halbinsel Sanggar (Soembawa). — Zeitschr. für Vulkanologie, vol. 4, 1918, pp. 189–192, pls. 34, 35, map.

UMBGROVE, J. H. F. Het genus Pellatispira in het indo-pacifische gebied (with summary in English). — Wetensch. Med., Mijn. Ned.-Ind., No. 10, 1928, pp. 1–60, 16 pls., 10 text figs.
Heterospira, a New Foraminiferal Genus from the Tertiary of Borneo. — Leidische Geol. Meded., vol. 8, 1936, pp. 115–159, 1 pl.
A Second Species of *Biplanispira* from the Eocene of Borneo. — l. c., vol. 10, 1938, pp. 82–89, 17 text figs.

VAUGHAN, T. W. Species of Lepidocyclina and Carpenteria from the Cayman Islands. — Quart. Journ. Geol. Soc., vol. 82, 1926, pp. 388–400, pls. 24–26.

VLERK, I. M. VAN DER. Een Overgangsvorm Tusschen Orthophragmina en Lepidocyclina uit het Tertiar van Java. — Verh. Geol.-Mijnb. Gen. Nederland en Kolonien, Geol. Ser., vol. 7, 1923, pp. 91–98, 1 pl.
Foraminiferen uit het Tertiar van Java. — Wetens. Med. Dienst Mijn. Ned. Ind., No. 1, 1924, pp. 1–13, pls. 3–5.
A Study of Tertiary Foraminifera from the "Tidoengsche Landen" (E. Borneo). — l. c., 1925, pp. 13–38, pls. 1–6.
The Genus Lepidocyclina in the Far East. — Eclog. geol. Helv., vol. 21, 1928, pp. 182–211, pls. 6–23, with 3 tables.
Grotte foraminiferen van N. O. Borneo (with summary in English). — Wetens. Med., No. 9, 1929, pp. 1–44, 7 pls.

VLERK, I. M. VAN DER, and J. H. F. UMBGROVE. Tertiaire Gidsforaminiferen van Nederlandsch Oost-Indië. — Wetens. Med. Dienst Mijn. Med. Ind., No. 6, 1927, pp. 1–31, 24 text figs., table.

WHIPPLE, G. L. Eocene Foraminifera. — Bull. 96, Bishop Mus., 1932, pp. 79–90, pls. 20–23.

YABE, H. Notes on *Operculina*-Rocks from Japan, with Remarks on "*Nummulites*" *cumingi* Carpenter. — Sci. Rep't, Tohoku Imp. Univ., ser. 2 (Geol.), vol. 4, No. 3, 1918, pp. 105–126, pl. 17. Notes on a *Carpenteria*-Limestone from B. N. Borneo. — l. c., vol. 5, No. 1, 1918, pp. 15–30, pls. 3–5. Notes on a *Lepidocyclina*-Limestone from Cebu. — l. c., vol. 5, No. 2, 1919, pp. 37–51, pls. 6, 7. Notes on Some Eocene Foraminifera. I. Four Arenaceous Foraminifera from the Eocene of Hahajima (Hillsborough Is.), Oga-Sawara (Bonin) Group. II. Notes on Two Foraminiferal Limestones from E. D. Borneo. III. Notes on *Pellatispira*, Boussac. — l. c., vol. 5, No. 4, 1921, pp. 97–108, pls. 16–20.

YABE, H., and K. ASANO. New Occurrence of *Rotaliatina* in the Pliocene of Java. — Journ. Geol. Soc. Japan, vol. 44, No. 523, 1937, pp. 39–41, text figs. 1–3.
Contribution to the Palaeontology of the Tertiary Formations of West Java. Part 1. Minute Foraminifera from the Neogene of West Java. Tohoku Imp. Univ., Sci. Rep'ts, ser. 2 (Geol.), vol. 19, 1937, pp. 87–126 (1–40), pls. XVII–XIX (I–III).

YABE, H., and S. HANZAWA. A Geological Problem Concerning the Raised Coral-Reefs of the Riukiu Islands and Taiwan; etc. — l. c., vol. 7, No. 2, 1925, pp. 29–56, pls. 5–10. Nummulitic Rocks of the Islands of Amakusa (Kyushu, Japan). — l. c., No. 3, 1925, pp. 73–82, pls. 18–22. I. A Foraminiferous Limestone, with a Questionable Fauna, from Klias Peninsula, British North Borneo. II. *Choffatella* Schlumberger and *Pseudocyclammina*, a New Genus of Arenaceous Foraminifera. III. Geological Age of *Orbitolina*-bearing Rocks of Japan. — l. c., vol. 9, No. 1, 1926, pp. 1–20, pls. 1–6.

Lepidocyclina from Naka-Kosaka, Province of Kozuke, Japan. — Jap. Journ. Geol. Geogr., vol. 1, No. 1, 1922, pp. 45–50, pls. 5–8. Foraminifera from the Natsukawa-Limestone, with a note on a new subgenus of *Polystomella*. — l. c., vol. 2, No. 4, 1923, pp. 95–100, text figs. A *Lepidocyclina*-Limestone from Sangkoelirang, D. E. Borneo. — l. c., vol. 3, 1924, pp. 71, 72, pls. 9–12, text fig. A *Lepidocyclina*-Limestone from Klias Peninsula, B. N. Borneo. — Verh. Geol.-Mijn. Gen. Ned. Kol., Geol. Ser., vol. 8, 1925, pp. 617–632, pls. 1–4.

AMERICA

APPLIN, E. R., A. E. ELLISOR and H. T. KNIKER. Subsurface Stratigraphy of the Coastal Plain of Texas and Louisiana. — Bull. Amer. Assoc. Petr. Geol., vol. 9, 1925, pp. 79–122, pl. 3, 1 text fig.

APPLIN, E. R., and L. JORDAN. Diagnostic Foraminifera from subsurface formations in Florida. — Journ. Pal., vol. 19, 1945, pp. 129–148, pls. 18–21, 2 text figs.

APPLIN, P. L., and E. R. APPLIN. Regional Subsurface Stratigraphy and Structure of Florida and Southern Georgia. — Bull. Amer. Assoc. Petr. Geol., vol. 28, 1944, pp. 1673–1753, 5 pls., 38 text figs.

BAGG, R. M. Foraminifera [Eocene of Maryland]. — Maryland Geol. Surv., Eocene, 1901, pp. 233–258, pls. 62–64.

Foraminifera [Miocene of Maryland]. — l. c., Miocene, 1904, pp. 460–483, pls. 131–133.

Miocene Foraminifera from the Monterey Shale of California, with a few species from the Tejon formation. — Bull. 268, U. S. Geol. Surv., 1905, pp. 1–78, pls. 1–11.

Pliocene and Pleistocene Foraminifera from Southern California. — Bull. 513, U. S. Geol. Surv., 1912, pp. 1–153, pls. 1–28.

BANDY, O. L. Eocene Foraminifera from Cape Blanco, Oregon. — Journ. Pal., vol. 18, 1944, pp. 366–377, pls. 60–62.

BARBAT, W. F., and F. E. VON ESTORFF. Lower Miocene Foraminifera from the Southern San Joaquin Valley. — l.c., vol. 7, 1933, pp. 164–174, pl. 23.

BARBAT, W. F., and F. L. JOHNSON. Stratigraphy and Foraminifera of the Reef Ridge Shale, Upper Miocene, California. — l. c., vol. 8, 1934, pp. 3–17, pl. 1.

BARKER, R. W. Three Species of Larger Tertiary Foraminifera from S. W. Ecuador. — Geol. Mag., vol. 69, 1932, pp. 277–281, pl. 16, text figs.

Larger Foraminifera from the Eocene of Santa Elena Peninsula, Ecuador. — l. c., pp. 302–310, pls. 21, 22, 4 text figs.

BARKER, R. W., and T. F. GRIMSDALE. Studies of Mexican Fossil Foraminifera. — Ann. Mag. Nat. Hist., ser. 10, vol. 19, 1937, pp. 161–178, pls. V–IX.

BECK, R. S. Eocene Foraminifera from Cowlitz River, Lewis County, Washington. — Journ. Pal., vol. 17, 1943, pp. 584–614, pls. 98–109, text figs. 1–4.

VAN BELLEN, R. C., J. F. C. DE WITT PUYT, A. C. RUTGERS, and J. VAN SOEST. Smaller Foraminifera from the Lower Oligocene of Cuba. — Proc. Ned. Akad. Wetensch., Amsterdam, vol. 44, No. 9, 1941, pp. 1–8, pl., table.

Bergquist, H. R. Scott County Fossils; Jackson Foraminifera and Ostracoda. — Bull. 49, Mississippi State Geol. Survey, 1942, pp. 1–146, pls. 1–11.

Bermudez, P. J. Nuevas especies de Foraminíferos del Eoceno de Cuba. — Mem. Soc. Cubana Hist. Nat., vol. 11, 1937, pp. 137–150, pls. 16–19.
Notas sobre *Hantkenina brevispina* Cushman. — l. c., pp. 151, 152, pl. 19 (pt.).
Estudio micropaleontologico de dos formaciones eocénicas de las cercanías de la Habana, Cuba. — l. c., pp. 153–180.
Foraminíferos pequeños de las margas eocénicas de Guanajay, Provincia Pinar del Río, Cuba. — l. c., pp. 319–346, 1 map.
Nuevas especies de Foraminíferos del Eoceno de las cercanías de Guanajay. — l. c., pp. 237–248, pls. 20, 21.
Foraminíferos pequeños de las margas eocénicas de Guanajay, Provincia del Río, Cuba, 2nd Part. — l. c., vol. 12, 1938, pp. 1–26.
Foraminíferos de la Fauna de Jicotea (Eoceno Medio), Provincia Santa Clara, Cuba. — l. c., pp. 91–96.
Flintina clenchi, un Nuevo Foraminífero de Puerto Plata, Santo Domingo. — l. c., vol. 13, 1939, pp. 199–200, pl. 30.

Berry, E. W. The Foraminifera of the Restin Shale of Northwest Peru. — Eclog. geol. Helv., vol. 21, 1928, pp. 130–135, 6 text figs.
The Smaller Foraminifera of the Middle Lobitos Shale of Northwestern Peru. — l. c., pp. 390–405, with 27 text figs.
Asterodiscocyclina, a new Subgenus of *Orthophragmina*. — l. c., pp. 405–407, pl. 33.
Two new species of "Orthophragmina" from Calita Sal, Peru. — Journ. Washington Acad. Sci., vol. 19, 1929, pp. 142–145, text figs. 1–4.
The Larger Foraminifera of the Talara Shale of Northwestern Peru. — l. c., vol. 22, 1932, pp. 1–9, text figs. 1–11.
The Foraminifera of the Heath Formation of Northwestern Peru, South America. — Eclog. geol. Helv., vol. 25, 1932, pp. 25–31, 2 pls.

Caudri, C. M. B. The Larger Foraminifera from San Juan de los Morros, State of Guarico, Venezuela. — Bull. Amer. Pal., vol. 28, No. 114, 1944, pp. 1–54, pls. 1–5, 2 text figs., chart.

Clark, W. B. Foraminifera [Pleistocene]. — Maryland Geol. Surv., Pliocene and Pleistocene, 1906, pp. 214–216, pl. 66.

Cole, W. S. A Foraminiferal Fauna from the Guayabal Formation in Mexico. — Bull. Amer. Pal., vol. 14, No. 51, 1927, pp. 1–46, pls. 1–5.
A Foraminiferal Fauna from the Chapapote Formation in Mexico. — l. c., No. 53, 1928, pp. 1–32, pls. 1–4.
The Pliocene and Pleistocene Foraminifera of Florida. — Bull. 6, Florida State Geol. Surv., 1931, pp. 1–79, pls. 1–7, table.
Oligocene Orbitoids from near Duncan Church, Washington County, Florida. — Journ. Pal., vol. 8, 1934, pp. 21–28, pls. 3, 4.
Stratigraphy and Micropaleontology of Two Deep Wells in Florida. — Florida Dept. Conservation, Geol. Bull. 16, 1938, pp. 1–73, pls. 1–12, text figs. 1–3.
Stratigraphic and Paleontologic Studies of Wells in Florida. — l. c., Geol. Bull. 19, 1941, pp. 1–91, pls. 1–18, 4 text figs.
Stratigraphic and Paleontologic Studies of Wells in Florida. — No. 2. — l. c., Geol. Bull. 20, 1942, pp. 1–89, pls. 1–16, text figs. 1–3.
Lockhartia in Cuba. — Journ. Pal., vol. 16, 1942, pp. 640–642, pl. 92.
Stratigraphic and Paleontologic Studies of Wells in Florida. — No. 3. — Florida Dept. Conservation, Geol. Bull. 26, 1944, pp. 1–168, pls. 1–29, text figs. 1–5.
Stratigraphic and Paleontologic Studies of Wells in Florida. — No. 4. — l. c., Geol. Bull. 28, 1945, pp. 1–160, pls. 1–22.

Cole, W. S., and R. Gillespie. Some Small Foraminifera from the Meson Formation of Mexico. — Bull. Amer. Pal., vol. 15, No. 57b, 1930, pp. 123–137 (1–15), pls. 18–21 (1–4).

Cole, W. S., and G. M. Ponton. The Foraminifera of the Marianna Limestone of Florida. — Bull. 5, Florida State Geol. Surv., 1930, pp. 19–69, pls. 5–11.

New Species of *Fabularia*, *Asterocyclina*, and *Lepidocyclina* from the Florida Eocene. — Amer. Midland Nat., vol. 15, No. 2, 1934, pp. 138–147, pls. 1, 2.

Cooper, C. L. Smaller Foraminifera from the Porters Creek Formation (Paleocene) of Illinois. — Journ. Pal., vol. 18, 1944, pp. 343–354, pls. 54, 55, 2 text figs.

Coryell, H. N., and J. R. Embich. The Tranquilla Shale (Upper Eocene) of Panama and its Foraminiferal Fauna. — l. c., vol. 11, 1937, pp. 289–305, pls. 41–43, text fig. 1.

Coryell, H. N., and R. W. Mossman. Foraminifera from the Charco Azul Formation, Pliocene, of Panama. — l. c., vol. 16, 1942, pp. 233–246, pl. 36.

Coryell, H. N., and F. C. Rivero. A Miocene Microfauna of Haiti. — l. c., vol. 14, 1940, pp. 324–344, pls. 41–44.

Cushman, J. A. Orbitoid Foraminifera of the Genus *Orthophragmina* from Georgia and Florida. — U. S. Geol. Surv., Prof. Paper 108-G, 1917, pp. 115–124, pls. 40–44.

The Smaller Fossil Foraminifera of the Panama Canal Zone. — Bull. 103, U. S. Nat. Mus., 1918, pp. 45–87, pls. 19–33.

The Larger Fossil Foraminifera of the Panama Canal Zone. — l. c., pp. 89–102, pls. 34–45.

Some Pliocene and Miocene Foraminifera of the Coastal Plain of the United States. — Bull. 676, U. S. Geol. Surv., 1918, pp. 1–100, pls. 1–31.

Fossil Foraminifera from the West Indies. — Publ. 291, Carnegie Inst. Washington, 1919, pp. 21–71, pls. 1–15, 8 text figs.

The American Species of *Orthophragmina* and *Lepidocyclina*. — U. S. Geol. Surv., Prof. Paper 125-D, 1920, pp. 39–105, pls. 7–35.

Lower Miocene Foraminifera of Florida. — l. c., 128-B, 1920, pp. 67–74, pl. 11.

American Species of *Operculina* and *Heterostegina*. — l. c., 128-E, 1921, pp. 125–142, 5 pls.

The Byram Calcareous Marl of Mississippi and Its Foraminifera. — l. c., 129-E, 1922, pp. 87–122, pls. 14–28.

The Foraminifera of the Mint Spring Calcareous Marl Member of the Marianna Limestone. — l. c., 129-F, 1922, pp. 123–152, pls. 29–35.

The Foraminifera of the Vicksburg Group. — l. c., 133, 1923, pp. 11–71, pls. 1–8.

An Eocene Fauna from the Moctezuma River. — Bull. Amer. Assoc. Petr. Geol., vol. 9, 1925, pp. 298–303, pls. 6–8.

Eocene Foraminifera from the Cocoa Sand of Alabama. — Contr. Cushman Lab. Foram. Res., vol. 1, 1925, pp. 65–68, pl. 10.

Foraminifera of the Typical Monterey of California. — l. c., vol. 2, 1926, pp. 53–66, pls. 7–9.

Some Pliocene Bolivinas from California. — l. c., vol. 2, 1926, pp. 40–47, pl. 6.

Additional Foraminifera from the Upper Eocene of Alabama. — l. c., vol. 4, 1928, pp. 73–79, pl. 10.

Cycloloculina in the Western Hemisphere. — l. c., vol. 5, 1929, pp. 4, 5, pl. 1.

New Foraminifera from Trinidad. — l. c., pp. 6–17, pls. 2, 3.

Notes on the Foraminifera of the Byram Marl. — l. c., pp. 40–48, pls. 7, 8.

An American *Virgulina* Related to *V. pertusa* Reuss. — l. c., pp. 53, 54, pl. 9.

Pliocene Lagenas from California. — l. c., pp. 67–72, pl. 11.

A Late Tertiary Fauna of Venezuela and Other Related Regions. — l. c., pp. 77–101, pls. 12–14.
Fossil Species of *Hastigerinella*. — l. c., vol. 6, 1930, pp. 17–19, pl. 3.
The Foraminifera of the Choctawhatchee Formation of Florida. — Bull. 4, Florida State Geol. Surv., 1930, pp. 1–89, pls. 1–12.
Post-Cretaceous Occurrence of *Gümbelina* with a Description of a New Species. — Contr. Cushman Lab. Foram. Res., vol. 9, 1933, pp. 64–69, pl. 7 (pt.).
New Species of Foraminifera from the Lower Oligocene of Mississippi. — l. c., vol. 11, 1935, pp. 25–39, pls. 4, 5.
Some New Foraminifera from the Late Tertiary of Georges Bank. — l. c., pp. 77–83, pl. 12.
Upper Eocene Foraminifera of the Southeastern United States. — U. S. Geol. Survey, Prof. Paper 181, 1935, pp. I–II, 1–88, pls. 1–23, 3 distribution charts.
Eocene Foraminifera from Submarine Cores off the Eastern Coast of North America. — Contr. Cushman Lab. Foram. Res., vol. 15, 1939, pp. 49–76, pls. 9 (pt)–12.
Midway Foraminifera from Alabama. — l. c., vol. 16, 1940, pp. 51–73, pls. 9–12.
Some Fossil Foraminifera from Alaska. — l. c., vol. 17, 1941, pp. 33–38, pl. 9.
A Foraminiferal Fauna of the Wilcox Eocene, Bashi Formation, from near Yellow Bluff, Alabama. — Amer. Journ. Sci., vol. 242, 1944, pp. 7–18, pls. 1, 2.
A Paleocene Foraminiferal Fauna from the Coal Bluff Marl Member of the Naheola Formation of Alabama. — Contr. Cushman Lab. Foram. Res., vol. 20, 1944, pp. 29–50, pls. 5–7, 8 (part).
A Foraminiferal Fauna from the Twiggs Clay of Georgia. — l. c., vol. 21, 1945, pp. 1–11, pls. 1, 2.
A Rich Foraminiferal Fauna from the Cocoa Sand of Alabama. — Special Publ. 16, Cushman Lab. Foram. Res., 1946, 40 pp., 8 pls.

CUSHMAN, J. A., and B. C. ADAMS. New Late Tertiary Bolivinas from California. — l. c., vol. 11, 1935, pp. 16–20, pl. 3 (pt.).

CUSHMAN, J. A., and E. R. APPLIN. Texas Jackson Foraminifera. — Bull. Amer. Assoc. Petr. Geol., vol. 10, 1926, pp. 154–189, pls. 5–10.
The Foraminifera of the Type Locality of the Yegua Formation of Texas. — Contr. Cushman Lab. Foram. Res., vol. 19, 1943, pp. 28–46, pls. 7, 8.

CUSHMAN, J. A., and W. F. BARBAT. Notes on Some Arenaceous Foraminifera from the Temblor Formation of California. — l. c., vol. 8, 1932, pp. 29–40, pls. 4, 5.

CUSHMAN, J. A., and J. D. BARKSDALE. Eocene Foraminifera from Martinez, California. — Contr. Dept. Geol. Stanford Univ., vol. 1, 1930, pp. 55–73, pls. 11, 12, map.

CUSHMAN, J. A., and P. J. BERMUDEZ. The Foraminiferal Genus *Amphimorphina* in the Eocene of Cuba. — Contr. Cushman Lab. Foram. Res., vol. 12, 1936, pp. 1–3, pl. 1 (pt.).
New Genera and Species of Foraminifera from the Eocene of Cuba. — l. c., pp. 27–38, pls. 5, 6.
Additional New Species of Foraminifera and a New Genus from the Eocene of Cuba. — l. c., pp. 55–63, pls. 10, 11.
Further New Species of Foraminifera from the Eocene of Cuba. — l. c., vol. 13, 1937, pp. 1–29, pls. 1, 2, 3 (pt.).
Additional New Species of Eocene Foraminifera from Cuba. — l. c., pp. 106–110, pl. 16.

CUSHMAN, J. A., and E. D. CAHILL. Miocene Foraminifera of the Coastal Plain of the Eastern United States. — U. S. Geol. Surv., Prof. Paper 175-A, 1933, pp. 1–50, pls. 1–13, table.

Cushman, J. A., and W. S. Cole. Pleistocene Foraminifera from Maryland. — Contr. Cushman Lab. Foram. Res., vol. 6, 1930, pp. 94–100, pl. 13.

Cushman, J. A., and A. N. Dusenbury, Jr. Eocene Foraminifera of the Poway Conglomerate of California. — l. c., vol. 10, 1934, pp. 51–65, pls. 7–9.

Cushman, J. A., and A. C. Ellisor. Some New Tertiary Foraminifera from Texas. — l. c., vol. 7, 1931, pp. 51–58, pl. 7.

Additional New Eocene Foraminifera. — l. c., vol. 8, 1932, pp. 40–43, pl. 6.

Two New Texas Foraminifera. — l. c., vol. 9, 1933, pp. 95, 96, pl. 10 (pt.).

New Species of Foraminifera from the Oligocene and Miocene. — l. c., vol. 15, 1939, pp. 1–14, pls. 1, 2.

The Foraminiferal Fauna of the Anahuac Formation. — Journ. Pal., vol. 19, 1945, pp. 545–572, pls. 71–78.

Cushman, J. A., and P. G. Edwards. The Described American Eocene Species of *Uvigerina*. — Contr. Cushman Lab. Foram. Res., vol. 13, 1937, pp. 74–87, pls. 11, 12.

Cushman, J. A., and D. L. Frizzell. Foraminifera from the Type Area of the Lincoln Formation (Oligocene) of Washington State. — l. c., vol. 19, 1943, pp. 79–89, pls. 14, 15.

Cushman, J. A., and E. W. Galliher. Additional New Foraminifera from the Miocene of California. — l. c., vol. 10, 1934, pp. 24–26, pl. 4 (pt.).

Cushman, J. A., and J. B. Garrett, Jr. New Species of *Triloculina* from the Claiborne of Louisiana. — l. c., pp. 65–70, pl. 9 (pt.).

Three New Rotaliform Foraminifera from the Lower Oligocene and Upper Eocene of Alabama. — l. c., vol. 14, 1938, pp. 62–66, pl. 11 (pt.).

Eocene Foraminifera of Wilcox Age from Woods Bluff, Alabama. — l. c., vol. 15, 1939, pp. 77–89, pls. 13–15.

Cushman, J. A., and P. P. Goudkoff. A New Species of *Pulvinulinella* from the California Miocene. — l. c., vol. 14, 1938, pp. 1, 2, pl. 1 (pt.).

Cushman, J. A., and U. S. Grant, IV. Late Tertiary and Quarternary Elphidiums of the West Coast of North America. — Trans. San Diego Soc. Nat. Hist., vol. 5, 1927, pp. 69–82, pls. 7, 8.

Cushman, J. A., and H. B. Gray. A Foraminiferal Fauna from the Pliocene of Timms Point, California. — Special Publ. 19, Cushman Lab. Foram. Res., 1946, pp. 1–46, pls. 1–8.

Cushman, J. A., and G. D. Hanna. Foraminifera from the Eocene near Coalinga, California. — Proc. Calif. Acad. Sci., ser. 4, vol. 16, 1927, pp. 205–229, pls. 13, 14.

Cushman, J. A., and M. A. Hanna. Foraminifera from the Eocene near San Diego, California. — Trans. San Diego Soc. Nat. Hist., vol. 5, 1927, pp. 45–64, pls. 4–6.

Cushman, J. A., and H. D. Hedberg. A New Genus of Foraminifera from the Miocene of Venezuela. — Contr. Cushman Lab. Foram. Res., vol. 11, 1935, pp. 13–16, pl. 3 (pt.).

Cushman, J. A., L. G. Henbest, and K. E. Lohman. Notes on a Core-sample from the Atlantic Ocean Bottom Southeast of New York City. — Bull. Geol. Soc. Amer., vol. 48, 1937, pp. 1297–1306, 1 pl.

Cushman, J. A., and S. M. Herrick. The Foraminifera of the Type Locality of the McBean Formation. — Contr. Cushman Lab. Foram. Res., vol. 21, 1945, pp. 55–73, pls. 9–11.

Cushman, J. A., and H. D. Hobson. A Foraminiferal Faunule from the Type San Lorenzo Formation, Santa Cruz County, California. — l. c., vol. 11, 1935, pp. 53–64, pls. 8, 9.

Cushman, J. A., and D. D. Hughes. Some Later Tertiary Cassidulinas of California. — l. c., vol. 1, 1925, pp. 11–16, pl. 2.

Cushman, J. A., and P. W. Jarvis. Miocene Foraminifera from Buff Bay, Jamaica. — Journ. Pal., vol. 4, 1930, pp. 353–368, pls. 32–34.

Some New Eocene Foraminifera from Jamaica. — Contr. Cushman Lab. Foram. Res., vol. 7, 1931, pp. 75–78, pl. 10.

Some Interesting New Uniserial Foraminifera from Trinidad. — l. c., vol. 10, 1934, pp. 71–75, pl. 10.

Three New Foraminifera from the Miocene, Bowden marl, of Jamaica. — l. c., vol. 12, 1936, pp. 3–5, pl. 1 (pt.).

Cushman, J. A., and R. M. Kleinpell. New and Unrecorded Foraminifera from the California Miocene. — l. c., vol. 10, 1934, pp. 1–23, pls. 1–4.

Cushman, J. A., and B. Laiming. Miocene Foraminifera from Los Sauces Creek, Ventura County, California. — Journ. Pal., vol. 5, 1931, pp. 79–120, pls. 9–14, 5 text figs.

Cushman, J. A., and L. W. Le Roy. A Microfauna from the Vaqueros Formation, Lower Miocene, Simi Valley, Ventura County, California. — l. c., vol. 12, 1938, pp. 117–126, pl. 22, text figs. 1–3.

Cushman, J. A., and W. McGlamery. Oligocene Foraminifera from Choctaw Bluff, Alabama — U. S. Geol. Survey, Prof. Paper 189-D, 1938, pp. 101–119, pls. 24–28.

New Species of Foraminifera from the Lower Oligocene of Alabama. — Contr. Cushman Lab. Foram. Res., vol. 15, 1939, pp. 45–49, pl. 9 (pt.).

Oligocene Foraminifera near Millry, Alabama. — l. c., Prof. Paper 197-B, 1942, pp. 65–84, pls. 4–7.

Cushman, J. A., and J. H. McMasters. Middle Eocene Foraminifera from the Llajas Formation, Ventura County, California. — Journ. Pal., vol. 10, 1936, pp. 497–517, pls. 74–77, text figs. 1–4.

Cushman, J. A., and F. L. Parker. Miocene Foraminifera from the Temblor of the East Side of the San Joaquin Valley, California. — Contr. Cushman Lab. Foram. Res., vol. 7, 1931, pp. 1–16, pls. 1, 2.

Some American Eocene Buliminas. — l. c., vol. 12, 1936, pp. 39–45, pls. 7, 8 (pt.).

Cushman, J. A., and G. M. Ponton. Some Interesting New Foraminifera from the Miocene of Florida. — l. c., vol. 8, 1932, pp. 1–4, pl. 1.

An Eocene Foraminiferal Fauna of Wilcox Age from Alabama. — l. c., pp. 51–72, pls. 7–9.

Foraminifera of the Upper, Middle and part of the Lower Miocene of Florida. — Bull. 9, Florida State Geol. Surv., 1932, pp. 1–147, pls. 1–17.

Cushman, J. A., and H. H. Renz. Eocene, Midway, Foraminifera from Soldado Rock, Trinidad. — Contr. Cushman Lab. Foram. Res., vol. 18, 1942, pp. 1–14, pls. 1–3.

Eocene Foraminifera of the Navet and Hospital Hill Formations of Trinidad, B. W. I. — Special Publ. 24, Cushman Lab. Foram. Res., 1948, pp. 1–42, pls. 1–8.

Cushman, J. A., and H. G. Schenck. Two Foraminiferal Faunules from the Oregon Tertiary. — Univ. Calif. Publ. Bull. Dept. Geol. Sci., vol. 17, 1928, pp. 305–324, pls. 42–45.

Cushman, J. A., and S. S. Siegfus. Foraminifera from the Type Area of the Kreyenhagen Shale of California. — Trans. San Diego Soc. Nat. Hist., vol. 9, No. 34, 1942, pp. 385–426, pls. 14–19, diagram, table.

CUSHMAN, J. A., and R. R. SIMONSON. Foraminifera from the Tumey Formation, Fresno County, California. — Journ. Pal., vol. 18, 1944, pp. 186–203, pls. 30–34, 5 text figs.

CUSHMAN, J. A., and R. M. STAINFORTH. The Foraminifera of the Cipero Marl Formation of Trinidad, British West Indies. — Special Publ. 14, Cushman Lab. Foram. Res., 1945, 91 pp., 16 pls., 2 charts.

CUSHMAN, J. A., and R. E. and K. C. STEWART. Tertiary Foraminifera from Humboldt County, California. A Preliminary Survey of the Fauna. — Trans. San Diego Soc. Nat. Hist., vol. 6, 1930, pp. 41–94, pls. 1–8, chart.

Five Papers on Foraminifera from the Tertiary of Western Oregon. — Bull. 36, Oregon Dept. Geol. & Min. Industries, 1947 (1948), pp. 1–111, pls. 1–13.

CUSHMAN, J. A., and B. STONE. An Eocene Foraminiferal Fauna from the Chira Shale of Peru. — Special Publ. 20, Cushman Lab. Foram. Res., 1947, pp. 1–27, pls. 1–4.

CUSHMAN, J. A., and N. L. THOMAS. Abundant Foraminifera of the East Texas Greensands. — Journ. Pal., vol. 3, 1929, pp. 176–184, pls. 23, 24.

Common Foraminifera of the East Texas Greensands. — l. c., vol. 4, 1930, pp. 33–41, pls. 3, 4.

CUSHMAN, J. A., and R. TODD. The Foraminifera of the Type Locality of the Naheola Formation. — Contr. Cushman Lab. Foram. Res., vol. 18, 1942, pp. 23–46, pls. 5–8.

A Foraminiferal Fauna from the Lisbon Formation of Alabama. — l. c., vol. 21, 1945, pp. 11–21, pls. 3, 4.

Foraminifera of the Type Locality of the Moodys Marl Member of the Jackson Formation of Mississippi. — l. c., vol. 21, 1945, pp. 79–105, pls. 13–16.

Miocene Foraminifera from Buff Bay, Jamaica. — Special Publ. 15, Cushman Lab. Foram. Res., 1945, 85 pp., 12 pls.

A Foraminiferal Fauna from the Paleocene of Arkansas. — Contr. Cushman Lab. Foram. Res., vol. 22, 1946, pp. 45–65, pls. 7–11.

A Foraminiferal Fauna from the Byram Marl at its Type Locality. — l. c., pp. 76–102, pls. 13 (part), 14–16.

A Foraminiferal Fauna from Amchitka Island, Alaska. — l. c., vol. 23, 1947, pp. 60–72, pls. 14 (part), 15, 16.

Foraminifera from the Red Bluff-Yazoo Section at Red Bluff, Mississippi. — l. c., vol. 24, 1948, pp. 1–12, pls. 1, 2 (part).

DAWSON, G. M. Notes on the Post-pliocene Geology of Canada. — Canad. Nat., vol. 6, 1872, pp. 19–42, 166–187, 241–259, 369–416, pl. 3.

DETLING, M. R. Foraminifera of the Coos Bay Lower Tertiary, Coos County, Oregon. — Journ. Pal., vol. 20, 1946, pp. 348–361, pls. 46–51, 2 text figs.

DOUVILLÉ, H. Les Orbitoides de l'Île de la Trinité. — C. R. Acad. Sci., vol. 161, 1915, pp. 87–92; vol. 164, 1917, pp. 841–847, 6 text figs.

DURHAM, J. W. *Operculina* in the Lower Tertiary of Washington. — Journ. Pal., vol. 11, 1937, p. 366, 1 text fig.

ELLIS, A. D., JR. Significant Foraminifera from the Chickasawhay Beds of Wayne County, Mississippi. — l. c., vol. 13, 1939, pp. 423, 424, pl. 48.

ELLISOR, A. C. Jackson Group of Formations in Texas with Notes on Frio and Vicksburg. — Bull. Amer. Assoc. Petr. Geol., vol. 17, 1933, pp. 1293–1350, 7 pls., 8 text figs.

FRANKLIN, E. S. Microfauna from the Carapita Formation of Venezuela. — Journ. Pal., vol. 18, 1944, pp. 301–319, pls. 44–48.

GALLOWAY, J. J. Notes on the Genus *Polylepidina*, and a New Species. — Journ. Pal., vol. 1, 1928, pp. 299–303, pl. 51, text figs.

GALLOWAY, J. J., and C. E. HEMINWAY. The Tertiary Foraminifera of Porto Rico. — Sci. Survey of Porto Rico and the Virgin Ids., vol. 3, pt. 4, 1941, pp. 275–491, pls. 1–36, map.

GALLOWAY, J. J., and M. MORREY. A Lower Tertiary Foraminiferal Fauna from Manta, Ecuador. — Bull. Amer. Pal., vol. 15, No. 55, 1929, pp. 1–56, pls. 1–6.

GALLOWAY, J. J., and S. G. WISSLER. Pleistocene Foraminifera from the Lomita Quarry, Palos Verdes Hills, California. — Journ. Pal., vol. 1, 1927, pp. 35–87, pls. 7–12, 2 tables.

GARRETT, J. B., JR. Occurrence of *Nonionella cockfieldensis* at Claiborne, Alabama. — l. c., vol. 10, 1936, pp. 785, 786.

The Hackberry Assemblage. — An Interesting Foraminiferal Fauna of Post-Vicksburg Age from Deep Wells in the Gulf Coast. — l. c., vol. 12, 1938, pp. 309–317, pl. 40, figs. 1, 2.

Some middle Tertiary smaller Foraminifera from subsurface beds of Jefferson County, Texas. — l. c., vol. 13, 1939, pp. 575–579, pls. 65, 66.

Some Miocene Foraminifera from subsurface strata of coastal Texas. — l. c., vol. 16, 1942, pp. 461–463, pl. 70.

GARRETT, J. B., JR., and A. D. ELLIS, JR. Distinctive Foraminifera of the Genus *Marginulina* from Middle Tertiary Beds of the Gulf Coast. — l. c., vol. 11, 1937, pp. 629–633, pl. 86.

GORTER, N. E., and I. M. VAN DER VLERK. Larger Foraminifera from Central Falcon (Venezuela). — Leidsche Geol. Med., Deel IV, 1932, pp. 94–122, pls. 11–17.

GRAVELL, D. W. Tertiary Larger Foraminifera of Venezuela. — Smithsonian Misc. Coll., vol. 89, No. 11 (Publ. 3223), Dec. 9, 1933, pp. 1–44, pls. 1–6.

GRAVELL, D. W., and M. A. HANNA. Larger Foraminifera from the Moody's Branch Marl, Jackson Eocene, of Texas, Louisiana, and Mississippi. — Journ. Pal., vol. 9, 1935, pp. 327–340, pls. 29–32, 1 text fig. (map).

The *Lepidocyclina texana* Horizon in the *Heterostegina* Zone, Upper Oligocene, of Texas and Louisiana. — l. c., vol. 11, 1937, pp. 517–529, pls. 60–65.

Subsurface Tertiary Zones of Correlation through Mississippi, Alabama and Florida. — Bull. Amer. Assoc. Petr. Geol., vol. 22, 1938, pp. 984–1013, pls. 1–7, 5 text figs.

New Larger Foraminifera from the Claiborne of Mississippi. — Journ. Pal., vol. 14, 1940, pp. 412–416, pl. 57.

GUPPY, R. J. L. Observations on some of the Foraminifera of the Oceanic Rocks of Trinidad. — Geol. Mag., dec. 5, vol. 1, 1904, pp. 241–250, pls. 8, 9.

HADLEY, W. H., JR. Some Tertiary Foraminifera from the North Coast of Cuba. — Bull. Amer. Pal., vol. 20, No. 70A, 1934, pp. 1–40, pls. 1–5.

Seven New Species of Foraminifera from the Tertiary of the Gulf Coast. — l. c., vol. 22, No. 74, 1935, pp. 1–10, pl. 1.

HANNA, G. D., and C. C. CHURCH. Notes on *Marginulina vacavillensis* (Hanna). — Journ. Pal., vol. 11, 1937, pp. 530, 531.

HANNA, G. D., and M. A. HANNA. Foraminifera from the Eocene of Cowlitz River, Lewis County, Washington. — Univ. Washington Publ. in Geol., vol. 1, No. 4, 1924, pp. 57–62, pl. 13.

HANNA, M. A. Wilcox Eocene Production at Segno Field, Polk County, and Cleveland Field, Liberty County, Texas. — Bull. Amer. Assoc. Petr. Geol., vol. 22, 1938, pp. 1274–77.

Hanzawa, S. Notes on Some Interesting Cretaceous and Tertiary Foraminifera from the West Indies. — Journ. Pal., vol. 11, 1937, pp. 110–117, pls. 20, 21.

Hedberg, H. D. Foraminifera of the Middle Tertiary Carapita Formation of Northeastern Venezuela. — l. c., pp. 661–697, pls. 90–92, text figs.

Hermes, J. J. Geology and Paleontology of East Camaguey and West Oriente, Cuba. — Geogr. Geol. Mededeel., Physiogr.-Geol. Reeks, ser. II, No. 7, 1945, pp. 1–75, pls. 1–5, tables, map.

Hodson, H. K. Foraminifera from Venezuela and Trinidad. — Bull. Amer. Pal., vol. 12, No. 47, 1926, pp. 1–46, pls. 1–8.

Hoffmeister, W. S., and C. T. Berry. A New Genus of Foraminifera from the Miocene of Venezuela and Trinidad. — Journ. Pal., vol. 11, 1937, pp. 29, 30, pl. 5 (pt.).

Howe, H. V. The Genus Bolivinella in the Oligocene of Mississippi. — Journ. Pal., vol. 4, 1930, pp. 263–267, pl. 21.

Distinctive New Species of Foraminifera from the Oligocene of Mississippi. — l. c., pp. 327–331, pl. 27.

Louisiana Cook Mountain Eocene Foraminifera. — State of Louisiana, Department of Conservation, Geol. Bull. 14, 1939, pp. 1–122, pls. 1–14, charts.

Howe, H. V., and S. M. McDonald. Two New Species of the Foraminiferal Genus *Marginulina* from the Sorrento Oil Field, Louisiana. — l. c., Bull. 13, 1938, pp. 209–211, pl. 1.

Howe, H. V., and W. E. Wallace. Foraminifera of the Jackson Eocene at Danville Landing on the Ouachita, Catahoula Parish, Louisiana. — Louisiana Geol. Bull. No. 2, 1932, pp. 1–118, pls. 1–15, 2 text figs.

Apertural Characteristics of the Genus *Hantkenina*, with Description of a new Species. — Journ. Pal., vol. 8, 1934, pp. 35–37, pl. 5, figs. 13–17.

Israelsky, M. C. Tentative Foraminiferal Zonation of Subsurface Claiborne of Texas and Louisiana. — Bull. Amer. Assoc. Petr. Geol., vol. 19, 1935, pp. 689–695, 5 text figs.

Notes on Some Foraminifera from Marysville Buttes, California. — Proc. 6th Pac. Sci. Congress, 1939, pp. 569–595, pls. 1–7.

Keijzer, F. G. Outline of the Geology of the Eastern Part of the Province of Oriente, Cuba (E of 76° WL), with Notes on the Geology of other Parts of the Island. — Geogr. Geol. Mededeel., Physiogr.-Geol. Reeks, ser. II, No. 6, 1945, pp. 1–238, pls. 1–11, figs. 1–34, map.

Kleinpell, R. M. Miocene Stratigraphy of California. — Amer. Assoc. Petr. Geol., 1938, pp. i–vi, 1–450, pls. I–XXII, text figs. 1–13, tables and maps.

Kline, V. H. Clay County Fossils; Midway Foraminifera and Ostracoda. — Bull. 53, Mississippi State Geol. Survey, 1943, pp. 1–98, pls. 1–8.

Koch, R. Tertiärer Foraminiferenkalk von der Insel Curacao (Niederländisch West-Indien). — Eclog. geol. Helv., vol. 21, 1928, pp. 51–56, pl. 3, 1 text fig.

Lalicker, C. G., and P. J. Bermudez. Some Foraminifera of the Family Textulariidae from the Eocene of Cuba. — Journ. Pal., vol. 12, 1938, pp. 170–172, pl. 28 (pt.).

Martin, L. T. Eocene Foraminifera from the Type Lodo Formation, Fresno County, California. — Stanford Univ. Publ., Univ. Ser., Geol. Sci., vol. 3, No. 3, 1943, pp. 91–125, pls. 5–9, 3 text figs., 2 tables.

Moberg, M. W. New Species of Coskinolina and Dictyoconus (?) from Florida. — 19th Ann. Rep't Florida Geol. Surv., 1928, pp. 166–175, pls. 3–5.

Mornhinveg, A. R., and J. B. Garrett, Jr. Study of Vicksburg Group at Vicksburg, Mississippi. — Bull. Amer. Assoc. Petr. Geol., vol. 19, pp. 1645–1667, 5 figs.

Natland, M. L. New Species of Foraminifera from off the West Coast of North America and from the Later Tertiary of the Los Angeles Basin. — Bull. Scripps Inst. Oceanography, Tech. Ser., vol. 4, 1938, pp. 137–164, pls. 3–7.

Nuttall, W. L. F. Tertiary Foraminifera from the Naparima Region of Trinidad (British West Indies). — Quart. Journ. Geol. Soc., vol. 84, 1928, pp. 57–112, pls. 3–8, text figs. 1–13.

Notes on the Tertiary Foraminifera of Southern Mexico. — Journ. Pal., vol. 2, 1928, pp. 372–376, pl. 50.

Eocene Foraminifera from Mexico. — l. c., vol. 4, 1930, pp. 271–293, pls. 23–25.

Lower Oligocene Foraminifera from Mexico. — l. c., vol. 6, 1932, pp. 3–35, pls. 1–9.

Two Species of *Miogypsina* from the Oligocene of Mexico. — l. c., vol. 7, 1933, pp. 175–177, pl. 24.

Upper Eocene Foraminifera from Venezuela. — l. c., vol. 9, 1935, pp. 121–131, pls. 14, 15.

Ovey, C. D. Notes on the Foraminifera from the Fossiliferous Tuffs of Roche Bluff. — Phil. Trans., Roy. Soc. London, vol. 229, No. 557, 1938, p. 81.

Palmer, D. K. The Foraminiferal Genus *Gümbelina* in the Tertiary of Cuba. — Mem. Soc. Cubana Hist. Nat., vol. 8, 1934, pp. 73–76, 8 text figs.

Some Large Fossil Foraminifera from Cuba. — l. c., pp. 235–264, pls. 12–16, 19 text figs.

New Genera and Species of Cuban Oligocene Foraminifera. — l. c., vol. 10, 1936, pp. 123–128, pl. 5, text figs. 1–3.

Planulina alavensis, A New Cuban Oligocene Foraminifer. — l. c., vol. 12, 1938, pp. 345, 346, text figs. A–C.

Foraminifera of the Upper Oligocene Cojimar Formation of Cuba (issued in 5 parts). — l. c., vol. 14, No. 1, 1940, pp. 19–35; No. 2, pp. 113–132, pls. 17, 18; No. 4, pp. 277–304, pls. 51–53; vol. 15, No. 2, 1941, pp. 181–200, pls. 15–17; No. 3, pp. 281–306, pls. 28–31.

Notes on the Foraminifera from Bowden, Jamaica. — Bull. Amer. Pal., vol. 29, No. 115, 1945, pp. 1–82, pls. 1, 2.

Palmer, D. K., and P. J. Bermudez. Late Tertiary Foraminifera from the Matanzas Bay region, Cuba. — Mem. Soc. Cubana Hist. Nat., vol. 9, No. 4, 1936, pp. 237–258, pls. 20–22.

An Oligocene Foraminiferal Fauna from Cuba. — l. c., vol. 10, 1936, pp. 227–316, pls. 13–20.

Parker, F. L., and P. J. Bermudez. Eocene Species of the Genera *Bulimina* and *Buliminella* from Cuba. — Journ. Pal., vol. 11, 1937, pp. 513–516, pls. 58, 59.

Pijpers, P. J. Geology and Paleontology of Bonaire (D. W. I.). — 4to, *Utrecht*, 1933, pp. i–vi, 1–103, 2 pls., 157 text figs., map.

Plummer, H. J. Foraminifera of the Midway Formation in Texas. — Bull. 2644, Univ. Texas, 1926 (1927), pp. 1–206, 15 pls., 13 text figs., table.

Rutten, M. G. Larger Foraminifera of Northern Santa Clara Province, Cuba. — Journ. Pal., vol. 9, 1935, pp. 527–545, pls. 59–62.

Stadnichenko, M. M. The Foraminifera and Ostracoda of the Marine Yegua of the Type Sections. — Journ. Pal., vol. 1, 1927, pp. 221–243, pls. 38, 39.

Stewart, R. E. and K. C. Post-Miocene Foraminifera from the Ventura Quadrangle, Ventura County, California. Twelve New Species and Varieties from the Pliocene. — l. c., vol. 4, 1930, pp. 60–72, pls. 8, 9.

Taliaferro, N. L., and H. G. Schenck. *Lepidocyclina* in California. — Amer. Journ. Sci., vol. 25, 1933, pp. 74–80, 4 text figs.

THALMANN, H. E. Das Vorkommen der Gattung *Miogypsina* Sacco 1893 in Ost-Mexiko. — Eclog. geol. Helv., vol. 25, 1932, pp. 282–286.

Mitteloligozän in der Umgebung von Tampico (Mexiko). — Geologischen Rundschau, Bd. 25, 1934, pp. 325–329.

Lepidocyclina canellei Lemoine & R. Douvillé im Oligocän von Tabasco (Mexiko). — Centralblatt für Min., Geol., und Pal., Jahrg. 1934, Abt. B, 1934, pp. 446–448.

THIADENS, A. A. Cretaceous and Tertiary Foraminifera from Southern Santa Clara Province, Cuba. — Journ. Pal., vol. 11, 1937, pp. 91–109, pls. 15–19, text figs. 1–3.

TODD, J. U. On *Lepidocyclina* (*Lepidocyclina*) *atascaderensis* Berry from the Atascadero Limestone (Eocene) of N. W. Peru. — Geol. Mag., vol. 70, No. 830, 1933, pp. 347–351, pl. XVIII.

TODD, J. U., and R. W. BARKER. Tertiary Orbitoids from Northwestern Peru. — Geol. Mag., vol. 69, 1932, pp. 529–543, pls. 39–42, 7 text figs.

TOLMACHOFF, I. P. A Miocene Microfauna and Flora from the Atrato River, Colombia, South America. — Annals Carnegie Museum, vol. 23, 1934, pp. 275–356, pls. XXXIX–XLIV, 1 map.

TOULMIN, L. D. Eocene smaller Foraminifera from the Salt Mountain limestone of Alabama. — Journ. Pal., vol. 15, 1941, pp. 567–611, pls. 78–82, 4 text figs.

VAN DE GEYN, W. A. E., and I. M. VAN DER VLERK. A Monograph on the Orbitoididae, Occurring in the Tertiary of America, compiled in connection with an examination of a collection of Larger Foraminifera from Trinidad. — Leidsche Geol. Med., Deel 7, 1935, pp. 221–272, 9 pls.

VAUGHAN, T. W. Studies of the Larger Tertiary Foraminifera from Tropical and Subtropical America. — Proc. Nat. Acad. Sci., vol. 9, 1923, pp. 253–257.

Larger Foraminifera of the Genus Lepidocyclina related to Lepidocyclina mantelli. — Proc. U. S. Nat. Mus., vol. 71, Art. 8, 1927, pp. 1–5, pls. 1–4.

Species of Large Arenaceous and Orbitoidal Foraminifera from the Tertiary Deposits of Jamaica. — Journ. Pal., vol. 1, 1928, pp. 277–298, pls. 43–50.

Yaberinella jamaicensis, A New Genus and Species of Arenaceous Foraminifera. — Journ. Pal., vol. 2, 1928, pp. 7–12, pls. 4, 5.

New Species of Operculina and Discocyclina from the Ocala Limestone. — 19th Ann. Rep't Florida Geol. Surv., 1928, pp. 155–165, pls. 1, 2.

Studies of Orbitoidal Foraminifera: The Subgenus Polylepidina of Lepidocyclina and Orbitocyclina, a New Genus. — Proc. Nat. Acad. Sci., vol. 15, 1929, pp. 288–295, 1 pl.

Descriptions of New Species of Foraminifera of the Genus Discocyclina from the Eocene of Mexico. — Proc. U. S. Nat. Mus., vol. 76, Art. 3, 1929, pp. 1–18, pls. 1–7.

Actinosiphon semmesi, A New Genus and Species of Orbitoidal Foraminifera, and *Pseudorbitoides trechmanni* H. Douvillé. — Journ. Pal., vol. 3, 1929, pp. 163–169, pl. 21.

Additional New Species of Tertiary Larger Foraminifera from Jamaica. — l. c., pp. 375–383, pls. 39–41.

A Note on *Lepidocyclina hilli* Cushman. — l. c., vol. 5, 1931, pp. 41, 42.

American Species of the Genus *Dictyoconus*. — l. c., vol. 6, 1932, pp. 94–99, pl. 14.

American Paleocene and Eocene Larger Foraminifera. — Geol. Soc. Amer., Mem. 9, pt. I, 1945, pp. 1–175, pls. 1–46.

VAUGHAN, T. W., and W. S. COLE. New Tertiary Foraminifera of the Genera *Operculina* and *Operculinoides* from North America and the West Indies. — Proc. U. S. Nat. Mus., vol. 83, No. 2996, 1936, pp. 487–496, pls. 35–38.

Triplalepidina veracruziana, a New Genus and Species of Orbitoidal Foraminifera from the Eocene of Mexico. — Journ. Pal., vol. 12, 1938, pp. 167–169, pl. 27.

WEINZIERL, L. L., and E. R. APPLIN. The Claiborne Formation on the Coastal Domes. — l. c., vol. 3, 1929, pp. 384–410, pls. 42–44.

WOODRING, W. P. Middle Eocene Foraminifera of the Genus *Dictyoconus* from the Republic of Haiti. — Journ. Washington Acad. Sci., vol. 12, 1922, pp. 244–247.

Some New Eocene Foraminifera of the Genus *Dictyoconus*. — App. I, pp. 608–610, pls. 9, 13, in W. P. Woodring, J. S. Brown and W. S. Burbank, Geology of the Republic of Haiti, 1924.

Upper Eocene Orbitoid Foraminifera from the Western Santa Ynez Range, California, and Their Stratigraphic Significance. — Trans. San Diego Soc. Nat. Hist., vol. 4, 1930, pp. 145–170, pls. 13–17.

WOODWARD, A., and B. W. THOMAS. On the Foraminifera of the Boulder-clay taken from a well-shaft twenty-two feet deep, Meeker County, Central Minnesota. — Thirteenth Ann. Rep't Geol. Nat. Hist. Surv., Minnesota, 1884 (1885), pp. 164–177, pls. 3, 4.

IV. CRETACEOUS

BRITISH ISLES

BURROWS, H. W., C. D. SHERBORN and G. BAILEY. The Foraminifera of the Red Chalk of Yorkshire, Norfolk and Lincolnshire. — Journ. Roy. Micr. Soc., 1890, pp. 549–566, pls. 10, 11.

CHAPMAN, F. The Foraminifera of the Gault of Folkestone. Pts. 1–10. — l. c., 1891–1898.

Some New Forms of Hyaline Foraminifera from the Gault. — Geol. Mag., 1892, pp. 52–54, pl. 2.

Microzoa from the Phosphatic Chalk of Taplow. — Quart. Journ. Geol. Soc., vol. 48, 1892, pp. 514–518, pl. 15.

The Bargate Beds of Surrey and their Microscopic Contents. — l. c., vol. 50, 1894, pp. 677–730, pls. 33, 34.

Foraminifera from the Cambridge Greensand. — Ann. Mag. Nat. Hist., ser. 7, vol. 3, 1899, pp. 48–66, text figs. 1–5; pp. 302–317, text figs. 1–3.

EARLAND, ARTHUR. Chalk: Its Riddles and Some Possible Solutions. — Trans. Hertfordshire Nat. Hist. Soc., vol. 21, pt. 1, 1939, pp. 6–37, pls. 1, 2.

ELEY, H. Geology in the Garden, or the Fossils in the Flint Pebbles. — 8vo, *London*, 1859, pls.

HERON-ALLEN, E. Prolegomena towards the Study of the Chalk Foraminifera, etc. — 8vo, *London*, 1894, pp. 1–36, text figs.

SHERLOCK, R. L. The Foraminifera of the Speeton Clay of Yorkshire. — Geol. Mag., dec. 6, vol. 1, 1914, pp. 216–222, 255–265, 289–296, pls. 18, 19.

WRIGHT, J. A List of the Cretaceous Microzoa of the North of Ireland. — Rep't Proc. Belfast Nat. Field Club, 1873–74, App., 1875, pp. 73–99, pls. 2, 3.

Cretaceous Foraminifera of Keady Hill, Co. Derry. — l. c., 1885–86 (1886), App., pp. 327–332, pl. 27.

EASTERN AND CENTRAL EUROPE

BEISSEL, I. Die Foraminiferen der Aachener Kreide. — Abhandl. kön. Preuss. Geol. Landes., Neue Folge, Heft 3, 1891, pp. 1–78, Atlas of 16 pls.

BROTZEN, F. Foraminiferen aus dem Schwedischen untersten Senon von Eriksdal in Schonen. — Sveriges Geologiska Undersökning, Ser. C, No. 396, vol. 30, No. 3, 1936, pp. 1–206, pls. 1–14, 69 text figs.

Die Foraminiferen in Sven Nilssons Petrificata Suecana 1827. — Geol. Fören. Förhandl., vol. 59, 1937, pp. 59–76, pl. II, text figs. 1–6.

De Geologiska Resultaten från Borrningarna vid Höllviken. Preliminär rapport, Del I: Kritan. — Sveriges Geologiska Undersökning, Ser. C, No. 465, 1944 (1945), pp. 1–65, pls. 1, 2, text figs. 1–10.

CORNUEL, J. Description de nouveaux fossiles microscopiques du terrain crétacé inférieur du départment de la Haute-Marne. — Mém. Soc. Géol. France, ser. 2, vol. 3, 1848, pp. 241–263, pls. 3, 4.

EGGER, J. G. Der Bau der Orbitolinen und verwandter Formen. — Abhandl. kön. bay. Akad. Wiss. München, Cl. II, vol. 21, 1902, pp. 577–600, pls. 1–6.

Mikrofauna der Kreideschichten des Westlichen bayerischen Waldes und des Gebietes um Regensburg. — Ber. Nat. Ver. Passau, 1907, pp. 1–75, pls. 1–9.

Foraminiferen der Seewener Kreideschichten. — Sitz. Bay. Akad. Wiss. München, 1909, pp. 1–52, pls. 1–6.

Ostrakoden und Foraminiferen des Eybrunner Kreidemergels in der Umgebung von Regensburg. — Bericht Nat. Ver. Regensburg, 1907–09 (1910), pp. 1–48, pls. 1–6.

EHRENBERG, C. G. Mikrogeologie. Das Wirken des unsichtbaren kleinen Lebens auf der Erde. — 2 vols. fol., *Leipzig*, 1854, 40 pls.

FRANKE, A. Die Foraminiferen der Kreideformation des Münsterschen Beckens. — Verh. Nat. Ver. preuss. Rheinlande und Westfalens, Jahrg. 69, 1912, pp. 255–285, pl. 6.

Die Foraminiferen und Ostracoden des Emschers, besonders von Obereving und Derne nördlich Dortmund. — Zeitschr. Deutsch. Geol. Ges., vol. 66, 1914, pp. 428–443, pl. 27.

Die Foraminiferen der pommerschen Kreide. — Abhandl. Geol. Pal. Instit. Univ. Greifswald, vol. 6, 1925, pp. 1–96, pls. 1–8.

Die Foraminiferen der Aachener Kreide, Engänzungen und Berichtigung zu dem gleichnamigen Buche mit Atlas von Ignaz Beissel, Herausgegeben von E. Holzapfel 1891. — Jahrb. Preuss. Geol. Landes für 1927, vol. 48, 1927, pp. 667–698.

Die Foraminiferen der Oberen Kreide Nord- und Mitteldeutschlands. — Abhandl. Preuss. Geol. Landes., vol. 111, 1928, pp. 1–207, pls. 1–18, 2 text figs.

GRZYBOWSKI, J. Otwornice warstw Inoceramowych okolicy Gorlic. — Rozpr. Akad. Um. mat. przyr., 1901, pp. 1–71, pls. 7, 8.

HECHT, F. E. Standard-Gliederung der Nordwest-deutschen Unterkreide nach Foraminiferen. — Senckenbergischen Naturforschenden Gesellschaft, Abhandl. 443, 1938, pp. 1–42, pls. 1–24, tables and charts.

HOFKER, J. Die Foraminiferen aus dem Senon Limburgens. I–XIII. — Nat. Maan., Nat. Gen. Limburg, Pts. 1–13, 1926–1932.

JEDLITSCHKA, H. Beitrag zur Kenntnis der Mikrofauna der subbeskidischen Schichten. — Mitth. nat. Ver. Troppau, C. S. R., 1935, Sep., pp. 1–18, pl., figs. 1–19.

MARSSON, T. Die Foraminiferen der Weissen Schreibkreide der Inseln Rügen. — Mitth. nat. Ver. Neu-Vorpommern und Rügen, Jahrb. 10, 1878, pp. 115–196, pls. 1–5.

ORBIGNY, A. D. D'. Mémoire sur les Foraminifères de la Craie blanche du Bassin de Paris. — Mém. Soc. Géol. France, vol. 4, 1840, pp. 1–51, pls. 1–4.

PERNER, J. Foraminifery Ceskeho Cenomanu. — Ceska Akad. Cis. Frantiska Josefa, vol. 16, 1891 (1892), pp. 49–65, pls. 1–10.

REUSS, A. E. Die Foraminiferen und Entomostraceen des Kreidemergels von Lemburg. — Haidinger's Nat. Abhandl., vol. 4, 1851, pp. 17–52, pls. 2–6.

Die Foraminiferen des westphälischen Kreideformation. — Sitz. Akad. Wiss. Wien, vol. 40, 1860, pp. 147–238, pls. 1–13.

Die Foraminiferen des norddeutschen Hils und Gault. — l. c., vol. 46, 1862 (1863), pp. 5–100, pls. 1–13.

RICHTER, K. Horizontbestimmung von Ober-Kreidegeschieben mittels Foraminiferenstatistik. — "Frankfurter Beiträge zur Geschiebeforschung" Beiheft zur Zeitschrift für Geschiebeforschung, 1935, pp. 20–28, 3 text figs., and table.

SCHIJFSMA, E. The Foraminifera from the Hervian (Campanian) of Southern Limburg. — Mededeel. Geol. Stichting, ser. C–V, No. 7, 1946, pp. 1–174, pls. 1–10.

SCHLUMBERGER, C., and P. CHOFFAT. Note sur le Genre *Spirocyclina* Munier-Chalmas et quelques autres genres du même auteur. — Bull. Soc. Géol. France, ser. 4, vol. 4, 1904, pp. 358–368, pls. 9, 10.

SCHMITT, W. Tonmergelgeschiebe aus den Gault. — Zeitschr. für Geschiebeforschung, Bd. V., Heft 3, 1929, pp. 125–128, pl. 5.

SCHNETZER, R. Foraminiferen des Betzensteiner Kreidekalks. Studien über die fränkischen albüberdeckende Kreide III. — Centralblatt für Min., etc., Jahrg. 1934, Abt. B, No. 2, pp. 86–95, 9 text figs.

STORM, H. Zur Kenntnis der Foraminiferenfauna in Oberturon und Emscher der Böhmischen Kreideformation. — Lotos, Prag, 77, 1929, pp. 39–54, figs. 1–14.

Zur Stratigraphischen Stellung der Oberturon- und Emschermergel in der Umgebung von Leitmeritz. — Firgenwald, 4 Jahrgang, 1931, pp. 3–29, 1 pl.

SOUTHERN EUROPE

ARNI, P. Foraminiferen des Senons und Untereocäns im Prätigauflysch. — Beiträge Geol. Karte der Schweiz., n. ser., pt. 65, 1933, pp. I–VIII, 1–18, pls. 1–5.

Siderolites heracleae im Maestrichtien des thessalischen Pindos. — Eclog. geol. Helv., vol. 27, 1933, pp. 105–109, pl. V.

COLOM, G. Las margas Rojas con Rosalinas del senoniense de Velez-Blanco (Prov. de Almeria). — Bull. Inst. Cat. Hist. Nat., vol. 31, No. 1, 1931, pp. 1–8, 2 pls., 20 text figs.

Contribucion al Conocimiento de las Facies Lito-paleontologicas del Cretacico de la Baleares. — Geol. des Pays Catalans, vol. 3, No. 2 (Pt. V), 1934, pp. 1–11, pl. I, 2 text figs.

SCHLUMBERGER, C. Note sur quelques foraminifères nouveaux ou peu connus du crétacé d'Espagne. — Bull. Soc. Géol. France, ser. 3, vol. 27, 1899, pp. 456–465, pls. 8–11.

SILVESTRI, A. Orbitoidi Cretace nell'Eocene della Brianza. — Mem. Pont. Accad. Nuovi Lincei, ser. 2, vol. 5, 1919, pp. 31–107, pl. 1, text figs.

Fossili cretacei della contrada Calcasacco presso Termini-Imerese (Palermo). — Pal. Ital., vol. 14, 1908, pp. 121–170, pls. 17–20; 2d part — l. c., vol. 18, 1912, pp. 29–56, pls. 6, 7.

INDO-PACIFIC

CHAPMAN, F. Monograph of the Foraminifera and Ostracoda of the Gingin Chalk. — Bull. Geol. Surv. W. Australia, No. 72, 1917, pp. 1–81, pls. 1–14.

The Cretaceous and Tertiary Foraminifera of New Zealand, with an Appendix on the Ostracoda. — New Zealand Geol. Surv., Pal. Bull. 11, 1926, pp. 1–119, pls. 1–22.

Foraminifera and Ostracoda from the Upper Cretaceous of Need's Camp, Buffalo River, Cape Province. — Ann. So. African Mus., vol. 12, 1916, pp. 107–118, pls. 14, 15.

CRESPIN, I. Upper Cretaceous Foraminifera from the Northwest Basin, Western Australia. — Journ. Pal., vol. 12, 1938, pp. 391–395, 1 text fig. (map).

NAUMANN, E., and M. NEUMAYR. Zur Geologie und Paläontologie von Japan. — Denkschr. Akad. Wiss. Wien, vol. 57, 1890, pp. 1–42, pls. 1–5.

SILVESTRI, A. Revisione di Foraminiferi Preterziarii del Sud-ouest di Sumatra. — Riv. Ital. Pal., Anno 38, 1932, pp. 75–107, pls. 2–4.

AMERICA

ALBRITTON, C. C., JR. Upper Jurassic and Lower Cretaceous Foraminifera from the Malone Mountains, Trans Pecos Texas. — Journ. Pal., vol. 11, 1937, pp. 19–23, pl. 4 (pt.).

Stratigraphy and Structure of the Malone Mountains, Texas. — Bull. Geol. Soc. America, vol. 49, 1938, pp. 1747–1806, 9 pls., 5 text figs.

ALBRITTON, C. C., JR., and F. B. PHLEGER, JR. Foraminiferal Zonation of Certain Upper Cretaceous Clays of Texas. — Journ. Pal., vol. 11, 1937, pp. 347–354, 3 text figs.

ALEXANDER, C. I., and J. P. SMITH. Foraminifera of the Genera *Flabellammina* and *Frankeina* from the Cretaceous of Texas. — Journ. Pal., vol. 6, 1932, pp. 299–311, pls. 45–47, text figs. 1, 2.

BAGG, R. M. The Cretaceous Foraminifera of New Jersey. — Bull. 88, U. S. Geol. Surv., 1898, pp. 1–89, pls. 1–6.

BERMUDEZ, P. J. Nueva Especie de Bulimina del Cretacico Superior Cubano. — Mem. Soc. Cubana Hist. Nat., vol. 12, 1938, pp. 89–90, text figs.

Nueva Especie de Seabrookia del Cretacico Superior Cubano. — l. c., pp. 163–165, text figs. 1–3.

CARMAN, K. W. Some Foraminifera from the Niobrara and Benton Formations of Wyoming. — Journ. Pal., vol. 3, 1929, pp. 309–315, pl. 34.

CARSEY, D. O. Foraminifera of the Cretaceous of Central Texas. — Univ. Texas Bull. 2612, 1926, pp. 1–56, pls. 1–8.

CUSHMAN, J. A. The Foraminifera of the Velasco Shale of the Tampico Embayment. — Bull. Amer. Assoc. Petr. Geol., vol. 10, 1926, pp. 581–612, pls. 15–21.

Some Foraminifera from the Mendez Shale of Eastern Mexico. — Contr. Cushman Lab. Foram. Res., vol. 2, 1926, pp. 16–26, pls. 2, 3.

Some Characteristic Mexican Fossil Foraminifera. — Journ. Pal., vol. 1, 1927, pp. 147–172, pls. 23–28.

New and Interesting Foraminifera from Mexico and Texas. — Contr. Cushman Lab. Foram. Res., vol. 3, 1927, pp. 111–117, pls. 22, 23.

Some Foraminifera from the Cretaceous of Canada. — Trans. Roy. Soc. Canada, Sec. IV, 1927, pp. 127–132, pl. 1.

American Upper Cretaceous Species of Bolivina and Related Species. — Contr. Cushman Lab. Foram. Res., vol. 2, 1927, pp. 85–91, pls. 11, 12.
The American Cretaceous Foraminifera Figured by Ehrenberg. — Journ. Pal., vol. 1, 1928, pp. 213–217, pls. 34–36.
A Peculiar *Clavulina* from the Upper Cretaceous of Texas. — Contr. Cushman Lab. Foram. Res., vol. 4, 1928, pp. 61, 62, pl. 8.
A Cretaceous *Cyclammina* from California. — l. c., p. 80, pl. 9.
Some Species of *Siphogenerinoides* from the Cretaceous of Venezuela. — l. c., pp. 55–59, pl. 9.
Notes on Upper Cretaceous Species of *Vaginulina, Flabellina* and *Frondicularia* from Texas and Arkansas. — l. c., vol. 6, 1930, pp. 25–38, pls. 4, 5.
Cretaceous Foraminifera from Antigua, B. W. I. — l. c., vol. 7, 1931, pp. 33–46, pls. 5, 6.
Hastigerinella and Other Interesting Foraminifera from the Upper Cretaceous of Texas. — l. c., pp. 83–90, pl. 11.
A Preliminary Report on the Foraminifera of Tennessee. — Bull. 41, Tenn. Geol. Div., 1931, pp. 1–62, pls. 1–13.
The Foraminifera of the Saratoga Chalk. — Journ. Pal., vol. 5, 1931, pp. 297–315, pls. 34–36.
The Foraminifera of the Annona Chalk. — l. c., vol. 6, 1932, pp. 330–345, pls. 50, 51.
New American Cretaceous Foraminifera. — Contr. Cushman Lab. Foram. Res., vol. 9, 1933, pp. 49–64, pls. 5–7.
Notes on Some American Cretaceous Flabellinas. — l. c., vol. 11, 1935, pp. 83–89, pl. 13.
Notes on Some American Cretaceous Frondicularias. — l. c., vol. 12, 1936, pp. 11–22, pls. 3, 4.
Geology and Paleontology of the Georges Bank Canyons. Part IV. Cretaceous and Late Tertiary Foraminifera. — Bull. Geol. Soc. Amer., vol. 47, 1936, pp. 413–440, pls. 1–5.
Some American Cretaceous Species of *Ellipsonodosaria* and *Chrysalogonium*. — Contr. Cushman Lab. Foram. Res., vol. 12, 1936, pp. 51–55, pl. 9.
Cretaceous Foraminifera of the Family Chilostomellidae. — l. c., pp. 71–78, pl. 13 (pt.).
Some Notes on Cretaceous Species of *Marginulina*. — l. c., vol. 13, 1937, pp. 91–99, pls. 1, 2.
A Few New Species of American Cretaceous Foraminifera. — l. c., pp. 100–105, pl. 15.
Additional New Species of American Cretaceous Foraminifera. — l. c., vol. 14, 1938, pp. 31–50, pls. 5–8.
Some New Species of Rotaliform Foraminifera from the American Cretaceous. — l. c., pp. 66–71, pl. 11 (pt.), 12.
New American Cretaceous Foraminifera. — l. c., vol. 15, 1939, pp. 89–93, pl. 16.
Upper Cretaceous Foraminifera of the Gulf Coastal Region of the United States and Adjacent Areas. — U. S. Geol. Survey Prof. Paper 206, 1946, pp. 1–241, pls. 1–66.
A Foraminiferal Fauna from the Santa Anita Formation of Venezuela. — Contr. Cushman Lab. Foram. Res., vol. 23, 1947, pp. 1–18, pls. 1–4.

Cushman, J. A., and C. I. Alexander. *Frankeina*, A New Genus of Arenaceous Foraminifera. — Contr. Cushman Lab. Foram. Res., vol. 5, 1929, pp. 61, 62, pl. 10.
Some Vaginulinas and Other Foraminifera from the Lower Cretaceous of Texas. — l. c., vol. 6, 1930, pp. 1–10, pl. 3.

Cushman, J. A., and A. S. Campbell. A New *Spiroplectoides* from the Cretaceous of California. — l. c., vol. 10, 1934, pp. 70, 71, pl. 9 (pt.).

Cretaceous Foraminifera from the Moreno Shale of California. — l. c., vol. 11, 1935, pp. 65–73, pls. 10, 11.

A New *Siphogenerinoides* from California. — l. c., vol. 12, 1936, pp. 91, 92, pl. 13 (pt.).

Cushman, J. A., and C. C. Church. Some Upper Cretaceous Foraminifera from near Coalinga, California. — Proc. Calif. Acad. Sci., ser. 4, vol. 18, 1929, pp. 497–530, pls. 36–41.

Cushman, J. A., and W. H. Deaderick. Cretaceous Foraminifera from the Marlbrook marl of Arkansas. — Journ. Pal., vol. 18, 1944, pp. 328–342, pls. 50–53.

Cushman, J. A., and P. P. Goudkoff. Some Foraminifera from the Upper Cretaceous of California. — Contr. Cushman Lab. Foram. Res., vol. 20, 1944, pp. 53–64, pls. 9, 10.

Cushman, J. A., and H. D. Hedberg. Notes on Some Foraminifera from Venezuela and Colombia. — l. c., vol. 6, 1930, pp. 64–69, pl. 9.

Cushman, J. A., and P. W. Jarvis. Cretaceous Foraminifera from Trinidad. — l. c., vol. 4, 1928, pp. 84–103, pls. 12–14.

Upper Cretaceous Foraminifera from Trinidad. — Proc. U. S. Nat. Mus., vol. 80, Art. 14, 1932, pp. 1–60, pls. 1–16.

Cushman, J. A., and F. L. Parker. Some American Cretaceous Buliminas. — Contr. Cushman Lab. Foram. Res., vol. 11, 1935, pp. 96–101, pl. 15.

Notes on Some Cretaceous Species of *Buliminella* and *Neobulimina*. — l. c., vol. 12, 1936, pp. 5–10, pl. 2.

Cushman, J. A., and H. H. Renz. The Foraminiferal Fauna of the Lizard Springs Formation of Trinidad, British West Indies. — Special Publ. 18, Cushman Lab. Foram. Res., 1946, pp. 1–48, pls. 1–8, 1 map.

Cushman, J. A., and J. A. Waters. Some Arenaceous Foraminifera from the Upper Cretaceous of Texas. — Contr. Cushman Lab. Foram. Res., vol. 2, 1927, pp. 81–85, pl. 10.

Some Arenaceous Foraminifera from the Taylor Marl of Texas. — l. c., vol. 5, 1929, pp. 63–66, pl. 10.

Cushman, J. A., and R. T. D. Wickenden. The Development of *Hantkenina* in the Cretaceous with a Description of a New Species. — l. c., vol. 6, 1930, pp. 39–43, pl. 6.

Ellis, B. F. *Gallowayina browni*, A New Genus and Species of Orbitoid from Cuba, with Notes on the American Occurrence of *Omphalocyclus macropora*. — Amer. Mus. Novitates, No. 568, 1932, pp. 1–8, 9 text figs.

Frizzell, D. L. Upper Cretaceous Foraminifera from northwestern Peru. — Journ. Pal., vol. 17, 1943, pp. 331–353, pls. 55–57, text figs. 1, 2.

Galloway, J. J., and M. Morrey. Late Cretaceous Foraminifera from Tobasco, Mexico. — l. c., vol. 5, 1931, pp. 329–354, pls. 37–40.

Gravell, D. W. The Genus Orbitoides in America, with Description of a New Species from Cuba. — l. c., vol. 4, 1930, pp. 268–270, pl. 22.

Hanzawa, S. Notes on Some Interesting Cretaceous and Tertiary Foraminifera from the West Indies. — l. c., vol. 11, 1937, pp. 110–117, pls. 20, 21.

Hedberg, H. D. Stratigraphy of the Rio Querecual Section of Northeastern Venezuela. — Bull. Geol. Soc. Amer., vol. 48, 1937, pp. 1971–2024, 9 pls., 2 text figs.

Keller, B. M. Correlation of the Upper Cretaceous Deposits in Eastern Mexico and in the Western Caucasus. — Bull. Acad. Sci., U. S. S. R., 1937, pp. 825–838.

LOEBLICH, A. R., JR. Foraminifera from the Type Pepper Shale of Texas. — Journ. Pal., vol. 20, 1946, pp. 130–139, pl. 22, text figs. 1–3.

LOETTERLE, G. J. The Micropaleontology of the Niobrara Formation in Kansas, Nebraska, and South Dakota. — Nebraska Geol. Survey, Ser. 2, Bull. 12, 1937, pp. 1–73, pls. I–XI, text figs. 1–3.

LOZO, F. E., JR. Biostratigraphic Relations of Some North Texas Trinity and Fredericksburg (Comanchean) Foraminifera. — Amer. Midland Nat., vol. 31, No. 3, 1944, pp. 513–582, 5 pls., 22 text figs.

MOREMAN, W. L. Fossil Zones of the Eagle Ford of North Texas. — Journ. Pal., vol. 1, 1927, pp. 89–101, pls. 13–16 [pp. 98–100, pl. 16].

MORROW, A. L. Foraminifera and Ostracoda from the Upper Cretaceous of Kansas. — l .c., vol. 8, 1934, pp. 186–205, pls. 29–31.

NAUSS, A. W. Cretaceous Microfossils of the Vermilion area, Alberta. — l. c., vol. 21, 1947, pp. 329–343, pls. 48, 49, text figs. 1–3.

PALMER, D. K. The Upper Cretaceous Age of the Orbitoidal Genus *Gallowayina* Ellis. — l. c., vol. 8, 1934, pp. 68–70.

PLUMMER, H. J. Some Cretaceous Foraminifera in Texas. — Univ. Texas Bull. 3101, 1931, pp. 109–203, pls. 8–15.

Gaudryinella, A New Foraminiferal Genus. — Amer. Midland Nat., vol. 12, 1931, pp. 341, 342, text figs. *a*, *b*.

Ammobaculoides, A New Foraminiferal Genus. — l. c., vol. 13, 1932, pp. 86–88, text figs. 1*a–d*.

SAMPLE, C. H. *Cribratina*, A New Genus of Foraminifera from the Comanchean of Texas. — l. c., pp. 319–322, pl. 30.

SANDIDGE, J. R. Significant Foraminifera from the Ripley Formation of Alabama. — l. c., pp. 190–202, pl. 19.

Fossil Foraminifera from the Cretaceous, Ripley Formation, of Alabama. — l. c., pp. 312–318, pl. 29.

Additional Foraminifera from the Ripley Formation in Alabama. — l. c., pp. 333–377, pls. 31–33.

Foraminifera from the Ripley Formation of Western Alabama. — Journ. Pal., vol. 6, 1932, pp. 265–287, pls. 41–44.

SILVESTRI, A. Revisione di Orbitoline Nordamericane e Nuova Località di Chapmanine. — Mem. Pont. Accad. Sci. Nuovi Lincei, vol. 16, 1932, pp. 371–394, pls. 1, 2.

TAPPAN, H. Foraminifera from the Grayson Formation of Northern Texas. — Journ. Pal., vol. 14, 1940, pp. 93–126, pls. 14–19.

Foraminifera from the Duck Creek Formation of Oklahoma and Texas. — l. c., vol. 17, 1943, pp. 476–517, pls. 77–83.

THIADENS, A. A. Cretaceous and Tertiary Foraminifera from Southern Santa Clara Province, Cuba. — l. c., vol. 11, 1937, pp. 91–109, pls. 15–19, text figs. 1–3.

TYRRELL, J. B. Foraminifera and Radiolaria from the Cretaceous of Manitoba. — Trans. Roy. Soc. Canada, vol. 8, Sect. IV, 1890, pp. 111–115.

VANDERPOOL, H. C. Upper Trinity Microfossils from Southern Oklahoma. — Journ. Pal., vol. 7, 1933, pp. 406–411, pl. 49.

VAUGHAN, T. W. Species of *Orbitocyclina*, a Genus of American Orbitoid Foraminifera from the Upper Cretaceous of Mexico and Louisiana. — l. c., vol, 3, 1929, pp. 170–175, pl. 22.

A Note on *Orbitoides browni* (Ellis) Vaughan. — l. c., vol. 8, 1934, pp. 70–72.

Vaughan, T. W., and W. S. Cole. Cretaceous Orbitoidal Foraminifera from the Gulf States and Central America. — Proc. Nat. Acad. Sci., vol. 18, 1932, pp. 611–616, pls. 1, 2.

Vieux, D. G. New Foraminifera from the Denton formation in northern Texas. — Journ. Pal., vol. 15, 1941, pp. 624–628, pl. 85.

Voorwijk, G. H. Foraminifera from the upper cretaceous of Habana, Cuba. — Proc. Roy. Acad. Amsterdam, vol. 40, 1937, pp. 190–198, pl. III, 49 text figs.

Weller, S. A Report on the Cretaceous Palaeontology of New Jersey. — Geol. Surv. New Jersey, Pal., vol. 4, 1907, pp. 189–265, pls. 1–4.

White, M. P. Some Index Foraminifera of the Tampico Embayment Area of Mexico. — Journ. Pal., vol. 2, 1928, pp. 177–215, pls. 27–29; pp. 280–317, pls. 38–42; vol. 3, 1929, pp. 30–58, pls. 4, 5.

Wickenden, R. T. D. New Species of Foraminifera from the Upper Cretaceous of the Prairie Provinces. — Trans. Roy. Soc. Canada, 3rd ser., vol. 26, Sect. IV, 1932, pp. 85–91, pl. 1.

A Useful Foraminifera Horizon in the Alberta Shale of Southern Alberta. — Journ. Pal., vol. 6, 1932, pp. 203–207, pl. 29.

ASIA

Brotzen, F. Foraminiferen aus dem Senon Palästinas. — Zeitschr. Deutschen Palästina-Vereins, vol. 57, 1934, pp. 28–72, pls. 1–4.

Schnetzer, R. Nummuliten und Orbitolinen aus dem Gilboagebirge in Palästina. — Centralblatt für Min., etc., Jahrg. 1934, Abt. B, No. 2, pp. 19–27, 4 text figs.

V. JURASSIC

Albritton, C. C., Jr. Upper Jurassic and Lower Cretaceous Foraminifera from the Malone Mountains, Trans Pecos Texas. — Journ. Pal., vol. 11, 1937, pp. 19–23, pl. 4 (pt.).

Bartenstein, H., and E. Brand. Mikro-paläontologische Untersuchungen zur Stratigraphie des nordwest-deutschen Lias und Doggers. — Abhandl. Senck. Nat. Ges., No. 439, 1937, pp. 1–224, pls. 1–20, text figs. 1–20.

Berthelin, G. Foraminefères du Lias Moyen de la Vendée. — Revue et Mag. Zool., 1879, pp. 24–41, 1 pl.

Mémoire sur les Foraminifères fossiles de l'étage Albien de Moncley (Doubs.) — Mém. Soc. Géol. France, ser. 3, vol. 1, No. 5, 1880, pp. 1–84, pls. 24–27.

Bornemann, J. G. Ueber die Liasformation in der Umgegend von Göttingen, und ihre organischen Einschlüsse. — 8vo, *Berlin*, 1854, 77 pp., 1 map and 3 pls.

Brady, H. B. Synopsis of the Foraminifera of the Upper and Middle Lias of Somersetshire. In Moore, On the Middle and Upper Lias of the Southwest of England. — Proc. Somerset Arch. Nat. Hist. Soc., vol. 13, 1865–66 (1867), pp. 220–230, pls. 1–3.

Brückmann, R. Die Foraminiferen des Litauisch-Kurischen Jura. — Schrift. Ges. Königsb., vol. 45, 1904, pp. 1–36, pls. 1–4.

Burbach, O. Beiträge zur Kenntniss der Foraminiferen des mittleren Lias von grossen Seeberg bei Gotha. I. Die Gattung *Frondicularia*, Defr. — Zeitchr. Naturw. Halle, vol. 59, 1886, pp. 30–53, pls. 1, 2; II. Die Miliolideen. — l. c., pp. 493–502, pl. 5.

Chapman, F. On Some Foraminifera and Ostracoda from Jurassic (Lower Oolite) Strata, near Geraldton, Western Australia. — Proc. Roy. Soc., Victoria, vol. 16, 1904, pp. 185–206, pls. 22, 23.

Colom, G. Estudios Litologicos sobre el Jurasico de Mallorca. — Géol. Méditerranée Occid., vol. 3, No. 4, pt. 5, 1935, pp. 1–17, pls. I, II, text figs. 1–4.

Cushman, J. A. Note sur quelques Foraminifères Jurassiques d'Auberville (Calvados). — Bull. Soc. Linn. Normandie, ser. 8, vol. 2, 1929, pp. 132–135, pl. 4.

Crick, W. D., and C. D. Sherborn. On Some Liassic Foraminifera from Northamptonshire. — Journ. Northamptonshire Nat. Hist. Soc., vol. 4, 1891, pp. 208–214, pl. 6; pp. 67–72, pl. 7.

Deecke, W. Die Foraminiferenfauna der zone des Stephanoceras Humphriesianum in Unter Elsass. — Abhandl. Geol. Special-Karte Elsass-Lothringen, vol. 4, pt. 1, 1884, pp. 1–68, pls. 1, 2.

Les Foraminifères de l'Oxfordien des environs de Montbéliard (Doubs). — Mém. Soc. émul. Montbéliard, ser. 3, vol. 16, 1886, pp. 1–47, pls. 1, 2.

Franke, A. Die Foraminiferen des deutschen Lias. — Abhandl. Preuss. Geol. Landes., Heft 169, 1936, 138 pp., 12 pls., 2 text figs.

Haeusler, R. Monographie der Foraminiferenfauna der Schweizerschen transversarius-zone. — Abhandl. Schweiz. Pal. Ges., vol. 17, 1890, pp. 1–134, pls. 1–15.

Hanzawa, S. On the Occurrence of *Acervulina*, an Encrusting Form of Foraminifera in the Jurassic Torinosu Limestone from the Kwanto Mountainland, Central Japan. — Journ. Geol. Soc. Japan., vol. 46, No. 547, April 20, 1939, pp. 201–203, pl. 12.

Macfadyen, W. A. Jurassic Foraminifera. — From the Geology and Palaeontology of British Somaliland, Part II. The Mesozoic Palaeontology of British Somaliland,, London, 1935, pp. 7–20, 1 pl.

d'Orbigny's Lias Foraminifera. — Journ. Roy. Micr. Soc., vol. 56, 1936, pp. 147–153, pl. I, text fig. 1.

Foraminifera from the Green Ammonite Beds, Lower Lias, of Dorset. — Phil. Trans. Roy. Soc. London, ser. B, Biol. Sci., No. 576, vol. 231, 1941, pp. 1–73, pls 1–4, text figs. 1–6.

Paalzow, R. Beiträge zur Kenntniss der Foraminiferenfauna des Unteren Weissen Jura in Süddeutschland. — Abhandl. Nat. Geol. Nürnberg, vol. 19, 1917, pp. 1–48, pls. 41–47.

Die Foraminiferen der Parkinsoni-Mergel von Heidenheim am Hahnenkamm. — l. c., vol. 22, 1922, pp. 1–35, pls. 1–4.

Die Foraminiferen aus den *Transversarius*-Schichten und *Impressa*-Tonen der nordöstlichen Schwäbischen Alb. — Jahreshefte Ver. vat. Nat. Württemberg, 88 Jahrg, 1932, pp. 81–142, pls. 4–11.

Sandidge, J. Foraminifera from the Jurassic in Montana. — Amer. Midland Nat., vol. 14, 1933, pp. 174–185, 1 pl., text figs.

Schwager, C. Beitrag zur Kenntniss der Mikroskopischen Fauna jurassichen Schichten. — Württemberg, Nat. Jahreshefte, vol. 21, 1865, pp. 82–151, pls. 2–7.

Terquem, O. Recherches sur les Foraminifères du Lias. — Series of Six Mémoirs, in Mém. Acad. Imp. Metz, 1858–1866, with 22 pls. Four Mémoirs, sur les Foraminifères du Système Oolithique. *Metz*, 1868–74; *Paris*, 1883, with 45 pls.

Recherches sur les Foraminifères du Bajocien de la Moselle. — Bull. Soc. Géol. France, ser. 3, vol. 4, 1876, pp. 477–500, pls. 15–17.

Les Foraminifères et les Ostracodes du Fuller's Earth (Zone à Ammonites Parkinsoni), des environs de Varsovie. — Mém. Soc. Géol. France, ser. 3, vol. 4, pt. 2, 1886, pp. 1–112, 12 pls.

Terquem, O., and G. Berthelin. Étude microscopique des Marnes du Lias Moyen d'Ersey-lès-Nancy, Zone inférieur de l'assise à Ammonites margaritatus. — l. c., ser. 2, vol. 10, Mém. 3, 1875, pp. 1–126, pls. 11–20.

TOBLER, A. Über Pseudocyclammina and Choffatella in Schweizerischen Juragebirge. — Eclog. geol. Helv., vol. 21, 1928, pp. 212–216, pl. 24.

WICHER, C. A. Mikrofaunen aus Jura und Kreide insbesondere Nordwestdeutschlands. 1. Teil: Lias *a–e*. — Preuss. Geol. Landes., Neue Folge, Heft 193, 1938, pp. 1–16, 27 pls., 4 text figs.

WICKENDEN, R. T. D. Jurassic Foraminifera from Wells in Alberta and Saskatchewan. — Trans. Roy. Soc. Canada, ser. 34, Sect. IV, vol. 27, 1933, pp. 157–170, pls. 1, 2.

VI. TRIASSIC

CHAPMAN, F. On Rhaetic Foraminifera from Wedmore, in Somerset. — Ann. Mag. Nat. Hist., ser. 6, vol. 16, 1895, pp. 307–329, pls. 11, 12.

On Some Microzoa from the Wianamatta Shales, New South Wales. — Rec. Geol. Surv. N. S. Wales, vol. 8, 1909, pp. 1–5, pl. 54.

GÜMBEL, C. W. Ueber Foraminiferen, Ostracoden, und Mikroskopische Thier-Ueberreste in den St. Cassianer und Raibler Schichten. — Jahrb. k. k. Geol. Reichsanst., vol. 19, 1869, pp. 175–186, pls. 5, 6.

JONES, T. R., and W. K. PARKER. On Some Fossil Foraminifera from Chellaston, near Derby. — Quart. Journ. Geol. Soc., vol. 16, 1860, pp. 452–458, pls. 19, 20.

SCHWAGER, C., in Dittmar. Die Contorta Zone. — 8vo, *München*, 1864, pp. 198–201, pl. 3.

VII. PALEOZOIC

BAGG, R. M. The Foraminifera of the Bonaventure Cherts of Gaspé. — New York State Bull. 219, 220 — Fifteenth Report of the Director, 1918 (1921), pp. 1–60, pls. 1–6.

BEEDE, J. W., and H. T. KNIKER. Species of the Genus *Schwagerina* and Their Stratigraphic Significance. — Bull. 2433, Univ. Texas, 1924, pp. 1–96, 9 pls.

BERRY, E. W. Distribution of the Fusulinidae. — Pan-Amer. Geol., vol. 56, 1931, pp. 181–187, 1 map.

BRADY, H. B. On *Saccammina Carteri*, a New Foraminifer from the Carboniferous Limestone of Northumberland. — Ann. Mag. Nat. Hist., ser. 4, vol. 7, 1871, pp. 177–184, pl. 12.

On *Archaediscus Karreri*, a new Type of Carboniferous Foraminifera. — l. c., vol. 12, 1873, pp. 286–290, pl. 11.

On a True Carboniferous Nummulite. — l. c., vol. 13, 1874, pp. 222–230, pl. 12.

A Monograph of Carboniferous and Permian Foraminifera (the genus Fusulina excepted). — Palaeont. Soc., vol. 30, 1876, pp. 1–166, pls. 1–12.

Notes on a Group of Russian Fusulinae. — Ann. Mag. Nat. Hist., ser. 4, vol. 18, 1876, pp. 414–422, pl. 18.

BRAND, E. Über Foraminiferen im Zechstein der Wetterau. — Senckenbergiana, Bd. 19, 1937, pp. 375–380, 1 text fig.

CHAPMAN, F. Foraminifera from an Upper Cambrian Horizon in the Malverns, together with a Note on some of the Earliest-known Foraminifera. — Quart. Journ. Geol. Soc., vol. 56, 1900, pp. 257–263, pl. 15.

On Some Fossils of Wenlock Age from Mulde, near Klinteberg, Gotland. — Ann. Mag. Nat. Hist., ser. 7, vol. 7, 1901, pp. 141–160, pl. 3.

Devonian Foraminifera; Tamworth District, New South Wales. — Proc. Linn. Soc. N. S. Wales, vol. 43, 1918, pp. 385–391, pls. 39–41.

Report on Fossils from an Upper Cambrian Horizon at Loyola, near Mansfield, Australia. — Bull. 46, Geol. Surv. Victoria, App. 1, 1923, pp. 34–46, pl. 13.

Correlation of Carboniferous and Permian Rocks of Australia and New Zealand. I. Note on Our Present Kowledge of the Permian Foraminifera of Western Australia. — Rep't Australian and New Zealand Assoc. Adv. Sci., vol. 21, 1933, pp. 453, 454.

Some Palaeozoic Fossils from Victoria. — Proc. Roy. Soc. Victoria, vol. 45, 1933, pp. 245–248, pl. XI.

Chapman, F., and W. J. Parr. On the Discovery of Fusulinid Foraminifera in the Upper Palaeozoic of North-West Australia; With a Note on a New Bivalve. — Victorian Naturalist, vol. 53, 1937, pp. 175–179, pl. XVI.

Chen, S. Fusulinidae of South China. Part I. — Palaeontologia Sinica, ser. B, vol. 4, fasc. 2, 1934, pp. 1–185, pls. I–XVI, 36 graphs.

A New Species of Fusulinidae from the Meitien Limestone. — Bull. Geol. Soc. China, vol. 13, 1934, pp. 237–242, pl. 1.

Colani, M. Nouvelle Contribution à l'étude des Fusulinides de l'extrêmeorient. — Mém. Serv. Géol. de l'Indochine, vol. 11, 1924.

Cooper, C. L. Upper Kinkaid (Mississippian) Microfauna from Johnson County, Illinois. — Journ. Pal., vol. 21, 1947, pp. 81–94, pls. 19–23, 1 text fig.

Crespin, I. Foraminifera in the Permian Rocks of Australia. — Commonwealth of Australia, Bureau of Mineral Resources, Geology and Geophysics, Bull. 15 (Pal. Ser. 5), 1947, pp. 1–31, pls. 1, 2, fig. 1 (map), tables.

Crespin, I., and W. J. Parr. Arenaceous Foraminifera from the Permian Rocks of New South Wales. — Journ. Proc. Roy. Soc. New South Wales, vol. 74, 1940, pp. 300–311, pls. 12, 13.

Cushman, J. A. The Microspheric and Megalospheric Forms of *Apterrinella grahamensis*. — Contr. Cushman Lab. Foram. Res., vol. 4, 1928, pp. 68, 69, pl. 9.

Notes on Early Paleozoic Foraminifera. — l. c., vol. 6, 1930, pp. 43, 44, pl. 6.

Paleozoic Foraminifera, their Relationships to Modern Faunas, and to their Environment. — Journ. Pal., vol. 9, 1935, pp. 284–287.

Cushman, J. A., and M. A. Stainbrook. Some Foraminifera from the Devonian of Iowa. — Contr. Cushman Lab. Foram. Res., vol. 19, 1943, pp. 73–79, pl. 13.

Cushman, J. A., and J. A. Waters. Pennsylvanian Foraminifera from Michigan. — l. c., vol. 3, 1927, pp. 107–110, pl. 22.

Arenaceous Palaeozoic Foraminifera from Texas. — l. c., pp. 146–153, pls. 26, 27.

Some Foraminifera from the Pennsylvanian and Permian of Texas. — l. c., vol. 4, 1928, pp. 31–55, pls. 4–7.

Upper Paleozoic Foraminifera from Sutton County, Texas. — Journ. Pal., vol. 2, 1928, pp. 358–371, pls. 47–49.

Additional Cisco Foraminifera from Texas. — Contr. Cushman Lab. Foram. Res., vol. 4, 1928, pp. 62–67, pl. 8.

Hyperamminoides, A New Name for *Hyperamminella* Cushman and Waters. — l. c., p. 112.

Foraminifera of the Cisco Group of Texas. — Univ. Texas Bull. 3019, 1930, pp. 22–81, pls. 2–12.

Deprat, J. Étude des Fusulinides de Chine et d'Indochine et classification des calcaires à Fusulines. — Mém. Serv. Géol. de l'Indochine, vol. 1, pt. 3, 1912, pp. 1–63, pls. 1–9, text figs. 1–30.

Les Fusulinides des calcaires carboniferiens et permiens du Tonkin, du Laos et du Nord-Annam. — l. c., vol. 2, pt. 1, 1913, pp. 1–76, pls. 1–10, text figs. 1–25.

Étude comparative des Fusulinides d'Akasaka (Japan) et des Fusulinides de Chine et d'Indochine. — l. c., vol. 3, pt. 1, 1914, pp. 1–45, pls. 1–8, text figs. 1–8.

Les Fusulinides des calcaires carboniferiens et permiens du Tonkin, du Laos et du Nord-Annam. — l. c., vol. 4, pt. 1, 1915, pp. 1–30, pls. 1–3, text figs. 1–11.

Douvillé, H. Sur la structure du test dans les Fusulines. — Compt. Rend., Acad. Sci., Paris, vol. 143, 1906, pp. 258–261.

Les calcaires à Fusulines de l'Indochine. — Bull. Soc. Géol. France, ser. 4, vol. 6 1907, pp. 576–587, pls. 17, 18.

Dunbar, C. O. *Neoschwagerina* in the Permian Faunas of British Columbia. — Trans. Roy. Soc. Canada, ser. 3, vol. 26, Sect. IV, 1932, pp. 45–49, pl. 1.

Stratigraphic Significance of the Fusulinids of the Lower Productus Limestone of the Salt Range. — Rec. Geol. Survey India, 1933, pt. 4, pp. 405–413, pl. 22.

Permian Fusulines from Central America. — Journ. Pal., vol. 13, 1939, pp. 344–348, pls. 35, 36.

Permian Fusulines from Sonora. — Bull. Geol. Soc. Amer., vol. 50, 1939, pp. 1745–1760, pls. 1–4.

Dunbar, C. O., and G. E. Condra. The Fusulinidae of the Pennsylvanian System in Nebraska. — Nebraska Geol. Surv., 2d Ser. Bull. 2, 1927, pp. 1–135, pls. 1–15, text figs. 1–13.

Dunbar, C. O., and L. G. Henbest. The Fusulinid Genera *Fusulina, Fusulinella,* and *Wedekindella.* — Amer. Journ. Sci., vol. 20, 1930, pp. 357–365.

Pennsylvanian Fusulinidae of Illinois. — Illinois State Geol. Survey, Bull. 67, 1942, pp. 1–218, pls. 1–23, text figs. A, 1–13.

Dunbar, C. O., and N. D. Newell. Marine Early Permian of the Central Andes and its Fusuline Faunas. — Amer. Journ. Sci., vol. 244, 1946, pp. 377–402; 457–491, pls. 1–12, text figs, 1–3.

Dunbar, C. O., and J. Skinner. New Fusulinid Genera from the Permian of West Texas. — l. c., vol. 22, 1931, pp. 252–268, 3 pls.

Dunbar, C. O., J. W. Skinner and R. E. King. Dimorphism in Permian Fusulines. — Univ. Texas Bull. 3501, 1936, pp. 173–190, pls. 1–3, text fig. 30.

Dunn, P. H. Silurian Foraminifera of the Mississippi Basin. — Journ. Pal., vol. 16, 1942, pp. 317–342, pls. 42–44.

Dyhrenfurth, G. Monographie der Fusulinen, Teil 2. Die Fusulinen von Darwas. — Palaeontographica, vol. 56, 1909, pp. 137–176, pls. 13–16, 10 text figs.

Eisenack, A. Neue Mikrofossilien des baltischen Silurs. II. (Foraminiferen, Hydrozoen, Chitinozoen u. a.). — Pal. Zeitschr., vol. 14, 1932, pp. 257–277, pls. 11, 12, text figs. 1–13.

Neue Mikrofossilien des baltischen Silurs. IV. Foraminiferen. — l. c., vol. 19, 1937, pp. 233–243, pls. 15, 16, text figs. 8–22.

Galloway, J. J., and B. H. Harlton. Some Pennsylvanian Foraminifera of Oklahoma, with Special Reference to the Genus *Orobias.* — Journ. Pal., vol. 2, 1928, pp. 338–357, pls. 45, 46.

Endothyranella, A Genus of Carboniferous Foraminifera. — l. c., vol. 4, 1930, pp. 24–28.

Galloway, J. J., and C. Ryniker. Foraminifera from the Atoka Formation of Oklahoma. — Oklahoma Geol. Surv. Circular No. 21, 1930, pp. 1–37, pls. 1–5.

Galloway, J. J., and L. E. Spock. Pennsylvanian Foraminifera from Mongolia. — American Museum Novitates, No. 658, 1933, pp. 1–7, figs. 1, 2.

GIRTY, G. H. On the names of American Fusulinas. — Journ. Geol., vol. 22, 1914, pp. 237–242.

GUBLER, J. La valeur stratigraphique des Fusulinidés du Permien. — Comptes Rendus Géol. Soc. France, séance du 22 Janvier, 1934, pp. 381–383.

Les Fusulinidés du Permien de l'Indochine. — Mém. Soc. Géol. France, n. ser., T. 11, fasc. 4, Mem. No. 26, 1935, pp. 1–171, pls. I–VIII, text figs. 1–54.

HANZAWA, S. On a *Neoschwagerina*-Limestone from Okinawa-jima, the Riukiu (Loochoo) Islands. — Jap. Journ. Geol. Geogr., vol. 10, 1933, pp. 107–110, pl. VII, map.

Stratigraphical Distributions of the Genera *Pseudoschwagerina* and *Paraschwagerina* in Japan with Descriptions of Two New Species of *Pseudoschwagerina* from the Kitakami Mountainland, Northeastern Japan. — l. c., vol. 16, 1938, pp. 65–73, pl. IV.

An Aberrant Type of the Fusulinidae from the Kitakami Mountainland, Northeastern Japan. — Proc. Imperial Acad., vol. 14, 1938, pp. 255–259, text figs. 1–16.

HARLTON, B. H. Some Pennsylvanian Foraminifera of the Glenn Formation of Southern Oklahoma. — Journ. Pal., vol. 1, 1927, pp. 15–27, pls. 1–5.

Pennsylvanian Foraminifera of Oklahoma and Texas. — l. c., 1928, pp. 305–310, pls. 52, 53.

Some Pennsylvanian Ostracoda and Foraminifera from Southern Oklahoma. A Correction. — l. c., vol. 3, 1929, p. 308.

Micropaleontology of the Pennsylvanian John Valley Shale of the Ouachita-Mountains, Oklahoma, and Its Relationship to the Mississippian Caney Shale. — l. c., vol. 7, 1933, pp. 3–29, pls. 1–7.

HAYDEN, H. H. Fusulinidae from Afghanistan — Records Geol. Surv. India, vol. 38, 1909, pp. 230–256, pls. 17–22.

HENBEST, L. G. Fusulinellas from the Stonefort Limestone Member of the Tradewater Formation. — Journ. Pal., vol. 2, 1928, pp. 70–85, pls. 9, 10, text figs.

Nanicella, a new genus of Devonian Foraminifera. — Journ. Washington Acad. Sci., vol. 25, 1935, pp. 34, 35.

HOWCHIN, W. Additions to the Knowledge of the Carboniferous Foraminifera. — Journ. Roy. Micr. Soc., 1888, pp. 533–545, pls. 8, 9.

Carboniferous Foraminifera of Western Australia, with Descriptions of New Species. — Trans. Roy. Soc. So. Australia, vol. 19, 1895, pp. 194–198, pl. 10.

HOWELL, B. F., and J. R. SANDIDGE. Cambrian Foraminifera: Chapman's Species from the Dolgelly Beds of England. — Bull. Wagner Free Inst. Sci., Philadelphia, vol. 8, 1933, pp. 59–62, pls. 1, 2.

HUZIMOTU, H. Some Foraminiferous Fossils from the Kôten Series of Zidô Coal-Field, Tyosen. — Journ. Geol. Soc. Japan, vol. 45, No. 533, 1938, pp. 271–276, pl. 8 (1).

Science Reports of the Tokyo Bunrika Daigaku, Section C, No. 2. Stratigraphical and Palaeontological Studies of the Titibu System of the Kwanto Mountainland, Part 2. — Palaeontology, vol. 1, 1936, pp. 29–125, pls. I–XXVI.

Some Fusulinids from Kawanobori-Mura, Kyusyu, Japan. — Jap. Journ. Geol. Geogr., vol. 14, 1937, pp. 117–125, pls. VII, VIII.

IRELAND, H. A. Devonian and Silurian Foraminifera from Oklahoma. — Journ. Pal., vol. 13, 1939, pp. 190–202, text figs.

KAHLER, F. Fusulinidae. — Paleontographica, vol. 79, 1933, pp. 168–172.

LANGE, E. Eine Mittelpermische Fauna von Guguk Bulat (Padanger Oberland, Sumatra). — Verhandl. Geol.-Mijn. Gen. Ned. Kol., Geol. Ser., vol. 7, 1925, pp. 213–295, pls. 1–5, 10 text figs.

LEE, J. S. Fusulinidae of North China. — Paleontologia Sinica, Geol. Surv. China, ser. B, vol. 4, fasc. 1, 1927.

LEE, J. S., S. CHEN and S. CHU. Huanglung Limestone and Its Fauna. — Mem. Nat. Research Inst. Geol. (China), No. 9, 1930, pp. 85–144, pls. 1–15.

LEE, W., C. O. NICKELL, J. S. WILLIAMS and L. G. HENBEST. Stratigraphic and Paleontologic Studies of the Pennsylvanian and Permian Rocks in North-Central Texas. — Univ. Texas Publ. No. 3801, 1938, pp. 1–252.

LIEBUS, A. Die Fauna des deutschen Unterkarbons. 3. Teil. Die Foraminiferen. — Preuss. Geol. Landes., Neue Folge, Heft 141, 1932, pp. 133–175, pls. 9, 10.

MAITRE, D. LE. Observations sur les Algues et las Foraminifères des calcaires dévoniens. — Ann. Soc. Géol. du Nord, vol. 55, 1930, pp. 42–50, pl. 3.

Foraminifères des terrains dévoniens de Bartine (Turquie). — l. c., vol. 56, 1931, pp. 1–8, pl. 1.

MERCHANT, F. E., and R. P. KEROHER. Some fusulinids from the Missouri Series of Kansas. — Journ. Pal., vol. 13, 1939, pp. 594–614, pl. 69, 4 text figs.

MILLER, A. K., and A. M. CARMER. Devonian Foraminifera from Iowa. — l. c., vol. 7, 1933, pp. 423–431, pl. 50, figs. 10, 11.

MÖLLER, V. VON. Ueber Fusulinen und ähnliche Foraminiferenformen des russischen Kohlenkalks. — Neues Jahrb. f. Min., etc., Jahrb. 1877, pp. 138–146.

Die Spiral-gewundenen Foraminiferen des russischen Kohlenkalks. — Mém. Acad. Imp. Sci. St.-Pétersbourg, ser. 7, vol. 25, No. 9, 1878, pp. 1–147, pls. 1–15, text figs.

Die Foraminiferen des russischen Kohlenkalks. — l. c., vol. 27, No. 5, 1879, pp. 1–131, pls. 1–7.

MOREMAN, W. L. Arenaceous Foraminifera from Ordovician and Silurian Limestones of Oklahoma. — Journ. Pal., vol. 4, 1930, pp. 42–59, pls. 5–7.

Arenaceous Foraminifera from the Lower Paleozoic Rocks of Oklahoma. — l. c., vol. 7, 1933, pp. 393–397, pl. 47.

NEWELL, N. D. Some Mid-Pennsylvanian Invertebrates from Kansas and Oklahoma: I. Fusulinidae, Brachiopoda. — l. c., vol. 8, 1934, pp. 422–432, pls. 52–55.

NEWELL, N. D., and R. P. KEROHER. The Fusulinid, *Wedekindellina*, in Mid-Pennsylvanian Rocks of Kansas and Missouri. — l. c., vol. 11, 1937, pp. 698–705, pl. 93, 4 text figs.

NEWTON, R. B. On *Fusulina* and Other Organisms in a Partially Calcareous Quartzite from near the Malayan-Siamese Frontier. — Ann. Mag. Nat. Hist., ser. 9, vol. 17, 1926, pp. 49–64, pls. 2, 3.

OZAWA, Y. Preliminary Notes on the Classification of the Family Fusulinidae. — Journ. Geol. Soc. Tokyo, vol. 29, 1922.

On some species of *Fusulina* from Honan, China. — Jap. Journ. Geol. Geogr., vol. 2, 1923, pp. 35–39, pl. 5.

A Brief Critical Revision of the *Fusulina* Species Recently Described, with Additional Studies on Japanese Fusulinae. — Journ. Geol. Soc., vol. 32, 1925, pp. 19–27, pls. 9, 10.

On the Classification of Fusulinidae. — Journ. Coll. Sci., Tokyo Imp. Univ., vol. 45, art. 4, 1925, pp. 1–26, pls. 1–4, text figs. 1–3.

Paleontological and Stratigraphical Studies on the Permo-Carboniferous Limestone of Nagato. Part II. Paleontology. — l. c., art. 6, 1925, pp. 1–90, pls. 1–14.

Stratigraphical Studies of the *Fusulina* Limestone of Akasaka, Province of Mino. — l. c., section II, vol. 2, 1927, pp. 121–164, pls. 34–46.

A New Genus, *Depratella*, and its Relation to *Endothyra*. — Contr. Cushman Lab. Foram. Res., vol. 4, 1928, pp. 9–11.

PAALZOW, R. Die Foraminiferen im Zechstein des östlichen Thüringen. — Jahrb. Preuss. Geol. Landes. für 1935, vol. 56, 1935, pp. 26–45, pls. 3–5.

PARR, W. J. Foraminifera and a tubicolous Worm from the Permian of the North-West Division of Western Australia. — Journ. Roy. Soc. Western Australia, vol. 27, 1940–41 (1942), pp. 97–115, 2 pls.

PLUMMER, H. J. Calcareous Foraminifera in the Brownwood Shale near Bridgeport, Texas. — Univ. Texas Bull. 3019, 1930, pp. 5–21, pl. 1.

Smaller Foraminifera in the Marble Falls, Smithwick, and Lower Strawn Strata around the Llano Uplift in Texas. — l. c., Publ. 4401, 1945, pp. 209–271, pls. 15–17.

RAUSER-CERNOUSSOVA, D. Rugosofusulina, a new Genus of Fusulinids. — Studies in Micropaleontology, Moscow Univ., vol. 1, fasc. 1, 1937, pp. 9–26, pls. I–III, 2 text figs.

ROTH, R., and J. SKINNER. The Fauna of the McCoy Formation, Pennsylvanian, of Colorado. — Journ. Pal., vol. 4, 1930, pp. 332–352, pls. 28–31.

SCHELLWIEN, E. Die Fauna des karnischen Fusulinenkalks, I. — Paleontographica, vol. 39, 1892.

Die Fauna des karnischen Fusulinenkalks, II. — Paleontographica, vol. 44, 1897 (1898), pp. 237–282, pls. 17–24, text figs. 1–7.

SCHWAGER, C. Carbonische Foraminiferen aus China und Japan. In Richtofen's China. — 4to, *Berlin*, vol. 4, 1883. "Beitr. zur Paläont. von China," pp. 106–159, pls. 15–18.

SILVESTRI, A. Sulle cosiddette Schwagerine della Valle del Sosio (Palermo). — Boll. Soc. Geol. Ital., vol. 51, 1932, pp. 253–264, pl. 8.

SKINNER, J. W. New Permo-Pennsylvanian Fusulinidae from Northern Oklahoma. — Journ. Pal., vol. 5, 1931, pp. 16–22, pls. 3, 4.

SPANDEL, E. Die Foraminiferen des deutschen Zechsteines (vorläufige Mitteilung) und ein zweifelhaftes mikroskopisches Fossil ebendaher. — Verl. Inst., General-Anzeigers, Nürnberg, 1898, pp. 1–15, 11 text figs.

Die Foraminiferen des Permo-Carbon von Hooser, Kansas, Nord Amerika. — Festschr. Nat. ges. Nürnberg, 1901, pp. 175–194, 10 text figs.

STAFF, H. VON. Monographie der Fusulinen, Teil I, Die Fusulinen des russischarktischen Meeresgebietes. — Paleontographica, vol. 55, 1908, pp. 145–194, pls. 13–20.

A. Zur Entwicklung der Fusuliniden. — Centralb. f. Min., etc., No. 22, 1908, pp. 691–703.

B. Ueber Schalenverschmalzungen und Dimorphismus bei Fusulinen. — Sitz. Ges. Nat. Freunde, Berlin, 1908, pp. 217–237.

Beiträge zur Kenntnis der Fusuliniden. — Neues Jahrb. für Min., vol. 27, 1909, pp. 461–508, pls. 7, 8, 16 text figs.

Die Anatomie und Physiologie der Fusulinen. — Zoologica, Heft 58, 1910, pp. 1–93, pls. 1, 2, text figs. 1–62.

Monographie der Fusulinen, Teil III, Die Fusulinen (Schellwienien) Nordamerikas. — Paleontographica, vol. 59, 1912, pp. 157–192, pls. 15–20.

STAFF, H. VON, and R. WEDEKIND. Der Oberkarbone Foraminiferen-sapropelit Spitzbergens. — Bull. Geol. Inst. Upsala, vol. 10, 1910, pp. 81–123, pls. 2–4.

STEINMANN, G. Mikroskopische Thierreste aus dem deutschen Kohlenkalke (Foraminiferen und Spongien). — Zeitschr. Deutsch. Geol. Ges., vol. 32, 1880, pp. 394–400, pl. 19.

STEWART, G. A., and L. LAMPE. Foraminifera from the Middle Devonian Bone Beds of Ohio. — Journ. Pal., vol. 21, 1947, pp. 529–536, pls. 78,79.

STEWART, G. A., and R. R. PRIDDY. Arenaceous Foraminifera from the Niagaran rocks of Ohio and Indiana. — l. c., vol. 15, 1941, pp. 366–375, pl. 54.

THOMAS, A. O. Late Devonian Foraminifera from Iowa. — l. c., vol. 5, 1931, pp. 40, 41, pl. 7.

THOMAS, N. L. New Early Fusulinids from Texas. — Univ. Texas Bull. 3101, 1931, pp. 27–33, pl. 1.

THOMPSON, M. L. The Fusulinids of the Des Moines Series of Iowa. — Univ. Iowa Studies in Natural History, vol. 16, No. 4, N. S. No. 284, 1934, pp. 273–332, pls. 20–23.

The Fusulinid Genus *Staffella* in America. — Journ. Pal., vol. 9, 1935, pp. 111–120, pl. 13.

Fusulinids from the Lower Pennsylvanian Atoka and Boggy Formations of Oklahoma. — l. c., pp. 291–306, pl. 26.

The fusulinid genus *Yangchienia* Lee. — Eclogae geologicae Helvetiae, vol. 28, No. 2, 1935, pp. 511–517, pl. XVII.

Nagatoella, a New Genus of Permian Fusulinids. — Journ. Geol. Soc. Japan, vol. 43, No. 510, 1936, pp. 195–202, pl. 12 (2).

Fusulinids from the Black Hills and Adjacent Areas in Wyoming. — Journ. Pal., vol. 10, 1936, pp. 95–113, pls. 13–16.

Pennsylvanian Fusulinids from Ohio. — l. c., pp. 673–683, pls. 90, 91.

The Fusulinid Genus *Verbeekina*. — l. c., pp. 193–201, pl. 24.

Lower Permian Fusulinids from Sumatra. — l. c., pp. 587–592, figs. 1–13.

Fusulinids of the Subfamily Schubertellinae. — l. c., vol. 11, 1937, pp. 118–125, pl. 22.

New Genera of Pennsylvanian Fusulinids. — Amer. Journ. Sci., vol. 240, 1942, pp. 403–420, pls. 1–3.

Pennsylvanian Rocks and Fusulines of East Utah and Northwest Colorado Correlated with Kansas Section. — State Geol. Survey of Kansas, Bull. 60, 1945 Reports of Studies, pt. 2, 1945, pp. 17–84, pls. 1–6, text figs. 1–11.

Permian Fusulinids from Afghanistan. — Journ. Pal., vol. 20, 1946, pp. 140–157, pls. 23–26, text fig. 1.

THOMPSON, M. L., and C. L. FOSTER. Middle Permian Fusulinids from Szechuan, China. — l. c., vol. 11, 1937, pp. 126–144, pls. 23–25.

THOMPSON, M. L., and A. K. MILLER. *Schwagerina* from the Western Edge of the Red Basin, China. — l. c., vol. 9, 1935, pp. 647–652, pl. 79.

The Permian of southernmost Mexico and its fusulinid faunas. — l. c., vol. 18, 1944, pp. 481–504, pls. 79–84.

THOMPSON, M. L., and H. E. WHEELER. Permian fusulinids from British Columbia, Washington, and Oregon. — l. c., vol. 16, 1942, pp. 700–711, pls. 105–109, 2 text figs.

WARTHIN, A. D., JR. Micropaleontology of the Wetumka, Wewoka, and Holdenville Formations. — Bull. 53, Oklahoma Geol. Surv., 1930, pp. 1–94, pls. 1–7, and map.

WATERS, J. A. A Group of Foraminifera from the Dornick Hills Formation of the Ardmore Basin. — Journ. Pal., vol. 1, 1927, pp. 129–133, pl. 22.

A Group of Foraminifera from the Canyon Division of the Pennsylvanian Formation in Texas. — l. c., vol. 1, 1928, pp. 271–275, pl. 42.

WHITE, M. P. Some Texas Fusulinidae. — Univ. Texas Bull. 3211, 1932, pp. 1–104, pls. 1–10, text figs. 1–3.

Some Fusulinid Problems. — Journ. Pal., vol. 10, 1936, pp. 123–133, pls. 18–20.

WOOD, A. The Supposed Cambrian Foraminifera from the Malverns. — Quart. Journ. Geol. Soc. London. vol. 102, 1947, pp. 447–460, pls. 26–28.

YABE, H. A Contribution to the Genus Fusulina, with Notes on a Fusulina-Limestone from Korea. — Journ. Coll. Sci. Imp. Univ., vol. 21, art. 5, 1906, pp. 1–36, pls. 1–3.

Structurproblem der Fusulinenschale. — Beitr. z. Geol. u. Pal. Oesterreich-Ungarns, vol. 23, 1910, pp. 273–282.

YABE, H., and S. HANZAWA. Tentative Classification of the Foraminifera of the Fusulinidae. — Proc. Imp. Acad. Japan, vol. 8, No. 2, 1932, pp. 40–44.

Foraminifera? Remains from Ordovician Limestone of Manchuria. — l. c., vol. 11, 1935, pp. 55–57, text figs. 1–3.

VIII. MORPHOLOGY, TECHNIQUE, ETC.

ADAMS, B. C. Distribution of Foraminifera of the Genus *Bolivina* in Canada de Aliso, Ventura County, California. — Amer. Journ. Sci., vol. 237, 1939, pp. 500–511.

ASANO, K. On the Japanese Species of *Elphidium* and Its Allied Genera. — Journ. Geol. Soc. Japan, vol. 45, 1938, pp. 581–591, pl. 14 (3).

On the Japanese Species of *Nonion* and Its Allied Genera. — l. c., pp. 592–599, pl. 15 (4).

On the Japanese Species of *Bolivina* and Its Allied Genera. — l. c., pp. 600–609, pl. 16 (5).

On the Japanese Species of *Uvigerina* and Its Allied Genera. — l. c., pp. 609–18, pl. 17 (6).

BAKX, L. A. J. De Genera Fasciolites en Neoalveolina in het Indo-Pacifische Gebied. — Verhandl. Geol.-Mijn. Gen. Ned. Kolonien, Geol. Ser., Deel 9, 1932, pp. 205–266, pls. 1–4, text figs.

BONTE, A. Foraminifères à structure organique conservée. — Annales Protistologie, vol. 5, 1936, pp. 139–149, pl. 5, text figs. 1–24.

BROECK, E. VANDEN. Étude sur le Dimorphisme des Foraminifères et des Nummulites en particulier. — Bull. Soc. Belge Géol., vol. 7, 1893, pp. 6–41.

CARPENTER, W. B. On the Microscopic Structure of Nummulina, Orbitolites, and Orbitoides. — Quart. Journ. Geol. Soc., vol. 6, 1849, pp. 21–39, pls. 3–8.

Researches in the Foraminifera. Pt. I. Containing general Introduction and Monograph of the Genus Orbitolites. — Phil. Trans., vol. 146, 1856, pp. 181–236, pls. 4–9; Pt. II. On the Genera Orbiculina, Alveolina, Cycloclypeus, and Heterostegina. — l. c., pp. 547–569, pls. 28–31; Pt. III. On the Genera Peneroplis, Operculina, and Amphistegina. — l. c., vol. 149, 1860, pp. 1–41, pls. 1–6; Fourth and concluding Series Contains Polystomella, Calcarina, Tinoporus, Carpenteria, and summary. — l. c., vol. 150, 1861, pp. 535–594, pls. 17–22. Supplemental Memoir. On the Abyssal type of the genus Orbitolites; a Study in the Theory of Descent. — l. c., vol. 174, 1883, pp. 551–573, pls. 37, 38.

Report on the specimens of the genus Orbitolites collected by H. M. S. *Challenger* during the years 1873–1876. — Rep. Voy. *Challenger,* Zoology, vol. 7, pt. 21, 1883, pp. 1–47, pls. 1–8.

On the Structure of Orbitolites. — Journ. Quekett Micr. Club, ser. 2, 1885, pp. 91–104, figs.

CARPENTER, W. B., and H. B. BRADY. Description of Parkeria and Loftusia, two gigantic types of arenaceous Foraminifera. — Phil. Trans., 1869, pp. 721–754, pls. 72–80.

CARPENTER, W. B., W. K. PARKER and T. R. JONES. Introduction to the Study of the Foraminifera. — Ray Soc., *London*, 1862, 319 pp., 22 pls.

CARSON, C. M. A Method of Concentrating Foraminifera. — Journ. Pal., vol. 7, 1933, p. 439.

CAUDRI, C. M. B. De Foraminiferen-Fauna van eenige *Cycloclypeus*-houdende Gesteenten van Java. — Verhandl. Geol.-Mijn. Gen. Nederland en Koloniën, Geol. Ser., Deel 9, 1932, pp. 171–204, pls. 1–3.

CHAPMAN, F. On Dimorphism in the Recent Foraminifer, Alveolina boscii Defr. sp. — Journ. Roy. Micr. Soc., 1908, pp. 151–153, pls. 2, 3.

The Importance of Foraminifera in Modern Geological Work. — Micr. Soc. Victoria, vol. 8, 1938, pp. 24–30.

COLE, W. S. Internal Structure of some Floridian Foraminifera. — Bull. Amer. Pal., vol. 31, No. 126, 1947, pp. 1–30, pls. 1–5, text fig. 1, 1 table.

CORDINI, I. R. Algunas ideas para la manipulacion de foraminiferos. — Rev. Contr. Est. Ciencias Naturales, vol. 1, 1935, pp. 20–27, 2 pls., 2 text figs.

COSIJN, A. J. Statistical Studies on the Phylogeny of Some Foraminifera. *Cycloclypeus* and *Lepidocyclina* from Spain, *Globorotalia* from the East-Indies. — Leiden, 1938, 70 pp., 5 pls., 12 text figs.

On the Phylogeny of the Embryonic Apparatus of Some Foraminifera. — Leidsche Geol. Med., vol. 13, pt. 1, 1942, pp. 140–171, text figs., tables.

CULLISON, J. S. A Suitable Tray for Comparative Examination of Minute Opaque Objects under the Binocular Microscope. — Journ. Pal., vol. 8, 1934, p. 247.

CUSHMAN, J. A. Apertural Characters in the Lagenidae. — Contr. Cushman Lab. Foram. Res., vol. 4, 1928, pp. 22–25, pl. 3.

Fistulose Species of *Gaudryina* and *Heterostomella*. — l. c., pp. 107–112, pl. 16.

The Term "Arenaceous Foraminifera" and Its Meaning. — l. c., vol. 5, 1929, pp. 25–27.

The Microspheric and Megalospheric Forms of *Valvulina oviedoiana* d'Orbigny. — l. c., vol. 7, 1931, pp. 17, 18, pl. 3.

The Megalospheric and Microspheric Forms of *Frondicularia sagittula* van den Broeck and their Bearing on Specific Descriptions. — l. c., vol. 19, 1943, pp. 25, 26, pls. 5, 6 (part).

CUSHMAN, J. A., and R. TODD. Statistical Studies of Some Bolivinas. — l. c., vol. 17, 1941, pp. 29–31, pls. 5–8.

CUSHMAN, J. A., and W. C. WARNER. A Preliminary Study of the Structure of the Test in the So-called Porcellanous Foraminifera. — l. c., vol. 16, 1940, pp. 24–26, pl. 4.

DAVIES, L. M. An Early *Dictyoconus*, and the Genus *Orbitolina*: their Contemporaneity, Structural Distinction, and Respective Natural Allies. — Trans. Roy. Soc. Edinburgh, vol. 59, pt. III (N. 29), 1939, pp. 773–790, pls. I, II.

DETTMER, F. Über das Variieren der Foraminiferengattung *Frondicularia* Defr. — Neues Jahrb. für Min., 1911, pp. 149–159, pl. 12.

DREYER, F. Die Principien der Gerüstbildung bei Rhizopoden, Spongien und Echinodermen. — Zeitschr. Nat. Jena, vol. 26, 1892, pp. 294–296, 5 pls.; pp. 297–468, 10 pls.

Peneroplis. Eine Studie zur biologischen Morphologie und zur Speciesfrage. — Leipzig, 1898, 119 pp., 5 pls., 25 text figs.

DRIVER, H. L. An Aid in Disintegrating Samples for Micro-Organic Study. — Journ. Pal., vol. 1, 1928, pp. 253, 254.

DOUVILLÉ, H. Sur la Structure des Orbitolines. — Bull. Soc. Géol. France, ser. 4, vol. 4, 1905, pp. 653–661, pl. 17.

DUNBAR, C. O. The Geologic and Biologic Significance of the Evolution of the Fusulinidae. — Trans. New York Acad. Sci., ser. 2, vol. 7, No. 3, 1945, pp. 57–60.

EARLAND, A. Some Notes on Selective Building. — Trans. Hertfordshire Nat. Hist. Soc., vol. 21, 1939, pp. 106–113.

FRANKE, A. Die Präparation von Foraminiferen und anderen mikroskopischen Tierresten. — In Keilhack, Praktische Geologie, vol. 2, pp. 509–576.

Die Trennung der Mikrofossilien aus sandigen Schlämmrückständen mit Tetrachlorkohlenstoff. — Zeitschr. für Ges., Bd. 6, Heft 4, 1930, pp. 162–164.

Ein einfacher Auslesetisch für Mikrofossilien. — Senckenbergiana, Bd. 17, No. ½, 1935, pp. 87–89, text figs.

Sammeln, Präparieren und Aufbewahren von Mikrofossilien. — l. c., No. ¾, 1935, pp. 124–137, figs. 1–6 in text.

A simple apparatus for sorting microfossils. — Journ. Pal., vol. 13, 1939, pp. 225–227, 2 text figs.

GUBLER, J. Structure et Sécrétion du Test des Fusulinidés. — Annales de Protistologie, vol. 4, 1934, separate, pp. 1–24, 15 text figs.

HAYASAKA, I. A Twinned or Double Fossil Shell of *Rotalia.* — Trans. Pal. Soc. Japan, No. 7, 1936 (Journ. Geol. Soc. Japan, vol. 43, No. 508, 1936), pp. 5–7, text figs. 1*a–c.*

HECHT, F. Arbeitsweisen der Mikropaläontologie. — Senkenbergiana, vol. 15, 1933, pp. 346–362, 11 text figs.

Einfache Geräte zum Fotografieren von Mikrofossilien insbesondere Foraminiferen. — l. c., vol. 16, 1934, pp. 65–77, text figs. 1–11.

Eine neue Verteilungszelle zum Aufbewahren von Mikrofossilien, vornehmlich Foraminiferen. — l. c., pp. 152–155, 2 text figs.

HENBEST, L. G. The Use of Selective Stains in Paleontology. — Journ. Pal., vol. 5, 1931, pp. 355–364.

A New Term for the Youthful Stage of Foraminiferal Shells. — Science, vol. 79, No. 2051, 1934, pp. 363, 364.

Keriothecal Wall-Structure in *Fusulina,* and Its Influence on Fusuline Classification. — Journ. Pal., vol. 11, 1937, pp. 212–230, pls. 34, 35.

HENSON, F. R. S. Stratigraphical Correlation by Small Foraminifera in Palestine and Adjoining Countries. — Geol. Mag., vol. 75, 1938, pp. 227–233.

HICKSON, S. J. On *Gypsina plana,* and on the Systematic Position of the Stromatoporoids. — Quart. Journ. Micr. Sci., vol. 76, pt. 3, 1934, pp. 433–480, pls. 26, 27, 13 text figs.

HODSON, F., and H. K. Short Cuts in Picking out and Sectioning Foraminifera. — Bull. Amer. Assoc. Petr. Geol., vol. 10, 1926, pp. 1173, 1174.

HOFKER, J. Preliminary Note on a Statistic Statement of Trimorphism in *Biloculina sarsi* Schlumberger. — Tijdschr. Ned. Dierkundige Vereeniging, ser. 3, vol. 2, 1931, pp. 179–184, 2 text figs.

Une analyse du foraminifère fossile *Orthophragmina advena,* Cushman. — Ann. Prot., vol. 3, 1932, pp. 209–215, pl. 20.

HUCKE, K. Über die Gewinnung von Mikrofossilien aus Geschieben. — Zeitschr. für Ges., Bd. 9, Heft 1, 1933, pp. 42–48, 3 text figs.

LACROIX, E. Du choix des Coccolithes par les Foraminifères Arénacés pour l'Édification de leur tests. — Compte rendu Congrès Lyon Assoc. Française Avan. Sci., 1926 (1927), pp. 418–421, text figs. 1–9.

Sur la texture du test de *Textularia sagittula* Defrance. — Compte rendus des séances Acad. Sci., vol. 184, 1927, pp. 1202, 1203.

Microtexture du test des Textularidae. — Bull. Inst. Océanographique, No. 582, 1931, pp. 1–18, 11 text figs.

Le pseudomorphisme chez les Textularidae. — l. c., No. 622, May 20, 1933, pls. 1–12, 10 text figs.

Le Calvez, J. Embryons à cinq loges de *Planorbulina mediterranensis* (d'Orb.) et trimorphisme de cette espèce. — Bull. Soc. Zool. France, vol. 59, 1934, pp. 284–290, 4 text figs.

Liebus, A. The Variability of *Vulvulina pennatula* Batsch. — Journ. Pal., vol. 6, 1932, pp. 208–210, text figs. 1–8.

Lister, J. J. On the Dimorphism of the English species of *Nummulites*, and the Size of the Megalosphere in Relation to that of the Microspheric and Megalospheric Tests in This Genus. — Proc. Roy. Soc., vol. B 76, 1905, pp. 298–319, pls. 3–5.

Munier-Chalmas and C. Schlumberger. Note sur les Miliolidées trematophorées. — Bull. Soc. Géol. France, ser. 3, vol. 13, 1885, pp. 273–323, pls. 13, 14, 14 bis, 45 text figs.

Nicol, D. New West American Species of the Foraminiferal Genus *Elphidium*. — Journ. Pal., vol. 18, 1944, pp. 172–185, pl. 29, 7 text figs.

Nuttall, W. L. F. Micropaléontologie Appliquée aux parallélisations géologiques en Amérique. — Congrès International des Mines, de la Métallurgie et de la Géologie appliquée. VII, Session Paris 20–26, Oct. 1935, pp. 413–418.

Plummer, H. J. Structure of *Ceratobulimina*. — Amer. Midland Nat., vol. 17, 1936, pp. 460–463, 1 pl.

Reichel, M. Sur la structure des Alveolines. — Eclog. geol. Helv., vol. 24, 1931, pp. 289–303, pls. 13–18.

Rhumbler, L. Der Aggregatzustand und die physikalischen Besonderheiten des lebenden Zellinhaltes. — Zeitschr. allgem. Phys., pt. 1, vol. 1, 1902, pp. 279–388, 31 text figs.; pt. 2, vol. 2, pp. 183–340, 1 pl., 80 text figs.

Die Doppelschalen von Orbitolites und anderer Foraminiferen von entwicklungsmechanischen standpunkt aus betrachtet. — Arch. Prot., vol. 1, 1902, pp. 193–296, pls. 7, 8, 17 text figs.

Rijsinge, C. P. I. van. Description of Some Foraminifera of a Boring near Bunde (Dutch South-Limburg) with a Discussion of the Theories of Trimorphism and Dimorphism in Foraminifera. — 8 vo, pp. 1–112, pls. 1–4, text figs., 1 chart, 1932. Press of L. Gerretsen, Den Haag.

Rutten, M. G. Über Stolonen bei *Lepidorbitoides socialis* (Leymerie). — "De Ingenieur in Nederlandsch-Indië," IV. Mijnbouw Geol., Jaarg. III, 1936, pp. 82–84, text figs. 1, 2.

On an interseptal Canal-system in the foraminiferal Species *Discocyclina papyracea* (Boubée). — Proc. Roy. Acad. Amsterdam, vol. 39, 1936, pp. 413–418, pl. I, text figs. 1–7.

Scheffen, W. De Fylogenetische Beteekenis der Bestanddeelen van het Foraminiferen skelet. — De Mijningenieur, No. 11, 1931, pp. 206–208, 7 text figs.

Zur Morphologie und Morphogenese der "Lepidocyclinen." — Pal. Zeitschr., vol. 14, 1932, pp. 233–256, pls. 9, 10, text figs. 1–6.

Die besonderen Vorteile der Transparent-Zelle bei der Untersuchung von Klein- und Gross-Foraminiferen. — Senckenbergiana, vol. 19, 1937, pp. 193–200, text figs. 1–5.

Neuere Methoden und Erfolge der Micro-paläontologie. — Oel und Kohle vereinigt mit Erdoel und Teer, 13 Jahrg., Heft 21, 1937, pp. 483–486, 11 text figs.

SCHENCK, H. G., and B. C. ADAMS. Operations of Commercial Micropaleontologic Laboratories. — Journ. Pal., vol. 17, 1943, pp. 554–583, pl. 97, 13 text figs.

SCHENCK, H. G., and S. E. AGUERREVERE. Morphologic Nomenclature of Orbitoidal Foraminifera. — Amer. Journ. Sci., ser. 5, vol. 11, 1926, pp. 251–256, 3 text figs.

SCHLUMBERGER, C. Première Note sur les Orbitoides. — Bull. Soc. Géol. France, ser. 4, vol. 1, 1901, pp. 459–467, pls. 7–9. Deuxième Note sur les Orbitoides. — l. c., vol. 2, 1902, pp. 255–261, pls. 6–8, 3 text figs. Troisième Note sur les Orbitoides. — l. c., vol. 3, 1903, pp. 273–289, pls. 8–12, 6 text figs. Quatrième Note sur les Orbitoides. — l. c., vol. 4, 1904, pp. 119–135, pls. 3–6, 6 text figs.

SCHULTZE, M. S. Ueber den Organismus der Polythalamien (Foraminiferen) nebst Bemerkungen über die Rhizopoden im Allgemeinen. — 4to, *Leipzig*, 1854, 68 pp., 7 pls.

SECRIST, M. H. Technique for the Recovery of Paleozoic Arenaceous Foraminifera. — Journ. Pal., vol. 8, 1934, pp. 245, 246.

SILVESTRI, A. Osservazione critiche sul genere *Baculogypsina* Sacco. — Atti Accad. Pont. Lincei, vol. 58, 1905, pp. 65–82, 8 text figs.

Considerazioni palaeontologiche e morphologiche sui generi *Operculina*, *Heterostegina*, *Cycloclypeus*. — Boll. Soc. Geol. Ital., vol. 26, 1907, pp. 29–62, pl. 2.

Ortostilia e flessostilia nei Rhizopodi reticolari. — Atti Pont. Accad. Nuovi Lincei, 1919, pp. 30–70, text figs.

Lo stipite delle Ellissoforme e le sue affinita. — Mem. Pont. Accad. Nuovi Lincei, ser. 2, vol. 6, 1923, pp. 231–270, pl. 1.

Sul Genere "Lepidorbitoides" A. Silvestri e di un Suo Nuovo giacimento. — l. c., vol. 10, 1927, pp. 109–140, pl. 1.

STAFF, H. VON. Die Anatomie und Physiologie der Fusulinen. — Zoologica, Bd. 22, Heft 58, Lief. 6, 1910, pp. 1–93, 2 pls., 62 text figs.

TAN, S. H. Zur Theorie des Trimorphismus und zum Initialpolymorphismus der Foraminiferen. — Natuurk. Tijdschrift 3e Afl. van Deel 95, 1935, pp. 171–188.

Die Peri-embryonalen Aquatorialkammern bei einigen Orbitoididen. — "De Ingenieur in Nederlandsch-Indië," IV. Mijnbouw en Geologie, 2 de Jaargang, No. 12, 1935, pp. 113–126, pl. 1.

TODD, J. U. Metagenesis in *Lepidocyclina* from the Eocene of Peru. — Geol. Mag., vol. 70, 1933, pp. 393–396.

VAN DER VLERK, I. M. The Task of the Oil Paleontologist. — Geologie & Mijnbouw, No. 2, 1933, 4 pp., 4 text figs., and chart.

VAUGHAN, T. W. The Biogeographic Relations of the Orbitoid Foraminifera. — Proc. Nat. Acad. Sci., vol. 19, 1933, pp. 922–938.

IX. LIVING ANIMAL

CHAPMAN, F. Notes on the Appearance of Some Foraminifera in the Living Condition, from the Challenger Collection. — Proc. Roy. Soc. Edinburgh, 1899–1901, pp. 391–395, pls. 1–3.

CUSHMAN, J. A. Observations on Living Specimens of *Iridia diaphana*, a Species of Foraminifera. — Proc. U. S. Nat. Mus., vol. 57, 1920, pp. 153–158, pls. 19–21.

The Interrelation of Foraminifera and Algae. — Journ. Washington Acad. Sci., vol. 20, 1930, pp. 395, 396.

HERON-ALLEN, E. Contributions to the Study of the Bionomics and Reproductive Processes of the Foraminifera. — Philos. Trans., vol. 206, 1915, pp. 227–279, pls. 13–18.

The Further and Final Researches of Joseph Jackson Lister upon the Reproductive Processes of *Polystomella crispa* (Linné). (An unpublished paper completed and edited from his notebooks.) — Smithsonian Misc. Coll., vol. 82, No. 9, 1930, pp. 1–11, pls. 1–7.

HERON-ALLEN, E., and A. EARLAND. An Experimental Study of the Foraminiferal Species *Verneuilina polystropha* (Reuss), and some others, being a Contribution to a Discussion "On the Origin, Evolution, and Transmission of Biological Characters." — Proc. Roy. Irish Acad., vol. 35, 1920, pp. 153–177, pls. 16–18, text figs.

HOFKER, J. On Heterogamy in Foraminifera. — Tijdschr. Ned. Dierk. Vereen, ser. 2, vol. 19, 1925, pp. 68–70.

Der Generationswechsel von *Rotalia beccarii*, var. *flevensis*, nov. var. — Zeitschr. Zellforsch. mikr. Anat., vol. 10, 1930, pp. 756–768, 1 text fig.

JEPPS, M. W. Contribution to the Study of *Gromia oviformis* Dujardin. — Quart. Journ. Micr. Sci., vol. 70, n. ser., 1926, pp. 701–719, pls. 37–39.

Studies on *Polystomella* Lamarck. — Journ. Marine Biol. Assoc. United Kingdom, vol. 25, 1942, pp. 607–666, pls. 4, 5, 10 text figs.

KOFOID, C. A. *An Interpretation of the conflicting views as to the life cycle of the Foraminifera.* — Science, vol. 79, No. 2054, 1934, p. 436.

LE CALVEZ, J. Les gamètes de quelques Foraminifères. — Comptes Rendus des Séances de l'Académie des Sciences, vol. 201, Dec. 23, 1935, pp. 1505–7 (1–3), text figs. 1–3.

Modifications du test des Foraminifères pélagiques en rapport avec la reproduction: *Orbulina universa* d'Orb. et *Tretomphalus bulloides* d'Orb. — Ann. Prot., vol. 5, 1936, pp. 125–133, text figs. 1–8.

Observations sur le Genre *Iridia.* — Archiv. Zool. Exper. Gen., vol. 78, 1936, pp. 115–131, pl. 1, text figs. I–VII.

Processus schizogoniques chez le Foraminifère *Planorbulina mediterranensis* d'Orb. — Comptes rendus Séances Acad. Sci., vol. 204, Jan. 11, 1937, pp. 147–149, text figs. 1–4.

Les Chromosomes spiraux de la première mitose schizogonique du Foraminifère *Patellina corrugata* Will. — l. c., vol. 205, Nov. 29, 1937, pp. 1106–1108, 6 text figs.

LISTER, J. J. Contributions to the Life-History of the Foraminifera. — Stud. Mar. Lab., vol. 6, 1894, pp. 108–180, pls. 5–8.

MYERS, E. H. Multiple Tests in the Foraminifera. — Proc. Nat. Acad. Sci., vol. 19, 1933, pp. 893–899.

The Life History of *Patellina corrugata* Williamson, a Foraminifer. — Bull. Scripps Inst. Oceanography, Tech. Ser., vol. 3, 1935, pp. 355–392, pls. 10–16, 1 text fig.

Morphogenesis of the Test and the Biological Significance of Dimorphism in the Foraminifer *Patellina corrugata* Williamson. — l. c., pp. 393–404, 1 text fig.

The Life-Cycle of *Spirillina vivipara* Ehrenberg, with notes on Morphogenesis, Systematics and Distribution of the Foraminifera. — Journ. Roy. Micr. Soc., vol. 56, 1936, pp. 120–146, pls. 1–3.

The Present State of our Knowledge Concerning the Life Cycle of the Foraminifera. — Proc. Nat. Acad. Sci., vol. 24, 1938, pp. 10–17, text figs. A–C.

A Quantitative Study of the Productivity of the Foraminifera in the Sea. — Proc. Amer. Philos. Soc., vol. 85, No. 4, 1942, pp. 325–342, 1 pl., 6 text figs.

Life Activities of Foraminifera in Relation to Marine Ecology. — l. c., vol. 86, No. 3, 1943, pp. 439–458, 1 pl., 7 text figs.

Biology, Ecology, and Morphogenesis of a Pelagic Foraminifer. — Stanford Univ. Publ., Univ. Ser., Biol. Sci., vol. 9, No. 1, 1943, pp. 1–30, pls. 1–4.

Rhumbler, L. Beiträge zur Kenntniss der Rhizopoden. II. Saccammina sphaerica M. Sars. — Zeitschr. Wiss. Zool., vol. 57, 1893, pt. 1, pp. 433–586, pls. 21–24; pt. 2, pp. 587–617, pl. 25.

Winter, F. W. Zur Kenntnis der Thalamophoren. I. Untersuchung über Peneroplis pertusus (Forskal). — Arch. Prot., vol. 10, 1907, pp. 1–113, pls. 1, 2, text figs.

X. FORAMINIFERA AS RELATED TO OCEANOGRAPHY

Chapman, F. The Foraminifera of the Funafuti Boring. — Ann. Mag. Nat. Hist., ser. 11, vol. 11, 1944, pp. 98–110.

Crickmay, G. W., H. S. Ladd, and J. E. Hoffmeister. Shallow-water *Globigerina* Sediments. — Bull. Geol. Soc. Amer., vol. 52, 1941, pp. 79–106, pls. 1, 2, 4 text figs.

Cushman, J. A. A Study of the Foraminifera Contained in Cores from the Bartlett Deep. — Amer. Journ. Sci., vol. 239, 1941, pp. 128–147, pls. 1–6, text figs. 1–10.

Cushman, J. A., and L. G. Henbest. Geology and Biology of North Atlantic Deep-Sea Cores between Newfoundland and Ireland. Part 2, Foraminifera. — U. S. Geol. Survey Prof. Paper 196–A, 1940, pp. 35–56, pls. 8–10, text figs. (charts) 11–21.

Ladd, H. S. *Globigerina* Beds as Depth Indicators in the Tertiary Sediments of Fiji. — Science, n. ser., vol. 83, No. 2152, 1936, pp. 301, 302.

Myers, E. H. Biological Evidence as to the Rate at which Tests of Foraminifera are Contributed to Marine Sediments. — Journ. Pal., vol. 16, 1942, pp. 397, 398, 1 text fig.

Natland, M. L. The Temperature and Depth-Distribution of Some Recent and Fossil Foraminifera in the Southern California Region. — Bull. Scripps Inst. Oceanography, Tech. Ser., vol. 3, 1933, pp. 225–230, table.

Norton, R. D. Ecologic Relations of Some Foraminifera. — l. c., vol. 2, 1930, pp. 331–388.

Parker, F. L. Foraminifera of the Continental Shelf from the Gulf of Maine to Maryland. — Bull. Mus. Comp. Zoöl., vol. 100, No. 2, 1948, pp. 213–241, pls. 1–7, text figs. 1–4, tables 1–10.

Phleger, F. B., Jr. Foraminifera of Submarine Cores from the Continental Slope. — Bull. Geol. Soc. Amer., vol. 50, 1939, pp. 1395–1422, pls. 1–3, 4 text figs.

Foraminifera of Submarine Cores from the Continental Slope, Part 2. — l. c., vol. 53, 1942, pp. 1073–1097, pls. 1–3, text figs. 1–6.

Vertical Distribution of Pelagic Foraminifera. — Amer. Journ. Sci., vol. 243, 1945, pp. 377–383, text figs. 1, 2.

Foraminifera of three submarine cores from the Tyrrhenian Sea (with a foreword by Hans Pettersson). — Meddelanden från Oceanografiska Institutet I Göteborg. 13. (Göteborgs Kungl. Vetenskaps- och Vitterhets-Samhälles Handlinger. Sjätte Följden. ser. B, Band 5, No. 5), 1947, pp. 1–19, 2 tables.

Phleger, F. B., Jr., and W. A. Hamilton. Foraminifera of Two Submarine Cores from the North Atlantic Basin. — Bull. Geol. Soc. Amer., vol. 57, 1946, pp. 951–966, pl. 1, tables 1–3, figs. 1–3.

Schenck, H. G. The Biostratigraphic Aspect of Micropaleontology. — Journ. Pal., vol. 2, 1928, pp. 158–165, pl. 26.

Stubbings, H. G. The Marine Deposits of the Arabian Sea; an Investigation into their Distribution and Biology. — British Museum (Nat. Hist.), John Murray Expedition 1933–34, Sci. Rept., vol. 3, No. 2, 1939, pp. 31–158, 8 pls., 5 text figs., 20 tables.

Stratification of Biological Remains in Marine Deposits. — l. c., No. 3, 1939, pp. 159–192, 4 text figs., 8 tables.

Stubbs, S. A. Studies of Foraminifera from seven Stations in the Vicinity of Biscayne Bay. — Proc. Florida Acad. Sci., vol. 4, 1939 (1940), pp. 225–230, 2 tables.

Vaughan, T. W. The Biogeographic Relations of the Orbitoid Foraminifera. — Proc. Nat. Acad. Sci., vol. 19, 1933, pp. 922–938.

Ecology of Modern Marine Organisms with Reference to Paleogeography. — Bull. Geol. Soc. Amer., vol. 51, 1940, pp. 433–468, 8 text figs.

XI. CLASSIFICATION AND GENERAL

Abrard, R. Description d'une nouvelle variété de Nummulite. — Mém. Soc. Sci. Nat. Maroc., vol. 37, 1933, pp. 58, 59, pl. 12, figs. 6–9.

Barker, R. W. Some Notes on the Genus *Helicolepidina* Tobler. — Journ. Pal., vol. 8, 1934, pp. 344–351, pl. 47, text figs. 1*a–e*.

Species of the Foraminiferal Family Camerinidae in the Tertiary and Cretaceous of Mexico. — Proc. U. S. Nat. Mus., vol. 86, No. 3052, 1939, pp. 305–330, pls. 11–22.

Bermudez, P. J. Un Genero Y Especie Nueva de Foraminiferos Vivientes de Cuba. — Mem. Soc. Cubana Hist. Nat., vol. 8, 1934, pp. 83–86, 3 text figs.

Foraminíferos del Género *Recurvoides*, descripción de una Especie Nueva. — l. c., vol. 13, 1939, pp. 57–62, pl. 5.

Nuevo género y especies nuevas de Foraminíferos. — l. c., pp. 247–252, pl. 33.

Brady, H. B., W. K. Parker and T. R. Jones. A Monograph of the Genus *Polymorphina*. — Trans. Linn. Soc., London, vol. 27, 1870, pp. 197–253, pls. 39–42, text figs.

Brotzen, F. Die Foraminiferengattung *Gavelinella* nov. gen. und die Systematik der Rotaliiformes. — Sveriges Geologiska Undersökning, Ser. C, No. 451, 1942, pp. 1–60, pl. 1, 18 text figs.

Bütschli, O., in Bronn, Klassen und Ordnungen des Thier-Reichs, Wissenschaftlich dargestellt in Wort und Bild. Bd. 1. Protozoa. Neu bearbeitet. — 8vo, *Leipzig* and *Heidelberg*. Lief. 1–7 (1880); Lief. 8, 9 (1881).

Chapman, F. The Foraminifera. — 8vo, Longmans, Green & Co., 1902, 354 pp., 14 pls., 42 text figs.

Chapman, F., and W. J. Parr. A Classification of the Foraminifera. — Proc. Roy. Soc. Victoria, vol. 49, Pt. 1 (New Ser.), 1936, pp. 139–151.

Colom, G. Estudios sobre las Calpionelas. — Bol. Soc. Española Hist. Nat., vol. 34, 1934, pp. 379–388, pls. XXX–XXXII, text figs. 1, 2.

Las especies de la familia *Peneroplidae* actuales y fósiles de las Baleares. — l. c., vol. 35, 1935, pp. 83–102, pls. VI–XVI, 5 text figs.

Cushman, J. A. An Introduction to the Morphology and Classification of the Foraminifera. — Smithsonian Misc. Coll., vol. 77, No. 4, 1925, pp. 1–77, pls. 1–16, 11 text figs.

Some Palaeontologic Evidence Bearing on a Classification of the Foraminifera. — Amer. Journ. Sci., vol. 13, 1927, pp. 53–56.

Phylogenetic Studies of the Foraminifera. Part I. — l. c., vol. 13, 1927, pp. 315–326, text figs. Part II. — l. c., vol. 14, 1927, pp. 317–324, 24 figs.

An Outline of a Re-Classification of the Foraminifera. — Contr. Cushman Lab. Foram. Res., vol. 3, 1927, pp. 1–105, pls. 1–22.

Foraminifera of the Genus *Siphonina* and Related Genera. — Proc. U. S. Nat. Mus., vol. 72, Art. 20, 1927, pp. 1–15, pls. 1–4.

Notes on Foraminifera in the Collection of Ehrenberg. — Journ. Washington Acad. Sci., vol. 17, 1927, pp. 487–491.

Some Notes on the Genus *Ceratobulimina*. — Contr. Cushman Lab. Foram. Res., vol. 3, 1927, pp. 171–179, pls. 29, 30.

Epistomina elegans (d'Orbigny) and *E. partschiana* (d'Orbigny). — l. c., pp. 180–187, pls. 31, 32.

Additional Genera of the Foraminifera. — l. c., vol. 4, 1928, pp. 1–8, pl. 1.

Foraminifera, Their Classification and Economic Use. — l. c., Special Publ. No. 1, 1928, pp. 1–401, pls. 1–59, text figs. and charts.

The Genus *Bolivinella* and Its Species. — l. c., vol. 5, 1929, pp. 28–34, pl. 5.

The Genus *Trimosina* and Its Relationships to Other Genera of the Foraminifera. — Journ. Washington Acad. Sci., vol. 19, 1929, pp. 155–159, 3 text figs.

The Development and Generic Position of *Sagrina* (?) *tessellata* H. B. Brady. — l. c., pp. 337–339, text figs. 1–5.

Notes on the Genus *Virgulina*. — Contr. Cushman Lab. Foram. Res., vol. 8, 1932, pp. 7–23, pls. 2, 3.

The Genus *Vulvulina* and Its Species. — l. c., pp. 75–85, pl. 10.

Textularia and Related Forms from the Cretaceous. — l. c., pp. 86–97, pl. 11.

The Relationships of *Textulariella* and Description of a New Species. — l. c., pp. 97, 98, pl. 11.

Some New Foraminiferal Genera. — l. c., vol. 9, 1933, pp. 32–38, pls. 3, 4.

Relationships and Geologic Distribution of the Genera of the Valvulinidae. — l. c., pp. 38–44, table.

Two New Genera, *Pernerina* and *Hagenowella*, and Their Relationships to Other Genera of the Valvulinidae. — Amer. Journ. Sci., vol. 26, 1933, pp. 19–26, pls. 1, 2.

Foraminifera, Their Classification and Economic Use. Second Edition. Revised and Enlarged. — Cushman Lab. Foram. Res., Special Publ. No. 4, 1933, pp. i–viii, 1–349, pls. 1–31, text figs., charts, maps.

An Illustrated Key to the Genera of the Foraminifera. — l. c., No. 5, 1933, pages, pls. 1–40.

Notes on the Genus *Tretomphalus*, with Descriptions of Some New Species and a New Genus, *Pyropilus*. — Contr. Cushman Lab. Foram. Res., vol. 10, 1934, pp. 79–101, pls. 11–13.

The Relationships of *Ungulatella*, with Descriptions of Additional Species. — l. c., pp. 101–104, pl. 13 (pt.).

Some New Species of *Nonion*. — l. c., vol. 12, 1936, pp. 63–69, pl. 12.

Some New Species of *Elphidium* and Related Genera. — l. c., pp. 78–89, pls. 13 (pt.), 14, 15.

A Monograph of the Foraminiferal Family Verneuilinidae. — Special Publ. No. 7, Cushman Lab. Foram. Res., 1937, pp. i-xiii, 1–157, pls. 1–20.

A Monograph of the Foraminiferal Family Valvulinidae. — l. c., No. 8, pp. i–xiii, 1–210, pls. 1–24.

A Monograph of the Subfamily Virgulininae of the Foraminiferal Family Buliminidae. — l. c., No. 9, pp. i–xv, 1–228, pls. 1–24.

Cretaceous Species of *Gümbelina* and Related Genera. — Contr. Cushman Lab. Foram. Res., vol. 14, 1938, pp. 2–28, pl. 1 (pt.), 2–4.

A Monograph of the Foraminiferal Family Nonionidae. — U. S. Geol. Survey Prof. Paper 191, 1939, pp. 1–100, pls. 1–20.

Cushman, J. A., and P. G. Edwards. *Astrononion*, a New Genus of the Foraminifera and Its Species. — Contr. Cushman Lab. Foram. Res., vol. 13, 1937, pp. 29–36, pl. 3 (pt.).

Notes on the Early Described Eocene Species of *Uvigerina* and Some New Species. — l. c., pp. 54–61, pl. 8.

Notes on the Oligocene Species of *Uvigerina* and *Angulogerina*. — l. c., vol. 14, 1938, pp. 74–89, pls. 13 (pt.), 14, 15.

Notes on Early Described Miocene Species of *Uvigerina*. — l. c., vol. 15, 1939, pp. 33–40, pl. 8.

Cushman, J. A., and S. Hanzawa. Notes on Some of the Species Referred to *Vertebralina* and *Articulina*, and a New Genus of *Nodobaculariella*. — l. c., vol. 13, 1937, pp. 41–46, pl. 5 (pt.).

Cushman, J. A., and L. W. Le Roy. *Cribrolinoides*, a New Genus of the Foraminifera, its Development and Relationships. — l. c., vol. 15, 1939, pp. 15–19, pls. 3, 4.

Cushman, J. A., and Y. Ozawa. An Outline of a Revision of the Polymorphinidae. — l. c., vol. 4, 1928, pp. 13–21, pl. 2.

A Revision of Polymorphinidae. — Jap. Journ. Geol. Geogr., vol. 6, 1929, pp. 79–83, text figs. 1, 2.

A Monograph of the Foraminiferal Family Polymorphinidae, Recent and Fossil. — Proc. U. S. Nat. Mus., vol. 77, Art. 6, 1930, pp. 1–185, pls. 1–40, text figs. 1, 2.

Cushman, J. A., and F. L. Parker. Bulimina and Related Foraminiferal Genera. — U. S. Geol. Survey Prof. Paper 210–D, 1947, pp. 55–176, pls. 15–30.

Defrance, J. L. M. Articles on various Genera of Foraminifera in Dictionnaire des Sciences Naturelles. — 8vo, *Strassburg*, 1816–1830.

Douvillé, H. Essai d'une revision des Orbitolites. — Bull. Soc. Géol. France, ser. 4, vol. 2, 1902, pp. 289–306, pls. 9, 10, 7 text figs.

Evolution et enchainments des Foraminifères. — l. c., vol. 6, 1907, pp. 588–602, pl. 18.

Revision des Lepidocyclines. — Mém. Soc. Géol. France, n. ser., vol. 1, Mém. 2, 1924, pp. 1–49, pls. 5, 6, 48 text figs.; vol. 2, Mém. 2, 1925, pp. 51–115, pls. 3–7, 82 text figs.

Ehrenberg, C. G. Microgeologische Studien über das kleinste Leben der Meeres-Tiefgründe aller Zonen und dessen geologischen Einfluss. — Abhandl. k. Akad. Wiss. Berlin, 1872 (1873), pp. 131–397, pls. 1–12, map.

Eimer, G. H. T., and C. Fickert. Die Artbildung und Verwandtschaft bei den Foraminiferen. Entwurf einer natürlichen Eintheilung derselben. — Zeitschr. Wiss. Zool., vol. 65, 1899, pp. 599–708, 45 text figs.

Fichtel, L. von, and J. P. C. von Moll. Testacea microscopica alaique minuta ex generibus Argonauta et Nautilus ad naturam delineata et descripta. — 4to, *Wien*, 1798, 24 pls.; second ed., 1803.

Folin, L. Les Bathysiphons; premières pages d'une monographie du Genre. — Actes Soc. Linn. Bordeaux, ser. 4, vol. 10, livr. 5, 1887, pp. 271–291; livr. 6, 1888, pls. 5–8.

Galloway, J. J. A Manual of Foraminifera. — 4to, Bloomington, Ind., 1933, pp. i–xii, 1–483, 42 pls.

Glaessner, M. F. Die Entfaltung der Foraminiferenfamilie Buliminidae. — Problems of Paleontology, Moscow Univ., vols. 2, 3, 1937, pp. 411–423, 2 text figs.

On a New Family of Foraminifera. — Studies in Micropaleontology, vol. 1, 1937, pp. 19–29, pls. I, II.

Principles of Micropaleontology. — Melbourne Univ. Press, 1945, pp. i–xvi, 1–296, pls. 1–14, tables 1–7, text figs. 1–64.

HERON-ALLEN, E. Alcide d'Orbigny, His Life and His Work. — Journ. Roy. Micr. Soc., 1917, pp. 1–105, pls. 1–13.

HERON-ALLEN, E., and A. EARLAND. On the Pegididae, a New Family of Foraminifera. — l. c., vol. 48, 1928, pp. 283–299, 3 pls.

HOFKER, J. On *Faujasina* d'Orbigny. — Contr. Cushman Lab. Foram. Res., vol. 4, 1928, pp. 80, 81, pl. 11.

HUCKE, K. Ein Beitrag zur Phylogenie der Thalamophoren. — Arch. Prot., vol. 9, 1907, pp. 33–52, 2 text figs.

JEDLITSCHKA, H. Über *Candorbulina,* eine neue Foraminiferen-Gattung, und zwei neue *Candeina-Arten.* — Verhandl. Nat. Ver. Brunn, 65 J. 1934 (1933), pp. 17–26, text figs. 1–23.

LACROIX, E. Sur une texture méconnue de la coquille de diverses Massilines des mers tropicales. — Bull. Instit. Océanographique, No. 750, 1938, pp. 1–8, text figs. 1–4.

LAMARCK, J. B. P. A. M. Système des Animaux sans vertèbres, ou tableau général des classes, des ordres et des genres de ces animaux. — 8vo, *Paris,* 1801, pp. 100–103.

Histoire naturelle des Animaux sans Vertèbres. — 8vo, *Paris,* vol. 2, 1816, pp. 193–197; vol. 7, 1822, pp. 580–632.

LE CALVEZ, J. Un Foraminifères géant *Bathysiphon filiformis* G. O. Sars. — Archives de Zoologie Expérimentale et Générale, vol. 79, 1937, pp. 82–88, text figs. I, II.

Recherches sur les Foraminifères. 1. Développement et reproduction. — l. c., vol. 80, 1938, pp. 163–333, pls. II–VII, text figs. 1–26.

LE CALVEZ, Y. Révision des Foraminifères Lutétiens du Bassin de Paris, I. Miliolidae. — Mémoires pour servir à l'explication de la carte géologique détaillée de la France, Paris, 1947, pp. 1–41, pls. 1–4.

LIEBUS, A. Zur Stammesgeschichte der Foraminiferen. — Pal. Zeitschr., vol. 10, 1928, pp. 130–135, text figs. 1–15.

LISTER, J. J. The Foraminifera. — In Lankester, "A Treatise on Zoology," pt. 1, fasc. 2, 1903, pp. 47–149, 59 text figs.

MONTFORT, P. D. Conchyliologie systématique et classification méthodique des Coquilles, 2 vols. — 8vo, *Paris,* 1808–1810, figs.

NEUMAYR, M. Die natürlichen Verwandschaftsverhältnisse der schalentragenden Foraminiferen. — Sitz. Akad. Wiss. Wien, vol. 95, 1887, pp. 156–186, table.

ORBIGNY, A. D. D'. Tableau Méthodique de la Classe des Céphalopodes. — Ann. Sci. Nat., vol. 7, 1826, pp. 245–314, pls. 10–17.

Modèles de Céphalopodes microscopiques vivans et fossiles, etc. — *Paris,* 1826.

OZAWA, Y. On the Classification of the Fusulinidae. — Journ. Coll. Sci. Imp. Univ. Tokyo, vol. 45, art. 4, 1925, pp. 1–26, pls. 1–4.

PALMER, D. K. Cuban Foraminifera of the Family Valvulinidae. — Mem. Soc. Cubana Hist. Nat., vol. 12, 1938, pp. 281–301, pls. 19–23.

REICHEL, M. Étude sur les Alvéolines. — Mem. Soc. pal. Suisse, vols. 57 (pt.), 59 (pt.), 1936–1937, pp. 1–147, pls. I–XI, 29 text figs.

REUSS, A. E. Entwurf einer systematischen Zusammenstellung der Foraminiferen. — Sitz. Akad. Wiss. Wien, vol. 44, 1861, pp. 355–396.

Die Foraminiferen-Familie der Lagenideen. — l. c., vol. 46, 1862 (1863), pp. 303–342, pls. 1–7.

RHUMBLER, L. Entwurf eines natürlichen Systems der Thalamophoren. — Nachr. k. Ges. Wiss. Göttingen, Math. Phys. Kl., 1895, pp. 51–98.

Ueber die phylogenetisch abfallende Schalen-Ontogenie der Foraminiferen und deren Erklärung. — Verh. deutsch. Zool. Ges., 1897, pp. 162–192, 21 text figs.

RIJSINGE, C. P. I. VAN. Some Remarks on *Dictyoconoides*, Nuttall (= *Conulites*, Carter = *Rotalia*, Lamarck). — Ann. Mag. Nat. Hist., ser. 10, vol. 5, 1930, pp. 116–135, pls. 5, 6, 12 text figs.

RUTTEN, M. G. Zur Einführung geographischer Rassenkreise bei fossilen Foraminiferen. Antwort an Hans E. Thalmann. — Pal. Zeitschr., vol. 17, 1935, pp. 257–262.

SCHUBERT, R. J. Beiträge zu einer natürlicher Systematik der Foraminiferen. — Neues Jahrb. für Min., vol. 25, 1907, pp. 233–260, text fig.

Palaeontologische Daten zur Stammesgeschichte der Protozoen. — Pal. Zeitschr., vol. 3, 1920, pp. 129–188.

SENN, A. Die stratigraphische Verbreitung der tertiären Orbitoiden, mit spezieller Berücksichtigung ihres Vorkommens in Nord-Venezuela und Nord-Marokko. — Eclogae geologicae Helvetiae, vol. 28, No. 1, 1935, pp. 51–113.

Nachtrag zu: Die stratigraphische Verbreitung der tertiären Orbitoiden. — l. c., pp. 369–373, pls. VIII, IX.

SILVESTRI, A. Osservazioni su fossili Nummulitici. — Riv. Ital. Pal., Anno 35, 1929, pp. 1–21, pls. 1–3.

Sul modo do presentarsi delle Alveoline Eoceniche nei loro giacimenti primari. — Mem. Pont. Accad. Sci. Nuovi Lincei, ser. 2, vol. 12, 1929, pp. 465–492, 3 pls., 6 text figs.

Sul genre *Chapmanina* e sulla *Alveolina maiellana*, n. sp. — Boll. Soc. Geol. Ital., vol. 50, 1931, pp. 63–73, pl. 1.

Sul Modo di Presentarsi di Alveoline Eoceniche in loro giacimento secondario. — Mem. Pont. Accad. Sci. Nuovi Lincei, vol. 15, 1931, pp. 203–231, pl. 1, text figs. 1–3.

TAN, S. H. Vindplaatsen van *Globotruncana* Cushman in West-Borneo. — Natuurk. Tijdschrift le Alf. van Deel 96, 1936, pp. 14–18, text figs. 1–8.

On *Polylepidina*, *Orbitocyclina* and *Lepidorbitoides*. — "De Ingenieur in Nederlandsch-Indië," IV. Mijnbouw Geol., Jaargang VI, No. 5, May 1939, pp. 53–84, pls. I, II.

THALMANN, H. E. Die Foraminiferen-Gattung *Hantkenina* Cushman 1924 und ihre regionalstratigraphische Verbreitung. — Eclog. geol. Helv., vol. 25, 1932, pp. 287–292.

Zwei neue Vertreter der Foraminiferen-Gattung *Rotalia* Lamarck 1804: *R. cubana* nom. nov. und *R. trispinosa* nom. nov. — vol. 26, 1933, pp. 248–251, pl. XII.

Die Regional-stratigraphische Verbreitung der oberkretazischen Foraminiferen-Gattung *Globotruncana* Cushman, 1927. — Eclogae geologicae Helvetiae, Bd. 27, 1934, pp. 413–428, 1 text fig.

VAUGHAN, T. W. Studies of American Species of Foraminifera of the Genus *Lepidocyclina*. — Smithsonian Misc. Coll., vol. 89, No. 10 (Publ. 3222), 1933, pp. 1–53, pls. 1–32.

VIENNOT, P. Sur la valeur paleontologique et stratigraphique d'*Orbitolina subconcava* Leymerie. — Comptes Rendus, Soc. Géol. France, No. 6, 1929, pp. 75–77.

WENGEN, W. A. Phylogenetic Considerations of the Nummulinidae. — Proc. Fourth Dutch East Indian Congress of Natural Science, Weltvreden (Java), Sep. 22–26, 1926, Geographic-Geologic Section, 1927, pp. 448–466, 3 charts.

XII. NOMENCLATURE

BARKER, R. W. On *Camerina petri* M. G. Rutten and *Nummulites striatoreticulatus* L. Rutten. — Geol. Mag., vol. 75, 1938, pp. 49–51, pl. III.

CHAPMAN, F., W. HOWCHIN, and W. J. PARR. A Revision of the Nomenclature of the Permian Foraminifera of New South Wales. — Proc. Roy. Soc. Victoria, vol. 47 (n. ser.), pt. 1, 1934, pp. 175–189, 5 text figs.

CORYELL, H. N. *Textularia hockleyensis*, var. *malkinae* Coryell and Embich, A New Name for *Textularia hockleyensis*, var. *panamensis* Coryell and Embich. — Journ. Pal., vol. 11, 1937, p. 714.

CUSHMAN, J. A. Some Notes on the Early Foraminiferal Genera Erected before 1808. — Contr. Cushman Lab. Foram. Res., vol. 3, 1927, pp. 122–126, pl. 24.
Notes on the Collection of Defrance. — l. c., pp. 141–145, pl. 28.
The Work of Fichtel and Moll and of Montfort. — l. c., pp. 168–171.
The Designation of Some Genotypes in the Foraminifera. — l. c., pp. 188–190.
Notes on Foraminifera in the Collection of Ehrenberg. — Journ. Washington Acad. Sci., vol. 17, 1927, pp. 487–491.
On *Rotalia beccarii* (Linné). — Contr. Cushman Lab. Foram. Res., vol. 4, 1928, pp. 103–107, pl. 15.
On *Quinqueloculina seminula* (Linné). — l. c., vol. 5, 1929, pp. 59, 60, pl. 9.
Planulina ariminensis d'Orbigny and *P. wuellerstorfi* (Schwager). — l. c., pp. 102–105, pl. 15.
On *Uvigerina pigmea* d'Orbigny. — l. c., vol. 6, 1930, pp. 62, 63, pl. 9.
A Résumé of New Genera of the Foraminifera Erected Since Early 1928. — l. c., pp. 73–94, pls. 10–12.
Some Notes on the Genus *Flabellinella* Schubert. — l. c., vol. 7, 1931, pp. 16, 17, pl. 3.
Notes on the Foraminifera Described by Batsch in 1791. — l. c., pp. 62–72, pls. 8, 9.
Some Notes on d'Orbigny's Models. — l. c., vol. 9, 1933, pp. 70–73.
The Generic Position of "*Cornuspira cretacea* Reuss." — l. c., vol. 10, 1934, pp. 44–47.
Notes on the Genus *Spiroplectoides* and Its Species. — l. c., pp. 37–44, pl. 6 (pt.).
Some New Names in the Foraminifera. — l. c., vol. 14, 1938, pp. 28, 29.
Marginulina texasensis Cushman, A New Name. — l. c., p. 95.

CUSHMAN, J. A., and D. H. LEAVITT. On *Elphidium macellum* (Fichtel and Moll), *E. striato-punctatum* (Fichtel and Moll) and *E. crispum* (Linné). — l. c., vol. 5, 1929, pp. 18–22, pl. 4.

CUSHMAN, J. A., and F. L. PARKER. Notes on Some of the Earlier Species Originally Described as *Bulimina*. — l. c., vol. 10, 1934, pp. 27–36, pls. 5, 6.
The Recent Species of *Bulimina* Named by d'Orbigny in 1826. — l. c., vol. 14, 1938, pp. 90–94, pl. 16.
Bulimina macilenta Cushman and Parker, a New Name. — l. c., vol. 15, 1939, pp. 93, 94.

DUNBAR, C. O., and L. G. HENBEST. The Fusulinid Genera *Fusulina*, *Fusulinella* and *Wedekindella*. — Amer. Journ. Sci., vol. 20, 1930, pp. 357–364, text fig.

DUNBAR, C. O., and J. W. SKINNER. *Schwagerina* versus *Pseudoschwagerina* and *Paraschwagerina*. — Journ. Pal., vol. 10, 1936, pp. 83–91, pls. 10, 11.

FORNASINI, C. Il *Nautilus Legumen* di Linneo e la *Vaginulina elegans* di d'Orbigny. — Boll. Soc. Geol. Ital., vol. 5, 1886, pp. 25–30, pl. 1.
Foraminiferi illustrati da Soldani e citati delgi Autori. — l. c., pp. 131–254.

(Many of the tracings of the originals of d'Orbigny's Planches inédités, 1826, were published by Fornasini in a series of papers and are given here. They afford an excellent means of knowledge of d'Orbigny's early species.) (Papers in Mem. Accad. Sci. Istit. Bologna.) Indice ragionato delle Rotaliine fossili d'Italia spettanti ai generi *Truncatulina, Planorbulina, Anomalina, Rotalia* e *Discorbina.* — ser. 5, vol. 7, 1898, pp. 239–290, text figs. Globigerine adriatiche. — 1899, pp. 575–586, pls. 1–4. Le Polystomelline fossili d'Italia. — pp. 639–660, 5 text figs. Intorno ad alcuni esemplari di Foraminiferi adriatici, — vol. 8, 1900, pp. 357–402, 50 text figs. Intorno a la nomenclatura di alcuni nodosaridi neogenici italiani. — vol. 9, 1901, pp. 45–76, 27 text figs. Sinnosi metodica dei Foraminiferi sin qui rinvenuti nella sabbia del Lido di Rimini. — vol. 10, 1902, pp. 1–68, 63 text figs. Illustrazione di Specie orbignyane di Foraminiferi institute nel 1826. — ser. 6, vol. 1, 1904, pp. 3–17, pls. 1–4. Illustrazioni di specie orbignyane di Miliolidi institute nel 1826. — vol. 2, 1905, pp. 59–70, 4 pls. Illustrazione di specie orbignyane di Rotalidi institute nel 1826. — vol. 3, 1906, pp. 61–70, 4 pls. Illustrazione di specie orbignyane di Nodosaridi, Rotalidi e d'altri Foraminiferi, institute, nel 1826. — vol. 5, 1908, pp. 41–54, 3 pls. (Also in Rend. Accad. Sci. Istit. Bologna) Sopra alcune Specie di "*Globigerina*" institute da d'Orbigny nel 1826. — vol. 7, 1903, pp. 139–142, 1 pl. Le otto pretese specie di "*Amphistegina*" institute da d'Orbigny nel 1826. — pp. 142–145, 1 pl. (Also in Pal. Ital.) Le Globigerine fossili d'Italia. — vol. 4, 1899, pp. 203–216, 5 figs. (Also in Bull. Soc. Geol. Ital.) Le Polimorfine e le Uvigerine fossili d'Italia. — vol. 19, 1900, pp. 132–172, 7 figs. Le Bulimine e le Cassiduline fossili d'Italia. — vol. 20, 1901, pp. 159–214, 5 figs. Illustrazione di specie orbignyane di Nummulitidi institute nel 1826. — vol. 22, 1903, pp. 395–398, 1 pl. (Also in Riv. Ital. Pal.) La *Clavulina cylindrica* di A. D. d'Orbigny. — vol. 3, 1897, pp. 13, 14, figs. La "*Biloculina alata*" di A. D. d'Orbigny. — vol. 5, 1899, pp. 23, 24, 3 figs. Intorno ad alcune specie di "*Textilaria*" institute da d'Orbigny nel 1826. — vol. 7, 1901, pp. 1–3, 1 pl. Intorno ad alcune specie di "*Polymorphina*" institute da d'Orbigny nel 1826. — vol. 8, 1902, pp. 1–13, 9 figs. Sopra tre specie di "*Textilaria*" del pliocene italiano institute da d'Orbigny nel 1826. — pp. 44–47, 3 figs.

Galloway, J. J. A Revision of the Family Orbitoididae. — Journ. Pal., vol. 2, 1928, pp. 45–69, text figs.

Garrett, J. B. Use of the name *Marginulina mexicana.* — l. c., vol. 13, 1939, p. 622.

Glaessner, M. F. Die Foraminiferengattungen *Pseudotextularia* und *Amphimorphina.* — Problems of Paleontology, vol. 1, 1936, pp. 95–134, pls. I, II, text figs. 1, 2.

Henbest, L. G. The Species *Endothyra baileyi* (Hall). — Contr. Cushman Lab. Foram. Res., vol. 7, 1931, pp. 90–93, pls. 11, 12.

Heron-Allen, E. The Genus *Keramosphaera,* Brady. — Journ. Roy. Micr. Soc., vol. 56, 1936, pp. 113–119, pl. I.

Howe, H. V. The Foraminiferal Genus *Palmula* Isaac Lea, 1833. — Journ. Pal., vol. 10, 1936, pp. 415, 416, text figs. 1, 2.

Jedlitschka, H. Revision der Foraminiferen-gattungen *Siphonodosaria, Nodogenerina, Sagrinnodosaria.* — Verhandl. Nat. Ver. Brunn, vol. 66, 1935, pp. 61–72, 3 text figs.

Koch, R. E. Namensänderung einiger Tertiär-Foraminiferen aus Niederländisch Ost-Indien. — Eclogae geol. Helvetiae, vol. 28, 1935, pp. 557, 558.

Lacroix, E. *Textularia sagittula* ou *Spiroplecta Wrightii* (?) — Bull. Inst. Océanographique, No. 532, 1929, pp. 1–12, text figs. 1–12.

Revision du genre *Massilina.* — l. c., No. 754, 1938, pp. 1–11, text figs. 1–9.

MACFADYEN, W. A. Modern Studies of the Foraminifera. — Nature, vol. 141, 1938, pp. 750–751.

On *Ophthalmidium*, and Two New Names for Recent Foraminifera of the Family *Ophthalmidiidae*. — Journ. Roy. Micr. Soc., ser. 3, vol. 59, 1939, pp. 162–169, 3 text figs.

MACFADYEN, W. A., and E. J. ANDRÉ KENNY. On the Correct Writing, in Form and Gender, of the Names of the Foraminifera. — Journ. Roy. Micr. Soc., vol. 54, 1934, pp. 177–181.

OVEY, C. D. Difficulties in Establishing Relationships in the Foraminifera. — Proc. Geol. Assoc., vol. 49, 1938, pp. 160–170, pls. 8, 9, text figs. 29–32.

OZAWA, Y. On *Guttulina lactea* (Walker and Jacob), *Polymorphina burdigalensis* d'Orbigny and *Pyrulina gutta* d'Orbigny. — Contr. Cushman Lab. Foram. Res., vol. 5, 1929, pp. 34–39, pl. 6.

PARKER, W. K., and T. R. JONES. On the Nomenclature of the Foraminifera. I. On the Species enumerated by Linnaeus and Gmelin. — Ann. Mag. Nat. Hist., ser. 3, vol. 3, 1859, pp. 474–482. II. On the Species enumerated by Walker and Montagu. — l. c., vol. 4, 1859, pp. 333–351. III. The Species enumerated by Von Fichtel and Von Moll. — l. c., vol. 5, 1860, pp. 98–116, 174–183. IV. The Species enumerated by Lamarck. — l. c., pp. 285–298, 466–477; vol. 6, 1860, pp. 29–40. V. The Foraminifera enumerated by Denys de Montfort. — l. c., pp. 337–347. VI. *Alveolina.* — l. c., vol. 8, 1861, pp. 161–168. VII. *Operculina* and *Nummulina.* — l. c., pp. 229–238. VIII. *Textularia.* — l. c., vol. 9, 1863, pp. 91–98, figs. IX. The Species enumerated by De Blainville and Defrance. — l. c., vol. 12, 1863, pp. 200–219. X. The Species enumerated by d'Orbigny in the Annales des Sciences Naturelles, vol. vii, 1826. — l. c., pp. 429–441. XV. The Species figured by Ehrenberg. — l. c., vol. 9, 1872, pp. 211–230, 280–303; vol. 10, 1872, pp. 184–200, 253–271; App., pp. 453–457. PARKER, W. K., T. R. JONES and H. B. BRADY. XI. The Species enumerated by Batsch in 1791. — l. c., vol. 15, 1865, pp. 225–232. XII. The Species enumerated by d'Orbigny in the Annales des Sciences Naturelles, vol. vii, 1826. (3) The Species illustrated by Modèles — l. c., vol. 16, 1865, pp. 15–41, pls. 1–3. XIV. (4) The Species founded upon the Figures in Soldani's Testaceographia ac Zoophytographia. — l. c., ser. 4, vol. 8, 1871, pp. 145–179, 238–266, pls. 8–12. JONES, T. R., W. K. PARKER, and J. W. KIRKBY. XIII. The Permian *Trochammina pusilla* and its Allies. — l. c., ser. 4, vol. 4, 1869, pp. 386–392, pl. 13.

SCHENCK, H. G., and D. L. FRIZZELL. Subgeneric Nomenclature in Foraminifera. — Amer. Journ. Sci., vol. 31, 1936, pp. 464–466.

SILVESTRI, A. Intorno all' *Alveolina melo* d'Orbigny (1846). — Riv. Ital. Pal., Anno 34, 1928, pp. 17–44, pls. 1–4, 1 text fig.

Protozoi cretacei ricordati e figurati da B. Faujas de Saint-Fond. — Atti Pont. Accad. Sci. Nuovi Lincei, Anno 82, 1929, pp. 325–343, 1 pl., 9 text figs.

Sulla Validita del Genere "*Fusulina*" Fischer. — Boll. Soc. Geol. Ital., vol. 54, 1935, pp. 203–219, pl. 9, text figs. 1–3.

THALMANN, H. E. Nomenclator (Um- und Neubenennungen) zu den Tafeln 1 bis 115 in H. B. Brady's Werk über die Foraminiferen der Challenger-Expedition, London, 1884. — Eclog. geol. Helv., vol. 25, 1932, pp. 293–312.

Validité du nom générique "*Globotruncana* Cushman, 1927." — Comptes Rendus Séances Soc. Géol. France, Séance, 6 Nov. 1933, No. 13, pp. 200, 201.

Nachtrag zum Nomenclator zu Brady's Tafelband der Foraminiferen der "Challenger"-Expedition. — Eclogae geologicae Helvetiae, vol. 26, 1933, pp. 251–255.

Mitteilungen über Foraminiferen I. — l. c., vol. 27, 1934, pp. 428–440, 1 plate, 5 text figs.

Mitteilungen über Foraminiferen II. — l. c., vol. 28, 1935, pp. 592–606, 2 text figs.

Mitteilungen über Foraminiferen III. 9, Über *Polystomella bolivinoides* Schubert, 1911. 10, Über das Genus *Staffia* Schubert, 1911. 11, Weitere Nomina mutata in Brady's Werk über die Foraminiferen der "Challenger"-Expedition (1884). 12, Zwei Nomina Conservanda: *Nummulites* Lamarck, 1801, una *Cristellaria* Lamarck, 1812. 13, Notizen zur Systematik der Gattung *Uvigerina* d'Orbigny, 1826. 14, Bemerkungen zu den Gattungen *Vaginulinopsis* Silvestri, 1904, *Marginulinopsis* Silvestri, 1904 und *Hemicristellaria* Stache, 1864. — l. c., vol. 30, 1937, pp. 337–356, pls. XXI–XXIII.

Mitteilungen über Foraminiferen IV. 16, Bemerkungen zur Frage des Vorkommens kretazischer Nummuliten. 17, Wert und Bedeutung morphogenetischer Untersuchungen an Gross-Foraminiferen für die Stratigraphie. 18, Stratigraphische Verbreitung der Familien und Genera der Foraminiferen. 19, Foraminiferen-Statistik. — l. c., vol. 31, 1938, pp. 327–344.

Foraminiferal Homonyms. — Amer. Midland Nat., vol. 28, 1942, pp. 457–462.

Nomina Bradyana Mutata. — l. c., pp. 463, 464.

THOMPSON, M. L. The Genotype of *Fusulina*, S. S. — Amer. Journ. Sci., vol. 32, 1936, pp. 287–291.

UMBGROVE, J. H. F. A New Name for the Foraminiferal Genus *Heterospira.* — Leidsche Geologische Mededeelingen, vol. 8, 1937 (Jan. 15, 1937), p. 155.

VAUGHAN, T. W. Notes on the Types of *Lepidocyclina mantelli* (Morton) Gümbel and on Topotypes of *Nummulites floridanus* Conrad. — Proc. Acad. Nat. Sci. Philadelphia, vol. 79, 1927, pp. 299–303.

A Note on the Names *Cyclosiphon* Ehrenberg, 1856, and *Lepidocyclina* Gümbel, 1868. — Journ. Pal., vol. 3, 1929, pp. 28, 29.

XIII. BIBLIOGRAPHY

BARKER, R. W. Micropaleontology in Mexico with Special Reference to the Tampico Embayment. — Bull. Amer. Assoc. Petr. Geol., vol. 20, 1936, pp. 433–456, text figs. 1, 2.

BEUTLER, K. Paläontologisch-Stratigraphische und Zoologisch-Systematische Literatur über marine Foraminiferen fossil und rezent bis Ende 1910. — 8vo, *München*, pp. 1–144.

CUSHMAN, J. A. A Bibliography of American Foraminifera. — Special Publ. No. 3, Cushman Lab. Foram. Res., 1932, pp. 1–40.

ELLIS, B. F. and A. R. MESSINA. A catalogue of Foraminifera. — American Museum of Natural History, New York.

JAWORSKI, E. Bibliographia Palaeontologica für die Jahre 1914–1926 by E. Jaworski. Erste Lieferung. Allgemein Paläontologie. Spezielle Paläozoologie: Protozoa, Coelenterata, Echinodermata, Molluscoides, Mollusca (ausser Cephalopoda). (Published by Max Weg, Leipzig, Germany. Price: 18 Marks.) The part on foraminifera takes up pages 53–77 with papers numbered 781–1152 arranged under various groupings.

LIEBUS, A. Fossilium Catalogus, I: Animalia. Editus a W. Quenstedt. Pars 49: A. Liebus. Bibliographia foraminiferum recentium et fossilium II. (1911–1930). — W. Junk, Berlin, 1931, pp. 1–36.

LIEBUS, A., and H. E. THALMANN. Fossilium Catalogus. I. Animalia. Editus a W. Quenstedt. Pars 59: Bibliographia foraminiferum recentium et fossilium I (–1910). 1933, pp. 1–179; Pars 60: III (1911–1930, Supplementum). 1933, pp. 1–28.

RUTTEN, L. M. R. Bibliography of West Indian Geology. — Geog. Geol. Med., Phisiografisch-geologische Reeks. No. 16, 1938, pp. 1–103.

SHERBORN, C. D. A Bibliography of the Foraminifera, Recent and Fossil, from 1565 to 1888. — *London*, 1888.

An Index to the Genera and Species of the Foraminifera. Pts. 1, 2. — Smithsonian Misc. Coll., Publ. No. 856, 1893, 1896.

SILVESTRI, A. Bibliografia delle Fusulinidi. — Mem. Pont. Accad. Sci., Nuovi Lincei, vol. 17, 1933, pp. 523–554.

THALMANN, H. E. A series of papers entitled "Bibliography and Index to New Genera, Species, and Varieties of Foraminifera" appearing at intervals in Journ. Pal., as follows:

Year 1931. — Journ. Pal., vol. 7, 1933, pp. 343–355; vol. 8, 1934, pp. 238–244.
Year 1932. — vol. 8, 1934, pp. 356–387.
Year 1933. — vol. 9, 1935, pp. 715–743.
Year 1934. — vol. 10, 1936, pp. 294–322.
Year 1935. — vol. 12, 1938, pp. 177–208.
Year 1936. — vol. 13, 1939, pp. 425–465.
Years 1937 and 1938. — vol. 15, 1941, pp. 629–690.
Year 1939. — vol. 16, 1942, pp. 489–520.
Year 1940. — vol. 17, 1943, pp. 388–408.
Year 1941. — vol. 18, 1944, pp. 387–404.
Year 1942. — vol. 19, 1945, pp. 396–410.
Year 1943. — vol. 19, 1945, pp. 648–656.
Year 1944. — vol. 20, 1946, pp. 172–183.
Year 1945 (with supplements). — vol. 20, 1946, pp. 591–619; vol. 21, 1947, pp. 278–281; 355–395.

Foraminifera. — Fortschritte der Paläontologie, Bd. 1, 1937, pp. 66–82.

TOUTKOWSKY, P. Index bibliographique de la litterature sur les Foraminifères vivants et fossiles (1888–1898). — Mém. Soc. Nat. Kiev, vol. 16, 1898, pp. 137–240.

WINTER, F. W. Foraminifera (Testacea reticulosa) für 1891–1895. — Arch. Nat., Jahrb. 1901, Bd. 2, Heft 3, 1901, pp. 37–146; 1896–1900. — l. c., Jahrg. 1905, Bd. 2, Heft 3, 1905, pp. 1–78; 1901–1905. — l. c., Jahrg. 1908, Bd. 2, Heft 3, 1908, pp. 1–61, suppl. 1–8.

WOODWARD, A. The Bibliography of the Foraminifera, Recent and Fossil, including Eozoön and Receptaculites. 1565–Jan. 1, 1886. — Fourteenth Ann. Rep't Geol. Nat. Hist. Surv. Minnesota, 1885 (1886), pp. 167–311.

AN ILLUSTRATED KEY TO THE GENERA OF THE FORAMINIFERA

AN ILLUSTRATED KEY TO THE GENERA OF THE FORAMINIFERA

THE KEYS AND PLATES in this Key section should be found useful by those studying the foraminifera or making use of them for economic purposes. The section contains, first, a key to the families of the foraminifera; second, a key to the genera under each family; and, finally, a series of plates illustrating the various genera. After the family name in the first key will be found references to the plate or plates in this section where the various genera of the family are illustrated. At the head of each key to a family will be found a reference to the page of the text where the family is treated as well as references to the plates, in both text and Key, where various genera of the family are illustrated. After the genus name in the later keys will be found a reference to the page of the text where the more detailed description of the genus is given, together with data concerning its geologic range, etc.

The keys are intended only to give clues to the families and genera, since abbreviated keys, especially with the foraminifera, cannot give the full characters. It is expected that the plates will prove more useful than the keys. The explanations of the plates give the age and locality of fossil specimens, and usually the magnification. The black backgrounds make the specimens appear much as they do under a microscope when seen against a black slide or tray. Figures are given of both fossil and recent forms where space is available, and often of other species than those illustrated in the text. Text and Key plates should be used together.

I. A KEY TO THE FAMILIES OF THE FORAMINIFERA

I. Test wanting or of thin chitinous material, living forms in fresh or brackish water or in the ocean. .... Family 1. *Allogromiidae.*
(On Plate 8 in the text will be found figures of some of the genera of this family.)

II. Test wholly or in part arenaceous.
- A. Test single chambered or rarely an irregular group of similar chambers loosely attached.
  - 1. Test with a central chamber and usually two or more arms.
    Family 2. *Astrorhizidae.* Key, pls. 1, 42.
  - 2. Test without a central chamber, elongate, open at both ends.
    Family 3. *Rhizamminidae.* Key, pls. 1, 42.
  - 3. Test a chamber, or rarely a series of similar chambers loosely attached, with normally a single opening.
    Family 4. *Saccamminidae.* Key, pls. 2, 42.
- B. Test two chambered, a proloculum and long undivided tubular second chamber.

1. Test with the second chamber simple or branching, not coiled.
Family 5. *Hyperamminidae.* Key, pls. 3, 42.
2. Test with the second tubular chamber usually coiled at least in the young.
a. Test of arenaceous material with much cement, usually yellowish- or reddish-brown. ...Family 7. *Ammodiscidae.* Key, pls. 4, 42.
b. Test of siliceous material, second chamber partially divided.
Family 16. *Silicinidae.* Key, pls. 13, 46.

C. Test typically many chambered.
1. Test with all the chambers in a rectilinear series.
Family 6. *Reophacidae.* Key, pls. 3, 42.
2. Test planispirally coiled at least in the young.
a. Axis of coiling short, many uncoiled forms.
Family 8. *Lituolidae.* Key, pls. 4–6, 40, 42, 43.
b. Axis of coiling usually long, all close coiled.
(1). Interior not labyrinthic, late Paleozoic.
(a). Septa folded, except in most primitive genera; without external aperture, but with a tunnel.
Family 12. *Fusulinidae.* Key, pls. 10, 11, 44, 45.
(b). Septa not folded; septula may subdivide chambers; aperture a row of round foramina; without tunnel.
Family 13. *Neoschwagerinidae.* Key, pl. 12.
(2). Interior labyrinthic, Cretaceous.
Family 14. *Loftusiidae.* Key, pl. 13.
3. Test typically biserial at least in the young of the microspheric form.
Family 9. *Textulariidae.* Key, pls. 7, 43, 46.
4. Test typically triserial at least in the young of the microspheric form, aperture usually without a tooth.
Family 10. *Verneuilinidae.* Key, pls. 7, 8, 43.
5. Test typically multiserial at least in the young of the microspheric form, aperture typically with a tooth.
Family 11. *Valvulinidae.* Key, pls. 5, 8, 9, 43, 46.
6. Test with whole body labyrinthic, large, flattened or cylindrical.
Family 15. *Neusinidae.* Key, pl. 13.
7. Test trochoid at least in the young.
a. Mostly free and typically trochoid throughout.
Family 20. *Trochamminidae.* Key, pls. 18, 40, 47, 48.
b. Attached, young trochoid, later stages variously formed.
Family 21. *Placopsilinidae.* Key, pls. 19, 48.
c. Free, conical, mostly of large size.
Family 22. *Orbitolinidae.* Key, pl. 19.
8. Test coiled in varying planes, wall imperforate, with arenaceous portion only on the exterior.
Family 17. *Miliolidae* (in part). Key, pls. 14, 15, 46, 47.

III. Test calcareous, imperforate, porcellanous.
A. Test with the chambers coiled in varying planes, at least in the young, aperture large, toothed.
Family 17. *Miliolidae* (in part). Key, pls. 14, 15, 46, 47.
B. Test trochoid. ....Family 19. *Fischerinidae.* Key, pl. 17.
C. Test planispiral, at least in the early stages.
1. The axis very short, chambers usually simple.
Family 18. *Ophthalmidiidae.* Key, pls. 16, 17, 47.

2. The axis short, test typically compressed and often discoid, chambers mostly with many chamberlets.
   Family 27. *Peneroplidae.* Key, pls. 24, 25, 40, 49.
3. The axis typically elongate, chamberlets developed.
   Family 28. *Alveolinellidae.* Key, pls. 25, 49, 50.

D. Test globular, apertures small, not toothed.
   Family 29. *Keramosphaeridae.* Key, pl. 25.

IV. Test calcareous, perforate.

A. Test vitreous with a glassy lustre, aperture typically radiate, not trochoid.
1. Test planispirally coiled or becoming straight, or single chambered.
   Family 23. *Lagenidae.* Key, pls. 20, 21, 48.
2. Test biserial or elongate spiral.
   Family 24. *Polymorphinidae.* Key, pls. 22, 48.

B. Test not vitreous, aperture not radiate.
1. Test planispiral, occasionally trochoid, then usually with retral processes along the suture lines, septa single, no canal system.
   Family 25. *Nonionidae.* Key, pls. 23, 48, 49.
2. Test planispiral, at least in the young, generally lenticular, septa double, canal system in higher forms.
   Family 26. *Camerinidae.* Key, pls. 23, 49.
3. Test at least in the microspheric form of most genera biserial, aperture usually large, without teeth.
   Family 30. *Heterohelicidae.* Key, pls. 26, 50.
4. Test planispiral, with elongate spines and lobed aperture.
   Family 41. *Hantkeninidae.* Key, pls. 35, 54.
5. Test typically with an internal tube, elongate.
   a. Aperture generally loop-shaped or cribrate.
      Family 31. *Buliminidae.* Key, pls. 21, 26, 27, 28, 50, 51.
   b. Aperture narrow, curved, with an overhanging portion.
      Family 32. *Ellipsoidinidae.* Key, pls. 28, 51.
6. Test of inflated chambers, in opposed pairs.
   Family 34. *Pegidiidae.* Key, pl. 31.
7. Test trochoid, at least in the young of the microspheric form, usually coarsely perforate, when lenticular, with equatorial and lateral chambers.
   a. Test trochoid throughout, simple, aperture ventral.
      (1). No alternating supplementary chambers on the ventral side.
         Family 33. *Rotaliidae.* Key, pls. 29–31, 51, 52.
      (2). Alternating supplementary chambers on the ventral side.
         Family 35. *Amphisteginidae.* Key, pls. 31, 52.
   b. Test trochoid and aperture ventral at least in the early stages.
      (1). With supplementary material and large spines independent of the chambers. ....Family 36. *Calcarinidae.* Key, pl. 32.
      (2). With later chambers in annular series or globose with multiple apertures but not covering the earlier ones.
         Family 37. *Cymbaloporidae.* Key, pls. 32, 40, 52.
      (3). With later chambers mostly somewhat biserial, aperture elongate, in the axis of coiling.
         Family 38. *Cassidulinidae.* Key, pls. 33, 53.
      (4). With later chambers becoming involute, very few making up the exterior in the adult, aperture typically elongate, semicircular, in a few species circular.
         Family 39. *Chilostomellidae.* Key, pls. 33, 53.

(5). With chambers mostly finely spinose and wall cancellated, adapted for pelagic life, globular forms with the last chamber completely involute, apertures umbilicate or along the sutures. ....Family 40. *Globigerinidae.* Key, pls. 34, 35, 53.

(6). Early chambers globigerine, later ones spreading and compressed. ..Family 42. *Globorotaliidae.* Key, pls. 35, 54.

c. Test trochoid at least in the young, aperture peripheral or becoming dorsal.

(1). Mostly attached, dorsal side usually flattened.
Family 43. *Anomalinidae.* Key, pls. 36, 54.

(2). Later chambers in annular series.
Family 44. *Planorbulinidae.* Key, pls. 37, 54.

d. Test trochoid in the very young, later growing upward.

(1). Typically attached, later chambers in a loose spiral, aperture often with a neck and lip.
Family 45. *Rupertiidae.* Key, pls. 37, 54.

(2). Typically free in adult, later chambers in an irregular rounded mass. ....Family 46. *Victoriellidae.* Key, pl. 37.

(3). Later chambers in masses or branching, highly colored.
Family 47. *Homotremidae.* Key, pl. 37.

e. Test trochoid in the very young of the microspheric form, chambers becoming annular later, with definite equatorial chambers covered on either side by lateral chambers or appressed laminae, often with pillars.

(1). Equatorial chambers arcuate, hexagonal or modified hexagonal in plan without a canal system.
Family 48. *Orbitoididae.* Key, pls. 38–41, 54, 55.

(2). Equatorial chambers rectangular or faintly hexagonal in plan, a canal system present.
Family 49. *Discocyclinidae.* Key, pls. 38, 39, 41, 55.

(3). Equatorial chambers diamond-shaped, rhomboid or a modification of these forms with a spiral canal and interseptal canals. ....Family 50. *Miogypsinidae.* Key, pls. 39, 41, 55.

II. Keys to the Genera of the Various Families

Family 1. ALLOGROMIIDAE, p. 65, pl. 8.

The members of this family are so rare and so little known that a key to the genera is not given.

Family 2. ASTRORHIZIDAE, p. 70; Key, pls. 1, 42.

I. Test with large, rigid arms beyond the general periphery.
  A. Arms usually short and thick, central chamber large.
    1. Wall thick, loosely cemented. ....*Astrorhiza*, p. 70
    2. Wall very thin, firmly cemented. ....*Pseudastrorhiza*, p. 71.
  B. Arms usually long, central chamber small. ....*Rhabdammina*, p. 71.

II. Test without rigid arms, usually flexible and short or wanting.
  A. Test generally spherical.
    1. With usually two arms, test large. ....*Pelosphaera*, p. 72.
    2. With several short arms, test small. ....*Astrammina*, p. 72.
    3. Without distinct arms, wall usually thick. ....*Crithionina*, p. 71.
  B. Test generally flat and thin.
    1. Test entirely arenaceous. ....*Masonella*, p. 71.

2. Test of central part thin and chitinous. .... *Vanhoeffenella*, p. 72.
C. Test irregularly polygonal. .... *Ordovicina*, p. 72.
D. Test fusiform. .... *Amphitremoidea*, p. 72.

Family 3. RHIZAMMINIDAE, p. 73; Key, pls. 1, 42.

I. Test cylindrical, not branching.
A. Wall of sand grains and sponge spicules.
1. Exterior showing the spicules. .... *Marsipella*, p. 73.
2. Exterior smooth, wall thick, spicules mostly on the interior.
Bathysiphon, p. 73.
B. Wall of fine arenaceous material. .... *Hippocrepinella*, p. 74.
C. Wall of coarse arenaceous material. .... *Arenosiphon*, p. 74.
II. Test usually branching, wall of foraminiferal tests and chitin.
Rhizammina, p. 73.

Family 4. SACCAMMINIDAE, p. 74; Key, pls. 2, 42.

I. Test without a definite aperture, usually free.
A. Wall thick, of coarse material.
1. Test a single chamber.
a. Test globular.
(1). Wall of agglutinated material. ...... *Psammosphaera*, p. 75.
(2). Wall largely chitinous. .... *Pilalla*, p. 75.
b. Test elongate.
(1). Test regular in form. .... *Stegnammina*, p. 77.
(2). Test irregular or branching. .... *Raibosammina*, p. 77.
c. Test plano-convex. .... *Causia*, p. 75.
d. Test horn-shaped. .... *Ceratammina*, p. 77.
2. Test of several chambers.
a. Chambers in an elongate series. .... *Psammophax*, p. 77.
b. Chambers in an irregular series.
(1). Chambers numerous, small. .... *Sorosphaera*, p. 76.
(2). Chambers very few, large. .... *Arenosphaera*, p. 76.
B. Wall thick, of fine amorphous material. .... *Storthosphaera*, p. 77.
C. Wall thin, mostly chitin and fine sand.
1. Test rounded or irregular. .... *Blastammina*, p. 76.
2. Test angular. .... *Thekammina*, p. 76.
D. Wall thin, of coarse sand grains. .... *Shidelerella*, p. 76.

II. Test usually with a definite aperture.
A. Test free, usually with brownish cement.
1. Wall of coarse material.
a. Aperture terminal, usually with a neck.
(1). Test globular. .... *Saccammina*, p. 78.
(2). Test flask-shaped or pyriform; neck without a lip.
(a). Wall with little chitin, thickly encrusted.
Proteonina, p. 78.
(b). Wall largely chitin, thinly encrusted.
Lagenammina, p. 78.
(3). Test flask-shaped, with a flaring lip. ...... *Lagunculina*, p. 79.
b. Aperture terminal, without a neck, and at the broad end of the test.
Urnulina, p. 79.
c. Aperture not terminal.
(1). Curved, in a depression. .... *Millettella*, p. 79.
(2). Rounded, not in a depression. .... *Marsupulina*, p. 79.

(3). Rounded, with a thickened rim. ...... *Placentammina*, p. 79.
2. Wall of fine material.
a. With two apertures close to one another. .. *Ammosphaeroides*, p. 80.
b. With numerous remote apertures, usually with necks.
(1). Short nipple-like protuberances. .... *Thurammina*, p. 80.
(2). Numerous elongate necks. .... *Armorella*, p. 81.
c. Interior labyrinthic. .... *Thuramminoides*, p. 80.
3. Wall entirely chitinous. .... *Leptodermella*, p. 80.
B. Test free, usually whitish.
1. Aperture rounded.
a. Wall of material without definite arrangement, spicules few.
Pelosina, p. 81.
b. Wall largely of spicules, definitely arranged. .... *Technitella*, p. 81.
2. Aperture elongate, narrow. .... *Pilulina*, p. 82.
3. Apertures at ends of test. .... *Croneisella*, p. 82.
C. Test attached.
1. Test circular, low.
a. Single chambered.
(1). Aperture peripheral. .... *Webbinella*, p. 82.
(2). Aperture central. .... *Colonammina*, p. 83.
(3). Aperture indistinct. .... *Amphifenestrella*, p. 83.
b. Divided into numerous parts. .... *Kerionammina*, p. 83.
c. Several polygonal chambers. .... *Urnula*, p. 84.
2. Test irregular.
a. Aperture peripheral.
(1). Chamber large, irregular. .... *Iridia*, p. 82.
(2). Chamber finely tubular. .... *Diffusilina*, p. 84.
3. Test subglobular.
a. Apertures at ends, near base. .... *Tholosina*, p. 83.
b. Apertures on upper side, in a depression. .... *Verrucina*, p. 83.

Family 5. HYPERAMMINIDAE, p. 84; Key, pls. 3, 42.

I. Test free, unbranched.
A. Elongated chamber cylindrical.
1. Test chitinous, with a few sand grains. .... *Nubeculariella*, p. 85.
2. Test arenaceous.
a. Test single, usually straight. .... *Hyperammina*, p. 85.
b. Tests in masses, twisted. .... *Normanina*, p. 85.
B. Elongated chamber tapering.
1. Coarsely arenaceous with little cement. .... *Jaculella*, p. 85.
2. Finely arenaceous with much cement.
a. Cement ferruginous, aperture often lobed. .... *Hippocrepina*, p. 85.
b. Cement siliceous, aperture not lobed. .... *Hyperamminoides*, p. 86.
3. Of fine calcareous granules. .... *Earlandia*, p. 86.

II. Test free, branched.
A. Chamber irregular, tubular portion very small. .... *Chitinodendron*, p. 86.
B. Chamber globular, tubular portion large. .... *Saccorhiza*, p. 86.

III. Test usually attached, branching.
A. Test dichotomously branching.
1. Wholly attached. .... *Sagenina*, p. 88.
2. Attached by proloculum only, tubular part erect.
Psammatodendron, p. 88.
B. Test irregularly branching, largely erect.

1. Test mostly arenaceous, without spicules.
 a. Arborescent. ....*Dendrophrya*, p. 87.
 b. Branching from a bulbous base. ....*Saccodendron*, p. 87.
2. Test fusiform or tapering, with many spicules at the outer end.
 a. Simple or slightly branching. ....*Haliphysema*, p. 87.
 b. A complex arborescent form. ....*Dendronina*, p. 87.
C. Test a mass of anastomosing tubes, arenaceous. ....*Syringammina*, p. 88.
D. Test inside other foraminiferal tests, chitinous.
 1. Test simply branched and winding. ....*Ophiotuba*, p. 88.
 2. Test anastomosing. ....*Dendrotuba*, p. 88.

Family 6. REOPHACIDAE, p. 89; Key, pls. 3, 42.

I. Chambers irregular in size and arrangement.
 A. Wall chitinous, in the interior of other foraminiferal tests.
 Hospitella, p. 90.
 B. Wall mostly finely arenaceous, apertures at the ends of tubular necks.
 Aschemonella, p. 89.
 C. Wall largely of chitin with scattered arenaceous fragments.
 Kalamopsis, p. 90.

II. Chambers usually in a regular rectilinear series, not enveloped.
 A. Wall mostly arenaceous.
 1. Interior simple.
 a. Aperture round, terminal.
 (1). Test of coarsely agglutinated material. ....*Reophax*, p. 90.
 (2). Test finely arenaceous with much cement. *Hormosina*, p. 91.
 (3). Test usually with an outer coating. ......*Nodosinella*, p. 91.
 b. Aperture elongate, in a depression. ....*Sulcophax*, p. 91.
 c. Aperture multiple. ....*Polychasmina*, p. 91.
 2. Interior labyrinthic. ....*Haplostiche*, p. 92.
 B. Wall mostly chitinous.
 1. With some arenaceous material, aperture elongate. *Turriclavula*, p. 92.
 2. Wholly chitinous, aperture rounded. ....*Nodellum*, p. 92.

III. Chambers enveloping previous ones in the series.
 A. Free, a rectilinear series. ....*Sphaerammina*, p. 92.
 B. Adherent, an eccentric series. ....*Ammosphaerulina*, p. 92.

Family 7. AMMODISCIDAE, p. 93, pl. 9; Key, pls. 4, 42.

I. Test free.
 A. Test completely coiled throughout.
 1. Planispiral, at least in the young, not labyrinthic.
 a. Planispiral throughout. ....*Ammodiscus*, p. 94.
 b. Planispiral in the young, later coils partially covering one side.
 Hemidiscus, p. 94.
 c. Planispiral in the young, last coil double. ....*Bifurcammina*, p. 97.
 2. Planispiral, labyrinthic. ....*Discammina*, p. 98.
 3. Conical spiral, at least in the young.
 a. Wall arenaceous, with much cement.
 (1). Conical spiral throughout. ....*Turritellella*, p. 96.
 (2). Conical spiral in the young, later nearly planispiral.
 Ammodiscoides, p. 96.
 b. Wall chitinous. ....*Spirillinoides*, p. 96.
 4. Irregularly winding.
 a. Undivided.

(1). Mostly in an irregular ball. .... *Glomospira*, p. 96.
(2). Later portion nearly planispiral. .... *Glomospirella*, p. 97.
b. Divided into irregular chambers. .... *Ammoflintina*, p. 97.
B. Test partially uncoiled.
1. Tubular chamber not compressed. .... *Lituotuba*, p. 97.
2. Tubular chamber compressed, complanate. .... *Psammonyx*, p. 97.
II. Test attached.
A. Early portion coiled, not elongate spiral.
1. Later portion irregularly winding. .... *Tolypammina*, p. 98.
2. Later portion bending back and forth. .... *Ammovertella*, p. 98.
B. Early portion coiled, elongate spiral. .... *Trepeilopsis*, p. 99.
C. Early portion with a large proloculum, tubular chamber nearly straight.
1. Wall chitinous. .... *Xenotheka*, p. 99.
2. Wall arenaceous. .... *Ammolagena*, p. 99.

Family 8. LITUOLIDAE, p. 99, pl. 10; Key, pls. 4–6, 40, 42, 43.

I. Test of simple chambers, not labyrinthic.
A. Wall mostly of agglutinated material.
1. Close coiled throughout.
a. Aperture simple.
(1). Test not at all involute. .... *Trochamminoides*, p. 101.
(2). More or less involute, planispiral throughout.
(a). Wall not alveolar.
(i). Aperture at base of apertural face.
Haplophragmoides, p. 102.
(ii). Aperture in the apertural face.
Labrospira, p. 103.
(b). Wall alveolar. .... *Alveolophragmium*, p. 102.
(3). More or less involute, later part not planispiral.
Recurvoides, p. 102.
b. Aperture, several at base of apertural face. .. *Cribrostomoides*, p. 102.
2. Later portion uncoiled.
a. Aperture simple.
(1). Test compressed.
(a). Test evolute, much compressed, later chambers not frondicularian.
(i). Chambers distinct from exterior.
Ammomarginulina, p. 103.
(ii). Chambers indistinct from exterior.
Ammoscalaria, p. 103.
(b). Test mostly involute, later chambers frondicularian.
Flabellammina, p. 103.
(2). Test not greatly compressed.
(a). Adult chambers rounded in section.
Ammobaculites, p. 103.
(b). Adult chambers triangular in section
(i). Early portion distinctly coiled. ... *Frankeina*, p. 104.
(ii). Uniserial throughout. .... *Triplasia*, p. 106.
b. Aperture multiple.
(1). Uncoiled portion very small, linear. .... *Cribrospirella*, p. 106.
(2). Uncoiled portion large, linear. ..... *Haplophragmium*, p. 106.
(3). Uncoiling laterally. .... *Ammoastuta*, p. 106.
B. Wall agglutinated but consisting largely of cement.
1. Close coiled throughout.

a. Aperture simple at base of apertural face. ...... *Endothyra*, p. 107.
b. Aperture slit-like.
(1). Numerous, regular. .... *Bradyina*, p. 107.
(2). Few, irregular. .... *Glyphostomella*, p. 108.
c. Aperture cribrate. .... *Cribrospira*, p. 108.
2. Later portion uncoiled.
a. Aperture simple. .... *Endothyranella*, p. 108.
b. Aperture cribrate. .... *Septammina*, p. 109.
3. Chambers alternating, biserial. .... *Biseriammina*, p. 109.

II. Test with labyrinthic chambers.
A. Close coiled throughout, not annular.
1. Only slightly compressed. .... *Cyclammina*, p. 109.
2. Strongly compressed.
a. Interior of chambers only slightly divided. ..... *Choffatella*, p. 110.
b. Interior of chambers complex.
(1). Divided by branching divisions. .... *Dictyopsella*, p. 110.
(2). Divided by criss-cross divisions. .... *Yaberinella*, p. 110.
B. Later chambers uncoiled.
1. Irregular in shape, interior with partitions. .... *Lituola*, p. 110.
2. Regular in shape, interior simply labyrinthic. *Pseudocyclammina*, p. 110.
C. Later chambers annular.
1. Chambers in a single plane, not thickened at the edge.
a. Simply divided. .... *Cyclolina*, p. 111.
b. Irregularly divided. .... *Spirocyclina*, p. 111.
2. Test greatly thickened at the edge. .... *Orbitopsella*, p. 111.
3. Chambers in a double plane. .... *Cyclopsinella*, p. 111.

Family 9. TEXTULARIIDAE, p. 112, pl. 11; Key, pls. 7, 43, 46.

I. A large portion of the early test planispiral, later biserial or uniserial.
A. Adult biserial.
1. Chambers undivided. .... *Spiroplectammina*, p. 113.
2. Chambers partially divided. .... *Septigerina*, p. 114.
B. Adult uniserial, fan shaped.
1. Apertures in a single row. .... *Ammospirata*, p. 114.
2. Apertures in a double row. .... *Semitextularia*, p. 114.
C. Adult uniserial, narrow.
1. Uniserial chambers few, compressed. .... *Ammobaculoides*, p. 114.
2. Uniserial chambers numerous, rounded. .... *Spiroplectella*, p. 114.

II. Test mostly biserial, no uniserial stage.
A. Attached. .... *Textularioides*, p. 116.
B. Free.
1. Aperture simple.
a. Closely biserial throughout.
(1). Aperture arched.
(a). Wall arenaceous. .... *Textularia*, p. 115.
(b). Wall siliceous. .... *Silicotextulina*, p. 115.
(2). Aperture tubular. .... *Siphotextularia*, p. 116.
b. Becoming loosely biserial. .... *Haeuslerella*, p. 116.
c. Chambers tending to extend backwards. .... *Pseudopalmula*, p. 115.
2. Aperture cribrate. .... *Cribostomum*, p. 120.

III. Test with definite biserial and uniserial stages.
  A. Aperture simple.
    1. Not much compressed, aperture rounded. .... *Bigenerina*, p. 116.
    2. Much compressed, aperture elongate. .... *Vulvulina*, p. 117.
  B. Aperture multiple.
    1. Apertures several, rounded. .... *Climacammina*, p. 120.
    2. Apertures two, elliptical. .... *Deckerella*, p. 120.

IV. Test uniserial throughout, at least in the megalospheric form.
  A. Aperture simple.
    1. Not strongly compressed. .... *Monogenerina*, p. 117.
    2. Early stages at least compressed.
      a. Sides strongly concave. .... *Geinitzina*, p. 118.
      b. Sides flattened or convex. .... *Spandelina*, p. 117.
  B. Aperture cribrate. .... *Cribrogenerina*, p. 121.

Family 10. VERNEUILINIDAE, p. 121, pl. 11; Key, pls. 7, 8, 43.

I. Triserial with no biserial stage.
  A. Triserial throughout.
    1. Aperture arched. .... *Verneuilina*, p. 122.
    2. Aperture elongate, rounded. .... *Flourensina*, p. 123.
    3. Aperture with a neck. .... *Barbourinella*, p. 123.
  B. Triserial except last chamber, which is terminal. .... *Tritaxia*, p. 122.

II. Triserial, becoming biserial.
  A. Aperture at inner margin of chamber. .... *Gaudryina*, p. 123.
  B. Aperture in apertural face connecting with inner margin. *Migros*, p. 125.
  C. Aperture terminal, with short neck.
    1. Later portion regular.
      a. Angles of test often fistulose. .... *Heterostomella*, p. 127.
      b. Angles of test not fistulose. .... *Bermudezina*, p. 125.
    2. Later portion very irregular. .... *Rudigaudryina*, p. 127.

III. Triserial, biserial, then uniserial.
  A. Biserial stage short and irregular.
    1. Uniserial portion short.
      a. Triserial stage distinct. .... *Gaudryinella*, p. 125.
      b. Triserial stage indistinct. .... *Pseudogaudryinella*, p. 126.
    2. Uniserial portion long.
      a. Triangular in section. .... *Clavulinoides*, p. 126.
      b. Rounded in section. .... *Pseudoclavulina*, p. 126.
  B. Biserial stage long and regular. .... *Spiroplectinata*, p. 125.

Family 11. VALVULINIDAE, p. 127, pl. 12; Key, pls. 5, 8, 9, 43, 46.

I. Triserial in the early stages.
  A. Triserial throughout.
    1. Aperture with a flattened tooth. .... *Valvulina*, p. 129.
    2. Aperture without a tooth. .... *Eggerellina*, p. 129.
    3. Aperture cribrate. .... *Pseudogoësella*, p. 130.
  B. Later chambers uniserial. .... *Clavulina*, p. 130.
  C. Later chambers more than three in a whorl.
    1. Aperture with several openings. .... *Cribrobulimina*, p. 130.
    2. Aperture simple. .... *Arenobulimina*, p. 131.

II. Early stages with more than three chambers.
- A. Chambers undivided.
  - 1. Adult with more than three chambers to a whorl.
    - a. With five or six chambers to a whorl. ...... *Valvulammina*, p. 131.
    - b. With three or four chambers to a whorl. .... *Makarskiana*, p. 131.
  - 2. Adult triserial.
    - a. Aperture simple.
      - (1). Last chambers not strongly embracing. .... *Eggerella*, p. 131.
      - (2). Last chambers strongly embracing. .... *Eggerina*, p. 132.
    - b. Aperture cribrate. .... *Chrysalidina*, p. 132.
  - 3. Adult biserial.
    - a. Aperture at base of apertural face.
      - (1). Apertural face concave. .... *Marssonella*, p. 134.
      - (2). Apertural face convex, interior undivided. ... *Dorothia*, p. 136.
      - (3). Apertural face convex, interior subdivided.
        - (a). Early, 5-chambered stage persisting. ... *Matanzia*, p. 136.
        - (b). Early, 5-chambered stage wanting. ... *Cubanina*, p. 136.
    - b. Aperture in apertural face, rounded, no neck. .... *Plectina*, p. 137.
    - c. Aperture in apertural face, with neck. .... *Karreriella*, p. 138.
  - 4. Adult uniserial.
    - a. Aperture large, rounded, depressed. .... *Goësella*, p. 137.
    - b. Aperture cribrate. .... *Cribrogoësella*, p. 137.
    - c. Aperture with or without a short neck.
      - (1). Biserial stage present. .... *Schenckiella*, p. 138.
      - (2). Biserial stage wanting. .... *Martinottiella*, p. 138.
- B. Chambers with interior pillars.
  - 1. Aperture simple.
    - a. Rounded.
      - (1). Not uncoiled. .... *Ataxophragmium*, p. 139.
      - (2). Becoming uncoiled. .... *Orbignyna*, p. 140.
    - b. Becoming flattened. .... *Pernerina*, p. 141.
  - 2. Aperture multiple, cribrate.
    - a. With pillars only.
      - (1). With vertical pillars. .... *Lituonella*, p. 140.
      - (2). With radiating pillars. .... *Coprolithina*, p. 140.
    - b. With pillars and peripheral chamberlets.
      - (1). Peripheral chamberlets not subdivided. ... *Coskinolina*, p. 141.
      - (2). Peripheral chambers divided into cellules.
        - (a). Test conical. .... *Dictyoconus*, p. 141.
        - (b). Test compressed. .... *Gunteria*, p. 142.
- C. Chambers divided but without pillars.
  - 1. Adult with four or more chambers to a whorl. ... *Hagenowella*, p. 139.
  - 2. Adult typically biserial.
    - a. Test not greatly compressed.
      - (1). Test conical. .... *Textulariella*, p. 134.
      - (2). Test plano-convex. .... *Pseudorbitolina*, p. 135.
    - b. Test much compressed.
      - (1). Test cuneiform, wedge-shaped.
        - (a). Aperture a series of rounded openings. *Cuneolina*, p. 135.
        - (b). Aperture a series of elongate openings. *Cuneolinella*, p. 135.
      - (2). Test disciform. .... *Dicyclina*, p. 135.
  - 3. Adult typically uniserial.
    - a. Interior labyrinthic, aperture multiple. .... *Liebusella*, p. 139.

b. Interior with radiating partitions.
(1). Aperture simple. .... *Tritaxilina*, p. 138.
(2). Aperture multiple. .... *Coskinolinoides*, p. 141.
c. With pillar-like structures below the aperture. . . *Camagueyia*, p. 139.

Family 12. FUSULINIDAE, p. 146, pl. 13; Key, pls. 10, 11, 44, 45.

I. Walls of test not alveolar.
A. Axis of coiling short.
1. Periphery rounded, shell subspherical.
a. Inner whorls subspherical like the outer.
(1). Chomata heavy; wall thin, of three layers or with a fourth in outer whorls. .... *Pseudostaffella*, p. 148.
(2). Chomata thin; wall of two layers only. .... *Pisolina*, p. 149.
b. Inner whorls narrow and nautiliform. .... *Sphaerulina*, p. 150.
2. Periphery rounded, shells lenticular or nautiliform.
a. Small (less than 2 mm. in diameter). .... *Nummulostegina*, p. 149.
b. Large (over 5 mm. in diameter); axis not over half the equatorial diameter. .... *Nankinella*, p. 152.
c. Medium to large; axis more than half the equatorial diameter. *Staffella*, p. 147.
3. Periphery narrow and acute or sharply rounded.
a. Minute, narrow, widely umbilicate, outer whorls partly evolute. *Millerella*, p. 149.
b. Small, fully involute, periphery acutely angular in outer whorls. *Ozawainella*, p. 151.
B. Axis of coiling the greatest diameter, shells fusiform to subcylindrical.
1. Pseudostaffelloid in early whorls changing to fusiform in later whorls. *Leëlla*, p. 150.
2. Pseudostaffelloid in early whorls, then changing axis and becoming elongate. .... *Rauserella*, p. 151.
3. Fusiform in early whorls, but in the last becoming evolute and flaring. *Codonofusiella*, p. 154.
4. Fusiform throughout (except for endothyroid juvenarium of microspheric shells).
a. Spiral wall composed of two layers: tectum and diaphanotheca.
(1). Septa deeply folded. .... *Boultonia*, p. 153.
(2). Septa plane or slightly folded; chomata of moderate size. *Schubertella*, p. 155.
(3). Septa plane or slightly folded; chomata very massive. *Yangchienia*, p. 155.
b. Spiral wall composed of three layers: tectum, and inner and outer tectoria. .... *Fusiella*, p. 153.
c. Spiral wall composed of four layers: tectum, diaphanotheca, and inner and outer tectoria.
(1). Septa plane or only gently folded across the middle of the shell, more deeply folded in the end zones; chomata massive.
(a). Microspheric shells with endothyroid juvenaria predominate .... *Eoschubertella*, p. 155.
(b). Megalospheric shells with fusulinoid juvenaria predominate. .... *Fusulinella*, p. 152.
(2). Septa deeply folded throughout; chomata massive to slight. *Fusulina*, p. 156.
(3). Septa plane (except rarely near the ends of the outer whorls);

chomata low but wide, grading laterally into the tectorium; axis of shell more or less filled with epitheca.
Wedekindellina, p. 154.

d. Spiral wall thin and homogeneous (presumably degenerate); shells of moderately large size, with deeply folded septa and without chomata.
   (1). Wall coated by a very thin dark film on inside as well as outside; no mural pores visible. .... *Gallowainella*, p. 162.
   (2). Wall with distinct pores. .... *Quasifusulina*, p. 163.

II. Wall of test alveolar.
   A. Septa more or less folded and with a median tunnel; but septula are lacking.
      1. Equatorial expansion is abruptly accelerated during ontogeny, leaving a tightly coiled juvenarium clearly set from strongly inflated adult whorls.
         a. Whorls of juvenarium like *Triticites*; septa only slightly to moderately folded.
            (1). Wall and septa rather thin. ... *Pseudoschwagerina* s. s., p. 160.
            (2). Wall rather thick; septa thickened by epitheca; septal pores very coarse. .... Subgenus *Zellia*, p. 160.
         b. Whorls of juvenarium like *Schwagerina*; septa deeply and regularly folded throughout. .... *Paraschwagerina*, p. 161.
      2. Equatorial expansion not abruptly changed during growth.
         a. Chomata well developed.
            (1). Septa slightly to moderately folded across the middle of the shell, more deeply folded in the end zones.
            Triticites, p. 157.
         b. Chomata obsolete.
            (1). Septal folds regular and very deep even to the spiral wall, appearing pillar-like in axial sections.
            Paleofusulina, p. 163.
            (2). Septal folds deep and regular, appearing as septal loops in axial sections.
               (a). Septal folds entire, extending to the floor of the volution.
                  (i). Wall normally arched. ..*Schwagerina*, p. 157.
                  (ii). Wall dimpled so as to appear corrugated in sections. ..Subgenus *Rugofusulina*, p. 159.
               (b). Septal folds curled up at their tips to leave cuniculi connecting the chambers; basal sutures of the septa run around the shell.
                  (i). A single median tunnel is present.
                  Parafusulina, p. 161.
                  (ii). Supplementary tunnels are paired on opposite sides of the median tunnel.
                  Polydiexodina, p. 162.
      3. Shell evolute and uncoiled after the first two or three volutions.
      Nipponitella, p. 162.

Family 13. NEOSCHWAGERINIDAE, p. 164; Key, pl. 12.

A. Septula lacking.
   1. Shell spheroidal, with a slit-like median tunnel; parachomata appear late in ontogeny. .... *Eoverbeekina*, p. 165.

2. Shell spheroidal; without median tunnel, parachomata lacking in the early whorls. . . . . *Verbeekina*, p. 165.
3. Shell melon-shaped; parachomata present throughout. *Misellina*, p. 166.
4. Shell subspherical and laterally compressed; parachomata present throughout. . . . . *Brevaxina*, p. 166.

B. Septula present.
1. Septula limited to one set running spirally. . . . . *Cancellina*, p. 167.
2. Septula of two sets, one spiral and the other meridional.
a. Septa alveolar to their free margin. . . . . . . *Neoschwagerina*, p. 167.
b. Inner whorls as in *Neoschwagerina;* outer as in *Yabeina*. *Colania*, p. 168.
c. Lower edges of septula with lamellae thickened and fused into a solid plate. . . . . *Yabeina*, p. 168.

II. Shell wall thin and compact; not alveolar.
A. Septula lacking. . . . . *Pseudodoliolina*, p. 166.
B. Septula present.
1. Septula of uniform length. . . . . *Sumatrina*, p. 169.
2. Septula of two or more sets of unequal length. . . . . *Lepidolina*, p. 169.

Family 14. LOFTUSIIDAE, p. 170; Key, pl. 13.

Represented by a single genus. . . . . *Loftusia*, p. 170.

Family 15. NEUSINIDAE, p. 170; Key, pl. 13.

I. Test much compressed.
A. Fan-shaped from a pointed beginning. . . . . *Neusina*, p. 171.
B. Radiating from a center. . . . . *Jullienella*, p. 171.

II. Test generally cylindrical.
A. Not branched, only partially if at all labyrinthic. . . . . *Protobotellina*, p. 171.
B. With short branches or none. . . . . *Botellina*, p. 171.
C. Widely branching in one plane. . . . . *Schizammina*, p. 172.

Family 16. SILICINIDAE, p. 172; Key, pls. 13, 46.

I. Test planispiral throughout.
A. Chambers not a half coil in length.
1. Wall siliceous.
a. Early coils usually covered. . . . . *Silicina*, p. 173.
b. Early coils usually visible. . . . . *Involutina*, p. 173.
2. Wall partly calcareous. . . . . *Problematina*, p. 173.
B. Chambers a half coil in length.
1. Mostly involute. . . . . *Rzehakina*, p. 173.
2. Mostly evolute. . . . . *Spirolocammina*, p. 174.

II. Test planispiral only in the young.
A. Adult sigmoid in end view. . . . . *Silicosigmoilina*, p. 174.
B. Adult milioline in end view. . . . . *Miliammina*, p. 174.

Family 17. MILIOLIDAE, p. 174, pl. 14; Key, pls. 14, 15, 46, 47.

I. Test not divided into chambers, irregularly winding. . . *Agathammina*, p. 177.

II. Test chambered.
A. Test not reaching a triloculine stage.
1. Quinqueloculine throughout.
a. Aperture simple, with a simple tooth. . . . . *Quinqueloculina*, p. 177.
b. Aperture with a series of short teeth. . . . . *Dentostomina*, p. 178.

c. Aperture cribrate.
(1). Entirely calcareous. .... *Miliola*, p. 182.
(2). Exterior arenaceous. .... *Schlumbergerina*, p. 184.
2. Later chambers of various shapes.
a. Later chambers two to a coil, laterally spreading.
(1). Aperture simple.
(a). Planispiral chambers few.
(i). Wall with tubules. .... *Pseudomassilina*, p. 178.
(ii). Wall without tubules.
(A). Wall entirely calareous. ... *Massilina*, p. 178.
(B). Wall arenaceous on exterior.
Proemassilina, p. 178.
(b). Planispiral chambers many.
(i). Interior undivided. .... *Spiroloculina*, p. 178.
(ii). Interior subdivided. .... *Riveroina*, p. 179.
(c). Later chambers sigmoid. .... *Sigmoilina*, p. 179.
(d). Sigmoid chambers followed by spiroloculine stage.
Spirosigmoilina, p. 183.
(2). Aperture cribrate.
(a). Entirely calcareous. .... *Heterillina*, p. 183.
(b). Exterior arenaceous. .... *Ammomassilina*, p. 184.
b. Later chambers more than two to a coil.
(1). Aperture simple, with a broad tooth. *Nummoloculina*, p. 179.
(2). Aperture cribrate.
(a). Adult planispiral. .... *Hauerina*, p. 183.
(b). Adult uncoiling. .... *Raadshoovenia*, p. 183.
c. Later chambers uniserial.
(1). Quinqueloculine stage prominent, uniserial chambers few.
Articulina, p. 180.
(2). Quinqueloculine stage reduced, uniserial chambers many.
(a). Chambers indistinct, without a tooth. *Tubinella*, p. 182.
(b). Chambers indistinct, dentate. .... *Nubeculina*, p. 182.
(c). Aperture cribrate. .... *Poroarticulina*, p. 180.
B. Test reaching a triloculine stage but not a biloculine one.
1. Triloculine throughout.
a. Wall simple, aperture with simple tooth.
(1). Aperture with narrow or bifid tooth. .... *Triloculina*, p. 184.
(2). Aperture with broad flat tooth. .... *Miliolinella*, p. 177.
b. Wall alveolate, aperture cribrate. .... *Austrotrillina*, p. 185.
2. Later stage planispiral, not uncoiled.
a. Two chambers in adult coil. .... *Cribrolinoides*, p. 184.
b. Three chambers in adult coil. .... *Flintina*, p. 185.
3. Later stage uncoiling, becoming uniserial. ...... *Ptychomiliola*, p. 182.
C. Test reaching a biloculine stage.
1. Adult biloculine.
a. Aperture with broad bifid tooth, interior simple.
(1). Completely involute throughout. .... *Pyrgo*, p. 185.
(2). Later stages evolute, spiroloculine. .... *Flintia*, p. 186.
b. Aperture with broad flat tooth, interior simple. *Biloculinella*, p. 186.
c. Aperture of several elongate openings. .... *Pyrgoella*, p. 185.
d. Aperture cribrate.
(1). Interior undivided. .... *Cribropyrgo*, p. 186.
(2). Interior labyrinthic. .... *Fabularia*, p. 186.
2. Adult exterior mostly formed by the last-formed chamber.

a. Penultimate chamber showing as a small basal area.
(1). Aperture simple, radiate, test elongate. ....*Nevillina*, p. 187.
(2). Aperture complex, radiate, test broadly ovate. *Idalina*, p. 187.
b. Penultimate chamber not visible.
(1). Test subglobular. ....*Periloculina*, p. 187.
(2). Test strongly compressed. ....*Lacazina*, p. 187.

Family 18. OPHTHALMIDIIDAE, p. 188, pl. 15; Key, pls. 16, 17, 47.

I. Test not divided into a series of chambers.
A. Test free.
1. Planispiral throughout.
a. Evolute. ....*Cornuspira*, p. 189.
b. Involute. ....*Vidalina*, p. 189.
2. Early stages not entirely planispiral.
a. Close coiled.
(1). Later coils inflated, not entirely involute. *Gordiospira*, p. 193.
(2). Later coils not inflated, entirely involute. *Hemigordius*, p. 192.
(3). Later coils in zigzag bends. ....*Meandrospira*, p. 193.
b. Uncoiled in adult. ....*Orthovertella*, p. 193.
3. Later chambers variously formed.
a. Adult uncoiled, tubular. ....*Rectocornuspira*, p. 190.
b. Adult fan-shaped. ....*Cornuspiroides*, p. 192.
c. Adult with long peripheral extensions. ......*Cornuspirella*, p. 192.
B. Test typically attached.
1. Adult uncoiled, tube without definite zigzag bends.
a. Tubular chambers simple. ....*Apterrinella*, p. 194.
b. Tubular chambers branched.
(1). Tubular chambers undivided. ....*Cornuspiramia*, p. 192.
(2). Tubular chambers divided. ....*Rhizonubecula*, p. 199.
2. Adult tubular with definite zigzag bends.
a. Bends long and irregular. ....*Calcitornella*, p. 193.
b. Bends becoming obsolescent in adult.
(1). Test narrow, elongate. ....*Calcivertella*, p. 194.
(2). Test planispirally coiled. ....*Plummerinella*, p. 194.

II. Test divided into a series of chambers, or degenerate single chambered.
A. No chambers a half coil in length.
1. Test free, adults chambers uniserial.
a. Test compressed, aperture somewhat lateral, narrow.
(1). Early chambers planispiral. ....*Nodobaculariella*, p. 195.
(2). Early chambers in a trochoid spiral. .....*Vertebralina*, p. 196.
b. Test not compressed, aperture terminal, rounded.
Nodophthalmidium, p. 195.
2. Test attached or becoming free.
a. Early chambers coiled, later irregular.
(1). Irregularly coiled in adult, in one plane. ..*Nubecularia*, p. 199.
(2). Adult forming a solid mass. ....*Sinzowella*, p. 200.
(3). Adult uniserial. ....*Nubeculinella*, p. 199.
b. Early chambers not definitely coiled.
(1). Several chambers in adult, free.
(a). Test branched. ....*Calcituba*, p. 200.
(b). Test with irregular inflated chambers. ..*Parrina*, p. 200.
(2). A single chamber in adult, attached. .....*Squamulina*, p. 200.
B. At least some of the chambers a half coil in length.

1. Test with a thin plate between the coils.
   a. Adult with two chambers in a coil. ..... *Spiropthalmidium*, p. 196.
   b. Adult usually with more than two chambers in a coil. *Ophthalmidium*, p. 196.
   c. Adult with annular chambers and chamberlets. *Discospirina*, p. 197.
2. Test without a thin plate between the coils.
   a. Much compressed. .... *Wiesnerella*, p. 198.
   b. Somewhat compressed. .... *Glomulina*, p. 200.
   c. Becoming uniserial. .... *Meandroloculina*, p. 199.
   d. Globular. .... *Ophthalmina*, p. 197.

C. Test with several chambers to a coil.
1. Adult with three chambers to a coil.
   a. Aperture simple. .... *Planispirinella*, p. 198.
   b. Aperture cribrate. .... *Trisegmentina*, p. 198.
2. Adult with more than three chambers to a coil.
   a. Several to a coil, partially involute.
      (1). Aperture simple. .... *Planispirina*, p. 197.
      (2). Aperture cribrate. .... *Polysegmentina*, p. 197.
   b. Evolute, chambers spreading, becoming annular. *Renulina*, p. 198.

Family 19. FISCHERINIDAE, p. 201, pl. 15; Key, pl. 17.

Represented by a single genus. .... *Fischerina*, p. 201.

Family 20. TROCHAMMINIDAE, p. 201; Key, pls. 18, 40, 47, 48.

I. Test with a single layer.
A. Test typically trochoid, not involute.
1. Test arenaceous.
   a. Chambers simple.
      (1). Aperture terminal. .... *Conotrochammina*, p. 202.
      (2). Aperture at ventral margin. .... *Trochammina*, p. 202.
      (3). Aperture in ventral face.
         (a). Test free. .... *Trochamminita*, p. 202.
         (b). Test attached. .... *Trochamminella*, p. 203.
   b. Chambers subdivided. .... *Remanica*, p. 203.
2. Test of cement with distinct small bodies.
   a. Small bodies fusiform, aperture single. .... *Carterina*, p. 204.
   b. Small bodies irregular, apertures in pairs. .... *Entzia*, p. 203.
3. Test of matted spicules above, chitin below. ..... *Rotaliammina*, p. 203.

B. Test irregularly spiral with globose chambers, not involute. *Globotextularia*, p. 204.
C. Test becoming biserial in adult. .... *Mooreinella*, p. 204.
D. Test becoming uniserial in adult. .... *Ammocibicides*, p. 204.
E. Test with adult chambers more or less involute.
1. Test coarsely arenaceous.
   a. Aperture at umbilical border. .... *Ammosphaeroidina*, p. 205.
   b. Aperture terminal, test elongate. .... *Nouria*, p. 205.
2. Test finely arenaceous with much cement, aperture at base or in face of last-formed chamber. .... *Cystammina*, p. 205.

II. Test with two distinct layers.
A. Test trochoid, not symmetrical from above.
1. Chambers not labyrinthic. .... *Globivalvulina*, p. 206.
2. Chambers labyrinthic, irregular. .... *Ruditaxis*, p. 206.

B. Test symmetrical from above.

1. Chambers in fours in adult. .... *Tetrataxis*, p. 206.
2. Chambers numerous in adult. .... *Polytaxis*, p. 206.

Family 21. PLACOPSILINIDAE, p. 207; Key, pls. 19, 48.

I. Chambers simple, not labyrinthic.
  A. Early chambers close coiled, later uncoiled in a linear series, aperture simple.
    1. Adult chambers in a single series. .... *Placopsilina*, p. 207.
    2. Adult chambers in a double series. .... *Placopsilinella*, p. 208.
  B. Test spreading, apertures numerous. .... *Bdelloidina*, p. 208.
  C. Test elongate, becoming uniserial. .... *Acruliammina*, p. 208.

II. Chambers labyrinthic.
  A. Aperture simple.
    1. Aperture a crescentic slit. .... *Haddonia*, p. 208.
    2. Aperture rounded, often with a neck. .... *Stacheia*, p. 209.
  B. Aperture cribrate, test cylindrical. .... *Polyphragma*, p. 209.
  C. Aperture a ring of pores, test cylindrical. .... *Stylolina*, p. 209.
  D. Aperture a few irregular pores, test cylindrical. .... *Adhaerentia*, p. 209.

Family 22. ORBITOLINIDAE, p. 209; Key, pl. 19.

I. Chambers undivided. .... *Howchinia*, p. 210.

II. Chambers divided, with cellules at the exterior.
  A. Entirely trochoid. .... *Valvulinella*, p. 210.
  B. Chambers discoid or annular, complex. .... *Orbitolina*, p. 210.

Family 23. LAGENIDAE, p. 211, pl. 16; Key, pls. 20, 21, 48.

I. Test mostly close coiled.
  A. Aperture rounded, or the median slit much enlarged.
    1. Adult planispiral. .... *Robulus*, p. 213.
    2. Adult tending to become trochoid. .... *Darbyella*, p. 213.
  B. Aperture evenly radiate.
    1. Sides typically convex. .... *Lenticulina*, p. 214.
    2. Sides typically flattened or concave. .... *Planularia*, p. 214.

II. Test coiled in early stages at least of microspheric form.
  A. Test much compressed.
    1. Later chambers not chevron-shaped.
      a. Aperture radiate.
        (1). Later chambers extending obliquely back on one side. *Vaginulina*, p. 218.
        (2). Later chambers inflated, rectilinear. .... *Amphicoryne*, p. 218.
      b. Aperture elliptical. .... *Lingulina*, p. 218.
    2. Later chambers chevron-shaped.
      a. With a definite coiled stage in both forms. .... *Palmula*, p. 219.
      b. A definite coiled stage in microspheric form only.
        (1). A biserial stage present. .... *Kyphopyxa*, p. 220.
        (2). No biserial stage.
          (a). Early chambers like Vaginulina. ... *Flabellinella*, p. 220.
          (b). Early and later chambers chevron-shaped. *Frondicularia*, p. 219.
      c. No early coiled stage apparent.
        (1). Biserial in the adult. .... *Dyofrondicularia*, p. 219.
        (2). Uniserial in the adult. .... *Parafrondicularia*, p. 219.

B. Test not much compressed.
  1. Test rounded in section in adult. .... *Marginulina*, p. 214.
  2. Test triangular in section in adult. .... *Saracenaria*, p. 218.

III. Test not coiled.
  A. Chambers several.
    1. Not involute.
      a. Sutures oblique, aperture at peripheral angle. .... *Dentalina*, p. 215.
      b. Sutures directly transverse, aperture terminal, central.
        (1). Aperture radiate. .... *Nodosaria*, p. 215.
        (2). Aperture rounded, not radiate.
          (a). Triangular in section. .... *Tristix*, p. 215.
          (b). Quadrangular in section. .... *Quadratina*, p. 216.
        (3). Aperture becoming divided into pores.
          Chrysalogonium, p. 216.
    2. Involute. .... *Pseudoglandulina*, p. 216.
  B. Chambers single. .... *Lagena*, p. 221.

Family 24. POLYMORPHINIDAE, p. 221, pls. 17, 18; Key, pls. 22, 48.

I. Test in the adult biserial or uniserial.
  A. Test biserial in the adult.
    1. Test much compressed in section.
      a. Sigmoid only in the early stages, later not embracing.
        (1). Regularly biserial in adult. .... *Polymorphina*, p. 229.
        (2). Irregularly biserial, uncoiling. ...... *Polymorphinella*, p. 229.
      b. Sigmoid throughout.
        (1). Chambers extending to the base. .... *Sigmoidella*, p. 229.
        (2). Later chambers not reaching the base. *Sigmomorphina*, p. 229.
    2. Test rounded in section.
      a. Early chambers in threes, chambers elongate. ..... *Pyrulina*, p. 228.
      b. Early chambers in fives, chambers short.
        Pseudopolymorphina, p. 228.
      c. Early chambers irregularly spiral, chambers inflated, loosely biserial. .... *Paleopolymorphina*, p. 224.
  B. Test uniserial in the adult.
    1. Early stages triserial. .... *Dimorphina*, p. 226.
    2. Early stages biserial. .... *Glandulina*, p. 228.
    3. Early stages coiled. .... *Polymorphinoides*, p. 230.

II. Test in adult with more than two chambers in a whorl.
  A. Test typically with four or five chambers.
    1. Five chambers to a whorl throughout. .... *Guttulina*, p. 224.
    2. Four chambers to a whorl. .... *Quadrulina*, p. 224.
    3. Early chambers five to a whorl, later one terminal.
      Pseudopolymorphinoides, p. 224.
  B. Test typically with three chambers, globular. .... *Globulina*, p. 226.
  C. Test loosely spiral. .... *Eoguttulina*, p. 224.

III. Test irregular.
  A. Attached in a linear series with definite chambers. .... *Bullopora*, p. 230.
  B. Attached or free, chambers tubular, indefinite. .... *Ramulina*, p. 230.

Family 25. NONIONIDAE, p. 232, pl. 19; Key, pls. 23, 48, 49.

I. Adult test planispiral at least in the young, bilaterally symmetrical.
  A. Close coiled throughout, no supplementary chambers.
    1. Aperture simple, sutures simple.

a. Aperture at the margin of the apertural face. ...... *Nonion*, p. 232.
b. Aperture rounded, in the apertural face. ...... *Paranonion*, p. 233.
2. Apertures several, sutures with bridges. .... *Elphidium*, p. 234.
3. Apertures several, sutures with double row of pores. *Elphidiella*, p. 236.
4. Apertures several, with supplementary ones in the apertural face. *Cribroelphidium*, p. 234.

B. Close coiled throughout, supplementary chambers present. *Astrononion*, p. 233.

C. Adult uncoiled. .... *Ozawaia*, p. 236.

II. Adult test somewhat trochoid.
A. Last-formed chambers extending toward and often covering the umbilicus at one side.
1. Without retral processes. .... *Nonionella*, p. 233.
2. With retral processes. .... *Elphidioides*, p. 234.
B. Test in the adult typically trochoid, planoconvex.
1. Dorsally convex, ventrally flattened. .... *Polystomellina*, p. 236.
2. Dorsally flattened, ventrally convex. .... *Faujasina*, p. 236.
C. Test biconvex. .... *Notorotalia*, p. 236.

Family 26. CAMERINIDAE, p. 237, pl. 20; Key, pls. 23, 49.

I. Test with long undivided second chamber. .... *Archaediscus*, p. 237.

II. Test with numerous chambers, not divided into chamberlets.
A. Test involute throughout.
1. No secondary skeleton nor canal system. .... *Nummulostegina*, p. 237.
2. Secondary skeleton and canal system. .... *Camerina*, p. 238.
3. Supplementary skeleton of pillars only. .... *Miscellanea*, p. 238.
B. Test evolute, not complanate. .... *Assilina*, p. 239.
C. Test complanate.
1. Young lenticular, later complanate. .... *Operculinella*, p. 239.
2. Complanate throughout. .... *Operculina*, p. 239.
3. Becoming involute in the adult. .... *Operculinoides*, p. 238.

III. Test with chamberlets.
A. Adult without annular chambers.
1. Early chambers simple, later divided. .... *Heterostegina*, p. 240.
2. Nearly all chambers divided. .... *Spiroclypeus*, p. 240.
B. Adult with annular chambers.
1. Early chambers mostly spiral, adult annular. ... *Heteroclypeus*, p. 240.
2. Nearly all chambers annular. .... *Cycloclypeus*, p. 242.

Family 27. PENEROPLIDAE, p. 242; Key, pls. 24, 25, 49.

I. Chambers simple, without chamberlets.
A. Test more or less compressed, mostly coiled.
1. Aperture elongate, slit-like, or a linear series of rounded openings. *Peneroplis*, p. 243.
2. Aperture dendritic. .... *Dendritina*, p. 243.
B. Test in adult uncoiled, aperture terminal, simple.
1. Wall thick, aperture without neck. .... *Spirolina*, p. 243.
2. Wall thin, aperture with neck. .... *Monalysidium*, p. 244.
C. Test in adult uncoiled, aperture multiple. .... *Taberina*, p. 244.

II. Chambers divided into chamberlets.
A. Annular chambers not developed. .... *Fallotia*, p. 246.
B. Annular chambers only in adult if at all. .... *Archaias*, p. 244.

C. Annular chambers very early developed.
  1. Chamberlets in a single plane. .... *Sorites*, p. 245.
  2. Chamberlets in two planes. .... *Amphisorus*, p. 245.
  3. Chamberlets complex in the central portion.
    a. Sides of test without thickened imperforate layers.
      (1). Adjacent chamberlets in each annular chamber connecting. *Marginopora*, p. 245.
      (2). Adjacent chamberlets in each annular chamber not connecting. .... *Orbitolites*, p. 246.
    b. Sides of test with thickened imperforate layers. *Opertorbitolites*, p. 246.

Family 28. ALVEOLINELLIDAE, p. 248; Key, pls. 25, 49, 50.

I. No preseptal canal developed.
  A. Number of chamberlets the same as the number of openings on the apertural face.
    1. Apertures in a single row throughout. .... *Ovalveolina*, p. 249.
    2. Apertures in several rows toward the poles. ..... *Praealveolina*, p. 249.
  B. Number of chamberlets smaller than number of openings.
    1. No postseptal canal developed.
      a. A single row of accessory openings. .... *Subalveolina*, p. 249.
      b. More than one row of accessory openings. .... *Bullalveolina*, p. 250.
    2. A postseptal canal developed.
      a. Globular. .... *Borelis*, p. 250.
      b. Fusiform.
        (1). Early coils high. .... *Flosculina*, p. 250.
        (2). Early coils low. .... *Boreloides*, p. 250.
II. A preseptal canal developed.
  A. Two layers of chamberlets. .... *Flosculinella*, p. 251.
  B. More than two layers of chamberlets. .... *Alveolinella*, p. 251.

Family 29. KERAMOSPHAERIDAE, p. 251; Key, pl. 25.

This family is represented by the single genus. .... *Keramosphaera*, p. 251.

Family 30. HETEROHELICIDAE, p. 252, pl. 21; Key, pls. 26, 50.

I. Coiled stage in the young prominent.
  A. Biserial chambers few, increasing in size as added. .... *Heterohelix*, p. 253.
  B. Biserial chambers many, of nearly uniform size. ...... *Bolivinopsis*, p. 253.
  C. Biserial stage absent, adult uniserial. .... *Nodoplanulis*, p. 253.
II. Biserial stage in the young prominent.
  A. Biserial throughout.
    1. Chambers more or less globular. .... *Gümbelina*, p. 254.
    2. Chambers compressed.
      a. Periphery much thickened. .... *Bolivinoides*, p. 257.
      b. Periphery concave, with sharp keels.
        (1). Aperture at the inner margin of the chamber. *Bolivinita*, p. 257.
        (2). Aperture terminal. .... *Bolivinitella*, p. 258.
      c. Periphery sharp, chambers elongate. .... *Bolivinella*, p. 258.
  B. Later chambers variously arranged.
    1. Uniserial.
      a. Test not compressed.
        (1). Chambers regular, aperture terminal. *Rectogümbelina*, p. 256.

(2). Chambers irregular, aperture somewhat lateral. ..... *Tubitextularia*, p. 256.
b. Test compressed. ..... *Plectofrondicularia*, p. 258.
2. Multiserial.
a. Test in later development spiral. ..... *Pseudotextularia*, p. 256.
b. Test in later development spiral, later with numerous globular chambers in one plane. ..... *Planoglobulina*, p. 257
c. Test in later development with a fan-shaped series directly after the biserial stage. ..... *Ventilabrella*, p. 257.
3. Triserial at least in part.
a. Irregularly biserial in adult. ..... *Eouvigerina*, p. 259.
b. Completely biserial in adult. ..... *Zeauvigerina*, p. 259.

III. Early stages triserial.
A. Triserial throughout.
1. Chambers angled, aperture terminal. ..... *Pseudouvigerina*, p. 260.
2. Chambers rounded, aperture at base of chamber. ..... *Gümbelitria*, p. 256.
B. Triserial in the young, later uniserial. ..... *Siphogenerinoides*, p. 260.
C. Triserial in the young, later multiserial. ..... *Gümbelitriella*, p. 256.

IV. Test wholly uniserial.
A. Early portion compressed, aperture without a lip.
1. Early portion concave. ..... *Amphimorphina*, p. 259.
2. Early portion not concave. ..... *Nodomorphina*, p. 259.
B. Not compressed, aperture with a lip. ..... *Nodogenerina*, p. 260.

Family 31. BULIMINIDAE, p. 260, pl. 22; Key, pls. 21, 26–28, 50, 51.

I. Test an elongate spiral, the spiral suture prominent.
A. Test a proloculum and long tubular second chamber. *Terebralina*, p. 262.
B. Test of numerous chambers.
1. Usually more than three to a whorl in adult.
a. Aperture with a broad base. ..... *Turrilina*, p. 264.
b. Aperture comma-shaped.
(1). Chambers in a single series. ..... *Buliminella*, p. 264.
(2). Chambers in a double series.
(a). Chambers of one series much smaller. ..... *Pseudobulimina*, p. 266.
(b). Chambers of both series nearly equal. ..... *Robertina*, p. 265.
c. Aperture rounded, subterminal. ..... *Elongobula*, p. 266.
d. Aperture rounded, terminal. ..... *Buliminellita*, p. 264.
2. Later chambers becoming uniserial.
a. Aperture rounded. ..... *Buliminoides*, p. 265.
b. Aperture comma-shaped. ..... *Ungulatella*, p. 265.

II. Test triserial, at least in the early stages, or thoughout.
A. Aperture comma-shaped.
1. Chambers subglobular, test rounded in section.
a. Strongly involute at base. ..... *Globobulimina*, p. 267.
b. Not strongly involute, triserial throughout.
(1). Aperture comma-shaped. ..... *Bulimina*, p. 266.
(2). Apertures along the sutures. ..... *Delosina*, p. 276.
c. Not strongly involute, biserial in adult.
(1). Aperture with a distinct tooth. ..... *Neobulimina*, p. 267.
(2). Aperture without a distinct tooth. ..... *Virgulopsis*, p. 267.
2. Chambers angled, test triangular in section. ..... *Reussella*, p. 271.

B. Aperture of fine pores.
  1. Triserial throughout. .... *Pseudochrysalidina*, p. 272.
  2. Young triserial, adult uniserial, angular. .... *Chrysalidinella*, p. 273.
  3. Young triserial, adult fan-shaped, spreading. .... *Pavonina*, p. 272.
C. Aperture with collar, open at side, with a tooth. ...... *Uvigerinella*, p. 273.
D. Aperture with cylindrical neck and phialine lip.
  1. Triserial throughout.
    a. Chambers distinctly angled in section.
      (1). Chambers with prominent spines. .... *Trimosina*, p. 272.
      (2). Chambers without prominent spines. ... *Angulogerina*, p. 275.
    b. Chambers rounded.
      (1). Wall without tubuli. .... *Uvigerina*, p. 273.
      (2). Wall with tubuli. .... *Tritubulogenerina*, p. 271.
  2. Later stages biserial.
    a. Aperture terminal. .... *Hopkinsina*, p. 274.
    b. Aperture at base of chamber. .... *Mimosina*, p. 272.
  3. Later stages uniserial.
    a. Chambers simple, angled. .... *Trifarina*, p. 275.
    b. Chambers simple, rounded.
      (1). Early chambers triserial or biserial.
        (a). Early chambers triserial. .... *Rectuvigerina*, p. 273.
        (b). Early chambers mostly biserial. .. *Siphogenerina*, p. 274.
      (2). Early chambers uniserial in megalospheric form.
        Unicosiphonia, p. 274.
C. Chambers subdivided. .... *Schubertia*, p. 270.

III. Test uniserial throughout.
  A. Early stages angled, later rounded. .... *Dentalinopsis*, p. 275.
  B. Rounded throughout, aperture with a neck and phialine lip.
    Siphonodosaria, p. 274.

IV. Test a single chamber, with internal tube, aperture terminal.
  Entosolenia, p. 267.

V. Test biserial in adult or at least in early stages.
  A. Adult biserial, young an elongate spiral or twisted.
    1. Aperture comma-shaped, not in median line. ...... *Virgulina*, p. 268.
    2. Aperture in median line. .... *Bolivina*, p. 268.
      a. Aperture elongate in the apertural face.
        (1). Aperture low, semi-lunate. .... *Suggrunda*, p. 269.
        (2). Aperture rounded, with a lip. ...... *Bitubulogenerina*, p. 270.
  B. Test becoming uniserial in adult, aperture terminal.
    1. Last chambers becoming uniserial, aperture broader at one end.
      Loxostomum, p. 269.
    2. Typically uniserial in adult.
      a. Aperture single, terminal.
        (1). Adult somewhat compressed, neck slightly developed.
          (a). Aperture rounded. .... *Rectobolivina*, p. 270.
          (b). Aperture elongate. .... *Bifarinella*, p. 269.
          (c). Aperture double. .... *Geminaricta*, p. 270.
        (2). Adult not compressed, neck and lip strongly developed.
          Bifarina, p. 269.
      b. Aperture cribrate or a circle of pores. ...... *Tubulogenerina*, p. 271

Family 32. ELLIPSOIDINIDAE, p. 276, pl. 23; Key, pls. 28, 51.

I. Test of numerous chambers.
  A. Chambers biserial throughout. .... *Pleurostomella*, p. 277
  B. Early chambers biserial, later uniserial.
    1. Completely involute in the adult. .... *Ellipsobulimina*, p. 278.
    2. Not completely involute.
      a. Later chambers forming an elongate test. ...... *Nodosarella*, p. 278.
      b. Later chambers forming a glanduline test. *Ellipsopleurostomella*, p. 277.
  C. All chambers uniserial.
    1. Not involute.
      a. Test not compressed.
        (1). Aperture semi-elliptical. .... *Ellipsonodosaria*, p. 278.
        (2). Aperture with a large, flat tooth. .... *Pleurostomellina*, p. 277.
        (3). Aperture without a tooth. .... *Dentalinoides*, p. 277.
      b. Test compressed, aperture elliptical. ...... *Ellipsolingulina*, p. 280.
    2. Partially involute.
      a. Test not compressed.
        (1). Aperture semi-elliptical. .... *Ellipsoglandulina*, p. 280.
        (2). Aperture multiple. .... *Pinaria*, p. 280.
      b. Test somewhat compressed, aperture elliptical. *Gonatosphaera*, p. 280.
    3. Completely involute. .... *Ellipsoidina*, p. 280.
II. Test monothalamous. .... *Parafissurina*, p. 281.

Family 33. ROTALIIDAE, p. 281, pl. 24; Key, pls. 29–31, 51, 52.

I. Test tubular, undivided.
  A. Test planispiral. .... *Spirillina*, p. 283.
  B. Test trochoid.
    1. Ventral side open, umbilicate. .... *Turrispirillina*, p. 283.
    2. Ventral side closed.
      a. Without ventral pillars.
        (1). Ventral surface not lobed.
          (a). Ventral surface with simple plate. *Conicospirillina*, p. 284.
          (b). Ventral surface with lattice plate. .. *Coscinoconus*, p. 284.
        (2). Ventral side lobed. .... *Paalzowella*, p. 284.
      b. With ventral pillars. .... *Trocholina*, p. 284.
II. Test of numerous chambers, usually two to a whorl or annular.
  A. Adult with four chambers in a whorl. .... *Conorbina*, p. 286.
  B. Adult with two chambers to a whorl.
    1. Chambers not subdivided.
      a. High spired, periphery thickened. .... *Patellinella*, p. 285.
      b. Low spired, periphery not thickened. .... *Patellinoides*, p. 285.
    2. Chambers subdivided. .... *Patellina*, p. 285.
  C. Adult with annular chambers. .... *Annulopatellina*, p. 285.
III. Test of numerous chambers, several in a whorl.
  A. Umbilicus usually open, without pillars or plug.
    1. Ventral side with a prolongation of the chamber over or toward the umbilicus.
      a. Test typically planoconvex, umbilicus deep.
        (1). Umbilicus open.
          (a). Smooth and polished ventrally. .... *Lamarckina*, p. 288

(b). Usually radially striate ventrally. . . *Heronallenia*, p. 289.
(2). Chambers projecting to umbilicus. . . . . *Discorbis*, p. 286.
(3). With supplementary ventral aperture.
(a). Chambers undivided. . . . . *Discorbinella*, p. 288.
(b). Chambers partially divided. . . . . *Torresina*, p. 288.
b. Test strongly biconvex. . . . . *Valvulineria*, p. 289.
2. Ventral side without prolongations across the umbilicus.
a. Test in a low spire.
(1). Sutures not raised and ornate.
(a). Umbilicus small.
(i). Aperture confined to the middle of the ventral border. *Gyroidina*, p. 290.
(ii). Aperture connecting with the umbilicus. *Pseudovalvulineria*, p. 290.
(b). Umbilicus large. . . . . *Gavellinella*, p. 289.
(2). Sutures raised and ornate. . . . . *Stensiöina*, p. 290.
b. Test in a high spire. . . . . *Rotaliatina*, p. 291.
B. Umbilicus usually filled or covered.
1. With distinct umbilical plug or pillars.
a. Test not greatly spreading without intercalary whorls.
(1). Ventral pillars few.
(a). Spire low. . . . . *Rotalia*, p. 293.
(b). Spire high. . . . . *Asanoina*, p. 294.
(2). Ventral pillars many.
(a). Spire low. . . . . *Lockhartia*, p. 294.
(b). Spire high. . . . . *Sakesaria*, p. 294.
(3). Without ventral pillars. . . . . *Rotorbinella*, p. 294.
b. Test greatly spreading, with intercalary whorls. *Dictyoconoides*, p. 295.
2. Without a distinct umbilical plug or pillars.
a. Apertural face with a distinct, thin, clear plate.
(1). Aperture simple.
(a). Test not involute dorsally. . . . . *Cancris*, p. 297.
(b). Test strongly involute dorsally. . . . . *Baggina*, p. 298.
(c). With a distinct spire. . . . . *Baggatella*, p. 298.
(2). Aperture cribrate. *Neocribrella*, p. 298.
b. Apertural face without a distinct, thin, clear plate.
(1). Without supplementary apertures.
(a). Aperture without a neck.
(i). Close coiled.
(A). Dorsal and ventral sides distinct. *Eponides*, p. 291.
(B). Dorsal and ventral sides similar. *Crespinella*, p. 292.
(ii). Adult spreading and irregular. *Planopulvinulina*, p. 293.
(iii). Adult uncoiled.
(A). Plano-convex. . . . . *Coleites*, p. 293.
(B). Biconvex. . . . . *Rectoeponides*, p. 293.
(b). Aperture with a neck and lip.
(i). Test globular. . . . . *Siphoninoides*, p. 297.
(ii). Test biconvex, adult close coiled. *Siphonina*, p. 296.
(iii). Test biconvex, adult uncoiled. *Siphoninella*, p. 297.
(iv). Test biconvex, adult biserial. . . . *Siphonides*, p. 297.

(2). With supplementary apertures but without supplementary chambers.
(a). Supplementary apertures dorsal only. *Epistomaria*, p. 296.
(b). Supplementary apertures ventral only.
(i). Near peripheral margin. ...... *Epistomina*, p. 295.
(ii). At ventral margin. .... *Parrella*, p. 291.
(iii). In apertural face. .... *Poroeponides*, p. 295.
(iv). In umbilical area. .... *Discorinopsis*, p. 296.
(c). Supplementary apertures on both sides. *Mississippina*, p. 296.
(3). With supplementary chambers.
(a). Aperture simple. .... *Rotalidium*, p. 292.
(b). Aperture with a large tooth. .... *Eponidella*, p. 292.

Family 34. PEGIDIIDAE, p. 298; Key, pl. 31.

I. Test elongate, not spherical.
A. Chambers subglobular.
1. Outer surface smooth. .... *Physalidia*, p. 300.
2. Surface rugose. .... *Rugidia*, p. 300.
B. Chambers somewhat compressed. .... *Pegidia*, p. 299.
II. Test spherical. .... *Sphaeridia*, p. 300.

Family 35. AMPHISTEGINIDAE, p. 300, pl. 25; Key, pls. 31, 52.

I. Dorsal sutures simple, curved.
A. Suplementary chambers not reaching the periphery. .. *Asterigerina*, p. 301.
B. Supplementary chambers becoming peripheral.
1. Supplementary chambers undivided. .... *Helicostegina*, p. 301.
2. Supplementary chambers divided. .... *Eoconuloides*, p. 301.
II. Dorsal sutures with a distinct angle. .... *Amphistegina*, p. 302.

Family 36. CALCARINIDAE, p. 302, pl. 25; Key, pl. 32.

I. Test with large blunt spines.
A. Spines mostly peripheral in one plane.
1. Later chambers distinct. .... *Calcarina*, p. 303.
2. Later chambers indistinct. .... *Siderolites*, p. 303.
B. Spines in varying planes.
1. Surface without bosses and connecting rods. *Baculogypsinoides*, p. 303.
2. Surface with bosses and connecting rods. ...... *Baculogypsina*, p. 304.
II. Test without large stout spines.
A. Large vacuolar openings in section at side. .... *Arnaudiella*, p. 304.
B. No large vacuolar openings.
1. Spiral not visible from exterior. .... *Siderina*, p. 307.
2. Spiral visible from exterior.
a. Chambers in a single plane. .... *Pellatispira*, p. 306.
b. Chambers in two planes. .... *Biplanispira*, p. 307.

Family 37. CYMBALOPORIDAE, p. 307, pl. 25; Key, pls. 32, 40, 52.

I. Test with final globular chamber. .... *Tretomphalus*, p. 310.
II. Test without final globular chamber.
A. Umbilicus open.
1. Chambers nearly horizontal, test conical. .... *Cymbalopora*, p. 308.
2. Chambers nearly vertical, test compressed. .... *Cymbaloporella*, p. 309.

B. Umbilicus not open.
   1. Covered by a thin plate. .... *Cymbaloporetta*, p. 308.
   2. Filled by secondary shell material.
      a. Filling, a porous mass, without definite chambers. *Halkyardia*, p. 309.
      b. Filling, a series of chambers in definite layers. *Chapmanina*, p. 310.
      c. Filling, a series of irregular chambers. .... *Pyropilus*, p. 309.

Family 38. CASSIDULINIDAE, p. 311, pl. 26; Key, pls. 33, 53.

I. Test trochoid throughout.
   A. Aperture umbilical, covered by a thin plate. ...... *Ceratobulimina*, p. 312.
   B. Aperture not umbilical, without a plate, and nearly in the plane of coiling. .... *Pseudoparrella*, p. 312.
   C. Aperture along the ventral margin with a vertical fold near the periphery. .... *Alabamina*, p. 313.

II. Test biserial, at least in the early stages.
   A. Close coiled throughout.
      1. Supplementary chambers large.
         a. Chambers of rather uniform shape. .... *Cassidulina*, p. 313.
         b. Chambers becoming very elongate. .... *Cassidulinella*, p. 313.
         c. With supplementary apertures along the sutures. *Stichocassidulina*, p. 314.
      2. Supplementary chambers small.
         a. Single aperture. .... *Cerobertina*, p. 314.
         b. Double apertures. .... *Cushmanella*, p. 314.
   B. Uncoiling in the adult.
      1. Chambers irregular, sutures not depressed.
         a. Test rounded in section in adult. .... *Orthoplecta*, p. 315.
         b. Test triangular in section in adult. .... *Epistominoides*, p. 314.
      2. Chambers regular, sutures usually depressed.
         a. Chambers rounded, test not spinose. ...... *Cassidulinoides*, p. 315.
         b. Chambers triangular in section, often spinose. *Ehrenbergina*, p. 315.

Family 39. CHILOSTOMELLIDAE, p. 315, pl. 26; Key, pls. 33, 53.

I. Test trochoid throughout.
   A. Three chambers to a whorl. .... *Allomorphina*, p. 318.
   B. Four chambers to a whorl. .... *Quadrimorphina*, p. 318.
   C. Five or more chambers to a whorl. .... *Rotamorphina*, p. 318.

II. Test trochoid in the young, later planispiral.
   A. Two chambers in each coil.
      1. Aperture at the side of the test.
         a. Aperture narrow. .... *Chilostomella*, p. 319.
         b. Aperture rounded. .... *Chilostomelloides*, p. 319.
      2. Aperture terminal. .... *Seabrookia*, p. 319.
   B. More than two chambers in each coil.
      1. Chambers high.
         a. Aperture simple, narrow, low. .... *Allomorphinella*, p. 320.
         b. Aperture fimbriate. .... *Chilostomellina*, p. 320.
      2. Chambers low.
         a. Aperture broad, low, chambers regular. .... *Pullenia*, p. 320.
         b. Aperture narrow, semi-elliptical, chambers irregularly involute. *Sphaeroidina*, p. 321.
         c. Aperture cribrate. .... *Cribropullenia*, p. 321.

Family 40. GLOBIGERINIDAE, p. 321, pl. 27; Key, pls. 34, 35, 53.

I. Test trochoid throughout.
  A. Not involute.
    1. Aperture single, ventral. .... *Globigerina*, p. 322.
    2. Apertures several.
      a. Apertures small, all along the sutures. .... *Candeina*, p. 327.
      b. Apertures large, one umbilicate, others along the sutures. *Globigerinoides*, p. 322.
  B. Involute.
    1. Aperture not in a deep depression. .... *Pulleniatina*, p. 326.
    2. Aperture in a deep depression. .... *Sphaeroidinella*, p. 326.

II. Test not trochoid throughout.
  A. Test in the adult planispiral.
    1. Wall cancellated, spines fine and numerous.
      a. Chambers globular. .... *Globigerinella*, p. 323.
      b. Chambers with lateral prolongations toward the previous whorl. *Globigerinelloides*, p. 323.
    2. Wall rather smooth, spines coarse, flat, few. .... *Hastigerina*, p. 323.
  B. Test in the adult irregularly spiral, spines at the ends of the chambers. *Hastigerinella*, p. 324.
  C. Test in the adult with the last chamber spherical and completely enclosing the earlier ones.
    1. Aperture single. .... *Orbulina*, p. 326.
    2. Aperture a ring of small pores. .... *Candorbulina*, p. 327.
  D. Test with supplementary chambers over the sutures. *Globigerinatella*, p. 323.

Family 41. HANTKENINIDAE, p. 327; Key, pls. 35, 54.

I. Test with low aperture at base, often several spines to a chamber. *Schackoina*, p. 328.

II. Test with aperture running into the face, one spine only to a chamber.
  A. Aperture single. .... *Hantkenina*, p. 328.
  B. Apertures multiple. .... *Cribrohantkenina*, p. 329.

Family 42. GLOBOROTALIIDAE, p. 329, pl. 27; Key, pls. 35, 54.

I. Test trochoid throughout.
  A. Periphery truncate, usually with a double keel. .... *Globotruncana*, p. 329.
  B. Periphery acute or rounded, with a single keel.
    1. Aperture single.
      a. Ventral side slightly convex. .... *Globorotalia*, p. 330.
      b. Ventral side strongly convex or conical. ..... *Globorotalites*, p. 330.
    2. Aperture double. .... *Rotalipora*, p. 330.
    3. Apertures multiple. .... *Cribrogloborotalia*, p. 330.

II. Test becoming annular.
  A. A single layer of chambers. .... *Cycloloculina*, p. 331.
  B. Chamberlets on the flattened surface. .... *Sherbornina*, p. 331.

Family 43. ANOMALINIDAE, p. 331, pl. 28; Key, pls. 36, 54.

I. Test nearly symmetrical.
  A. Test more or less involute.
    1. Aperture usually median in the adult, at the base of the chamber.
      a. Test usually completely involute. .... *Anomalina*, p. 332.
      b. Test becoming somewhat evolute. .... *Planomalina*, p. 333.

2. Aperture median, elongate, in the terminal face. *Palmerinella*, p. 333.
3. Aperture extending over onto the dorsal side. . . *Anomalinoides*, p. 333.
4. Aperture completely dorsal. . . . . *Buningia*, p. 336.
5. Aperture extending over onto the ventral side. . . . . *Boldia*, p. 333.
6. With a supplementary peripheral aperture. . . . . . *Anomalinella*, p. 335.

B. Test little if at all involute, much compressed.
1. Without a broad keel.
a. Aperture single at inner margin. . . . . *Planulina*, p. 334.
b. With a second aperture at periphery. . . . . *Kelyphistoma*, p. 334.
2. With a broad, thin keel. . . . . *Laticarinina*, p. 334.

C. Test nearly involute. . . . . *Cibicidoides*, p. 335.

II. Test strongly plano-convex.
A. Test close coiled.
1. Aperture nearly peripheral. . . . . *Ruttenia*, p. 334.
2. Aperture extending to dorsal side. . . . . *Cibicides*, p. 335.

B. Test with later chambers annular.
1. Apertures peripheral, tubular. . . . . *Annulocibicides*, p. 338.
2. Apertures, pores scattered over surface. . . . . *Cyclocibicides*, p. 339.

C. Test with later chambers very irregular, aperture single, large, with lip. *Cibicidella*, p. 339.

D. Test with later chambers biserial. . . . . *Dyocibicides*, p. 338.

E. Test with later chambers uniserial.
1. Aperture single.
a. Trochoid stage well developed. . . . . *Stichocibicides*, p. 336.
b. Trochoid stage largely wanting. . . . . *Webbina*, p. 339.
2. Apertures numerous. . . . . *Rectocibicides*, p. 338.

III. Test subcylindrical, aperture rounded. . . . . *Vagocibicides*, p. 338.

Family 44. PLANORBULINIDAE, p. 339, pl. 29; Key, pls. 37, 54.

I. Test plano-convex or globular, not bilaterally symmetrical.
A. Chambers definitely in one plane.
1. Chambers closely grouped. . . . . *Planorbulina*, p. 340.
2. Later chambers forming a reticulate network. *Planorbulinoides*, p. 342.

B. Chambers piled up irregularly or in a spherical mass.
1. Chambers few, hemispherical, test compressed and attached. *Acervulina*, p. 343.
2. Chambers numerous, compressed, test globular, free. . . *Gypsina*, p. 343.

II. Test bilaterally symmetrical.
A. Sides papillate but not greatly thickened. . . . . *Planorbulinella*, p. 342.
B. Sides with a very thick secondary mass. . . . . *Linderina*, p. 342.
C. Sides without a thick secondary mass. . . . . *Eoannularia*, p. 343.

Family 45. RUPERTIIDAE, p. 343; Key, pls. 37, 54.

I. Test in a loose spiral throughout. . . . . *Rupertia*, p. 344.

II. Test spiral in young, later irregular.
A. Adult test high. . . . . *Carpenteria*, p. 344.
B. Adult test low, scale-like. . . . . *Neocarpenteria*, p. 345.

Family 46. VICTORIELLIDAE, p. 345; Key, pl. 37.

I. Aperture single.
A. Slit-like at base of chamber, without lip. . . . . *Eorupertia*, p. 345.
B. Triangular or arched, with a lip. *Victoriella*, p. 345.

II. Aperture of several pores. . . . . *Hofkerina*, p. 346.

Family 47. HOMOTREMIDAE, p. 346; Key, pl. 37.

I. Surface solid with large foramina, no fine pores.
  A. Foramina with perforated plate. .... *Homotrema*, p. 346.
  B. Foramina open, without perforated plate. .... *Sporadotrema*, p. 347.
II. Surface with open foramina and numerous fine pores ...... *Miniacina*, p. 347.

Family 48. ORBITOIDIDAE, p. 347, pls. 30, 31; Key, pls. 38–41, 54, 55.

KEY TO THE SUBFAMILIES OF ORBITOIDIDAE HERE RECOGNIZED

I. Equatorial zone with radial plates .... PSEUDORBITOIDINAE, p. 352.
II. Equatorial zone without radial plates.
  A. Embryonic apparatus of megalospheric forms, either with a thick outer wall or in the form of a trochoid spire so that all of the embryonic chambers are not in the equatorial plane. ...... ORBITOIDINAE, p. 353.
  B. Embryonic apparatus of megalospheric forms with a thin outer wall and bilocular initial chambers, except in teratological individuals.
    a. Without peri-embryonic chambers, or with thin-walled peri-embryonic chambers. .... LEPIDOCYCLINAE, p. 357.
    b. With peri-embryonic chambers with thick distal walls, or as a row of spirally arranged larger chambers from the nucleoconch. HELICOLEPIDINAE, p. 365.

KEYS TO THE GENERA AND SUBGENERA ACCORDING TO SUBFAMILIES

Pseudorbitoidinae

I. Normal equatorial chambers to the periphery of the test. *Pseudorbitoides*, p. 353.
II. Normal equatorial chambers replaced peripherally by rectangular chamberlets. .... *Vaughanina*, p. 352.

Orbitoidinae

I. Test biconcave. Lateral chambers absent over center. . . *Omphalocyclus*, p. 353.
II. Test spherical. Lateral chambers present over the center, arranged in an irregularly concentric manner around the nucleoconch. *Torreina*, p. 354.
III. Test asymmetrically lenticular. Chambers of megalospheric nucleoconch not in equatorial plane, trochoid in shape. .... *Clypeorbis*, p. 355.
IV. Test lenticular.
  A. Embryonic chambers four or less in number, or, teratologically, with more than four irregularly arranged chambers, in either case surrounded by a common thick wall. .... *Orbitoides*, p. 355.
  B. Embryonic chambers bilocular in equatorial sections, trilocular in vertical sections. .... *Pseudolepidina*, p. 356.

Lepidocyclinae

I. Megalospheric nucleoconch small in relation to equatorial chambers. *Lepidorbitoides*, p. 357.

(Subgenera)

  A. Test lenticular, not stellate. .... *Lepidorbitoides*, p. 357.
  B. Test stellate, with or without raised ribs. .... *Asterorbis*, p. 358.
II. Megalospheric nucleoconch large in relation to equatorial chambers.
  A. Equatorial chambers with a medianly situated radial stolon in curved outer wall. .... *Actinosiphon*, p. 358.
  B. Equatorial chambers without a medianly situated radial stolon in curved outer wall.

1. Equatorial layer at periphery not divided into three parts by a wedge-shaped layer of clear shell material. . . *Lepidocyclina*, p. 359.

(Subgenera)

a. Nucleoconch composed of two or more chambers of subequal size, other chambers of smaller size but larger than the equatorial chambers. Outer walls of the equatorial chambers arcuate, lateral walls inwardly convergent; four-stolon system. . . . . *Polylepidina*, p. 360.

b. Nucleoconch composed of two large subequal chambers, separated by a straight wall with variable peri-embryonic chambers; or teratologically one large chamber with smaller chambers around its periphery; four-stolon system. *Pliolepidina*, p. 361.

c. Nucleoconch composed of two subequal chambers, separated by a straight wall; six- or eight-stolon system. *Lepidocyclina*, p. 362.

d. Nucleoconch composed of two chambers, one of which partly embraces the other; six-stolon system. . . *Nephrolepidina*, p. 362.

e. Nucleoconch composed of two chambers, one of which completely embraces the other except at the place of attachment of the inner chamber; six-stolon system. . . . . *Eulepidina*, p. 363.

f. Nucleoconch multilocular, composed of a large central chamber, surrounded by five to ten smaller chambers; eight-stolon system. . . . . *Multilepidina*, p. 364.

2. Outer portion of the equatorial layer divided into three parts by a wedge-shaped layer of clear shell material which separates layer of small equatorial chambers; at least four stolons in each equatorial chamber. . . . . *Triplalepidina*, p. 364.

Helicolepidinae

I. Nucleoconch forming a trochoid spiral with thick outer wall. *Eulinderina*, p. 365.

II. Nucleoconch bilocular, with a row of spirally arranged larger chambers from the nucleoconch. . . . . *Helicolepidina*, p. 366.

Family 49. DISCOCYCLINIDAE, p. 367, pl. 30; Key, pls. 38, 39, 41, 55.

I. Radial chamber walls alternate in position in adjacent annuli. *Discocyclina*, p. 368.

A. Test circular in plan.

(Subgenera)

1. Without costae. . . . . *Discocyclina*, p. 368.
2. With radiating costae. . . . . *Aktinocyclina*, p. 369.

B. Test angular in plan, usually with projecting costae. *Asterocyclina*, p. 369.

II. Radial chamber walls in adjacent annuli aligned. *Pseudophragmina*, p. 371.

(Subgenera)

A. Radial chamber walls complete. . . . . *Proporocyclina*, p. 371.

B. Radial chamber walls incomplete.

1. Distal part degenerate, in places represented by a row of granules. *Pseudophragmina*, p. 371.

2. Walls absent or indistinct.
   a. Test without rays. .... *Athecocyclina*, p. 371.
   b. Test with rays. .... *Asterophragmina*, p. 372.

Family 50. MIOGYPSINIDAE, p. 372, pl. 31; Key, pls. 39, 41, 55.

I. Lateral chambers present. .... *Miogypsina*, p. 373.

(Subgenera)

A. Initial chambers near periphery with peri-embryonic chambers. *Miogypsina*, p. 373.

B. Initial chambers part of subcentral spiral. .... *Miolepidocyclina*, p. 374.

II. Lateral chambers absent. .... *Miogypsinoides*, p. 376.

PLATES 1–55

PLATE 1.

Family 2. ASTRORHIZIDAE, p. 70.

FIG.

1. *Astrorhiza arenaria* Norman. Recent. × 5. (After H. B. Brady). p. 70.
2, 3. *Astrorhiza granulosa* H. B. Brady. Recent. × 5. (After H. B. Brady). 3, Section showing interior. .... p. 70.
4. *Pseudastrorhiza silurica* Eisenack. Silurian, Germany. × 50. (After Eisenack). .... p. 71.
5. *Pelosphaera cornuta* Heron-Allen and Earland. Recent. × 9. (After Heron-Allen and Earland). .... p. 72.
6. *Astrammina rara* Rhumbler. Recent. × 15. (After Wiesner). ..... p. 72.
7, 8. *Crithionina mamilla* Goës. × 15. (After Goës). 7, Exterior. 8, Section. .... p. 71.
9. *Rhabdammina abyssorum* M. Sars. Recent. × 8. .... p. 71.
10, 11. *Rhabdammina linearis* H. B. Brady. Recent. × 10. (After H. B. Brady). 10, Exterior. 11, Longitudinal section. .... p. 71.
12. *Rhabdammina irregularis* W. B. Carpenter. Recent. × 10. (After H. B. Brady). .... p. 71.
13. *Masonella planulata* H. B. Brady. Recent. × 6. (After H. B. Brady). *a*, broad face; *b*, edge view. .... p. 71.
14. *Vanhoeffenella gaussi* Rhumbler. Recent. × 50. (After Wiesner). p. 72.

Family 3. RHIZAMMINIDAE, p. 73.

15. *Bathysiphon rufescens* Cushman. Recent. × 12. .... p. 73.
16, 17. *Bathysiphon filiformis* M. Sars. Recent. (After H. B. Brady). 16, Transverse section, × 32. 17, Longitudinal section, × 100. ... p. 73.
18, 19. *Rhizammina algaeformis* H. B. Brady. Recent. (After H. B. Brady). 18, Exterior, × 3. 19, Section, × 25. .... p. 73.
20. *Hippocrepinella hirudinea* Heron-Allen and Earland. Recent. × 15. (After Heron-Allen and Earland). .... p. 74.
21. *Hippocrepinella alba* Heron-Allen and Earland. Recent. × 15. (After Heron-Allen and Earland). .... p. 74.
22. *Marsipella elongata* Norman. Recent. × 50. (After Norman). .... p. 73.

Plate 1

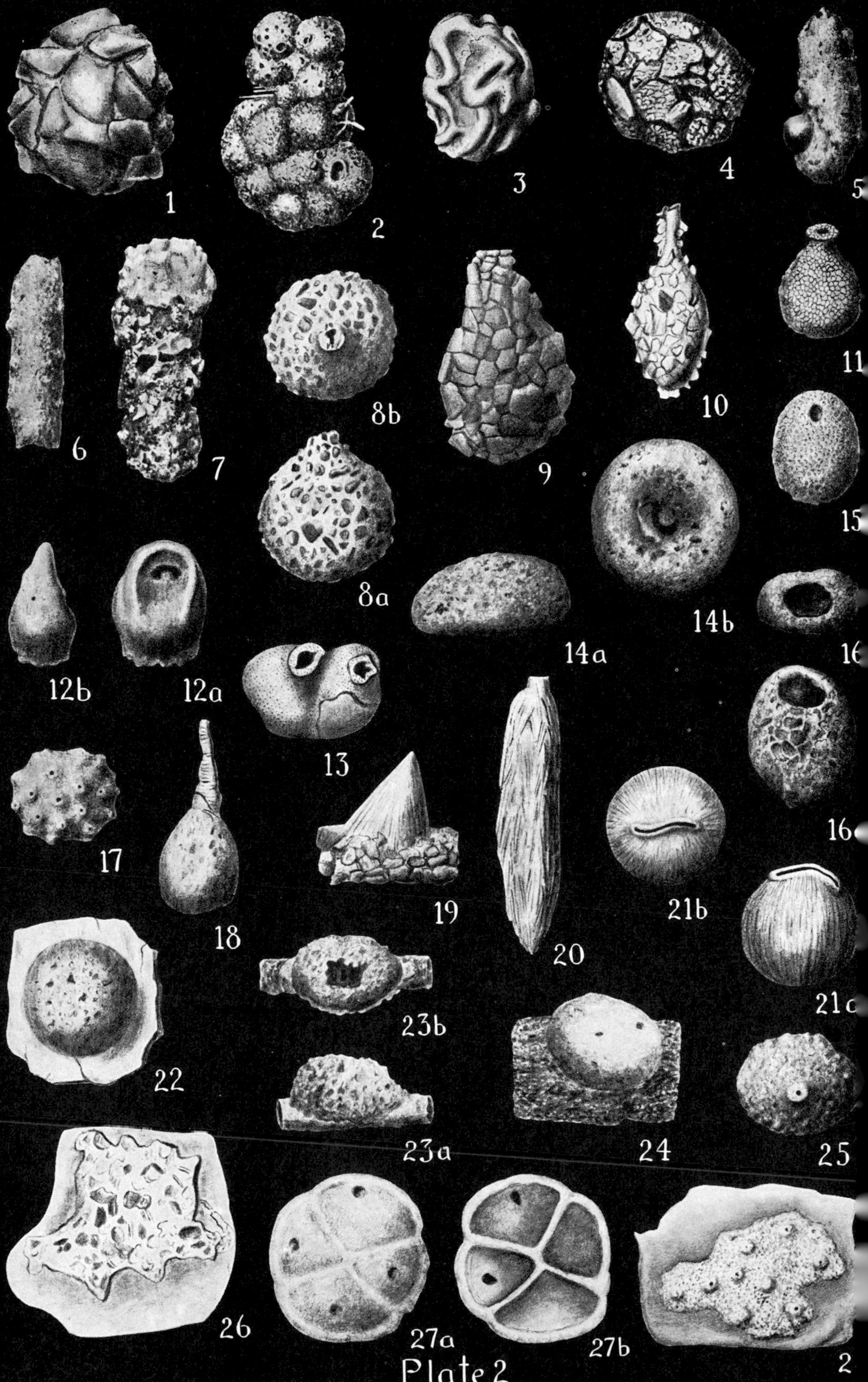
1
2
3
4
5
6
7
8b
8a
9
10
11
12b
12a
13
14a
14b
15
16
16
17
18
19
20
21b
21a
22
23b
23a
24
25
26
27a
27b
2

Plate 2

PLATE 2.

Family 4. SACCAMMINIDAE, p. 74.

FIG.

1. *Psammosphaera fusca* Schulze. Recent. × 50. p. 75.
2. *Sorosphaera confusa* H. B. Brady. Recent. × 7. (After H. B. Brady). p. 76.
3. *Storthosphaera albida* Schulze. Recent. × 50. p. 77.
4. *Blastammina polyedra* Eisenack. Silurian, Germany. × 50. (After Eisenack). p. 76.
5. *Raibosammina mica* Moreman. Ordovician, Oklahoma. × 35. (After Moreman). p. 77.
6. *Stegnammina cylindrica* Moreman. Silurian, Oklahoma. × 35. (After Moreman). p. 77.
7. *Psammophax consociata* Rhumbler. Recent. × 10. (After Wiesner). p. 77.
8. *Saccammina sphaerica* M. Sars. Recent. × 7. (After H. B. Brady). *a*, side view; *b*, apertural view. p. 78.
9. *Proteonina difflugiformis* H. B. Brady. Cretaceous, Texas. × 50. p. 78.
10. *Lagenammina laguncula* Rhumbler. Recent. (After Rhumbler). p. 78.
11. *Laguncula urnula* (Gruber). Recent. (After Gruber). p. 79.
12. *Millettella pleurostomelloides* (Millett). Recent. × 60. (After Millett). *a*, apertural view; *b*, side view. p. 79.
13. *Ammosphaeroides distoma* Cushman. Recent. × 25. (After Cushman). p. 80.
14. *Leptodermella arenata* (Cushman). Miocene, Florida. × 60, *a*, side view; *b*, apertural view. p. 80.
15. *Marsupulina schultzei* Rhumbler. Recent. × 65. (After Rhumbler). p. 79.
16. *Urnulina compressa* Cushman. Miocene, Florida. × 60, *a*, side view; *b*, apertural view. p. 79.
17. *Thurammina papillata* H. B. Brady. Recent. × 18. (After H. B. Brady). p. 80.
18. *Pelosina rotundata* H. B. Brady. Recent. × 20. (After H. B. Brady). p. 81.
19. *Rhaphidoscene conica* Vaughan Jennings. Recent. × 8. (After Vaughan Jennings). Not a foraminifer but the young form of a sponge. See Earland, Journ. Roy. Micr. Soc., ser. 3, vol. 64, 1944 (1946), pp. 26–30.
20. *Technitella legumen* Norman. Recent. × 85. (After Norman). p. 81.
21. *Pilulina jeffreysi* W. B. Carpenter. Recent. × 6. (After H. B. Brady). *a*, side view; *b*, apertural view. p. 82.
22. *Webbinella hemisphaerica* Jones, Parker and H. B. Brady. Recent. × 10. (After H. B. Brady). p. 82.
23. *Verrucina rudis* Goës. Recent. × 7. (After Goës). *a*, side view; *b*, apertural view. p. 83.
24. *Tholosina bulla* (H. B. Brady). Recent. × 10. p. 83.
25. *Colonammina verruca* Moreman. Silurian, Oklahoma. × 50. (After Moreman). p. 83.
26. *Iridia diaphana* Heron-Allen and Earland. Recent. × 8. (After Heron-Allen and Earland). p. 82.
27. *Urnula arctica* Cushman. Recent. × 60. *a*, dorsal side; *b*, ventral attached side. p. 84.
28. *Diffusilina humilis* Heron-Allen and Earland. Recent. × 20. (After Heron-Allen and Earland). p. 84.

PLATE 3.

Family 5. HYPERAMMINIDAE, p. 84.

FIG.

1. *Hyperammina bulbosa* Cushman and Waters. Pennsylvanian, Texas. × 60. .... p. 85.
2. *Hyperammina elongata* H. B. Brady. Cretaceous, Trinidad. × 40. p. 85.
3. *Saccorhiza ramosa* (H. B. Brady). Recent. × 10. (After H. B. Brady). .... p. 86.
4. *Jaculella acuta* H. B. Brady. Recent. × 6. (After H. B. Brady). ... p. 85.
5. *Nubeculariella birulai* Awerinzew. Recent. (After Awerinzew). .. p. 85.
6. *Earlandia perparva* Plummer. Pennsylvanian, Texas. × 25. (After Plummer). .... p. 86.

7, 8. *Hyperamminoides elegans* (Cushman and Waters). Pennsylvanian, Texas. × 35. 7, Megalospheric. 8, Microspheric. .... p. 86.

9. *Hippocrepina indivisa* Parker. Recent. × 40. *a*, side view; *b*, apertural view. .... p. 85.
10. *Dendronina arborescens* Heron-Allen and Earland. Recent. × 7. (After Heron-Allen and Earland). .... p. 87.
11. *Psammatodendron arborescens* Norman. Recent. × 15. (After H. B. Brady). .... p. 88.
12. *Dendrophrya erecta* Str. Wright. Recent. × 15. (After H. B. Brady). p. 87.
13. *Normanina conferta* (Norman). Recent. (After Norman). *a*, colony; *b*, individual test. .... p. 85.
14. *Haliphysema tumanowiczii* Bowerbank. Recent. (After Bowerbank). .... p. 87.
15. *Sagenina frondescens* (H. B. Brady). Recent. × 6. (After H. B. Brady). .... p. 88.
16. *Syringammina fragillissima* H. B. Brady. Recent. (After H. B. Brady). *a*, from side, × ⅔; *b*, section, × 5. .... p. 88.

Family 6. REOPHACIDAE, p. 89.

17. *Aschemonella scabra* H. B. Brady. Recent. × 10. (After H. B. Brady). .... p. 89.
18. *Hospitella fulva* Rhumbler. Recent. (After Rhumbler). .... p. 90.
19. *Kalamopsis vaillanti* deFolin. Recent. (After deFolin). .... p. 90.

20–22. *Hormosina globulifera* H. B. Brady. Recent. × 5. (After H. B. Brady). 20, 21, Megalospheric forms. 22, Microspheric form. p. 91.

23. *Sulcophax claviformis* Rhumbler. Recent. × 40. (After Wiesner). *a*, side view; *b*, apertural view. .... p. 91.
24. *Turriclavula interjecta* Rhumbler. Recent. (After Rhumbler). .... p. 92.

25, 26. *Haplostiche texana* (Conrad). Lower Cretaceous, Texas. × 6. (After Plummer). *a*, side view; *b*, apertural view. 26, Longitudinal section. .... p. 92.

27. *Reophax asper* Cushman and Waters. Pennsylvanian, Texas. × 50. (After Cushman and Waters). .... p. 90.
28. *Reophax texana* Cushman and Waters. Cretaceous, Texas. × 15. p. 90.
29. *Nodosinella perelegans* Plummer. Pennsylvanian, Texas. × 50. (After Plummer). .... p. 91.
30. *Nodosinella glennensis* Harlton. Pennsylvanian, Texas. × 75. (After Cushman and Waters). .... p. 91.
31. *Nodellum membranaceum* (H. B. Brady). Recent. × 50. (After H. B. Brady). .... p. 92.
32. *Ammosphaerulina adhaerens* Cushman. Recent. × 20. .... p. 92.
33. *Sphaerammina ovalis* Cushman. Recent. × 10. .... p. 92.

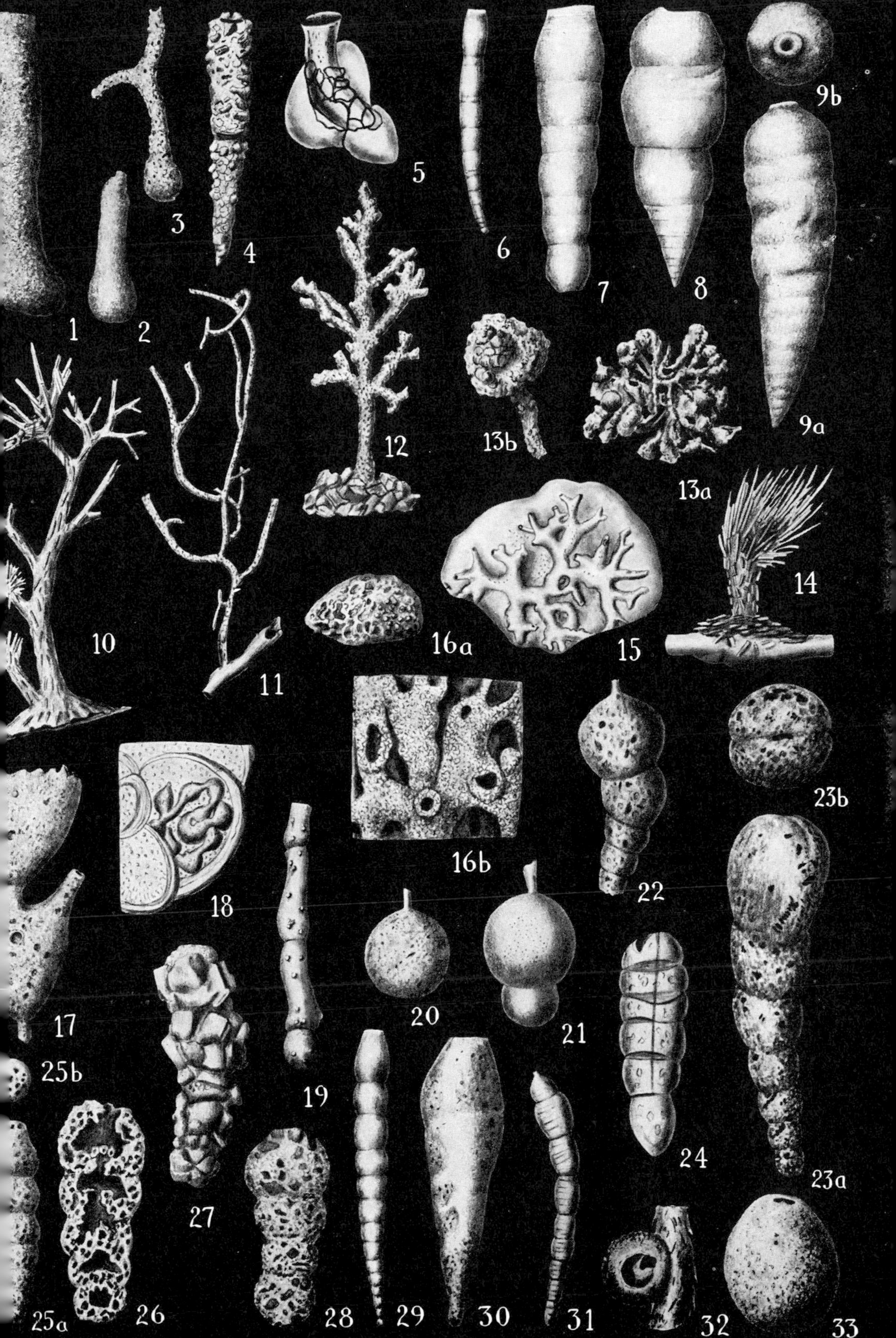
1
2
3
4
5
6
7
8
9a
9b
10
11
12
13a
13b
14
15
16a
16b
17
18
19
20
21
22
23a
23b
24
25a
25b
26
27
28
29
30
31
32
33

Plate 3

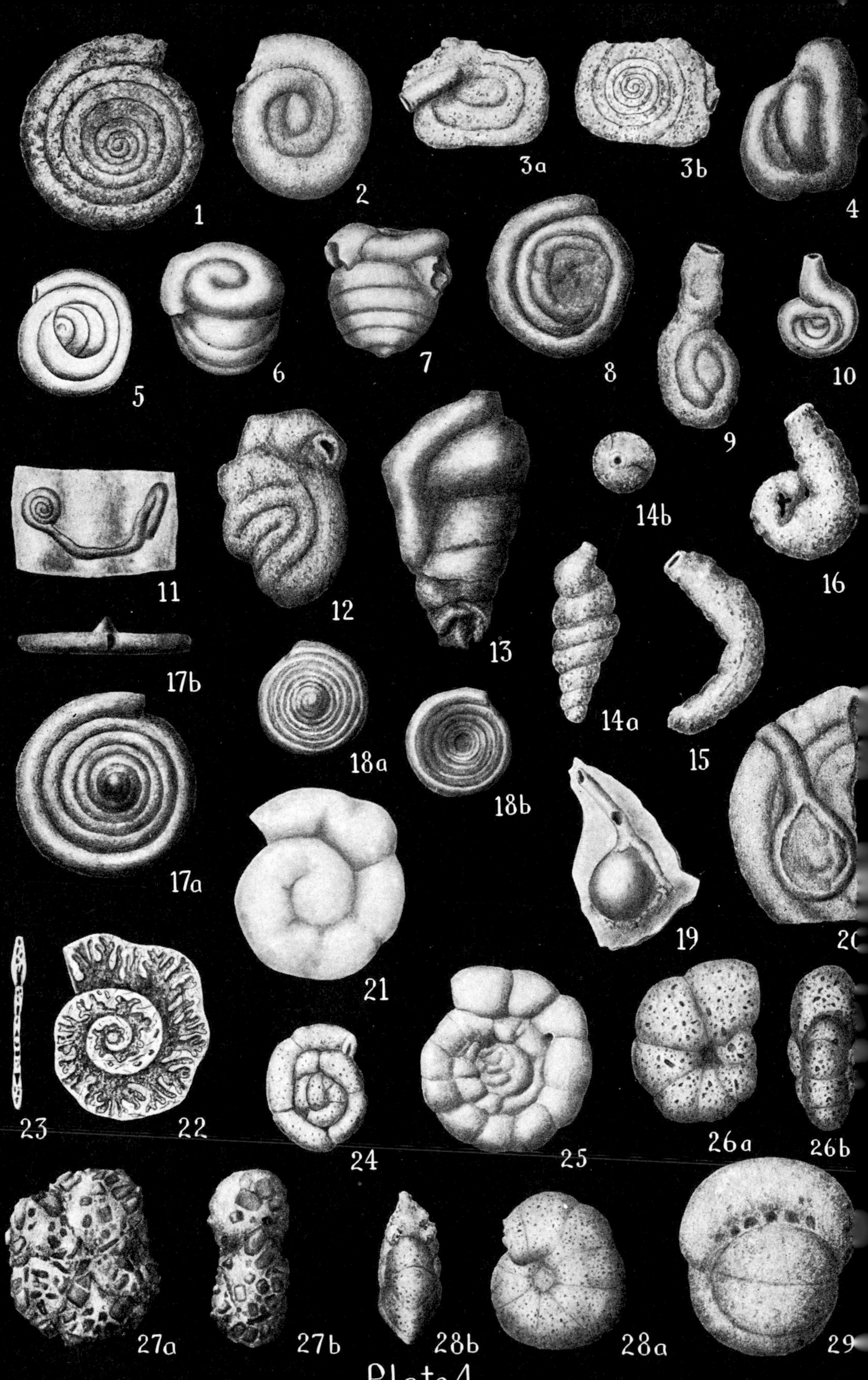
1
2
3a
3b
4
5
6
7
8
9
10
11
12
13
14a
14b
15
16
17a
17b
18a
18b
19
20
21
22
23
24
25
26a
26b
27a
27b
28a
28b
29

Plate 4

PLATE 4.

Family 7. AMMODISCIDAE, p. 93.

1. *Ammodiscus semiconstrictus* Waters, var. *regularis* Waters. Pennsylvanian, Texas. × 90. (After Cushman and Waters). .... p. 94.
2. *Ammodiscus pennyi* Cushman and Jarvis. Cretaceous, Trinidad. × 35. (After Cushman and Jarvis). .... p. 94.
3. *Hemidiscus perversus* (Sidebottom). Recent. × 35. (After Sidebottom). *a*, *b*, opposite sides. .... p. 94.
4. *Glomospira simplex* Harlton. Pennsylvanian, Texas. × 90. (After Cushman and Waters). .... p. 96.
5. *Glomospira gordialis* (Jones and Parker). Jurassic, Switzerland. (After Haeusler). .... p. 96.

6, 7. *Glomospira charoides* (Jones and Parker), var. *corona* Cushman and Jarvis. Cretaceous, Trinidad. × 60. (After Cushman and Jarvis). .... p. 96.

8. *Glomospira diversa* Cushman and Waters. Pennsylvanian, Texas. × 90. (After Cushman and Waters). .... p. 96.
9. *Lituotuba lituiformis* (H. B. Brady). Cretaceous, Trinidad. × 35. (After Cushman and Jarvis). .... p. 97.
10. *Lituotuba exserta* Moreman. Silurian, Oklahoma. × 50. (After Moreman). .... p. 97.
11. *Tolypammina delicatula* Cushman and Waters. Pennsylvanian, Texas. × 50. (After Cushman and Waters). .... p. 98.
12. *Ammovertella inclusa* (Cushman and Waters). Pennsylvanian, Texas. × 50. (After Cushman and Waters). .... p. 98.
13. *Trepeilopsis grandis* (Cushman and Waters). Pennsylvanian, Texas. × 25. (After Cushman and Waters). .... p. 99.
14. *Turritellella spirans* Cushman and Waters. Pennsylvanian, Texas. × 100. (After Cushman and Waters). *a*, side view; *b*, apertural view. .... p. 96.

15, 16. *Psammonyx vulcanicus* Döderlein. Recent. (After Rhumbler). 15, Megalospheric. 16, Microspheric. .... p. 97.

17. *Ammodiscoides conica* Cushman and Waters. Pennsylvanian, Texas. × 90. (After Cushman and Waters). *a*, from side; *b*, from edge. .... p. 96.
18. *Ammodiscoides turbinatus* Cushman, Cretaceous, Trinidad. × 30. (After Cushman and Jarvis). *a*, *b*, opposite sides. .... p. 96.
19. *Ammolagena contorta* Waters. Pennsylvanian, Oklahoma. × 20. .. p. 99.
20. *Ammolagena clavata* (Jones and Parker). Cretaceous, Trinidad. × 30. (After Cushman and Jarvis). .... p. 99.

21–23. *Discammina fallax* Lacroix. Recent. × 25. (After Lacroix). 21, Side view. 22, 23, Sections. .... p. 98.

Family 8. LITUOLIDAE, p. 99.

24, 25. *Trochamminoides proteus* (Karrer). Recent. × 20. 24 (After H. B. Brady). .... p. 101.

26. *Haplophragmoides canariensis* (d'Orbigny). Recent. × 60. (After Cushman and Parker). *a*, side view; *b*, apertural view. .... p. 102.
27. *Haplophragmoides rugosa* Cushman and Waters. Cretaceous, Tennessee. × 70. *a*, side view; *b*, apertural view. .... p. 102.
28. *Haplophragmoides glabra* Cushman and Waters. Cretaceous, Texas. × 70. *a*, side view; *b*, apertural view. .... p. 102.
29. *Cribrostomoides trinitatensis* Cushman and Jarvis. Cretaceous, Trinidad. × 60. (After Cushman and Jarvis). .... p. 102.

PLATE 5.

Family 8. LITUOLIDAE, p. 99.

FIG.

1. *Endothyra bowmani* Phillips. Pennsylvanian, England. × 35. (After H. B. Brady). *a*, side view; *b*, apertural view. .... p. 107.
2. *Endothyra media* Waters. Pennsylvanian, Texas. × 40. (After Cushman and Waters). *a*, side view; *b*, apertural view. .... p. 107.
3. *Bradyina nautiliformis* Möller. Pennsylvanian, Russia. × 8. (After Möller). *a*, side view; *b*, apertural view. .... p. 107.

4–6. *Glyphostomella triloculina* (Cushman and Waters). Pennsylvanian, Texas. × 35. (After Cushman and Waters). 4, 5, Young stages. 6, Adult. .... p. 108.

7. *Cribrospira panderi* Möller. Pennsylvanian, Russia. × 15. (After Möller). *a*, side view; *b*, apertural view. .... p. 108.
8. *Cribrospira grande* (Reuss). Cretaceous, Austria. (After Reuss). *a*, side view; *b*, apertural view. .... p. 108.

Family 11. VALVULINIDAE, p. 127.

9. *Orbignyna ovata* Hagenow. Cretaceous, Germany. × 15. (After Marsson). *a*, side view; *b*, apertural view. .... p. 140.

Family 8. LITUOLIDAE, p. 99.

10. *Ammobaculites coprolithiforme* (Schwager). Cretaceous, Trinidad. × 40. (After Cushman and Jarvis). .... p. 103.
11. *Ammobaculites stenomeca* Cushman and Waters. Pennsylvanian, Texas. × 60. (After Cushman and Waters). *a*, side view; *b*, apertural view. .... p. 103.
12. *Ammobaculites taylorensis* Cushman and Waters. Cretaceous, Texas. × 25. .... p. 103.
13. *Ammomarginulina ensis* Wiesner. Recent. × 65. (After Wiesner). p. 103.
14. *Ammomarginulina foliacea* (H. B. Brady). Recent. × 30. (After H. B. Brady). *a*, side view; *b*, edge view. .... p. 103.
15. *Endothyranella powersi* (Harlton). Pennsylvanian, Oklahoma. × 20. (After Harlton). *a*, side view; *b*, apertural view. .... p. 108.
16. *Flabellammina saratogaensis* Cushman. Cretaceous, Arkansas. × 18. (After Cushman). *a*, side view; *b*, apertural view. .... p. 103.
17. *Frankeina goodlandensis* Cushman and Alexander. Cretaceous, Texas. × 50. (After Cushman and Alexander). *a*, *b*, side views; *c*, apertural view. .... p. 104.
18. *Haplophragmium taylorensis* Cushman and Waters. Cretaceous, Texas. × 25. (After Cushman and Waters). *a*, side view; *b*, section; *c*, apertural view. .... p. 106.
19. *Haplophragmium aequale* Roemer. Cretaceous, Germany. (After Reuss). *a*, side view, *b*, apertural view. .... p. 106.
20. *Septammina bradyi* Cushman. Pennsylvanian, England. × 15. (After H. B. Brady). *a*, side view; *b*, apertural view. .... p. 109

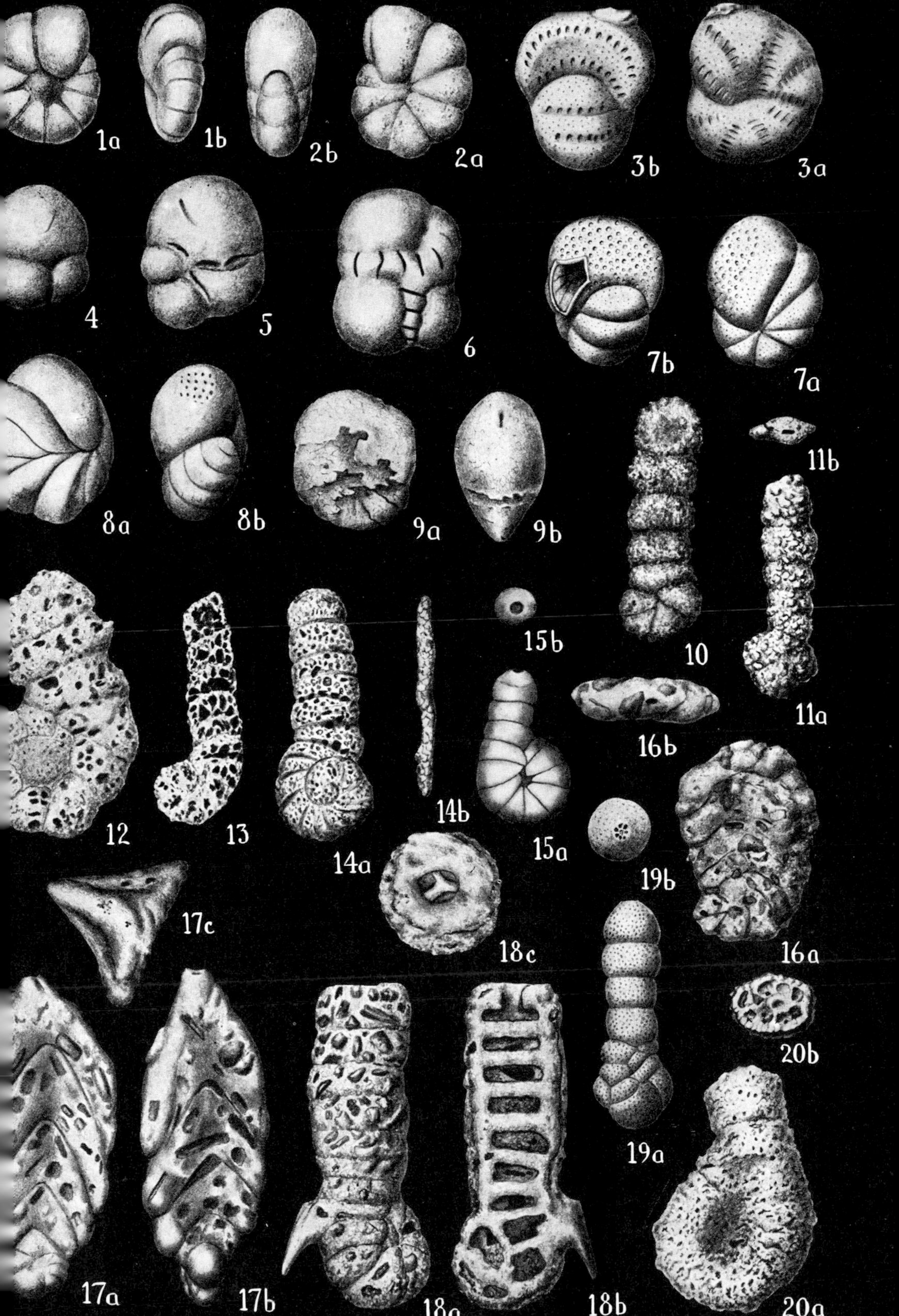
1a
1b
2b
2a
3b
3a
4
5
6
7b
7a
8a
8b
9a
9b
11b
10
11a
12
13
14b
14a
15b
15a
16b
16a
17c
18c
19b
19a
20b
17a
17b
18a
18b
20a

Plate 5

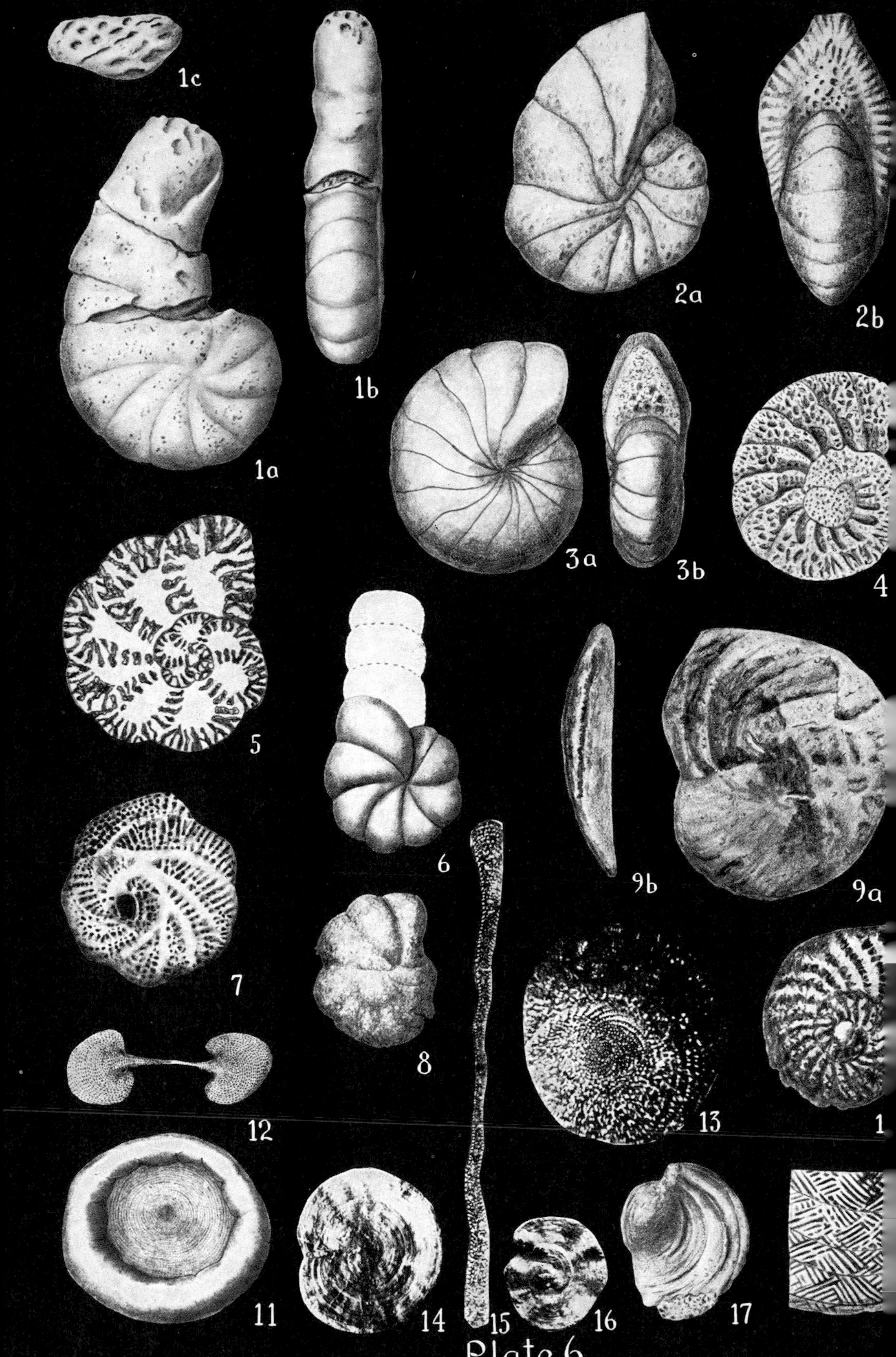

Plate 6

PLATE 6.

Family 8. LITUOLIDAE, p. 99.

FIG.

1. *Lituola taylorensis* Cushman and Waters. Cretaceous, Texas. × 10. (After Cushman and Waters). *a,* side view; *b,* edge view; *c,* apertural view. .... p. 110.
2. *Cyclammina elegans* Cushman and Jarvis. Cretaceous, Trinidad. × 20. (After Cushman and Jarvis). *a,* side view; *b,* apertural view. .... p. 109.

3, 4. *Cyclammina cancellata* H. B. Brady. Recent. × 10. (After H. B. Brady). 3 *a,* side view; *b,* apertural view. 4, Section showing labyrinthic interior. .... p. 109.

5, 6. *Pseudocyclammina lituus* (Yokoyama). Cretaceous, Japan. 5, Side view (After Yokoyama). 6, Section (After Yabe and Hanzawa). p. 110.

7. *Dictyopsella kiliani* Schlumberger. Cretaceous, Spain. × 15. (After Schlumberger). Section. .... p. 110.

8. *Dictyopsella chalmasi* Schlumberger. Cretaceous, Spain. × 10. (After Schlumberger). .... p. 110.

9, 10. *Choffatella decipiens* Schlumberger. Cretaceous, Portugal. (After Schlumberger). 9 *a,* side view; *b,* edge view. 10, Section. .... p. 110.

11, 12. *Orbitopsella praecursor* (Gümbel). Jurassic, Italy. × 8. (After Gümbel). 11, Side view. 12, Section. .... p. 111.

13–16. *Spirocyclina choffati* Munier-Chalmas. Cretaceous, France. (After Schlumberger and Choffat). 13, 15, Sections, × 8. 14, 16, Exteriors, × 2. .... p. 111.

17, 18. *Yaberinella jamaicensis* Vaughan. Eocene, Jamaica. (After Vaughan). 17, Exterior, × 4. 18, Part of section showing labyrinthic interior, × 15. .... p. 110.

PLATE 7.

Family 9. TEXTULARIIDAE, p. 112.

FIG.

1. *Spiroplectammina castensis* Cushman and Waters. Pennsylvanian, Texas. × 50. (After Cushman and Waters). *a*, side view; *b*, apertural view. .... p. 113.
2. *Spiroplectammina biformis* (Parker and Jones). Recent. × 90. (After Heron-Allen and Earland). .... p. 113.
3. *Spiroplectammina scotti* Cushman and Alexander. Lower Cretaceous, Texas. × 35. (After Cushman and Alexander). *a*, side view; *b*, apertural view. .... p. 113.
4. *Ammobaculoides navarroensis* Plummer. Cretaceous, Texas. × 50. *a*, side view; *b*, apertural view. .... p. 114.
5. *Textularia grahamensis* Cushman and Waters. Pennsylvanian, Texas. × 50. (After Cushman and Waters). *a*, front view; *b*, edge view. .... p. 115.
6. *Textularia ripleyensis* W. Berry. Cretaceous, Tennessee. × 65. *a*, front view; *b*, apertural view. .... p. 115.
7. *Cribrostomum bradyi* Möller. Carboniferous, Russia. × 25. (After Möller). .... p. 120.
8. *Deckerella clavata* Cushman and Waters. Pennsylvanian, Texas. × 12. (After Cushman and Waters). *a*, side view; *b*, apertural view. .... p. 120.
9. *Textularioides inflata* Cushman. Recent. × 12. (After Cushman). *a*, side view; *b*, apertural view. .... p. 116.
10. *Bigenerina ciscoensis* Cushman and Waters. Pennsylvanian, Texas. × 50. (After Cushman and Waters). *a*, side view; *b*, apertural view. .... p. 116.
11. *Bigenerina wintoni* Cushman and Alexander. Lower Cretaceous, Texas. × 35. (After Cushman and Alexander). *a*, side view; *b*, apertural view. .... p. 116.
12. *Climacammina cylindrica* Cushman and Waters. Pennsylvanian, Texas. × 30. (After Cushman and Waters). *a*, side view; *b*, apertural view. .... p. 120.
13. *Vulvulina advena* Cushman. Eocene, Alabama. × 30. .... p. 117.
14. *Vulvulina colei* Cushman. Eocene, Mexico. × 20. .... p. 117.
15. *Monogenerina atava* Spandel. Permian, Kansas. × 50. (After Spandel). Section. .... p. 117.
16. *Ammospirata mexicana* (Cushman). Oligocene, Mexico. × 35. (After Cushman). .... p. 114.
17. *Cribrogenerina*. Idealized section. .... p. 121.
18. *Spandelina excavata* Cushman and Waters. Permian, Texas. × 50. (After Cushman and Waters). *a*, side view; *b*, apertural view. p. 117.
19. *Geinitzina ciscoensis* Cushman and Waters. Pennsylvanian, Texas. × 50. (After Cushman and Waters). *a*, side view; *b*, apertural view. .... p. 118.

Family 10. VERNEUILINIDAE, p. 121.

20. *Spiroplectinata annectens* (Parker and Jones). Lower Cretaceous, England. × 40. (After Chapman). *a*, side view; *b*, edge view. p. 125.
21. *Verneuilina schizea* Cushman and Alexander. Lower Cretaceous, Texas. × 50. (After Cushman and Alexander). *a*, side view; *b*, apertural view. .... p. 122.
22. *Verneuilina polystropha* Reuss. Cretaceous, Trinidad. × 60. (After Cushman and Jarvis). *a*, side view; *b*, apertural view. .... p. 122.
23. *Tritaxia pyramidata* Reuss. Cretaceous, Trinidad. × 25. (After Cushman and Jarvis). *a*, side view; *b*, apertural view. .... p. 122.
24. *Gaudryina triangularis* Cushman. Miocene, California. × 30. *a*, side view; *b*, edge view. .... p. 123.
25. *Heterostomella foveolata* Marsson. Cretaceous, Texas. × 60. *a*, side view; *b*, apertural view. .... p. 127.
26. *Gaudryinella delrioensis* Plummer. Lower Cretaceous, Texas. × 35. (After Plummer). .... p. 125.

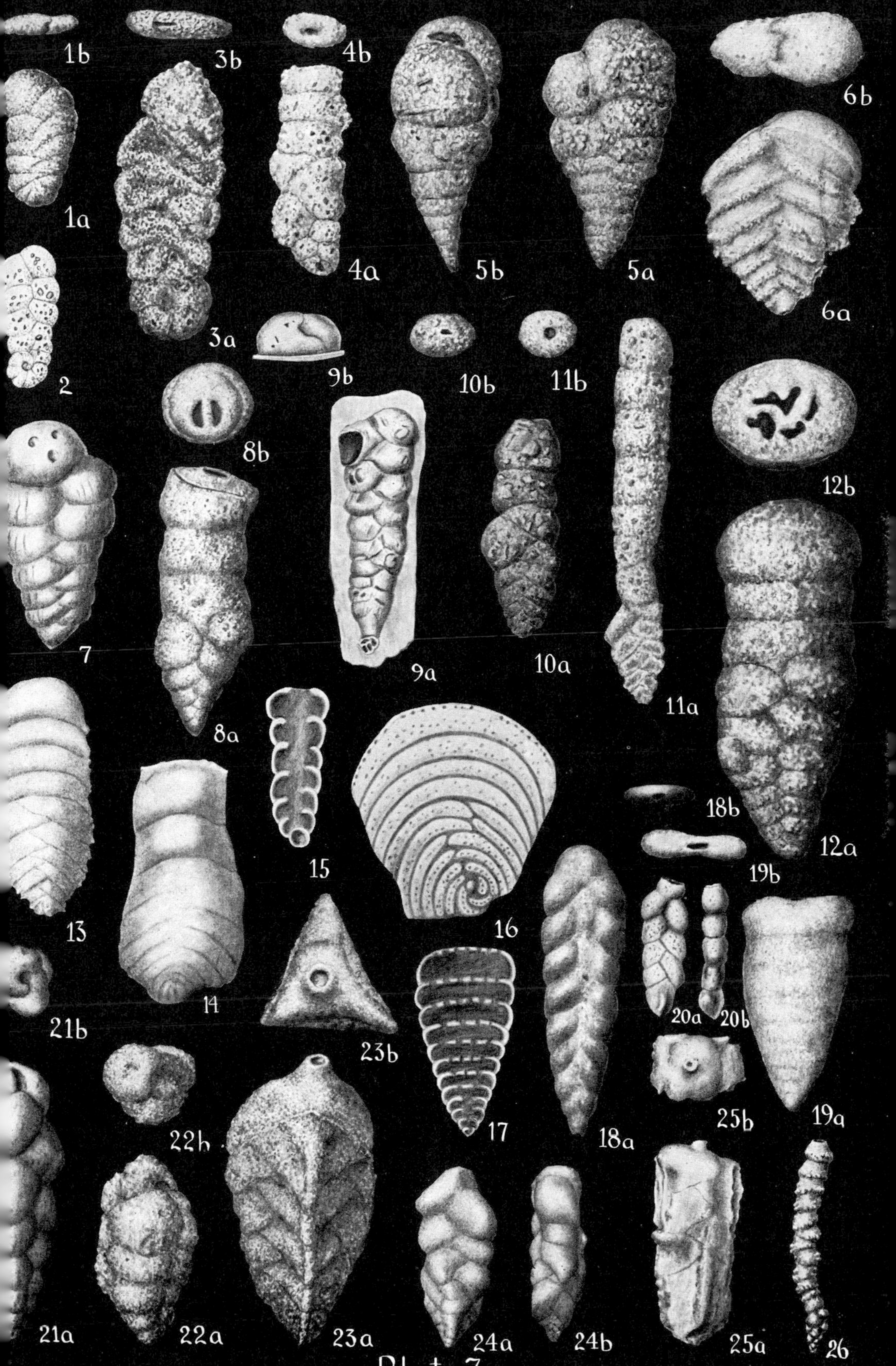
1b
3b
4b
6b
1a
4a
5b
5a
6a
3a
2
9b
10b
11b
8b
12b
7
9a
10a
11a
8a
15
16
18b
12a
19b
13
14
20a
20b
21b
23b
17
25b
19a
22h
18a
21a
22a
23a
24a
24b
25a
26

Plate 7

1b
2b
3
4
6
9b
1a
2a
5b
8
7
9c
5a
12b
10a
11b
13b
14b
10b
16b
11a
12a
13a
14a
17b
16a
18b
20
19
18a
21
22a
17a
23b
23a
24b
24a
25a

Plate 8

PLATE 8.

Family 11. VALVULINIDAE, p. 127.

FIG.

1. *Valvulina triangularis* d'Orbigny. Eocene, France. (After d'Orbigny's Model). *a*, side view; *b*, apertural view. .... p. 129.

Family 10. VERNEUILINIDAE, p. 121.

2. *Clavulinoides insignis* (Plummer). Cretaceous, Texas. × 20, *a*, side view; *b*, apertural view. .... p. 126.

Family 11. VALVULINIDAE, p. 127.

3. *Clavulina parisiensis* d'Orbigny. Eocene, France. (After d'Orbigny's Model). .... p. 130.

4, 5. *Arenobulimina preslii* (Reuss). Cretaceous, Bohemia. 4 (After Reuss). 5 *a*, side view; *b*, apertural view. .... p. 131.

6–8. *Martinottiella communis* (d'Orbigny). Topotypes. Recent, Rimini, Italy. × 20. 6, Microspheric. 7, Megalospheric. 8, Section near base. .... p. 138.

9. *Eggerella bradyi* (Cushman). Recent. × 50. *a*, side view; *b*, apertural view; *c*, basal view. .... p. 131.

10. *Dorothia bulletta* (Carsey). Cretaceous, Texas. × 60. *a*, side view; *b*, basal view. .... p. 136.

11. *Plectina ruthenica* (Reuss). Cretaceous, Galicia. (After Reuss). *a*, side view; *b*, apertural view. .... p. 137.

12. *Goësella rotundata* (Cushman). Recent. × 18. *a*, side view; *b*, apertural view. .... p. 137.

13, 14. *Karreriella siphonella* (Reuss). Oligocene, Germany. (After Reuss). *a*, *a*, side views; *b*, *b*, apertural views. 13, Young microspheric form. 14, Adult megalospheric form. .... p. 138.

15. *Schenckiella primaeva* (Cushman). Recent. × 18. *a*, side view; *b*, apertural view. .... p. 138.

16. *Hagenowella gibbosa* (d'Orbigny). Cretaceous, Texas. × 30. *a*, side view; *b*, section. .... p. 139.

17. *Valvulammina globulosa* (d'Orbigny). Eocene, France. × 20. *a*, dorsal view; *b*, ventral view. .... p. 131.

18–21. *Liebusella soldanii* (Jones and Parker). Recent. × 20. 18 (After H. B. Brady). *a*, side view; *b*, apertural view. 19, Section of early stage. 20, Section of later stage. 21 (After H. B. Brady). Longitudinal section. .... p. 139.

22. *Tritaxilina caperata* (H. B. Brady). Recent. × 10. (After H. B. Brady). *a*, side view; *b*, apertural view. .... p. 138.

23. *Marssonella oxycona* (Reuss). Cretaceous, Texas. × 25. *a*, side view; *b*, apertural view. .... p. 134.

24, 25. *Textulariella cretosa* Cushman. Cretaceous, England. × 15. 24 *a*, side view; *b*, end view showing interior. 25 *a*, side view; *b*, apertural view. .... p. 134.

PLATE 9.

Family 11. VALVULINIDAE, p. 127.

FIG.

1–3. *Cuneolina conica* d'Orbigny. Cretaceous, Spain. × 15. (After Schlumberger). 1, Side view. 2, 3, Sections. .... p. 135.

4. *Cuneolina angusta* Cushman. Recent. × 15. Section near base of microspheric form. .... p. 135.

5, 6. *Dicyclina schlumbergeri* Munier-Chalmas. Cretaceous, France. (After Schlumberger and Choffat). Idealized sections. .... p. 135.

7–9. *Ataxophragmium variabile* (d'Orbigny). Cretaceous, France. (After d'Orbigny). .... p. 139.

10, 11. *Pernerina depressa* (Perner). Cretaceous, Bohemia. × 25. 10 *a*, *c*, opposite sides; *b*, edge view. 11, Section showing pillars of interior. .... p. 141.

12. *Chrysalidina gradata* d'Orbigny. Cretaceous, France. (After d'Orbigny). *a*, side view; *b*, apertural view. .... p. 132.

13. *Cribrobulimina polystoma* (Parker and Jones). Pliocene, Australia. × 20. .... p. 130.

14–16. *Lituonella roberti* Schlumberger. Eocene, France. × 5. (After Schlumberger). 14, 15, Side views. 16, Apertural view. .... p. 140.

17, 18. *Lituonella douvilléi* Davies. Eocene, India. (After Davies). 17, Side view, × 8. 18, Section showing pillars, × 50. .... p. 140.

19, 20. *Lituonella liburnica* Schubert. Eocene, Dalmatia. × 10. (After Schubert). 19, Side view. 20, Apertural view. .... p. 140.

21, 22. *Coskinolina balsilliei* Davies. Eocene, India. (After Davies). 21, Side view, × 8. 22, Section showing pillars and outer chamberlets, × 50. .... p. 141.

23, 24. *Coskinolina liburnica* Stache. Eocene, Dalmatia. × 7. (After Schubert). 23, Side view. 24, Apertural view. .... p. 141.

25–27. *Dictyoconus indicus* Davies. Eocene, India. (After Davies). 25, Side view, × 8. 26, Section, × 12. 27, Section showing pillars, outer chamberlets and cellules, × 50. .... p. 141.

28–30. *Gunteria floridana* Cushman and Ponton. Eocene, Florida. × 40. (After Cushman and Ponton). 28 Exterior. 29, Eroded specimen showing interior structure. 30, Apertural view of young specimen. .... p. 142.

1
2
3
7
8
9
4
5
6
10a
10b
11
10c
13
12a
12b
15
16
14
21
26
24
17
20
19
25
23
18
22
27
28
30
29

Plate 9

1
te
d
t
2
3
5
4
t
te
9
6
8
7
10
13
11
14
15
12
16
18
21
17
tu
ch
22
19
tu
ch
24
20
23

Plate 10

PLATE 10.

Family 12. FUSULINIDAE, p. 146.

FIG.

1. Diagram of the structure of a primitive fusuline wall. In the last few chambers there are only the two primary layers, (1) the thin dark tectum (*t*) on the outside, and (2) the clear calcite layer, diaphanotheca (*d*), on the inside. In preceding chambers a secondary deposit of grey color coats the wall forming the tectorium (*te*), which, in all but the outer volution, forms a layer on both inside and outside of the wall.

2, 3. *Staffella sphaerica* (Möller) [non Abich ?]. (After Möller). 3, specimen weathered at the middle so as to show the tunnel. .... p. 147.

4. *Pseudostaffella.* Axial section. (After Lee and Chen). .... p. 148.

5, 6. *Fusulinella bocki* Möller, the latter a weathered specimen with the spirotheca abraded to show the edges of the simple septa and part of the tunnel. (After Möller). .... p. 152.

7. *Fusulinella itoi* Ozawa, axial section. (After Ozawa). .... p. 152.

8. *Schubertella transitoria* Staff and Wedekind. Copy of original illustration. .... p. 155.

9. *Fusiella typica* Lee and Chen. A sagittal section showing the thin wall to consist of tectum (*t*), coated on both inside and out by tectoria (*te*). (After Lee and Chen). .... p. 153.

10, 11. *Fusulina* sp. nov., from the Lonsdale limestone in Illinois, showing the deep septal folding characteristic of this genus. .... p. 156.

12. *Fusulina cylindrica* Fischer. (After Möller). .... p. 156.

13–15. *Wedekindellina euthysepta* (Henbest), axial section and two exterior views. .... p. 154.

16–20. *Triticites ventricosus* (Meek and Hayden); 16, 17, juvenile and adult specimens showing antetheca; 18, sagittal section; 19, a weathered and exfoliated specimen showing the tunnel (*tu*), the edges of the septa, and the chomata (*ch*); 20, axial slice showing the tunnel (*tu*) and chomata (*ch*). .... p. 157.

21–23. *Schwagerina huecoensis* (Dunbar and Skinner); 21, external view showing the antetheca; 22, tangential slice showing the tunnel in the penultimate whorl, and the plan of the septal folds; 23, axial section showing the septal loops and the lack of chomata. p. 157.

24. *Paleofusulina prisca* Deprat, showing the pillar-like sections of the septal folds. (After Deprat). .... p. 163.

PLATE 11.

Family 12. FUSULINIDAE, p. 146.

FIG.

1, 2. *Paraschwagerina yabei* (Staff); 1, external view and 2, sagittal section showing the tightly coiled juvenarium and the sudden inflation. .... p. 161.

3, 6, 7. *Parafusulina wordensis* Dunbar and Skinner; 3, a drawing of a fragment to show the shape of the septal folds; 6, fragment of a silicified specimen with the wall of the outer volution and part of the septa broken away, showing the complex septal folds and the spirally directed basal margins of the septa (*s*); 7, nearly half of a similar specimen. .... p. 161.

4, 5, 8. *Parafusulina kattaensis* (Schwager); 4, external view showing the deeply folded antetheca and the lack of an external aperture; 5, axial slice showing the tunnel (*t*) and the open arches (*a*) at the base of the septal folds; 8, tangential slice which for a distance on each side of the middle cuts near the floor of the penultimate whorl, showing the cuniculi. .... p. 161.

9–12. *Polydiexodina shumardi* Dunbar and Skinner; 9, axial section showing 4 pairs of accessory tunnels (t_1–t_4); 10, sagittal section showing the close coiling and an irregularly shaped proloculum; 11, a tangential slice showing the accessory tunnels (*t*) in section in the outer whorls and in plan in the inner whorl; 12, another tangential slice, more greatly enlarged. .... p. 162.

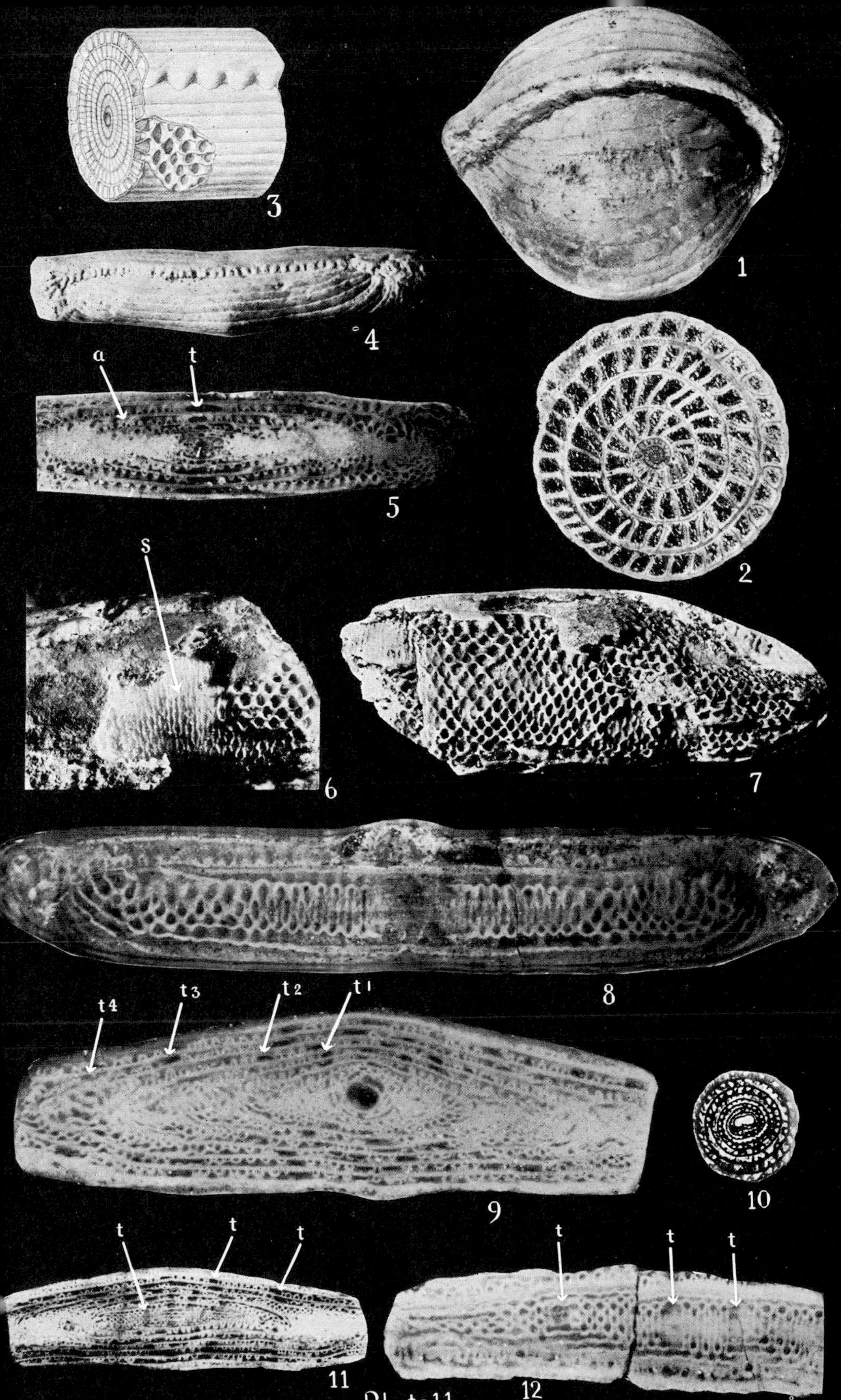

Plate 11

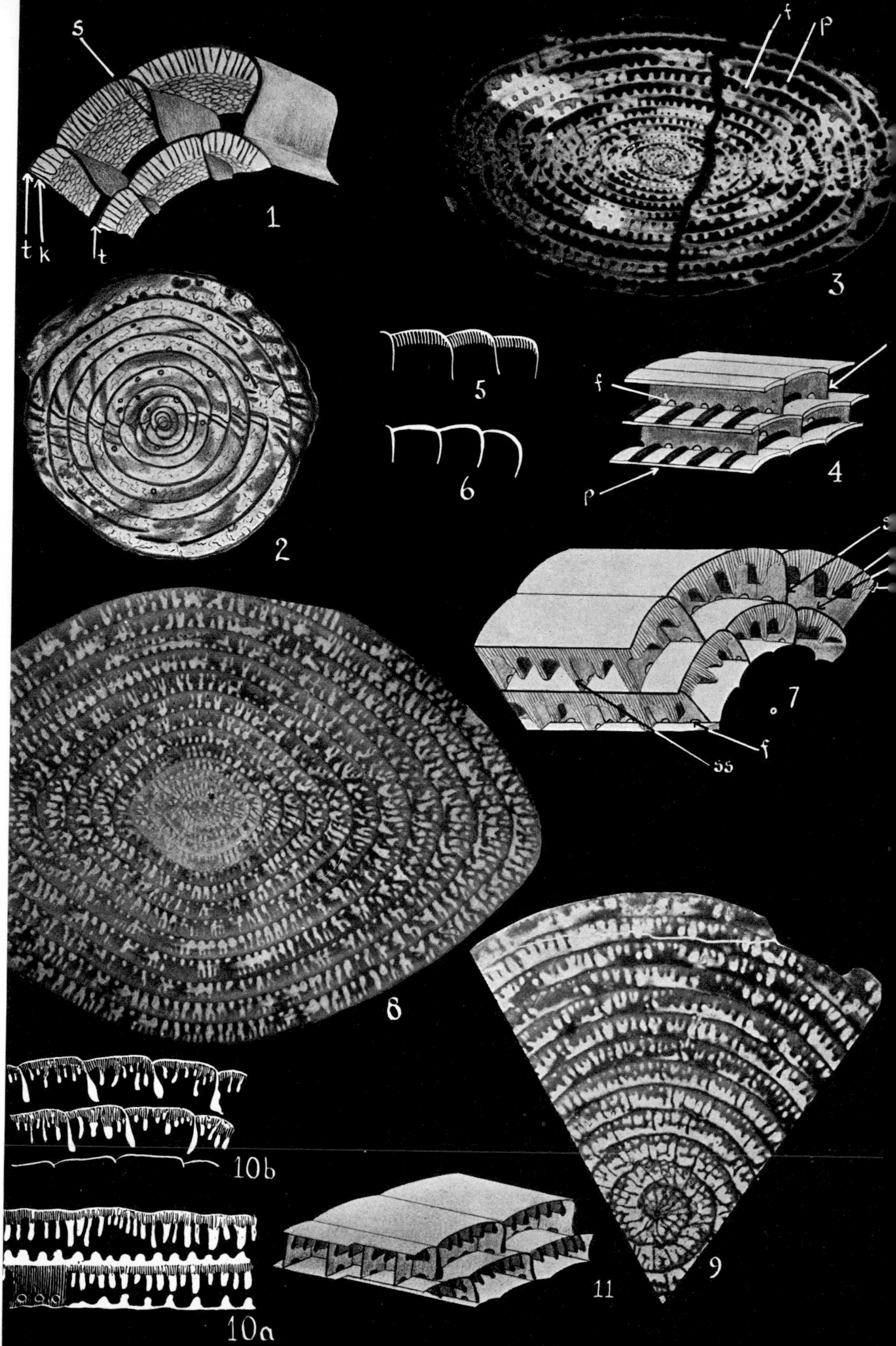

Plate 12

PLATE 12.

Family 13. NEOSCHWAGERINIDAE, p. 164.

FIG.

1. Diagram to show the structure of the wall and the septa in the Neoschwagerinidae and in the Schwagerininae of the family Fusulinidae: *k*, keriotheca; *t*, tectum; *s*, septum.
2. *Verbeeckina verbeecki* (Geinitz). Axial section (After Deprat, showing the feeble appearance of the parachomata. .... p. 165.
3. *Misellina lepida* (Schwager). Axial section, showing the parachomata (*p*) and in places part of a septum with basal foramina (*f*). .... p. 166.
4. Diagram of part of two volutions to show the structure of *Misellina*: *f*, foramina; *p*, parachomata; *s*, septum. .... p. 166.
5. Section of the wall of *Misellina*, showing the keriotheca. .... p. 166.
6. Section of the wall of *Pseudodoliolina*, showing the thin compact wall. p. 166.
7. Diagram to show the shell structure of *Neoschwagerina*: *f*, foramen at the base of a septum; *k*, keriotheca; *s*, septum; *sa*, axial septulum; *ss*, spiral septulum; *t*, tectum. .... p. 167.

8, 9. *Yabeina globosa kotoi* Ozawa. Axial section and sector of a transverse but eccentric section. .... p. 168.

10. Diagram of the wall strucure of *Yabeina*: *a*, part of axial section; *b*, part of a sagittal section. (After Deprat). .... p. 168.
11. Diagram to show the shell structure of *Sumatrina*. .... p. 169.

PLATE 13.

Family 14. LOFTUSIIDAE, p. 170.

FIG.

1\. *Loftusia persica* H. B. Brady. Upper Cretaceous, Persia. × 1. (After H. B. Brady). p. 170.

Family 15. NEUSINIDAE, p. 170.

2–5\. *Jullienella foetida* Schlumberger. Recent. (After Schlumberger). 2, 3, Exterior, × 2/3. 4, Longitudinal section of tip. 5, Transverse section. p. 171

6–8\. *Neusina agassizi* Goës. Recent. 6, 7, (After Goës). 6, Complete specimen, × 1/4. 7, Transverse section, × 8. 8, Specimen showing coiled young, × 2/3 (After Cushman). p. 171.

9\. *Schizammina labyrinthica* Heron-Allen and Earland. Recent. × 2. (After Heron-Allen and Earland). p. 172.

10–12\. *Protobotellina cylindrica* Heron-Allen and Earland. Recent. (After Heron-Allen and Earland). 10, Exterior, × 2. 11, Longitudinal section. 12, Transverse section, × 8. p. 171.

13, 14\. *Botellina pinnata* Pearcey. Recent. (After Pearcey). 13, Exterior, × 1. 14, Section, × 3. p. 171.

Family 16. SILICINIDAE, p. 172.

15\. *Silicina limitata* (Terquem). Jurassic, France, × 25. (After Terquem). *a*, side view; *b*, apertural view. p. 173

16, 17\. *Involutina silicea* (Terquem). Jurassic, France. × 8. (After Terquem). 16, side view; 17, section. p. 173.

18\. *Problematina deslongchampsi* (Terquem). Jurassic, France. × 50. (After Terquem). *a*, side view; *b*, apertural view. p. 173.

19\. *Rzehakina epigona* (Rzehak), var. *lata* Cushman and Jarvis. Cretaceous, Trinidad. × 30. (After Cushman and Jarvis). *a*, side view; *b*, apertural view. p. 173

20\. *Rzehakina epigona* (Rzehak). Cretaceous, Moravia. (After Rzehak). Section. p. 173

21, 22\. *Silicosigmoilina californica* Cushman and Church. Cretaceous, California. × 40. (After Cushman and Church). 21, Side view. 22, Transverse section. p. 174.

23, 24\. *Miliammina oblonga* (Chapman). Recent. × 35. (After Heron-Allen and Earland). 23, Section. 24 *a*, side view; *b*, apertural view. p. 174.

25\. *Miliammina lata* Heron-Allen and Earland. Recent. × 35. (After Heron-Allen and Earland). *a*, side view; *b*, apertural view ... p. 174.

26\. *Miliammina circularis* Heron-Allen and Earland. Recent. × 25. (After Heron-Allen and Earland). *a*, side view; *b*, apertural view. p. 174

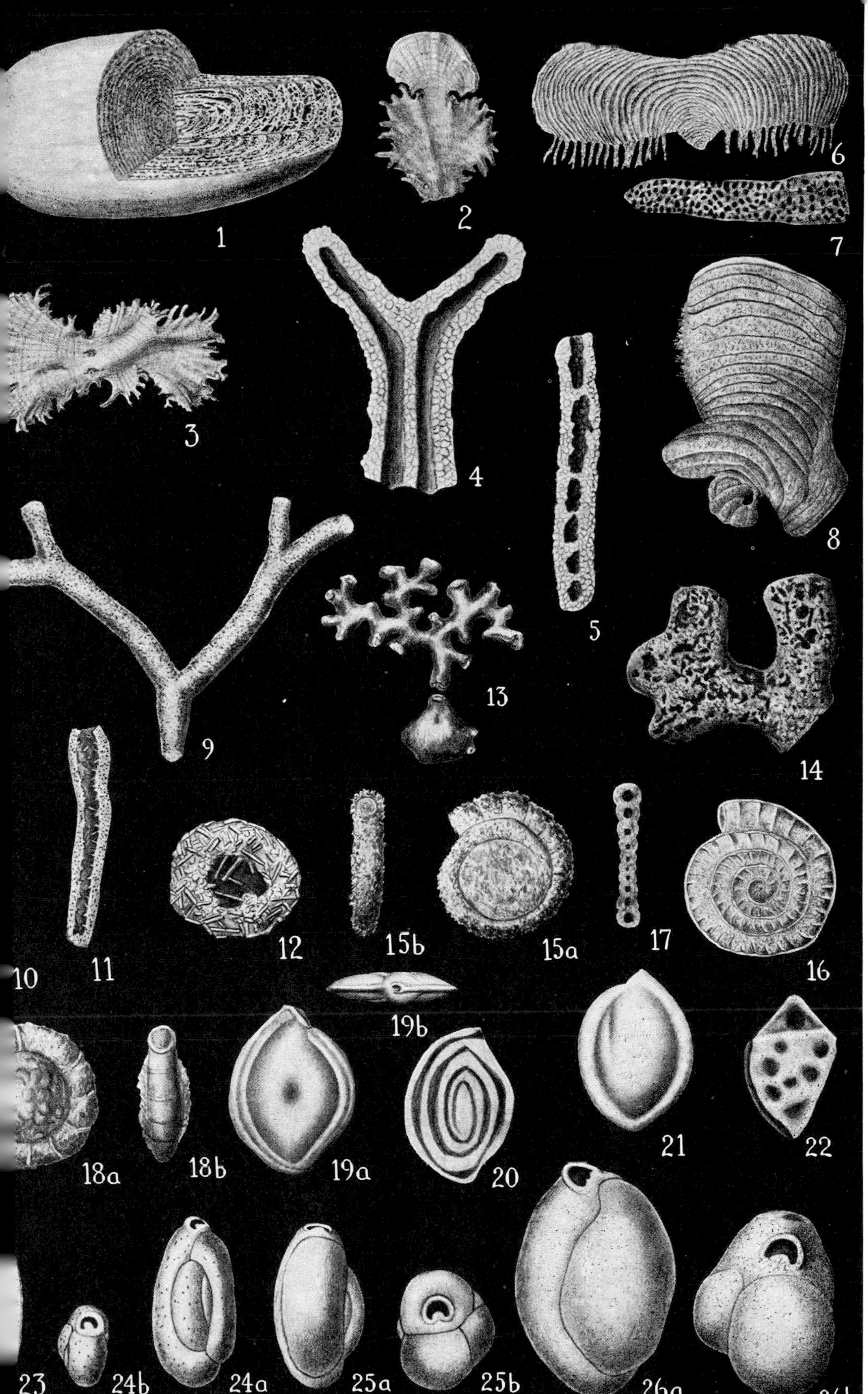
1
2
6
7
3
4
5
8
9
13
14
10
11
12
15b
15a
17
16
19b
18a
18b
19a
20
21
22
23
24b
24a
25a
25b
26a
26b

Plate 13

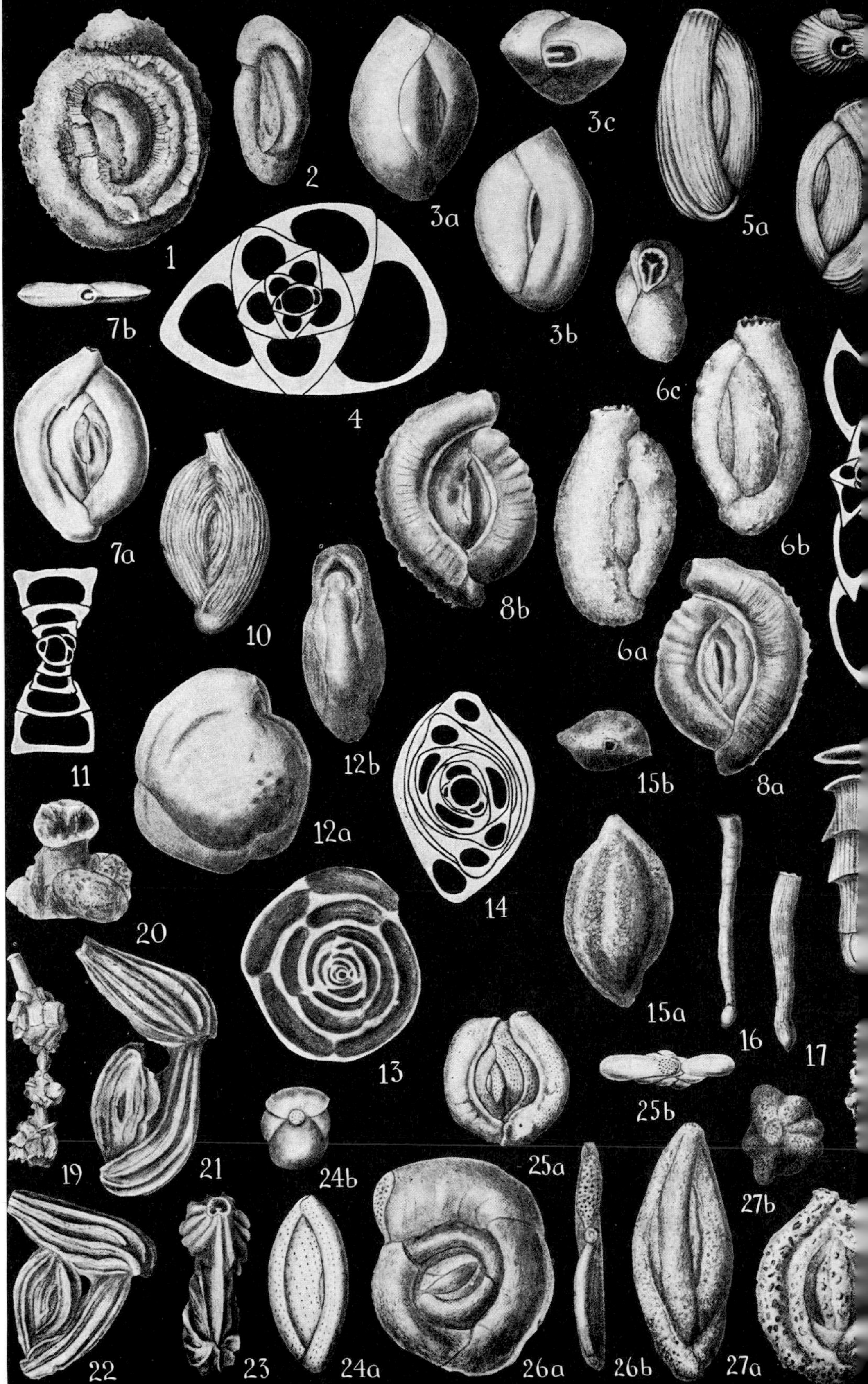

1
2
3a
3c
3b
5a
4
7b
6c
7a
6b
8b
10
6a
11
12b
15b
8a
12a
14
20
15a
13
16
17
25b
24b
25a
19
21
27b
22
23
24a
26a
26b
27a

Plate 14

PLATE 14.

Family 17. MILIOLIDAE, p. 174.

FIG.

1, 2. *Agathammina pusilla* (Geinitz). (After H. B. Brady). 1, Carboniferous, Great Britain, × 35. 2, Permian, × 15. .... p. 177.

3, 4. *Quinqueloculina seminula* (Linné). Recent. 3, × 40. (After H. B. Brady). *a*, *b*, opposite sides; *c*, apertural view. 4, Transverse section, × 65. (After Schlumberger). .... p. 177.

5. *Quinqueloculina costata* d'Orbigny. Recent. × 50. (After Cushman and Valentine). *a*, *b*, opposite sides; *c*, apertural view. .... p. 177.

6. *Dentostomina bermudiana* Carman. Recent. × 20. (After Carman). *a*, *b*, opposite sides; *c*, apertural view. .... p. 178.

7. *Massilina pratti* Cushman and Ellisor. Eocene, Texas. × 40. (After Cushman and Ellisor). *a*, side view; *b*, apertural view. .... p. 178.

8, 9. *Massilina secans* (d'Orbigny). Recent. (After Schlumberger). 8 *a*, *b*, opposite sides, × 12. 9, Section, × 20. .... p. 178.

10. *Spiroloculina antillarum* d'Orbigny. Recent. × 50. .... p. 178.

11. *Spiroloculina depressa* d'Orbigny. Recent. × 50. (After Martinotti). Section. .... p. 178.

12, 13. *Nummoloculina contraria* (d'Orbigny). Recent. × 15. (After H. B. Brady). 12 *a*, side view; *b*, apertural view. 13, Section. ..... p. 179.

14. *Sigmoilina sigmoidea* (H. B. Brady). Recent. × 50. (After Schlumberger). Transverse section. .... p. 179.

15. *Sigmoilina celata* (Costa). Recent. × 35. (After H. B. Brady). *a*, side view; *b*, apertural view. .... p. 179.

16. *Tubinella inornata* (H. B. Brady). Recent. × 20. (After H. B. Brady). .... p. 182.

17. *Tubinella funalis* (H. B. Brady). Recent. × 20. (After H. B. Brady). .... p. 182.

18. *Articulina sagra* d'Orbigny. Recent. × 35. (After H. B. Brady). *a*, side view; *b*, apertural view. .... p. 180.

19, 20. *Nubeculina divaricata* (H. B. Brady). Recent. × 40. (After H. B. Brady). 19, Side view. 20, Enlarged view of aperture showing everted lip and teeth. .... p. 182.

21–23. *Ptychomiliola separans* (H. B. Brady). Recent. × 12. (After H. B. Brady). 23, Apertural view. .... p. 182.

24. *Miliola saxorum* Lamarck. Eocene, France. (After Terquem). *a*, side view; *b*, apertural view. .... p. 182.

25. *Heterillina guespellensis* Schlumberger. Eocene, France. (After Schlumberger). *a*, side view; *b*, apertural view. .... p. 183.

26. *Hauerina bradyi* Cushman. Recent. × 35. (After H. B. Brady). *a*, side view; *b*, apertural view. .... p. 183.

27. *Schlumbergerina alveoliniformis* (H. B. Brady). Recent. × 15. (After H. B. Brady). *a*, side view; *b*, apertural view. .... p. 184.

28. *Ammomassilina alveoliniformis* (Millett). Recent. × 20. (After Millett). *a*, side view; *b*, apertural view. .... p. 184.

PLATE 15.

Family 17. MILIOLIDAE, p. 174.

FIG.

1. *Triloculina trigonula* (Lamarck). Recent. × 55. *a*, side view; *b*, apertural view. p. 184
2. *Triloculina laevigata* d'Orbigny. Recent. × 65. (After Schlumberger). Transverse section. p. 184.
3. *Triloculina bicarinata* d'Orbigny. Recent. × 40. *a*, *b*, opposite sides; *c*, apertural view. p. 184.
4. *Austrotrillina howchini* (Schlumberger). Eocene, France. × 35. (After Schlumberger). Section. p. 185.
5. *Flintina bradyana* Cushman. Recent. × 20. (After H. B. Brady). *a*, *b*, opposite sides; *c*, apertural view. p. 185.
6. *Miliolinella oblonga* (Montagu). Recent. × 65. (After Cushman and Valentine). *a*, *b*, opposite sides; *c*, apertural view. p. 177.

7, 8. *Pyrgo sarsi* (Schlumberger). Recent. (After Schlumberger). 7, Section, × 20. 8 *a*, side view; *b*, edge view; *c*, apertural view, × 12. p. 185.

9. *Pyrgo murrhina* (Schwager). Recent. × 40. *a*, side view; *b*, apertural view. p. 185.

10. *Pyrgo denticulata* (H. B. Brady), var. *striolata* (H. B. Brady). Recent. × 40. *a*, side view; *b*, apertural view. p. 185.

11. *Biloculinella globula* (Bornemann). Recent. × 40. *a*, front view; *b*, apertural view. p. 186.

12, 13. *Fabularia discolithes* Defrance. Eocene, France. (After Schlumberger). 12 *a*, side view; *b*, apertural view. × 35. 13, Transverse section, × 30. p. 186.

14. *Flintia robusta* (H. B. Brady). Recent. × 13. (After H. B. Brady). *a*, side view; *b*, apertural view. p. 186.

15, 16. *Nevillina coronata* (Millett). Recent. × 18. (After Sidebottom). 15, Interior. 16 *a*, side view; *b*, apertural view. p. 187.

17. *Idalina antiqua* (d'Orbigny). Upper Cretaceous, France. × 8. (After Munier-Chalmas and Schlumberger). *a*, side view; *b*, apertural view. p. 187.

18, 19. *Periloculina zittelli* Munier-Chalmas and Schlumberger. Upper Cretaceous, France. (After Munier-Chalmas and Schlumberger). 18 *a*, side view; *b*, apertural view. × 12. 19, Section, × 16. p. 187.

20. *Lacazina compressa* (d'Orbigny). Upper Cretaceous, France. × 3. (After Munier-Chalmas and Schlumberger). *a*, side view; *b*, edge view. p. 187.

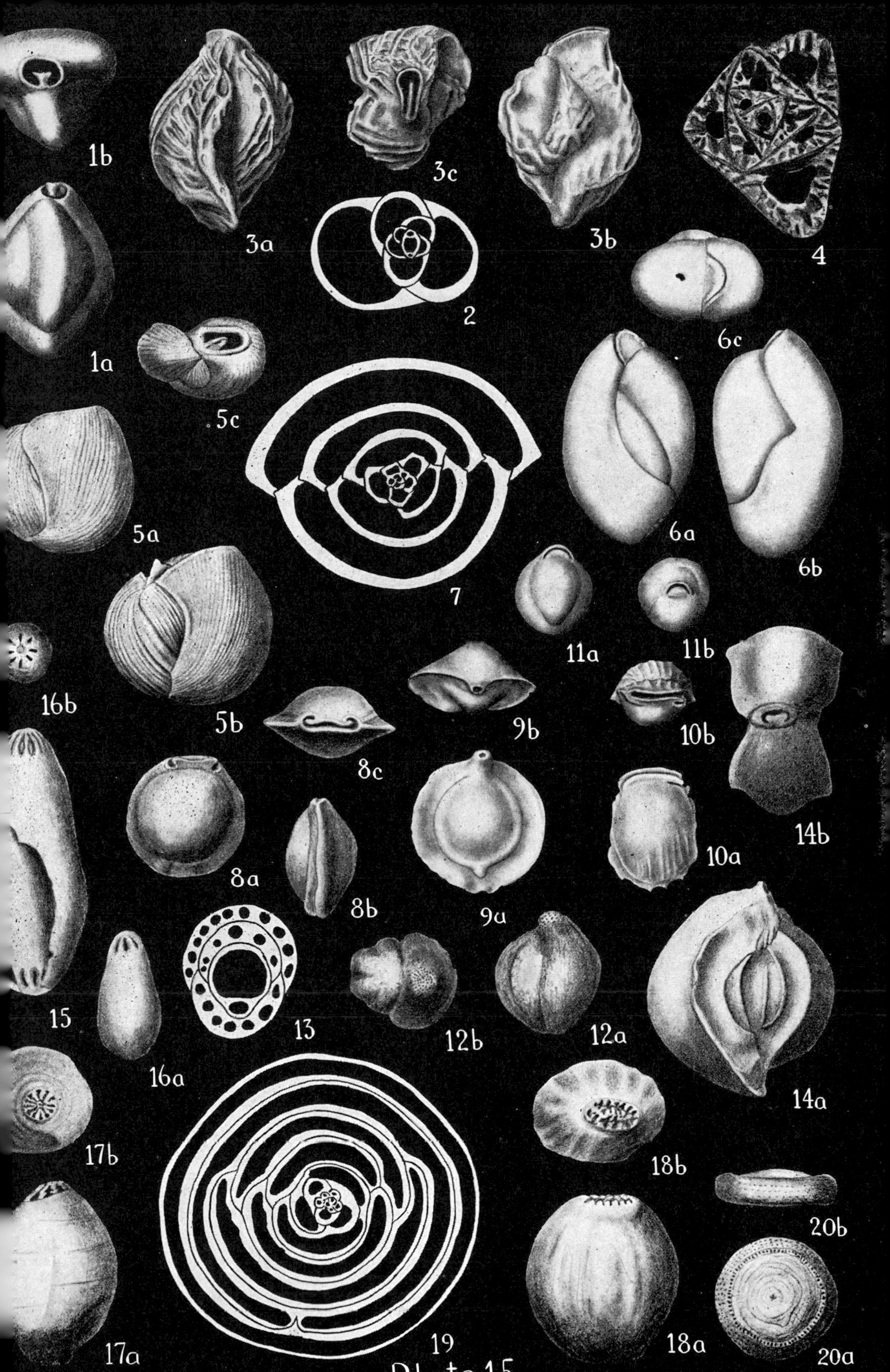
1b
3c
3a
3b
4
2
1a
6c
5c
5a
6a
6b
7
11a
11b
16b
5b
9b
10b
8c
14b
8a
10a
8b
9a
15
16a
13
12b
12a
14a
17b
18b
20b
19
17a
18a
20a

Plate 15

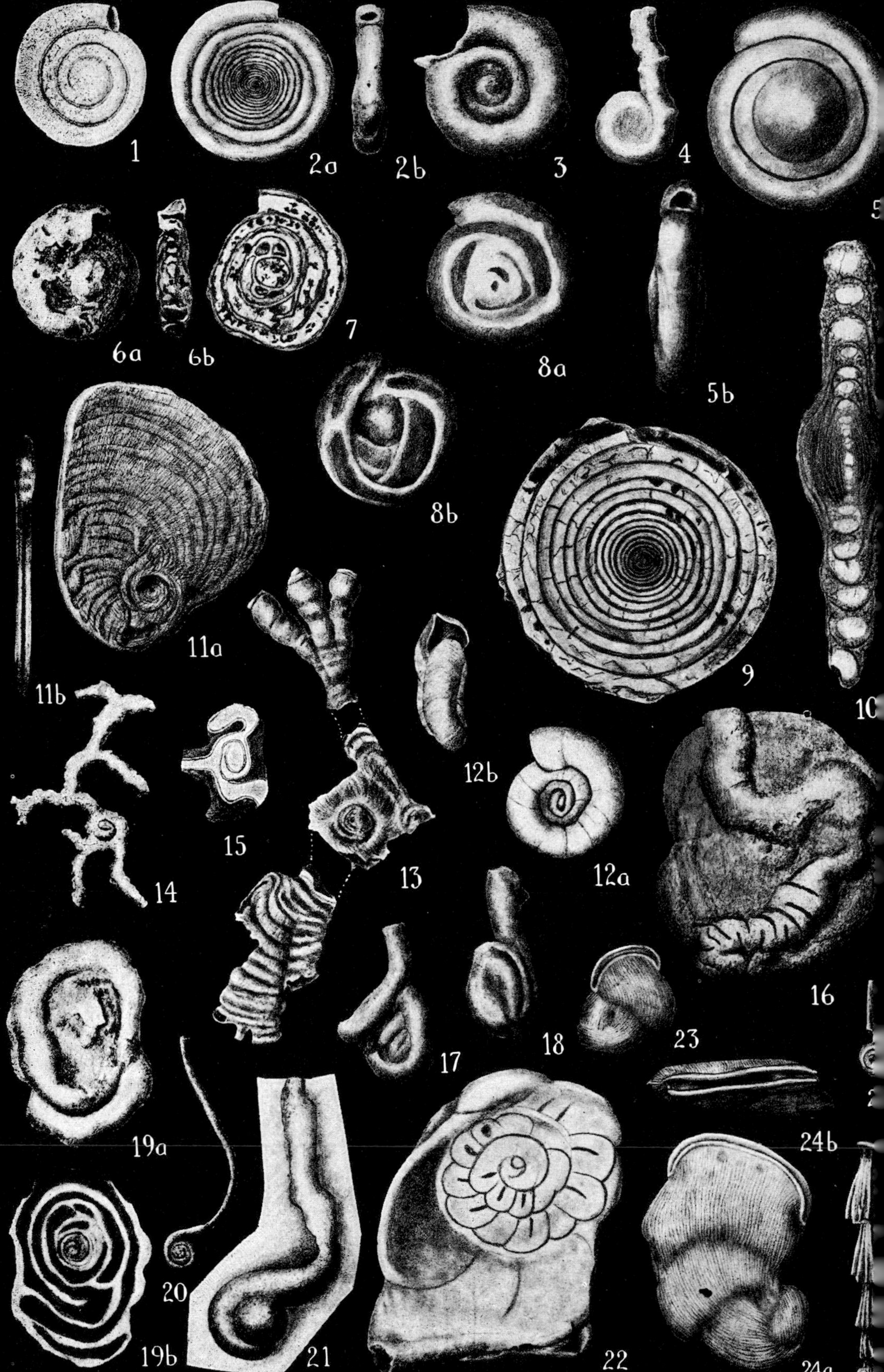
1
2a
2b
3
4
5
6a
6b
7
8a
5b
8b
9
10
11a
11b
12b
12a
13
14
15
16
17
18
23
24b
19a
19b
20
21
22
24a

Plate 16

PLATE 16.

Family 18. OPHTHALMIDIIDAE, p. 188.

FIG.

1. *Cornuspira planorbis* Schultze. Recent. × 65. (After Schultze). p. 189.
2. *Cornuspira involvens* Reuss. Recent. × 20. (After H. B. Brady). *a*, side view; *b*, apertural view. .... p. 189.
3. *Cornuspira thompsoni* Cushman and Waters. Pennsylvanian, Texas. × 90. (After Cushman and Waters). .... p. 189.
4. *Rectocornuspira lituiformis* Warthin. Pennsylvanian, Oklahoma. × 25. (After Warthin). .... p. 190.
5. *Hemigordius harltoni* Cushman and Waters. Pennsylvanian, Texas. × 90. (After Cushman and Waters). *a*, side view; *b*, apertural view. .... p. 192.

6, 7. *Hemigordius schlumbergeri* (Howchin). Carboniferous, Australia. (After Howchin). 6 *a*, side view; *b*, apertural view. 7, Section. p. 192.

8. *Hemigordius calcareus* Cushman and Waters. Pennsylvanian, Texas. × 70. (After Cushman and Waters). *a*, *b*, opposite sides. ..... p. 192.

9, 10. *Vidalina hispanica* Schlumberger. Cretaceous, Spain. (After Schlumberger). 9, Horizontal section, × 30. 10, Vertical section, × 55. p. 189.

11. *Cornuspiroides striolata* (H. B. Brady). Recent. × 3. (After H. B. Brady). *a*, side view; *b*, edge view. .... p. 192.
12. *Gordiospira fragilis* Heron-Allen and Earland. Recent. × 20. (After Heron-Allen and Earland). *a*, side view; *b*, apertural view. ... p. 193.
13. *Cornuspirella diffusa* (Heron-Allen and Earland). Recent. × 13. (After Heron-Allen and Earland). .... p. 192.

14, 15. *Cornuspiramia antillarum* (Cushman). Recent. 14, Adult. 15, Early stage. .... p. 192.

16. *Calcivertella adherens* Cushman and Waters. Pennsylvanian, Texas. × 40. (After Cushman and Waters). .... p. 194.

17, 18. *Orthovertella protea* Cushman and Waters. Pennsylvanian, Texas. × 90. (After Cushman and Waters). .... p. 193.

19. *Calcitornella heathi* Cushman and Waters. Pennsylvanian, Texas. × 60. (After Cushman and Waters). *a*, outer side; *b*, attached side. .... p. 193.

20, 21. *Apterrinella grahamensis* (Harlton). Pennsylvanian, Texas. × 55. (After Cushman and Waters). 20, Microspheric. 21, Megalospheric. .... p. 194.

22. *Plummerinella complexa* Cushman and Waters. Pennsylvanian, Texas. × 75. (After Cushman and Waters). .... p. 194.

23, 24. *Vertebralina striata* d'Orbigny. Recent. × 40. 23, Young. 24, Adult, *a*, side view; *b*, apertural view. .... p. 196.

25. 26. *Nodophthalmidium milletti* (Cushman). Recent. 25, Adult, × 40. 26, Early portion of same specimen, × 75. .... p. 195.

PLATE 17

Family 18. OPHTHALMIDIIDAE, p. 188.

FIG.

1, 2. *Ophthalmidium inconstans* H. B. Brady. Recent. (After H. B. Brady). 1, Adult, ×20. 2, Young, ×35. .... p. 196.

3, 4. *Planispirinella exigua* (H. B. Brady). Recent. ×30. (After H. B. Brady). 3 *a*, *b*, opposite sides; *c*, apertural view. 4, Section ... p. 198.

5, 6. *Planispirina communis* Seguenza. ×12. 5, Pliocene, Italy (After Seguenza), *a*, side view; *b*, apertural view. 6, Recent (After H. B. Brady), Section. .... p. 197.

7. *Spirophthalmidium acutimargo* (H. B. Brady). Recent. ×35. (After H. B. Brady). *a*, side view; *b*, apertural view. .... p. 196.

8. *Discospirina tenuissima* (W. B. Carpenter). Recent. ×8. (After H. B. Brady). *a*, from flat side; *b*, from edge. .... p. 197.

9, 10. *Wiesnerella auriculata* (Egger). Recent. ×100. 9 *a*, *b*, opposite sides; *c*, apertural view. 10, Section. .... p. 198.

11, 12. *Renulina opercularia* Lamarck. Eocene, France. ×35. 11, Exterior of topotype. 12, Section. .... p. 198.

13. *Nubecularia lucifuga* Defrance. Recent. ×12. (After H. B. Brady). *a*, outer side; *b*, attached side. .... p. 199.

14. *Trisegmentina sidebottomi* Cushman. Recent. ×50. (After Sidebottom). *a*, side view; *b*, apertural view. .... p. 198.

15. *Nubeculinella bigoti* Cushman. Jurassic, France. ×80. .... p. 199.

16, 17. *Parrina bradyi* (Millett). Recent. 16 (After Millett), ×35. 17, ×55. .... p. 200.

18–20. *Sinzowella deformis* (Karrer and Sinzow). Miocene, Russia. (After Karrer and Sinzow). 18, Section of young. 19, Section of adult. 20, Exterior, with apertures. .... p. 200.

21. *Calcituba polymorpha* Roboz. Recent. (After Roboz). .... p. 200.

22. *Squamulina laevis* Schultze. Recent. ×50. (After Schultze). ..... p. 200.

Family 19. FISCHERINIDAE, p. 201.

23. *Fischerina helix* Heron-Allen and Earland. Recent. ×65. (After Heron-Allen and Earland). *a*, dorsal view; *b*, apertural view. p. 201.

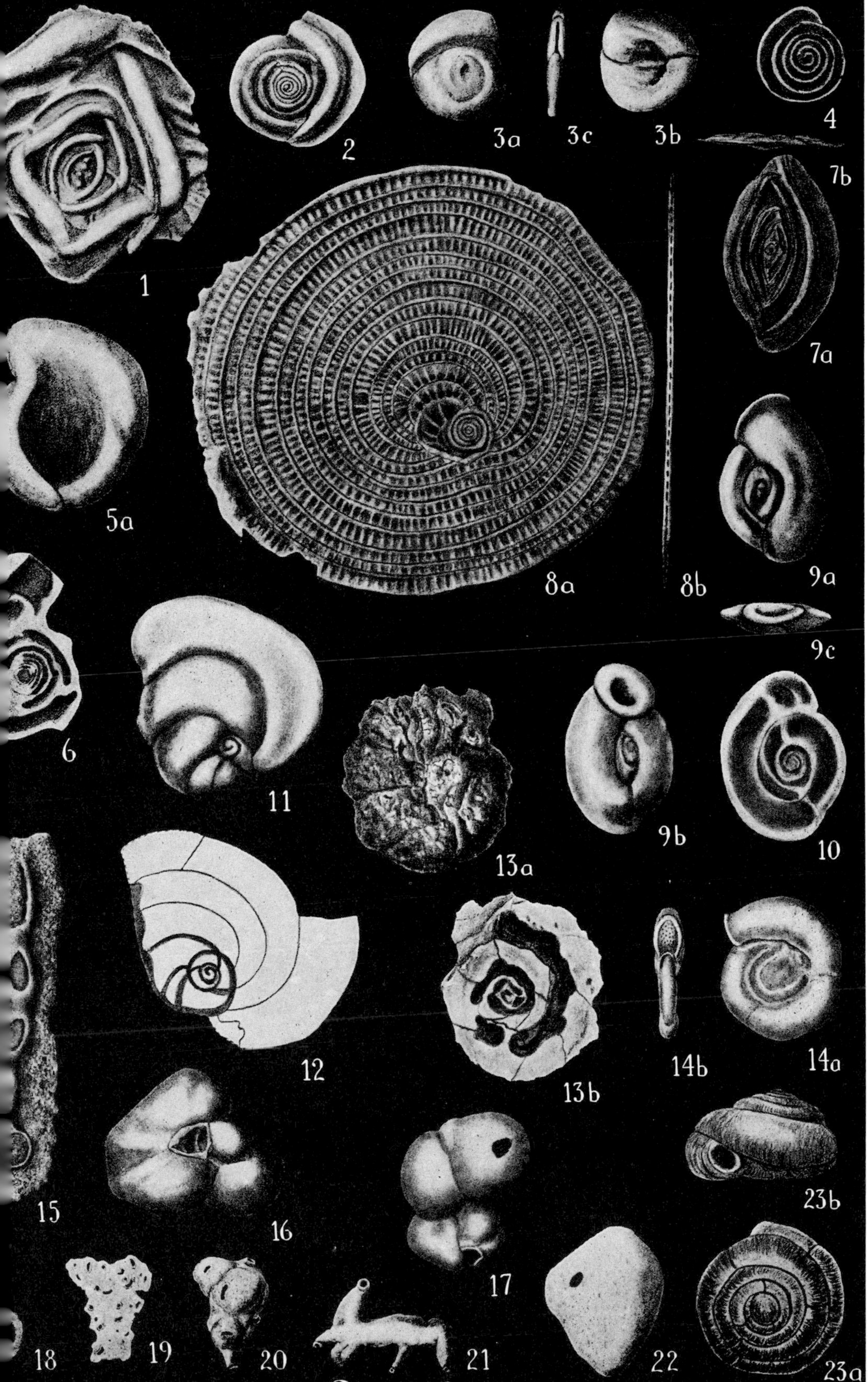
1
2
3a
3c
3b
4
7b
7a
5a
8a
8b
9a
9c
6
11
13a
9b
10
12
13b
14b
14a
23b
15
16
17
18
19
20
21
22
23a

Plate 17

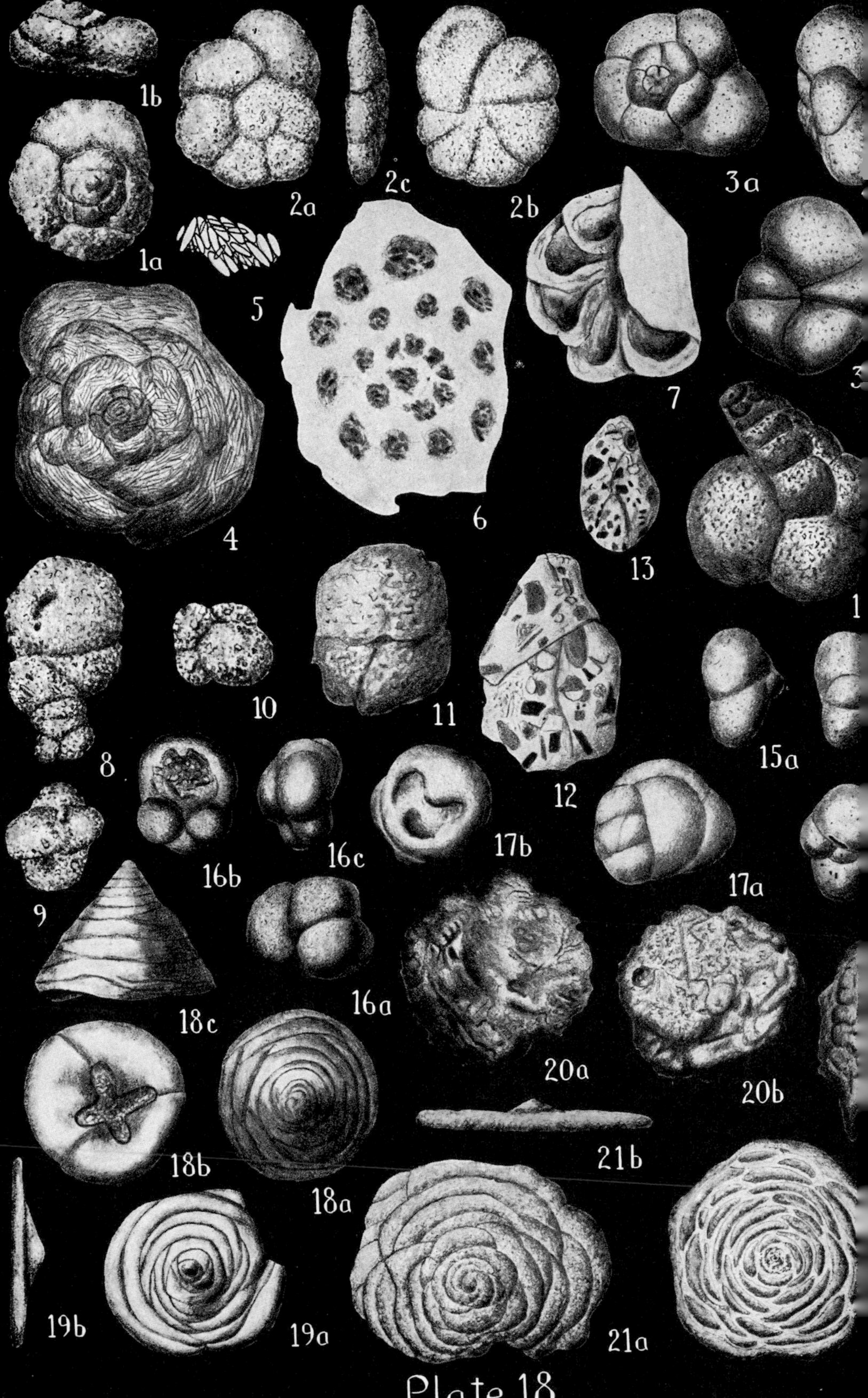
1a
1b
2a
2b
2c
3a
4
5
6
7
8
9
10
11
12
13
15a
16a
16b
16c
17a
17b
18a
18b
18c
19a
19b
20a
20b
21a
21b

Plate 18

PLATE 18

Family 20. TROCHAMMINIDAE, p. 201.

FIG.

1. *Trochammina arenosa* Cushman and Waters. Pennsylvania, Texas. × 75. (After Cushman and Waters). *a*, dorsal view; *b*, peripheral view. ..... p. 202.
2. *Trochammina diagonis* Carsey. Cretaceous, Texas. × 50. *a*, dorsal view; *b*, ventral view; *c*, peripheral view. ..... p. 202.
3. *Trochammina inflata* (Montagu). Recent. × 35. (After H. B. Brady). *a*, dorsal view; *b*, ventral view; *c*, peripheral view. p. 202.

4, 5. *Carterina spiculotesta* (Carter). Recent. (After H. B. Brady). 4, Dorsal view, × 35. 5, Enlarged surface. ..... p. 204.

6, 7. *Rotaliammina mayori* Cushman. Recent. × 65. 6, Dorsal view. 7, Ventral view. ..... p. 203.

8–10. *Mooreinella biserialis* Cushman and Waters. Pennsylvanian, Texas. × 25. (After Cushman and Waters). 8, Adult. 9, 10, Opposite sides of young trochoid stage. ..... p. 204.

11. *Ammosphaeroidina sphaeroidiniformis* (H. B. Brady). Recent. × 12. p. 205.

12, 13. *Nouria polymorphinoides* Heron-Allen and Earland. Recent. (After Heron-Allen and Earland). ..... p. 205.

14. *Globotextularia anceps* (H. B. Brady). Recent. × 12. (After H. B. Brady). ..... p. 204.

15. *Cystammina pauciloculata* (H. B. Brady). Recent. × 35. (After H. B. Brady). *a*, dorsal view; *b*, ventral view; *c*, peripheral view. ..... p. 205.

16. *Globivalvulina bulloides* (H. B. Brady). Carboniferous, Great Britain. × 35. (After H. B. Brady). *a*, dorsal view; *b*, ventral view; *c*, peripheral view. ..... p. 206.

17. *Globivalvulina biserialis* Cushman and Waters. Pennsylvanian, Texas. × 50. (After Cushman and Waters). *a*, dorsal view; *b*, ventral view. ..... p. 206.

18. *Tetrataxis palaeotrochus* Ehrenberg. Carboniferous, Great Britain. × 30. (After H. B. Brady). *a*, dorsal view; *b*, ventral view; *c*, peripheral view. ..... p. 206.

19. *Tetrataxis scutella* Cushman and Waters. Pennsylvanian, Texas. × 55. (After Cushman and Waters). *a*, dorsal view; *b*, peripheral view. ..... p. 206.

20. *Ruditaxis rudis* (H. B. Brady). Carboniferous, Great Britain. × 35. (After H. B. Brady). *a*, dorsal view; *b*, ventral view; *c*, peripheral view. ..... p. 206.

21. *Polytaxis multiloculata* (Cushman and Waters). Pennsylvanian, Texas. × 30. (After Cushman and Waters). *a*, dorsal view; *b*, peripheral view. ..... p. 206.

22. *Polytaxis laheei* Cushman and Waters. Pennsylvanian, Texas. × 20. (After Cushman and Waters). ..... p. 206.

PLATE 19

Family 21. PLACOPSILINIDAE, p. 207.

FIG.

1. *Placopsilina redoakensis* (Galloway and Harlton). Pennsylvanian, Texas. × 40. (After Galloway and Ryniker). .... p. 207.
2. *Placopsilina ciscoensis* Cushman and Waters. Pennsylvanian, Texas. × 60. (After Cushman and Waters). .... p. 207.
3. *Placopsilina cenomana* d'Orbigny. Cretaceous, Austria. (After Reuss). *a, b,* opposite sides. .... p. 207.

4, 5. *Bdelloidina aggregata* Carter. Recent. (After H. B. Brady). 4, Exterior, × 10. 5, Interior, × 20. .... p. 208.

6, 7. *Haddonia torresiensis* Chapman. Recent. (After Chapman). ...... p. 208.

8–10. *Polyphragma cribrosum* Reuss. Cretaceous. 8, Side view. 10, Apertural view. (After Reuss). 9, Branching form. (After Perner). p. 209.

11, 12. *Stylolina lapugyensis* Karrer. Miocene, Hungary. (After Karrer). *a, a,* side views; *b, b,* showing apertures. .... p. 209.

13–15. *Stacheia marginulinoides* H. B. Brady. Carboniferous, Great Britain. (After H. B. Brady). 13 *a,* side view; *b,* apertural view. × 35. 14, Side view, × 35. 15, Section showing subdivision of the chambers, × 25. .... p. 209.

Family 22. ORBITOLINIDAE, p. 209.

16–18. *Howchinia bradyana* (Howchin). Carboniferous, England. × 35. (After Howchin). 16 *a,* side view; *b,* basal view. 17, Side view. 18, Vertical section. .... p. 210.

19–21. *Valvulinella youngi* (H. B. Brady). Carboniferous, Great Britain. × 30. (After H. B. Brady). 19, Vertical section. 20, Transverse section. 21 *a,* dorsal view; *b,* side view. .... p. 210.

22. *Orbitolina texana* Roemer. Lower Cretaceous, Texas. × 3. (After Carsey). *a,* dorsal view; *b,* ventral view; *c,* peripheral view. .... p. 210.

23, 24. *Orbitolina concava* (Lamarck). Lower Cretaceous, Germany. (After Egger). 23, Vertical section. 24, Horizontal section. .... p. 210.

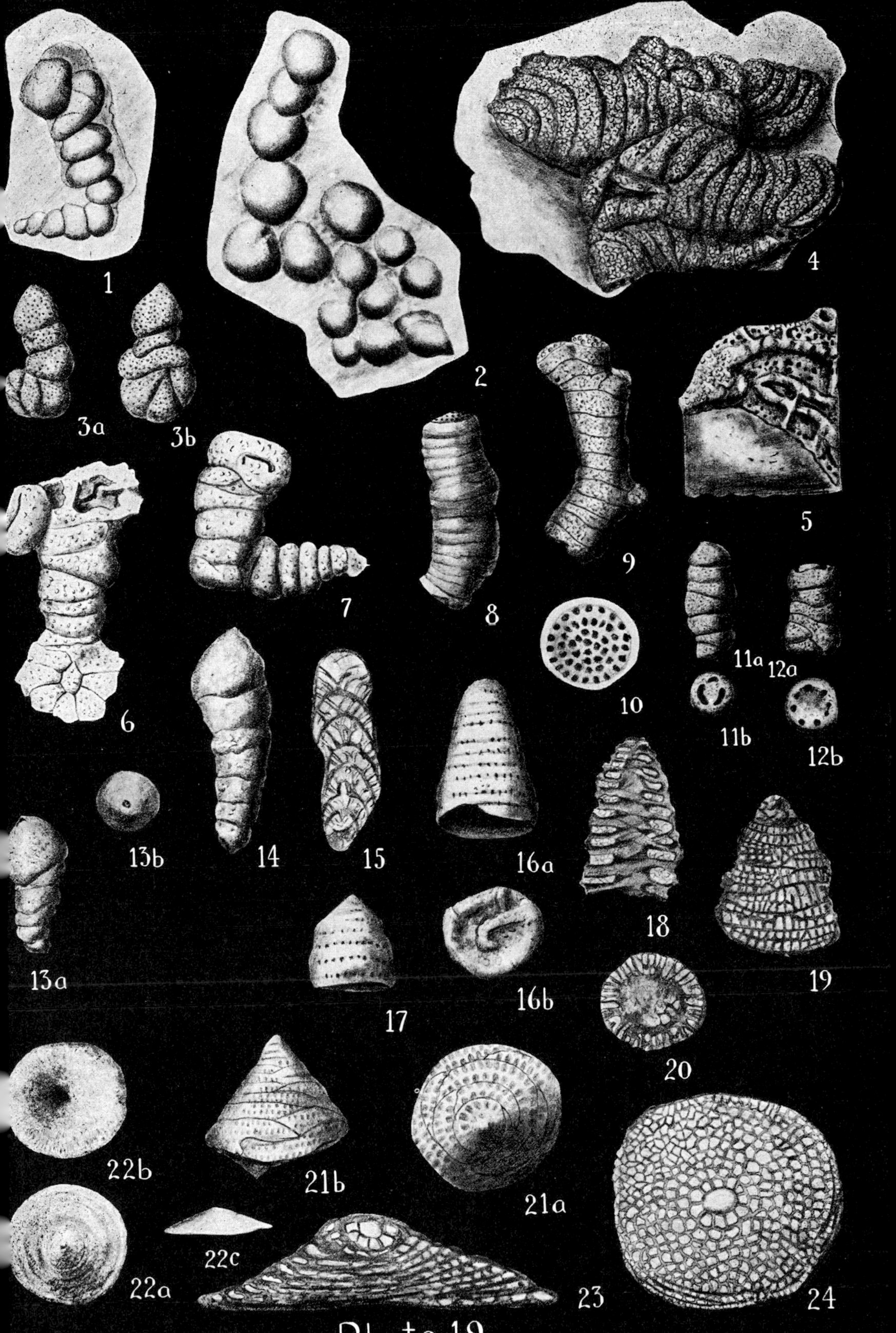
1
2
3a
3b
4
5
6
7
8
9
10
11a
11b
12a
12b
13a
13b
14
15
16a
16b
17
18
19
20
21a
21b
22a
22b
22c
23
24

Plate 19

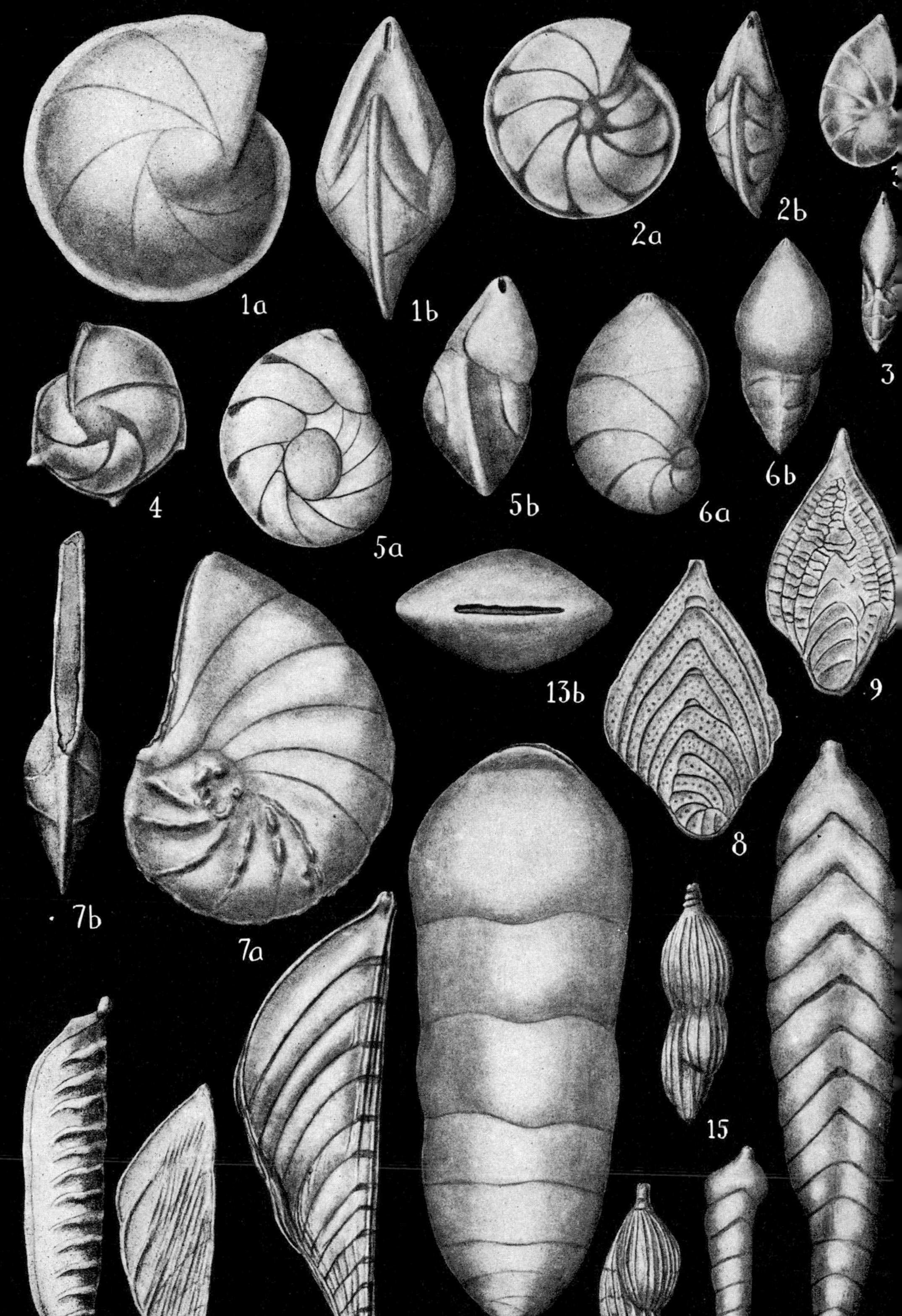

Plate 20

PLATE 20.

Family 23. LAGENIDAE, p. 211.

FIG.

1. *Robulus macrodiscus* (Reuss). Cretaceous, Trinidad. × 40. (After Cushman and Jarvis). *a*, side view; *b*, apertural view. .... p. 213.
2. *Robulus stephensoni* Cushman. Cretaceous, Tennessee. × 30. *a*, side view; *b*, apertural view. .... p. 213.
3. *Robulus mayi* Cushman and Parker. Miocene, California. × 35. (After Cushman and Parker). *a*, side view; *b*, apertural view. p. 213.
4. *Robulus calcar* (Linné). Miocene, Venezuela. × 33 .... p. 213.
5. *Darbyella danvillensis* Howe and Wallace. Eocene, Louisiana. × 35. (After Howe and Wallace). *a*, side view; *b*, apertural view. p. 213.
6. *Lenticulina navicula* (d'Orbigny). Cretaceous, Trinidad. × 40. (After Cushman and Jarvis). *a*, side view; *b*, apertural view. p. 214.
7. *Planularia advena* Cushman Jarvis. Cretaceous, Trinidad. × 15. (After Cushman and Jarvis). *a*, side view; *b*, apertural view. p. 214.
8. *Palmula rugosa* (d'Orbigny). Cretaceous, Arkansas. × 40. ...... p. 219.
9. *Palmula reticulata* (Reuss). Cretaceous, Trinidad. × 50. (After Cushman and Jarvis). .... p. 219.
10. *Vaginulina legumen* (Linné). Recent, Adriatic. (After Gaultieri). p. 218.
11. *Vaginulina kochii* Roemer, var. *striolata* Reuss. Lower Cretaceous, Texas. × 30. .... p. 218.
12. *Vaginulina webbervillensis* Carsey. Cretaceous, Texas. × 20. ..... p. 218.
13. *Lingulina wilcoxensis* Cushman and Ponton. Eocene, Alabama. × 20. (After Cushman and Ponton). *a*, side view; *b*, apertural view. .... p. 218.

14, 15. *Amphicoryne falx* (Jones and Parker). Recent. × 42–50. (After H. B. Brady). .... p. 218.

16, 17. *Flabellinella zittelliana* (Egger). Cretaceous, Bavaria. × 25. 16, Young. 17, Adult. .... p. 220.

PLATE 21.

Family 23. LAGENIDAE, p. 211.

FIG.

1, 2. *Kyphopyxa christneri* (Carsey). Cretaceous, Texas. $\times$ 30. .... p. 220.
3. *Frondicularia goldfussi* Reuss. Cretaceous, Texas. $\times$ 18. .... p. 219.
4. *Frondicularia archiaciana* d'Orbigny. Cretaceous, Texas. $\times$ 18. ..... p. 219.
5. *Frondicularia gracilis* Franke. Cretaceous, Texas. $\times$ 18. .... p. 219.
6. *Marginulina dubia* Neugeboren. Miocene, California. $\times$ 70. (After Cushman and Laiming). .... p. 214.
7. *Marginulina bullata* Reuss. Cretaceous, Texas. $\times$ 50. *a*, side view; *b*, apertural view. .... p. 214.
8, 9. *Saracenaria italica* Defrance. 8, Recent, $\times$ 10. (After H. B. Brady). *a*, side view; *b*, apertural view. 9, Miocene, Venezuela, $\times$ 25. p. 218.
10. *Dentalina pauperata* d'Orbigny. Miocene, Venezuela. $\times$ 35. ...... p. 215.
11. *Dentalina communis* d'Orbigny. Eocene, Alabama. $\times$ 35. (After Cushman and Ponton). .... p. 215.
2–14. *Nodosaria affinis* Reuss. Cretaceous, Tennessee. 12, 13, Megalospheric, $\times$ 35. 14, Microspheric, $\times$ 30. .... p. 215.
15. *Chrysalogonium cretaceum* Cushman and Church. Cretaceous, California. $\times$ 40. (After Cushman and Church). *a*, side view; *b*, apertural view. .... p. 216.
16. *Pseudoglandulina conica* (Neugeboren). Eocene, California. $\times$ 50. (After Cushman and Barksdale). .... p. 216.
17. *Pseudoglandulina parallela* (Marsson). Cretaceous, Trinidad. $\times$ 40. (After Cushman and Jarvis). .... p. 216.
18. *Lagena perlucida* (Montagu). Recent. $\times$ 70. *a*, side view; *b*, apertural view. .... p. 221.
24. *Lagena spiralis* H. B. Brady. Recent. $\times$ 70. .... p. 221.

Family 31. BULIMINIDAE, p. 260.

19. *Entosolenia orbignyana* (Seguenza). Cretaceous, Tennessee. $\times$ 100. *a*, side view; *b*, peripheral view. .... p. 267.
20. *Entosolenia hexagona* Williamson, Cretaceous, Tennessee. $\times$ 70. .. p. 267.
21. *Entosolenia auriculata* (H. B. Brady), var. *duplicata* (Sidebottom). Recent. $\times$ 70. *a*, side view; *b*, apertural view. .... p. 267.
22. *Entosolenia formosa* (Schwager). Recent. $\times$ 70. .... p. 267
23. *Entosolenia radiato-marginata* (Parker and Jones). Recent. $\times$ 70. .. p. 267.
25. *Entosolenia orbignyana* (Seguenza), var. *clathrata* (H. B. Brady). Recent. $\times$ 70. .... p. 267.

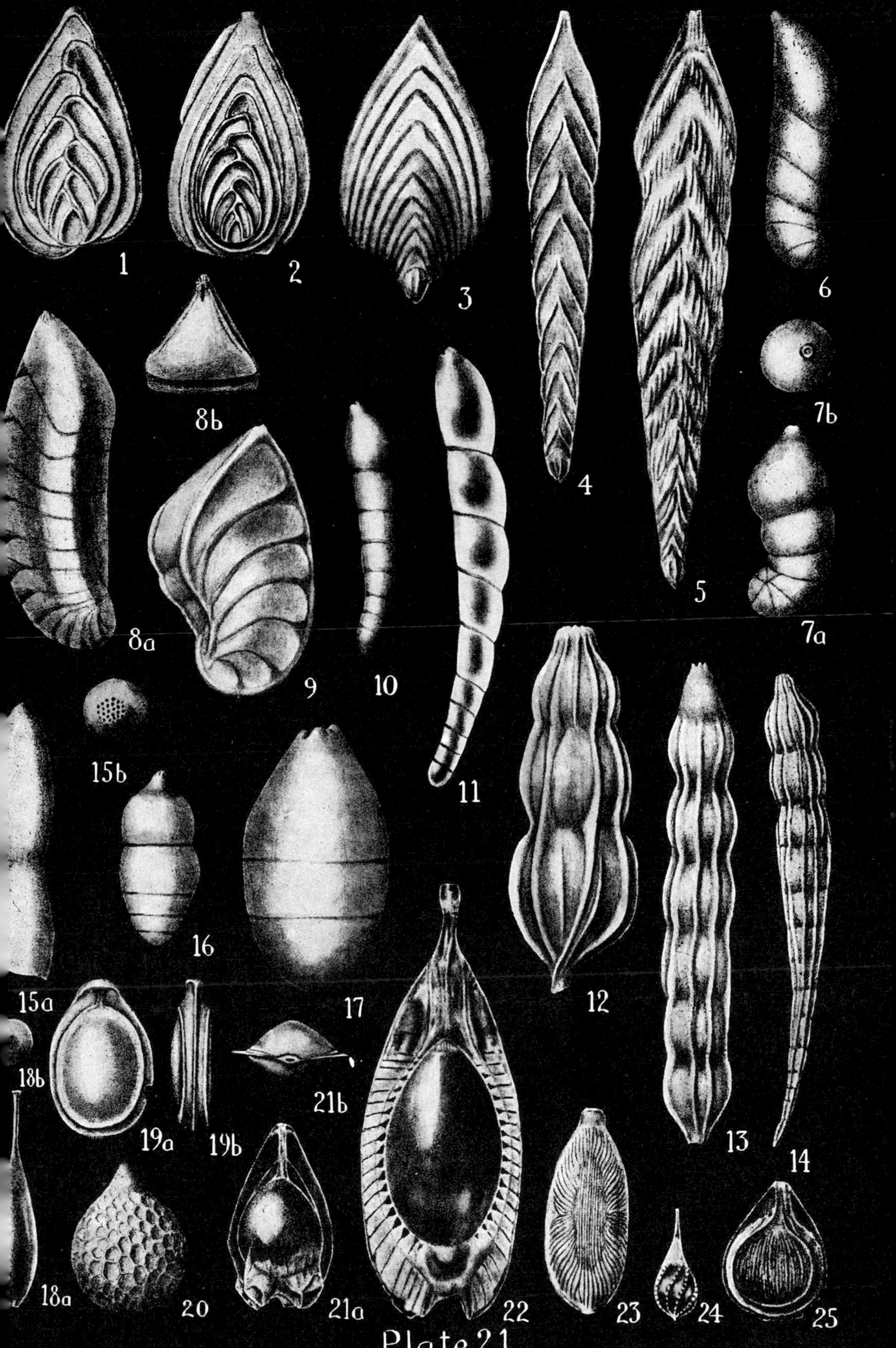
1
2
3
4
5
6
7a
7b
8a
8b
9
10
11
12
13
14
15a
15b
16
17
18a
18b
19a
19b
20
21a
21b
22
23
24
25

Plate 21

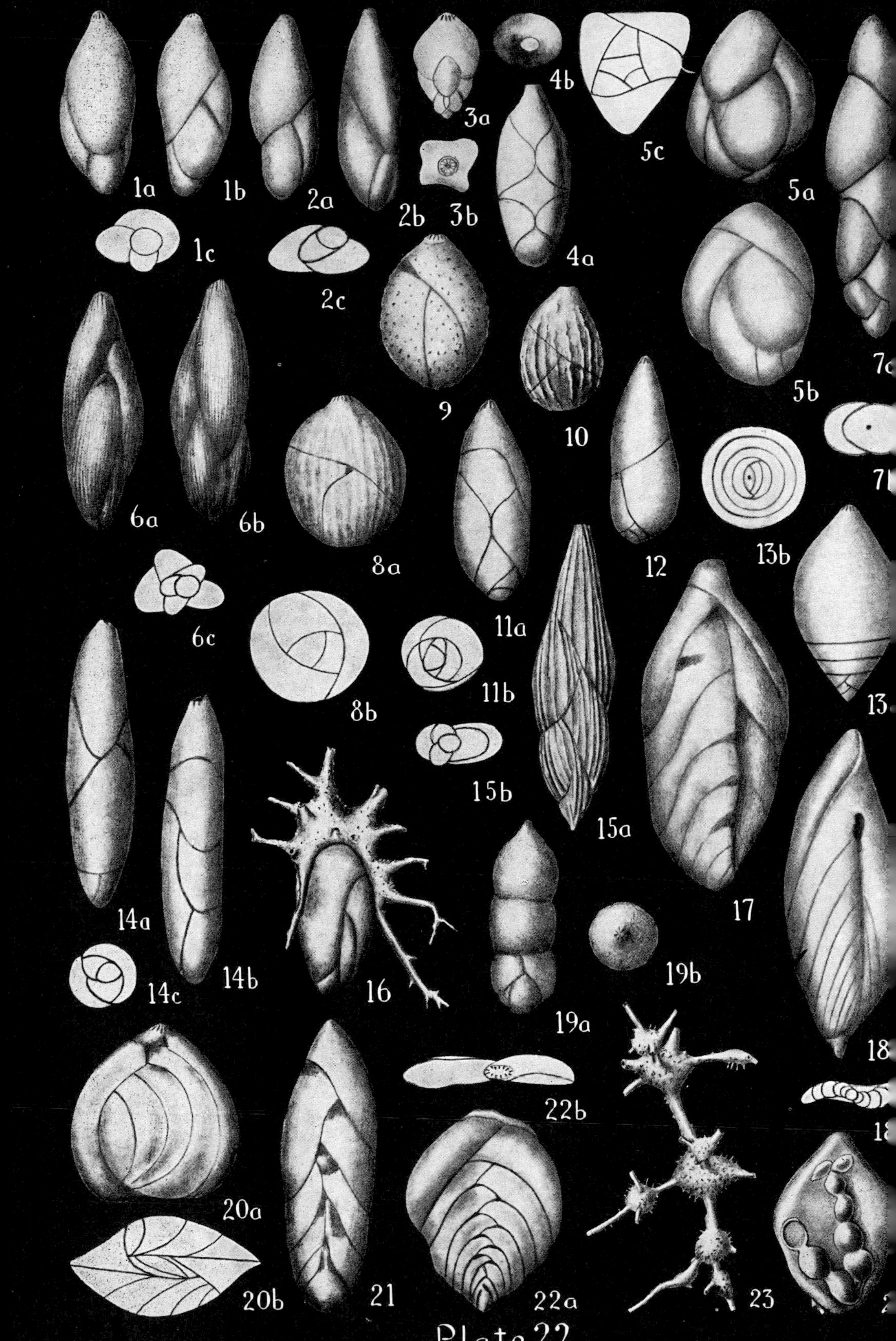
1a
1b
2a
2b
3a
3b
4a
4b
5a
5b
5c
1c
2c
6a
6b
6c
7c
8a
8b
9
10
11a
11b
12
13b
14a
14b
14c
15a
15b
16
17
19a
19b
20a
20b
21
22a
22b
23

Plate 22

PLATE 22.

Family 24. POLYMORPHINIDAE, p. 221.

FIG.

1. *Eoguttulina anglica* Cushman and Ozawa. Cretaceous, England. × 75. (After Cushman and Ozawa). *a, b,* side views; *c,* basal view. .... p. 224
2. *Eoguttulina polygona* (Terquem). Jurassic, England. × 80. (After Cushman and Ozawa). *a, b,* side views; *c,* basal view. .... p. 224.
3. *Quadrulina rhabdogonioides* (Chapman). Cretaceous, England. × 40. (After Chapman). *a,* side view; *b,* apertural view. .... p. 224.
4. *Quadrulina lagenalis* (Terquem). Jurassic, France. (After Terquem). *a,* side view; *b,* apertural view. .... p. 224.
5. *Guttulina problema* d'Orbigny. Miocene, France. × 35. (After Cushman and Ozawa). *a, b,* side views; *c,* basal view. .... p. 224.
6. *Guttulina pulchella* d'Orbigny. Recent. × 45. (After Cushman and Ozawa). *a, b,* side views; *c,* basal view. .... p. 224.
7. *Paleopolymorphina pleurostomelloides* (Franke). Cretaceous, Germany. × 65. (After Cushman and Ozawa). *a,* side view; *b,* apertural view. .... p. 224.
8. *Globulina gibba* d'Orbigny, var. *myristiformis* (Williamson). Recent. × 65. (After Cushman and Ozawa). *a,* side view; *b,* basal view. .... p. 226.
9. *Globulina inaequalis* Reuss, var. *caribaea* d'Orbigny. Recent. × 80. (After Cushman and Ozawa). .... p. 226.
10. *Globulina gibba* d'Orbigny, var. *striata* Egger. Miocene, Germany. × 80. (After Cushman and Ozawa). .... p. 226.

11, 12. *Pyrulina gutta* d'Orbigny. Eocene, Belgium. × 45. (After Cushman and Ozawa). 11 *a,* side view; *b,* basal view. .... p. 228.

13. *Glandulina laevigata* d'Orbigny. Miocene, Austria. × 45. (After Cushman and Ozawa). *a,* side view; *b,* basal view. .... p. 228.
14. *Pyrulina cylindroides* (Roemer). Cretaceous, Germany. × 45. (After Cushman and Ozawa). *a, b,* side views; *c,* basal view. .... p. 228.
15. *Pseudopolymorphina rutila* (Cushman). Oligocene, Mississippi. × 65. (After Cushman and Ozawa). *a,* side view; *b,* basal view. p. 228.
16. *Sigmomorphina semitecta* (Reuss), var. *terquemiana* (Fornasini). Recent. × 35. (After Cushman and Ozawa). Fistulose form. p. 229.
17. *Sigmomorphina pseudoregularis* Cushman and Thomas. Eocene, Texas. × 25. (After Cushman and Ozawa). .... p. 229.
18. *Sigmomorphina frondiculariformis* (Galloway and Wissler). Pliocene, California. × 35. (After Cushman and Ozawa). *a,* side view; *b,* basal view. .... p. 229.
19. *Dimorphina tuberosa* d'Orbigny. Recent. (After d'Orbigny). *a,* side view; *b,* apertural view. .... p. 226.
20. *Sigmoidella elegantissima* (Parker and Jones). Recent. × 25. (After Cushman and Ozawa). *a,* side view; *b,* basal view. .... p. 229.
21. *Polymorphina charlottensis* Cushman. Pliocene, California. × 25. (After Cushman and Ozawa). .... p. 229.
22. *Polymorphina complanata* d'Orbigny. Miocene, Austria. × 35. (After Cushman and Ozawa). *a,* side view; *b,* apertural view. p. 229.
23. *Ramulina globulifera* H. B. Brady. Recent. × 20. (After H. B. Brady). .... p. 230.
24. *Bullopora laevis* (Sollas). Cretaceous, New Jersey. × 2. (After Bagg). .... p. 230.

PLATE 23.

Family 25. NONIONIDAE, p. 232.

FIG.

1. *Nonion pompilioides* (Fichtel and Moll). Recent. × 80. *a*, side view; *b*, apertural view. .... p. 232.
2. *Nonion labradoricum* (Dawson). Recent. × 60. *a*, side view; *b*, apertural view. .... p. 232.
3. *Nonionella miocenica* Cushman. Miocene, California. × 50. *a*, dorsal view; *b*, ventral view; *c*, apertural view. .... p. 233.
4. *Nonionella turgida* (Williamson). Recent. × 100. *a*, dorsal view; *b*, ventral view; *c*, apertural view. .... p. 233.
5. *Elphidium lessonii* (d'Orbigny). Recent. × 60. .... p. 234.
6. *Elphidella arctica* (Parker and Jones). Recent. × 50. .... p. 236.
7. *Polystomellina discorbinoides* Yabe and Hanzawa. Pliocene, Japan. × 40. (After Yabe and Hanzawa). *a*, dorsal view; *b*, ventral view; *c*, peripheral view. .... p. 236.
8. *Faujasina carinata* d'Orbigny. Cretaceous, Holland. (After d'Orbigny). *a*, dorsal view; *b*, ventral view; *c*, peripheral view. .... p. 236.

9, 10. *Ozawaia tongaensis* Cushman. Recent. × 40. 10 *a*, side view; *b*, apertural view. .... p. 236.

Family 26. CAMERINIDAE, p. 237.

11, 12. *Archaediscus karreri* H. B. Brady. Carboniferous, England. × 25. (After H. B. Brady). 11 *a*, side view; *b*, peripheral view. 12, Section. .... p. 237.

13. *Nummulostegina velibitana* Schubert. Carboniferous, Dalmatia. (After Schubert). *a*, side view; *b*, apertural view. .... p. 237.

14, 15. *Camerina elegans* (Sowerby). Eocene, England. × 6. (After Jones). 14, Exterior. 15, Section. .... p. 238.

16. *Assilina undata* d'Orbigny. Eocene, France. (After d'Orbigny). *a*, side view; *b*, peripheral view. .... p. 239.

17. *Operculina granulosa* Leymerie. Recent. × 8. (After H. B. Brady). p. 239.

18, 19. *Operculinella cumingii* (W. B. Carpenter). Recent. (After W. B. Carpenter). 18, Young. 19, Adult. *a*, *a*, side views; *b*, *b*, peripheral views. .... p. 239.

20. *Heterostegina depressa* d'Orbigny. Recent. × 8. (After H. B. Brady). .... p. 240.

21. *Cycloclypeus guembelianus* H. B. Brady. Recent. × 20. .... p. 242.

22. *Spiroclypeus*. Idealized figure. (After Van der Vlerk and Umbgrove). .... p. 240.

1a
1b
2a
2b
3a
3c
3b
4a
4c
4b
5
6
7c
7b
7a
8c
8b
8a
9
10b
10a
11a
11b
12
13b
13a
14
15
16a
16b
17
18a
18b
19b
19a
20
21
22

Plate 23

Plate 24

PLATE 24.

Family 27. PENEROPLIDAE, p. 242.

FIG.

1. *Peneroplis planatus* (Fichtel and Moll). Recent. × 35. (After H. B. Brady). *a*, side view; *b*, apertural view. .... p. 243.
2. *Peneroplis proteus* d'Orbigny. Recent. × 45. .... p. 243.
3. *Dendritina arbuscula* d'Orbigny. Miocene, France. (After d'Orbigny's Model). *a*, side view; *b*, apertural view. .... p. 243.
4. *Spirolina arietinus* (Batsch). Recent. × 60. .... p. 243.
5. *Monalysidium sollasi* Chapman. Recent. × 35. (After Chapman). p. 244.
6. *Monalysidium politum* Chapman. Recent. × 100. .... p. 244.

7, 8. *Archaias angulatus* (Fichtel and Moll). Recent. × 20. 7 *a*, side view; *b*, apertural view. (After H. B. Brady). 8, Eroded specimen showing chamberlets. From photograph). .... p. 244.

9. *Sorites marginalis* (Lamarck). Recent. × 20. (From photograph). p. 245.
10. *Sorites orbiculus* (Forskal). Recent. (After Ehrenberg). *a*, side view; *b*, peripheral view. .... p. 245.
11. *Sorites*. Idealized median section. (After H. Douvillé). .... p. 245.

12–14. *Amphisorus hemprichii* Ehrenberg. Recent. 12 (After Ehrenberg). 13, 14, × 20. (From photographs). .... p. 245.

15–17. *Marginopora vertebralis* Blainville. Recent. 15, 16, × 12. (After H. B. Brady). 15 *a*, side view; *b*, peripheral view. 16, Section. 17, Portion of idealized section. (After W. B. Carpenter). .... p. 245.

PLATE 25.

Family 27. PENEROPLIDAE, p. 242.

FIG.

1\. *Fallotia jacquoti* H. Douvillé. Upper Cretaceous, France. × 3. (After H. Douvillé). p. 246.

2\. *Orbitolites*. Idealized median section. (After H. Douvillé). p. 246.

3, 4. *Meandropsina vidali* Schlumberger. (After Schlumberger). 3, Surface, × 3. 4, Portion of surface, enlarged. p. 247.

5\. *Opertorbitolites douvilléi* Nuttall. Eocene, India. × 10. (After Nuttall). Section. p. 246.

6, 7. *Praesorites moureti* H. Douvillé. Cretaceous, France. × 2. (After H. Douvillé). p. 247.

15\. *Craterites rectus* Heron-Allen and Earland. Recent. × 20. (After Heron-Allen and Earland). *a*, side view; *b*, from above. p. 246.

Family 28. ALVEOLINELLIDAE, p. 248.

8–10. *Borelis melo* (Fichtel and Moll). Recent. × 35. (After H. B. Brady). 8 *a*, side view; *b*, end view. 9, Transverse section. 10, Longitudinal section. p. 250.

11, 12. *Flosculina globosa* Leymerie. Eocene, India. × 3. (After Nuttall). 11, Transverse section. 12, Longitudinal section. p. 250.

13, 14. *Alveolinella boscii* (Defrance). Recent. × 15. (After H. B. Brady). 13 *a*, side view; *b*, end view. 14, Transverse section. p. 251.

Family 29. KERAMOSPHAERIDAE, p. 251.

16, 17. *Keramosphaera murrayi* H. B. Brady. Recent. (After H. B. Brady). 16, Exterior, × 15. 17, Section, × 65. p. 251.

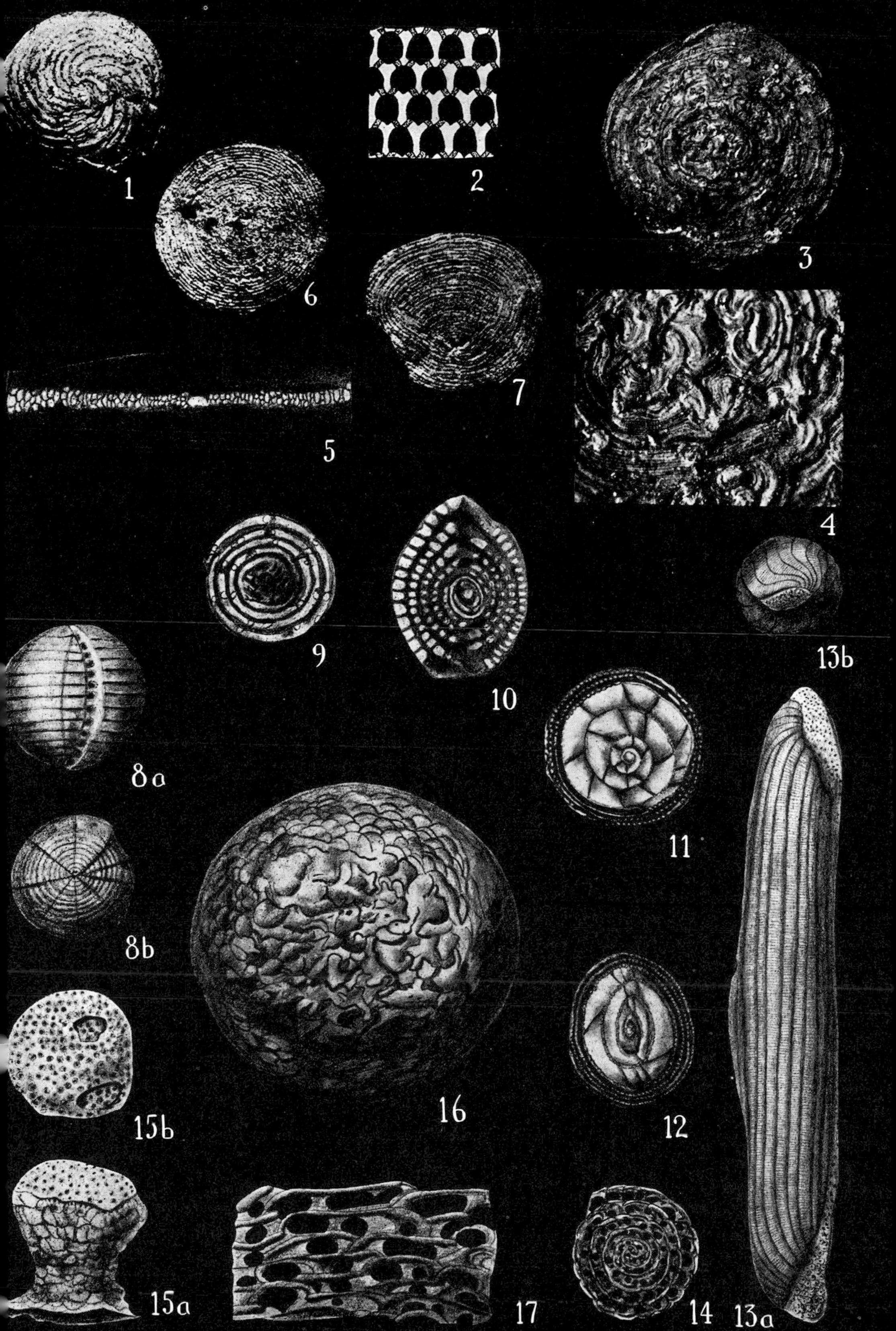

Plate 25

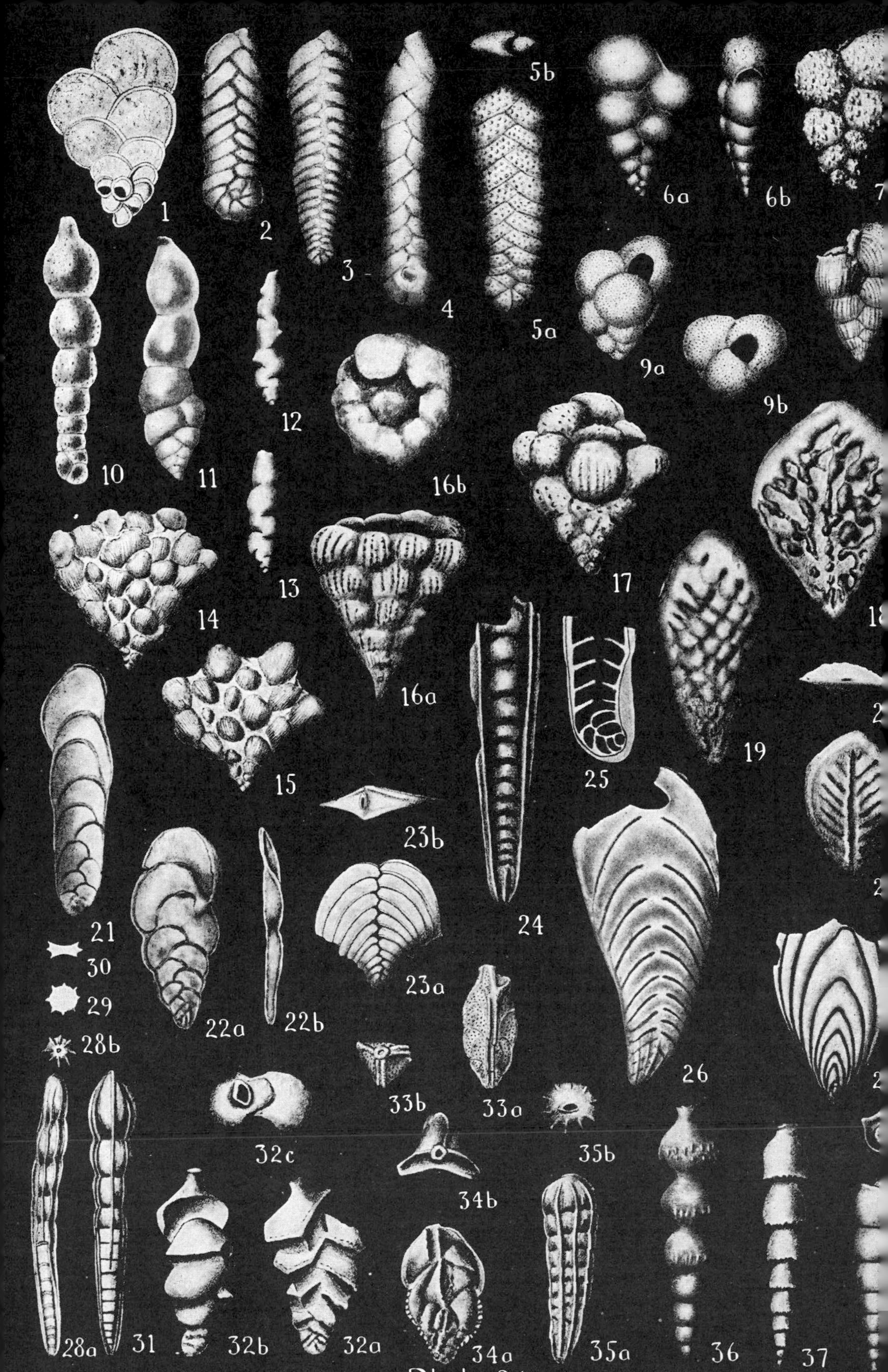
1
2
3
4
5a
5b
6a
6b
9a
9b
10
11
12
13
14
15
16a
16b
17
19
21
22a
22b
23a
23b
24
25
26
28a
28b
29
30
31
32a
32b
32c
33a
33b
34a
34b
35a
35b
36
37

Plate 26

PLATE 26.

Family 30. HETEROHELICIDAE, p. 252.

FIG.

1. *Heterohelix americana* Ehrenberg. Cretaceous. (After Ehrenberg). p. 253.

2, 3. *Bolivinopsis clotho* (Grzybowski). Cretaceous, Trinidad. × 75. (After Cushman and Jarvis). 2, Megalospheric form. 3, Microspheric form. p. 253.

4. *Bolivinopsis rosula* (Ehrenberg). Cretaceous, Tennessee. × 120. p. 253.

5. *Bolivinopsis papillata* (Cushman). Cretaceous, Tennessee. × 120. *a*, front view; *b*, apertural view. p. 253.

6. *Gümbelina globulosa* (Ehrenberg). Cretaceous, Tennessee. × 120. *a*, side view; *b*, peripheral view. p. 254.

7. *Gümbelina spinifera* Cushman. Cretaceous, Tennessee. × 120. p. 254.

8. *Gümbelina striata* (Ehrenberg). Cretaceous, Tennessee. × 120. p. 254.

9. *Gümbelitria cretacea* Cushman. Cretaceous, Texas. × 100. *a*, side view; *b*, apertural view. p. 256.

10. *Rectogümbelina texana* Cushman. Cretaceous, Texas. × 160. p. 256.

11. *Rectogümbelina cretacea* Cushman. Cretaceous, Arkansas. × 145. p. 256.

12, 13. *Tubitextularia bohemica* (Sulc). Cretaceous, Bohemia. (After Sulc). p. 256.

14, 15. *Ventilabrella eggeri* Cushman. Cretaceous, Texas. × 50. p. 257.

16. *Pseudotextularia fruticosa* (Egger). Cretaceous, Austria. (After Rzehak). *a*, side view; *b*, apertural view. p. 256.

17. *Planoglobulina acervulinoides* (Egger). Cretaceous, Bavaria. (After Egger). p. 257.

18. *Bolivinoides decorata* (Jones). Cretaceous, Texas. × 50. p. 257.

19. *Bolivinoides decorata* (Jones), var. *delicatula* Cushman. Cretaceous, Trinidad. × 75. (After Cushman and Jarvis). p. 257.

20. *Bolivinoides draco* (Marsson). Cretaceous, Germany. (After Marsson). *a*, side view; *b*, apertural view. p. 257.

21. *Bolivinitella eleyi* (Cushman). Cretaceous, Texas. × 60. p. 258.

22. *Bolivinita planata* Cushman. Cretaceous, Texas. × 50. *a*, side view; *b*, peripheral view. p. 257.

23. *Bolivinella folia* (Parker and Jones). Recent. × 40. (After H. B. Brady). *a*, side view; *b*, apertural view. p. 258.

24, 25. *Plectofrondicularia californica* Cushman and Stewart. Pliocene, California. (After Cushman and Stewart). 24, Exterior, × 35. 25, Section of microspheric young, × 50. p. 258.

26. *Plectofrondicularia miocenica* Cushman. Miocene, California. × 70. (After Cushman and Laiming). p. 258.

27. *Plectofrondicularia vaughani* Cushman. Miocene, Jamaica. × 30. (After Cushman and Jarvis). p. 258.

28–30. *Amphimorphina haueriana* Neugeboren. Miocene, Austria. (After Neugeboren). 28 *a*, side view; *b*, apertural view. 29, Section near apertural end. 30, Section near base. p. 259.

31. *Nodomorphina compressiuscula* (Neugeboren). Miocene, Austria. (After Neugeboren). p. 259.

32. *Eouvigerina americana* Cushman. Cretaceous, Texas. × 75. *a*, front view; *b*, side view; *c*, apertural view. p. 259.

34. *Pseudouvigerina plummerae* Cushman. Cretaceous, Texas. × 65. *a*, side view; *b*, apertural view. p. 260.

35. *Siphogenerinoides plummeri* (Cushman). Cretaceous, Texas. × 40. *a*, side view; *b*, apertural view. p. 260.

36. *Nodogenerina lepidula* (Schwager). Pliocene, California. × 65. (After Cushman, Stewart and Stewart). p. 260.

37. *Nodogenerina bradyi* Cushman. Recent. × 50. (After H. B. Brady). p. 260.

38. *Nodogenerina advena* Cushman and Laiming. Miocene, California. (After Cushman and Laiming). *a*, side view; *b*, apertural view. p. 260.

Family 31. BULMINIDAE, p. 260.

33. *Angulogerina wilcoxensis* (Cushman and Ponton.) Eocene, Alabama. × 60. (After Cushman and Ponton). *a*, side view; *b*, apertural view. p. 275.

PLATE 27.

Family 31. BULIMINIDAE, p. 260.

FIG.

1. *Terebralina regularis* Terquem. Jurassic, France. × 50. (After Terquem). .... p. 262.
2. *Turrilina andreaei* Cushman. Oligocene, France. × 65. (After Andreae). *a, b*, side views. .... p. 264.
3. *Turrilina alsatica* Andreae. Oligocene, France. × 65. (After Andreae). *a, b*, side views; *c*, apertural view. .... p. 264.
4. *Buliminella elegantissima* (d'Orbigny). Recent. × 135. (After d'Orbigny). *a, b*, side views. .... p. 264.
5. *Buliminella carseyae* Plummer. Cretaceous, Texas. × 50. *a*, side view; *b*, apertural view. .... p. 264.

6, 7. *Buliminoides williamsoniana* (H. B. Brady). Recent. × 55. (After H. B. Brady). 6 *a, b*, side views. 7 *a*, side view; *b*, apertural view. .... p. 265.

8. *Ungulatella pacifica* Cushman. Recent. × 60. *a*, front view; *b*, side view; *c*, apertural view. .... p. 265.
9. *Robertina charlottensis* (Cushman). Recent. × 25. *a, b*, opposite sides. .... p. 265.
10. *Robertina arctica* d'Orbigny. Recent. (After d'Orbigny). *a, b*, opposite sides. .... p. 265.
11. *Bulimina marginata* d'Orbigny. Recent. (After d'Orbigny). *a, b*, opposite sides. .... p. 266.
12. *Bulimina aculeata* d'Orbigny. Recent. × 40. (After H. B. Brady). p. 266.
13. *Bulimina buchiana* d'Orbigny. Recent. × 55. (After H. B. Brady). p. 266.
14. *Bulimina kickapooensis* Cole, var. *pingua* Cushman and Parker. Cretaceous, Texas. × 50. *a*, side view; *b*, apertural view. .... p. 266.
15. *Neobulimina canadensis* Cushman and Wickenden. Cretaceous, Tennessee. × 150. *a, b*, side views; *c*, apertural view. .... p. 267.
16. *Globobulimina pacifica* Cushman. Recent. × 50. *a*, side view; *b*, apertural view. .... p. 267.

17, 18. *Entosolenia globosa* (Reuss). Recent. (After Williamson). 17 *a*, side view; *b*, apertural view. 18, Longitudinal section showing entosolenian tube. .... p. 267.

19. *Entosolenia staphyllearia* (Schwager). Recent. × 30. (After H. B. Brady). .... p. 267.
20. *Entosolenia orbignyana* (Seguenza). Recent. (After Williamson). .. p. 267.
21. *Virgulina squammosa* d'Orbigny. Pliocene, Italy. (From d'Orbigny's Model). *a*, front view; *b*, apertural view. .... p. 268.
22. *Virgulina* sp. (?). Recent. (After Rhumbler). Showing internal spiral tube. .... p. 268.
23. *Virgulina* (*Virgulinella*) *gunteri* Cushman. Miocene, Florida. × 40. p. 268.
24. *Virgulina* (*Virgulinella*) *miocenica* Cushman and Ponton. Miocene, Florida. × 50. (After Cushman and Ponton). .... p. 268.
25. *Bolivina plicata* d'Orbigny. Recent. (After d'Orbigny). .... p. 268.
26. *Bolivina hantkeniana* H. B. Brady. Recent. × 40. (After H. B. Brady). .... p. 268.
27. *Bolivina subangularis* H. B. Brady. Recent. × 40. (After H. B. Brady). *a*, front view; *b*, apertural view. .... p. 268.
28. *Bolivina incrassata* Reuss. Cretaceous, Tennessee. × 70. .... p. 268.
29. *Rectobolivina bifrons* (H. B. Brady). Recent. × 35. (After H. B. Brady). *a*, front view; *b*, apertural view. .... p. 270.
30. *Loxostomum subrostratum* Ehrenberg. Cretaceous. (After Ehrenberg). .... p. 269.
31. *Loxostomum karrerianum* (H. B. Brady). Recent. × 40. (After H. B. Brady). *a*, front view; *b*, apertural view. .... p. 269.
32. *Loxostomum plaitum* (Carsey). Cretaceous, Texas. × 70. .... p. 269.

33, 34. *Tubulogenerina mooraboolensis* (Cushman). Miocene, Australia. × 75. (After Heron-Allen and Earland). 33, Exterior. 34, Diagrammatic longitudinal section. .... p. 271.

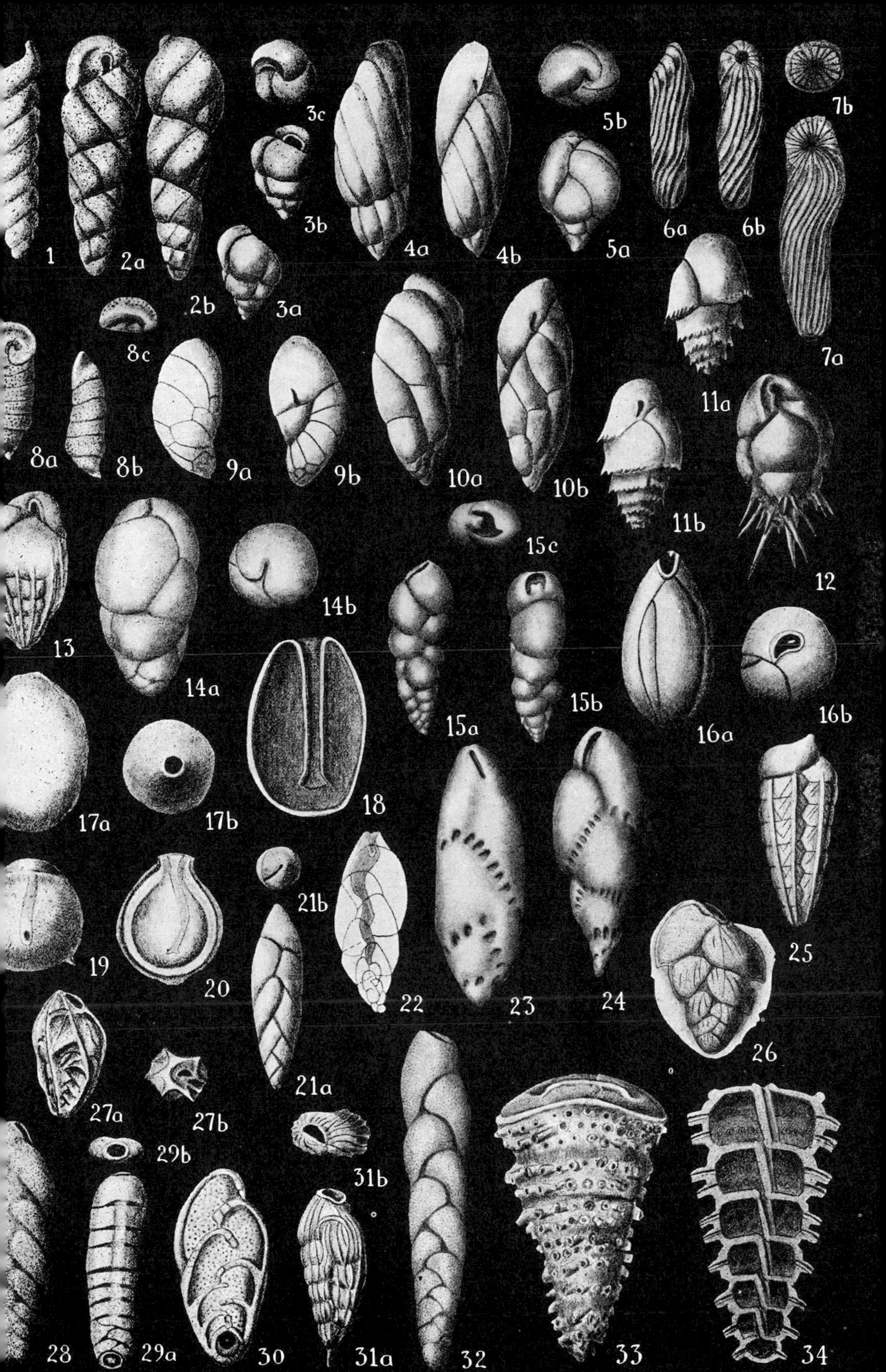
1
2a
2b
3a
3b
3c
4a
4b
5a
5b
6a
6b
7a
7b
8a
8b
8c
9a
9b
10a
10b
11a
11b
12
13
14a
14b
15a
15b
15c
16a
16b
17a
17b
18
19
20
21a
21b
22
23
24
25
26
27a
27b
28
29a
29b
30
31a
31b
32
33
34

Plate 27

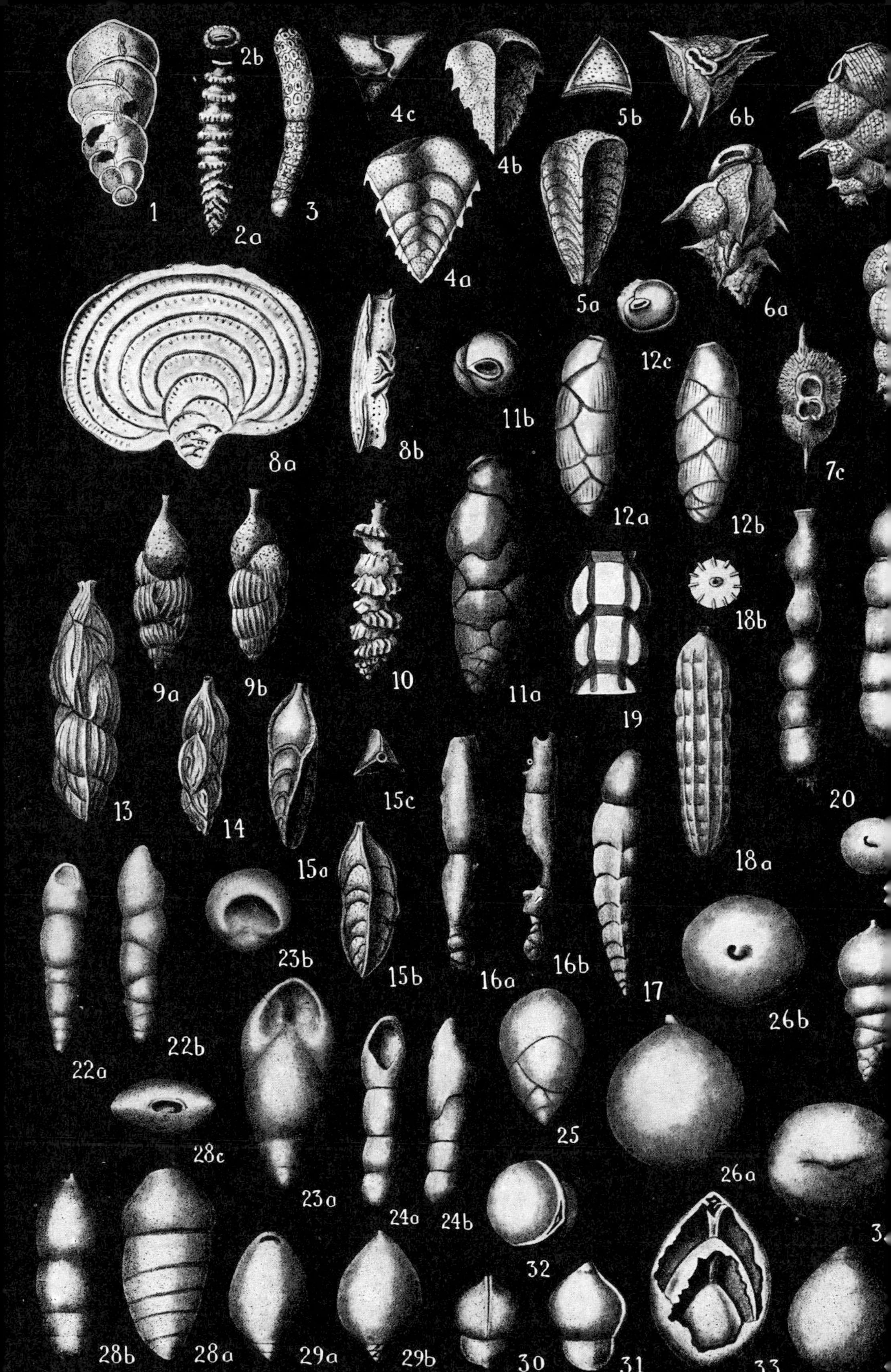

1
2a
2b
3
4a
4b
4c
5a
5b
6a
6b
7c
8a
8b
9a
9b
10
11a
11b
12a
12b
12c
13
14
15a
15b
15c
16a
16b
17
18a
18b
19
20
22a
22b
23a
23b
24a
24b
25
26a
26b
28a
28b
28c
29a
29b
30
31
32
33

Plate 28

PLATE 28.

Family 31. BULIMINIDAE, p. 260.

FIG.

1. *Bifarina saxipara* (Ehrenberg). Cretaceous. (After Ehrenberg). ... p. 269.
2. *Bifarina fimbriata* (Millett). Recent. × 28. (After Millett). *a*, front view; *b*, apertural view. ... p. 269.
3. *Schubertia tessellata* (H. B. Brady). Recent. × 150. ... p. 270.
4. *Reussella spinulosa* (Reuss). Miocene, Austria. (After Reuss). *a*, *b*, side views; *c*, apertural view. ... p. 271.
5. *Chrysalidinella dimorpha* (H. B. Brady). Recent. × 50. (After H. B. Brady). *a*, side view; *b*, apertural view. ... p. 273.
6. *Trimosina milletti* Cushman. Recent. × 60. (After Millett). *a*, side view; *b*, apertural view. ... p. 272.
7. *Mimosina hystrix* Millett. Recent. × 50. (After Millett). *a*, front view; *b*, side view; *c*, apertural view. ... p. 272.
8. *Pavonina flabelliformis* d'Orbigny. Recent. × 40. (After Parr). *a*, front view, Pliocene, Australia; *b*, basal view, Recent. ... p. 272.
9. *Uvigerina pigmea* d'Orbigny. Pliocene, Italy. (After d'Orbigny). *a*, *b*, side views. ... p. 273.
10. *Uvigerina porrecta* H. B. Brady. Recent. × 50. (After H. B. Brady). p. 273.
11. *Uvigerinella californica* Cushman. Miocene, California. × 50. *a*, side view; *b*, apertural view. ... p. 273.
12. *Hopkinsina danvillensis* Howe and Wallace. Eocene, Louisiana. × 55. (After Howe and Wallace). *a*, *b*, side views; *c*, apertural view. ... p. 274.

13, 14. *Angulogerina angulosa* (Williamson). Recent. 13, × 40. (After H. B. Brady). 14 (After Williamson). ... p. 275.

15. *Trifarina bradyi* Cushman. Recent. × 50. (After H. B. Brady). *a*, *b*, side views; *c*, apertural view. ... p. 275.

17. *Dentalinopsis subtriquetra* Reuss. Cretaceous, Germany. (After Reuss). ... p. 275.

18, 19. *Siphogenerina raphanus* (Parker and Jones). Recent. 18, Exterior (After Parker and Jones). 19, Section. ... p. 274.

20, 21. *Siphonodosaria abyssorum* (H. B. Brady). Recent. × 15. (After H. B. Brady). ... p. 274.

Family 23. LAGENIDAE, p. 211.

16. *Sporadogenerina flintii* Cushman. Recent. × 20. *a*, *b*, side views. p. 220.

Family 32. ELLIPSOIDINIDAE, p. 276.

22. *Pleurostomella subnodosa* Reuss. Cretaceous, Texas. × 35. *a*, *b*, side views. ... p. 277.
23. *Pleurostomella clavata* Cushman. Cretaceous, Trinidad. × 40. (After Cushman and Jarvis). *a*, front view; *b*, apertural view. p. 277.
24. *Pleurostomellina barroisi* (Berthelin). Cretaceous. × 120. (After Berthelin). *a*, *b*, side views. ... p. 277.
25. *Ellipsopleurostomella curta* Cushman. Cretaceous, Trinidad. × 40. (After Cushman and Jarvis). ... p. 277.
26. *Ellipsobulimina* sp. Cretaceous, California. × 40. (After Cushman and Church). *a*, side view; *b*, apertural view. ... p. 278.
27. *Nodosarella coalingensis* Cushman and Church. Cretaceous, California. × 30. (After Cushman and Church). *a*, front view; *b*, apertural view. ... p. 278.
28. *Ellipsolingulina impressa* (Terquem). Eocene, France. (After Terquem). *a*, front view; *b*, side view; *c*, apertural view. ... p. 280.
29. *Ellipsoglandulina velascoensis* Cushman. Cretaceous, Mexico. × 50. *a*, front view; *b*, side view. ... p. 280.

30–32. *Gonatosphaera prolata* Guppy. Miocene, Trinidad. (After Guppy). 30, Side view. 31, Front view. 32, Apertural view. ... p. 280.

33. *Ellipsoidina ellipsoides* Seguenza. Miocene, Italy. (After H. B. Brady). ... p. 280.

34, 35. *Parafissurina apiculata* (Reuss). Miocene, California. × 65. (After Cushman, Stewart and Stewart). 34, Front view. 35, Apertural view. ... p. 281.

PLATE 29.

Family 33. ROTALIIDAE, p. 281.

FIG.

1, 2. *Spirillina vivipara* Ehrenberg. Recent. 1, (After Ehrenberg). 2, × 65. (After H. B. Brady). *a*, side view; *b*, apertural view. .... p. 283.

3. *Spirillina vivipara* Ehrenberg, var. *revertens* Rhumbler. Recent. × 60. *a*, ventral view; *b*, peripheral view. .... p. 283.

4. *Spirillina inaequalis* H. B. Brady. Recent. × 40. *a*, dorsal view; *b*, ventral view; *c*, peripheral view. .... p. 283.

5. *Turrispirillina conoidea* (Paalzow). Jurassic, Germany. (After Paalzow). *a*, side view; *b*, ventral view. .... p. 283.

6. *Conicospirillina trochoides* (Berthelin). Jurassic, France. (After Berthelin). *a*, dorsal view; *b*, ventral view; *c*, side view. .... p. 284.

7–9. *Trocholina conica* (Schlumberger). Jurassic, Germany. (After Paalzow). 7, side view. 8, vertical section. 9, ventral view. ...... p. 284

10. *Paalzowella scalariformis* (Paalzow). Jurassic, Germany. (After Paalzow). *a*, dorsal view; *b*, ventral view; *c*, side view. ...... p. 284.

11, 12. *Patellinoides conica* Heron-Allen and Earland. Recent. × 125. (After Heron-Allen and Earland). 11*a* dorsal view; *b*, side view. 12, ventral view. .... p. 285.

13. *Patellina corrugata* Williamson. Recent. × 80. (After H. B. Brady). *a*, dorsal view; *b*, ventral view; *c*, side view. .... p. 285.

14. *Patellinella inconspicua* (H. B. Brady). Recent. × 50. (After H. B. Brady). *a*, dorsal view; *b*, ventral view; *c*, side view. .... p. 285.

15. *Annulopatellina annularis* (Parker and Jones). Recent. × 30. (After Parr). *a*, dorsal view; *b*, ventral view; *c*, side view. .... p. 285.

16. *Discorbis vesicularis* Lamarck. Eocene, France. × 20. (Topotype). *a*, dorsal view; *b*, ventral view; *c*, side view. .... p. 286.

17. *Discorbis nitida* (Williamson). Recent. × 120. (After Cushman and Parker). *a*, dorsal view; *b*, ventral view; *c*, side view. .... p. 286.

18. *Lamarckina erinacea* (Karrer). Miocene, Austria. (After Karrer). *a*, dorsal view; *b*, ventral view; *c*, side view. .... p. 288.

19. *Lamarckina wilcoxensis* Cushman. Eocene, Alabama. × 40. (After Cushman and Ponton). *a*, dorsal view; *b*, ventral view; *c*, side view. .... p. 288

20. *Lamarckina ripleyensis* Cushman. Cretaceous, Tennessee. × 90. *a*, dorsal view; *b*, ventral view; *c*, side view. .... p. 288.

21. *Heronallenia wilsoni* (Heron-Allen and Earland). Miocene, Australia. × 60. (After Heron-Allen and Earland). *a*, dorsal view; *b*, ventral view; *c*, side view. .... p. 289.

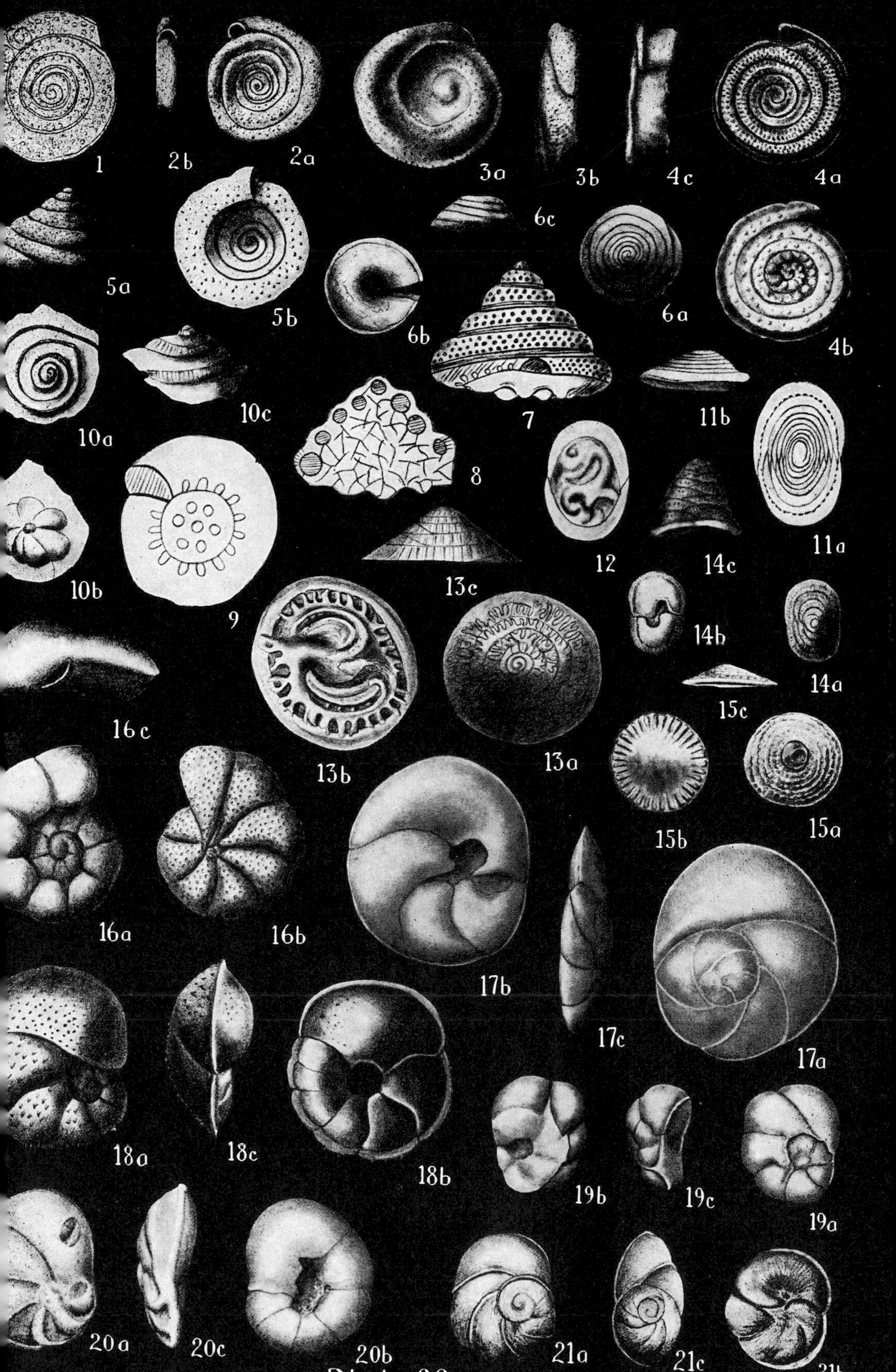
1
2b
2a
3a
3b
4c
4a
5a
6c
5b
6b
6a
4b
10a
10c
7
11b
8
9
10b
12
13c
14c
11a
14b
14a
16c
15c
13b
13a
15b
15a
16a
16b
17b
17c
17a
18a
18c
18b
19b
19c
19a
20a
20c
20b
21a
21c
21b

Plate 29

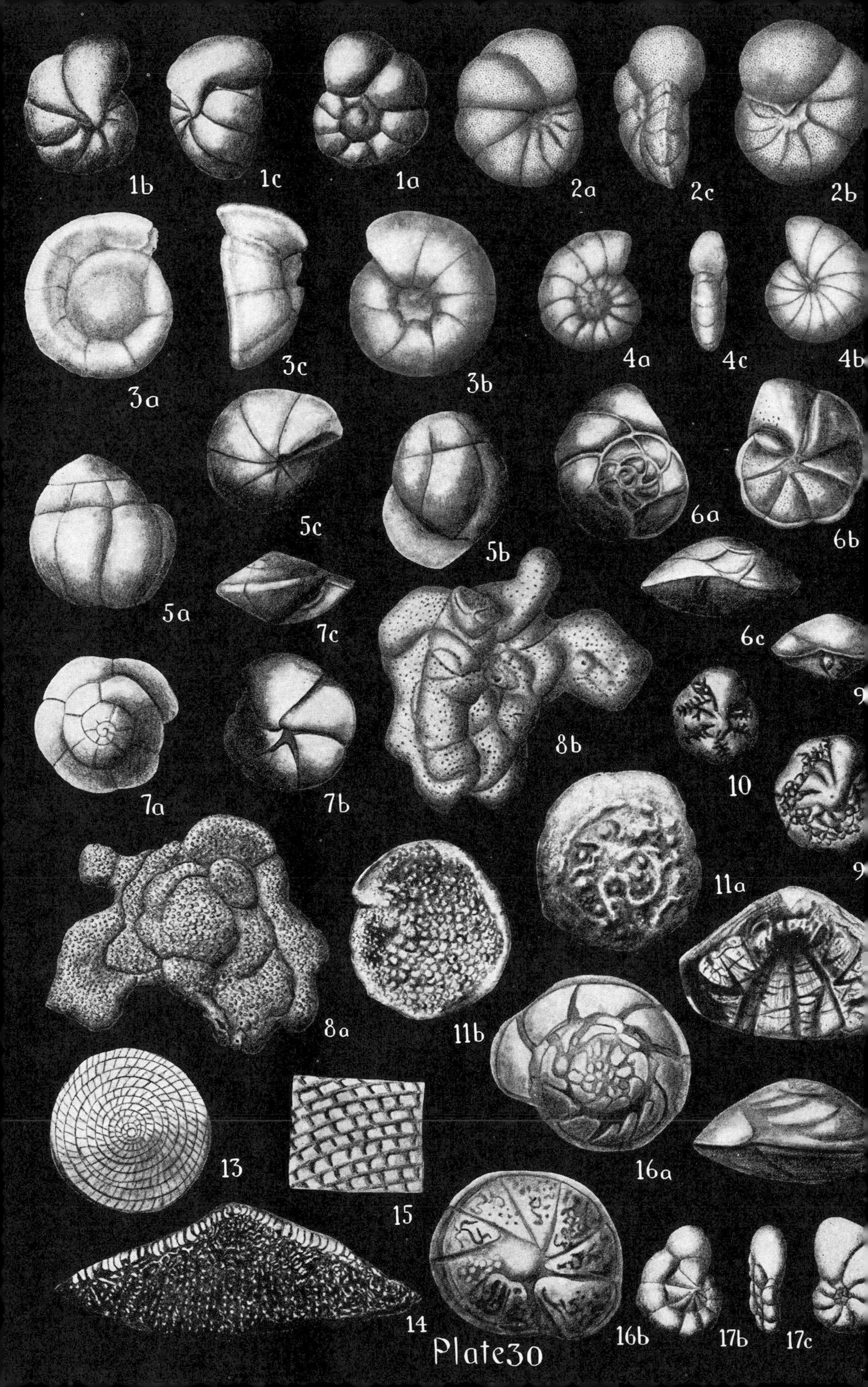
1b
1c
1a
2a
2c
2b
3a
3c
3b
4a
4c
4b
5c
5b
5a
6a
6b
7c
6c
7a
7b
8b
10
11a
8a
11b
13
15
16a
14
16b
17b
17c

Plate30

PLATE 30.

Family 33. ROTALIIDAE, p. 281.

FIG.

1. *Valvulineria californica* Cushman. Miocene, California. × 40. *a*, dorsal view; *b*, ventral view; *c*, side view. .... p. 289
2. *Valvulineria casitasensis* Cushman and Laiming. Miocene, California. (After Cushman and Laiming). *a*, dorsal view; *b*, ventral view; *c*, side view. .... p. 289.
3. *Gyroidina girardana* (Reuss). Cretaceous, Arkansas. × 40. *a*, dorsal view; *b*, ventral view; *c*, side view. .... p. 290.
4. *Gyroidina depressa* (Alth). Cretaceous, Trinidad. × 55. (After Cushman and Jarvis). *a*, dorsal view; *b*, ventral view; *c*, side view. .... p. 290.
5. *Rotaliatina mexicana* Cushman. Eocene, Mexico. × 40. *a*, dorsal view; *b*, ventral view; *c*, side view. .... p. 291.
6. *Eponides repandus* (Fichtel and Moll). Recent. × 20. (After H. B. Brady). *a*, dorsal view; *b*, ventral view; *c*, side view. .... p. 291.
7. *Eponides tenera* (H. B. Brady). Recent. × 50. (After H. B. Brady). *a*, dorsal view; *b*, ventral view; *c*, side view. .... p. 291.
8. *Planopulvinulina dispansa* (H. B. Brady). Recent. × 10. (After H. B. Brady). *a*, dorsal view; *b*, ventral view. .... p. 293.

9, 10. *Rotalia trochidiformis* Lamarck. Eocene, France. × 20. 9 *a*, ventral view; *b*, side view. .... p. 293.

11. *Lockhartia haimei* (Davies). Eocene, India. × 12. (After Davies). *a*, dorsal view; *b*, ventral view. .... p. 294.
12. *Lockhartia conditi* (Nuttall). Eocene, India. × 12. (After Davies). Section. .... p. 294.

13–15. *Dictyoconoides cooki* (Carter). Eocene, India. 13, Dorsal view (After Carter). 14, Vertical section (After Nuttall). 15, Portion of surface layer (After Nuttall). .... p. 295.

16. *Epistomina elegans* (d'Orbigny). Recent. × 25. (After H. B. Brady). *a*, dorsal view; *b*, ventral view; *c*, side view. .... p. 295.
17. *Epistomaria rimosa* (Parker and Jones). Eocene, France. × 15. (After Parker and Jones). *a*, dorsal view; *b*, ventral view; *c*, side view. .... p. 296.

PLATE 31.

Family 33. ROTALIIDAE, p. 281.

FIG.

1. *Mississippina monsouri* Howe. Lower Oligocene, Mississippi. × 60. (After Howe). *a*, dorsal view; *b*, ventral view; *c*, side view. p. 296.
2. *Siphonina tubulosa* Cushman. Recent. × 50. *a*, dorsal view; *b*, ventral view; *c*, side view. .... p. 296.
3. *Siphonina howei* Cushman. Eocene, Louisiana. × 65. *a*, dorsal view; *b*, ventral view. .... p. 296.
4. *Siphoninella claibornensis* Cushman and Howe. Eocene, Louisiana. × 65. (After Cushman and Howe). *a*, dorsal view; *b*, ventral view; *c*, apertural view. .... p. 297.
5. *Siphoninella soluta* (H. B. Brady). Recent. × 65. (After H. B. Brady). *a*, dorsal view; *b*, ventral view; *c*, apertural view. .... p. 297.
6. *Siphoninoides echinata* (H. B. Brady). Recent. × 65. (After H. B. Brady). *a*, *b*, side views; *c*, apertural view. .... p. 297.
7. *Baggina californica* Cushman. Miocene, California. × 40. *a*, dorsal view; *b*, ventral view; *c*, apertural view. .... p. 298.
8. *Cancris auriculus* (Fichtel and Moll). Recent. × 30. (After H. B. Brady). *a*, dorsal view; *b*, ventral view; *c*, side view. .... p. 297.
9. *Neocribrella globigerinoides* (Parker and Jones). Eocene, France. (After Parker and Jones). *a*, dorsal view; *b*, ventral view. ... p. 298.

Family 34. PEGIDIIDAE, p. 298.

10. *Pegidia dubia* (d'Orbigny). Recent. × 30. (After Heron-Allen and Earland). *a*, dorsal view; *b*, ventral view; *c*, side view. ...... p. 299.
11. *Sphaeridia papillata* Heron-Allen and Earland. Recent. × 40. (After Heron-Allen and Earland). .... p. 300.
12. *Physalidia simplex* Heron-Allen and Earland. Recent. × 40. (After Heron-Allen and Earland). .... p. 300.
13. *Rugidia corticata* (Heron-Allen and Earland). Recent. × 60. (After Heron-Allen and Earland). *a*, dorsal view; *b*, apertural view. p. 300.

Family 35. AMPHISTEGINIDAE, p. 300.

14. *Asterigerina bracteata* Cushman. Lower Oligocene, Mississippi. × 80. *a*, dorsal view; *b*, ventral view; *c*, side view. .... p. 301.
15. *Asterigerina carinata* d'Orbigny. Recent. × 30. (After d'Orbigny). *a*, dorsal view; *b*, ventral view; *c*, side view. .... p. 301.

16, 17. *Amphistegina lessonii* d'Orbigny. Recent. 16 (After d'Orbigny). 17, × 20. (After H. B. Brady). *a*, *a*, dorsal views; *b*, *b*, ventral views; *c*, *c*, side views. .... p. 302.

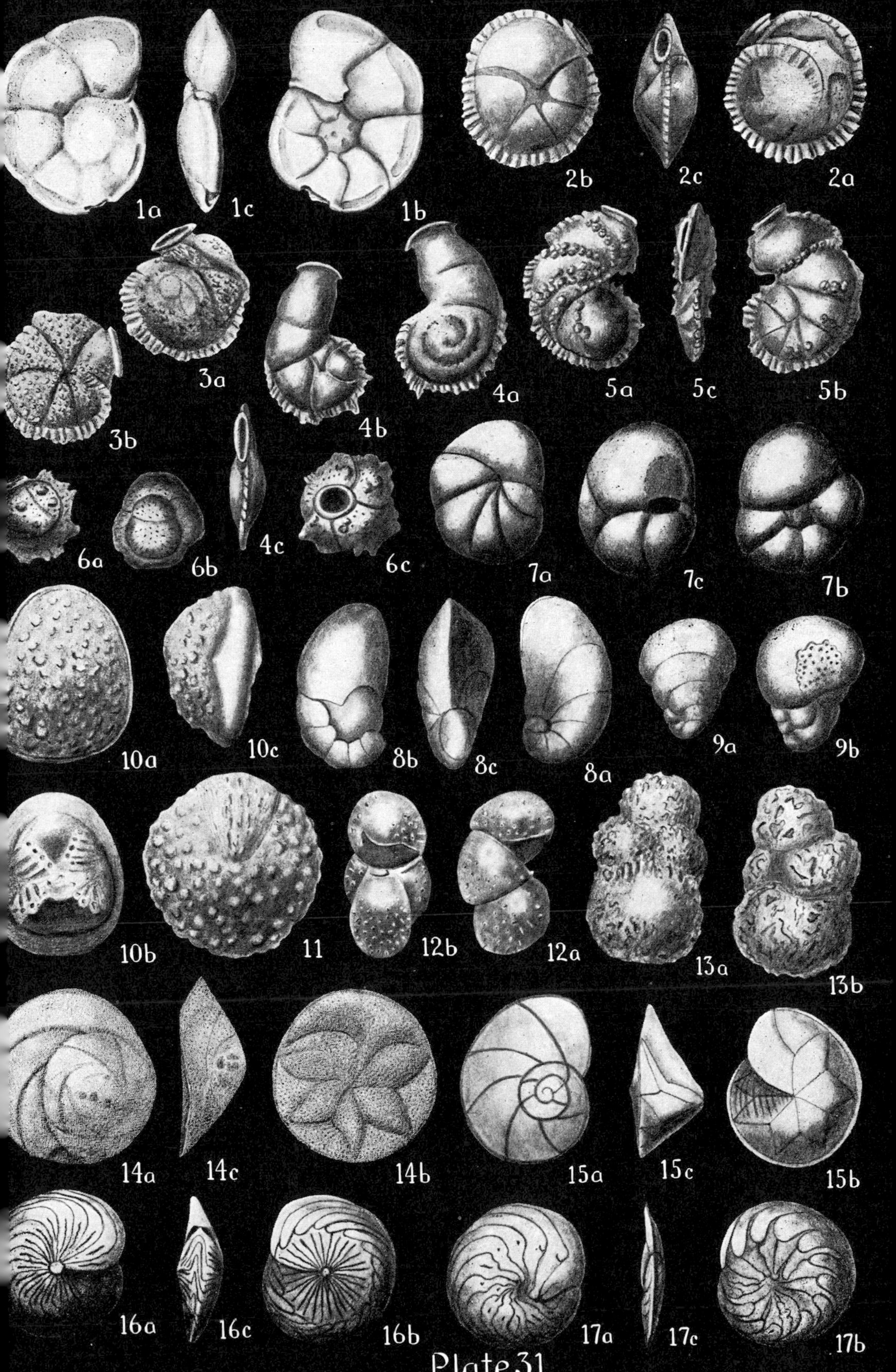
1a
1c
1b
2b
2c
2a
3a
3b
4b
4a
5a
5c
5b
6a
6b
4c
6c
7a
7c
7b
10a
10c
8b
8c
8a
9a
9b
10b
11
12b
12a
13a
13b
14a
14c
14b
15a
15c
15b
16a
16c
16b
17a
17c
17b

Plate 31

1a
1b
2
3
4
5
6
7
8
9b
9a
10
11
12
13b
14
15a
15b
15c
16
17
17
18
19a
19b
19c
20a
20b
21

Plate 32

PLATE 32.

Family 36. CALCARINIDAE, p. 302.

FIG.

1. *Calcarina spengleri* (Linné). Recent. × 20. (After H. B. Brady). *a*, dorsal view; *b*, ventral view. .... p. 303.

2, 3. *Siderolites calcitrapoides* Lamarck. Cretaceous, Holland. (After Hofker). 2, Exterior. 3, Section. .... p. 303.

4. *Baculogypsinoides spinosus* Yabe and Hanzawa. × 12. .... p. 303.

5, 6. *Baculogypsina sphaerulatus* (Parker and Jones). 5, × 35 (After W. B. Carpenter), Exterior of adult. 6, × 80 (After Hofker), Young. .... p. 304.

7, 8. *Arnaudiella grossouvrei* H. Douvillé. Upper Cretaceous, France. × 10. (After H. Douvillé). 7, Transverse section. 8, Detail of section showing pillars and vacuolar structure. .... p. 304.

9. *Pellatispira douvilléi* Boussac. Eocene, Italy. × 7. (After Boussac). *a*, equatorial section; *b*, transverse section. .... p. 306.

10. *Pellatispira madaraszi* (Hantken). Eocene, Austria. × 4. (After Boussac). Exterior. .... p. 306.

11, 12. *Siderina douvilléi* Abrard. Upper Cretaceous, France. (After Abrard). 11, Exterior. 12, Section. .... p. 307.

Family 37. CYMBALOPORIDAE, p. 307.

13, 14. *Cymbalopora radiata* Hagenow. Cretaceous, Holland. × 22. 13 *a*, dorsal view; *b*, ventral view. 14, Early portion with exterior eroded away by acid showing *Discorbis*-like young. .... p. 308.

15. *Cymbaloporetta squammosa* (d'Orbigny). Recent. × 35. (After H. B. Brady). *a*, dorsal view; *b*, ventral view; *c*, side view. ... p. 308.

16, 17. *Cymbaloporella tabellaeformis* (H. B. Brady). Recent. × 28. (After H. B. Brady). 16, Dorsal view. 17 *a*, ventral view; *b*, side view. p. 309.

18, 19. *Halkyardia minima* (Liebus). Eocene, France. × 25. (After Halkyard). 18, Vertical section. 19 *a*, dorsal view; *b*, ventral view; *c*, side view. .... p. 309.

20–22. *Tretomphalus bulloides* (d'Orbigny). Recent. × 40. 20, 21 (After H. B. Brady). 20 *a*, dorsal view; *b*, side view. 21, Ventral view. 22 (After Earland), showing inner "float" chamber. .... p. 310.

PLATE 33.

Family 38. CASSIDULINIDAE, p. 311.

FIG.

1. *Ceratobulimina cretacea* Cushman and Harris. Cretaceous, Texas. × 65. (After Cushman and Harris). *a*, dorsal view; *b*, ventral view; *c*, apertural view. .... p. 312.
2. *Pseudoparrella pacifica* (Cushman). Recent. × 35. *a*, dorsal view; *b*, peripheral view. .... p. 312.
3. *Cassidulina cretacea* Cushman. Cretaceous, Tennessee. × 90. *a*, *b*, opposite sides; *c*, apertural view. .... p. 313.
4. *Cassidulina laevigata* d'Orbigny. Recent. × 50. (After H. B. Brady). *a*, *b*, opposite sides; *c*, apertural view. .... p. 313.
5. *Cassidulinoides parkeriana* (H. B. Brady). Recent. ×50. (After H. B. Brady). *a*, *b*, side views; *c*, apertural view. .... p. 315.
6. *Orthoplecta clavata* H. B. Brady. Recent. × 65. (After H. B. Brady). p. 315.
7. *Ehrenbergina pacifica* Cushman. Recent. × 40. (After H. B. Brady). *a*, dorsal view; *b*, ventral view; *c*, apertural view. .... p. 315.

Family 39. CHILOSTOMELLIDAE, p. 315.

8. *Allomorphina trigona* Reuss. Recent. × 40. (After H. B. Brady). *a*, dorsal view; *b*, ventral view. .... p. 318.

9–11. *Chilostomella oolina* Schwager. Recent. × 35. (After H. B. Brady). 9 *a*, side view; *b*, apertural view. 10 *a*, dorsal view; *b*, side view. 11, Partially broken specimen. .... p. 319

12, 13. *Chilostomelloides oviformis* (Sherborn and Chapman). Eocene. × 12. (After Sherborn and Chapman). .... p. 319.

14, 15. *Seabrookia pellucida* H. B. Brady. Recent. × 135. (After H. B. Brady). 14 *a*, ventral view; *b*, apertural view. 15, By transmitted light. .... p. 319.

16. *Allomorphinella contraria* Reuss. Cretaceous, Galicia. (After Reuss). *a*, ventral view; *b*, apertural view; *c*, side view. .... p. 320.

17. *Chilostomellina fimbriata* Cushman. Recent. × 50. *a*, dorsal view; *b*, ventral view; *c*, apertural view. .... p. 320.

18. *Pullenia salisburyi* R. E. and K. C. Stewart. Pliocene, California. × 100. (After R. E. and K. C. Stewart). *a*, side view; *b*, apertural view. .... p. 320.

19. *Pullenia quinqueloba* Reuss. Recent. × 35. (After H. B. Brady). *a*, side view; *b*, apertural view. .... p. 320.

20. *Pullenia coryelli* White. Cretaceous, Texas. × 40. *a*, side view; *b*, apertural view. .... p. 320.

21, 22. *Sphaeroidina bulloides* d'Orbigny. 21, Pliocene, California, × 75 (After Cushman, Stewart and Stewart), *a*, side view; *b*, apertural view. 22 (After d'Orbigny's Model), *a*, side view; *b*, apertural view. .... p. 321.

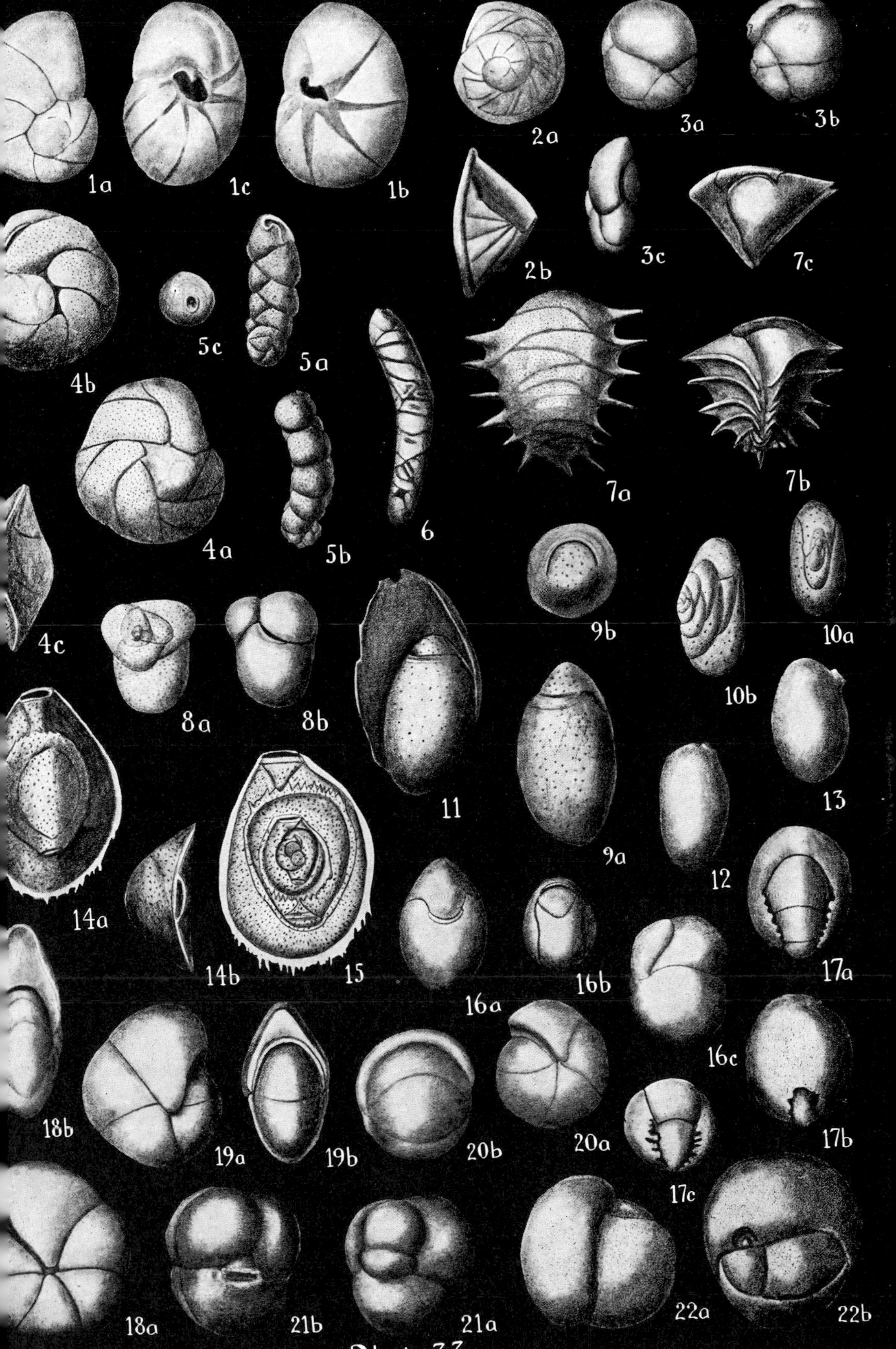
1a
1c
1b
2a
3a
3b
2b
3c
7c
4b
5c
5a
7a
7b
4a
5b
6
4c
8a
8b
9b
10a
10b
11
13
14a
14b
15
9a
12
16a
16b
17a
16c
18b
19a
19b
20b
20a
17b
17c
18a
21b
21a
22a
22b

Plate 33

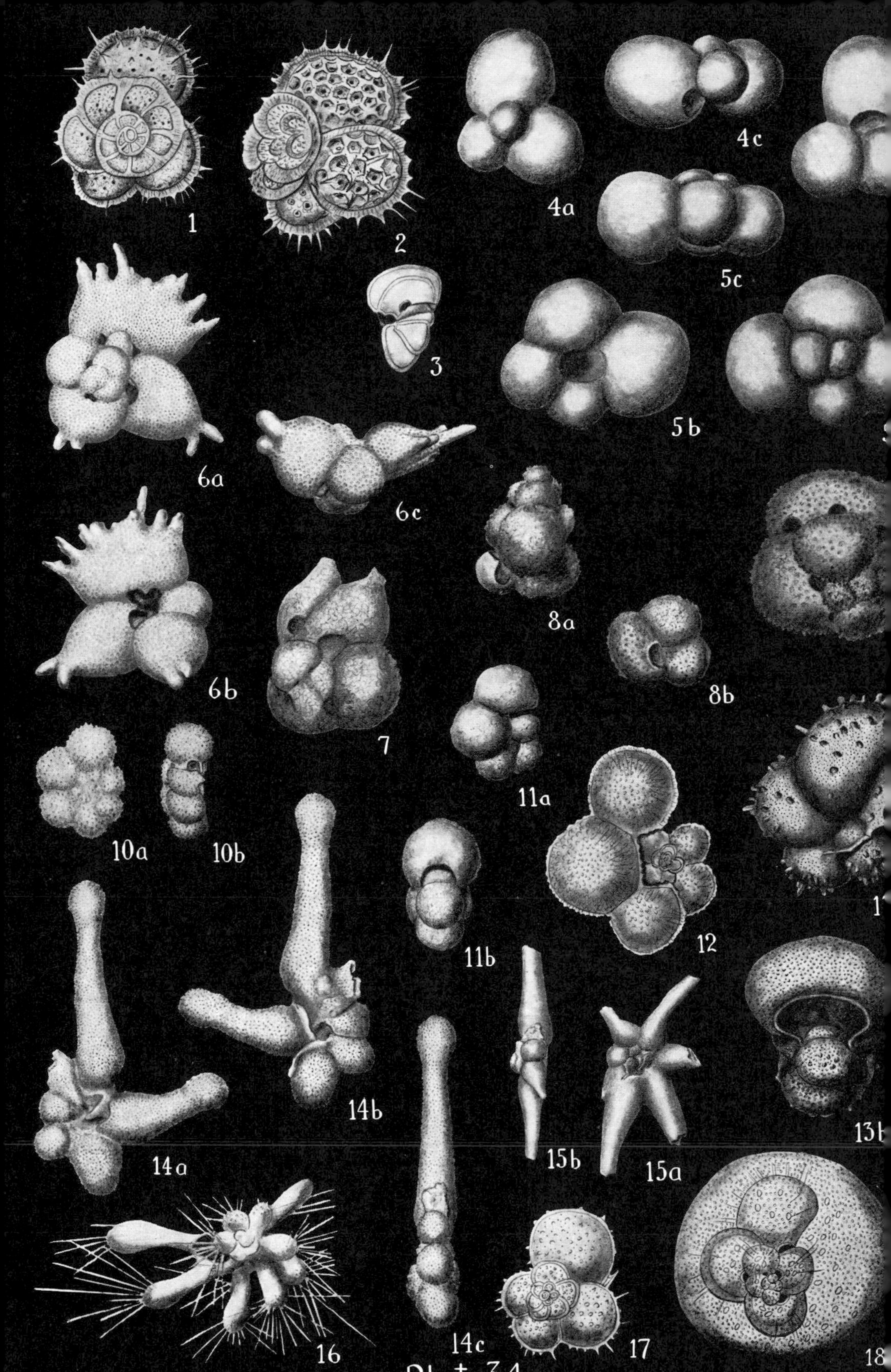
1
2
4a
4c
5c
3
5b
6a
6c
8a
6b
7
8b
10a
10b
11a
12
11b
14b
14a
15b
15a
13b
16
14c
17
18

Plate34

PLATE 34.

Family 40. GLOBIGERINIDAE, p. 321.

FIG.

1–3. *Globigerina triloba* Reuss. Recent. (After Rhumbler). 1, Dorsal view, showing *Discorbis*-like young of microspheric form. 2, Side view. 3, Side view of *Discorbis*-like young stage, smooth, without spines. .... p. 322.

4, 5. *Globigerina bulloides* d'Orbigny. (After d'Orbigny's Models). *a, a,* dorsal views; *b, b,* ventral views; *c, c,* side views. .... p. 322.

6. *Globigerinoides sacculifera* (H. B. Brady), var. *fistulosa* (Schubert). Recent. × 25. *a,* dorsal view; *b,* ventral view; *c,* side view. .... p. 322.

7. *Globigerinoides sacculifera* (H. B. Brady). Recent. × 35. (After H. B. Brady). .... p. 322.

8. *Globigerinoides rubra* (d'Orbigny). Recent. × 35. (After H. B. Brady). *a,* side view; *b,* ventral view. .... p. 322.

9. *Globigerinoides conglobata* (H. B. Brady). Recent. × 35. (After H. B. Brady). .... p. 322.

10. *Globigerinella aspera* (Ehrenberg). Cretaceous, Texas. × 50. *a,* side view; *b,* apertural view. .... p. 323.

11, 12. *Globigerinella acquilateralis* (H. B. Brady). Recent. 11, × 35 (After H. B. Brady), *a,* side view; *b,* apertural view. 12 (After Rhumbler), Showing smooth trochoid young. .... p. 323.

13. *Hastigerina pelagica* (d'Orbigny). Recent. × 35. (After H. B. Brady). *a,* side view; *b,* apertural view. .... p. 323.

14. *Hastigerinella watersi* Cushman. Cretaceous, Texas. × 60. *a, b,* opposite sides; *c,* peripheral view. .... p. 324.

15. *Hastigerinella alexanderi* Cushman. Cretaceous, Texas. × 60. *a,* side view; *b,* apertural view. .... p. 324.

16. *Hastigerinella digitata* (Rhumbler). Recent. (After Rhumbler). .... p. 324.

17, 18. *Orbulina universa* d'Orbigny. Recent. (After Rhumbler). 17, Early stage showing *Discorbis*-like young. 18, Adult globular chamber showing early *Globigerina* stage inside. .... p. 326.

PLATE 35.

Family 40. GLOBIGERINIDAE, p. 321.

FIG.

1–3. *Pulleniatina obliquiloculata* (Parker and Jones). Recent. 1, Exterior (After H. B. Brady). 2 (After Rhumbler), Viewed by transmitted light showing early included *Globigerina* stage. 3 (After H. B. Brady), Section. . . . . p. 326.

4–7. *Sphaeroidinella dehiscens* (Parker and Jones). Recent. 4, 5, × 35 (After H. B. Brady), Exteriors. 6, 7 (After Rhumbler). 6, Vertical section of wall. 7, Exterior. . . . . p. 326.

8. *Candeina nitida* d'Orbigny. Recent. × 35. (After H. B. Brady). *a*, dorsal view; *b*, ventral view; *c*, side view. . . . . p. 327.

Family 41. HANTKENINIDAE, p. 327.

9, 10. *Schackoina multispinata* (Cushman and Wickenden). Cretaceous, Canada. (After Cushman and Wickenden). 9, Adult, × 120. 10, Young, × 200. . . . . p. 328.

11. *Hantkenina mexicana* Cushman. Eocene, Mexico. × 50. *a*, side view; *b*, apertural view. . . . . p. 328.

12. *Hantkenina alabamensis* Cushman. Eocene, Alabama. × 40. . . . . . . p. 328.

Family 42. GLOBOROTALIIDAE, p. 329.

13. *Globorotalites micheliniana* (d'Orbigny). Cretaceous, Texas. × 40. *a*, dorsal view; *b*, ventral view; *c*, side view. . . . . p. 330.

14. *Globotruncana calcarata* Cushman. Cretaceous, Texas. × 50. *a*, dorsal view; *b*, ventral view; *c*, side view. . . . . p. 329.

15. *Globorotalia fimbriata* (H. B. Brady). Recent. × 25. (After H. B. Brady). *a*, dorsal view; *b*, ventral view. . . . . p. 330.

16. *Globorotalia tumida* (H. B. Brady). Recent. × 25. (After H. B. Brady). *a*, dorsal view; *b*, ventral view; *c*, side view. . . . . p. 330.

17. *Globotruncana arca* (Cushman). Cretaceous, Mexico. × 40. *a*, dorsal view; *b*, ventral view; *c*, side view. . . . . p. 329.

18. *Cycloloculina miocenica* Cushman and Ponton. Miocene, Florida. × 55. (After Cushman and Ponton). . . . . p. 331.

19–22. *Cycloloculina annulata* Heron-Allen and Earland. Eocene, England. (After Heron-Allen and Earland). 19, Early stage with chambers nearly annular, ×40. 20, Still earlier stage, × 32. 21, Detail of surface and spines, × 95. 22, Adult, × 32. . . . . p. 331.

23, 24. *Sherbornina atkinsoni* Chapman. Miocene, Tasmania. (After Chapman). 23, Portion of median section, × 16. 24, Surface view of adult, × 14. . . . . p. 331.

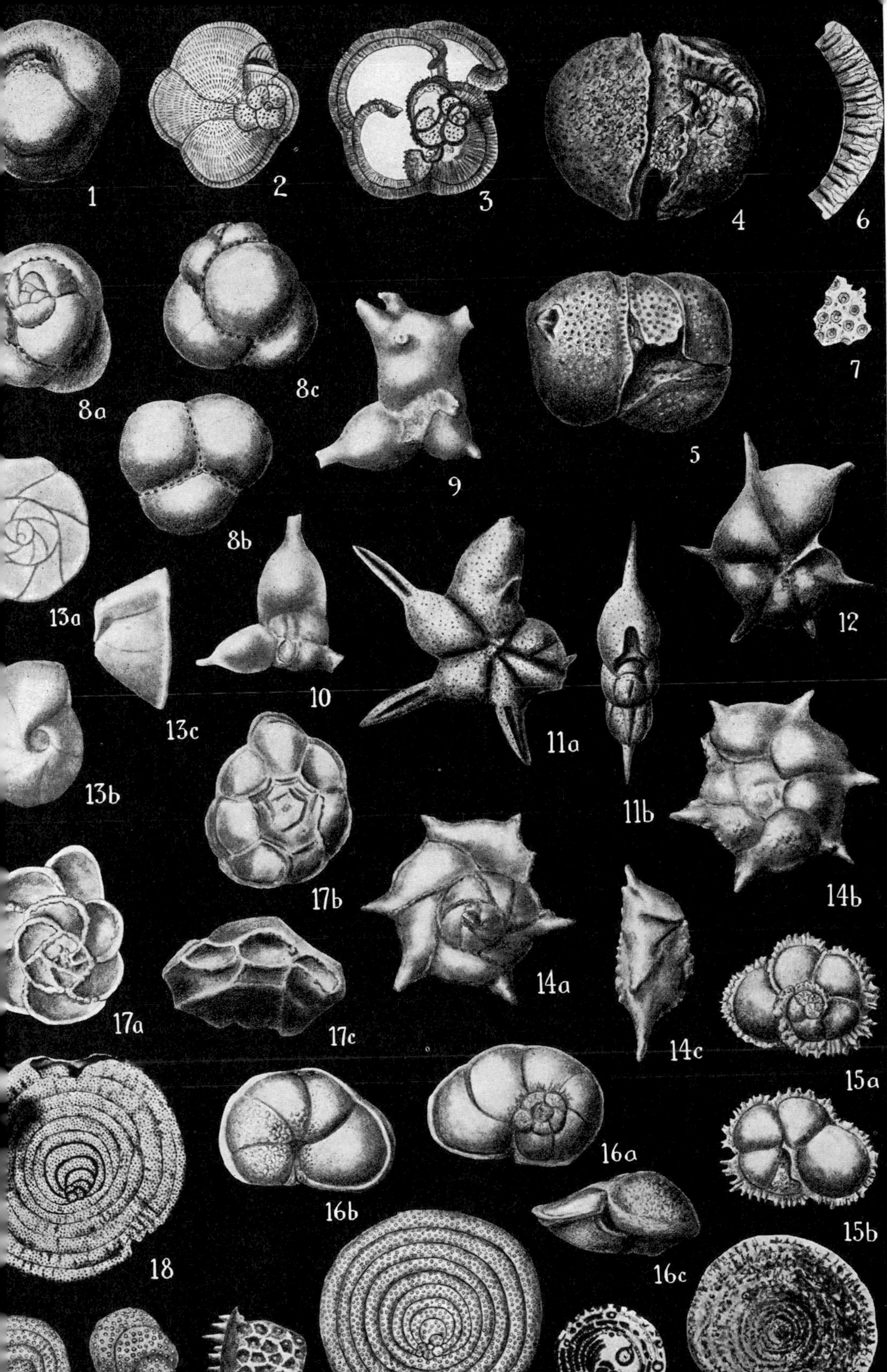
1
2
3
4
6
8a
8c
7
9
5
8b
13a
10
11a
11b
12
13c
13b
17b
14b
14a
17a
17c
14c
15a
18
16a
16b
15b
16c
19
20
21
22
23
24

Plate 35

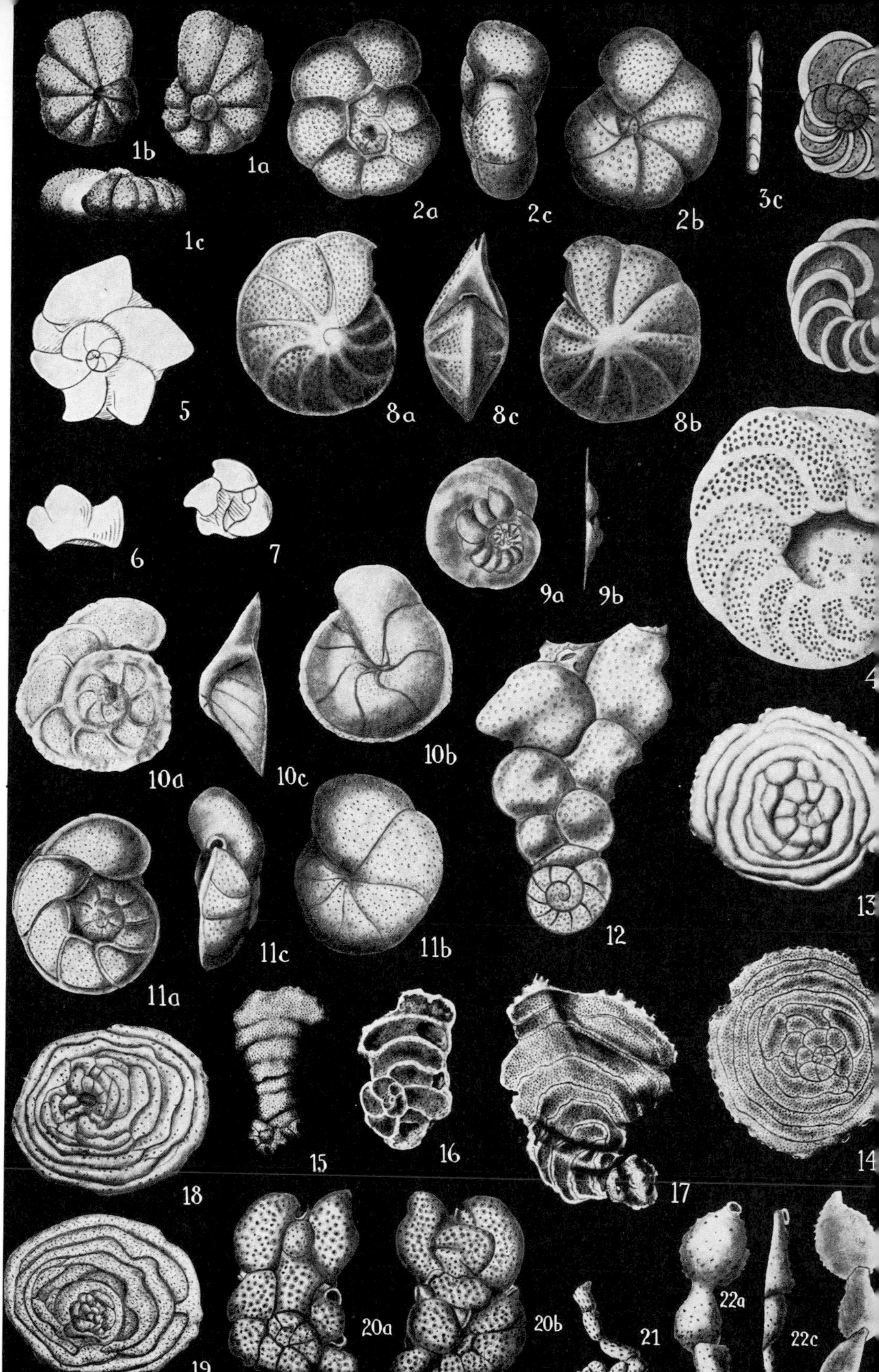
1b
1a
1c
2a
2c
2b
3c
5
8a
8c
8b
6
7
9a
9b
4
10a
10c
10b
12
13
11c
11b
11a
15
16
17
14
18
19
20a
20b
21
22a
22c

Plate36

PLATE 36.

Family 43. ANOMALINIDAE, p. 331.

FIG.

1. *Anomalina punctulata* d'Orbigny. Recent. (After d'Orbigny). *a*, dorsal view; *b*, ventral view; *c*, side view. .... p. 332.
2. *Anomalina grosserugosa* Gümbel (?). Recent. × 28. (After H. B. Brady). *a*, dorsal view; *b*, ventral view; *c*, side view. .... p. 332.
3. *Planulina ariminensis* d'Orbigny. Recent. (After d'Orbigny). *a*, dorsal view; *b*, ventral view; *c*, side view. .... p. 334.
4. *Planulina ornata* (d'Orbigny). Recent. × 35. .... p. 334.

5–7. *Ruttenia coronaeformis* (Pijpers). Eocene, Island of Bonaire. (After Pijpers). 5, Dorsal view. 6, Ventral view. 7, Side view. ..... p. 334.

8. *Anomalinella rostrata* (H. B. Brady). Recent. ×35. (After H. B. Brady). *a*, *b*, opposite sides; *c*, peripheral view. .... p. 335.
9. *Laticarinina pauperata* (Parker and Jones). Recent. × 20. (After Flint). *a*, side view; *b*, peripheral view. .... p. 334.
10. *Cibicides refulgens* Montfort. Recent. × 20. (After H. B. Brady). *a*, dorsal view; *b*, ventral view; *c*, peripheral view. .... p. 335.
11. *Cibicides lobatulus* (Walker and Jacob). Recent. × 40. (After H. B. Brady). *a*, dorsal view; *b*, ventral view; *c*, peripheral view. ... p. 335.
12. *Dyocibicides biserialis* Cushman and Valentine. Recent. × 30. (After Cushman and Valentine). .... p. 338.

13, 14. *Annulocibicides projectus* Cushman and Ponton. Miocene, Florida. × 40. (After Cushman and Ponton). 13, Upper side. 14, Attached side showing tubular apertures. .... p. 338.

15–17. *Rectocibicides miocenicus* Cushman and Ponton. Miocene, Florida. × 32. (After Cushman and Ponton). 15, Outer surface. 16, 17, Attached surfaces. 17, Showing tubular apertures. .... p. 338.

18, 19. *Cyclocibicides vermiculatus* (d'Orbigny). Recent. × 20. (After H. B. Brady). 18, 19, Opposite sides. .... p. 339.

20, 21. *Cibicidella variabilis* (d'Orbigny). Recent. 20, × 16 (After Sidebottom), *a*, *b*, opposite sides. 21 (After d'Orbigny), Specimen growing upward from attached surface. .... p. 339.

22. *Webbina rugosa* d'Orbigny. Recent. (After d'Orbigny). *a*, outer side; *b*, attached side; *c*, edge view. .... p. 339.

PLATE 37.

Family 44. PLANORBULINIDAE, p. 339.

FIG.

1, 2. *Planorbulina mediterranensis* d'Orbigny. Recent. 1 (After d'Orbigny), *a*, *b*, opposite sides; *c*, peripheral view. 2, × 28 (After H. B. Brady). .... p. 340.

3. *Planorbulinella larvata* (Parker and Jones). Recent. × 10. (After Parker and Jones). *a*, side view; *b*, peripheral view. .... p. 342.

4. *Planorbulinoides retinaculata* (Parker and Jones). Recent. × 10. (After Parker and Jones). .... p. 342.

5–7. *Linderina brugesi* Schlumberger. Eocene, France. (After Schlumberger). 5, Exterior, × 28. 6, Vertical section, × 25. 7, Details of vertical section, × 65. .... p. 342.

8–10. *Acervulina inhaerens* Schultze. Recent. 8, 9, × 25 (After H. B. Brady). 8 *a*, outer surface; *b*, peripheral view of compact form. 9, Spreading form. 10 (After Schultze). .... p. 343.

11. *Gypsina vesicularis* (Parker and Jones). Recent. × 20. (After H. B. Brady). .... p. 343.

Family 45. RUPERTIIDAE, p. 343.

12–15. *Rupertia stabilis* Wallich. Recent. × 28. (After H. B. Brady). 12, Young stage. 13, Longitudinal section. 14, Adult. 15, Transverse section of young stage. .... p. 344.

16. *Carpenteria proteiformis* Goës. Recent. × 10. .... p. 344.

17. *Carpenteria utricularis* Carter. Recent. × 10. (After H. B. Brady). Young specimen. .... p. 344.

18–20. *Carpenteria monticularis* Carter. Recent. (After H. B. Brady). 17, × 8, Outer surface. 18, × 20, From below. 19, × 20, From periphery. .... p. 344.

Family 46. VICTORIELLIDAE, p. 345.

21, 22. *Victoriella plecte* (Chapman). Miocene, Australia. × 10. (After Chapman and Crespin). 21, Horizontal section. 22, Vertical section. .... p. 345

23–25. *Eorupertia boninensis* (Yabe and Hanzawa). Eocene, Japan. (After Yabe and Hanzawa). 23, Exterior, × 8. 24, Vertical section, × 12. 25, Horizontal section, × 12. .... p. 345.

26. *Hofkerina semiornata* (Howchin). Oligocene, Australia. × 10. (After Chapman and Parr). .... p. 346.

Family 47. HOMOTREMIDAE, p. 346.

27–29. *Homotrema rubrum* (Lamarck). Recent. (After Hickson). 27, Exterior, × 3. 28, Section, × 65. 29, Surface, × 35. .... p. 346.

30–32. *Sporadotrema cylindricum* (Carter). Recent. (After Hickson). 30, Exterior, × 1½. 31, Section, × 65. 32, Surface, × 35. .... p. 347.

33–36. *Miniacina miniacea* (Pallas). Recent. 33–35 (After Hickson). 33, Exterior, × 3. 34, Section, × 65. 35, Surface, × 35. 36 (After Heron-Allen and Earland), Ventral view showing very young coiled stage. .... p. 347.

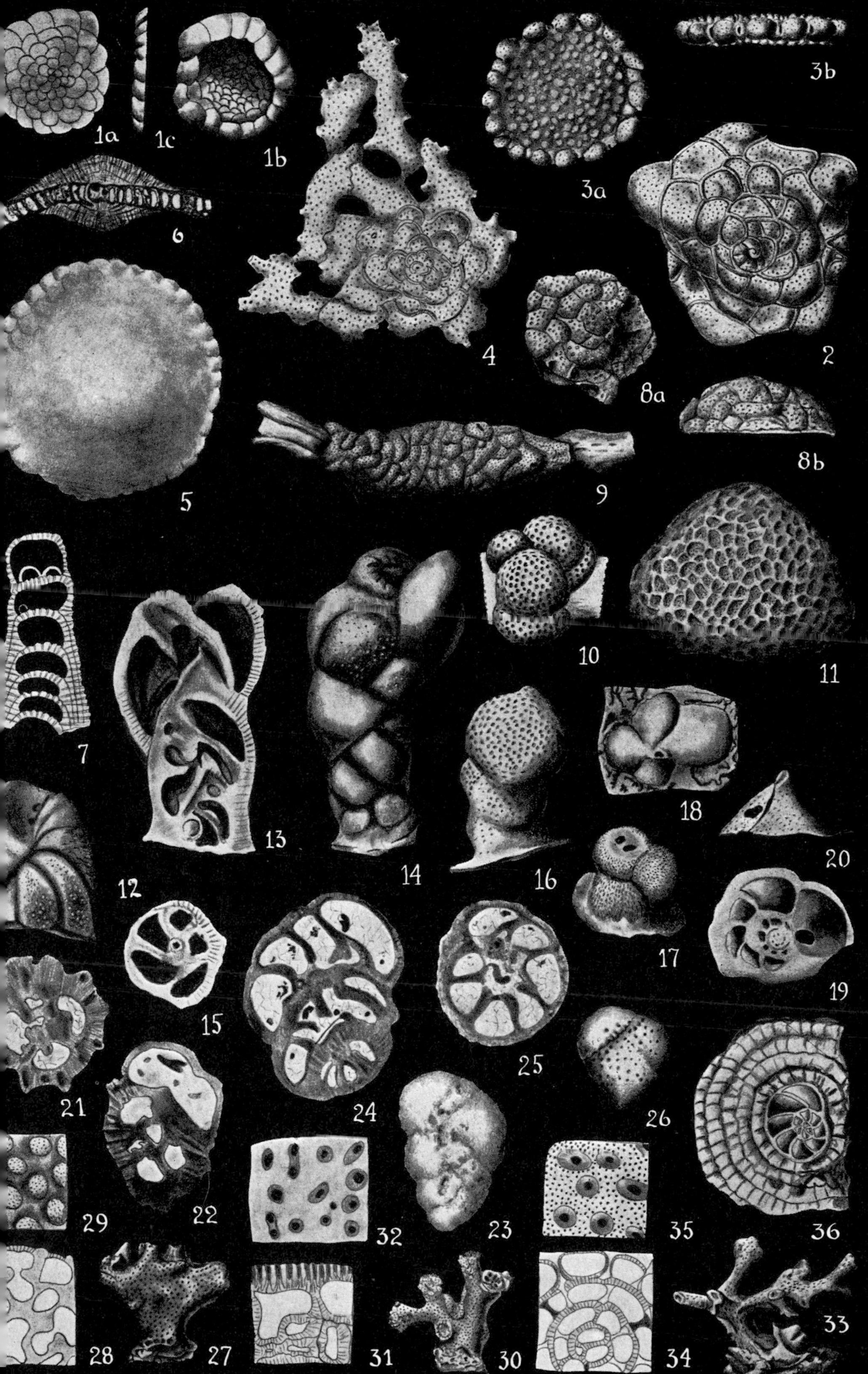
1a 1c 1b 3a 3b 2 4 6 5 8a 8b 9 7 13 14 10 11 16 18 20 12 17 19 15 21 24 25 26 22 23 29 32 35 36 28 27 31 30 34 33

Plate 37

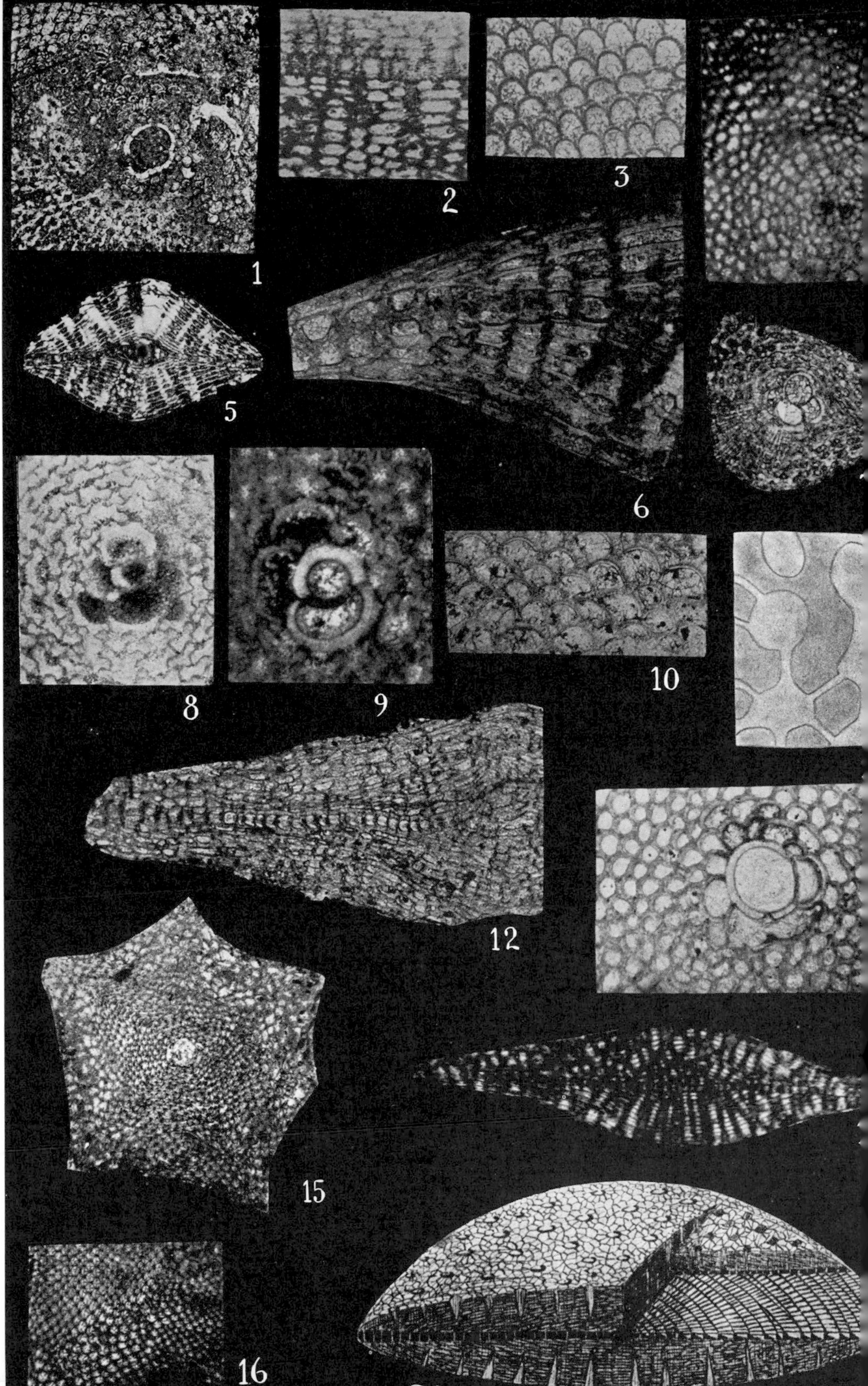

Plate 38

PLATE 38.

Family 48. ORBITOIDIDAE, p. 347.

Family 49. DISCOCYCLINIDAE, p. 367.

FIG.

1. *Orbitoides media* (d'Archiac). × 16. (After Vaughan). Showing quadrilocular embryonic apparatus surrounded by a common thick wall. .... p. 355.

2, 3. *Lepidorbitoides socialis* (Leymerie). × 80. (After H. Douvillé). 2, Portion of vertical section. 3, Portion of horizontal section. p. 357.

4, 6. *Pseudorbitoides trechmanni* H. Douvillé. 4, Section nearly in equatorial plane, × 30. 6, Vertical section of peripheral part of test, × 70. .... p. 353.

5, 7. *Pseudorbitoides israelskii* Vaughan and Cole. × 25. 5, Vertical section. 7, Section showing embryonic chambers. .... p. 353.

8–10. *Lepidorbitoides minima* (H. Douvillé). 8, 9, Embryonic chambers of two specimens, × 70. 10, Equatorial chambers near the periphery, × 18. .... p. 357.

11–13. *Actinosiphon semmesi* Vaughan. 11, Equatorial chambers, × about 300, to show the stoloniferous passages between chambers in the same radial series. 12, Vertical section, × 70. 13, Embryonic and equatorial chambers, × 70. .... p. 358.

14–16. *Lepidorbitoides* (*Asterorbis*) *rooki* Vaughan and Cole. × 25. 14, Vertical section. 15, Horizontal section passing through embryonic and equatorial chambers of a pentagonal specimen. 16, Oblique section passing through a part of the equatorial plane. p. 358.

17. *Discocyclina*. Diagrammatic figure to show structure. (After Van der Vlerk and Umbgrove). .... p. 368.

PLATE 39.

Family 48. ORBITOIDIDAE, p. 347.

Family 49. DISCOCYCLINIDAE, p. 367.

Family 50. MIOGYPSINIDAE, p. 372.

FIG.

1. *Pseudophragmina (Pseudophragmina) floridana* (Cushman). × 6. Exterior. .... p. 371.
2. *Asterocyclina georgiana* (Cushman). × 6. Exterior. .... p. 369.
3. *Lepidocyclina (Lepidocyclina) mortoni* Cushman. × 80. (After Vaughan). Part of horizontal section to show perforations in the chamber roofs and floors and details of the chamber walls. p. 362.
4. *Lepidocyclina.* Diagrammatic figure to show structure. (After Van der Vlerk and Umbgrove). .... p. 359.
5. *Lepidocyclina (Eulepidina) undosa* Cushman. × 16. (After Vaughan). Horizontal section through embryonic and equatorial chambers. .... p. 363.
6. *Lepidocyclina (Polylepidina) proteiformis* Vaughan. × 16. (After Vaughan). Section in median plane of megalospheric specimen. p. 360.
7. *Lepidocyclina (Eulepidina) favosa* Cushman. × 16. (After Vaughan). Horizontal section through embryonic and equatorial chambers. .... p. 363.
8. *Lepidocyclina (Lepidocyclina) parvula* Cushman. × 8. Vertical section. .... p. 362.
9. *Lepidocyclina (Polylepidina) proteiformis* Vaughan. Vertical section near periphery. .... p. 360.

10, 12. *Miogypsina (Miogypsina) cushmani* Vaughan. (After Vaughan). 10, Horizontal section to show embryonic and equatorial chambers, × 16. 12, Part of a horizontal section to show perforations in the roofs and floors of the chambers and details of the chamber walls, × 80. .... p. 373.

11. *Miogypsina.* Diagrammatic figure to show structure. (After Van der Vlerk and Umbgrove). .... p. 373.

13–15. *Omphalocyclus macroporus* (Lamarck). × 3. (After H. Douvillé). 13, Horizontal weathered section. 14, 15, Peripheral views. ... p. 353.

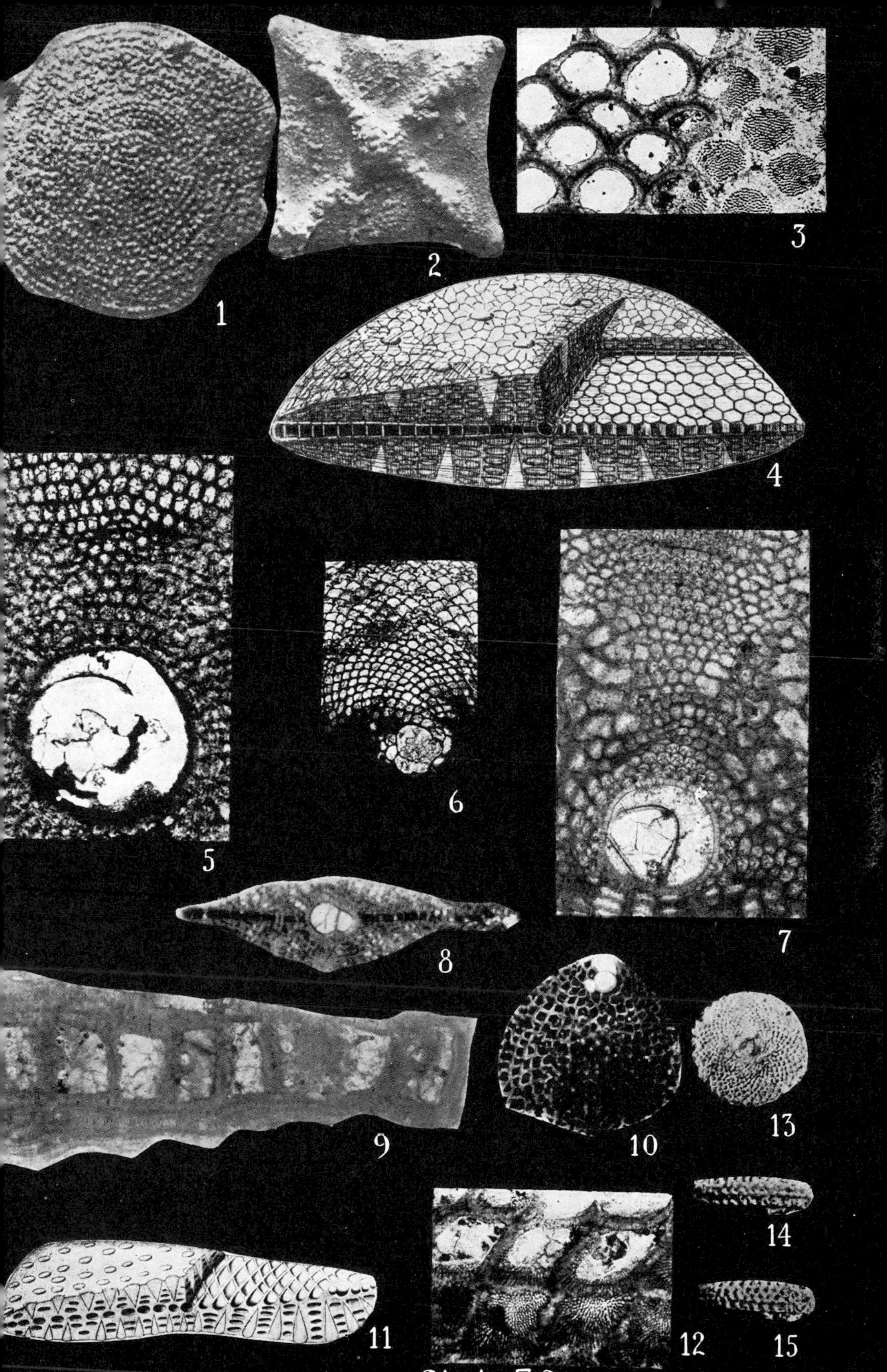

Plate 39

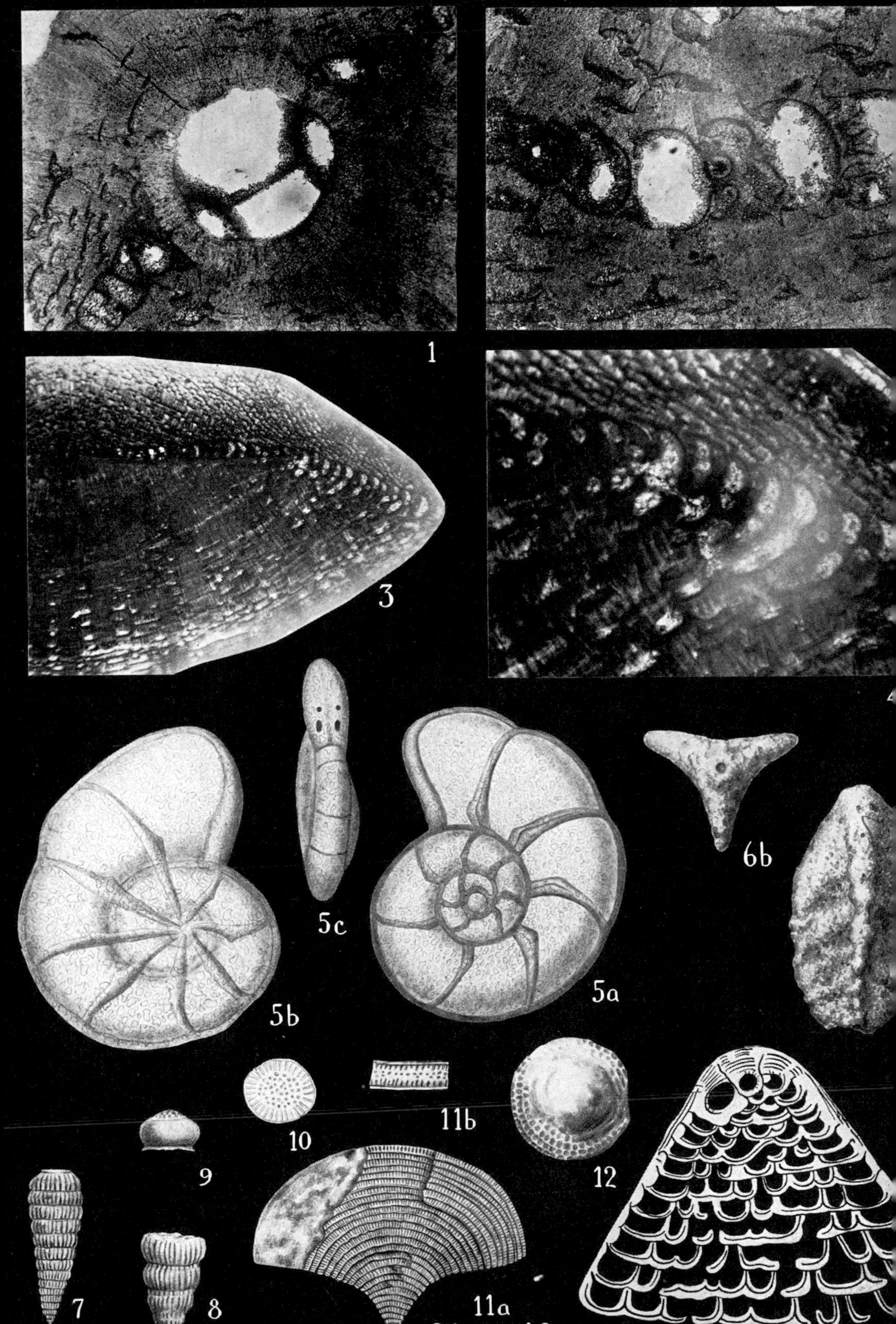

Plate 40

PLATE 40.

Family 48. ORBITOIDIDAE, p. 347.

FIG.

1, 2. *Orbitoides gensacicus* (Leymerie). From Maestricht, Limburg, Holland. 1, Vertical section through embryonic chambers, × 43. 2, Vertical section showing stoloniferous apertures, two of which are very distinct in the chamber wall of the middle chamber, × 65. .... p. 355.

3, 4. *Orbitoides browni* (Ellis) Vaughan. From Tarafa, Cuba. 3, Vertical section showing both equatorial and lateral chambers, × 15. 4, Vertical section showing stoloniferous apertures, × 40. ..... p. 355.

Family 20. TROCHAMMINIDAE, p. 201.

5. *Entzia tetrastomella* Daday. Recent, Hungary. (After Daday). *a*, dorsal view; *b*, ventral view; *c*, peripheral view. .... p. 203.

Family 8. LITUOLIDAE, p. 99.

6. *Triplasia murchisoni* Reuss. ×15. Topotype from Gosau, Austria. *a*, side view; *b*, apertural view. .... p. 106.

Family 27. PENEROPLIDAE (?), p. 242.

7. *Rhapydionina liburnica* (Stache). Eocene, Italy. (After Stache). ... p. 248.
8. *Rhapydionina liburnica* (Stache), var. *strangulata* (Stache). Eocene, Italy. (After Stache). .... p. 248.
9. *Rhapydionina liburnica* (Stache), var. *laevigata* (Stache). Eocene, Italy. (After Stache). Showing aperture. .... p. 248.
10. *Rhapydionina protocaenica* (Stache). Section. .... p. 248.
11. *Rhipidionina liburnica* (Stache). Eocene, Italy. (After Stache). *a*, side view; *b*, portion of periphery showing apertures. ...... p. 248.

Family 37. CYMBALOPORIDAE (?), p. 307.

12, 13. *Chapmanina gassinensis* (A. Silvestri). (After A. Silvestri). 12, Dorsal view, × 12. 13, Vertical section, × 40. .... p. 310.

PLATE 41.

Family 48. ORBITOIDIDAE, p. 347.

Family 49. DISCOCYCLINIDAE, p. 367.

Family 50. MIOGYPSINIDAE, p. 372.

FIG.

1–3. *Triplalepidina veracruziana* Vaughan and Cole. (After Vaughan and Cole). 1, Vertical section, × 15. 2, Equatorial section, × 15. 3, Portion of vertical section fig. 1, × 80, illustrating the threefold division of the equatorial zone near the periphery of the test. The shape and arrangement of the stoloniferous apertures of the lateral chambers are clearly indicated. .... p. 364.

4, 5. *Pseudolepidina trimera* Barker and Grimsdale. (After Barker and Grimsdale). 4, Equatorial section, × 20. 5, Vertical section, × 24, showing interlocking of the lateral zones to form the median zone which is composed of a double layer of chambers. .... p. 356.

6, 7. *Eulinderina guayabalensis* (Nuttall). (After Barker and Grimsdale). 6, Equatorial section, × 18, of a megalospheric individual to illustrate the embryonic apparatus and equatorial chambers. 7, Vertical section, × 18. .... p. 365.

8. *Torreina torrei* D. K. Palmer. (After D. K. Palmer). × 15. Showing embryonic apparatus and character of surrounding chambers. p. 354.

9. *Lepidocyclina* (*Multilepidina*) *irregularis* Hanzawa. (From unpublished photograph by Dr. Hanzawa). × 20. Showing multilocular embryonic apparatus and elongate, spatulate equatorial chambers. .... p. 364.

10. *Lepidocyclina* (*Pliolepidina*) *pustulosa* (H. Douvillé) forma *tobleri* (H. Douvillé), forma teratologica (After Vaughan and Cole). × 160. Decalcified Canada balsam preparation of an equatorial section to show the diagonal stolons, four connections to each chamber. Four-stolon system. .... p. 361.

11, 12. *Lepidocyclina* (*Lepidocyclina*) *mortoni* Cushman. (After Vaughan and Cole). 11, × 160. Decalcified Canada balsam preparation of part of an equatorial section to show diagonal and annular stolons. Six-stolon system. 12, × 160. Decalcified Canada balsam preparation of a vertical section along a cord to show fine tubules that perforate the roofs and floors; at the left, three apertures for stolons, and the annular stolons. .... pp. 359, 362.

13. *Discocyclina* (*Discocyclina*) *pratti* (Michelin). (After Vaughan). × 65. Equatorial chambers as seen in horizontal section, illustrating the annular stolons on the proximal sides of the radial chamber walls. .... p. 368.

14. *Pseudophragmina* (*Pseudophragmina*) *floridana* (Cushman). (After Vaughan). × 65. A small portion of an equatorial section, showing that the radial chamber walls are in alignment from one annulus to the next; they are wavy and interrupted, and the annular stolons are on the distal side. .... p. 371.

15. *Miogypsinoides formosensis* Yabe and Hanzawa. (After Cole). × 20. Vertical section, illustrating the appressed laminae on each side of the equatorial layer. .... p. 376.

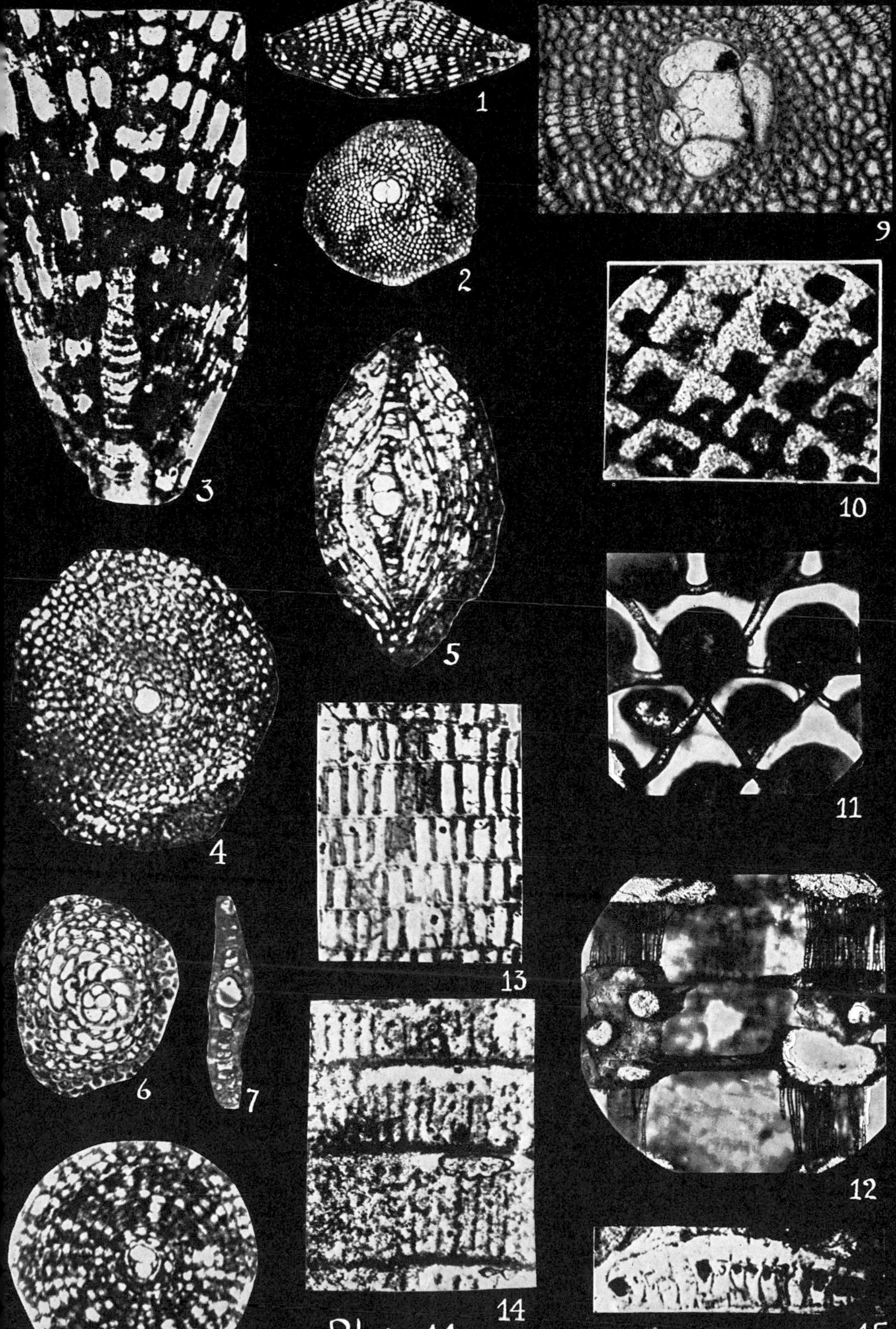

Plate 41

1
2
3
4
5
6
7
8
9
10
11
12
13c
13a
13b
14
15
16
17
18
19
20
21
22b
22a
23
24b
24a
25
26
27
28
29
30
31
32
33
34
35
36
37
38
Plate 42

PLATE 42.

Family 2. ASTRORHIZIDAE, p. 70.

FIG.

1. *Ordovicina oligostoma* Eisenack. Silurian. × 40. (After Eisenack). p. 72.
2. *Amphitremoidea citroniforma* Eisenack. Silurian. × 40. (After Eisenack). p. 72.

Family 3. RHIZAMMINIDAE, p. 73.

3. *Arenosiphon gigantea* Grubbs. Silurian. × 2. (After Grubbs). p. 74.

Family 4. SACCAMMINIDAE, p. 74.

4. *Pilalla exigua* Rhumbler. Recent. × 135. (After Rhumbler). p. 75.

5–7. *Causia injudicata* Rhumbler. Recent. (After Rhumbler). 5, 6, Dorsal views, × 130. 7, Outline of side view, × 95. p. 75.

8. *Shidelerella bicuspida* Dunn. Silurian. × 18. (After Dunn). p. 76.
9. *Thekammina quadrangularis* Dunn. Silurian. × 35. (After Dunn). p. 76.
10. *Arenosphaera perforata* Stschedrina. Recent, Arctic. × 40. (After Stschedrina). Three-chambered specimen with last chamber broken, showing interior. p. 76.

11, 12. *Ceratammina cornucopia* Ireland. Devonian. × 30. (After Ireland). p. 77.

13. *Placentammina placenta* (Grzybowski). Flysch, Hungary. (After Majzon). *a, b,* opposite sides; *c,* peripheral view. p. 79.
14. *Armorella sphaerica* Heron-Allen and Earland. Recent. × 20. (After Heron-Allen and Earland). p. 81

15, 16. *Thuramminoides sphaeroidalis* Plummer. Pennsylvanian, Texas. × 25. (After Plummer). 15, Labyrinthic interior of a specimen devoid of its test. 16, Common smooth test showing moderate secondary compression. p. 80.

17. *Croneisella typa* Dunn. Silurian. × 20. (After Dunn). p. 82.
18. *Amphifenestrella wiesneri* Rhumbler. Recent. × 40. (After Rhumbler). p. 83.
19. *Gastroammina williamsae* Dunn. Silurian. × 20. (After Dunn). p. 81.
22. *Kerionammina favus* Moreman. Ordovician. × 30. (After Moreman). *a,* dorsal side; *b,* ventral side. p. 83.

Family 5. HYPERAMMINIDAE, p. 84.

20. *Chitinodendron bacciferum* Eisenack. Silurian. × 30. (After Eisenack). p. 86.
21. *Saccodendron heronalleni* Rhumbler, var. *latericium* Rhumbler. Recent. × 26. (After Rhumbler). p. 87.

Family 6. REOPHACIDAE, p. 89.

23, 24. *Polychasmina pawpawensis* Loeblich and Tappan. Lower Cretaceous. × 4. (After Loeblich and Tappan). 23, 24 *a,* side views; 24 *b,* apertural view. p. 91.

Family 7. AMMODISCIDAE, p. 93.

25. *Glomospirella umbilicata* (Cushman and Waters). Pennsylvanian, Texas. × 25. (After Plummer). p. 97.

26, 27. *Spirillinoides circumcinctus* Rhumbler. Recent. × 120. (After Rhumbler). 26, Dorsal view. 27, Ventral view. p. 96.

28, 29. *Ammoflintina trihedra* Earland. Recent. × 55. (After Earland). p. 97.

30, 31. *Bifurcammina bifurca* Ireland. Silurian. × 25. (After Ireland). p. 97.

32. *Thalmannina nothi* Majzon. Flysch, Hungary. Approximately × 25. (After Majzon). p. 99.
33. *Xenotheka klinostoma* Eisenack. Silurian. × 30. (After Eisenack). p. 99.

Family 8. LITUOLIDAE, p. 99.

34–36. *Recurvoides contortus* Earland. Recent. × 27. (After Earland). 34, 36, Side views. 35, Apertural view. p. 102.

37, 38. *Alveolophragmium orbiculatum* Stschedrina, var. *ochotonensis* Stschedrina. Recent. (After Stschedrina). 37, Broken test showing interior. p. 102.

PLATE 43.

Family 8. LITUOLIDAE, p. 99.

FIG.

1. *Labrospira crassimargo* (Norman). Recent. × 18. (After Höglund). p. 103.
2. *Ammoscalaria tenuimargo* (H. B. Brady). Recent. × 15. (After Höglund). p. 103.

3, 4. *Cribrospirella difformis* (Lamarck). Upper Cretaceous, France. × 8. (After Marie). p. 106.

5. *Ammoastuta salsa* Cushman and Bronnimann. Recent. × 45. (After Cushman and Bronnimann). p. 106.

6, 7. *Biseriammina uralica* Tchernysheva. Mississippian. × 22. (After Tchernysheva). 6, Side view. 7, Apertural view. p. 109.

Family 9. TEXTULARIIDAE, p. 112.

8. *Morulaeplecta bulbosa* Höglund. Recent. × 60. (After Höglund). p. 113.

9, 10. *Septigerina dalmatica* Keijzer. Eocene, Dalmatia. × 20. (After Keijzer). 9 *a*, side view; *b*, apertural view. 10, Section. p. 114.

11. *Semitextularia thomasi* Miller and Carmer. Devonian. × 70. (After Miller and Carmer). *a*, front view; *b*, peripheral view; *c*, apertural view. p. 114.
12. *Haeuslerella pukeuriensis* Parr. Miocene. × 22. (After Parr). *a*, front view; *b*, side view of last two chambers, showing position of aperture. p. 116.
13. *Siphotextularia wairoana* Finlay. Tertiary. × 60. (After Finlay). *a*, front view; *b*, peripheral view of another specimen. p. 116.

14, 15. *Spiroplectella cylindroides* Earland. Recent. × 55. (After Earland). p. 114.

Family 10. VERNEUILINIDAE, p. 121.

16. *Gaudryina (Siphogaudryina) stephensoni* Cushman. Upper Cretaceous. × 45. *a*, front view; *b*, apertural view. p. 124.
17. *Gaudryina (Pseudogaudryina) atlantica* (Bailey). Recent. × 16. *a*, front view; *b*, apertural view. p. 124.
18. *Barbourinella atlantica* (Bermudez). Recent. × 28. (After Bermudez). *a*, front view; *b*, apertural view. p. 123.

19, 20. *Flourensina lemoinei* Marie. Cretaceous, France. × 20. (After Marie). 19, Front view. 20, Apertural view. p. 123.

21. *Digitina recurvata* Crespin and Parr. Permian. × 25. (After Crespin and Parr). *a*, front view; *b*, apertural view. p. 124.
22. *Migros medwayensis* (Parr). Tertiary, New Zealand. × 15. (After Parr). p. 125.
23. *Bermudezina cubensis* (D. K. Palmer and Bermudez). Oligocene, Cuba. × 38. (After D. K. Palmer and Bermudez). *a*, *b*, opposite sides; *c*, peripheral view. p. 125.
24. *Pseudoclavulina clavata* (Cushman). Upper Cretaceous, Texas. × 40. *a*, front view; *b*, apertural view. p. 126.
25. *Pseudogaudryinella capitosa* (Cushman). Upper Cretaceous, Mississippi. × 20. *a*, front view; *b*, side view; *c*, apertural view. p. 126.

26, 27. *Rudigaudryina inepta* Cushman and McCulloch. Recent. × 30. (After Cushman and McCulloch). p. 127.

Family 11. VALVULINIDAE, p. 127.

28. *Eggerellina brevis* (d'Orbigny), var. *conica* Marie. Upper Cretaceous, France. × 15. (After Marie). *a*, front view; *b*, side view; *c*, apertural view of another specimen. p. 129.

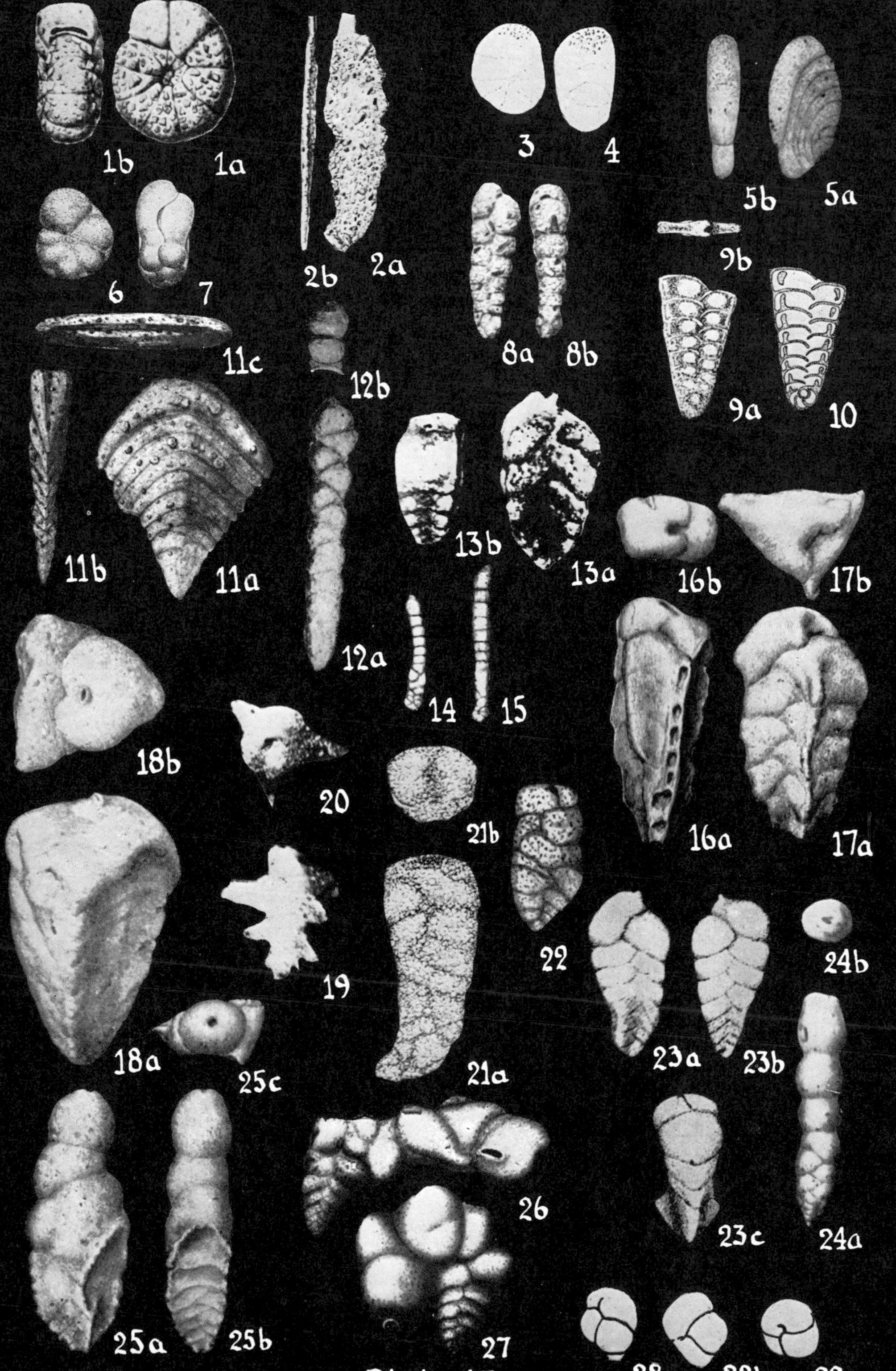
1b
1a
2b
2a
3
4
5b
5a
6
7
8a
8b
9b
9a
10
11c
12b
11b
11a
12a
13b
13a
16b
17b
14
15
18b
20
21b
16a
17a
22
19
21a
23a
23b
24b
18a
25c
26
23c
24a
25a
25b
27
28a
28b
28c

Plate 43

Plate 44

PLATE 44.

Family 12. FUSULINIDAE, p. 146.

FIG.

1. Diagram to illustrate evolutionary change from the fusulinellid to the schwagerinid type of wall. *A*, primitive (fusulinellid) wall, consisting of 4 layers (or zones) of differing opacity, the pores being simple, very minute and tubular; *B*, a primitive stage of the schwagerinid wall in which the tectoria are lacking and the pores are enlarging and commonly uniting except near the outer margin; *C*, an advanced schwagerinid wall in which the alveoli of the keriotheca are large and the tectum is distinctly set off and is perforate (compare with figure 18); *a*, alveolus; *d*, diaphanotheca; *i*, inner tectorium; *k*, keriotheca; *l*, lamella; *o*, outer tectorium; *p*, pore; *t*, tectum. (After Dunbar and Skinner).

2. Diagram to show the character of the wall in *Rugofusulina*. *A*, normal wall of *Schwagerina*; *B*, dimpled wall of early species of *Rugofusulina*; *C*, finely dimpled wall of later species of *Rugofusulina*. (After Rauser-Chernoussova). .... pp. 157, 159.

3, 4. *Codonofusiella paradoxica* Dunbar and Skinner; 3, sagittal section; 4, axial section. .... p. 154.

5. *Leëlla bellula* Dunbar and Skinner, axial section showing ontogenetic change in shape. .... p. 150.

6, 7. *Ozawainella angulata* (Colani), axial sections. (After Colani). .... p. 151.

8, 9. *Nankinella gigas* (Mansuy), lateral and peripheral view of weathered specimens. (After Mansuy). .... p. 152.

10. Cross section of one of the chomata in *Triticites ventricosus* (Meek and Hayden), showing its layered structure. The alveolar wall is shown below. .... p. 157.

11. Front view of a perfect specimen of *Schwagerina diversiformis* Dunbar and Skinner, showing a deeply and regularly folded antetheca. .... p. 157.

12–14. *Paraschwagerina gigantea* (White); 12, bit of a tangential slice showing the septa in plan; 13, axial section with broken ends; 14, sagittal section. .... p. 161.

15, 16. *Pseudoschwagerina uddeni* (Beede); 15, sagittal section showing the rapid inflation after about 1½ whorls; 16, axial section showing the simple septa. .... p. 160.

17, 18. *Nipponitella explicata* Hanzawa. Axial section of a specimen and greater enlargement of the coiled juvenarium of the same. Photo by Dunbar of the holotype. .... p. 162.

19. Greatly enlarged axial section of *Fusulinella llanoensis* Thomas, showing tectum (*t*), diaphanotheca (*d*), outer tectorium (*o*) and inner tectorium (*i*). Compare with figure 1*A*. .... p. 152.

20. Greatly enlarged bit of axial section of *Schwagerina gracilitatis* Dunbar and Skinner, showing the coarse alveoli (*a*) of the keriotheca, and the pores (*p*) in the tectum. .... p. 157.

PLATE 45.

Family 12. FUSULINIDAE, p. 146.

FIG.

1, 2. *Nummulostegina velibitana* Schubert. Permian (Schwagerinakalk) of Austria. × 10. (After type figures). .... p. 149.

3, 4. *Nankinella discoidea* (Lee). × 10. Permian (Chihsia limestone) of China. Copy of original illustrations of Lee. 3, Axial section. 4, Sagittal section. .... p. 152.

5, 6. *Staffella sphaerica* Abich. ×10. Permian of southern Armenia. From the type collection of Abich, after Likharev. 5, Axial section. 6, Saggital section. .... p. 147.

7, 8. *Millerella marblensis* Thompson. Lower Pennsylvanian (Marble Falls limestone), central Texas. (Holotype and paratype in axial section, after Thompson). 7, Holotype. × 100. 8, Paratype. × 50. .... p. 149.

9–11. *Rauserella erratica* Dunbar. High Permian, Mexico and Texas. 9, 10, Holotype axial section. × 10 and × 25. 11, Paratype. × 25. p. 151.

12, 13. *Dunbarinella ervinensis* Thompson. × 10. (After Thompson). Upper Pennsylvanian (Virgil series), Oklahoma. 12, Axial section of holotype. 13, Tangential section of a paratype. ...... p. 159.

14, 15. "*Chusenella ishanensis* Hsu." × 10. (After original figures by Hsu). Permian of China. 14, Sagittal section. 15, Axial section here designated as the holotype. .... p. 159.

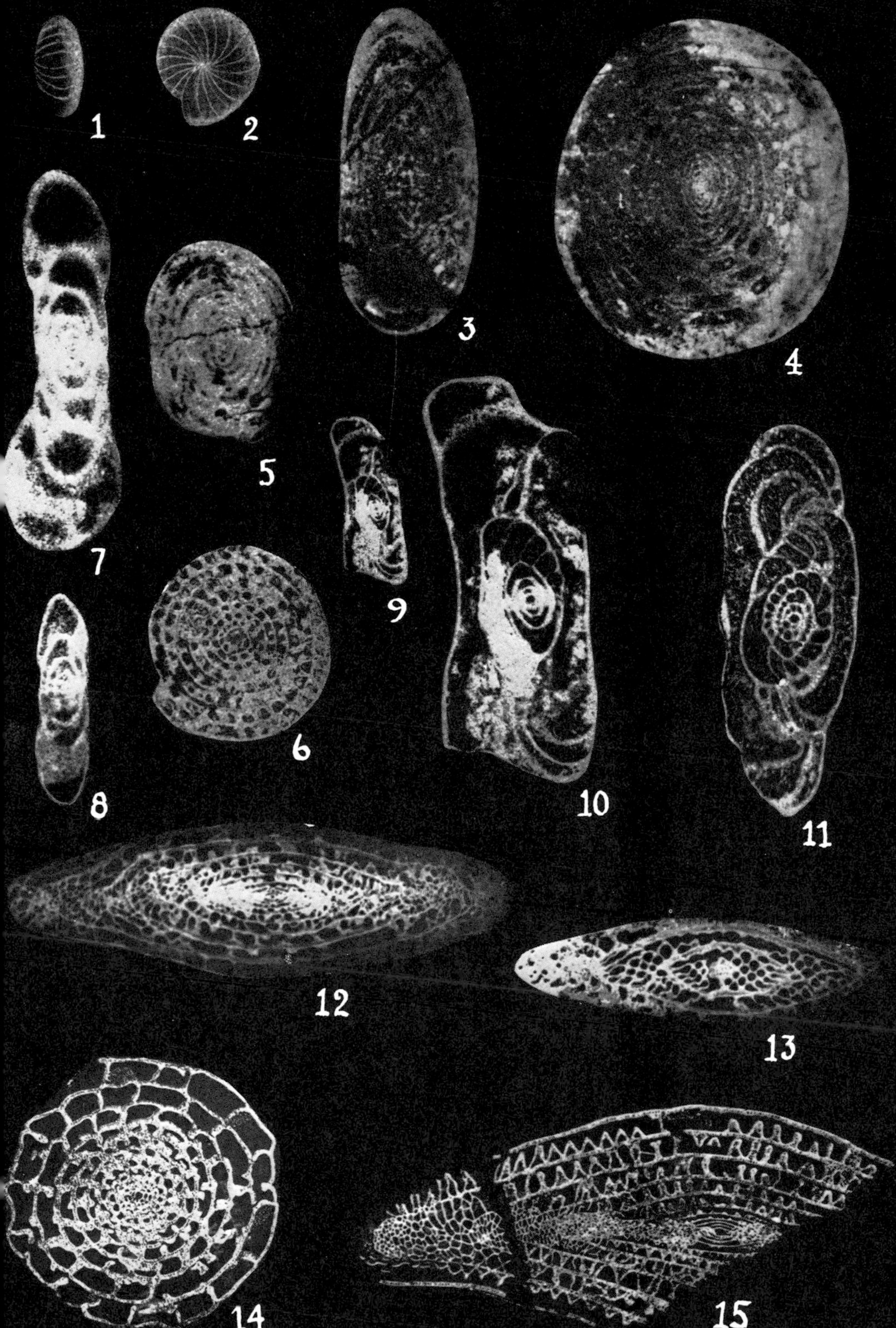

Plate 45

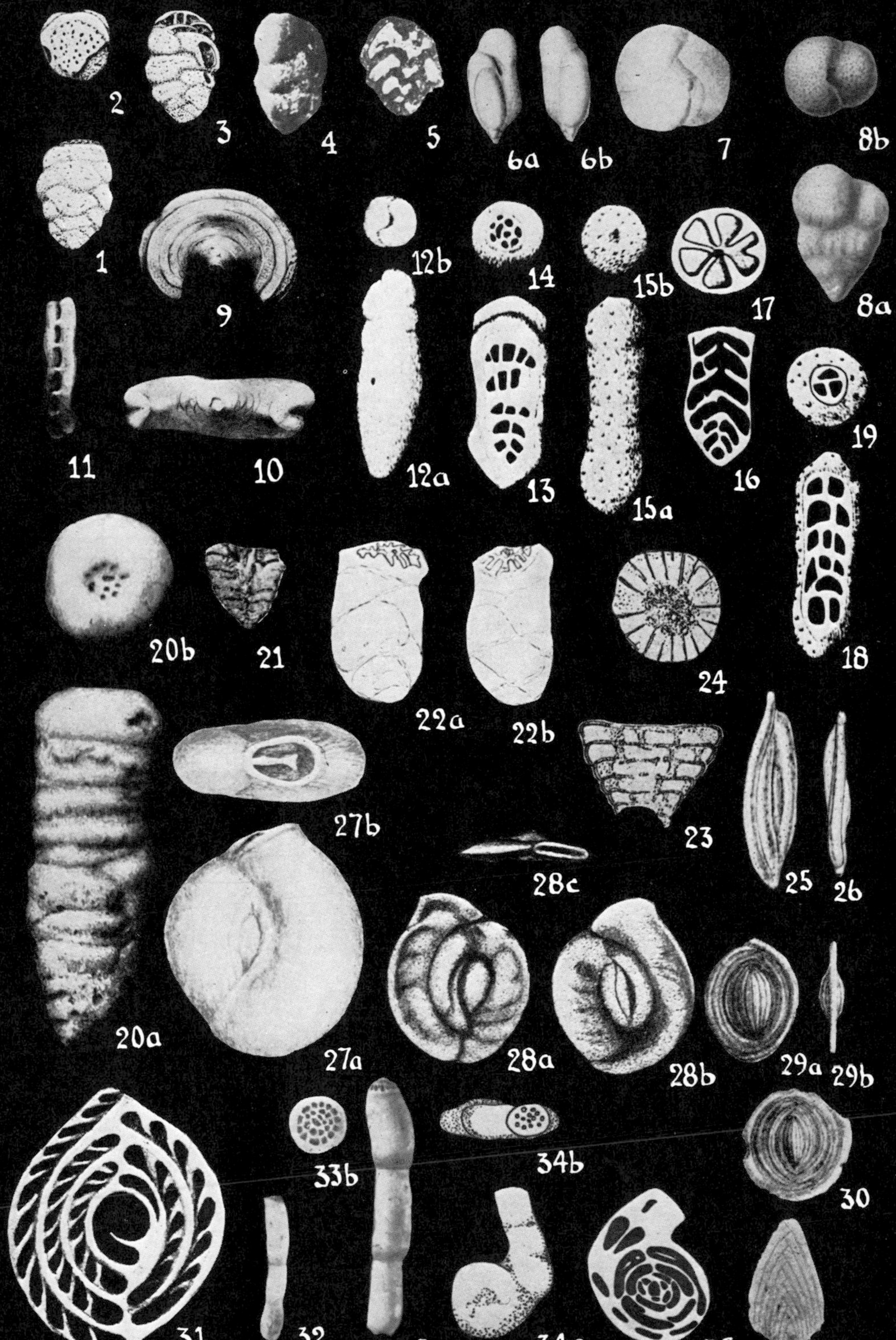
2
3
4
5
6a
6b
7
8b
1
9
12b
14
15b
17
8a
11
10
12a
13
15a
16
19
18
20b
21
22a
22b
24
20a
27b
23
28c
25
26
27a
28a
28b
29a
29b
33b
34b
30
31
32
33a
34a
35
36

Plate 46

PLATE 46.

Family 11. VALVULINIDAE, p. 127.

FIG.

1–3. *Pseudogoësella cubana* Keijzer. Eocene, Cuba. × 12. (After Keijzer). 1, 3, Opposite sides; 2, apertural view, all of same specimen. .... p. 130.

4, 5. *Makarskiana trochoidea* van Soest. Eocene, Dalmatia. × 12. (After van Soest). 4, Side view. 5, Section. .... p. 131.

6, 7. *Eggerina cylindrica* Toulmin. Eocene, Alabama. (After Toulmin). 6 *a*, *b*, opposite sides. × 25. 7, Basal view of another specimen. × 40. .... p. 132.

8. *Remesella mariae* Vasicek. Eocene, Moravia. × 22. (After Vasicek). *a*, front view; *b*, apertural view. .... p. 134.

9–11. *Cuneolinella lewisi* Cushman and Bermudez. Miocene, Dominican Republic. 9, 11, × 3. 10, × 5. (After Cushman and Bermudez). 9, Front view. 10, Apical view. 11, Apertural view. .... p. 135.

12–14. *Matanzia bermudezi* D. K. Palmer. Oligocene, Cuba. (After D. K. Palmer). 12, 13, × 20. 14, × 30. 12 *a*, front view; *b*, apertural view. 13, Tangential vertical section showing supplementary chamber partitions. 14, Section of initial end. .... p. 136.

15–19. *Cubanina alavensis* D. K. Palmer. Oligocene, Cuba. (After D. K. Palmer). 15, × 12. 16, 19, × 18. 17, × 22. 18, × 16. 15 *a*, front view; *b*, apertural view. 16, Vertical section through megalospheric form. 17, Uniserial chambers showing supplementary partitions. 18, Tangential vertical section showing numerous supplementary partitions. 19, Section of initial end showing the chambers immediately following the megalosphere. .... p. 136.

20. *Cribrogoësella robusta* (H. B. Brady). Recent. × 28. (After H. B. Brady). *a*, side view; *b*, apertural view. .... p. 137.

21. *Camagueyia perplexa* Cole and Bermudez. Eocene, Cuba. × 15. (After Cole and Bermudez). Axial section. .... p. 139.

22. *Coprolithina subcylindrica* Marie. Upper Cretaceous, France. × 10. (After Marie). *a*, *b*, side views at 90° from each other. .... p. 140.

23, 24. *Coskinolinoides texanus* Keijzer. Lower Cretaceous, Texas. × 50. (After Keijzer). 23, Vertical section. 24, Horizontal section. .... p. 141.

Family 16. SILICINIDAE, p. 172.

25, 26. *Spirolocammina tenuis* Earland. Recent. × 55. (After Earland). 25, Front view. 26, Peripheral view. .... p. 174.

Family 17. MILIOLIDAE, p. 174.

27. *Proemassilina rugosa* (Sidebottom). Recent. × 33. (After Sidebottom). *a*, front view; *b*, apertural view. .... p. 178.

28. *Pseudomassilina australis* (Cushman). Recent. × 26. (After Lacroix). *a*, *b*, opposite sides; *c*, apertural view. .... p. 178.

29, 30. *Spirosigmoilina tateana* (Howchin). Tertiary, Australia. 29, × 25. 30, × 22. (After Howchin). 29 *a*, 30, front views; 29 *b*, peripheral view. .... p. 183.

31. *Riveroina caribaea* Bermudez. Recent, Cuba. × 55. (After Bermudez). Horizontal section. .... p. 179.

32, 33. *Poroarticulina glabra* Cushman. Miocene, Rumania. × 32. (After Cushman). 32, Initial end. 33 *a*, apertural end; *b*, apertural view enlarged. .... p. 180.

34, 35. *Raadshoovenia guatemalensis* van den Bold. Eocene, Guatemala. × 20. (After van den Bold). 34 *a*, side view; *b*, apertural view. 35, Section through plane of coiling. .... p. 183.

Family 9. TEXTULARIIDAE, p. 112.

36. *Pseudopalmula palmuloides* Cushman and Stainbrook. Devonian, Iowa. × 40. (After Cushman and Stainbrook). .... p. 115.

PLATE 47.

Family 17. MILIOLIDAE, p. 174.

FIG.

1. *Cribrolinoides curta* (Cushman). Pliocene, Java. × 9. (After Cushman and LeRoy). *a*, side view; *b*, apertural view. .... p. 184.

2–4. *Pyrgoella sphaera* (d'Orbigny). 2, Recent. × 18. (After Flint). 3, 4, Pleistocene, California. × 27. (After Cushman and White). p. 185.

5, 6. *Cribropyrgo robusta* Cushman and Bermudez. Recent, off Cuba. × 10. (After Cushman and Bermudez). 5 *a*, side view; *b*, apertural view of another specimen. 6, Transverse section. .... p. 186.

Family 18. OPHTHALMIDIIDAE, p. 188.

7. *Meandrospira washitensis* Loeblich and Tappan. Lower Cretaceous. × 55. (After Loeblich and Tappan). *a*, *b*, opposite sides; *c*, peripheral view. .... p. 193.

8. *Carixia langi* Macfadyen. Lias, England. × 15. (After Macfadyen). Adherent to shale fragment. .... p. 195.

9. *Nodobaculariella atlantica* Cushman and Hanzawa. Recent. × 27. (After Cushman and Hanzawa). *a*, *b*, opposite sides; *c*, apertural view. .... p. 195.

10, 11. *Ophthalmina kilianensis* Rhumbler. Recent. × 45. (After Rhumbler). .... p. 197.

12, 13. *Polysegmentina circinata* (H. B. Brady). Recent. × 25. (After H. B. Brady). 12 *a*, side view; *b*, apertural view. .... p. 197.

14–16. *Meandroloculina bogatschovi* Bogdanowicz. Miocene. (After Bogdanowicz). 14, 15, × 25. 16, × 50. 14 *a*, front view; *b*, apertural view. 15, Vertical section. 16, Section showing proloculum and succeeding chambers. .... p. 199.

17, 18. *Rhizonubecula adherens* LaCalvez. Recent. (After LeCalvez). 17, Entire test, × 5. 18, Optical section of early chambers, × 75. . . p. 199.

19, 20. *Glomulina fistulescens* Rhumbler. Recent. × 45. (After Rhumbler). 19 *a*, *b*, opposite sides. 20, Balsam mount by transmitted light showing internal structure. .... p. 200.

Family 20. TROCHAMMINIDAE, p. 201.

21, 22. *Conotrochammina whangaia* Finlay. Upper Cretaceous, New Zealand. × 22. (After Finlay). 21, Peripheral view. 22, Dorsal view. .... p. 202.

23–25. *Trochamminella siphonifera* Cushman. Recent. × 25. 23, Attached form. 24, Dorsal view. 25, Ventral view. .... p. 203.

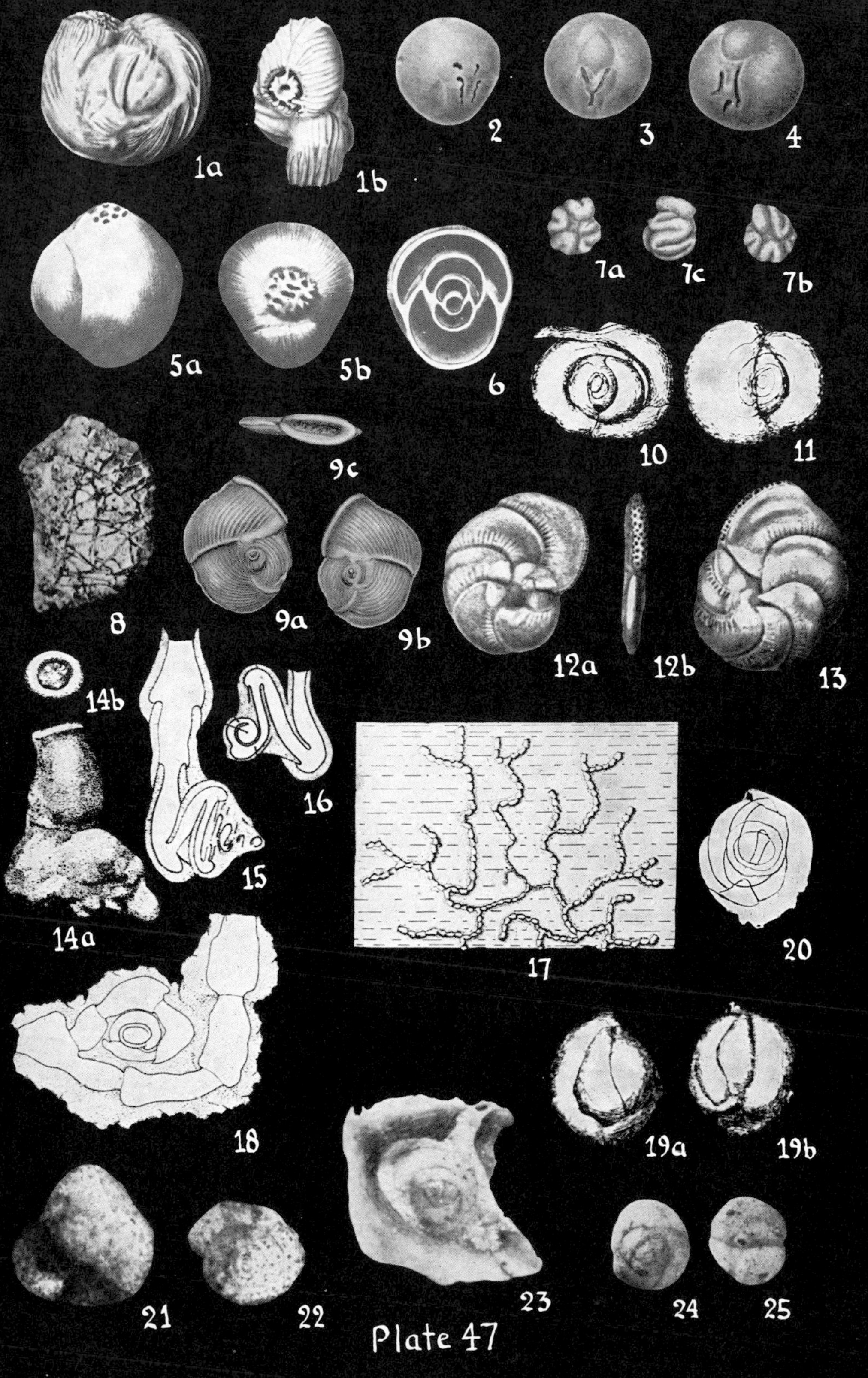
1a
1b
2
3
4
5a
5b
6
7a
7c
7b
10
11
9c
8
9a
9b
12a
12b
13
14b
16
15
14a
17
20
18
19a
19b
21
22
23
24
25

Plate 47

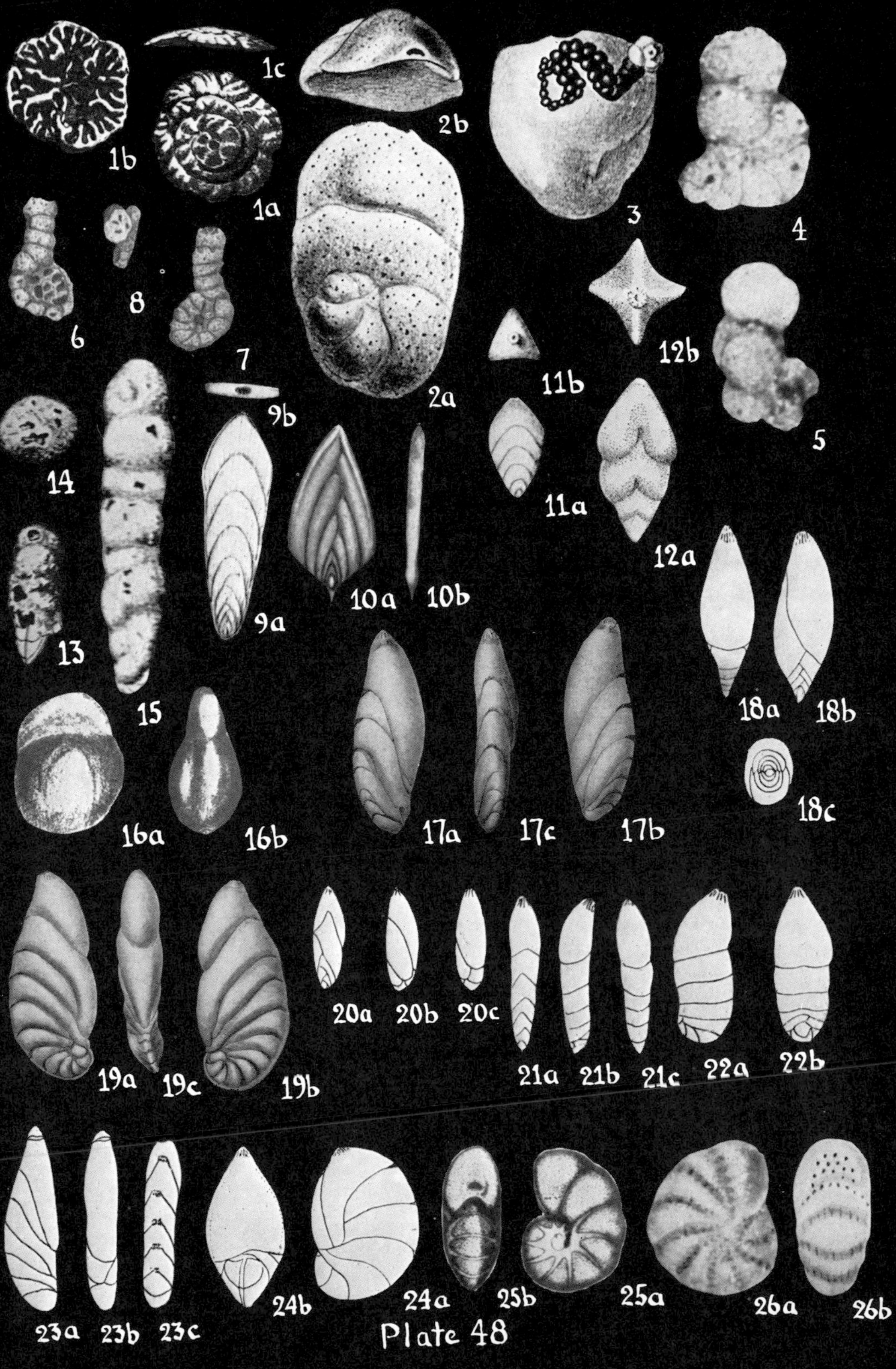
1c
1b
2b
1a
3
4
8
6
7
12b
11b
2a
5
9b
14
11a
10a
10b
9a
12a
13
15
18a
18b
18c
16a
16b
17a
17c
17b
20a
20b
20c
19a
19c
19b
21a
21b
21c
22a
22b
24b
24a
25b
25a
26a
26b
23a
23b
23c

Plate 48

PLATE 48.

Family 20. TROCHAMMINIDAE, p. 201.

FIG.

1. *Remaneica plicata* (Terquem). Recent. × 60. (After Rhumbler). *a*, dorsal view; *b*, ventral view; *c*, peripheral view. .... p. 203.
2. *Ammocibicides pontoni* Earland. Eocene, Alabama. × 55. (After Earland). *a*, dorsal view; *b*, apertural view. .... p. 204.
4, 5. *Trochamminita irregularis* Cushman and Bronnimann. Recent. × 70. (After Cushman and Bronnimann). .... p. 202.

Family 21. PLACOPSILINIDAE, p. 207.

3. *Placopsilinella aurantiaca* Earland. Recent. × 55. (After Earland). p. 208.
6–8. *Acruliammina longa* (Tappan). Lower Cretaceous. × 8. (After Loeblich and Tappan). 6, View showing attachment of coiled portion and free-growing uniserial portion. 7, 8, Dorsal and apertural views of same specimen. .... p. 208.
13–15. *Adhaerentia midwayensis* Plummer. Paleocene, Texas. × 15. (After Plummer). 13, View showing rounded aperture of early stage. 14, Apertural view. 15, Front view of adult specimen. .. p. 209.

Family 23. LAGENIDAE, p. 211.

9. *Parafrondicularia japonica* Asano. Pliocene, Japan. × 50. (After Asano). *a*, front view; *b*, apertural view. .... p. 219.
10. *Dyofrondicularia nipponica* Asano. Pliocene, Japan. × 40. (After Asano). *a*, front view; *b*, apertural view. .... p. 219.
11. *Tristix liasina* (Berthelin). Lias, England. × 100. (After Macfadyen). *a*, front view; *b*, apertural view. .... p. 215.
12. *Quadratina depressula* ten Dam. Lower Cretaceous, France. × 40. (After ten Dam). *a*, front view; *b*, apertural view. .... p. 216.

Family 24. POLYMORPHINIDAE, p. 221.

16. *Pseudopolymorphinoides limburgensis* van Bellen. Eocene, Netherlands. × 25. (After van Bellen). *a*, front view; *b*, peripheral view. .... p. 224.
17. *Polymorphinella vaginulinaeformis* Cushman and Hanzawa. Pleistocene, Ryukyu Islands. × 38. (After Cushman and Hanzawa). *a*, *b*, opposite sides; *c*, dorsal view. .... p. 229.
18. *Pyrulinoides acuminata* (d'Orbigny), forma *typica* Marie. Upper Cretaceous, France. × 35. (After Marie). *a*, dorsal view; *b*, left profile; *c*, projection of the sutures on the horizontal plan of the base. .... p. 228.
19. *Polymorphinoides spiralis* Cushman and Hanzawa. Pleistocene, Ryukyu Islands. × 25. (After Cushman and Hanzawa). *a*, *b*, opposite sides; *c*, ventral view. .... p. 230.
20. *Enantiomorphina lemoinei* Marie, forma *typica* Marie. Upper Cretaceous, France. × 35. (After Marie). *a*, dorsal view; *b*, right profile; *c*, front view. .... p. 231.
21. *Enantiodentalina communis* (d'Orbigny), forma *typica* Marie. Upper Cretaceous, France. × 25. (After Marie). *a*, dorsal view; *b*, left profile; *c*, front view. .... p. 231.
22. *Enantiomarginulina d'orbignyi* Marie. Upper Cretaceous, France. × 25. (After Marie). *a*, left profile; *b*, ventral view. .... p. 231.
23. *Enantiovaginulina recta* (d'Orbigny). Upper Cretaceous, France. × 25. (After Marie). *a*, right profile; *b*, ventral view; *c*, dorsal view. .... p. 231.
24. *Enantiocristellaria cayeuxi* Marie. Upper Cretaceous, France. × 25. (After Marie). *a*, left profile; *b*, ventral view. .... p. 231.

Family 25. NONIONIDAE, p. 232.

25. *Paranonion venezuelanum* Logue and Haas. Miocene, Venezuela. × 70. (After Logue and Haas). *a*, side view; *b*, apertural view. p. 233.
26. *Cribroelphidium vadescens* Cushman and Bronnimann. Recent. × 70. (After Cushman and Bronnimann). *a*, side view; *b*; apertural view. .... p. 234.

PLATE 49.

Family 25. NONIONIDAE, p. 232.

FIG.

1. *Hydromylina rutteni* de Witt Puyt. Eocene, Herzegovina. × 15. (After de Witt Puyt). *a*, *b*, opposite sides; *c*, peripheral view. . . p. 233.
2. *Elphidioides americanus* Cushman. Eocene, Georgia. × 60. (After Cushman). *a*, dorsal view; *b*, ventral view; *c*, apertural view. . . p. 234.
3. *Astrononion stelligerum* (d'Orbigny). Recent. × 75. (After d'Orbigny). *a*, side view; *b*, apertural view. .... p. 233.
4. *Elphidiella arctica* (Parker and Jones). Recent. × 37. *a*, side view; *b*, apertural view. .... p. 236.
5. *Notorotalia clathrata* (H. B. Brady). Recent. × 38. (After H. B. Brady). *a*, dorsal view; *b*, ventral view; *c*, peripheral view. .... p. 236.

Family 26. CAMERINIDAE, p. 237.

6. *Operculinoides wilcoxi* (Heilprin). Tertiary. × 7½. (After Cushman, 1921). .... p. 238.

7–9. *Miscellanea miscella* (d'Archiac and Haime). Eocene. (After Davies and Pinfold). 7, Equatorial section, × 15. 8, Meridian section, × 15. 9, Exterior, × 5. .... p. 238.

10. *Operculina* (*Sulcoperculina*) *dickersoni* (D. K. Palmer). Upper Cretaceous, Cuba. × 36. (After D. K. Palmer). *a*, side view; *b*, apertural view. .... p. 239.

Family 27. PENEROPLIDAE, p. 242.

11, 12. *Taberina cubana* Keijzer. Upper Cretaceous-Paleocene transitional beds, Cuba. × 15. (After Keijzer). 11, Median section, about perpendicular to axis of coiling. 12, Tangential section parallel with axis of coiling. .... p. 244.

13. *Fascispira colomi* A. Silvestri. Upper Cretaceous, Spain. × 9. (After A. Silvestri). .... p. 247.

Family 28. ALVEOLINELLIDAE, p. 248.

14. *Ovalveolina ovum* (d'Orbigny). .... p. 249.
15. *Praealveolina cretacea* (d'Archiac), var. *tenuis* Reichel. .... p. 249.
16. *Alveolinella quoyi* (d'Orbigny). .... p. 251.
17. *Subalveolina dordonica* Reichel. .... p. 249.

(Figures 14–17 copied from schematic figures of Reichel.)

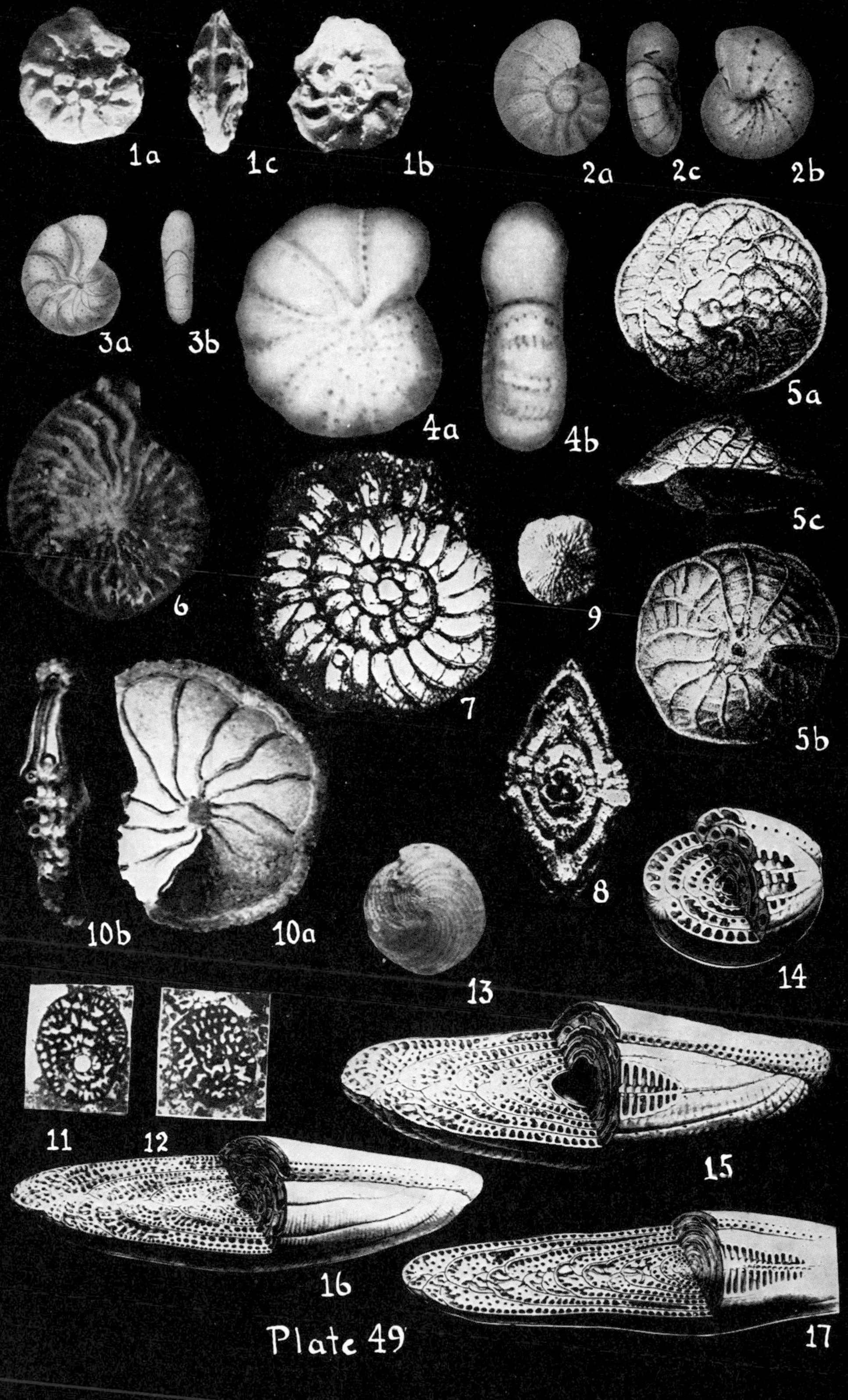
1a
1c
1b
2a
2c
2b
3a
3b
4a
4b
5a
5c
6
7
9
5b
8
10b
10a
13
14
11
12
15
16
17

Plate 49

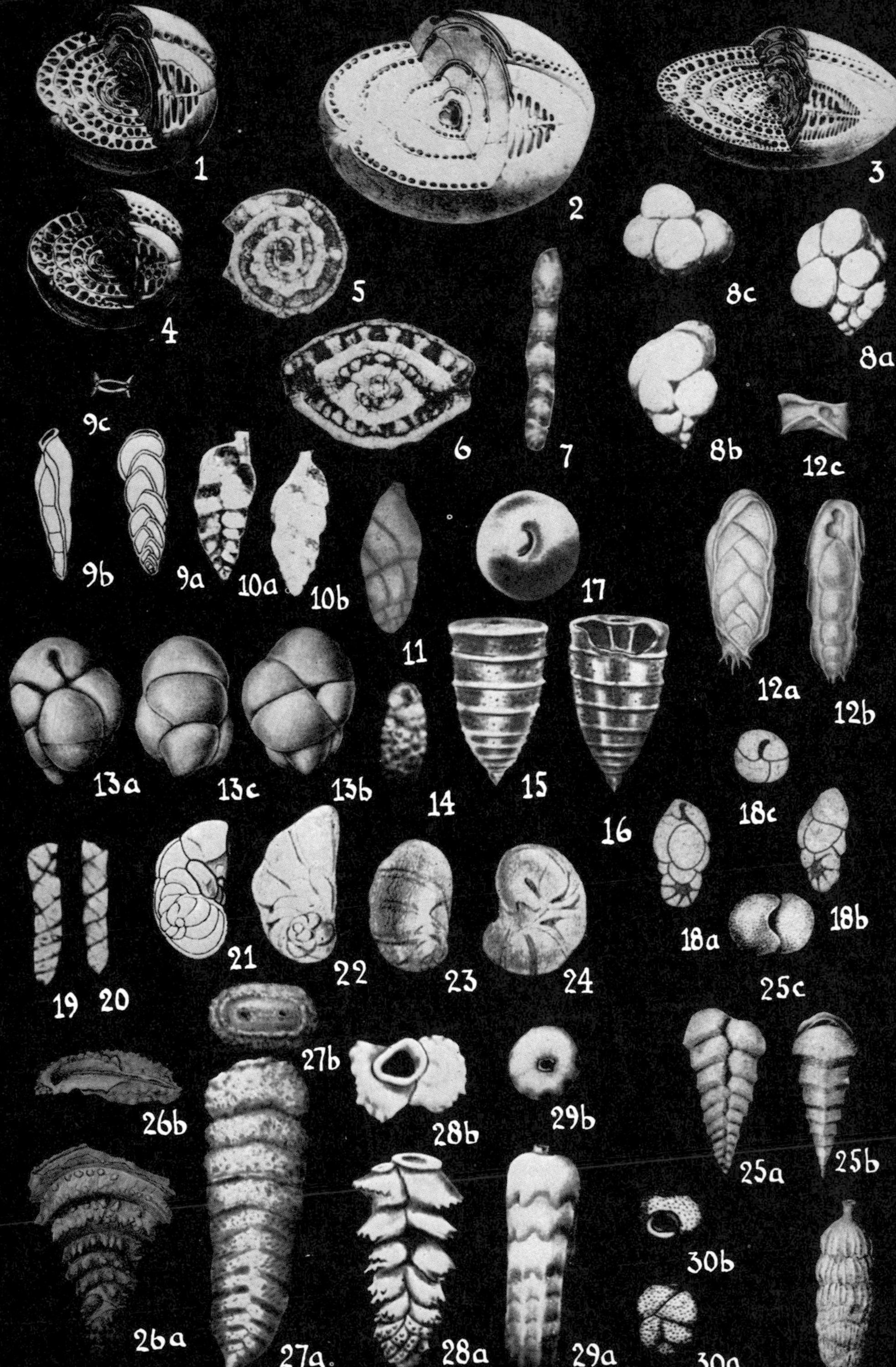
1
2
3
4
5
6
7
8a
8b
8c
9a
9b
9c
10a
10b
11
12a
12b
12c
13a
13b
13c
14
15
16
17
18a
18b
18c
19
20
21
22
23
24
25a
25b
25c
26a
26b
27a
27b
28a
28b
29a
29b
30a
30b
31

Plate 50

PLATE 50.

Family 28. ALVEOLINELLIDAE, p. 248.

FIG.

1. *Borelis melo* (Fichtel and Moll). .... p. 250.
2. *Flosculina globosa* (Leymerie). .... p. 250.
3. *Flosculinella bontangensis* Rutten. .... p. 251.
4. *Bullalveolina bulloides* (d'Orbigny). .... p. 250.
(Figures 1–4 copied from schematic figures of Reichel.)

5, 6. *Boreloides cubensis* Cole and Bermudez. Eocene, Cuba. × 20. (After Cole and Bermudez). 5, Transverse section. 6, Axial section. .... p. 250.

Family 30. HETEROHELICIDAE, p. 252.

7. *Nodoplanulis elongata* Hussey. Eocene, Louisiana. × 60. (After Hussey). .... p. 253.

8. *Gümbelitriella graysonensis* Tappan. Lower Cretaceous. × 75. (After Tappan). *a*, *b*, side views; *c*, apertural view. .... p. 256.

9. *Bolivinitella eleyi* (Cushman), forma *typica* Marie. Upper Cretaceous, France. × 35. (After Marie). *a*, front view; *b*, side view; *c*, apertural view. .... p. 258.

10. *Zeauvigerina zelandica* Finlay. Eocene, New Zealand. × 70. (After Finlay). *a*, side view; *b*, front view. .... p. 259.

12. *Bolivinita quadrilatera* (Schwager). Recent. × 32. (After Cushman, 1942). *a*, front view; *b*, side view; *c*, apertural view. .... p. 257.

Family 31. BULIMINIDAE, p. 260.

11. *Buliminellita mirifica* Cushman and Stainforth. Eocene, Ecuador. × 50. (After Cushman and Stainforth). .... p. 264.

13. "*Robertinoides normani* (Goës)." Recent. × 35. (After Höglund). *a*, front view; *b*, back view; *c*, side view. .... p. 266.

14. *Virgulopsis pustulata* Finlay. Miocene, New Zealand. × 50. (After Finlay). .... p. 267.

15–17. *Colomia cretacea* Cushman and Bermudez. Upper Cretaceous, Cuba. × 60. (After Cushman and Bermudez). 15, Side view. 16, Specimen broken away showing internal tubes. 17, Apertural view. .... p. 265.

18. *Lacosteina gouskovi* Marie. Upper Cretaceous, Morocco. × 50. (After Marie). *a*, front view; *b*, left profile; *c*, top view. .... p. 264.

19, 20. *Elongobula chattonensis* Finlay. Oligocene, New Zealand. × 45. (After Finlay). .... p. 266.

21–24. *Pseudobulimina chapmani* (Heron-Allen and Earland). × 30. (After Earland). 21, 22, Edge views. 23, Dorsal view. 24, Ventral view. .... p. 266.

25. *Suggrunda porosa* W. Hoffmeister and C. Berry. Miocene. × 60. (After W. Hoffmeister and C. Berry). *a*, front view; *b*, side view; *c*, apertural view. .... p. 269.

26. *Bifarinella ryukyuensis* Cushman and Hanzawa. Pleistocene, Ryukyu Islands. × 25. (After Cushman and Hanzawa). *a*, front view; *b*, end view. .... p. 269.

27. *Geminaricta virgata* (Cushman). Miocene, France. × 75. *a*, front view; *b*, apertural view. .... p. 270.

28. *Bitubulogenerina vicksburgensis* Howe. Oligocene, Mississippi. × 125. *a*, front view; *b*, apertural view. .... p. 270.

29. *Unicosiphonia crenulata* Cushman. Tertiary. × 38. *a*, front view; *b*, apertural view. .... p. 274.

30. *Tritubulogenerina mauricensis* Howe. Eocene, Louisiana. × 55. (After Howe). *a*, front view; *b*, apertural view. .... p. 271.

31. *Rectuvigerina multicostata* (Cushman and Jarvis). Tertiary. × 30. (After Cushman and Jarvis). .... p. 273.

PLATE 51.

Family 31. BULIMINIDAE, p. 260.

FIG.

1, 2. *Pseudochrysalidina floridana* Cole. Eocene, Florida. × 20. (After Cole). 1, Front view. 2, Apertural view. .... p. 272.

3, 4. *Washitella typica* Tappan. Lower Cretaceous. × 32. (After Tappan). .... p. 275.

5–7. *Delosina sutilis* Earland. Recent. × 35. (After Earland). 5, Exterior of microspheric form. 6, Longitudinal section. 7, Apertural view. .... p. 276.

Family 32. ELLIPSOIDINIDAE, p. 276.

8, 9. *Dentalinoides canulina* Marie. Upper Cretaceous, France. × 22. (After Marie). 8 *a*, 9, side views; 8 *b*, dorsal view of the last chamber showing the elliptical aperture, oblique with respect to the apex. .... p. 277.

10. *Pinaria heterosculpta* Bermudez. Eocene, Cuba. × 18. (After Bermudez). *a*, front view; *b*, apertural view. .... p. 280.

Family 33. ROTALIIDAE, p. 281.

11–13. *Coscinoconus alpinus* Leupold. Jurassic, Switzerland. × 18. (After Leupold and Bigler). 11, Peripheral view. 12, Dorsal view. 13, Ventral view. .... p. 284.

14. *Conorbina marginata* Brotzen. Upper Cretaceous, Sweden. × 75. (After Brotzen). *a*, dorsal view; *b*, ventral view; *c*, peripheral view. .... p. 286.

15. *Discorbinella montereyensis* Cushman and Martin. Recent. × 110. (After Cushman and Martin). *a*, dorsal view; *b*, ventral view; *c*, peripheral view. .... p. 288.

16. *Torresina haddoni* Parr. Recent. × 50. (After Parr). *a*, dorsal view; *b*, ventral view; *c*, peripheral view. .... p. 288.

17. *Earlmyersia punctulata* (d'Orbigny), var. *liliputana* Rhumbler. Recent. × 185. (After Rhumbler). *a*, dorsal view; *b*, ventral view; *c*, peripheral view. .... p. 288.

18. *Gavelinella pertusa* (Marsson). Upper Cretaceous, Rügen. (After Brotzen). *a*, dorsal view; *b*, oblique ventral view. .... p. 289.

19. *Parrella bengalensis* (Schwager). Pliocene, Kar Nicobar. × 20. (After Schwager). *a*, dorsal view; *b*, ventral view; *c*, peripheral view. .... p. 291.

20. *Gyroidinoides nitida* (Reuss). Upper Cretaceous, Bohemia. (After Brotzen). *a*, dorsal view; *b*, oblique ventral view. .... p. 290.

21–23. *Crespinella umbonifera* (Howchin and Parr). Miocene, Australia. 21, × 30. (After Parr). Horizontal section. 22, × 15. (After Howchin and Parr). *a*, side view; *b*, peripheral view. 23, × 20. (After Parr). Vertical section. .... p. 292.

24. *Pseudovalvulineria lorneiana* (d'Orbigny). Upper Cretaceous, France. (After Brotzen). *a*, dorsal view; *b*, oblique ventral view. .... p. 290.

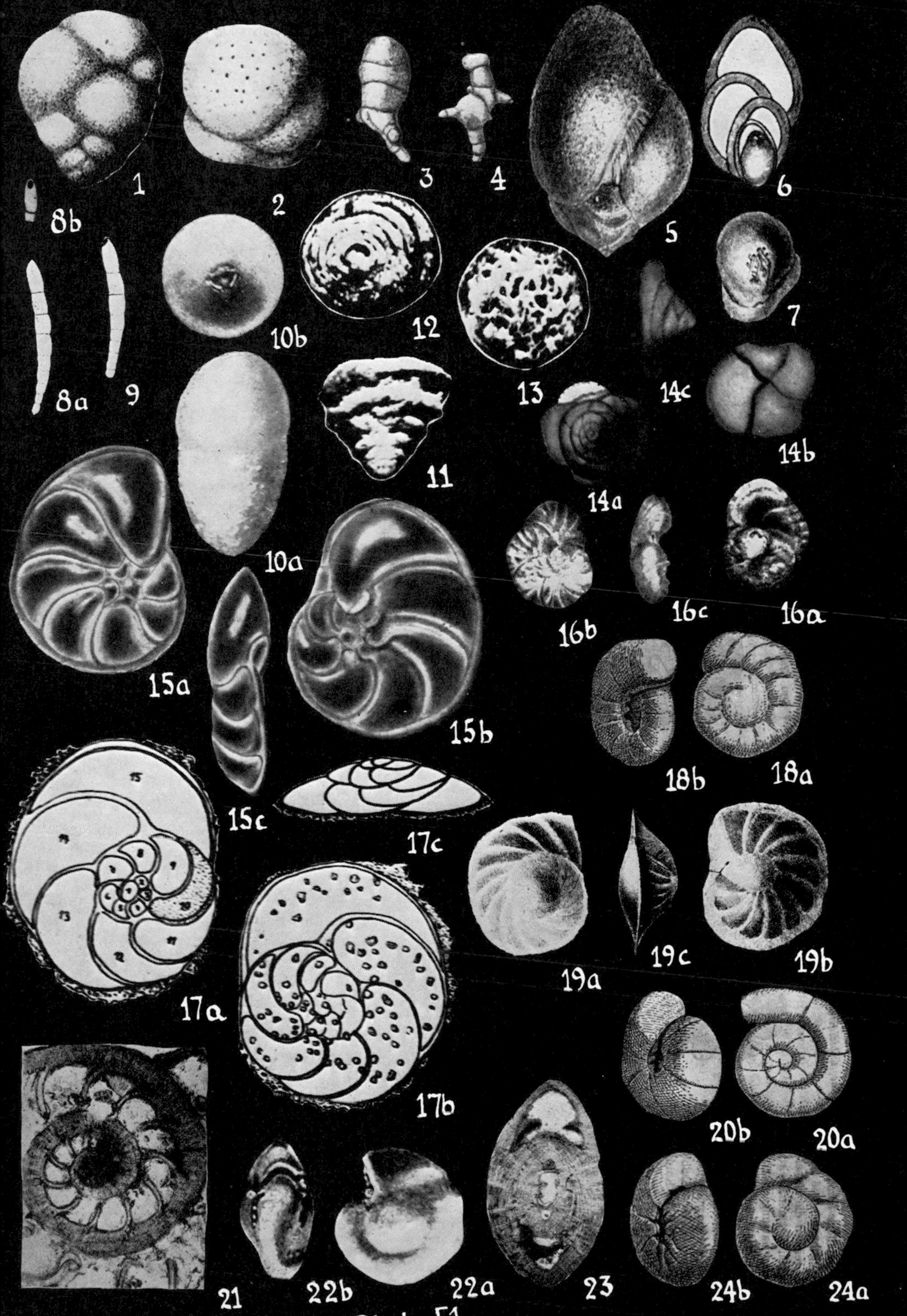
1
2
3
4
5
6
8b
8a
9
10b
10a
11
12
13
7
14c
14b
14a
15a
15b
15c
16b
16c
16a
18b
18a
17c
17a
17b
19a
19c
19b
20b
20a
21
22b
22a
23
24b
24a

Plate 51

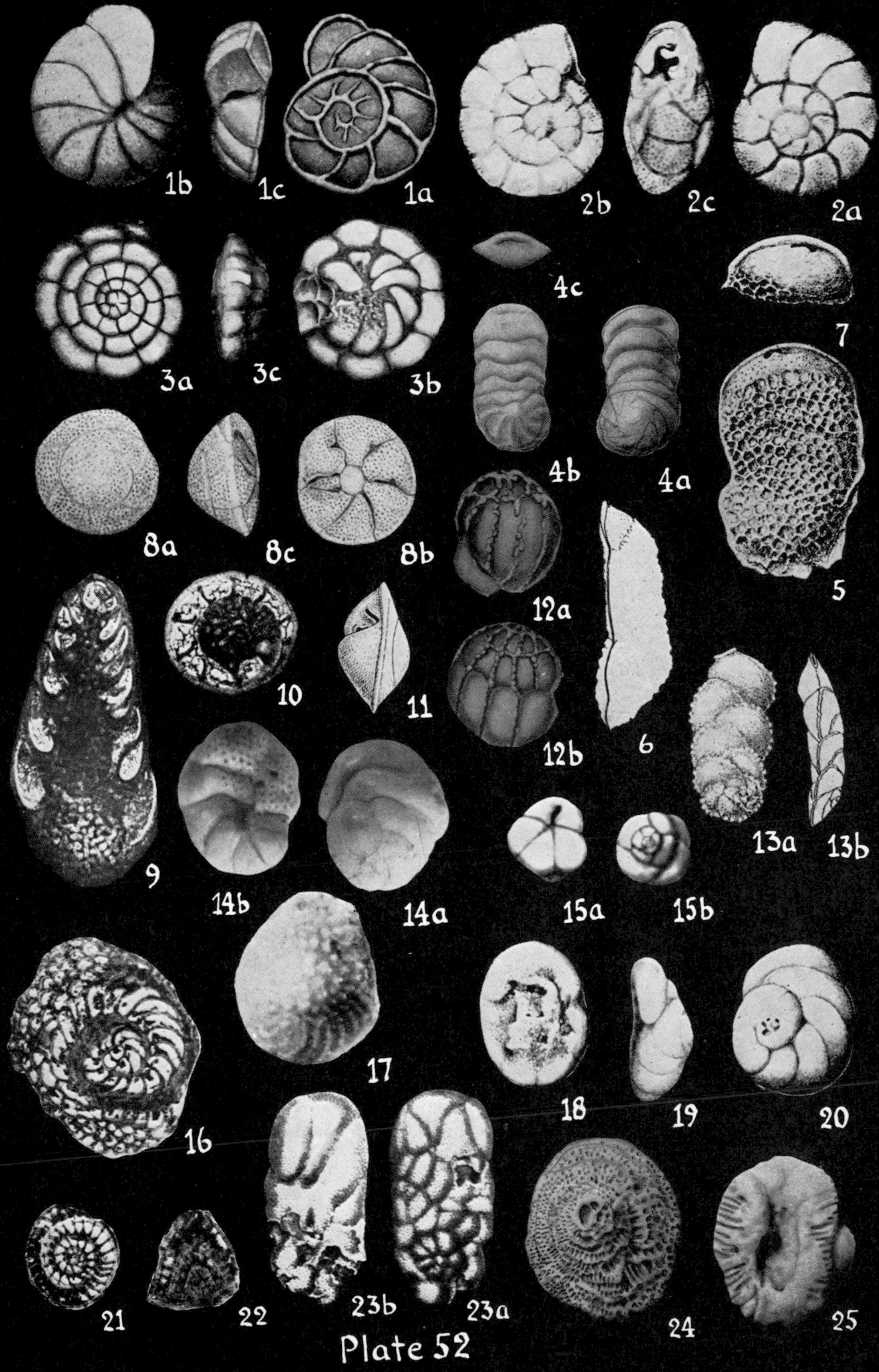
1b
1c
1a
2b
2c
2a
3a
3c
3b
4c
7
4b
4a
8a
8c
8b
12a
5
10
11
12b
6
13a
13b
9
14b
14a
15a
15b
17
18
19
20
16
21
22
23b
23a
24
25

Plate 52

PLATE 52.

Family 33. ROTALIIDAE, p. 281.

FIG.

1. *Stensiöina exsculpta* (Reuss). Cretaceous. × 75. (After Brotzen). *a*, dorsal view; *b*, ventral view; *c*, peripheral view. .... p. 290.
2. *Eponidella libertadensis* Cushman and Hedberg. Miocene, Venezuela. × 90. (After Cushman and Hedberg). *a*, dorsal view; *b*, ventral view; *c*, peripheral view. .... p. 292.
3. *Rotalidium pacificum* Asano. Recent. × 28. (After Asano). *a*, dorsal view; *b*, ventral view; *c*, peripheral view. .... p. 292.
4. *Rectoeponides cubensis* Cushman and Bermudez. Eocene, Cuba. × 25. (After Cushman and Bermudez). *a*, dorsal view; *b*, ventral view; *c*, apertural view. .... p. 293.

5–7. *Coleites reticulosus* (Plummer). Eocene, Texas. × 40. (After Plummer). 5, Ventral view of adult specimen. 6, Peripheral view. 7, Portion of ventral view showing aperture. .... p. 293.

8. *Rotorbinella colliculus* Bandy. Eocene, Oregon. × 45. (After Bandy). *a*, dorsal view; *b*, ventral view; *c*, peripheral view. ... p. 294.

9, 10. *Sakesaria cotteri* Davies. Eocene, India. × 12. (After Davies and Pinfold). 9, Longitudinal section. 10, Transverse section. ..... p. 294.

11. *Parrella lens* (Brotzen). "*Osangularia.*" Danian, Sweden. (After Brotzen). Peripheral view. .... p. 291.
12. *Asanoina globosa* (Yabe and Asano). Pliocene, Java. × 15. (After Yabe and Asano). *a*, *b*, opposite sides. .... p. 294.
13. *Siphonides biserialis* Feray. Eocene, Texas. × 120. (After Feray). *a*, dorsal view; *b*, peripheral view. .... p. 297.
14. *Poroeponides lateralis* (Terquem). Recent. × 22. *a*, dorsal view; *b*, ventral view. .... p. 295.
15. *Baggatella inconspicua* Howe. Eocene, Louisiana. × 75. (After Howe). *a*, apertural view; *b*, dorsal view. .... p. 298.

18–20. *Discorinopsis gunteri* Cole. Eocene, Florida. × 8. (After Cole). 18, 19, 20, Ventral, peripheral, and dorsal views of holotype. ...... p. 296.

Family 35. AMPHISTEGINIDAE, p. 300.

16. *Helicostegina dimorpha* Barker and Grimsdale. Eocene, Mexico. × 20. (After Barker and Grimsdale). Median section. .... p. 301.
17. *Helicostegina gyralis* Barker and Grimsdale. Eocene, Mexico. × 15. (After Barker and Grimsdale). Dorsal view. .... p. 301.

21, 22. *Eoconuloides wellsi* Cole and Bermudez. Eocene, Cuba. × 15. (After Cole and Bermudez). 21, Horizontal section. 22, Axial section. .... p. 301.

Family 37. CYMBALOPORIDAE, p. 307.

23. *Pyropilus rotundatus* Cushman. Recent. × 40. *a*, dorsal view; *b*, ventral view. .... p. 309.

24, 25. *Fabiania cubensis* (Cushman and Bermudez). Eocene, Cuba. × 21. (After Cushman and Bermudez). 24, Dorsal view. 25, Ventral view. .... p. 310.

PLATE 53.

Family 38. CASSIDULINIDAE, p. 311.

FIG.

1. *Roglicia sphaerica* van Bellen. Eocene, Yugoslavia. × 25. (After van Bellen). *a*, dorsal view; *b*, ventral view; *c*, peripheral view. p. 312.
2. *Alabamina wilcoxensis* Toulmin. Eocene, Alabama. × 45. (After Toulmin). *a*, dorsal view; *b*, ventral view; *c*, peripheral view. p. 313.
3. *Cassidulinella pliocenica* Natland. Pliocene, California. × 32. (After Natland). *a*, *b*, opposite sides; *c*, apertural view. .... p. 313.
4. *Stichocassidulina thalmanni* Stone. Eocene, Peru. × 22. (After Stone). Ventral view showing large aperture with toothlike plate and supplementary apertures along the sutures. .... p. 314.

5, 6. *Cerobertina bartrumi* Finlay. Miocene, New Zealand. × 45. (After Finlay). 5, Ventral view. 6, Dorsal view. .... p. 314.

7, 8. *Epistominoides wilcoxensis* (Cushman and Ponton). Eocene, Texas. × 50. (After Plummer). 7 *a*, dorsal view; *b*, ventral view; *c*, peripheral view. 8, Peripheral view with final chamber broken away showing internal structure. .... p. 314.

9–12. *Cushmanella brownii* (d'Orbigny). Recent. (After D. K. Palmer and Bermudez). 9, Side view, × 28. 10, Apertural view, × 28. 11, Broken chamber, apertural view, × 56. 12, Broken chambers showing internal structure, × 35. .... p. 314.

Family 39. CHILOSTOMELLIDAE, p. 315.

13. *Quadrimorphina allomorphinoides* (Reuss). Cretaceous. × 50. (After Reuss). *a*, dorsal view; *b*, ventral view; *c*, peripheral view. .... p. 318.

14. *Rotamorphina cushmani* Finlay. Upper Cretaceous, New Zealand. × 30. (After Finlay). *a*, dorsal view; *b*, ventral view of another specimen. .... p. 318.

15–17. *Cribropullenia marielensis* (D. K. Palmer). Oligocene, Cuba. × 65. (After D. K. Palmer). 15, Side view. 16, Apertural view. 17, Apertural view of broken specimen showing porous aperture. .. p. 321.

Family 40. GLOBIGERINIDAE, p. 321.

18–20. *Globigerinatella insueta* Cushman and Stainforth. Oligocene, Trinidad. × 50. (After Cushman and Stainforth). .... p. 323.

21–23. *Candorbulina universa* Jedlitschka. Miocene. × 12. (After Jedlitschka). 21, Broken specimen showing early chambers. 22, 23, Exteriors showing circular ring of pores. .... p. 327.

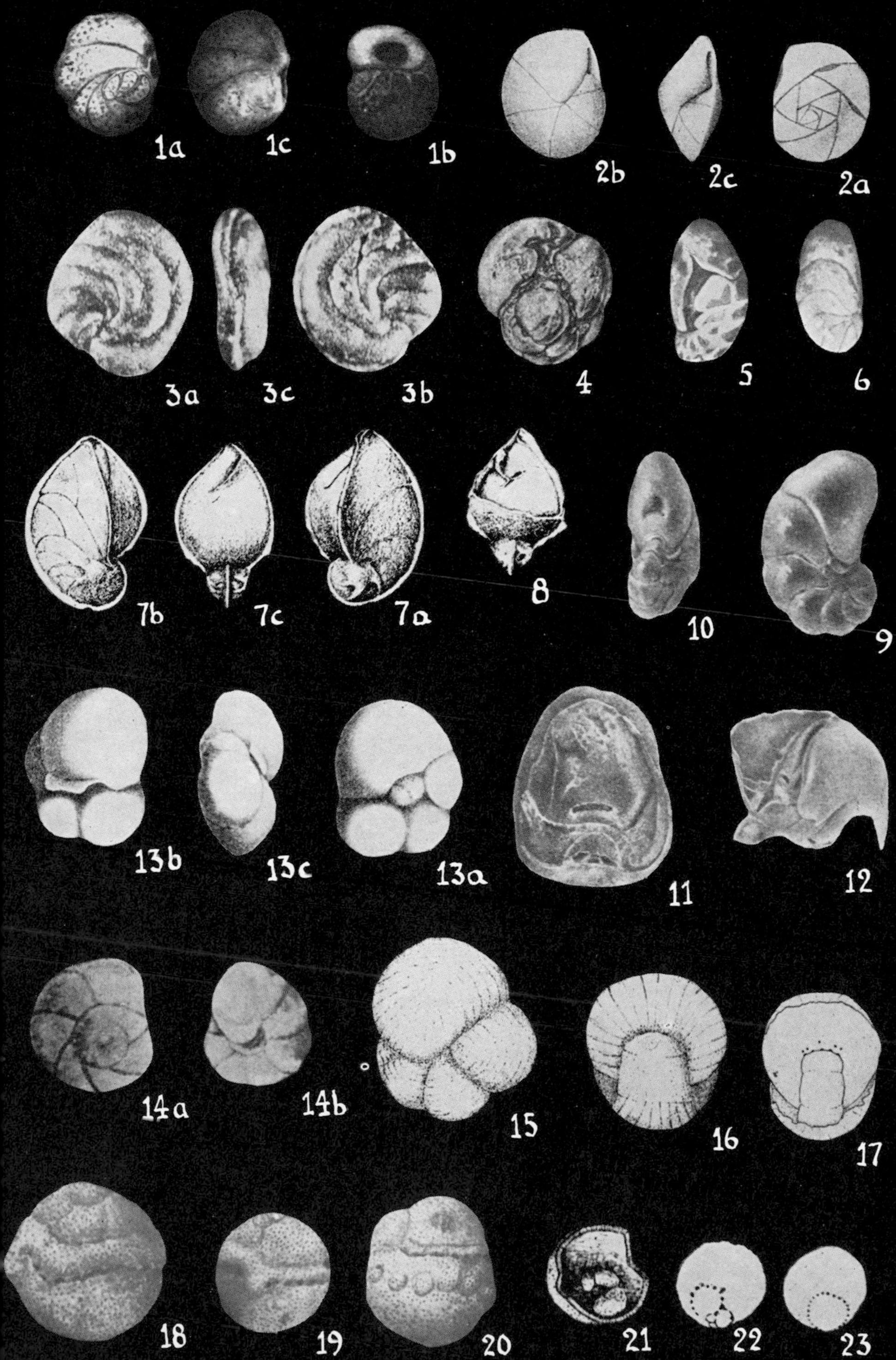
1a
1c
1b
2b
2c
2a
3a
3c
3b
4
5
6
7b
7c
7a
8
10
9
13b
13c
13a
11
12
14a
14b
15
16
17
18
19
20
21
22
23

Plate 53

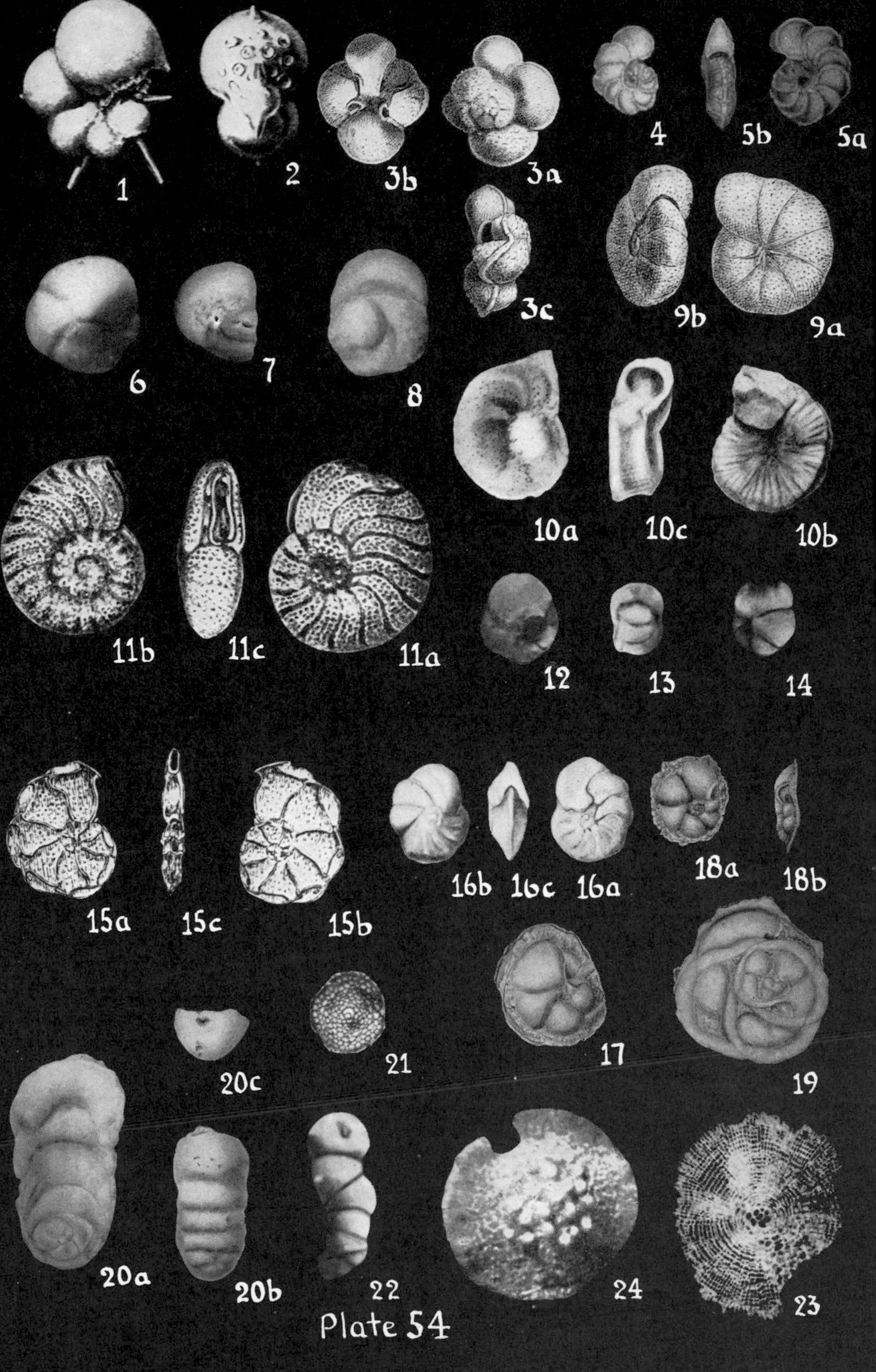
1
2
3b
3a
4
5b
5a
3c
9b
9a
6
7
8
10a
10c
10b
11b
11c
11a
12
13
14
15a
15c
15b
16b
16c
16a
18a
18b
17
19
20c
21
20a
20b
22
24
23

Plate 54

PLATE 54.

Family 41. HANTKENINIDAE, p. 327.

FIG.

1, 2. *Cribrohantkenina brevispina* (Cushman). Eocene, Cuba. × 45. (After Bermudez). 1, Side view. 2, Apertural view. .... p. 329.

Family 42. GLOBOROTALIIDAE, p. 329.

3. *Rotalipora turonica* Brotzen. Upper Cretaceous. × 28. (After Brotzen). *a*, dorsal view; *b*, ventral view; *c*, peripheral view. ..... p. 330.

6–8. *Cribrogloborotalia marielina* Cushman and Bermudez. Eocene, Cuba. × 23. (After Cushman and Bermudez). 6, Ventral view. 7, Peripheral view. 8, Dorsal view. .... p. 330.

Family 43. ANOMALINIDAE, p. 331.

4, 5. *Planomalina apsidostroba* Loeblich and Tappan. Lower Cretaceous. × 35. (After Loeblich and Tappan). 4, 5 *a*, side views; *b*, peripheral view. .... p. 333.

9. *Anomalinoides plummerae* Brotzen. Upper Cretaceous, Texas. (After Brotzen, after Plummer). *a*, dorsal view; *b*, oblique ventral view. .... p. 333.

10. *Boldia lobata* (Terquem). Eocene, Netherlands. × 50. (After van Bellen). *a*, dorsal view; *b*, ventral view; *c*, peripheral view. .... p. 333.

11. *Palmerinella palmerae* Bermudez. Recent. × 60. (After Bermudez). *a*, ventral view; *b*, dorsal view; *c*, apertural view. .... p. 333.

12–14. *Buningia creeki* Finlay. Miocene, New Zealand. × 35. (After Finlay). 12, Ventral view. 13, Peripheral view. 14, Dorsal view. .. p. 336.

15. *Kelyphistoma ampulloloculata* Keijzer. Tertiary, Cuba. × 35. (After Keijzer). *a*, ventral view; *b*, dorsal view; *c*, peripheral view. .... p. 334.

16. *Cibicidoides eriksdalensis* Brotzen. Upper Cretaceous, Sweden. × 45. (After Brotzen). *a*, dorsal view; *b*, ventral view; *c*, peripheral view. .... p. 335.

20. *Stichocibicides cubensis* Cushman and Bermudez. Eocene, Cuba. (After Cushman and Bermudez). *a*, dorsal view, × 30; *b*, *c*, ventral and apertural views of another specimen, × 18. .... p. 336.

22. *Vagocibicides maoria* Finlay. Tertiary, New Zealand. × 15. (After Finlay). .... p. 338.

Family 45. RUPERTIIDAE, p. 343.

17–19. *Neocarpenteria cubana* Cushman and Bermudez. Eocene, Cuba. × 25. (After Cushman and Bermudez). 17, 18 *a*, ventral views; *b*, peripheral view. 19, dorsal view. .... p. 345.

Family 44. PLANORBULINIDAE, p. 339.

21. *Eoannularia eocenica* Cole and Bermudez. Eocene, Cuba. × 15. (After Cole and Bermudez). Horizontal section of megalospheric individual. .... p. 343.

Family 48. ORBITOIDIDAE, p. 347.

23, 24. *Vaughanina cubensis* D. K. Palmer. Upper Cretaceous, Cuba. × 15. 23, Horizontal section (After Vaughan and Cole). 24, Exterior (After D. K. Palmer). .... p. 352.

PLATE 55.

Family 48. ORBITOIDIDAE, p. 347.

Family 49. DISCOCYCLINIDAE, p. 367.

Family 50. MIOGYPSINIDAE, p. 372.

FIG.

1, 2. *Miogypsina* (*Miolepidocyclina*) *burdigalensis* (Gümbel). Miocene. (After Brönnimann). 1, Vertical section. × 48. 2, Equatorial section. × 25. .... p. 374.

3, 4. *Pseudophragmina* (*Athecocyclina*) *cookei* (Vaughan). Eocene. (After Vaughan). 3, Equatorial section. × 15. 4, Vertical section. × 27. .... p. 371.

5. *Discocyclina* (*Aktinocyclina*) *radians* (d'Archiac). Eocene. (After van der Weijden). External view. × 5. .... p. 369.

6. *Miogypsinoides formosensis* Yabe and Hanzawa. Miocene. (After Yabe and Hanzawa). Tangential equatorial section through the nucleoconch. × 20. .... p. 376.

7. *Lepidocyclina* (*Pliolepidina*) *pustulosa* (H. Douvillé). (After Vaughan and Cole). Portion of equatorial section to illustrate normal embryonic chambers and equatorial chambers. × 20. .. p. 361.

8. *Pseudophragmina* (*Proporocyclina*) *perpusilla* (Vaughan). Eocene. (After Vaughan and Cole). Decalcified Canada balsam preparation of a portion of the equatorial layer to show the annular stolon on the distal side of the chambers and one radial stolon to a chamber. × 210. .... p. 371.

9. *Pseudophragmina* (*Asterophragmina*) *pagoda* Rao. Eocene. (After Rao). Equatorial section of a microspheric individual. × 45. .. p. 372.

10. *Discocyclina* (*Discocyclina*) *anconensis* Barker. Eocene. (After Vaughan). Portion of an equatorial preparation to illustrate canals in both the annular and radial walls with radial and vertical stolons. The black circles with light rims are vertical stolons. × 500. .... p. 368.

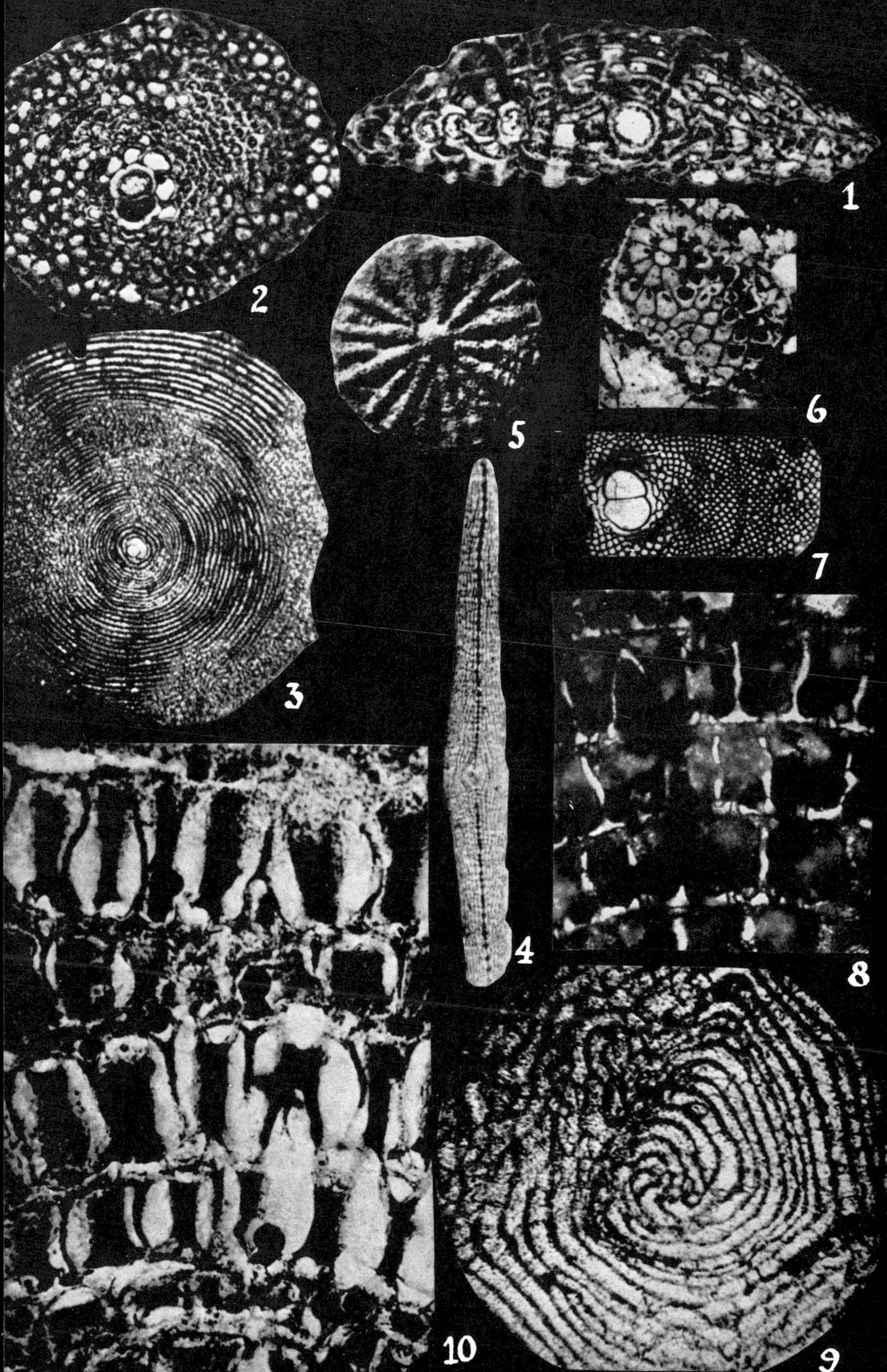

Plate 55

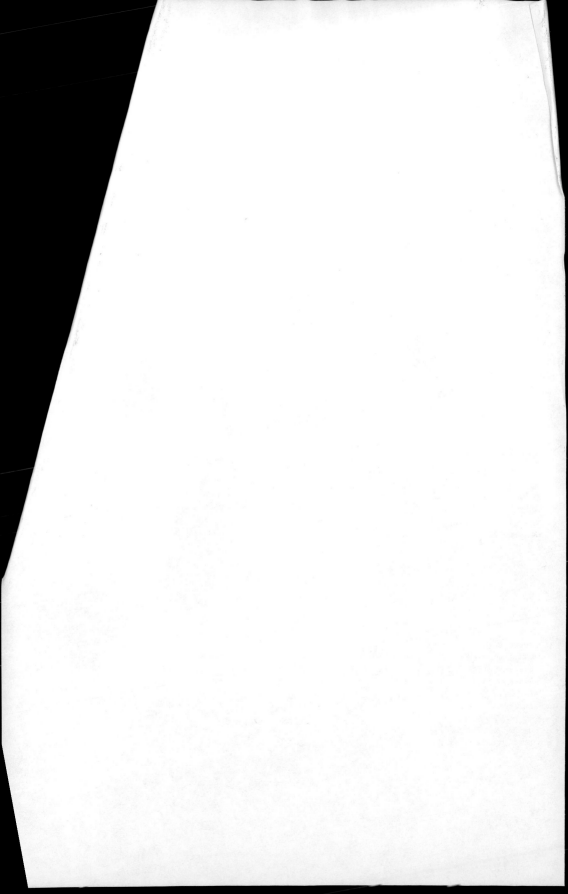

INDEX

INDEX

VALID GENERIC NAMES as used appear in boldface type; synonyms in italics; family and subfamily names, and genera which are only mentioned, in roman type, families being in capitals, and subfamilies in capitals and small capitals. Plate references are given, those preceded by K referring to the Key, and those without the K referring to the text. Page numbers are given in boldface type for the valid genera, and in italics for synonyms.